Studies in Logic

Mathematical Logic and Foundations

Volume 44

Meta-logical Investigations in Argumentation Networks

Studies in Logic Series Editor
Dov Gabbay   dov.gabbay@kcl.ac.uk

Meta-logical Investigations in Argumentation Networks

Dov M. Gabbay

ISBN 978-1-84890-103-2

College Publications
Scientific Director: Dov Gabbay
Managing Director: Jane Spurr
Department of Informatics
King's College London, Strand, London WC2R 2LS, UK

http://www.collegepublications.co.uk

Original cover design by Orchid Creative   www.orchidcreative.co.uk
Printed by Lightning Source, Milton Keynes, UK

Contents

Foreword to the Reader

I focussed my research more on argumentation networks and on deontic logic and law in late 2008 when I started working on Talmudic Logic. The Talmud is a vast body of legal debate and argumenation on every aspect of human behaviour. It is very logical and incredibly coherent. It is said that God gave Moses on Mount Sinai, not only the Ten Commandments, but also a body of laws and a body of rules of how to derive more laws. This led to the development of the Talmud and its way of reasoning and to our 20 volumes project of modelling Talmudic logic (we have 8 volumes so far).

Another reason for focussing on argumentation is my coming to forced retirement (2011). I figured that I need to focus my research because there is a lot to do and not so many years left to live.

So there I was in 2009, with a research agenda in argumentation, Talmudic logic and deontic logic and with a feeling that time is precious and that the right ideas do not come so easily. So I asked myself, what do I look for under these circumstances? Certainly it is a waste of effort from my point of view to take some technical problem and solve it, (for example the computational complexity of finding certain extensions). What I should look for are general principles of logic that instantitate over many areas, among them argumentation theory. These would give us new results and points of view in argumenation theory as well as give us new ideas for more principles to export from argumentation theory to other areas. It will also give us a better picture of what is going on in argumentation.

There is another dimension to this. Assuming we are God's creations in His own image, it is reasonable to assume that the Almighty formulated some general principles first and then proceeded to do his work.[2] So my job was to find some of these principles.

Looking at argumentation I was therefore looking for general questions. This book has this character to it. It follows principles rather than technical details. Of course details need to be worked out mainly to support and highlight the principles involved.

The following is a partial list of general prinicples I have been looking at during my 50 years in logic.

1. The interaction between object level and metalevel aspects in an (application) area.

---

[2] Some might think that this argument is doubtful given the mess this creation is in.

2. The handling of time and change.
3. The extent that algorithmic features of aspects in the area have declarative meaning.
4. The handling of loops in algorithms arising in the area.
5. Connections with other areas. How can simple ideas manifest themselves in different areas.
6. The role of preference ordering in the area.
7. the role of numerical–fuzzy evaluations in the area.
8. The equational approach to the area.
9. The move from the propositional case to the predicate case.
10. The idea of reactivity, the change in a system as a result of, and while, it is being used.
11. The declarative content of modelling in one area as seen in another area.
12. Uncertainty and probability in the area.
13. Combining (fibring) of systems.
14. Substitution of one system in another.
15. Instantiation of the abstract by the more concrete.
16. Algorithmic complexity.
17. Fallacies and errors.
18. Translations and reductions.
19. The theory of labelled deductive systems and the question of what is a logical system.

The above are some general principles and if, for example, you are a student looking to publish a new research paper, you can choose an area or subarea, choose a principle and try to write a paper about the two.

If you do it well and discover new ideas then you are a true reseacher with potential. If you fall into the easy way and do the obvious then you had better go and do something else! It is easy to do the obvious and, alas, there are many who do just that![3]

The material in this book looks at some of the principles listed above in the context of argumentation. You can tell by looking at chapter titles. The chapters are based on various papers of mine, including some joint with colleagues [7, 38, 39, 40, 81, 161, 174, 176, 183] There is a lot to be done and this is why in many places in the book you find me saying 'this is postponed to a future paper'.

Among the postponed research are two main serious topics:

1. The handling of time. Although we do have a chapter on modal and temporal argumentation networks, this is only the beginning. To do it right we will probably need another book. Some of the complexities involved have to do with the influence of the future on the past. Remember we create reality in law and such reality can go backwards. See our book [8]
2. The problem of instantiation. The debate about ASPIC is part of this, but this is a general problem. I would like to talk more about it because the debate in the community is very emotional.

---

[3] The student can look up the discussions and conclusion sections at the end of each chapter and see what topics are left by me for future papers. These are good topics and reading the current book gives the student a head start.

Take modal logic for example. Its semantics has possible worlds S and atomic formulas Q in the syntax and we give an arbitrary abstract assignment $h(t, q) = \top$ or $\bot$ for $t \in S$ and $q \in Q$. We may ask why is q true (or false) at t? we want specific information about the structure of t and the meaning of q which entails that q is true or false at t. This is instantiation. Those of us in argumenation theory who want to instantiate arguments and explain why and how one argument attacks another fall into the same camp as those in modal logic who want to instantiate worlds and propositions.

Acknowledgements

I am grateful to Jane Spurr for producing this book and for her continued support for almost 30 years.

Thanks also to Martin Caminada, Sanjay Modgil and Leon Van der Torre for their advice and suport and to the Argumentation community for welcoming me to their subject.

Research partially supported by the Israel Science Foundation Project 1321/10: Integrating Logic and Networks.

1

Logic and Argumentation Networks: A Manifesto

To the Reader

Dear Reader. Logic and its applications is a vast area spanning many communities with very little interaction between them. My view has always been that there is unity in Logic at least in all the areas of Logic attempting to model human reasoning and behaviour. In order to show that, I undertook to integrate the areas of discrete logical systems and network reasoning. These two areas have very little interaction between them. To be more specific, I first tried to integrate discrete Logic with the area of Argumentation Networks. In fact the integration is with the formal abstract theory of Argumentation, it being the closest to Formal Logic. Argumentation theory is a huge area in itself. So this opening chapter is unique in the sense that it does not survey what has been done but gives you an idea of what should be done!

This chapter introducess a research program centered around argumentation networks and offering several research directions for abstract argumentation networks, with a view of using such networks for integrating logics and network reasoning.

1.1 Logic and networks — a manifesto

In the past half century various formal tools have been proposed for the study of human behaviour in daily life. Such tools were developed in computer science, communication, artificial intelligence, language study, law, analytic philosophy, psychology and cognition, among others. Main among these tools are the formal logical systems (classical logic, non-monotonic logics, modal and temporal logics, etc, etc.) and various network models such as argumentation networks, neural networks, Bayesian networks, inheritance networks, and more. There is no unifying view for all these tools, and in fact they are developed by completely different international communities with very little common ground and communication and yet (see below) all of these features of human behaviour (logics and networks) do reside coherently in the individual human mind and enable him to function intelligently in his day-to-day activity.

There is some realisation among a few of these diverse communities that communication between them needs to take place and unifying principles are indeed sought.

Unfortunately not much is known and certainly no coherent and successful unifying view exists. The mission of this manifesto to provide such a view.

To explain what we have in mind, we start with a simple example.

Example 1.1 (The Messy Room). Mother goes into her teenage daughter's bedroom. Her instant impression is that it is a big mess. There is stuff scattered everywhere.

Mother's impression is that it is not characteristic of the girl to be like this.

What has happened?

Conjecture: The girl has boyfriend problems.

Further Analysis: Mother noticed a collapsed shelf. Did the girl smash it? Upon further observation, mother notices that the pattern of chaos shows that a shelf has collapsed because of excessive weight and scattered everything around, giving the impression of a big mess. But, actually, it is not a mess, it does make some (gravitational) sense.

There are several modes of reasoning:

1. Neural nets type of reasoning.
   She recognises the mess instantly, like we recognise a face.
2. Nonmonotonic deduction.
   Mother reasons from context and her knowledge of her daughter that the girl is not disorganised like this. She asks 'what happened?'.
3. Abduction/conjecture.
   She offers a reasonable explanation that the girl has boyfriend problems. This is common to that age.
4. She then applies a database AI deduction and recognises that the mess is due to gravity. This deduction is no longer a neural net impression. It is a careful calculation.
4*. Item (4) could have been a neural net impression.
   For example, a man who sees many shelf collapsing mess cases may recognise the pattern like it were a face (in which case it would be a neural net-like recognition).
2*. Item (2) could have been a Bayesian network.

Clearly all of these reasoning tools are working together in the mother's mind. Can we give a unified model? What does it look like in principle?

Furthermore, suppose both mother and father have seen the room. Father may reach different conclusions about the girl and demand some action. A dialogue, argumentation and negotiation between the parents will follow with a view to reaching a merged knowledge base and an agreed course of action.

The value of a unified model goes beyond just a unifying formal theory. Even if we take the view that each of these components model a different aspect of the human (constructed as a model for the purpose of installing on a computer or a robot) a unified theory can help extend their range of applicability and help integrate them better. But we hope for more. We hope that such a model built up carefully might give us a better insight on how people actually reason. Something of great interest to the philosopher, psychologist, linguist and cognitive scientist. A unified theory would be a better, sharper tool in their hands.

First let us list what systems (and communities studying them) are involved. We have:

A. **Networks**:
   neural nets, argumentation nets, Bayesian nets, fuzzy nets, biological predator-prey networks, transportation networks, flow networks, inheritance nets, mathematical graphs, Kripke models, Description logics, Electrical networks, legal jurisdiction nets, social networks, input-output nets, and more.

B. **Logics**:
   classical logic, modal and temporal logics, nonmonotonic logics, logic programming, Labelled deductive systems, and more.

C. **Mechanisms**:
   abduction, belief revision and merging, consistency/paraconsistency, meta-level *vs.* object level, and more.

D. **Metalevel principles**:
   fibring and combining systems, communication between systems, and more.

A quick analysis of this task immediately shows how huge the problem is. There is a lot of work to be done just in providing a unifying view between networks, let alone making a connection with logics. The rich variety of networks and the different research communities supporting them have different underlying assumptions, different ranges of applicability and different kinds of mathematics involved. How do we bring them together?

A simple working procedure seems successful. We first identify characteristic movements in each kind of network and then see whether we can generalise the other networks with similar features. Iterating this process will hopefully lead us to a general notion of network which can specialise to the various existing networks. Once we have that, we can try and see how this general network notion can unify with ordinary logics and other mechanisms.

This procedure has the advantage of extending and generalising each network we work with in a meaningful way. Thus each research area will benefit incrementally. Our understanding of characteristic movement existing in one source network (e.g. loops, feedback, aggregation) will be enhanced by developing its counterpart manifestations in other target networks, with the added benefit of meaningful generalisations in the target networks.

The task is not only a scientific problem but also a social problem.

The diverse communities of research in particular networks (e.g. Bayesian community, neural community, argumentation community, logic, etc.) are all immersed in their own areas and are not likely to respond to unifying theories. Our scientific strategy must be such that it is compatible with the social situation and is likely to generate enthusiasm and response from a significant group of researchers. This is why we proceed from the ground up as described above. Furthermore, it is fortunate that one type of network, the argumentation network is open to further generalisation and research in a natural and large scale way. Features existing in other networks can be naturally (though not easily) brought into argumentation networks and these features do have a natural meaning there. Furthermore the argumentation community itself is open-minded and is in fact desiring to expand their field. Many of their individual members have background in logic and will therefore easily see and respond to a general unifying theory with logic.

1.2 Introducing abstract argumentation networks

Let us begin with a motivating example, let us follow a familiar story.

Example 1.2 (The student and the exam). A girl student has one more exam to take for a degree. The course is Logic 101. She already has a job offer but she must get her degree to take up her position. She failed to show up to the exam and the university administration decided not to give her a degree. So her missing the exam turned out to be crucial to her getting a degree, since she was missing a mark in Logic for her total package of courses required for the degree.

We have the following arguments at play:
b = It is not possible to give a degree to the student because she is missing a grade in the Logic course. The student attacks b by giving two arguments
a_1 = I was in hospital at the time and could not come to the exam.
a_2 = My grade average is good and even if I were to come to the exam and get grade 0, I would still have had enough points for a degree.

The administration of the university countered the student's arguments as follows:
c = The student was not in the hospital at the time of the exam. Hospital records do not show any admittance of her name.
c obviously attacks a_1
d = According to the rules of the university all students must take a Logic examination. It may be true that had she taken the exam and failed, she would have had enough points to get a degree, but her obligation was to take the exam and this she did not do.

Argument d does not attack a_2 but it attacks the applicability of the attack from a_2 onto b.
The student counters with two more arguments
e = The university could not find a record of the student's admittance to the hospital because she was admitted to hospital under her married name, while at the university she is registered under her maiden name.

Argument e attacks argument c.
f = you cannot apply the dry letter of university rules to cases where a person's entire future is in the balance. The application of this rule in this case is not fair and is therefore not valid.

Argument f attacks the attack of d on the attack of a_2 on b.
Figure 1.1 displays the situation in this case.
It is clear that in this story the arguments $\{d, f, a_2, e, a_1\}$ are live and therefore the student will get her degree.

We now give the formal definitions, and give references.

Definition 1.3 (Higher level argumentation frames).

1. *An ordinary argumentation frame has the form* $\mathbf{A} = (S, R)$, *where S is the set of arguments and $R \subseteq S \times S$ is the attack relation.*
2. *A level $(0, 1)$ argumentation frame has the form* $\mathbf{A}^{0,1} = (S, \mathbb{R})$ *where S is the set of arguments and $\mathbb{R}$ is a set of pairs of the form (x, y), or $(z, (x, y))$, where $x, y, z \in S$. $(x, y) \in \mathbb{R}$ means that x attacks y and $(z, (x, y)) \in \mathbb{R}$ means that z attacks the attack (x, y).*
3. *Level $(0, n)$ argumentation frames are defined as follows*

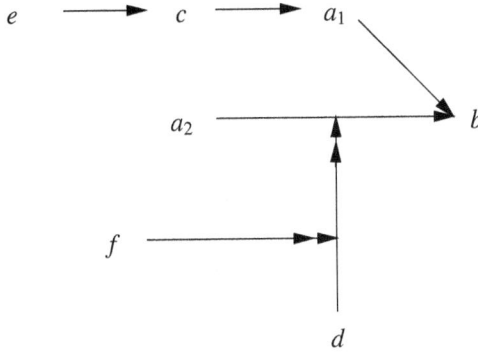

Fig. 1.1.

a) A pair $(x, y) \in S \times S$ is called a level $(0, 0)$ attack.
b) If $z \in S$ and α is level $(0, n)$ attack then (z, α) is a level $(0, n + 1)$ attack.
c) A level $(0, n)$ argumentation frame has the form $\mathbf{A}^{0,n} = (S, \mathbb{R})$ where $\mathbb{R}$ contains attacks β of level $(1, m)$ for $m \leq n$, including some elements of level $(0, n)$.

Definition 1.4 (Caminada labelling[1]).   Let $\mathbf{A} = (S, R)$ be ordinary argumentation frame. A function λ

$$\lambda : S \mapsto \{0, 1, ?\}$$

is called a complete Caminada labelling if it satisfies the following conditions:

1. $\lambda(x) = 1$ if for no y do we have yRx
2. If for some y, yRx and $\lambda(y) = 1$. Then $\lambda(x) = 0$
3. If for all y such that yRx we have $\lambda(y) = 0$ then $\lambda(x) = 1$.
4. If for all y such that yRx we have $\lambda(y) \neq 1$ and for some y such that yRx we have $\lambda(y) = ?$ then $\lambda(x) = ?$.
5. If $\lambda(x) = 1$ we say x is 'live' or 'on'. If $\lambda(x) = 0$, we say x is 'dead' or 'out'. If $\lambda(x) = ?$ we say x is 'undecided'.

Dear Reader. The above gives you an idea and formal definition of argumentation networks. To read more see Chapter 2 below and [294, 225]. Our task is to use abstract argumentation networks as a vehicle to showing how to integrate discrete logic and networks.

So our strategy is as follows.

1. Assume the current state of argumentation networks is S

---

[1] The term Caminada labelling seems to have gained ground in the literature, partly because of the extensive work of Caminada in this area. We should mention that Pollcok [1994], [283], was the first who gave a labelling-based formalisation of argumentation, while Verheij [1996], [328], gave the first labelling-based version of Dung's [114] stable semantics and Jakobovits & Vermeir [1999], [223] defined labelling-based versions of all of Dung's semantics (and define several other semantics of abstract argumentation networks).

2. Study other networks and identify features $F_1, F_2, \ldots$ existent in such networks but not in S.
3. Extend the theory of S to new S_F by incorporating and generalising all of these features in a meaningful way. This is not simple. It is real research which will be meaningful to the argumentation community. It will also generate further generalizations to be exported to other networks.
4. Emerge with a generalised theory of network, say $S_{\text{new}} \supset S_F$.
5. Specialise S_{new} into other networks (e.g. network T) to show that S_{new} is general enough to unify big chunks of other networks. Get, for example, T_{new}. Chances are that T_{new} will naturally and intuitively suggest new features to be exported back to S_{new} to form S_{new}^1.
6. Generalise further to a unified theory with logic, call it S_L.
7. Iterate the entire process to obtain $S_{\text{new}}^2, T_{\text{new}}^2, S_L^2, \ldots$ etc.

The following table lists features from other networks which can be imported into argumentation networks.

Network	Properties
Neural	Feedback loops. Numerical weights on arcs. Real number values, function approximation and learning emphasis.
Argumentation	Nodes attacking nodes. Also sometimes supporting nodes. Loops is an issue with currently no consensus. Logic programming way of thinking. Nodes have logical content.
Bayesian	Loops forbidden. Probabilistic approach.
Fuzzy	Real number values and aggregation.
Biological	Emphasis on loops and cycles through the loops.
Transportation	Emphasis on paths and costs through paths.
Flow	Emphasis on flows and counterflows through paths and nodes. Nodes are sources and Sinks.
Inheritance	Emphasis on persistence and exceptions.
Graphs	Mathematical theory. Counting of nodes, topology on connectivity. Pure mathematics point of view.
Kripke models	Propagation of values for evaluating formulas.
Description logic nets	Similar to Kripke models but with different emphasis. A fragment of predicate logic.
Electrical networks	Feedback loops. Equational approach. Network generates equations to be solved.
Legal jurisdiction network	Movability of data across jurisdiction.
Social networks	Information propagation.
Input-output networks	Nodes are logic processors.

To give the reader an idea of what S_F might look like, we present a diagram, Figure 1.2 of a generalised argumentation network and discuss its features.

This is a complex diagram with the following generalisation:

1. It has both attack and support. Each argument has strength and there is a rate of transmission. For example, node c (strength w) supports b and the transmission rate is e_3.
2. There is feedback from nodes to arcs $u : d$ attacks the attack arc from $x : a$ to $y : b$
3. Arcs can attack or support other nodes or arcs. E.g. the arc e_3 attacks the arc e_2 with strength e_6.
4. There are indirect attacks, for example, of a on c because c supports b and a attacks b.

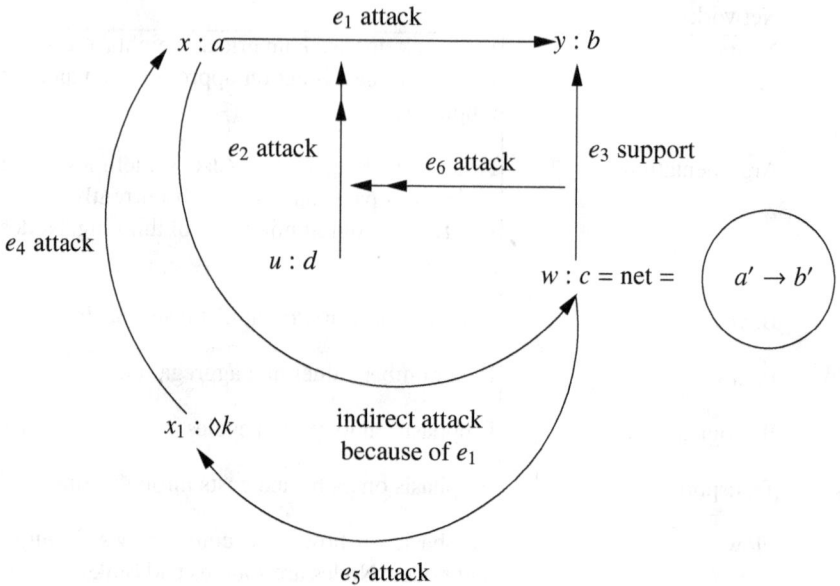

Fig. 1.2. Sample general network

5. There are loops, because c attacks $x_1 : \Diamond k$ which attacks $x : a$. If c is weakened then the attack of $x_1 : \Diamond k$ may succeed and hence $x : a$ fails. So by attacking $y : b$, the node $x : a$ opens itself to attack from $x_1 : \Diamond k$ because it weakens $w : c$.

6. The entire system can be temporal if the strength and transmissions are time dependent and the language of the arguments is temporal. The node $x_1 : \Diamond k$ is temporal because x_1 depends on time and $\Diamond k$ says there is the (future) possibility of argument k.

7. The net contains a subnet. c is the subnet $a' \to b'$. So we can also fibre nets within nets. This is also a problem of communication between networks.

8. We need algorithms to propagate values within the net to get the emerging winning arguments. We need theory of aggregating values and handling loops. We need to combine with temporal logic and change and we need to know how to substitute one net inside the other.

Connection with logic

We will present logic data as networks. So the basic semantical notions will be in the diagram: In Logic

$$\text{formula theory} \Vdash^{\text{evaluated}} \text{semantic model}$$

In our generalised S_L

$$\text{master network} \Vdash^{\text{evaluated}} \text{slave network}$$

This will include the logic case as a special case.

This is integrating with logic through the semantics. Proof theory can also be developed as rules of syntactical manipulation of networks. The framework for doing so is that of Labelled Deductive Systems [136].

To give the reader an idea of how we perceive logic as a network, consider the formula $p \wedge q \to r$. Its construction tree is Figure 1.3

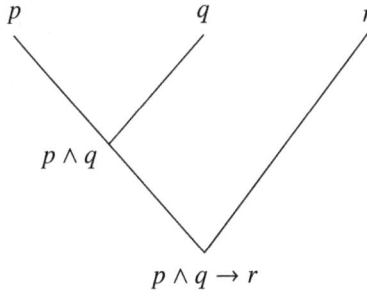

Fig. 1.3. Sample construction tree

If we give values $p = 1, q = 1, r = 0$ to the atoms, then we can propagate them down the tree. Our conclusion from this that a formula A is a network. Now examine how we evaluate $\Diamond q$ at a 3-node Kripke model, Figure 1.4.

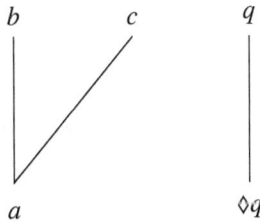

Fig. 1.4. Sample construction tree and a model

First we give joint values of 0 or 1 to pairs (x, q) (meaning $x \vDash q$). Then we jointly and inductively propagate the values down both networks.

So if α is a node in one network and E is a node in another network we define the value $\alpha \vDash E$ by recursive induction on the known values of the rest of the pairs $\alpha' \vDash E'$. The 'formulas' correspond to one network and the 'semantics' is the other network. This way we can define the notion of one network being true in another. $M \vDash N_1$.

Suppose we get that a semantics is a family of networks for evaluation. If we have $\forall M[M \vDash N_1 \to M \vDash N_2]$ we can develop rules for transforming N_1 into N_2. This is proof theory, LDS style, [136].

Sample problems to be addressed

The reader should bear in mind that our aim is to integrate general discrete logic with argumentation networks. This task is both scientific and social. It is not a job for one man but a goal for a community. We therefore need to show that both communities, the logic researchers and the argumentation researchers can benefit. So the way to proceed is to solve problems in both areas in such a way that the benefits of integration can be seen. The list of sample problems to be solved was chosen with this in mind:

1. Aggregation of values. A node is attacked and supported by other nodes. How do we aggregate? Joint attacks and disjunctive attacks. Nodes have strength and arcs have transmission values.
2. General theory of attack absorption and its relation to belief change.
3. Temporal dynamics . The effect of temporal change in each network. We need to add a temporal language to the net as well as make all parameters time dependent.
4. Contents and fibring of networks. Giving nodes a content, e.g. a theory, or another network etc.
5. Feedback, e.g. nodes attacking connections.
6. Handling loops.
7. Procedures and theories for propagating values in networks and extracting information from networks.
8. Metalevel, object level of networks, hierarchies.
9. Logic viewed as networks. A logical theory becomes a network.
10. Evaluating one network in another. One is the "theory" and the other is the semantical "model".
11. Proof theories for networks.

1.3 General view of argumentation networks

Having discussed our general manifesto in the previous section, let us outline our view of argumentation networks in general and the place of logic and fibring within this outline.

There are several ways of viewing and handling argumentation networks. Main among them are the logic programming approach, the classical (first-order or higher-order) logic approach and the algebraic equational approach.

For the purpose of fibring networks, the algebraic equational approach is the most convenient.

Let us take as our starting point a classical model (S, R) of a binary relation R on a non-empty set S. This is basically a directed graph. We can read xRy as an arrow from x to y: $x \rightarrow y$. If we view (S, R) as an argumentation network then xRy means x attacks y.

(S, R) in general can be a basis for many types of networks. To give it a more specific character, we need more annotations. Let us add the unary predicates $Q_0, \ldots, Q_n$. So $Q_i(x)$ is some property of x. Specific types of networks will have some axioms relating the relation R to the predicates $Q_0, \ldots, Q_n$.

For example, for argumentation networks we have three predicates Q_0, Q_1, Q_2, corresponding to the three values {*in, out, undecided*} in a Caminada labelling of arguments, and the relationships between them and R is governed by a theory Δ formalising the Caminada rules. See Chapter 2 for a survey and further results. See also [48], Besnard & Doutre [48], they characterise argumentation networks with propositional formulas and then find labellings with model checking and with satisfiability checking. See also [48], Grossi [207] has translated abstract argumentation networks into modal logic, while in Chapter 5 we define the "logical content" of an argumentation network using modal logic.

In the more general case, we have a theory $\Delta(R, Q_0, \ldots, Q_n)$ governing the relationship between R and $\{Q_i\}$. Any model of Δ will be a properly presented network of type Δ.

The kind of questions one can ask in this context is the existence of models of certain types, the existence and nature of models with maximal or minimal Q_i, and relationships between different models.

The mathematical answers we obtain for the above questions will have quality meaning in the context of the particular network we are studying. For example for argumentation network, Q_1 is the set of winning arguments and so the members of Q_1 give the 'logical content' of the network. For the general case the above questions are purely mathematical.

We can therefore employ general formal logical techniques to get answers to these questions.

For example, we can use second-order logic to ask for minimal models for Q_0. We write the wff

$$\theta(Q_0 \text{ minimal}) = \Delta(R, Q_0, \ldots, Q_n) \wedge \forall Q'_0, \ldots, Q'_n(\Delta(R, Q'_0, \ldots, Q'_n) \to Q_0 \subseteq Q'_0)$$

$(S, R) \vDash \theta$ says that out of all possible models for Q_i based on (S, R), our model is the one with $Q_0, \ldots, Q_n$ where with Q_0 minimal. To express this we need to quantify over subsets. There are also methods for eliminating second-order set quantifiers or for finding fix point solutions for them. See [155] These methods can also be profitably employed here, see Chapter 3.

First-order logic is not sufficient to express the situation in Figure 1.2. This is not because the figure contains several types of arrows (attack and support). The different types of arrows give rise to different binary relations. Also the annotations of nodes and arrows is not a problem. We can add parameters to the relations (so unary predicates become binary, binary relations become ternary, etc.). The problem is arrows leading to arrows.

For example in Figure 1.2 we have an arrow emanating from $c \to b$ going to the arrow emanating from d and attacking the arrow $a \to b$.

Writing this in full would yield a relation between 5 elements. This is too complicated to be natural, since these arrows can be iterated to any higher level. Thus a different approach is required if we are seriously dealing with complex networks as displayed in Figure 1.2.

The second approach is the equational algebraic approach. We regard the predicates $Q_i(x)$ as algebraic values attached to the nodes x. Let $x = a_i$ mean that $Q_i(x)$ holds. We now consider the axioms of Δ as a vehicle using R to formulate a system of equations on the algebra $\mathcal{A} = \{a_i\}$. Of course suitable operations on $\mathcal{A}$ need to be defined. These operations should arise from Δ and respect the declarative content of Δ. To see how this is done for the case of argumentation networks see the discussion in Chapter 7. It is not known what sufficient conditions we can put on a general network of type Δ to ensure that some algebraic equations can be extracted. This approach is better for networks with many levels of arrows attacking arrows. This is because the 'higher order' logic part gets processed into the equations and we just end up with complex equations on the nodes. So from this point of view the network is just an instrument for formulating equations.

The third approach is the logic programming approach. In fact in Dung's original paper [114], the words 'logic programming' are in the title. Consider the network in Figure 1.6. The nodes $x_1, \ldots, x_n$ are all the nodes attacking b. The Horn clause corresponding to it is

$$b \text{ if } \bigwedge_i \neg x_i$$

where $\neg$ is negation as failure

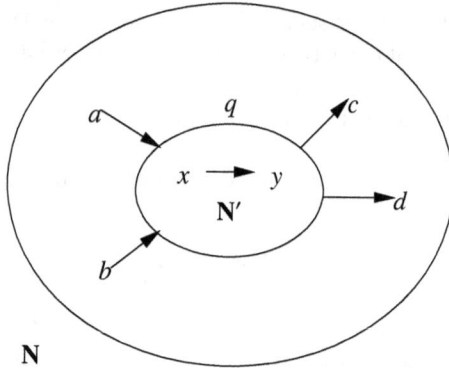

Fig. 1.5. Sample fibring of two networks

Given a network (S, R), we translate it into a set of Horn clauses by taking he clauses

$$y \text{ if } \bigwedge_{\{x \mid xRy\}} \neg x$$

for all $y \in S$.

This translation, originally presented in [161] (see Chapter 4), and further studied in [340] gives a logic program with two special properties.

1. Each literal is the head of at most one clause.
2. All literals in a body of clauses are negated.

Given a logic program with properties (1) and (2) we can regain the corresponding argumentation network by defining

xRy iff (definition) x appears in the body of the clause with head y.

The general question of how to find a corresponding argumentation network for any general logic program, for example

$$y \text{ if } a \wedge \neg b$$
$$y \text{ if } c \wedge \neg d$$

(which does not satisfy (1) and (2) above), is studied in [340]. We use the notion of a critical subset of an argumentation network introduced in [164], see also Chapter 7.

Given a general logic program, we can represent it as a classical model of the form $(S, R_1^{\pm}, \ldots, R_k^{\pm})$. The elements of S are the literals of the program. The relation xR_i^+y, xR_i^-y mean that the literal x appears positively (resp. negated) in the ith clause with head y.

So we have

clasue i with head y: y if $\bigwedge_{xR_i^+y} x \wedge \bigwedge_{zR_i^-y} z$

For extensive study of formal properties of argumentation networks see the references. This type of networks serves as a test case for integrating logic and networks.

1.4 Conclusion and overview of our approach to logic and networks

Given a network (S, R) with a node $x \in S$, we want to view it as a variable for which we can substitute values. This question is a logical question and can be asked of any system of any sort provided it allows for atomic elements. Thus we are trying to find connections between logic and networks by performing the logical substitution operation on networks.

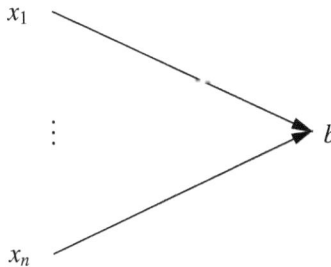

Fig. 1.6. Attack situation

The notion of fibring of networks arises when we substitute for x another network, see Figure 1.5. There are two options here

- General fibring: we substitute any other network in any network, e.g. substitute a neural network for a node x in argumentation network.
- Self fibring: substitute networks of the same type. So in our case substitute an argumentation netwrok for a node in another argumentation network.

We are faced with two immediate problems

- Give meaning to the substitution
- Generalise the notion of the network so that it is closed under substitution.

We discuss the meaning of the substitution and its properties in Chapter 7, Sections 2.1 and 2.2. In Chapter 7, Section 2.3 we generalise the notion of argumentation network so it is closed under substitution and we study its properties. We call these networks *higher-level networks*.

It turns out that the natural way of defining substitution differs from network to network. We examined this notion for Bayesian nets [337] and for Neural nets [149] and they are all different.

Another logical question which can be tested for networks when looked upon as logic , is what does it mean for one network to prove another, in other words we treat networks as if they were logical entities. Chapter 5 is a step in this direction. If we give the network a logical contents in some logic then we have an answer one network proves another if their logical content does so. This is however indirect, and we prefer manipulation rules which take us from one to the other.

To appreciate the difficulty, consider the connection we mentioned above between logic programs and argumentation networks. We can ask the same question of logic programs: what does it mean for one logic program (with nested negations as failure) to 'prove' another? There are many semantic interpretations for logic programs and the answer to this question is not clear. Would a solution for e.g. one of them (i.e. one of logic programs or argumentation networks) also work for the other through the translation? Or would the solution look unnatural? How about other networks like neural nets? Another question we can ask of a network is the following. If networks can be regarded as a form of logic, then we should be able to update, revise, change, expand and contract them in a natural logical way, as we do with logical theories. We need natural definitions for these concepts for networks. We are still working on that. What we have already accomplished though, is to make argumentation networks time dependent in a natural way, see Chapter 6.

We conclude here, as the reader can see, there is a lot to be done yet!

2

A Logical Account of Formal Argumentation

2.1 Introduction

Formal argumentation has become a popular approach for purposes varying from non-monotonic reasoning [25, 79], multi-agent communication [17] and reasoning in the semantic web [295]. Although some research on formal argumentation can be traced back to the early 1990s (like for instance the work of Vreeswijk [333] and of Simari and Loui [312]) the topic really started to take off with Dung's theory of abstract argumentation [114]. Here, arguments are seen as abstract entities (although they can be instantiated using approaches like [79] and [290]) among which an attack relationship is defined. The thus formed *argumentation framework* can be represented as a directed graph in which the arguments serve as nodes and the attack relation as the arrows.

Given such a graph, an interesting question is which sets of nodes can reasonably be accepted. Several criteria of acceptance have been stated, including grounded, complete, preferred and stable semantics [76], as well as more recent approaches like semi-stable semantics [74] and ideal semantics [116].

Despite its relative popularity, formal argumentation has been criticized for its lack of meta-theory [54]. Although quite extensive work has been done describing the properties of the various argument-based semantics [29, 28], what is lacking is a comprehensive way of expressing properties of argumentation using existing logical approaches.

In this chapter, we describe several alternative ways to express argumentation (and in particular argumentation semantics) other then the traditional extensions approach. In particular, we focus on argument labellings, classical logic and various forms of modal logic.

The remaining part of this chapter is structured as follows. First, in Section 2.2, we treat some basic concepts of abstract argumentation, including the traditional extensions approach. Then, in Section 2.3, we describe the labellings approach and show how this can be used an alternative way to describe the standard admissibility based argumentation semantics. Section 2.4 then describes how argumentation can be expressed in modal logic. In section 2.5 argumentation is described using classical logic, and in Section 2.6 we introduce an alternative approach for using modal logic to describe argumentation. We then round off with a discussion of related work in Section 2.8.

2.2 Argumentation preliminaries

An *argumentation framework* [114] consists of a set of arguments and an attack relation on these arguments. In order to simplify the discussion, we consider only finite argumentation frameworks.

Definition 2.1. *Let U be the universe of all possible arguments. An* argumentation framework *is a pair (S, R) where S is a finite subset of U and $R \subseteq S \times S$.*

We say that an argument A *attacks* an argument B iff $(A, B) \in R$. When convenient we write A *Attack* B to mean (A, B) in R.

An argumentation framework can be depicted as a directed graph in which the arguments are represented as nodes and the attack relation is represented as arrows. For instance, argumentation framework (S, R) where $S = \{A, B, C, D, E\}$ and $R = \{(A, B), (B, A), (B, C), (C, D), (D, E), (E, C)\}$ is represented in Figure 2.1.

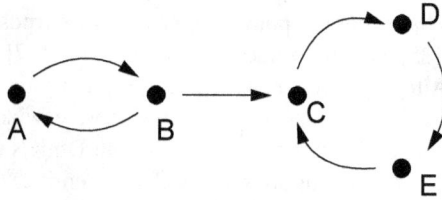

Fig. 2.1. An argumentation framework represented as a directed graph.

The shorthand notation A^+ and A^- stands for, respectively, the set of arguments attacked by argument A and the set of arguments that attack argument A. Likewise, if $\mathcal{A}rgs$ is a set of arguments, then we write $\mathcal{A}rgs^+$ for the set of arguments that are attacked by at least one argument in $\mathcal{A}rgs$, and $\mathcal{A}rgs^-$ for the set of arguments that attack at least one argument in $\mathcal{A}rgs$. In the definition below, $F(\mathcal{A}rgs)$ stands for the set of arguments that are acceptable in the sense of [114].

Definition 2.2 (defense / conflict-free). *Let (S, R) be an argumentation framework, $A \in S$ and $\mathcal{A}rgs \subseteq S$.*
We define A^+ as $\{B \mid A \, R \, B\}$ and $\mathcal{A}rgs^+$ as $\{B \mid A \, R \, B \text{ for some } A \in \mathcal{A}rgs\}$.
We define A^- as $\{B \mid B \, R \, A\}$ and $\mathcal{A}rgs^-$ as $\{B \mid B \, R \, A \text{ for some } A \in \mathcal{A}rgs\}$.
$\mathcal{A}rgs$ is conflict-free iff $\mathcal{A}rgs \cap \mathcal{A}rgs^+ = \emptyset$.
$\mathcal{A}rgs$ defends an argument A iff $A^- \subseteq \mathcal{A}rgs^+$.
We define the function $F : 2^S \to 2^S$ as
$F(\mathcal{A}rgs) = \{A \mid A \text{ is defended by } \mathcal{A}rgs\}$.

In the definition below, definitions of grounded, preferred and stable semantics are described in terms of complete semantics, which has the advantage of making the proofs in the remainder of this chapter more straightforward. These descriptions are not literally the same as the ones provided by Dung [114], but as was first stated in [75], these are in fact equivalent to Dung's original versions of grounded, preferred and stable semantics.

Definition 2.3 (acceptability semantics). *Let (S, R) be an argumentation framework and let $\mathcal{A}rgs \subseteq S$ be a conflict-free set of arguments.*

- $\mathcal{A}rgs$ is admissible *iff* $\mathcal{A}rgs \subseteq F(\mathcal{A}rgs)$.
- $\mathcal{A}rgs$ *is a* complete *extension iff* $\mathcal{A}rgs = F(\mathcal{A}rgs)$.
- $\mathcal{A}rgs$ *is a* grounded *extension iff* $\mathcal{A}rgs$ *is the minimal (w.r.t. set-inclusion) complete extension.*
- $\mathcal{A}rgs$ *is a* preferred *extension iff* $\mathcal{A}rgs$ *is a maximal (w.r.t. set-inclusion) complete extension.*
- $\mathcal{A}rgs$ *is a* stable *extension iff* $\mathcal{A}rgs$ *is a complete extension that attacks every argument in* $S \backslash \mathcal{A}rgs$.
- $\mathcal{A}rgs$ *is a* semi-stable *extension iff* $\mathcal{A}rgs$ *is a complete extension where* $\mathcal{A}rgs \cup \mathcal{A}rgs^+$ *is maximal (w.r.t. set-inclusion).*

As an example, in the argumentation framework of Figure 2.1 $\{B, D\}$ is a stable extension, $\{A\}$ is a preferred extension which is neither stable nor semi-stable, $\emptyset$ is the grounded extension, and $\{B\}$ is an admissible set which is not a complete extension.

It is known that for every argumentation framework, there exists at least one admissible set (the empty set), exactly one grounded extension, one or more complete extensions, one or more preferred extensions and zero or more stable extensions. Moreover, when the set of arguments in the argumentation framework is finite, as is assumed in the current chapter, there also exist one or more semi-stable extensions.

An overview of how the various extensions are related to each other is provided in Figure 2.2. The fact that every stable extension is also a semi-stable extension, and that every semi-stable extension is also a preferred extension was first stated in [74]. All other relations shown in Figure 2.2 have originally been stated in [114].

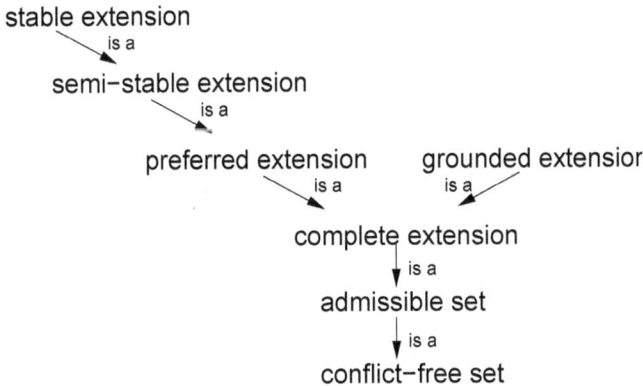

Fig. 2.2. An overview of argumentation semantics (extension based).

2.3 Argument labellings

The concepts of admissibility, as well as that of complete, grounded, preferred, stable or semi-stable semantics were originally stated in terms of sets of arguments. It is equally well possible, however, to express these concepts using argument labellings. This approach was pioneered by Pollock [284] and Jakobovits and Vermeir [223], and has more recently been extended by Caminada [75, 76], Vreeswijk [334] and Verheij [330]. The idea of a labelling is to associate with each argument exactly one label, which can either be in, out or undec. The label in indicates that the argument is explicitly accepted, the label out indicates that the argument is explicitly rejected, and the label undec indicates that the status of the argument is undecided, meaning that one abstains from an explicit judgment whether the argument is in or out.

Definition 2.4. *Let (S, R) be an argumentation framework. A* labelling *is a total function $\mathcal{L} : S \longrightarrow \{$in, out, undec$\}$.*

We write in($\mathcal{L}$) for $\{A \mid \mathcal{L}(A) = $ in$\}$, out($\mathcal{L}$) for $\{A \mid \mathcal{L}(A) = $ out$\}$ and undec($\mathcal{L}$) for $\{A \mid \mathcal{L}(A) = $ undec$\}$. Sometimes, we write a labelling $\mathcal{L}$ as a triple $(\mathcal{A}rgs_1, \mathcal{A}rgs_2, \mathcal{A}rgs_3)$ where $\mathcal{A}rgs_1 = $ in($\mathcal{L}$), $\mathcal{A}rgs_2 = $ out($\mathcal{L}$) and $\mathcal{A}rgs_3 = $ undec($\mathcal{L}$).

We distinguish three special kinds of labellings. The *all-in labelling* is a labelling that labels every argument in. The *all-out labelling* is a labelling that labels every argument out. The *all-undec labelling* is a labelling that labels every argument undec. We say that a labelling is *conflict-free* if no in-labelled argument attacks an (other or the same) in-labelled argument.

2.3.1 Complete labellings

We will now define the concept of a complete labelling and show its relationship with Dung's concept of a complete extension.

Definition 2.5. *Let (S, R) be an argumentation framework. A* complete labelling *is a labelling such that for every $A \in S$ it holds that:*

1. *if A is labelled* in *then all attackers of A are labelled* out
2. *if all attackers of A are labelled* out *then A is labelled* in
3. *if A is labelled* out *then A has a attacker that is labelled* in, *and*
4. *if A has a attacker that is labelled* in *then A is labelled* out

Conditions 1 and 2 essentially form a bi-implication ("A is labelled in iff all its attacker are labelled out"), just like conditions 3 and 4 ("A is labelled out iff it has a attacker that is labelled in").

It is also possible to characterize a complete labelling as a labelling without arguments that are illegally in, illegally out, or illegally undec.

Definition 2.6. *Let $\mathcal{L}$ be a labelling of argumentation framework (S, R) and let $A \in S$. We say that:*

1. *A is* illegally in *iff A is labelled* in *but not all its attackers are labelled* out
2. *A is* illegally out *iff A is labelled* out *but it does not have an attacker that is labelled* in
3. *A is* illegally undec *iff A is labelled* undec *but either all its attackers are labelled* out *or it has an attacker that is labelled* in.

We say that an argument is legally in *iff it is labelled* in *and is not illegally* in. *We say that an argument is* legally out *iff it is labelled* out *and is not illegally* out. *We say that an argument is* legally undec *iff it is labelled undec and is not illegally undec.*

Theorem 2.7. *Let* $\mathcal{L}$ *be a labelling of argumentation framework* (S, R). *It holds that* $\mathcal{L}$ *is a complete labelling iff*

- *every* in-*labelled argument is legally* in,
- *every* out-*labelled argument is legally* out, *and*
- *every* undec-*labelled argument is legally* undec.

Proof.
"$\Longrightarrow$": The fact that each in-labelled argument is legally in follows from point 1 of Definition 2.5. The fact that each out-labelled argument is legally out follows from point 3 of Definition 2.5. We will now prove that each undec-labelled argument is legally undec. Let A be an argument that is labelled undec. Then from point 2 of Definition 2.5 it follows that not all attackers of A are labelled out and from point 4 of Defintion 2.5 it follows that A does not have a attacker that is labelled in. Hence, A is legally undec.
"$\Longleftarrow$": Point 1 of Definition 2.5 follows from the fact that each in-labelled argument is legally in. Point 3 of Definition 2.5 follows from the fact that each out-labelled argument is legally out. Point 2 of Definition 2.5 can be proved as follows. Let A be an argument of which all attackers are labelled out. Then A cannot be labelled out (otherwise A would be illegally out) and A cannot be labelled undec (otherwise A would be illegally undec). Therefore, A can only be labelled in. Point 4 of Definition 2.5 can be proved as follows. Let A be an argument that has an attacker that is labelled in. Then A cannot be labelled in (otherwise A would be illegally in) and A cannot be labelled undec (otherwise A would be illegally undec). Hence, A can only be labelled out.

Using the results of Theorem 2.7 we can restate the concept of a complete labelling as follows.

Proposition 2.8. *Let* $\mathcal{L}$ *be a labelling of argumentation framework* (S, R). *It holds that* $\mathcal{L}$ *is a complete labelling iff for each argument A it holds that:*

- *if A is labelled* in *then all its attackers are labelled* out,
- *if A is labelled* out *then it has at least one attacker that is labelled* in, *and*
- *if A is labelled* undec *then it has at least one attacker that is labelled* undec *and it does not have an attacker that is labelled* in.

Lemma 2.9. *Let* $\mathcal{L}_1$ *and* $\mathcal{L}_2$ *be complete labellings of argumentation framework* $AF = (S, R)$. *It holds that:*

- $\text{in}(\mathcal{L}_1) \subseteq \text{in}(\mathcal{L}_2)$ *iff* $\text{out}(\mathcal{L}_1) \subseteq \text{out}(\mathcal{L}_2)$, *and*
- $\text{in}(\mathcal{L}_1) \subsetneq \text{in}(\mathcal{L}_2)$ *iff* $\text{out}(\mathcal{L}_1) \subsetneq \text{out}(\mathcal{L}_2)$

Proof.
We first prove that $\text{in}(\mathcal{L}_1) \subseteq \text{in}(\mathcal{L}_2)$ iff $\text{out}(\mathcal{L}_1) \subseteq \text{out}(\mathcal{L}_2)$.
"$\Longrightarrow$": Suppose $\text{in}(\mathcal{L}_1) \subseteq \text{in}(\mathcal{L}_2)$. Let $A \in \text{out}(\mathcal{L}_1)$. From point 3 of Definition 2.5 it then follows that A has an attacker (say B) that is labelled in by $\mathcal{L}_1$. That is,

$B \in \text{in}(\mathcal{L}_1)$. From the fact that $\text{in}(\mathcal{L}_1) \subseteq \text{in}(\mathcal{L}_2)$ it then follows that $B \in \text{in}(\mathcal{L}_2)$. From point 4 of Definition 2.5 it then follows that A is labelled out by $\mathcal{L}_2$. That is, $A \in \text{out}(\mathcal{L}_2)$.

"$\Longleftarrow$": Suppose $\text{out}(\mathcal{L}_1) \subseteq \text{out}(\mathcal{L}_2)$. Let $A \in \text{in}(\mathcal{L}_1)$. From point 1 of Definition 2.5 it then follows that each attacker of A must be labelled out by $\mathcal{L}_1$. From the fact that $\text{out}(\mathcal{L}_1) \subseteq \text{out}(\mathcal{L}_2)$ it then follows that each attacker of A is also labelled out by $\mathcal{L}_2$. From point 2 of Definition 2.5 it then follows that A is labelled in by $\mathcal{L}_2$. That is, $A \in \text{in}(\mathcal{L}_2)$.

We now prove that $\text{in}(\mathcal{L}_1) \subsetneq \text{in}(\mathcal{L}_2)$ $\text{out}(\mathcal{L}_1) \subsetneq \text{out}(\mathcal{L}_2)$.

"$\Longrightarrow$": Suppose $\text{in}(\mathcal{L}_1) \subsetneq \text{in}(\mathcal{L}_2)$. This means that $\text{in}(\mathcal{L}_1) \subseteq \text{in}(\mathcal{L}_2)$ and not $\text{in}(\mathcal{L}_2) \subseteq \text{in}(\mathcal{L}_1)$. It then follows that $\text{out}(\mathcal{L}_1) \subseteq \text{out}(\mathcal{L}_2)$ and not $\text{out}(\mathcal{L}_2) \subseteq \text{out}(\mathcal{L}_1)$. This means that $\text{out}(\mathcal{L}_1) \subsetneq \text{out}(\mathcal{L}_2)$.

"$\Longleftarrow$": Suppose $\text{out}(\mathcal{L}_1) \subsetneq \text{out}(\mathcal{L}_2)$. This means that $\text{out}(\mathcal{L}_1) \subseteq \text{out}(\mathcal{L}_2)$ and not $\text{out}(\mathcal{L}_2) \subseteq \text{out}(\mathcal{L}_1)$. It then follows that $\text{in}(\mathcal{L}_1) \subseteq \text{in}(\mathcal{L}_2)$ and not $\text{in}(\mathcal{L}_2) \subseteq \text{in}(\mathcal{L}_1)$. This means that $\text{in}(\mathcal{L}_1) \subsetneq \text{in}(\mathcal{L}_2)$.

Lemma 2.9 implies that a complete labelling is uniquely defined by the in-labelled part, as well as by the out-labelled part.

Lemma 2.10. *Let $\mathcal{L}_1$ and $\mathcal{L}_2$ be complete labellings of argumentation framework (S, R).*

1. *if $\text{in}(\mathcal{L}_1) = \text{in}(\mathcal{L}_2)$ then $\mathcal{L}_1 = \mathcal{L}_2$*
2. *if $\text{out}(\mathcal{L}_1) = \text{out}(\mathcal{L}_2)$ then $\mathcal{L}_1 = \mathcal{L}_2$*

Proof.

1. Suppose $\text{in}(\mathcal{L}_1) = \text{in}(\mathcal{L}_2)$. Then $\text{in}(\mathcal{L}_1) \subseteq \text{in}(\mathcal{L}_2)$ and $\text{in}(\mathcal{L}_2) \subseteq \text{in}(\mathcal{L}_1)$, so from Lemma 2.9 it follows that $\text{out}(\mathcal{L}_1) \subseteq \text{out}(\mathcal{L}_2)$ and $\text{out}(\mathcal{L}_2) \subseteq \text{out}(\mathcal{L}_1)$, so $\text{out}(\mathcal{L}_1) = \text{out}(\mathcal{L}_2)$. From the fact that $\text{in}(\mathcal{L}_1) = \text{in}(\mathcal{L}_2)$ and $\text{out}(\mathcal{L}_1) = \text{out}(\mathcal{L}_2)$ it then follows that $\text{undec}(\mathcal{L}_1) = \text{undec}(\mathcal{L}_2)$ so $\mathcal{L}_1 = \mathcal{L}_2$.
2. Suppose $\text{out}(\mathcal{L}_1) = \text{out}(\mathcal{L}_2)$. Then $\text{out}(\mathcal{L}_1) \subseteq \text{out}(\mathcal{L}_2)$ and $\text{out}(\mathcal{L}_2) \subseteq \text{out}(\mathcal{L}_1)$, so from Lemma 2.9 it follows that $\text{in}(\mathcal{L}_1) \subseteq \text{in}(\mathcal{L}_2)$ and $\text{in}(\mathcal{L}_2) \subseteq \text{in}(\mathcal{L}_1)$, so $\text{in}(\mathcal{L}_1) = \text{in}(\mathcal{L}_2)$. From the fact that $\text{in}(\mathcal{L}_1) = \text{in}(\mathcal{L}_2)$ and $\text{out}(\mathcal{L}_1) = \text{out}(\mathcal{L}_2)$ it then follows that $\text{undec}(\mathcal{L}_1) = \text{undec}(\mathcal{L}_2)$ so $\mathcal{L}_1 = \mathcal{L}_2$.

It turns out that there is a one-to-one relationship between complete labellings and complete extensions, and that it is relatively straightforward to convert a complete labelling to a complete extension and vice versa.

Definition 2.11. *Let $AF = (S, R)$ be an argumentation framework, clabellings be its set of all conflict-free labellings and csets be its set of all conflict-free sets.*

We define a function $\text{Ext2Lab}_{AF} : \text{csets} \rightarrow \text{clabellings}$ such that $\text{Ext2Lab}_{AF}(\mathcal{A}rgs) = \{(A, \text{in}) \mid A \in \mathcal{A}rgs\} \cup \{(A, \text{out} \mid A \in \mathcal{A}rgs^{+}\} \cup \{(A, \text{undec} \mid A \notin \mathcal{A}rgs \text{ and } A \notin \mathcal{A}rgs^{+}\}$.

We define a function $\text{Lab2Ext}_{AF} : \text{clabellings} \rightarrow \text{csets}$ such that $\text{Lab2Ext}_{AF}(\mathcal{L}) = \text{in}(\mathcal{L})$.

Sometimes, when the argumentation framework is either clear or not relevant, we write Ext2Lab and Lab2Ext instead of Ext2Lab_{AF} and Lab2Ext_{AF}.

Theorem 2.12. *Let $AF = (S, R)$ be an argumentation framework and let $\mathcal{L}$ be a complete labelling. Then $\text{Lab2Ext}_{AF}(\mathcal{L})$ is a complete extension.*

Proof. Let $\mathcal{A}rgs = \text{Lab2Ext}_{AF}(\mathcal{L})$. We now prove that $\mathcal{A}rgs$ is a fixpoint of F.

$\mathcal{A}rgs \subseteq F(\mathcal{A}rgs)$: Let $A \in \mathcal{A}rgs$. Then $\mathcal{L}(A) = \text{in}$. The fact that $\mathcal{L}$ is a complete labelling implies (point 1 of Definition 2.5) that each attacker B of A is labelled out. Point 3 of Definition 2.5 then implies that each such B has an attacker (say C) that is labelled in. From the definition of Lab2Ext_{AF} it then follows that $C \in \mathcal{A}rgs$. This means that for each attacker B of A, there is a $C \in \mathcal{A}rgs$ that attacks B. Therefore, $A \in F(\mathcal{A}rgs)$.

$F(\mathcal{A}rgs) \subseteq \mathcal{A}rgs$: Let $A \in F(\mathcal{A}rgs)$. Then each B that attacks A is attacked by some $C \in \mathcal{A}rgs$. From the definition of Lab2Ext_{AF} it follows that C is labelled in. The fact that $\mathcal{L}$ is a complete labelling then implies (point 4 of Definition 2.5) that each such B is labelled out, which then implies (point 2 of Definition 2.5) that A is labelled in. Therefore, by definition of Lab2Ext_{AF}, it holds that $A \in \mathcal{A}rgs$.

The fact that $\mathcal{A}rgs$ is a fixpoint of F, together with the fact that $\mathcal{A}rgs$ is conflict-free (which follows from the fact that each complete labelling is also a conflict-free labelling) implies that $\mathcal{A}rgs$ is a complete extension. $\square$

Theorem 2.13. *Let $AF = (S, R)$ be an argumentation framework and let $\mathcal{A}rgs$ be a complete extension. Then $\text{Ext2Lab}_{AF}(\mathcal{A}rgs)$ is a complete labelling.*

Proof. We first observe that $\text{Ext2Lab}_{AF}(\mathcal{A}rgs)$ is well-defined because the fact that $\mathcal{A}rgs$ is a complete extension implies that it is conflict-free. We now prove the four properties of Definition 2.5.

1. *"if A is labelled in then all attackers of A are labelled out"*
   Let A be an argument that is labelled in. From the definition of Ext2Lab_{AF} it then follows that $A \in \mathcal{A}rgs$. The fact that $\mathcal{A}rgs$ is a complete extension implies that it is an admissible set. That is, $\mathcal{A}rgs$ attacks every attacker of A. From the definition of Ext2Lab_{AF} it then follows that every attacker of A is labelled out.

2. *"if all attackers of A are labelled out then A is labelled in"*
   Let A be an argument such that every attacker of A is labelled out. From the definition of Ext2Lab_{AF} it then follows that every attacker of A is an element of $\mathcal{A}rgs^+$. This means that A is defended by $\mathcal{A}rgs$ ($A \in F(\mathcal{A}rgs)$). From the fact that $\mathcal{A}rgs$ is a complete extension it then follows that $A \in \mathcal{A}rgs$. From the definition of Ext2Lab_{AF} it then follows that A is labelled in.

3. *"if A is labelled out then A has an attacker that is labelled in"*
   Let A be an argument that is labelled out. From the definition of Ext2Lab_{AF} it then follows that $A \in \mathcal{A}rgs^+$, so there is an argument $B \in \mathcal{A}rgs$ that attacks A. From the definition of Ext2Lab_{AF} it follows that B is labelled in.

4. *"if A has an attacker that is labelled in then A is labelled out"*
   Let A be an argument that has an attacker (say B) that is labelled in. From the definition of Ext2Lab_{AF} it follows that $B \in \mathcal{A}rgs$, so $A \in \mathcal{A}rgs^+$. From the definition of Ext2Lab_{AF} it then follows that A is labelled out.

When the domain and range of Lab2Ext_{AF} are restricted to complete labellings and complete extensions, and the domain and range of Ext2Lab_{AF} are restricted to complete extensions and complete labellings, then the resulting functions (call them Lab2Ext_{AF}^c and Ext2Lab_{AF}^c) are bijective and each other's inverse.

Theorem 2.14. *Let $AF = (S, R)$ be an argumentation framework, cextensions its set of complete extensions and clabellings its set of complete labellings. Let $\text{Ext2Lab}_{AF}^c : cextensions \rightarrow clabellings$ be a function such that $\text{Ext2Lab}_{AF}^c(\mathcal{A}rgs) = \text{Ext2Lab}_{AF}(\mathcal{A}rgs)$ and $\text{Lab2Ext}_{AF}^c : clabellings \rightarrow cextensions$ be a function such that $\text{Lab2Ext}_{AF}^c(\mathcal{L}) = \text{Lab2Ext}_{AF}(\mathcal{L})$. The functions Ext2Lab_{AF}^c and Lab2Ext_{AF}^c are bijective and each other's inverse.*

Proof. As every function that has an inverse is bijective, we only need to prove that Lab2Ext_{AF}^c and Ext2Lab_{AF}^c are each other's inverses. That is $(\text{Lab2Ext}_{AF}^c)^{-1} = \text{Ext2Lab}_{AF}^c$ and $(\text{Ext2Lab}_{AF}^c)^{-1} = \text{Lab2Ext}_{AF}^c$. For this, we prove the following two things:

1. For every complete labelling $\mathcal{L}$ it holds that
   $\text{Ext2Lab}_{AF}^c(\text{Lab2Ext}_{AF}^c(\mathcal{L})) = \mathcal{L}$.
   Let $\mathcal{L}$ be a complete labelling of AF and let $A \in S$.
   If $\mathcal{L}(A) = $ in then $A \in \text{Lab2Ext}_{AF}^c(\mathcal{L})$, so $\text{Ext2Lab}_{AF}^c(\text{Lab2Ext}_{AF}^c(\mathcal{L}))(A) = $ in.
   If $\mathcal{L}(A) = $ out then A is attacked by $\text{Lab2Ext}_{AF}^c(\mathcal{L})$, so $\text{Ext2Lab}_{AF}^c(\text{Lab2Ext}_{AF}^c(\mathcal{L}))(A) = $ out.
   If $\mathcal{L}(A) = $ undec then $A \notin \text{Lab2Ext}_{AF}^c(\mathcal{L})$ and A is not attacked by $\text{Lab2Ext}_{AF}^c(\mathcal{L})$, so $\text{Ext2Lab}_{AF}^c(\text{Lab2Ext}_{AF}^c(\mathcal{L}))(A) = $ undec.
2. For every complete extension $\mathcal{A}rgs$ it holds that
   $\text{Lab2Ext}_{AF}^c(\text{Ext2Lab}_{AF}^c(\mathcal{A}rgs)) = \mathcal{A}rgs$.
   Let $\mathcal{A}rgs$ be a complete extension of AF. We now prove two things:
   a) $\text{Lab2Ext}_{AF}^c(\text{Ext2Lab}_{AF}^c(\mathcal{A}rgs)) \subseteq \mathcal{A}rgs$
      Let $A \in \text{Lab2Ext}_{AF}^c(\text{Ext2Lab}_{AF}^c(\mathcal{A}rgs))$. Then A is labelled in by $\text{Ext2Lab}_{AF}^c(\mathcal{A}rgs)$. Therefore $A \in \mathcal{A}rgs$.
   b) $\mathcal{A}rgs \in \text{Lab2Ext}_{AF}^c(\text{Ext2Lab}_{AF}^c(\mathcal{A}rgs)) \subseteq \mathcal{A}rgs$
      Let $A \in \mathcal{A}rgs$. Then A is labelled in by $\text{Ext2Lab}_{AF}^c(\mathcal{A}rgs)$. Therefore $A \in \text{Lab2Ext}_{AF}^c(\text{Ext2Lab}_{AF}^c(\mathcal{A}rgs))$.

From Theorem 2.13 it follows that complete labellings and complete extensions stand in a one-to-one relationship to each other. In essence, complete labellings and complete extensions are different ways of describing the same thing.

Given the one-to-one relationship between complete extensions and complete labellings, the next step would be to examine what kind of labellings are associated with stable, grounded, preferred and semi-stable extensions, all of which are essentially special forms of complete extensions. What would be the properties of the labellings associated with these types of extensions? This question is studied in the following sections.

2.3.2 Stable labellings

We start the discussion with examining the labellings that are associated with stable extensions. These labellings will be called *stable labellings*.

Definition 2.15. *Let AF = (S, R) be an argumentation framework. A stable labelling is a complete labelling* $\mathcal{L}$ *such that* Lab2Ext$_{AF}(\mathcal{L})$ *is a stable extension.*

The fact that for complete labellings and complete extensions, Lab2Ext and Ext2Lab are inverse functions implies that if $\mathcal{A}rgs$ is a stable extension, then Ext2Lab($\mathcal{A}rgs$) is a stable labelling.

Stable labellings can also be characterized as complete labellings without undec.

Theorem 2.16. *Let AF = (S, R) be an argumentation framework. The following statements are equivalent:*

1. $\mathcal{L}$ *is a complete labelling such that* undec($\mathcal{L}$) = $\emptyset$
2. $\mathcal{L}$ *is a stable labelling*

Proof.

from 1 to 2: Suppose $\mathcal{L}$ is a complete labelling such that undec($\mathcal{L}$) = $\emptyset$. Let $\mathcal{A}rgs$ be Lab2Ext$_{AF}(\mathcal{L})$. We now prove that $\mathcal{A}rgs$ is a stable extension (Definition 2.3). Let $A \in S\backslash\mathcal{A}rgs$. From the fact that $A \notin \mathcal{A}rgs$ it follows that A is not labelled in by $\mathcal{L}$. From the fact that undec($\mathcal{L}$) = $\emptyset$ it follows that A is not labelled undec either. Therefore, A is labelled out by $\mathcal{L}$. From the fact that $\mathcal{L}$ is a complete labelling, it follows that A is attacked by an argument (say B) that is labelled in. From the fact that B is labelled in it follows that $B \in \mathcal{A}rgs$. This means that A is attacked by $\mathcal{A}rgs$.

from 2 to 1: Suppose $\mathcal{L}$ is a stable labelling, meaning that $\mathcal{A}rgs$ = Lab2Ext$_{AF}(\mathcal{L})$ is a stable extension. We now prove that undec($\mathcal{L}$) = $\emptyset$. Let $A \in S$. We distinguish two cases:

1. $A \in \mathcal{A}rgs$. Then A is labelled in by $\mathcal{L}$.
2. $A \notin \mathcal{A}rgs$. Then from the fact that $\mathcal{A}rgs$ is a stable extension, it follows that A is attacked by $\mathcal{A}rgs$, so A is labelled out by $\mathcal{L}$.

In both cases, $A \notin$ undec($\mathcal{L}$). Since this folds for an arbitrary $A \in S$, it follows that undec($\mathcal{L}$) = $\emptyset$.

As an aside, one can also raise the question whether there exist labellings with empty in or empty out, and if so, what would be the meaning of these labellings. As for complete labellings with empty in it follows that these also have empty out, so each argument has to be labelled undec. Thus, a complete labelling with empty in is essentially the grounded labelling (see Definition 2.17 and Theorem 2.18) in an argumentation framework where each argument has at least one attacker.

As for complete labellings with empty out, it follows that each argument has to be labelled either in or undec. This means that no argument that is labelled in attacks any (other) argument. In essence, a complete labelling with empty out is the grounded labelling in an argumentation framework where each argument without attackers also does not attack any argument itself.

2.3.3 The grounded labelling

The next thing to be examined are the properties of the labelling associated with the grounded extension.

Definition 2.17. *Let AF = (S, R) be an argumentation framework. A grounded labelling is a complete labelling $\mathcal{L}$ such that* Lab2Ext$_{AF}(\mathcal{L})$ *is the grounded extension.*

The fact that the grounded extension is unique, and that for complete labellings and complete extensions, Lab2Ext and Ext2Lab are inverse functions, implies that the grounded labelling is unique, and that if $\mathcal{A}rgs$ is the grounded labelling, then Ext2Lab($\mathcal{A}rgs$) is the grounded extension.

The grounded labelling can be characterized as the complete labelling with minimal in, as the complete labelling with minimal out, or as the complete labelling maximal undec.

Theorem 2.18. *Let AF = (S, R) be an argumentation framework. The following statements are equivalent:*

1. *$\mathcal{L}$ is a complete labelling where* in($\mathcal{L}$) *is minimal (w.r.t. set inclusion) among all complete labellings*
2. *$\mathcal{L}$ is a complete labelling where* out($\mathcal{L}$) *is minimal (w.r.t. set inclusion) among all complete labellings*
3. *$\mathcal{L}$ is a complete labelling where* undec($\mathcal{L}$) *is maximal (w.r.t. set inclusion) among all complete labellings*
4. *$\mathcal{L}$ is the grounded labelling*

Proof.

from 1 to 2: Let $\mathcal{L}$ be a complete labelling where out($\mathcal{L}$) is not minimal. Then there exists a complete labelling $\mathcal{L}'$ with out($\mathcal{L}'$) $\subsetneq$ out($\mathcal{L}$). From Lemma 2.9 it then follows that in($\mathcal{L}'$) $\subsetneq$ in($\mathcal{L}$), so $\mathcal{L}$ is a labelling where in($\mathcal{L}$) is not minimal.

from 2 to 1: Let $\mathcal{L}$ be a complete labelling where in($\mathcal{L}$) is not minimal. Then there exists a complete labelling $\mathcal{L}'$ with in($\mathcal{L}'$) $\subsetneq$ in($\mathcal{L}$). From Lemma 2.9 it then follows that out($\mathcal{L}'$) $\subsetneq$ out($\mathcal{L}$), so $\mathcal{L}$ is a labelling where out($\mathcal{L}$) is not minimal.

from 1 to 4: Let $\mathcal{L}$ be a complete labelling where in($\mathcal{L}$) is minimal. Then Lab2Ext$_{AF}(\mathcal{L})$ is a minimal complete extension, which implies it is the grounded extension, so $\mathcal{L}$ is the grounded labelling.

from 4 to 1: Let $\mathcal{L}$ be the grounded labelling. Then Lab2Ext$_{AF}(\mathcal{L})$ is the grounded extension, which means it is the minimal complete extension. It then follows that in($\mathcal{L}$) is minimal.

from 1 to 3: Let $\mathcal{L}$ be a complete labelling where in($\mathcal{L}$) is minimal. Then Lab2Ext$_{AF}(\mathcal{L})$ is the grounded extension. Now suppose that undec($\mathcal{L}$) is not maximal. Then there exists a complete labelling $\mathcal{L}'$ with undec($\mathcal{L}$) $\subsetneq$ undec($\mathcal{L}'$). It holds that Lab2Ext$_{AF}(\mathcal{L}')$ is a complete extension, and from the fact that the grounded extension is a subset of each complete extension, it follows that Lab2Ext$_{AF}(\mathcal{L})$ $\subseteq$ Lab2Ext$_{AF}(\mathcal{L}')$, so in($\mathcal{L}$) $\subseteq$ in($\mathcal{L}'$). From Lemma 2.9 it then follows that out($\mathcal{L}$) $\subseteq$ out($\mathcal{L}'$). From the fact that in($\mathcal{L}$) $\subseteq$ in($\mathcal{L}'$) and out($\mathcal{L}$) $\subseteq$ out($\mathcal{L}'$) it then follows that undec($\mathcal{L}'$) $\subseteq$ undec($\mathcal{L}$). Contradiction.

from 3 to 1: Let $\mathcal{L}$ be a complete labelling where in($\mathcal{L}$) is not minimal. Then there exists a complete labelling $\mathcal{L}'$ with in($\mathcal{L}'$) $\subsetneq$ in($\mathcal{L}$). It then also follows (Lemma 2.9) that out($\mathcal{L}'$) $\subsetneq$ out($\mathcal{L}$), so undec($\mathcal{L}$) $\subsetneq$ undec($\mathcal{L}'$). Contradiction.

2.3.4 Preferred labellings

We now examine the properties of the labellings associated with preferred extensions.

Definition 2.19. *Let $AF = (S, R)$ be an argumentation framework. A preferred labelling is a complete labelling $\mathcal{L}$ such that* $\text{Lab2Ext}_{AF}(\mathcal{L})$ *is a preferred extension.*

The fact that for complete labellings and complete extensions, Lab2Ext and Ext2Lab are inverse functions implies that if $\mathcal{A}rgs$ is a preferred extension, then Ext2Lab($\mathcal{A}rgs$) is a preferred labelling.

Preferred labellings can be characterized as complete labellings with maximal in, or as complete labellings with maximal out.

Theorem 2.20. *Let $AF = (S, R)$ be an argumentation framework. The following statements are equivalent:*

1. *$\mathcal{L}$ is a complete labelling where* in($\mathcal{L}$) *is maximal (w.r.t. set inclusion) among all complete labellings of AF*
2. *$\mathcal{L}$ is a complete labelling where* out($\mathcal{L}$) *is maximal (w.r.t. set inclusion) among all complete labellings of AF*
3. *$\mathcal{L}$ is a preferred labelling*

Proof.

from 1 to 2: Let $\mathcal{L}$ be a complete labelling where out($\mathcal{L}$) is not maximal. Then there exists a complete labelling $\mathcal{L}'$ with out($\mathcal{L}$) $\subsetneq$ out($\mathcal{L}'$). From Lemma 2.9 it then follows that in($\mathcal{L}$) $\subsetneq$ in($\mathcal{L}'$), so $\mathcal{L}$ is a labelling where in($\mathcal{L}$) is not maximal.

from 2 to 1: Let $\mathcal{L}$ be a complete labelling where in($\mathcal{L}$) is not maximal. Then there exists a complete labelling $\mathcal{L}'$ with in($\mathcal{L}$) $\subsetneq$ in($\mathcal{L}'$). From Lemma 2.9 it then follows that out($\mathcal{L}$) $\subsetneq$ out($\mathcal{L}'$), so $\mathcal{L}$ is a labelling where out($\mathcal{L}$) is not maximal.

from 1 to 3: Let $\mathcal{L}$ be a complete labelling where in($\mathcal{L}$) is maximal. Then $\text{Lab2Ext}_{AF}(\mathcal{L})$ is a maximal complete extension, which implies it is a preferred extension, so $\mathcal{L}$ is a preferred labelling.

from 3 to 1: Let $\mathcal{L}$ be a preferred labelling. Then $\text{Lab2Ext}_{AF}(\mathcal{L})$ is a preferred extension, which means it is a maximal complete extension. It then follows that in($\mathcal{L}$) is maximal.

2.3.5 Semi-stable labellings

We now examine the properties of the labellings associated with semi-stable extensions.

Definition 2.21. *Let $AF = (S, R)$ be an argumentation framework. A semi-stable labelling is a complete labelling $\mathcal{L}$ such that* $\text{Lab2Ext}_{AF}(\mathcal{L})$ *is a semi-stable extension.*

The fact that for complete labellings and complete extensions, Lab2Ext and Ext2Lab are inverse functions implies that if $\mathcal{A}rgs$ is a semi-stable extension, then Ext2Lab($\mathcal{A}rgs$) is a semi-stable labelling.

Semi-stable labellings can be characterized as complete labellings with minimal undec.

Theorem 2.22. *Let $AF = (S, R)$ be an argumentation framework. The following statements are equivalent:*

1. *$\mathcal{L}$ is a complete labelling where $\mathrm{undec}(\mathcal{L})$ is minimal (w.r.t. set inclusion) among all complete labellings of AF*
2. *$\mathcal{L}$ is a semi-stable labelling*

Proof.

from 1 to 2: Suppose $\mathcal{L}$ is a complete labelling, but not a semi-stable labelling. Then $\mathrm{Lab2Ext}_{AF}(\mathcal{L})$ is a complete extension but not a semi-stable extension. This implies that $\mathrm{Lab2Ext}_{AF}(\mathcal{L}) \cup \mathrm{Lab2Ext}_{AF}(\mathcal{L})^+$ is not maximal. So there exists a complete extension $\mathcal{A}rgs'$ such that $\mathrm{Lab2Ext}_{AF}(\mathcal{L}) \cup \mathrm{Lab2Ext}_{AF}(\mathcal{L})^+ \subsetneq \mathcal{A}rgs' \cup \mathcal{A}rgs'^+$. Let $\mathcal{L}'$ be the labelling associated with $\mathcal{A}rgs'$ (that is: $\mathcal{L}' = \mathrm{Ext2Lab}_{AF}(\mathcal{A}rgs')$ and $\mathcal{A}rgs' = \mathrm{Lab2Ext}_{AF}(\mathcal{L}')$). It then follows that $\mathrm{Lab2Ext}_{AF}(\mathcal{L}) \cup \mathrm{Lab2Ext}_{AF}(\mathcal{L})^+ \subsetneq \mathrm{Lab2Ext}_{AF}(\mathcal{L}') \cup \mathrm{Lab2Ext}_{AF}(\mathcal{L}')^+$, so $\mathrm{in}(\mathcal{L}) \cup \mathrm{out}(\mathcal{L}) \subsetneq \mathrm{in}(\mathcal{L}') \cup \mathrm{out}(\mathcal{L}')$, so $\mathrm{undec}(\mathcal{L}') \subsetneq \mathrm{undec}(\mathcal{L})$. This means that $\mathcal{L}$ is not a labelling where $\mathrm{undec}(\mathcal{L})$ is minimal.

from 2 to 1: Suppose that $\mathcal{L}$ is a complete labelling where $\mathrm{undec}(\mathcal{L})$ is not minimal. This implies that there exists a complete labelling $\mathcal{L}'$ such that $\mathrm{undec}(\mathcal{L}') \subsetneq \mathrm{undec}(\mathcal{L})$. It then follows that $\mathrm{in}(\mathcal{L}) \cup \mathrm{out}(\mathcal{L}) \subsetneq \mathrm{in}(\mathcal{L}') \cup \mathrm{out}(\mathcal{L}')$. Let $\mathcal{A}rgs = \mathrm{Lab2Ext}_{AF}(\mathcal{L})$ and $\mathcal{A}rgs' = \mathrm{Lab2Ext}_{AF}(\mathcal{L}')$. It then follows that $\mathcal{A}rgs \cup \mathcal{A}rgs^+ \subsetneq \mathcal{A}rgs' \cup \mathcal{A}rgs'^+$, which means that $\mathcal{A}rgs$ is not a semi-stable extension, which implies that $\mathcal{L}$ is not a semi-stable labelling.

2.3.6 Roundup

The main results of the discussion until so far are summarized in Table 2.1. Notice that we have covered all combinations of minimal or maximal in, out or undec. Almost all combinations turn out to correspond to the traditional Dung-style semantics. The only exception are the labellings with minimal undec, which corresponds with a semantics not introduced in [114] but in [74].

restriction complete labellings	Dung-style semantics	linked by def. and th.
no restrictions	complete semantics	Def. 2.5 and Th. 2.7
empty undec	stable semantics	Def. 2.15 and Th. 2.16
maximal in	preferred semantics	Def. 2.19 and Th. 2.20
maximal out	preferred semantics	Def. 2.19 and Th. 2.20
maximal undec	grounded semantics	Def. 2.17 and Th. 2.18
minimal in	grounded semantics	Def. 2.17 and Th. 2.18
minimal out	grounded semantics	Def. 2.17 and Th. 2.18
minimal undec	semi-stable semantics	Def. 2.21 and Th. 2.22

Table 2.1. Argument labellings and Dung-style semantics.

Overall, one can see labellings as an alternative way to specify argumentation semantics. The essential rule is that an argument has to be accepted iff all its attackers are rejected, and an argument has to rejected iff it has at least one attacker that is accepted. Thus, argumentation can be explained without referring to things like admissibility or fixpoints of Dung's characteristic function. In a gunfight, one stays alive iff all attackers are dead, and one dies iff at least one attacker is still alive. Those who can understand this have basically understood what abstract argumentation is all about.

2.3.7 Admissible labellings

Until so far, we have modelled the concepts of complete, stable, grounded, preferred and semi-stable semantics in terms of labellings. This was done by *strengthening* the concept of complete labellings. Another route would be to *weaken* the concept of a complete labelling. An obvious way to do this would be to take only a subset of the four properties of Definition 2.5. However, these four properties are not completely independent. If one requires property 1 ("if A is labelled in then all attackers of A are labelled out") then it makes sense also to require property 3 ("if A is labelled out then A has an attacker that is labelled in"), and vice versa. If one requires each in label to make sense, and defines this in terms of out labels, then one should also require each out label to make sense (and vice versa). Together, properties 1 and 3 stand for admissibility.

Definition 2.23. *Let* (S, R) *be an argumentation framework. An* admissible labelling *is a labelling such that for every* $A \in S$ *it holds that:*

1. *if A is labelled* in *then all attackers of A are labelled* out
2. *if A is labelled* out *then A has an attacker that is labelled* in

Admissible labellings correspond to admissible sets, but the relationship is no longer one-to-one.

Theorem 2.24. *Let* $AF = (S, R)$ *be an argumentation framework. It holds that:*

1. *if* $Args$ *is an admissible set of AF then* $\text{Ext2Lab}_{AF}(Args)$ *is an admissible labelling of AF, and*
2. *if* $\mathcal{L}$ *is an admissible labelling of AF then* $\text{Lab2Ext}_{AF}(\mathcal{L})$ *is an admissible set of AF.*

Proof.

1. Let $Args$ be an admisssible set of AF, and let $\mathcal{L} = \text{Ext2Lab}_{AF}(Args)$. We now prove that $\mathcal{L}$ is an admissible labelling.
   a) Let A be an argument that is labelled in by $\mathcal{L}$. Let B be an arbitrary attacker of A. The fact that A is labelled in by $\mathcal{L}$ implies that $A \in Args$. The fact that $Args$ is an admissible set implies that $Args$ attacks B. That is, $B \in Args^+$, so B is labelled out by $\mathcal{L}$. Since this holds for any attacker B of A it follows that every attacker of A is labelled out by $\mathcal{L}$

b) Let A be an argument that is labelled out by $\mathcal{L}$. It then follows that $A \in \mathcal{A}rgs^+$, so there exists a $B \in \mathcal{A}rgs$ that attacks A. This B is labelled in by $\mathcal{L}$. This means that A has a attacker that is labelled in by $\mathcal{L}$.

2. Let $\mathcal{L}$ be an admissible labelling of AF, and let $\mathcal{A}rgs = \mathrm{Lab2Ext}_{AF}(\mathcal{L})$. We now prove that $\mathcal{A}rgs$ is an admissible set.

a) We first prove that $\mathcal{A}rgs$ is conflict-free. Suppose this is not the case. Then there exist $A, B \in \mathcal{A}rgs$ such that A attacks B. It then follows that both A and B are labelled in by $\mathcal{L}$. But this cannot be the case because the fact that $\mathcal{L}$ is an admissible labelling implies that all attackers of A (including B) are labelled out by $\mathcal{L}$.

b) We now prove that $\mathcal{A}rgs$ defends all its elements. Let $A \in \mathcal{A}rgs$. Then A is labelled in by $\mathcal{L}$. Let B be an arbitrary argument that attacks A. From the fact that $\mathcal{L}$ is an admissible labelling, it follows that B is labelled out by $\mathcal{L}$. From the fact that $\mathcal{L}$ is an admissible labelling it then also follows that B has an attacker C that is labelled in by $\mathcal{L}$, so $C \in \mathcal{A}rgs$. Therefore $\mathcal{A}rgs$ is self-defending.

Admissible labellings and admissible sets have a many-to-one relationship. That is, each admissible labelling is associated with exactly one admissible set, but each admissible set is associated with one or more admissible labellings. As an example, consider again the argumentation framework of Figure 2.1. Here, $(\{B\}, \{A, C\}, \{D, E\})$ and $(\{B\}, \{A\}, \{C, D, E\})$ are two admissible labellings associated with the same admissible set $\{B\}$.

For admissible labellings (say $\mathcal{L}_1$ and $\mathcal{L}_2$) it does *not* hold that if $\mathrm{in}(\mathcal{L}_1) \subseteq \mathrm{in}(\mathcal{L}_2)$ then $\mathrm{out}(\mathcal{L}_1) \subseteq \mathrm{out}(\mathcal{L}_2)$. As a counter example, consider again the argumentation framework of Figure 2.1, with $\mathcal{L}_1 = (\{B\}, \{A, C\}, \{D, E\})$ and $\mathcal{L}_2 = (\{B\}, \{A\}, \{C, D, E\})$. Similarly, it also does *not* hold that if $\mathrm{out}(\mathcal{L}_1) \subseteq \mathrm{out}(\mathcal{L}_2)$ then $\mathrm{in}(\mathcal{L}_1) \subseteq \mathrm{in}(\mathcal{L}_2)$. As a counter example, consider the argumentation framework $AF = (\{A, B, C, D\}, \{(A, C), (B, C), (C, D)\})$ with $\mathcal{L}_1 = (\{A, B, D\}, \{C\}, \emptyset)$ and $\mathcal{L}_2 = (\{A, D\}, \{C\}, \{B\})$.

Between the concept of an admissible labelling and a complete labelling, one can distinguish two intermediate forms.

Definition 2.25. *Let $AF = (S, R)$ be an argumentation framework. A JV-labelling is a labelling that satisfies:*

1. *if A is labelled* in *then all attackers of A are labelled* out
2. *if A is labelled* out *then A has an attacker that is labelled* in
3. *if A has an attacker that is labelled* in *then A is labelled* out

A VJ-labelling is a labelling that satisfies:

1. *if A is labelled* in *then all attackers of A are labelled* out
2. *if A is labelled* out *then A has an attacker that is labelled* in
3. *if all attackers of A are labelled* out *then A is labelled* in

In essence, an admissible labelling satisfies point 1 and 3 of a complete labelling (Definition 2.5), a JV-labelling satisfies point 1, 3 and 4, and a VJ-labelling satisfies point 1, 3 and 2. It immediately follows that every complete labelling is also a JV-labelling and a VJ-labelling, and that every JV-labelling or VJ-labelling is also an admissible labelling. We use the term JV-labelling, because these are quite close to a proposal of Jakobovits and Vermeir [223]. An important difference, however, is that in our approach each argument gets exactly one out of three possible labels (in, out or undec) whereas in Jakobovits and Vermeir's original proposal, there are four possibilities (either no label, single in, single out, or both in and out).[1]

It turns out that JV-labellings are uniquely identified by their in labelled part, whereas VJ-labellings are uniquely identified by their out labelled part.

Lemma 2.26. *Let* $AF = (S, R)$ *be an argumentation framework. Let* $\mathcal{L}_1$ *and* $\mathcal{L}_2$ *be JV-labellings of AF and let* $\mathcal{L}_3$ *and* $\mathcal{L}_4$ *be VJ-labellings of AF. It holds that:*

1. *if* $in(\mathcal{L}_1) \subseteq in(\mathcal{L}_2)$ *then* $out(\mathcal{L}_1) \subseteq out(\mathcal{L}_2)$
2. *if* $in(\mathcal{L}_1) \subsetneq in(\mathcal{L}_2)$ *then* $out(\mathcal{L}_1) \subsetneq out(\mathcal{L}_2)$
3. *if* $out(\mathcal{L}_3) \subseteq out(\mathcal{L}_4)$ *then* $in(\mathcal{L}_3) \subseteq in(\mathcal{L}_4)$
4. *if* $out(\mathcal{L}_3) \subsetneq out(\mathcal{L}_4)$ *then* $in(\mathcal{L}_3) \subsetneq in(\mathcal{L}_4)$

Proof. Similar to the proof of Lemma 2.9.

Lemma 2.27. *Let* $AF = (S, R)$ *be an argumentation framework. Let* $\mathcal{L}_1$ *and* $\mathcal{L}_2$ *be JV-labellings of AF and let* $\mathcal{L}_3$ *and* $\mathcal{L}_4$ *be VJ-labellings of AF. It holds that:*

1. *if* $in(\mathcal{L}_1) = in(\mathcal{L}_2)$ *then* $\mathcal{L}_1 = \mathcal{L}_2$
2. *if* $out(\mathcal{L}_3) = out(\mathcal{L}_4)$ *then* $\mathcal{L}_3 = \mathcal{L}_4$

Proof. Similar to the proof of Lemma 2.10.

Since JV-labellings are uniquely identified by their in labelled part, and the fact that an admissible set essentially specifies a set of in labelled arguments, it does not come as a surprise that there exists a one-to-one relation between admissible sets and JV-labellings.

Theorem 2.28. *Let* $AF = (S, R)$ *be an argumentation framework, asets be its set of admissible sets and JV labellings be its set of JV-labellings. Let* $\mathsf{Ext2Lab}_{AF}^{JV} :$ *asets* $\rightarrow$ *JV* $-$ *labellings be a function such that* $\mathsf{Ext2Lab}_{AF}^{JV}(\mathcal{A}rgs) = \mathsf{Ext2Lab}_{AF}(\mathcal{A}rgs)$ *and* $\mathsf{Lab2Ext}_{AF}^{JV} :$ *JV* $-$ *labellings* $\rightarrow$ *asets be a function such that* $\mathsf{Lab2Ext}_{AF}^{JV}(\mathcal{L}) = \mathsf{Lab2Ext}_{AF}(\mathcal{L})$. *The functions* $\mathsf{Ext2Lab}_{AF}^{JV}$ *and* $\mathsf{Lab2Ext}_{AF}^{JV}$ *are bijective and each other's inverse.*

Proof. Similar to the proof of Theorem 2.14.

A global overview of the relations between the various forms of labellings is provide in Figure 2.3.

---

[1] Another difference is that Jakobovits and Vermeir use the symbol "+" instead of in and the symbol "−" instead of out.

stable labelling
↘ is a

semi–stable labelling
↘ is a

preferred labelling   grounded labelling
↘ is a   is a ↗

complete labelling

is a ↗   ↘ is a

JV–admissible labelling   VJ–admissible labellii

is a ↘   ↗ is a

admissible labelling
↓ is a

conflict–free labelling

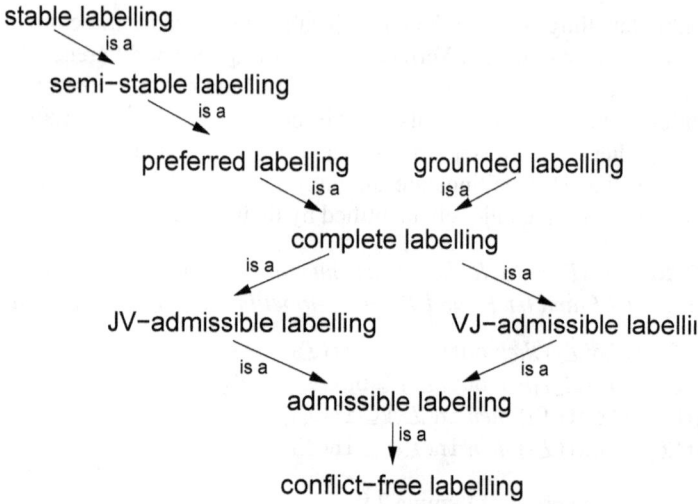

Fig. 2.3. An overview of argumentation semantics (labelling based).

2.4 Argumentation and modal logic

There are three main methods of introducing modal logic into argumentation theory: the metalevel approach, the object level approach and the mixed approach. We view an argumentation system as a combination of an argumentation framework $AF = (S, R)$ (where S is a set of atomic arguments and $R \subseteq S \times S$ is the attack relation) and a complete labelling $\mathcal{L}$ of AF. To us, the thus described argumentation system serves as the object level.

The metalevel approach talks about the argumentation framework from "above", using another language and logic. The metalevel language can be classical logic or modal logic. These are traditional metalevel languages and are used in this way in many areas. Classical logic can be full classical logic or the computational Horn clause logic programming language. When classical logic is used, one can think of the process as translation. There is a traditional translation of modal logic into classical logic and through this translation the metalevel approaches are related.

For the sake of clarity, we shall present all three metalevel versions in the appropriate sections: the classical logic one, the Horn clause logic programming one, and the modal logic one. They are related but are not the same; each one has its advantages and limitations.

2.4.1 Modal logic preliminaries

This subsection introduces some modal logic background needed for introducing our approaches.

The modal logic $\mathcal{K}$ is a propositional system with the modal operator $\Box$ and the atomic propositions $Q = \{q_1, q_2, q_3 \ldots\}$ and $\neg, \wedge, \vee, \rightarrow, \top$ and $\bot$. Models for $\mathcal{K}$ have the form $M = (S, R, h)$ where S is the set of possible worlds, $R \subseteq S \times S$, $S \neq \emptyset$ and h is the assignment function, giving to each atomic letter q a subset $h(q) \subseteq S$. Satisfaction is defined as follows for $t \in S$:

- $t \vDash q$ iff $t \in h(q)$, for atomic q
- $t \vDash A \wedge B$, $A \vee B$, $\neg A$, $A \to B$ are defined as usual
- $t \vDash \Box A$ iff for all s such that tRs we have $s \vDash A$
- A holds in (S, R, h) iff for all $t \in S$ we have $t \vDash A$.

$\mathcal{K}$ can be axiomatised as follows:

1. all substitution instances of truth functional tautologies
2. $\Box(A \to B) \to (\Box A \to \Box B)$
3. $\frac{\vdash A}{\vdash \Box A}$

It is complete for the class of all finite frames of the form (S, R) where $S \neq \emptyset$, S is finite, and $R \subseteq S \times S$.

2.4.2 Products of modal logics

We need the concept of an H-product of a fixed frame modal logic. To define this, we need to introduce the concept of a fixed frame modal logic as well as the concept of a product.

Definition 2.29. *A modal logic $\mathbb{L}$ is said to be a fixed frame modal logic (FF modal logic) if it can be characterised by a family of models of the form (S_0, R_0, h), $h \in H$, where (S_0, R_0) is a fixed frame and H is a set of assignments to this frame. We have $\mathbb{L} \vdash A$ iff A holds in all FF-models (S_0, R_0, h), $h \in H$.*

We can now define the cross product of the two modal logics. Let $\Box_1$ be a modality of modal logic $\mathbb{E}_1$ that is completely chacterised by a class of models of the form $\mathcal{M}_1 = \{(S_i^1, R_i^1, h_i^1)\}$, $i \in I_1$ and similarly for $\Box_2$, $\mathbb{E}_2$, $\mathcal{M}_2 = \{(S_j^2, R_j^2, h_j^2)\}$, $j \in I_2$. We define the flat-cross product of these two logics semantically, through their models. The language contains the two modalities $\{\Box_1, \Box_2\}$. The models have the form $(S^1 \times S^2, R_1 \cup R_2, h_{1,2})$ where (S^1, R^1, h^1) and (S^2, R^2, h^2) are any models of $\Box_1$ and $\Box_2$ respectively and $S^1 \times S^2$ is the product of the sets S^1 and S^2. We define R_1 and R_2 as follows.
$(x_1, x_2)R_1(y_1, x_2)$ iff $x_1 R^1 y_1$
$(x_1, x_2)R_2(x_1, y_2)$ iff $x_2 R^2 y_2$
We define $h_{1,2}$ by some boolean function of h_1 and h_2, for example
$(x_1, x_2) \in h_{1,2}(q)$ iff $x_1 \in h_1(q)$ or $x_2 \in h_2(q)$
or another definition
$(x_1, x_2) \in h_{1,2}(q)$ iff $x_1 \in h_1(q)$.
We have
$(x_1, x_2) \vDash \Box_1 A$ iff for all y_1, $x_1 R^1 y_1$ we have $(y_1, x_2) \vDash A$.
$(x_1, x_2) \vDash \Box_2 A$ iff for all y_2, $x_2 R^2 y_2$ we have $(x_1, y_2) \vDash A$.
The above definition defines a general flat product.

When we have a single FF-logic, with $\Box$ characterised by a fixed frame (S_0, R_0) and a family of assignments H, we can form the universal product of the logic with $\Box$ along the axis of H. We need a new modality for the H axis, which we denote by $\boxdot$.

Definition 2.30. *Let* (S_0, R_0) *be a frame for a fixed frame modal logic* $\mathbb{L}$ *with* $\square$. *Let H be a family of assignments* $h \in H$ *such that* (S_0, R_0, h) *is a model of* $\mathbb{L}$. *We form the universal H product of the frame as follows. We form the set* $\mathcal{M} = \{(S_0, R_0, h) \mid h \in H\}$. *We use a modality* $\boxdot$ *to move around* $\mathcal{M}$. *We then have two modalities:* $\square$ *and* $\boxdot$. *Satisfaction in* $\mathcal{M}$ *is defined as follows.*

$(t, h) \vDash q$ *iff* $t \in h(q)$, *for* $t \in S$, $h \in H$
$(t, h) \vDash \square A$ *iff for all* s, tRs *implies* $(s, h) \vDash A$
$(t, h) \vDash \boxdot A$ *iff for all* $h' \neq h$, $(t, h') \vDash A$
A holds in the model iff for all t, h, $(t, h) \vDash A$.

Figure 2.4 shows this is indeed a product.

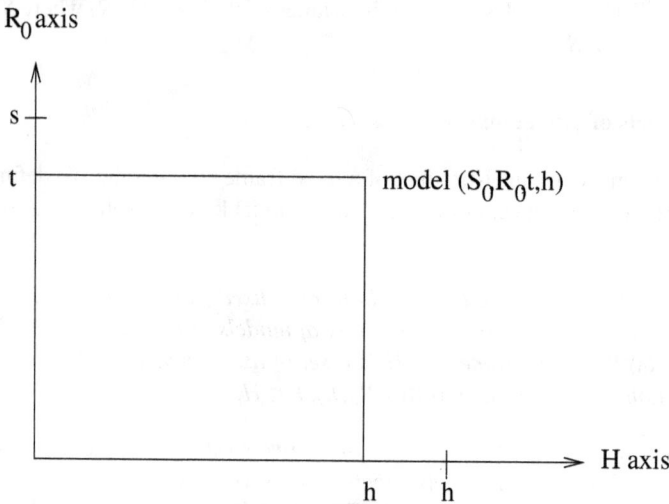

Fig. 2.4. $\square$ moves from t to s on the h vertical. $\boxdot$ moves from h to h' on the t horizontal.

See the book [144] for a wealth of material products.

2.4.3 Modal provability logic

Löb's logic for one modality $\square$ has the following axioms and rules.

1. axioms and rules of modal logic K
2. $\Diamond A \rightarrow \Diamond(A \wedge \square\neg A)$
3. $\square A \rightarrow \square\square A$

Axiom 2 says that if A is possible then it is possible for the last time. Axiom 3 says R is transitive.

It is complete for the class of all frames which are finite and acyclic (we can take all finite tree frames). The following holds.

Theorem 2.31 (fixed point theorem). *Let $\Psi(x)$ be a modal formula with the propositional variable x such that x is in the scope of a modality $\Box$. Then there is a formula ϕ which is a solution to the fixed point equation $x \leftrightarrow \Psi(x)$. Namely, we have $\vdash \phi \leftrightarrow \Psi(x/\phi)$.*

Proof. See Theorem 3.4 (page 464) of [313]. $\qquad\blacksquare$

We shall use an extension of this logic in our object level modal approach. The logic we use we call *LN*1; it is characterised by linear chains. The axiom we add for it is:

4. $\Diamond A \wedge \Diamond B \rightarrow \Diamond(A \wedge B) \vee \Diamond(A \wedge \Diamond B) \vee \Diamond(B \wedge \Diamond A)$

If we then also add the following axiom

5. $\Box\Box\Box\bot$

we get the logic of chains of length 3 (3-chains) of the form of Figure 2.5.

Fig. 2.5. A 3-chain.

The next step will be to connect these chains to labellings.

2.4.4 The metalevel approach

We use the modal logic $\mathcal{K}$ to describe the system of Definition 4. We can view (S, R) as a modal frame and view $\mathcal{L}$ as an assignment to three propositional atomic letters: q_1, q_0 and $q_?$, which we regard as constants. We have

$a \vDash q_1$ iff $\mathcal{L}(a) = \text{in}$,

$a \vDash q_0$ iff $\mathcal{L}(a) = \text{out}$, and

$a \vDash q_?$ iff $\mathcal{L}(a) = \text{undec}$.

The modality goes by the direction of "being attacked from", namely

$a \vDash \Box A$ iff for all y such that $y R a$ (that is, $(y, a) \in R$), we have $y \vDash A$.

We therefore must adopt the following axioms, written as $\Delta(q_1, q_0, q_?)$.

1. $\Box\bot \vee \Box q_0 \rightarrow q_1$ (if a is not attacked by any argument, i.e. $\Box\bot$ holds, or all attackets of a are out then a is in)
2. $\Diamond q_1 \rightarrow q_0$ (if a is attacked by an argument which is in then a is out)

3. $\Box(q_0 \vee q_?) \wedge \Diamond q_? \rightarrow q_?$ (if all the attackers of a are either out or undec and at least one attacker of a is undec then a is undec)
4. $\vdash^m (q_0 \vee q_1 \vee q_?)$, $m \geq 0$ (each argument has at least one label out, in or undec)
5. $\vdash^m (\neg(q_i \wedge q_j))$, $i \neq j$, $i, j \in \{0, 1, ?\}$, $m \geq 0$ (no argument has more than one label)

Let Δ be the above theory and let $AF = (S, R)$ be an argumentation framework and $\mathcal{L}$ be a labelling of AF. Then for any $a \in S$ it holds that $a \vDash \Delta(q_1, q_0, q_?)$, provided that
$a \vDash q_1$ iff $\mathcal{L}(a) = $ in
$a \vDash q_0$ iff $\mathcal{L}(a) = $ out
$a \vDash q_?$ iff $\mathcal{L}(a) = $ undec

Conversely if (S, R, h) is a modal model of Δ (i.e. for all $t \in S$, $t \vDash \Delta$) then it is an argumentation system with
$\mathcal{L}(a) = $ in iff $a \vDash q_1$
$\mathcal{L}(a) = $ out iff $a \vDash q_0$
$\mathcal{L}(a) = $ undec iff $a \vDash q_?$
This turns g into a complete labelling (see Theorem 2.7 and Proposition 2.8).

The above is may be a nice model but there is still not much we can do with it. Since the extensions are assignments to $q_0, q_1, q_?$ satisfying Δ we cannot directly talk about them, except for the grounded labelling. We shall see later how to deal with the other labellings, using circumscription. We shall see that to be able to fully understand our metalevel modal model and its options we should also introduce and compare with the classical logic metamodel. This we shall do in section 2.5.

Using the modality $\Box$ as above and results from [49], we can define extensions. Let E be a propositional letter. Then E defines a set of points in S. So according to our notation, $t \vDash \Box E$ iff for all s such that sRt we have $s \vDash E$.

So if E denotes a set of arguments, the $\Diamond E$ is the set of arguments attacked by E and $\Box \Diamond E$ is the set of arguments defended by E. This corresponds to Proposition 1 of [49]. We have:

1. E is stable iff $E = \neg \Diamond E = \Box \neg E$. This corresponds to Proposition 2 of [49].
2. a) E is conflict-free iff $E \rightarrow \neg \Diamond E$.
   b) E is admissible iff $E \rightarrow (\Box \Diamond E) \wedge (\Box \neg E)$ iff $E \rightarrow \Box(\neg E \wedge \Diamond E)$. This corresponds to Proposition 3 of [49].
3. E is a complete extension iff $E = \Box(\neg E \wedge \Diamond E)$. This corresponds to Proposition 5 of [49].

Using fixed points methods of Section 2.4.5 and reference [162], we can find the fixed point solutions for (1) (stable extensions) and (3) (complete extensions) for finite frames.

Altogether, the metalevel approach can be summarized as follows. Given $AF = (S, R)$

1. We view elements $a \in S$ as possible worlds and so AF becomes a model for modal logic $\mathcal{K}$ (with the modality $\Box$).
2. Labellings become assignments in the modal logic.
3. The frame of our modal logic is fixed, it is (S, R). What changes are the assignments defined by the labellings. This means we have what we call a fixed frame modal logic FF modality.
4. Properties of labellings are studied in a circumscription logic based on modal logic $\mathcal{K}$ (to be introduced in Section 2.5.

2.4.5 The object level approach

The previous view used modal logic to talk about argumentation. The now to be introduced object level approach will model argumentation from within. To introduce our point of view, let us ask what does an argumentation framework of the form $AF = (S, R)$ say to us? Here we view the arguments $a \in \mathscr{A}rgs$ as atoms in some logic. So let X be the logical content of the argumentation framework AF and labelling $\mathcal{L}$. Hence if a is in then $X \vdash a$ in some logic. If a is out then $X \vdash \neg a$ and if a is undec then we have neither. We might take the simple minded view and let $X_{\mathcal{L}} = \{a \mid \mathcal{L}(a) = \text{in}\} \cup \{\neg a \mid \mathcal{L}(a) = \text{out}\}$ and propose this as a solution. The problem, however, is that this is too simplistic because

1. It does not take into account the internal structure of AF (i.e. the attack relation)
2. It has an explicit dependence on $\mathcal{L}$

Our aim is therefore to introduce a more sophisticated approach.

We borrow the Löb modal provability logic and use a suitable extension of it, which we call $LN1$, and view the logical content of $(AF, \mathcal{L})$ as a formula $\mathcal{M}(AF, \mathcal{L})$ of $LN1$. $LN1$ is a fixed frame modal logic, the frame being a chain of 3 elements. We should therefore have (using $\boxminus$ as the modal provability operator)

1. $\mathcal{M}(AF, \mathcal{L}) \vdash \boxminus a$ if a is in
2. $\mathcal{M}(AF, \mathcal{L}) \vdash \boxminus \neg a$ if a is out
3. neither, if a is undec

We read $\boxminus$ as provability. Figure 2.6 shows how we find $\mathcal{M}(AF, \mathcal{L})$.

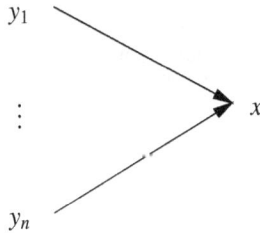

Fig. 2.6. Finding $\mathcal{M}(AF, \mathcal{L})$.

Let $x \in S$ and let $y_1, \ldots, y_n$ denote all arguments attacking it (i.e. $(y_i, x) \in R$). The labelling conditions basically say that $\mathcal{L}(x) = \text{in}$ iff for each attacker y_i it holds that $\mathcal{L}(y_i)$ is out. So if in means provable, we get $x \leftrightarrow \bigwedge_{y_i} \neg \boxminus \neg (\mathcal{M}(AF, \mathcal{L}) \wedge \neg y_i)$, so we have something like

$$\mathcal{M}(AF, \mathcal{L}) \leftrightarrow \bigwedge_{\substack{a \text{ not atttacked} \\ \text{by any } y}} \boxminus a \wedge [\bigwedge_{\substack{a \in S, x \\ \text{attacked by} \\ y_1, \ldots, y_n}} (x \leftrightarrow \bigwedge_{y_i} \neg \boxminus \neg (\mathcal{M}(AF, \mathcal{L}) \wedge \neg y_i))].$$

The exact solution formula for $M(AF, \mathcal{L})$ is given in the beginning of Section 2.6.

Thus, $M(AF, \mathcal{L})$ is obtained as a fixed point solution in a suitable modal provability logic. The fixpoint equation is generated by AF. All models of $M(AF)$ should give us all labellings of AF. A model of $M(AF)$ gives the atoms, the elements of S assignments and from the assignments we can tell which a is in, out or undec. This we will see in later sections.

Altogether, the object level approach can be summarised as follows. Let $AF = (S, R)$.

- We regard elements of S as atomic propositions in some modal provability logic $LN1$. The graph (S, R) generates a fixed point equation in the modal logic and the unique solution $M(AF)$ of this equation is a formula of modal logic representing the logical content of AF.
- The possible world models of $M(AF)$ are in one to one correspondence with all labellings $\mathcal{L}$ of AF.
- The logic we use, $LN1$ is a fixed frame modal logic.

2.4.6 The mixed approach

We saw that the metalevel approach regards arguments as possible worlds, while the object level approach regards them as propositions in a logic. The problem in both approaches is how to characterise extensions. We will now introduce a mixed approach. Consider the previous two approaches. In each of them, the fixed argumentation framework gives rise to a fixed frame modal logic. On the metalevel approach, the fixed frame is (S, R), and the set of assignments H is what all the possible labellings give us. On the object level approach the fixed frame is a chain of three elements (see Figure 2.5) and the assignments are all those assignments $\mathcal{L}$ that give us models of $M(AF)$. We shall see later that there is a one-to-one correspondance between the assignments which are models of $M(AF)$ and the labellings of AF.

In either case we get a fixed frame modal logic, one with $\square$ and the other with $\boxminus$. The mixed approach adds the modality $\boxdot$ to the existing modality and forms the universal products with $\{\square, \boxdot\}$ and with $\{\boxminus, \boxdot\}$ respectively.

To show how the mixed approach works, consider the semi-stable semantics of Table 2.1. For this we need to say that our labelling has a minimal undec. Let us explain the approach for the metalevel mixed case. Let g_{s-s} be the assignment arising from the semi-stable labelling. Then there is no other h for a model such that $h(q_2) \subsetneqq g_{s-s}(q_?)$. We can express this in the universal model.

The grounded extension for example is characterised by minimal in, so we obtain $\vDash q_1 \wedge \boxdot q_1$.

2.5 The metalevel approach

We saw that the metalevel approach uses modal logic $\mathcal{K}$ to talk about argumentation frameworks of the form $AF = (S, R)$. The arguments become the possible worlds of a Kripke model, the attack relation becomes the accessibility relation and an associated labelling $\mathcal{L}$ gives rise to and assignment. By varying $\mathcal{L}$ we get a fixed frame modal logic based on $\mathcal{K}$. To fully appreciate the advantages and limitations of this approach we need to compare it to two other approaches, the Logic Programming and the classical logic ones.

2.5.1 The classical logic metalevel approach

We start with classical logic with equality ("=") and three modal predicates Q_1, Q_0 and $Q_?$ and a binary relation R. Given an argumentation framework $AF = (S, R)$ and a labelling $\mathcal{L}$ we construct the associated model of classical logic. We take S as the domain, R as the relation R and use $\mathcal{L}$ to get the extensions of tree predicates Q_0, Q_1 and $Q_?$ as follows.

1. $a \in Q_1$ iff $\mathcal{L}$ labels a as in
2. $a \in Q_0$ iff $\mathcal{L}$ labels a as out
3. $a \in Q_?$ iff $\mathcal{L}$ labels a as undec

Consider the following classical theory $\Delta(R, Q_0, Q_1, Q_?)$.

1. $\forall x(Q_0(x) \lor Q_1(x) \lor Q_?(x))$
2. $\neg \exists x(Q_i(x) \land Q_j(x))$ for $i \neq j$, $i, j \in \{0, 1, ?\}$
3. $\forall y(\forall x(xRy \to Q_0(x)) \to Q_1(y))$
4. $\forall y(\exists x(xRy \land Q_1(x)) \to Q_0(y))$
5. $\forall y(\forall x(xRy \to (Q_0(x) \lor Q_?(x))) \land \exists x(xRy \land Q_?(x)) \to Q_?(y))$

Any model of Δ with domain D defines an argumentation framework with the set of arguments $S = D$, and the attack relation is R and $\mathcal{L}$ is what we obtain from the elements satisfying the respective predicates Q_0, Q_1 and $Q_?$. Notice that we are not using "=".

If we want to characterise any specific argumentation framework $AF = (S, R)$ we need equality and we need constant names for every element of S. We write the following additional axioms $\theta(AF)$.

6. $\forall x(\bigvee_{a \in S} x = \underline{a})$
7. $\bigwedge_{a,b \in S, a \neq b} \underline{u} \neq \underline{b}$
8. $\bigwedge_{a,b \in R} \underline{a} R \underline{b}$

We want to see how to characterise the different extensions of the argumentation frameworks obtained in classical logic as models of Δ. This means for example that we want to say that the predicate Q_1 is minimal in the model (in order to obtain the grounded extension) or for example that the predicate $Q_?$ is minimal to get the semistable extension, or that the predicate Q_1 is maximal to get the preferred extension. The concept "Q is maximal" or "Q is minimal" is not first order. We therefore need second order formula to express this, an approach that is know as predicate circumscription of John McCarthy; the subject has a very well developed theory.

So the theory we want is for example

$$\Delta_{semi-stable}(R, Q_0, Q_1, Q_?) = \Delta(R, Q_0, Q_1, Q_?) \land \text{ "}Q_? \text{ is minimal"}.$$

Any model of $\Delta_{semi-stable}$ yields an argumentation framework based on the domain of the model, with $\mathcal{L}$ (derived from Q_1, Q_1, $Q_?$) which is semi-stable.

Using circumscription we write

$$\Delta_{semi-stable} = \Delta(AF, Q_0, Q_1, Q_?) \wedge \forall Q_0 \forall Q_1 \forall Q_?^*((Q_?^* \subsetneq Q_?)$$
$$\rightarrow \neg \Delta(AF, Q_0, Q_1, Q_?^*))$$

where $X \subsetneq Y$ is defined as $\forall y(X(y) \rightarrow Y(y)) \wedge \exists z(Y(z) \wedge \neg X(z))$.

Similarly to maximise we use $\supseteq$ in the circumscription formulas for Q_1 and Q_1^*. According to our book [155] it is possible to eliminate such second order quantifiers under certain circumstance. We need to check whether this can be done in our case.

When we deal with a specific finite argumentation framework AF, our set of axioms is $\Delta(AF) = \Delta \cup \theta(AF)$. In this case characterising some of the extensions is easier. For one thing we can use provability to characterise some extensions and for others circumscription becomes first order.

The above axioms $\Delta(AF) = \Delta \cup \theta(AF)$ can immediately characterise the grounded extension of AF as the set of all x such that $\Delta(AF) \vdash Q_1(x)$.

The preferred semantics is characterised by maximal in, which implies that Q_1 is maximal, so $\neg Q_1$ is minimal. So let $\overline{Q_1}^{min} = \{x \mid \Delta(\mathcal{L}) \vdash \neg Q_1\}$ and add the axiom $\forall(Q_1(x) \leftrightarrow \overline{Q_1}^{min}(x))$. The problem with the above definition is that $Q_?^{min}$ and $\overline{Q_1}^{min}$ are defined using provability, which is outside the logic itself. To bring this in, we need to use circumscription as we did before. However, since for a given AF, S is finite, we can turn the second order quatlifier $\forall X$ into a big conjunction by enumerating all possible subsets $B \subseteq S$. We get $\forall X \Psi[y \in X]$ is replaced by $\bigwedge_{B \subseteq S} \Psi[\bigvee_{a \in B} y = \underline{a}]$. This makes circumscription first order for AF fixed.

2.5.2 The logic programming metalevel approach

The logic programming metalevel approach has been worked out in [340]. The approach is similar to the classical logic approach except that we use logic programming to represent the argumentation framework. The extensions correspond to the models of the corresponding logic program. For each argument x we write the clause $x \leftarrow \neg y_1, \ldots, \neg y_n$ ($n \geq 0$), where $\{y_1, \ldots, y_n\}$ is the set of all arguments attacking x. We get a resulting logic program satisfying

1. each atom is the head of exactly one clause
2. the bodies of clauses consist of weakly negated atoms only

For more information, we refer to [340] and to Chapter 4.

2.6 The modal provability object level approach

Let $AF = (S, R)$ be an argumentation framework. Let x be an argument whose set of attackers is $y_1, \ldots, y_n$, as was illustrated in Figure 2.6.

We construct the following modal formula in the modal logic $LN3$ of provability.

$$\mathcal{M}(AF) = G(\Box \bot \vee \bigwedge_{y \in S \text{ and } y \text{ has attackers } y_i} y \leftrightarrow \bigwedge_i \Diamond \neg y_i) \wedge$$
$$\bigwedge_{y \in S \text{ and } y \text{ is not attacked}} Gy$$

Here, GA stands for $A \wedge \Box A$. The logic $LN3$ has the following axioms.

1. all substitution instances of classical tautologies

2. all $K4$ axioms
   - $\Box(A \to B) \to (\Box A \to \Box B)$
   - $\Box A \to \Box\Box A$
   - $\dfrac{\vdash A}{\vdash \Box A}$
3. all substitution instances of Löb's axioms
   $$\Diamond A \to \Diamond(A \wedge \Box\neg A)$$
4. linearity axiom
   $$\Diamond A \wedge \Diamond B \to \Diamond(A \wedge B) \vee \Diamond(A \wedge \Diamond B) \vee \Diamond(B \wedge \Diamond A)$$
5. 3-chain axiom
   $$\Box\Box\Box\bot$$
6. axioms for atoms q only
   - $q \to \Box(\neg q \to \Box q)$
   - $\Box(\Box\bot \vee q) \leftrightarrow \Box q$
   - $\Box(\Box\bot \vee \neg q) \leftrightarrow \Box\neg q$

Lemma 2.32.

1. *The logic LN3 is complete for all 3-chain models whose assignment to atoms has one of the types of Table 2.2.*
2. *Any 3-chain model of LN1 allows for the atoms to have one of types 1, 0 or 2 assignments.*

Proof. The axioms of **LN1** force the following for atomic q in the chain $1 < 2 < 3$:

- $1 \vDash q$ and $2 \vDash q \Rightarrow 3 \vDash q$
- $1 \vDash\sim q$ and $2 \vDash\sim q \Rightarrow 3 \vDash\sim q$
- $1 \vDash q$ and $2 \vDash\sim q \Rightarrow 3 \vDash q$.

We need to prove that the option "$1 \vDash q$ and $2 \vDash\sim q$ and $3 \vDash\sim q$" cannot arise. Consider a point y such as y is being attacked. (If not then Gy is in $M(AF)$ and so y is of type 1.) We have $G(\Box\bot \vee (y \leftrightarrow \bigwedge \Diamond \sim y_i))$ holds in the model. Hence $y \leftrightarrow \bigwedge_i \Diamond \sim y_i$ holds at points 1 and 2. We distinguish several cases:

1. If for any j, if $y_j = \bot$ at 3 then $\Diamond \sim y_j$ is true at 1 and 2 and it plays no role in the conjunction. If for all y_j, $3 \vDash\sim y_j$ then $y = \top$ at 1 and 2 and y is of type 1.
2. Some y_j are $\top$ at 3. Then $\Diamond \sim y_j$ is $\bot$ at 2 and $y = \bot$ at 2.
3. We now check whether y_j is true or false at 2. If for some j, $y_j = \top$ at 2 then $\Diamond \sim y_j = \bot$ at 1 and $y = \bot$ at 1 and hence $y = \bot$ everywhere, and hence y is of type 0.

   If for all $j, y_j = \bot$ at 2 then $y = \top$ at 1 and we have that $y = \top$ at 1 and $y = \bot$ at 2 and we must have by the axiom 4.1 that $3 \vDash y$ and y is of type 3.

Theorem 2.33. *Let $AF = (S, R)$ be an argumentation framework and let $M(AF)$ be the modal sentence as defined. Then:*

1. *Any complete labelling $\mathcal{L}$ for AF gives rise to a model of $M(AF)$, where for any q, the correspondence is as in Table 2.2.*
   - *q is assigned type 1 assignment iff $\mathcal{L}(q) = 1$*
   - *q is assigned type 0 assignment iff $\mathcal{L}(q) = 0$*

world	type 1	type 0	type 2
3	⊤	⊥	⊤
2	⊤	⊥	⊥
1	⊤	⊥	⊤

Table 2.2. 3-chain models.

- q is assigned type 2 assignment iff $\mathcal{L}(q) = ?$
2. Let $M_1, M_2, M_3, \ldots$ be all the 3-chain models of $\mathcal{M}(AF)$. Then $\{M_i\}$ are all the complete labellings for AF, where these complete labellings are obtained as in Table 2.2.

Corollary 2.34. $\mathcal{M}(AF)$ characterises all the extensions of AF through the modal logic LN3.

In particular, we obtain the following theorem.

Theorem 2.35. x is in the grounded extension of AF iff $\mathcal{M}(AF) \vdash_{LN3} Gx$.

Proof. This holds because the grounded extension is the minimal complete extension [114].

Since Gx is in every model of $\mathcal{M}(AF)$ we cannot characterise other extensions, e.g. preferred extensions, in a similar way. We need the mixed approach with additional modalities.

See Chapter 5.

2.7 The equational approach

The Equational approach to classical logic has its conceptual roots in the 19th century following the algebraic equational approach to logic by George Boole [64], Louis Couturat [103] and Ernst Schroeder [301].

The equational algebraic approach was historically followed, in the first half of the 20th century by the Logical Truth (Tautologies) approach supported by giants such as G. Frege, D. Hilbert, B. Russell and L. Wittgenstein. In the second half of the twentieth Century the new current approach has emerged, which was to study logic through it consequence relations, as developed by A. Tarski, G. Gentzen, D. Scott and (for non-monotonic logic) D. Gabbay.

This section describes briefly the equational approach to argumentation. Full details in Chapter 10.

Let (S, R_A) be a Dung network. So $R_A \subseteq S^2$ is the attack relation. We are looking for a function $\mathbf{f} : S \mapsto [0, 1]$ assigning to each $a \in S$ a value of $0 \leq \mathbf{f}(a) \leq 1$ such that the following holds.

1. $(S, R_A, \mathbf{f})$ satisfies the following equations for some family of functions $\{\mathbf{h}_a\}, a \in S$:
   a) If a is not attacked (i.e. $\neg \exists x (x R_A a)$) then $\mathbf{f}(a) = 1$.

b) If $x_1, \ldots, x_n$ are all the attackers of a (i.e. $\bigwedge_{i=1}^{n} x_i R_A a \wedge \forall y(y R_A a \rightarrow \bigvee_{i=1}^{n} y = x_i)$) then we have that $\mathbf{f}(a) = \mathbf{h}_a(\mathbf{f}(x_1), \ldots, \mathbf{f}(x_n))$.

Let us take, for example the same $\mathbf{h}_a = \mathbf{h}$ for all a and let

$$\mathbf{h}(\mathbf{f}(x_1), \ldots, \mathbf{f}(x_n)) = \prod_{i=1}^{n}(1 - \mathbf{f}(x_i))$$

The above equation we shall call Eq_{inverse}. We shall define other possible equations later on.

Thus we get

Eq_{inverse} for the function $\mathbf{f}$:

$$\mathbf{f}(a) = \prod_{i=1}^{n}(1 - \mathbf{f}(x_i)).$$

2. For any Caminada labelling λ of (S, R_A), there exists an $(S, R_A, \mathbf{f})$ such that

$$E_{\mathbf{f}} = \begin{cases} \text{If } \lambda(a) = \text{in then } \mathbf{f}(a) = 1 \\ \text{If } \lambda(a) = \text{out then } \mathbf{f}(a) = 0 \\ \text{If } \lambda(a) = \text{undecided then don't care what } \mathbf{f}(a) \text{ is} \\ \quad \text{provided it satisfies the equations.} \end{cases}$$

Condition (1) above reads $\lambda(a) = \text{in as } \mathbf{f}(a) = 1$ and $\lambda(a) = \text{out as } \mathbf{f}(a) = 0$.

Therefore the equation

$$\mathbf{f}(a) = \prod_{i=1}^{n}(1 - \mathbf{f}(x_i))$$

ensures that:

If one of x_i (x_i are the attackers of a) is in then a is out.
If all the attackers are out then a is in.

The question is what happens with the undecided cases. Here we have condition (2).

Any Dung extension can have a corresponding function $\mathbf{f}$ which agrees with the "in" and "out", though may be also more specific about the undecided.

So if the Dung extension says I don't know, the function $\mathbf{f}$ can say whatever it wants, provided it satisfies the equations.

Note that we can have a different function $\mathbf{h}$. Time to give a formal definition.

Definition 2.36 (Possible equational systems). *Let (S, R_A) be a networks and let a be a node and let $x_1, \ldots, x_n$ be all of its attackers.*

We list below several possible equational systems, we write $Eq(\mathbf{f})$ to mean the equational system Eq applied to $\mathbf{f}$:

1. $Eq_{\text{inverse}}(\mathbf{f})$

$$\mathbf{f}(a) = \prod_{i}(1 - \mathbf{f}(x_i))$$

2. $Eq_{\text{geometrical}}(\mathbf{f})$

$$\mathbf{f}(a) = [\prod_i (1 - \mathbf{f}(x_i))]/[\prod_i (1 - \mathbf{f}(x_i)) + \prod_i x_i].$$

We call this equation $Eq_{\text{geometrical}}$ because it is connected to the projective geometry Cross Ratio, see our 2005 paper [37] (reproduced in Chapter 9).

3. $Eq_{\text{max}}(\mathbf{f})$

$$\mathbf{f}(a) = 1 - \max(\mathbf{f}(x_i)).$$

4. $Eq_{\text{suspect}}(\mathbf{f})$

We shall see the difference in the examples. In fact we shall see that this new function gives exactly the Caminada labelling.

Let us further introduce a fourth system of equations which we call $Eq_{\text{suspect}}(\mathbf{f})$:

$$\mathbf{h}_a(\mathbf{f}(x_1), \dots, \mathbf{f}(x_n)) = \prod_i (1 - \mathbf{f}(x_i)), \ \text{if } \neg aR_A a \text{ holds}$$

and

$$\mathbf{h}_a(\mathbf{f}(x_1), \dots, \mathbf{f}(x_n)) = \mathbf{f}(a) \prod_i (1 - \mathbf{f}(x_i)), \ \text{if } aR_A a \text{ holds}.$$

Example 2.37. Let us do an example using all four options for equations, namely $Eq_{\text{geometrical}}, Eq_{\text{inverse}}, Eq_{\text{max}}$ and Eq_{suspect}.

Consider Figure 2.7. We are looking for $\mathbf{f}$ solving the equations. Let $\mathbf{f}(a) = \alpha, \mathbf{f}(b) = \beta, \mathbf{f}(c) = \gamma$.

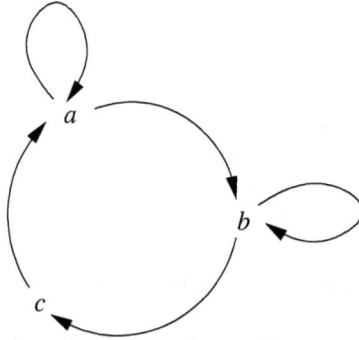

Fig. 2.7.

I We use $Eq_{inverse}$:

The equations are

1. $\alpha = (1 - \alpha)(1 - \gamma)$
2. $\beta = (1 - \beta)(1 - \alpha)$
3. $\gamma = 1 - \beta$

There are programs like Maple which can solve the equations of this sort and give all the solutions. We used one and got

$$\alpha = 1 - \frac{\sqrt{2}}{2}$$
$$\beta = \sqrt{2} - 1$$
$$\gamma = 2 - \sqrt{2}$$

The interest in this case is that we are getting all kinds of values which shows that these equations are sensitive to the nature of the loops involved!

II. We use Eq_{max}:

The equations are

1. $\alpha = 1 - \max(\alpha, \gamma)$
2. $\beta = 1 - \max(\beta, \alpha)$
3. $\gamma = 1 - \beta$

The only solution in this case is $\alpha = \beta = \gamma = \frac{1}{2}$.

III. We use $Eq_{suspect}$:

The equations are

1. $\alpha = \alpha(1 - \alpha) \cdot (1 - \gamma)$
2. $\beta = \beta(1 - \beta)(1 - \alpha)$
3. $\gamma = 1 - \beta$.

The solution is $\alpha = 0, \beta = 0, \gamma = 1$.

IV. We use $Eq_{geometrical}$.
The equations are:

1. $\alpha = \frac{(1-\alpha)(1-\gamma)}{(1-\alpha)(1-\gamma)+\alpha\gamma}$
2. $\beta = \frac{(1-\alpha)(1-\beta)}{(1-\alpha)(1-\beta)+\alpha\beta}$
3. $\gamma = 1 - \beta$.

The only solution is $\alpha = \beta = \gamma = \frac{1}{2}$.:

Example 2.38 (Comparing Eq_{max} and $Eq_{inverse}$). We shall show that these two equational systems may not yield the same extensions. the network is described in Figure 2.8.

Extensions according to Eq_{max}.
Let us compute the equations according to Eq_{max} and their possible solutions.
    The equations are (we write "x" instead of $\mathbf{f}(x)$):

1. $a = 1 - b$
2. $b = 1 - \max(a, b)$
3. $c = 1 - \max(b, e)$
4. $d = 1 - c$
5. $e = 1 - d.$

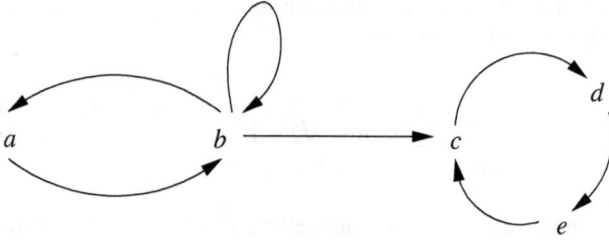

Fig. 2.8.

We get two extensions

1. $\{a\}, (a = 1, b = 0, c = d = e = \frac{1}{2})$
2. $\phi, (a = \frac{1}{2}, b - \frac{1}{2}, c = d = e = \frac{1}{2})$

Compare this result with Theorem 2.42 below.

We now deal with Figure 2.8 using Eq_{inverse}. The equations are:

1. $a = 1 - b$
2. $b = (1 - b)(1 - a)$
3. $c = (1 - b)(1 - e)$
4. $d = 1 - c$
5. $e = 1 - d.$

We can have only one extension

$$\{a\}, (a = 1, b = 0, c = d = e = \frac{1}{2}).$$

Let us now formally describe our equational approach. Conceptually the nodes and the Equations attached to them is the network and the solutions to the equations are the complete extensions, as we have seen in the examples of Section 1.

Definition 2.39 (Real equational networks).

1. An argumentation base is a pair (S, R_A) where $S \neq \varnothing$ is a finite set and $R_A \subseteq S^2$.
2. A real equation function in k variables $\{x_1, \ldots, x_k\}$ over the real interval $[0, 1]$ is a continuous function $\mathbf{h} : [0, 1]^k \mapsto [0, 1]$ such that
    a) $\mathbf{h}(0, \ldots, 0) = 1$
    b) $\mathbf{h}(x_1, \ldots, 1, \ldots, x_k) = 0$
    Sometimes we also have condition (c) below, as in ordinary Dung networks, but not always.

c) $\mathbf{h}(x_1, \ldots, x_k) = \mathbf{h}(y_1, \ldots, y_k)$ where $\{y_j\} = \{x_i\}$ are premutations of each other.

3. An equational argumentation network over $[0, 1]$ has the form $(S, R_A, \mathbf{h}_a)$, $a \in S$ where

   a) (S, R_A) is a base

   b) For each $a \in S$, $\mathbf{h}_a$ is a real equation function.

   c) If $\neg \exists y(y R_A a)$ then $\mathbf{h}_a \equiv 1$.

4. An extension is a function $\mathbf{f}$ from S into $[0, 1]$ such that the following holds:

   • $\mathbf{f}(a) = 1$ if $\neg \exists y(y R_a a)$

   • If $\{x_1, \ldots, x_k\}$ are all the elements in S such that $x_i R_A a$, then $\mathbf{h}_a$ is a k variable function and $\mathbf{f}(a) = \mathbf{h}_a(\mathbf{f}(x_1), \ldots, \mathbf{f}(x_k))$.

Theorem 2.40 (Existence theorem). *Let $(S, R_A, \mathbf{h}_a)$, $a \in S$ be a network as in Definition 2.39. Then there exists an extension function $\mathbf{f}$ satisfying (4) of Definition 2.39.*

Lemma 2.41. *Let (S, R_A) be a Dung argumentation network. Let $\lambda : S \mapsto \{in, out, undecided\}$ be a legitimate Caminada labelling, yielding an extension E_λ. Consider the functions $\mathbf{h}_a, a \in S$ as follows:*

1. $\mathbf{h}_a \equiv 0$ if $\lambda(a) = out$

2. $\mathbf{h}_a \equiv 1$ if $\lambda(a) = in$.

3. $\mathbf{h}_a$ arbitrary real equation function, otherwise.

Then there exists, by Theorem 2.40 an extension function $\mathbf{f}$ such that for all $a \in S$

$$\mathbf{f}(a) = \mathbf{h}_a(\mathbf{f}(x_1), \ldots, \mathbf{f}(x_k)),$$

where $\{x_i\}$ are all the nodes attacking a.

Note that we have argued in these examples that we get a good refinement of the undecided allocations.

To get exactly the Caminada labelling, we use the next theorem, Theorem 2.42.

Theorem 2.42 (Caminada labelling functions and Eq_{max}). *Consider the function*

$$\mathbf{h}_{max}(x_1, \ldots, x_n) = 1 - \max(x_1, \ldots, x_n).$$

This function is continuous in $[0, 1]^n \mapsto [0, 1]$ and therefore falls under Definition 2.39.

1. *Let $(S, R_A, \mathbf{h}_{max})$ be an equational network with $\mathbf{h}_{max}$ and let $\mathbf{f}$ be an extension, as in item 4 of Definition 2.39. Define a labelling $\lambda_\mathbf{f}$ dependent on $\mathbf{f}$ as follows*

$$\lambda_\mathbf{f}(a) = \begin{cases} in \ if \ \mathbf{f}(a) = 1 \\ out \ if \ \mathbf{f}(a) = 0 \\ undecided \ if \ 0 < \mathbf{f}(a) < 1 \end{cases}$$

Then $\lambda_\mathbf{f}$ is a proper Caminada extension of (S, R_A).

2. *Let λ be a Caminada extension for (S, R_A). Let $\mathbf{f}_\lambda$ be the real number function defined as follows*

$$\mathbf{f}_\lambda(a) = \begin{cases} 1 \ if \ \lambda(a) = in \\ 0 \ if \ \lambda(a) = out \\ \frac{1}{2} \ if \ \lambda(a) = undecided. \end{cases}$$

Then $\mathbf{f}_\lambda$ is a proper equational extension for $(S, R_A, \mathbf{h}_{max})$, i.e. $\mathbf{f}_\lambda$ solves the equations $\mathbf{f}_\lambda(a) = 1 - \max(\mathbf{f}_\lambda(x_1), \ldots, \mathbf{f}_\lambda(x_n))$ where x_i are all the attackers of a.

See Chapter 10.

2.8 Discussion

Grossi in [207] uses modal logic to represent argumentation frameworks. His approach is metalevel. We use (in Section 2.5.1) classical logic to talk about argumentation frameworks and use circumscription to define the various extensions. Grossi uses modal logic to describe the argumentation frameworks. He needs two modalities, one to go with the attack relation and one to go with the converse of the attack relation (like two temporal logic modalities). He also uses a universal modality and to get the extensions he needs μ-calculus on top. Our approach, on the other hand, is to use classical logic with circumscription to do the job. We would not be surprised if the approach of Grossi could simulate the relevant classical logic with circumscription, since all the ingredients are present. Note that our use of modal logic in section 2.6 is object level and is completely different from its use as a metalevel tool.

As was mentioned in Section 2.3 the approach of argument labellings can be traced back to Pollock, who in his 1995 book [284] described his OSCAR system in terms of labellings. As explained in [223], Pollock's approach essentially boils down to preferred semantics. The labelling approach of Jakobovits and Vermeir [223] is aimed not so much at describing Dung's original semantics but rather to defining additional semantics driven by what they perceive to be problems in Dung's original semantics. Caminada first described complete semantics in terms of labellings in [75] and showed how this can be used to provide labelling based descriptions of other semantics as well. This approach was then applied in [76] and [334] to provide labelling-based algorithms for computing particular argumentation sets and extensions.

Besnard and Doutre [49] examine how complete and preferred semantics can be expressed in terms of set theoretical equations, but do not provide a logical account of these semantics.

A connection between the current work and the topic of linear programming equations can be found in Chapter 7. We did use some of the equations of Chapter 7 in Section 2.4.4.

3

Annotation Approach to Argumentation

3.1 Introduction to annotation theories and related problems

In the current chapter we consider theories with vocabulary containing a number of binary and unary relation symbols. Binary relation symbols represent labeled edges of a graph and unary relations represent unique annotations of the graph's nodes. We call such theories *annotation theories*. They can be used in many applications, including the formalization of argumentation, approximate reasoning, semantics of logic programs, graph colouring, etc. Given an annotation theory, we address the following problems, where only finite models are considered:

- *satisfiability*: given a finite set D, is there a model for the theory with D as the underlying domain?
- *querying problem*: given a graph with edges represented by binary relations, are there annotations satisfying the theory?
- *specification of preferred models*: how to use circumscription on chosen annotations to specify preferred models and how to reduce second-order circumscription formulas to first-order or fixpoint logic?
- *model checking problem*: given a relational structure with edges and annotations, does it satisfy the circumscribed annotation theory?

These problems are of second-order nature. The methodology we use depends then on specifying them in the second-order logic and then on eliminating second-order quantifiers.[1] An application of quantifier elimination methods applied in this chapter, if successful, can result in:

- a formula of the first-order logic, validity of which (over finite models) is in LOGSPACE and therefore also in PTIME. Here we apply the DLS algorithm of [112], based on the Ackermann lemma (see Lemma [11]); also the SCAN algorithm of [135] can be used here;
- a formula of the fixpoint logic, validity of which (over finite models) is in PTIME. Here one can apply our Theorem 3.22, which substantially extends existing direct methods, including the theorem of Nonnengart and Szałas [272], theorem of Kachniarz and Szałas [226] and Ackermann's lemma [11] (quoted as Theorems 3.14, Theorem 3.13 and Lemma 3.15 in Section 3.5).

---

[1] For an overview of known second-order quantifier elimination techniques see [155].

Therefore Dung's argumentation theory, when successful, gives us also tractable algorithms for problems we address.

The main contribution of this chapter depends on providing a general second-order quantifier elimination result (Theorem 3.22), results for reducing circumscribed theories (Lemmas 3.27, 3.28), including annotation theories as well as results related to specific theories: Dung's argumentation theory[2] [114] (Section 3.6.3), a theory related to approximate reasoning (Section 3.6.4) and a theory formalizing a semantics of logic programs with negation, derived from considerations of [81], closely related to stable models [200, 1] (Section 3.6.5).

Complexity results are valid in the case of finite models only. On the other hand, the provided quantifier elimination techniques are not restricted to finite models nor to annotation theories. To make quantifier elimination possible, we define the notion of stratification of the considered theories, generalizing the corresponding idea known from logic programming. Theorem 3.22 works for any stratified theories. It is worth emphasizing that no existing up to now direct second-order quantifier elimination methods is successful for annotation theories (see the discussion provided in Section 3.6.1). Also resolution-based methods fail when axioms are recursive, which takes place in theories considered in this chapter.

The chapter is structured as follows. In Section 3.2 we recall some well-known notions used in the chapter. Section 3.3 introduces the concept of annotation theories and illustrates them by Dung's argumentation theory, a theory related to approximate reasoning as well as by a formalization of semantics of logic programs. In Section 3.4 we show that the satisfiability problem and the querying problem are NPTIME-complete by noticing that graph colorability can be formulated as an annotation theory. We also show that model checking problem for circumscribed annotation theories is co-NPTIME complete. Then, in Section 3.5, we provide second-order quantifier elimination results. Section 3.6 is devoted to elimination of second-order quantifiers in the context of circumscribed annotation theories. Finally, Section 3.7 concludes the chapter.

3.2 Preliminaries

3.2.1 Basics

Through the chapter we use the language of classical first- and second-order logic without function symbols.[3] We assume the standard first- and second-order semantics.

By a *literal* we understand a first-order formula of the form $R(\ldots)$ or $\neg R(\ldots)$, where R is a relation symbol. A formula A is in the *negation normal form* if it uses no propositional connectives other than $\neg, \vee, \wedge$ and the negation sign $\neg$ does not occur in A outside of literals. It is well-known that every classical first- and second- order formula can equivalently be transformed into a formula in negation normal form.

---

[2] Phan's argumentation theory is perhaps better known as "Dung's argumentation theory" and frequently cited using "Dung" as the author's family name. In fact, "Dung" is the first name and the family name is "Phan".

[3] We do not use function symbols as they do not appear in theories we deal with. Also, this allows us to use deductive databases [1, 121, 221] as a computational machinery.

Let $A(R)$ be a formula and $B(R)$ be a formula in the negation normal form equivalent to $A(R)$. An *occurrence of R is positive in $A(R)$*, if the corresponding occurrence of R in $B(R)$ is not preceded by $\neg$. An *occurrence of R is negative in $A(R)$*, if the corresponding occurrence of R in $B(R)$ is of the form $\neg R$. Formula $A(R)$ is *positive w.r.t. R* if all occurrences of R in A are positive. It is *negative w.r.t. R* if all occurrences of R in A are negative.[4]

Writing $A(\bar{X}, \bar{y})$, we mean that A contains variables $\bar{X}$ and $\bar{y}$, but we do not exclude other arguments.

If $M = \langle D, \bar{R} \rangle$ is a relational structure and v is a valuation of variables in D then we write $M, v \models A$ to denote that A is true in M under the valuation v. We write $M \models A$ to denote that A is true in M under all valuations of free variables occurring in A. If $\bar{R}$ is empty then we write $D \models A$ rather than $\langle D \rangle \models A$.

By $A(a,\ldots)^{\lambda \bar{x}_1.e_1,\ldots,\lambda \bar{x}_k.e_k}_{\lambda \bar{y}_1.f_1,\ldots,\lambda \bar{y}_k.f_k}$ we understand the expression obtained from A in such a way that for any $1 \le i \le k$, all occurrences of e_i of the form $e_i(\bar{a})$ are replaced by $f_i(\bar{y}_i)$, where $\bar{y}_i$ itself is replaced by $\bar{a}$. When λ-expression is a relation symbol applied to some arguments, say $P(\bar{z})$, then we write $P(\bar{z})$ rather than $\lambda \bar{z}.P(\bar{z})$. For example,

$$(P(s) \vee R(t))^{P(x),\ R(y)}_{\lambda u.(Q(a,u) \wedge Q(u,b)),\ S(z,z)} = (Q(a,s) \wedge Q(s,b)) \vee S(t,t).$$

3.2.2 Circumscription

In what follows we also use circumscription [247, 241, 112], which is basically a technique for minimizing chosen predicates with some other allowed to vary and all other fixed. Let us now formally define this concept.

Definition 3.1. *Let $\bar{P} = \langle P_1,\ldots,P_k \rangle, \bar{S} = \langle S_1,\ldots,S_m \rangle$ be disjoint tuples of relation symbols, and let $T(\bar{P}, \bar{S})$ be a first-order formula.*

The second-order circumscription *of $\bar{P}$ in $T(\bar{P}, \bar{S})$ with variable $\bar{S}$, written $\mathrm{Circ}_{\downarrow}(T; \bar{P}; \bar{S})$, is the second-order formula*

$$T(\bar{P}, \bar{S}) \wedge$$
$$\forall \bar{X} \forall \bar{Y} \Big\{ [T(\bar{P}, \bar{S})^{\bar{P},\bar{S}}_{\bar{X},\bar{Y}} \wedge \bigwedge_{i=1}^{k} \forall \bar{x}_i [X_i(\bar{x}_i) \to P_i(\bar{x}_i)]] \to \qquad (3.1)$$
$$\bigwedge_{i=1}^{k} \forall \bar{x}_i [P_i(\bar{x}_i) \to X_i(\bar{x}_i)] \Big\},$$

where $\bar{X}$ and $\bar{Y}$ are tuples of relational variables of the same arities as those in $\bar{P}$ and $\bar{S}$, respectively. □

We will also need a dual form of circumscription, where some predicates are maximized rather than minimized.

Definition 3.2. *Let $\bar{P}, \bar{S}$ and $T(\bar{P}, \bar{S})$ be as in Definition 3.1. The* dual second-order circumscription *of $\bar{P}$ in $T(\bar{P}, \bar{S})$ with variable $\bar{S}$, written $\mathrm{Circ}^{\uparrow}(T; \bar{P}; \bar{S})$, is the second-order formula*

---

[4] Observe that formula in which R does not occur is both positive and negative w.r.t. R.

$$T(\bar{P}, \bar{S}) \wedge$$

$$\forall \bar{X} \forall \bar{Y} \Big\{ [T(\bar{P}, \bar{S})^{\bar{P}, \bar{S}}_{\bar{X}, \bar{Y}} \wedge \bigwedge_{i=1}^{k} \forall \bar{x}_i [P_i(\bar{x}_i) \rightarrow X_i(\bar{x}_i)]] \rightarrow \qquad (3.2)$$

$$\bigwedge_{i=1}^{k} \forall \bar{x}_i [X_i(\bar{x}_i) \rightarrow P_i(\bar{x}_i)] \Big\}. \qquad \square$$

The class of all models of a theory T will be denoted by $mod(T)$. We assume that the class consists of relational structures of the form $M = \langle D^M, \langle R_i^M \rangle_{i \in I} \rangle$.

The semantics of circumscription is based on the concept of sub-models (see [240, 241]) defined as follows.

Definition 3.3. *Let $\bar{P}$, $\bar{S}$ and $T(\bar{P}, \bar{S})$ be as in Definitions 3.1 and 3.2. Let M and N be models of T. We say that M is a $(\bar{P}, \bar{S})$-submodel of N, written $M \leq^{(\bar{P}, \bar{S})} N$, iff*

1. $D^M = D^N$
2. $R^M = R^N$, *for any relation symbol R not in $\bar{P} \cup \bar{S}$*
3. $R^M \subseteq R^N$, *for any relation symbol R in $\bar{P}$.* $\qquad \square$

We write $M <^{(\bar{P}, \bar{S})} N$ when $M \leq^{(\bar{P}, \bar{S})} N$, but not $N \leq^{(\bar{P}, \bar{S})} M$. A model M of T is $(\bar{P}, \bar{S})$-*minimal* iff T has no model N such that $N <^{(\bar{P}, \bar{S})} M$. It is $(\bar{P}, \bar{S})$-*maximal* iff T has no model N such that $M <^{(\bar{P}, \bar{S})} N$. We also write $mod_{\downarrow}^{(\bar{P}, \bar{S})}(T)$ to denote the class of all $(\bar{P}, \bar{S})$-minimal models of T and $mod^{\uparrow(\bar{P}, \bar{S})}(T)$ to denote the class of all $(\bar{P}, \bar{S})$-maximal models of T. The semantics of circumscription is now given by

$$mod(Circ_{\downarrow}(T; \bar{P}; \bar{S})) = mod_{\downarrow}^{(\bar{P}, \bar{S})}(T) \qquad (3.3)$$

$$mod(Circ^{\uparrow}(T; \bar{P}; \bar{S})) = mod^{\uparrow(\bar{P}, \bar{S})}(T). \qquad (3.4)$$

3.2.3 Simultaneous fixpoints

We will also use the notion of simultaneous fixpoints (see, e.g., [1, 121, 221] for a detailed exposition of the theory of fixpoints and their applications as database queries).

Let $\bar{Q} = \langle Q_1, \ldots, Q_k \rangle$ be a tuple of relation symbols and $A_i(\bar{Q}, \bar{x}_i, \bar{y}_i)$, for $i = 1, \ldots, k$, be classical first-order formulas, where

- $\bar{x}_i$ and $\bar{y}_i$ are all free first-order variables of A_i
- the number of variables in $\bar{x}$ is k_i
- none of the x's is among the y's
- for $i = 1, \ldots, k$, Q_i is a k_i-argument relation symbol, whose all occurrences in $A_1, \ldots, A_k$ are positive.

Definition 3.4. *Under the above assumptions, the expression*

$$\textsc{Slfp} [Q_1(\bar{x}_1) \equiv A_1(\bar{Q}, \bar{x}_1, \bar{y}_1), \ldots, Q_k(\bar{x}_k) \equiv A_k(\bar{Q}, \bar{x}_k, \bar{y}_k)] \qquad (3.5)$$

is called the simultaneous least fixpoint *of $A_1, \ldots, A_k$, and the expression*

$$\textsc{Sgfp} [Q_1(\bar{x}_1) \equiv A_1(\bar{Q}, \bar{x}_1, \bar{y}_1), \ldots, Q_k(\bar{x}_k) \equiv A_k(\bar{Q}, \bar{x}_k, \bar{y}_k)] \qquad (3.6)$$

is called the simultaneous greatest fixpoint *of $A_1, \ldots, A_k$.* $\qquad \square$

Note that both SLFP and SGFP represent k-tuples of relations.

In the rest of the chapter we often abbreviate the formula in the scope of SLFP in (3.5) (and of SGFP in (3.6)) by $\bar{Q} \equiv \bar{A}$, formula (3.5) by SLFP$[\bar{Q} \equiv \bar{A}]$, and formula (3.6) by SGFP$[\bar{Q} \equiv \bar{A}]$.

Definition 3.5. *The semantics of* SLFP$[\bar{Q} \equiv \bar{A}]$ *is given by (the unique) tuple of relations $\bar{Q}$ satisfying* $\mathrm{Circ}_{\downarrow}(\bar{Q} \equiv \bar{A}; \bar{Q}; \emptyset)$, *and the semantics of* SGFP$[\bar{Q} \equiv \bar{A}]$ *is given by (the unique) tuple of relations $\bar{Q}$ satisfying* $\mathrm{Circ}^{\uparrow}(\bar{Q} \equiv \bar{A}; \bar{Q}; \emptyset)$. ☐

If $k = 1$ in formulas (3.5) and (3.6), then the simultaneous fixpoints reduce to the standard fixpoints. In such cases we write LFP$[Q(\bar{x}) \equiv A(Q, \bar{x}, \bar{y})]$ to stand for SLFP$[Q(\bar{x}) \equiv A(Q, \bar{x}, \bar{y})]$ and GFP$[Q(\bar{x}) \equiv A(Q, \bar{x}, \bar{y})]$ to stand for SGFP$[Q(\bar{x}) \equiv A(Q, \bar{x}, \bar{y})]$.

3.3 Annotation theories

3.3.1 Definition

We consider a directed graph or network, seen as a model of binary relations $\bar{R} = \langle R_i \rangle_{i=1,\ldots,m}$ ($m \geq 1$) on a finite set D. Relations in $\bar{R}$ represent *edges* of various kinds. D is called the *set of nodes*. If $m = 1$ then we omit the subscript and write R rather than R_1. We allow annotations on the nodes. The annotation is represented by unary predicates $\bar{Q} = \langle Q_j \rangle_{j=1,\ldots,n}$ ($n > 1$), where $Q_i(x)$ holds if node x is annotated by Q_i. Δ^{mn}-theories allow us to set requirements on annotations.

Definition 3.6. *Let $m \geq 1$ and $n > 1$. By an (m, n)-annotation theory, referred to as Δ^{mn}-theory, we understand any finite first-order theory $\Delta^{mn}(\bar{R}, \bar{Q})$ over the signature containing binary relation symbols $\bar{R} = \langle R_i \rangle_{i=1,\ldots,m}$ and unary relation symbols $\bar{Q} = \langle Q_j \rangle_{j=1,\ldots,n}$, where we assume that annotations in $\bar{Q}$ are unique, i.e., each Δ^{mn}-theory, in addition to specific axioms, contains also axioms:*

$$\sigma(n) \stackrel{\text{def}}{\equiv} \forall x\left[\bigvee_{1 \leq i \leq n} Q_i(x)\right] \text{ and } \pi(n) \stackrel{\text{def}}{\equiv} \bigwedge_{1 \leq i \neq j \leq n} \forall x\left[\neg Q_i(x) \vee \neg Q_j(x)\right]. \quad ☐$$

Observe that for each $1 \leq i \leq n$,

$$\sigma(n) \equiv \forall x\left[\left(\bigwedge_{1 \leq k \neq i \leq n}\neg Q_k(x)\right) \rightarrow Q_i(x)\right]. \tag{3.7}$$

The above simple observation is useful in specifying preferred models. In particular, it shows that annotations are strongly related to each other and minimizing/maximizing some of them usually requires varying all others. In is also useful in eliminating second-order quantifiers from circumscription axioms.

In the rest of the chapter we only allow a finite number of axioms. Any finite set of axioms can be represented by a single formula being the conjunction of its members. This restriction allows us to encode the considered problems by second-order formulas.

3.3.2 Example: Dung's argumentation theory

In argumentation theory the following Δ^{13}-theory, $\mathcal{A}(R, Q_1, Q_2, Q_3)$, is often considered (see Dung [115]):

$$\sigma(3) \wedge \pi(3) \tag{3.8}$$

$$\forall x[\forall y[R(y, x) \rightarrow Q_1(y)] \rightarrow Q_2(x)] \tag{3.9}$$

$$\forall x[\exists y[R(y, x) \wedge Q_2(y)] \rightarrow Q_1(x)] \tag{3.10}$$

$$\forall x[(\forall y[R(y, x) \rightarrow (Q_1(y) \vee Q_3(y))] \wedge \exists y[R(y, x) \wedge Q_3(y)]) \rightarrow Q_3(x)]. \tag{3.11}$$

The intended meaning of R, Q_1, Q_2, Q_3 is:

- elements of underlying models are arguments
- $R(x, y)$ means that argument x attacks argument y
- $Q_1(x)$ means that argument x is not-active/refuted, $Q_2(x)$ means that argument x is active and $Q_3(x)$ means that argument x is undecided.

Here one looks for minimal Q_2, maximal Q_2 or minimal Q_3, where in each case all relations, other than the minimized/maximized one, are allowed to vary (see Chapter 2). That is, we respectively consider $Circ_\downarrow(\mathcal{A}; Q_2; Q_1, Q_3)$ (see Section 3.6.3), $Circ^\uparrow(\mathcal{A}; Q_2; Q_1, Q_3)$ (see Section 3.6.3) and $Circ_\downarrow(\mathcal{A}; Q_3; Q_1, Q_2)$ (see Section 3.6.3).

3.3.3 Example: Theory related to approximate reasoning

In approximate reasoning one often uses a generalization of rough sets and relations [276], which depends on allowing arbitrary similarity relations, while in the rough set theory only equivalence relations are considered. Such generalized approximate reasoning has been shown useful in many application areas requiring the use of approximate knowledge structures (see, e.g., [111, 113]).

In order to formalize the fact that similarities should preserve properties of objects, we use the following Δ^{13}-theory, $\mathcal{R}(R, Q_1, Q_2, Q_3)$:

$$\sigma(3) \wedge \pi(3) \tag{3.12}$$

$$\forall x \forall y[(R(x, y) \wedge Q_1(x)) \rightarrow Q_1(y)] \tag{3.13}$$

$$\forall x \forall y[(R(x, y) \wedge Q_2(x)) \rightarrow Q_2(y)] \tag{3.14}$$

$$\forall x \forall y[(R(x, y) \wedge Q_3(x)) \rightarrow Q_3(y)]. \tag{3.15}$$

The intended meaning of R, Q_1, Q_2, Q_3 is:

- elements of underlying models are objects
- $R(x, y)$ means that object x is similar to object y
- $Q_1(x)$ means that x satisfies a given property, $Q_2(x)$ means that x does not satisfy the property and $Q_3(x)$ means that it is unknown whether x satisfies the property.

Here one often looks for simultaneous minimization of Q_1 and Q_3 with Q_2 allowed to vary, i.e., we consider $Circ_\downarrow(\mathcal{R}; Q_1, Q_3; Q_2)$ (see Section 3.6.4). This policy corresponds to the closed world assumption, where it is assumed that all positive facts are specified and all other facts should be considered false.

3.3.4 Example: Formalizing semantics for logic programs with negation

In order to provide a semantics for logic programs with negation allowed in the bodies of rules, we use a $\mathcal{A}^{23}$-theory, which we derive from considerations provided in [81].

Let P be a propositional logic program with negation as failure. Consider clauses of the form:

$$q : - \bigwedge_{i \in I} a_i, \bigwedge_{j \in J} \neg b_j, \tag{3.16}$$

where I, J are finite sets of indices.

We regard the atoms $q, \{a_i\}_{i \in I}, \{b_j\}_{j \in J}$ as elements in a classical model. The domain of the model, D_P, consists of all the atoms appearing in a given logic program.

We consider two binary relations R_+, R_- such that:

- for clauses of the form (3.16) we require that
  $R_+(a_i, q)$ and $R_-(b_j, q)$ hold ($i \in I, j \in J$) $\qquad(3.17)$
- for all other cases $R+, R_-$ are false.

Consider a logic program P satisfying the condition that

for any literal q, P contains at most one clause with q as its head. $\qquad(3.18)$

With such a program we can associate a model $\langle D_P, R_+, R_- \rangle$, where D_P is the set of atoms in P and R_+, R_- are defined as above. Conversely, with any finite domain model $\langle D_P, R_+, R_- \rangle$ with two binary relations R_+, R_- we can associate a logic program

$$y : - \bigwedge_{\{x | R_+(x,y)\}} x, \bigwedge_{\{x | R-(x,y)\}} \neg x. \tag{3.19}$$

Assume now that a program violates (3.18), i.e., contains several clauses with the same head,

$$q^k : - \bigwedge_{i \in I_k} a_{i_k}^k, \bigwedge_{j \in J_k} \neg b_{j_k}^k, \tag{3.20}$$

where $k \leq r$, for some $r \geq 1$.

We add r new propositions, $q_1^k, \ldots, q_r^k$ and we replace clauses $C_1, \ldots, C_r$ by clauses:

$$q_i^k : - \bigwedge_{i \in I_k} a_{i_k}^k, \bigwedge_{j \in J_k} \neg b_{j_k}^k, \quad q^* : - \bigwedge_{1 \leq k \leq r} \neg q_i^k, \quad q : - \neg q^*. \tag{3.21}$$

Now,

- q succeeds if q^* fails
- q loops if q^* loops
- q^* fails if at least one of $\neg q_i^k$ fails, i.e., if at least one of q_i^k succeeds, i.e., if at least one of bodies $\left[\bigwedge_{i \in I_k} a_{i_k}^k, \bigwedge_{j \in J_k} \neg b_{j_k}^k \right]$ succeeds.

Therefore, given a program P, one can write a program P' satisfying the assumption (3.18) as to unique heads.

We now formalize the semantics of logic programs with negation by a $\mathcal{A}^{23}$-theory $\mathcal{L}(R_+, R_-, Q_1, Q_2, Q_3)$, where

- elements of underlying models are atoms of a given logic program
- R_+, R_- are explained in (3.17)
- $Q_1(x)$ means that the computation of x fails
- $Q_2(x)$ means that the computation of x succeeds
- $Q_3(x)$ means that the computation of x loops.

The theory $\mathcal{L}$ consists of the following axioms:

$$\sigma(3) \wedge \pi(3) \tag{3.22}$$

$$\forall x[(\exists y[R_+(y,x) \wedge Q_1(y)] \vee \exists y[R_-(y,x) \wedge Q_2(y)]) \rightarrow Q_1(x)] \tag{3.23}$$

$$\forall x[(\forall y[R_+(y,x) \rightarrow Q_2(y)] \wedge \forall y[R_-(y,x) \rightarrow Q_1(y)]) \rightarrow Q_2(x)] \tag{3.24}$$

$$\forall x[(\forall y[R_+(y,x) \rightarrow (Q_2(y) \vee Q_3(y))] \wedge \forall y[R_-(y,x) \rightarrow (Q_1(y) \vee Q_3(y))] \\ \wedge \exists y[(R_-(y,x) \vee R_+(y,x)) \wedge Q_3(y)]) \rightarrow Q_3(x)]. \tag{3.25}$$

We look for models with minimal Q_2, where Q_1 and Q_3 are allowed to vary, which is expressed by $Circ_{\downarrow}(\mathcal{L}; Q_2; Q_1, Q_3)$ (see Section 3.6.5).

3.3.5 Some other applications

Observe that roles considered in description logics [26] can be represented as graphs whose edges correspond to roles. Annotations become necessary whenever one needs to uniquely identify nodes. In [218], Horrocks and Sattler discuss the need for annotations:

> "realistic ontologies typically contain references to named individuals within class descriptions. E.g., *Italians* might be described as persons who are citizens of *Italy*, where *Italy* is a named individual."

Yet another motivation for annotation theories is related to *nominals* which are a prominent feature of hybrid logics and their immediate ancestors, called modal logics with names [51, 204]. Consider, for example temporal reasoning. Once we refer to particular time points, e.g., by using dates ("it is going to be a board meeting on November 15th at 13:15"), we deal with unique annotations.

3.4 Complexity results

Consider first satisfiability checking and the querying problem.

Let $\mathcal{T}(R_1, \ldots, R_m, Q_1, \ldots, Q_n)$ be an arbitrary Δ^{mn}-theory. The *satisfiability problem* for $\mathcal{T}$ over a set of nodes D is expressed by

$$D \models \exists R_1 \ldots \exists R_m \exists Q_1 \ldots \exists Q_n[\mathcal{T}(R_1, \ldots, R_m, Q_1, \ldots, Q_n)]. \tag{3.26}$$

The *querying problem* assumes that a structure $\mathcal{M} = \langle D, R_1, \ldots, R_m \rangle$ is given and one asks whether

$$\mathcal{M} \models \exists Q_1 \ldots \exists Q_n[\mathcal{T}(R_1, \ldots, R_m, Q_1, \ldots, Q_n)]. \tag{3.27}$$

We start with the querying problem.

Theorem 3.7. *For $m \geq 1$ and $n \geq 3$, the querying problem for annotation theories is* NPTime-*complete.*

Proof. Let $\mathcal{T}(R_1, \ldots, R_m, Q_1, \ldots, Q_n)$ be an arbitrary Δ^{mn}-theory, where m, n are fixed natural numbers.

Given a finite model $M = \langle D, R_1, \ldots, R_m, Q_1, \ldots, Q_n \rangle$, one can check whether

$$M \models \mathcal{T}(R_1, \ldots, R_m, Q_1, \ldots, Q_n)$$

deterministically in time polynomial in the size of D (see [1, 121, 221]). So, given $\langle D, R_1, \ldots, R_m \rangle$, a nondeterministic polynomial time algorithm for the querying problem depends on guessing $Q_1, \ldots, Q_n$ and then accepting the result when the obtained model satisfies $\mathcal{T}(R_1, \ldots, R_m, Q_1, \ldots, Q_n)$. Of course, guessing $Q_1, \ldots, Q_n$ can be done in time linear in the size of D (recall that n is fixed). Thus the querying problem is in NPTIME.

To show NPTIME-completeness we consider the following Δ^{1n}-theory, denoted by $C(R, Q_1, \ldots, Q_n)$:

$$\sigma(n) \wedge \pi(n) \wedge \bigwedge_{1 \le i \le n} \forall x \forall y [R(x, y) \to (\neg Q_i(x) \vee \neg Q_i(y))]. \tag{3.28}$$

The theory C expresses the fact that a graph with edges represented by R can be colored using n colors $Q_1, Q_2, \ldots, Q_n$.

If only $\langle D, R \rangle$ is given, then checking n-colorability is expressed by:

$$\langle D, R \rangle \models \exists Q_1 \ldots \exists Q_n [(3.28)]. \tag{3.29}$$

It is well-known that this problem is NPTIME-complete already for $n = 3$.

By a similar proof we have the following theorem.

Theorem 3.8. *For $m \ge 1$ and $n \ge 3$, the satisfiability problem for annotation theories is NPTIME-complete.* □

The *model checking problem* for a Δ^{mn}-theory $\mathcal{T}$ assumes that a structure $M = \langle D, R_1, \ldots, R_m, Q_1, \ldots, Q_n \rangle$ is given and one asks whether

$$M \models Circ_\downarrow(\mathcal{T}, \bar{Q}', \bar{Q}'') \quad \text{(respectively, } M \models Circ^\uparrow(\mathcal{T}, \bar{Q}', \bar{Q}'')), \tag{3.30}$$

where $\bar{Q}', \bar{Q}''$ are chosen from $Q_1, \ldots, Q_n$.

Let us now show that model checking for circumscribed annotation theories is a co-NPTIME-complete problem. We adapt the proof given by Kolaitis and Papadimitriou (see pages 11-12 of [231]) for co-NPTIME-completeness of model checking for circumscription (Theorem 6 of [231]). We cannot use this result directly, as we want to show that co-NPTIME-completeness can be proved for circumscription on annotations, while the result of [231] applies circumscription to edges. For other results concerning the complexity of circumscription see [72, 229] and references there.

We need the following definition.

Definition 3.9. *We call a undirected graph* cubic *if all its nodes have degree three. A circuit of a graph is a closed path without repetitions of edges. A circuit is* long *if it contains at least twelve nodes. A graph is* simple *if it is a disjoint union of long circuits.*

We say that graph $G = \langle N, E \rangle$ is a subgraph *of graph $G' = \langle N', E' \rangle$ if $N = N'$ and $E \subseteq E'$. G is a* proper subgraph *of G' if it is a subgraph of G' and $E \neq E'$.* □

Of course, cubicity of a graph is a first-order property and can be expressed by a first-order formula $\rho(E)$ on edges.

Observe that simple graphs have all degrees two and there are no circuits of length eleven or less in them. Therefore, simplicity is also a first-order property and can be expressed by a first-order formula $\eta(E)$.

We have the following lemma.

Lemma 3.10 (Lemma 2 of Kolaitis and Papadimitriou [231]). *It is* NPTIME-*complete to check whether a cubic connected graph has a simple subgraph.*   □

Now we are in position to prove the announced complexity result.

Theorem 3.11. *There is an annotation theory and circumscriptive policy on annotations whose model checking is* co-NPTIME-*complete.*

Proof. Consider the following Δ^{12}-theory, denoted by $\mathcal{B}$ with axioms:

$$\sigma(2) \wedge \pi(2)$$
$$\rho(E) \vee \eta(E)$$
$$\eta(E'), \text{ where } E'(x, y) \stackrel{\text{def}}{\equiv} (Q_1(x) \wedge Q_1(y)).$$

Theory $\mathcal{B}$ states that graph $G = \langle N, E \rangle$ is either cubic or simple and that Q_1 "selects" a simple subgraph from G, while Q_2 annotates nodes outside of the selected subgraph. Consider $Circ_\downarrow(\mathcal{B}; Q_1; Q_2)$. It additionally says that there is no proper subgraph of G which is cubic or simple.

Consider a relational structure $M = \langle N, E, Q_1, Q_2 \rangle$ with $\langle N, E \rangle$ being a cubic graph. The question is whether

$$M \models Circ_\downarrow(\mathcal{B}; Q_1; Q_2). \tag{3.31}$$

As noted above, (3.31) holds when there is no proper subgraph of G which is cubic or simple. No proper subgraph of a cubic graph can be cubic, so (3.31) holds when no simple subgraph of G exists. By Lemma 3.10, checking existence of a simple subgraph of a given graph is NPTIME-complete. Therefore checking whether (3.31) holds is an co-NPTIME complete problem.

Remark 3.12. Observe that, in the light of Theorems 3.7, 3.8 and 3.11, the elimination of all second-order quantifiers from formulas (3.26), (3.27) and (3.30) is, in general, problematic. A successful elimination from formulas considered in proofs of Theorems 3.7, 3.8 and 3.11, would imply that PTIME =NPTIME.

The presence of both $\sigma(n)$ and $\pi(n)$ in Δ^{nn}-theories suggests that algorithmic quantifier elimination techniques depending on the syntactic shape of formulas require further simplifications of circumscribed formulas or certain syntactic restrictions as to the remaining axioms. To see the intuition, assume that the underlying database consisting of objects and edges is given. One can now translate a given annotation theory into a propositional theory with propositional variables corresponding to annotations. The resulting propositional theory is non-Schaefer [300]. Due to the Schaefer's dichotomy theorem, the satisfiability problem for theories containing axioms of that shape is NPTIME-complete. We can then strongly expect that second-order quantifier elimination methods depending on the shape of formulas cannot generally be applied here.

For a dichotomy theorem directly concerning circumscribed theories, also supporting this intuition, see [229]. □

3.5 Second-order quantifier elimination

3.5.1 Simultaneous elimination theorem

Let us start with a theorem allowing one to eliminate a number of second-order existential quantifiers at the same time. Theorem 3.13 is a special case of Theorem 3.22, but we formulate it separately for two reasons. First, it is useful in some applications which do not require the full strength of Theorem 3.22. Second, it considerably simplifies the proof of Theorem 3.22.

Theorem 3.13 (Kachniarz and Szałas [226]). *Let* $\bar{X} = X_1, \ldots, X_k$ *be distinct relation variables and* $C(\bar{X})$, $A_i(\bar{X}, \bar{x}_1, \ldots, \bar{x}_k, \bar{z})$ $(1 \leq i \leq k)$ *be classical first-order formulas, where the number of distinct variables in* $\bar{x}_i$ *is equal to the arity of* X_i *and* $A_i(\bar{X}, \ldots)$ *is positive w.r.t.* $\bar{X}$. *Then:*

- *if* $C(\bar{X})$ *is negative w.r.t.* $X_1, \ldots, X_k$ *then*

$$\exists X_1 \ldots \exists X_k \left\{ \bigwedge_{1 \leq i \leq k} \forall \bar{x}_i [A_i(\bar{X}, \bar{x}_1, \ldots, \bar{x}_k, \bar{z}) \rightarrow X_i(\bar{x}_i)] \wedge C(\bar{X}) \right\}$$

$$\parallel \qquad\qquad\qquad (3.32)$$

$$C(\bar{X})^{X_1(\bar{x}_1), \ldots, X_k(\bar{x}_k)}_{\lambda \bar{x}_1, \ldots, \bar{x}_k \mathrm{SLFP}\,[X_1(\bar{x}_1) \equiv A_1(\bar{X}, \ldots), \ldots, X_k(\bar{x}_k) \equiv A_k(\bar{X}, \ldots)]}.$$

- *if* $C(\bar{X})$ *is positive w.r.t.* $X_1, \ldots, X_k$ *then*

$$\exists X_1 \ldots \exists X_k \left\{ \bigwedge_{1 \leq i \leq k} \forall \bar{x}_i [X_i(\bar{x}_i) \rightarrow A_i(\bar{X}, \bar{x}_1, \ldots, \bar{x}_k, \bar{z})] \wedge C(\bar{X}) \right\}$$

$$\parallel \qquad\qquad\qquad (3.33)$$

$$C(\bar{X})^{X_1(\bar{x}_1), \ldots, X_k(\bar{x}_k)}_{\lambda \bar{x}_1, \ldots, \bar{x}_k \mathrm{SGFP}\,[X_1(\bar{x}_1) \equiv A_1(\bar{X}, \ldots), \ldots, X_k(\bar{x}_k) \equiv A_k(\bar{X}, \ldots)]}.$$

Proof. We prove (3.33). The proof of (3.32) is analogous. Let M be a relational structure and v be a valuation of variables.
($\rightarrow$) Assume that

$$M, v \models \exists X_1 \ldots \exists X_k \left\{ \bigwedge_{\leq i \leq k} \forall \bar{x}_i [X_i(\bar{x}_i) \rightarrow A_i(\bar{X}, \bar{x}_1, \ldots, \bar{x}_k, \bar{z})] \wedge C(\bar{X}) \right\}. \qquad (3.34)$$

Therefore, there exists a valuation V assigning to $X_1, \ldots, X_k$ relations over the domain of M such that

$$M, v, V \models \bigwedge_{1 \leq i \leq k} \forall \bar{x}_i [X_i(\bar{x}_i) \rightarrow A_i(\bar{X}, \bar{x}_1, \ldots, \bar{x}_k, \bar{z})] \wedge C(\bar{X}).$$

Note that the greatest (w.r.t. $\rightarrow$) $\bar{X}$ satisfying

$$\bigwedge_{1 \le i \le k} \forall \bar{x}_i [X_i(\bar{x}_i) \rightarrow A_i(\bar{X}, \bar{x}_1, \ldots, \bar{x}_k, \bar{z})]$$

is the greatest $\bar{X}$ satisfying $\bigwedge_{1 \le i \le k} \forall \bar{x}_i [X_i(\bar{x}_i) \equiv A_i(\bar{X}, \bar{x}_1, \ldots, \bar{x}_k, \bar{z})]$, which by Definition 3.5 and equation (3.4), is given by

$$\text{SGFP}\,[X_1(\bar{x}_1) \equiv A_1(\bar{X}, \ldots), \ldots, X_k(\bar{x}_k) \equiv A_k(\bar{X}, \ldots)].$$

Since $C(\bar{X})$ is positive in $X_1, \ldots, X_k$, it is also monotone in $X_1, \ldots, X_k$. Therefore we have that

$$M, v, V \models C(\bar{X})^{X_1(\bar{x}_1),\ldots,X_k(\bar{x}_k)}_{\lambda \bar{x}_1,\ldots,\bar{x}_k \text{SGFP}\,[X_1(\bar{x}_1) \equiv A_1(\bar{X},\ldots),\ldots,X_k(\bar{x}_k) \equiv A_k(\bar{X},\ldots)]}. \tag{3.35}$$

Relational variables $X_1, \ldots, X_k$ in (3.35) are bound by the simultaneous fixpoint operator SGFP, so V in (3.35) becomes redundant and we obtain

$$M, v \models C(\bar{X})^{X_1(\bar{x}_1),\ldots,X_k(\bar{x}_k)}_{\lambda \bar{x}_1,\ldots,\bar{x}_k \text{SGFP}\,[X_1(\bar{x}_1) \equiv A_1(\bar{X},\ldots),\ldots,X_k(\bar{x}_k) \equiv A_k(\bar{X},\ldots)]}. \tag{3.36}$$

($\leftarrow$) Assume now (3.36). It is easy to observe that $X_1, \ldots, X_k$ defined by respective coordinates of $\text{SGFP}\,[X_1(\bar{x}_1) \equiv A_1(\bar{X}, \ldots), \ldots, X_k(\bar{x}_k) \equiv A_k(\bar{X}, \ldots)]$ satisfy $M, v \models \bigwedge_{1 \le i \le k} \forall \bar{x}_i [X_i(\bar{x}_i) \rightarrow A_i(\bar{X}, \bar{x}_1, \ldots, \bar{x}_k, \bar{z})]$.

By (3.36), such $X_1, \ldots, X_k$ satisfy $M, v \models C(\bar{X})$. We have then indicated $X_1, \ldots, X_k$ for which $M, v \models \bigwedge_{1 \le i \le k} \forall \bar{x}_i [X_i(\bar{x}_i) \rightarrow A_i(\bar{X}, \bar{x}_1, \ldots, \bar{x}_k, \bar{z})] \wedge C(\bar{X})$. Therefore (3.34) holds, too.

3.5.2 Some consequences of the elimination theorem

We have two corollaries of Theorem 3.13, which we also use in further calculations.

The following theorem is a particular case of Theorem 3.13 when we eliminate a single relation variable.

Theorem 3.14 (Nonnengart and Szałas [272]). *Let X be a relation variable and $A(X, \bar{x}, \bar{z})$, $C(X)$ be a classical first-order formula, where the number of distinct variables in $\bar{x}$ is equal to the arity of X, $A(X, \bar{x}, \bar{z})$ is positive w.r.t. X. Then*

- *if $C(X)$ is negative w.r.t. X then*

$$\exists X \{\forall \bar{x}[A(X, \bar{x}, \bar{z}) \rightarrow X(\bar{x})] \wedge C(X)\} \equiv C(X)^{X(\bar{x})}_{\lambda \bar{x}.\text{LFP}\,[X(\bar{x}) \equiv A(X,\bar{x},\bar{z})]}. \tag{3.37}$$

- *if $C(X)$ is positive w.r.t. X then*

$$\exists X \{\forall \bar{x}[X(\bar{x}) \rightarrow A(X, \bar{x}, \bar{z})] \wedge C(X)\} \equiv C(X)^{X(\bar{x})}_{\lambda \bar{x}.\text{GFP}\,[X(\bar{x}) \equiv A(X,\bar{x},\bar{z})]}. \tag{3.38}$$

$\square$

The following lemma is a particular case of Theorem 3.14 (thus also of Theorem 3.13), when A contains no occurrences of the eliminated relational variable.

Lemma 3.15 (Ackermann [11]). *Let X be a relation variable and $A(\bar{x}, \bar{z})$, $C(X)$ be classical first-order formulas, where the number of distinct variables in $\bar{x}$ is equal to the arity of X. Let A contain no occurrences of X. Then*

- *if $C(X)$ is negative w.r.t. X then*

$$\exists X\{\forall \bar{x}[A(\bar{x}, \bar{z}) \rightarrow X(\bar{x})] \wedge C(X)\} \equiv C(X)^{X(\bar{x})}_{\lambda \bar{x}. A(\bar{x}, \bar{z})}. \tag{3.39}$$

- *if $C(X)$ is positive w.r.t. X then*

$$\exists X\{\forall \bar{x}[X(\bar{x}) \rightarrow A(\bar{x}, \bar{z})] \wedge C(X)\} \equiv C(X)^{X(\bar{x})}_{\lambda \bar{x}. A(\bar{x}, \bar{z})}. \tag{3.40}$$

$\square$

3.5.3 Strengthening the method

Observe that axioms of annotation theories are often formulated in the form of "rules". In considered examples of annotation theories, formulas (3.9)–(3.11), (3.13)–(3.15) and (3.23)–(3.25) have a form of rules. Also formula $\sigma(n)$ can be considered as a rule due to its form expressed as (3.7).

We shall strengthen the method formalized as Theorem 3.13 by using intuitions from the semantics of stratified logic programs and Datalog⁻, where recursion is not allowed to pass negation (cf. [1]). Namely, when we have rules with negated atoms in bodies, Theorem 3.13 is not applicable, but actually we can apply the theorem for separate strata and collect the results. The following example illustrates the idea.

Example 3.16. Consider the following second-order formula

$$\exists X \exists Y \{\forall x[\neg X(x) \wedge \neg Y(x)] \tag{3.41}$$

$$\forall x[\exists y[R(x, y) \vee X(y)] \rightarrow X(x)] \wedge \tag{3.42}$$

$$\forall x[(\exists y[R(x, y) \wedge Y(y)] \vee \neg X(x)) \rightarrow Y(x)]\}. \tag{3.43}$$

Theorem 3.13 cannot be applied due to the negative literal $\neg X(x)$ in (3.43). On the other hand, one can first "compute" X using (3.42) and then use its definition "computing" Y. Here we can even apply Theorem 3.14. The definition of X is given by:

$$\text{L\textsc{fp}}[X(x) \equiv \exists y[R(x, y) \vee X(y)]] \tag{3.44}$$

and, given (3.44) and (3.43), the definition of Y is given by

$$\text{L\textsc{fp}}[Y(x) \equiv \exists y[R(x, y) \wedge Y(y)] \vee \neg X(x)]. \tag{3.45}$$

Formula (3.41)–(3.41) is then equivalent to $\forall x[\neg X(x) \wedge \neg Y(x)]$, where X and Y are given by their definitions (3.44) and (3.45). $\square$

Other examples of applications of this method are provided in the next section.

In the rest of this section we formulate and prove the main second-order quantifier elimination result of this chapter, substantially extending Theorem 3.13. The strengthened version of the theorem can be formalized as follows.

Definition 3.17. *Let $A(P, \bar{T}, \bar{x})$ be a classical first-order formula positive w.r.t. P with $P, \bar{T}$ being its all relation symbols.*

A Pia *formula w.r.t. P is any formula of the form $\forall \bar{x}[A(P, \bar{T}, \bar{x}) \rightarrow P(\bar{x})]$.[5] By a* Pia *formula we understand a* Pia *formula w.r.t. P for some P.*

Dually, by an Aip *formula w.r.t. P we understand any formula of the form $\forall \bar{x}[P(\bar{x}) \rightarrow A(P, \bar{T}, \bar{x})]$.[6] By an* Aip *formula we understand an* Aip *formula w.r.t. P for some P.*

A rule is a Pia *formula or an* Aip *formula. In both cases the atom is called the* head *and the antecedent (consequent) is called the* body *of the rule. For a rule ρ, the head of ρ is denoted by head(ρ) and the body of ρ is denoted by body(ρ).*

By a Pia *set we understand any finite set of* Pia *formulas and by an* Aip *set we understand any finite set of* Aip *formulas. A set of rules is either a* Pia *set or an* Aip *set S such that any head of a rule of S appears in S in exactly one rule.[7]* □

Observe that the restriction as to uniqueness of heads' occurrences in rules is introduced solely to simplify presentation. Any set of Pia formulas and of Aip formulas can easily be transformed to a set satisfying this requirement. It suffices to rename variables and use the following tautologies:

$$(\forall \bar{x}[A(\bar{x})] \wedge \forall \bar{x}[B(\bar{x})]) \equiv \forall \bar{x}[A(\bar{x}) \wedge B(\bar{x})]$$
$$((A \rightarrow C) \wedge (B \rightarrow C)) \equiv ((A \vee B) \rightarrow C)$$
$$((A \rightarrow B) \wedge (A \rightarrow C)) \equiv (A \rightarrow (B \wedge C)).$$

The following definition generalizes the well-known definition of stratification of logic programs.

Definition 3.18. *A stratification of a set of rules S is a partition $S_1, \ldots, S_l$ of S such that there is a mapping δ from the set of heads appearing in S to $\{1, \ldots, l\}$, satisfying:*

- *all rules with the same head P are in the same partition $S_{\delta(P)}$*
- *if $(\forall \bar{x}[A(P, \bar{T}, \bar{x}) \rightarrow P(\bar{x})]) \in S$ (dually, if $(\forall \bar{x}[P(\bar{x}) \rightarrow A(P, \bar{T}, \bar{x})]) \in S$) and $P, \bar{T}$ are all relation symbols occurring in A, then for any T_i in $\bar{T}$:*
  - *if A is positive w.r.t. T_i then $\delta(T_i) \leq \delta(P)$*
  - *if there is a negative occurrence of T_i in A then $\delta(T_i) < \delta(P)$.*

Given a stratification $S_1, \ldots S_l$ of S, each S_i is called a stratum *of the stratification, and δ is called the* stratification mapping. □

We shall need an ordering on the set of heads of considered sets of rules, preserving the stratification mapping.

Definition 3.19. *Let r be the cardinality of a set of rules S stratifiable with a stratification mapping δ. A mapping $\gamma : \{1, \ldots, r\} \longrightarrow \{1, \ldots, r\}$ is a δ-order if it is one-to-one, onto and, for $1 \leq i \leq j \leq r$, satisfies the condition:*

$$\delta(head(rule \; \gamma(i))) \leq \delta(head(rule \; \gamma(j))). □$$

---

[5] The acronym Pia, introduced in [326], stands for "Positive antecedent Implies Atom".

[6] Aip stands for "Atom Implies Positive consequent".

[7] Let us emphasize that we do not consider sets containing both Pia and Aip formulas.

Let $\bar{Q} = \langle Q_1, \ldots, Q_k \rangle$ be a tuple of relation symbols and $A_i(\bar{Q}, \bar{x}_i, \bar{y}_i)$, for $i = 1, \ldots, k$, be classical first-order formulas, where

- $\bar{x}_i$ and $\bar{y}_i$ are all free first-order variables of A_i
- the number of variables in $\bar{x}$ is k_i
- none of the x's is among the y's
- for $i = 1, \ldots, k$, Q_i is a k_i-argument relation symbol
- the set of rules with bodies A_i and heads Q_i is stratifiable.

Definition 3.20. *Under the above assumptions, the expression*

$$\text{STLFP}[Q_1(\bar{x}_1) \equiv A_1(\bar{Q}, \bar{x}_1, \bar{y}_1), \ldots, Q_k(\bar{x}_k) \equiv A_k(\bar{Q}, \bar{x}_k, \bar{y}_k)] \qquad (3.46)$$

is called the stratified least fixpoint *of $A_1, \ldots, A_k$, and the expression*

$$\text{STGFP}[Q_1(\bar{x}_1) \equiv A_1(\bar{Q}, \bar{x}_1, \bar{y}_1), \ldots, Q_k(\bar{x}_k) \equiv A_k(\bar{Q}, \bar{x}_k, \bar{y}_k)] \qquad (3.47)$$

is called the stratified greatest fixpoint *of $A_1, \ldots, A_k$.* □

Definition 3.21. *Let $S_1, \ldots, S_l$ ($1 \le i \le l$) be strata of sets of rules considered in Definition 3.20. The semantics of $\text{STLFP}[\bar{Q} \equiv \bar{A}]$ is given by (the unique) tuple of relations $\bar{Q}$ given by $\text{SLFP}[S_1], \ldots, \text{SLFP}[S_l]$. The semantics of $\text{STGFP}[\bar{Q} \equiv \bar{A}]$ is given by (the unique) tuple of relations $\bar{Q}$ given by $\text{SGFP}[S_1], \ldots, \text{SGFP}[S_l]$.* □

We are now in position to formulate the elimination theorem. The intuition behind its proof is that one eliminates quantifiers starting from the first stratum and proceeds stratum by stratum until the last one. So, in the result, for each stratum there is a corresponding simultaneous fixpoint.

Theorem 3.22. *Let $\bar{X} = X_1, \ldots, X_k$ be distinct relation variables and $C(\bar{X})$, $A_i(\bar{X}, \bar{x}_1, \ldots, \bar{x}_k, \bar{z})$ ($1 \le i \le k$) be classical first-order formulas, where the number of distinct variables in $\bar{x}_i$ is equal to the arity of X_i. Then:*

- *if $\{\forall \bar{x}[A_i(\bar{X}, \ldots) \rightarrow X_i(\bar{x}_i)] \mid 1 \le i \le k\}$ is stratifiable with a stratification mapping δ, γ is a δ-order and $C(\bar{X})$ is negative w.r.t. $X_1, \ldots, X_k$ then:*

$$\exists X_1 \ldots \exists X_k \left\{ \bigwedge_{1 \le i \le k} \forall \bar{x}_i [A_i(\bar{X}, \bar{x}_1, \ldots, \bar{x}_k, \bar{z}) \rightarrow X_i(\bar{x}_i)] \wedge C(\bar{X}) \right\} \qquad (3.48)$$

$$\|$$

$$C(\bar{X})^{X_{\gamma(1)}(\bar{x}_{\gamma(1)}), \ldots, X_{\gamma(k)}(\bar{x}_{\gamma(k)})}_{\lambda \bar{x}_{\gamma(1)}, \ldots, \bar{x}_{\gamma(k)} \text{STLFP}[X_{\gamma(1)}(\bar{x}_{\gamma(1)}) \equiv A_{\gamma(1)}(\bar{X}, \ldots), \ldots, X_{\gamma(k)}(\bar{x}_{\gamma(k)}) \equiv A_{\gamma(k)}(\bar{X}, \ldots)]}.$$

- *if $\{\forall \bar{x}[X_i(\bar{x}_i) \rightarrow A_i(\bar{X}, \ldots)] \mid 1 \le i \le k\}$ is stratifiable with a stratification mapping δ, γ is a δ-order and $C(\bar{X})$ is positive w.r.t. $X_1, \ldots, X_k$ then:*

$$\exists X_1 \ldots \exists X_k \left\{ \bigwedge_{1 \le i \le k} \forall \bar{x}_i [X_i(\bar{x}_i) \rightarrow A_i(\bar{X}, \bar{x}_1, \ldots, \bar{x}_k, \bar{z})] \wedge C(\bar{X}) \right\} \qquad (3.49)$$

$$\|$$

$$C(\bar{X})^{X_{\gamma(1)}(\bar{x}_{\gamma(1)}), \ldots, X_{\gamma(k)}(\bar{x}_{\gamma(k)})}_{\lambda \bar{x}_{\gamma(1)}, \ldots, \bar{x}_{\gamma(k)} \text{STGFP}[X_{\gamma(1)}(\bar{x}_{\gamma(1)}) \equiv A_{\gamma(1)}(\bar{X}, \ldots), \ldots, X_{\gamma(k)}(\bar{x}_{\gamma(k)}) \equiv A_{\gamma(k)}(\bar{X}, \ldots)]}.$$

Proof. We prove the theorem by induction on the number of strata $l \geq 1$.

The case of a single stratum, $l = 1$, is formulated and proved as Theorem 3.13.

Assume that the theorem holds for sets of rules with $l > 1$ strata. Let S a PIA set with $l+1$ strata, $S_1, \ldots, S_l, S_{l+1}$, consisting of k rules. First, reorder existential second-order quantifiers in (3.48) (respectively, (3.49)) according to γ from right to left, so that the resulting sequence of quantifiers is $\exists\, head(\text{rule } \gamma(k)) \ldots \exists\, head(\text{rule } \gamma(1))$.

Existential second-order quantifiers binding heads of stratum $l + 1$ are the outermost ones. Use the inductive assumption to eliminate all quantifiers except those binding heads of stratum $l + 1$. In the resulting formula substitute any occurrence of a fixpoint operator of the form $\text{SLFP}\,[\bar{R} \equiv \bar{U}]$ by a new relation symbol, say N applied to the same arguments as $\text{SLFP}\,[\bar{R} \equiv \bar{U}]$. Next, add to the theory definitions of the new symbols as the respective fixpoints by using equivalences of the form:

$$\forall x[N(\bar{x}) \equiv \text{SLFP}\,[\bar{R} \equiv \bar{U}]]. \tag{3.50}$$

Now we apply Theorem 3.13, replace N's by fixpoints according to respective definitions (3.50) and finally remove those definitions.

Remark 3.23.

1. As noted in the above proof, Theorem 3.13 is a corollary of Theorem 3.22 in the case when there is a stratification consisting of a single stratum.
2. Observe that Theorem 3.22 can be extended to higher-order contexts along the lines of the elimination theorem of Gabbay and Szałas proved in [154].   □

3.6 Reducing circumscription formulas in annotation theories

3.6.1 Discussion

We have already indicated in Remark 3.12 that axioms of the form $\sigma(n) \wedge \pi(n)$, appearing in $\varLambda^{mn}$-theories, make second-order quantifier techniques presented in Section 3.5 rarely applicable.

In all examples of annotation theories we consider, formulas contain recursive clauses, which excludes the possibility of eliminating all second-order quantifiers using the lemma of Ackermann (i.e., Lemma 3.15). The resulting formulas are then at least formulas of the fixpoint logic. This makes the SCAN algorithm [135] inapplicable, too.

So a candidate could be the Theorem 3.13, which to our best knowledge provides the strongest second-order quantifier elimination method that could be applied here.[8] As we show below, this theorem is not directly applicable, too. The argument is based on van Benthem's result [326]. To present the result we need the following definition.

Definition 3.24. *A first-order formula* $A(P, \bar{T})$ *has the* intersection property *w.r.t.* P *iff in any relational structure* M*, whenever* $M, P_i \models A(P, \bar{T})$ *for all predicates in a family* $\{P_i \mid i \in I\}$*,* $A(P, \bar{T})$ *also holds for their intersection, i.e., we have that* $M, \bigcap_{i \in I} P_i \models$

$A(P, \bar{T})$.   □

---

[8] Except, of course, for our Theorem 3.22.

The following theorem has been proved by van Benthem in [326].

Theorem 3.25 (van Benthem [326]). *The following are equivalent for all first-order formulas $A(P, \bar{T})$:*

1. *$A(P, \bar{T})$ has the intersection property w.r.t. P;*
2. *There is a PIA formula equivalent to $A(P, \bar{T})$.*   □

We now have the following corollary.

Corollary 3.26. *For $n > 1$ and $m \geq 1$, no Δ^{mn}-theory theory is equivalent to a conjunction of PIA formulas.*

Proof. Let $n > 1$. Then the formula $\sigma(n) \wedge \pi(n)$ of Δ^{mn}-theory expresses the fact that every node of a graph is uniquely annotated. The intersection of two different annotations is then inconsistent.

The above result shows that the method based on Theorem 3.13 is not directly applicable in the case of the first form (3.32) of formulas required there.[9]

This shows that the method introduced in Section 3.5.3, based on Theorem 3.22 is indeed substantial.

3.6.2 Useful simplifications

As already discussed, axioms $\sigma(n) \wedge \pi(n)$ of Δ^{mn}-theories are a source of difficulties in applying second-order quantifier techniques. The following two lemmas allow us to simplify the formulas. Namely, in the case of minimization one can remove $\pi(n)$ (together with other conjuncts negative w.r.t. minimized relations) from the second-order part of circumscription formula (3.1). Dually, in the case of maximization one can remove $\sigma(n)$ (together with other conjuncts positive w.r.t. maximized relations) from the second-order part of circumscription formula (3.2).

Below we first consider circumscribed theories without varied predicates, so theories are denoted by $T(\bar{P})$ rather than $T(\bar{P}, \bar{S})$, as $\bar{S} = \emptyset$.

Lemma 3.27. *Let $T(\bar{P})$ be a theory. Assume that T is of the form of conjunction $T_{\pm}(\bar{P}) \wedge T_{+}(\bar{P})$, where $T_{+}(\bar{P})$ is positive w.r.t. all relation symbols in $\bar{P}$. Then $\text{Circ}^{\uparrow}(T; \bar{P}, \emptyset)$ is equivalent to*

$$T(\bar{P}) \wedge \forall \bar{X} \left\{ [T_{\pm}(\bar{P})_{\bar{X}}^{\bar{P}} \wedge \bigwedge_{i=1}^{k} \forall \bar{x}_i [P_i(\bar{x}_i) \rightarrow X_i(\bar{x}_i)] \,] \rightarrow \right.$$

$$\left. \bigwedge_{i=1}^{k} \forall \bar{x}_i [X_i(\bar{x}_i) \rightarrow P_i(\bar{x}_i)] \right\}. \tag{3.51}$$

Proof. [10] Since $T_{+}(\bar{P})$ positive w.r.t. all relation symbols in $\bar{P}$, it is also monotone w.r.t. all relation symbols in $\bar{P}$. We then have that for any relational structure M and valuation v:

---

[9] Similar argument applies to the form required in (3.33). This can be seen by considering the contraposition of the implication and respectively replace $\neg X_i$'s by X_i's.

[10] The proof is similar to the one given by Lifschitz [241] for proving a reduction result for separated formulas, but we deal with a more general context here.

$$M, v \models \left[T_+(\bar{P}) \wedge \bigwedge_{i=1}^{k} \forall \bar{x}_i [P_i(\bar{x}_i) \rightarrow X_i(\bar{x}_i)] \right] \rightarrow T_+(\bar{P})_{\bar{X}}^{\bar{P}}.$$

Thus, in the presence of the conjunct $T(\bar{P})$, formula (3.51) obtained from (3.2) by removing $T_+(\bar{P})_{\bar{X}}^{\bar{P}}$ is equivalent to (3.2).

Lemma 3.28. *Let $T(\bar{P})$ be a theory. Assume that T is of the form of conjunction $T_{\pm}(\bar{P}) \wedge T_-(\bar{P})$, where $T_-(\bar{P})$ is negative w.r.t. all relation symbols in $\bar{P}$. Then $\mathrm{Circ}_{\downarrow}(T; \bar{P}; \emptyset)$ is equivalent to*

$$T(\bar{P}) \wedge \forall \bar{X} \Big\{ [T_{\pm}(\bar{P})_{\bar{X}}^{\bar{P}} \wedge \bigwedge_{i=1}^{k} \forall \bar{x}_i [X_i(\bar{x}_i) \rightarrow P_i(\bar{x}_i)] \,] \rightarrow \tag{3.52}$$
$$\bigwedge_{i=1}^{k} \forall \bar{x}_i [P_i(\bar{x}_i) \rightarrow X_i(\bar{x}_i)] \Big\}.$$

Proof. Similar to the proof of Lemma 3.27, by observing that contraposition of all implications $\forall \bar{x}_i [X_i(\bar{x}_i) \rightarrow P_i(\bar{x}_i)]$ together with the monotonicity of T_- w.r.t. $\neg P_i$ imply the result.

In order to deal with varied predicates we use the following observation, allowing one to eliminate varied predicates.

Proposition 3.29 (Lifschitz [239]). *The circumscription $\mathrm{Circ}_{\downarrow}(T(\bar{P}, \bar{S}); \bar{P}; \bar{S})$ is equivalent to $T(\bar{P}, \bar{S}) \wedge \mathrm{Circ}_{\downarrow}(\exists \bar{Y} [T(\bar{P}, \bar{S})_{\bar{Y}}^{\bar{S}}]; \bar{P}; \emptyset)$.* □

Similarly, we have analogous proposition for the dual form of circumscription.

Proposition 3.30. *The circumscription $\mathrm{Circ}^{\uparrow}(T(\bar{P}, \bar{S}); \bar{P}; \bar{S})$ is equivalent to $T(\bar{P}, \bar{S}) \wedge \mathrm{Circ}^{\uparrow}(\exists \bar{Y} [T(\bar{P}, \bar{S})_{\bar{Y}}^{\bar{S}}]; \bar{P}; \emptyset)$.* □

The following proposition, known as the *purity deletion principle*, is sometimes useful.

Proposition 3.31 (Szałas [320]). *Let A be a classical first-order formula of the form $Q_1 x_1 .... Q_r x_r [A_1 \wedge ... \wedge A_q]$, where $Q_1, \ldots, Q_r \in \{\exists, \forall\}$ and each $A_1, ..., A_q$ containing an occurrence of P is of the form $(B \vee P(\bar{z}))$ with B being any first-order formula, possibly containing arbitrary occurrences of P. Then the formula $\exists P[A]$ is equivalent to $Q_1 x_1 .... Q_r x_r [A_{i_1} \wedge ... \wedge A_{i_s}]$, where $i_1, ..., i_s \in \{1, ..., q\}$ and $A_{i_1}, ..., A_{i_s}$ are all conjuncts that do not contain occurrences of P (the empty conjunction is, by convention, TRUE).*

The same holds when each $A_1, ..., A_q$ containing an occurrence of P is of the form $(B \vee \neg P(\bar{z}))$. □

3.6.3 Reducing circumscription in Dung's argumentation theory

In this section we consider Dung's theory $\mathcal{A}(R, Q_1, Q_2, Q_3)$ specified in Section 3.3.2. We show that all circumscriptive policies considered there are reducible to fixpoint logic.

Minimization on Q_2

Consider the circumscription formula $Circ_\downarrow(\mathcal{A}(R, Q_1, Q_2, Q_3); Q_2; Q_1, Q_3)$:

$$\mathcal{A}(R, Q_1, Q_2, Q_3)\wedge \tag{3.53}$$

$$\forall X_1 \forall X_2 \forall X_3 [(\mathcal{A}(R, X_1, X_2, X_3) \wedge \forall x[X_2(x) \rightarrow Q_2(x)]) \rightarrow \\ \forall x[Q_2(x) \rightarrow X_2(x)]]. \tag{3.54}$$

We focus on (3.54), which is equivalent to

$$\neg \exists X_1 \exists X_2 \exists X_3 \{\forall x[X_1(x) \vee X_2(x) \vee X_3(x)] \wedge$$

$$\forall x[\neg X_1(x) \vee \neg X_2(x)] \wedge \forall x[\neg X_1(x) \vee \neg X_3(x)] \wedge \forall x[\neg X_2(x) \vee \neg X_3(x)] \wedge$$

$$\forall x[\forall y[R(y, x) \rightarrow X_1(y)] \rightarrow X_2(x)] \wedge$$

$$\forall x[\exists y[R(y, x) \wedge X_2(y)] \rightarrow X_1(x)] \wedge$$

$$\forall x[(\forall y[R(y, x) \rightarrow (X_1(y) \vee X_3(y))] \wedge \exists y[R(y, x) \wedge X_3(y)]) \rightarrow X_3(x)] \wedge$$

$$\forall x[X_2(x) \rightarrow Q_2(x)] \wedge \exists z[Q_2(z) \wedge \neg X_2(z)]]\}.$$

Using (3.7) and minor transformations we obtain

$$\neg \exists X_1 \exists X_2 \exists X_3 \{\forall x[X_2(x) \rightarrow Q_2(x)] \wedge \exists z[Q_2(z) \wedge \neg X_2(z)] \wedge \tag{3.55}$$

$$\forall x[\neg X_1(x) \vee \neg X_2(x)] \wedge \forall x[\neg X_1(x) \vee \neg X_3(x)] \wedge \forall x[\neg X_2(x) \vee \neg X_3(x)] \wedge \tag{3.56}$$

$$\forall x[\forall y[R(y, x) \rightarrow X_1(y)] \rightarrow X_2(x)] \wedge \tag{3.57}$$

$$\forall x[\exists y[R(y, x) \wedge X_2(y)] \rightarrow X_1(x)] \wedge \tag{3.58}$$

$$\forall x[(\forall y[R(y, x) \rightarrow (X_1(y) \vee X_3(y))] \wedge \exists y[R(y, x) \wedge X_3(y)] \vee \\ (\neg X_1(x) \wedge \neg X_2(x))) \rightarrow X_3(x)]\}. \tag{3.59}$$

We have two strata: $\{(3.57), (3.58)\}$ and $\{(3.59)\}$. Using Theorem 3.22, we first simultaneously eliminate $\exists X_1 \exists X_2$ and then $\exists X_3$.

The elimination of $\exists X_1 \exists X_2$ provides us with the following definition of X_1 and X_2:

$$\textsc{Slfp}[X_1(x) \equiv \exists y[R(y, x) \wedge X_2(y)], \; X_2(x) \equiv \forall y[R(y, x) \rightarrow X_1(y)]]. \tag{3.60}$$

The elimination of $\exists X_3$ provides us with the following definition of X_3:

$$\textsc{Lfp}[X_3(x) \equiv \forall y[R(y, x) \rightarrow (X_1(y) \vee X_3(y))] \wedge \exists y[R(y, x) \wedge X_3(y)] \\ \vee (\neg X_1(x) \wedge \neg X_2(x))]. \tag{3.61}$$

The result of elimination is then

$$\neg \{\forall x[X_2(x) \rightarrow Q_2(x)] \wedge \exists z[Q_2(z) \wedge \neg X_2(z)] \wedge \\ \forall x[\neg X_1(x) \vee \neg X_2(x)] \wedge \forall x[\neg X_1(x) \vee \neg X_3(x)] \wedge \forall x[\neg X_2(x) \vee \neg X_3(x)]\} \tag{3.62}$$

with X_1, X_2, X_3 respectively substituted by definitions given by formulas (3.60) and (3.61).

Formula (3.62) can then be presented in a more readable form as:

$$\left. \begin{array}{l} \forall x[X_2(x) \rightarrow Q_2(x)] \wedge \forall x[\neg X_1(x) \vee \neg X_2(x)] \wedge \\ \forall x[\neg X_1(x) \vee \neg X_3(x)] \wedge \forall x[\neg X_2(x) \vee \neg X_3(x)] \end{array} \right\} \rightarrow \forall z[Q_2(z) \rightarrow X_2(z)].$$

Since the result is a fixpoint formula, we have the following corollary.

Corollary 3.32. *Model checking problem for the theory $\mathcal{A}(R, Q_1, Q_2, Q_3)$ with circumscriptive policy expressed by $Circ_\downarrow(\mathcal{A}(R, Q_1, Q_2, Q_3); Q_2; Q_1, Q_3)$ is in* PTIME *in the size of the structure.* □

Maximization on Q_2

Consider the dual circumscription formula $Circ^\uparrow(\mathcal{A}(R, Q_1, Q_2, Q_3); Q_2; Q_1, Q_3)$:

$$\mathcal{A}(R, Q_1, Q_2, Q_3) \wedge \tag{3.63}$$

$$\forall X_1 \forall X_2 \forall X_3 [(\mathcal{A}(R, X_1, X_2, X_3) \wedge \forall x [Q_2(x) \to X_2(x)]) \to$$
$$\forall x [X_2(x) \to Q_2(x)]]. \tag{3.64}$$

We focus on (3.64), which is equivalent to

$$\neg \exists X_1 \exists X_2 \exists X_3 \{\forall x [X_1(x) \vee X_2(x) \vee X_3(x)] \wedge$$
$$\forall x [\neg X_1(x) \vee \neg X_2(x)] \wedge \forall x [\neg X_1(x) \vee \neg X_3(x)] \wedge \forall x [\neg X_2(x) \vee \neg X_3(x)] \wedge$$
$$\forall x [\forall y [R(y, x) \to X_1(y)] \to X_2(x)] \wedge$$
$$\forall x [\exists y [R(y, x) \wedge X_2(y)] \to X_1(x)] \wedge$$
$$\forall x [(\forall y [R(y, x) \to (X_1(y) \vee X_3(y))] \wedge \exists y [R(y, x) \wedge X_3(y)]) \to X_3(x)] \wedge$$
$$\forall x [Q_2(x) \to X_2(x)] \wedge \exists z [X_2(z) \wedge \neg Q_2(z)]]\}$$

and further to

$$\neg \exists z \exists X_1 \exists X_2 \exists X_3 \{\neg Q_2(z) \wedge \forall x [\neg X_1(x) \vee \neg X_2(x)] \wedge \tag{3.65}$$

$$\forall x [\neg X_1(x) \vee \neg X_3(x)] \wedge \forall x [\neg X_2(x) \vee \neg X_3(x)] \wedge \tag{3.66}$$

$$\forall x [(\forall y [R(y, x) \to X_1(y)] \vee Q_2(x) \vee x = z) \to X_2(x)] \wedge \tag{3.67}$$

$$\forall x [\exists y [R(y, x) \wedge X_2(y)] \to X_1(x)] \wedge \tag{3.68}$$

$$\forall x [(\forall y [R(y, x) \to (X_1(y) \vee X_3(y))] \wedge \exists y [R(y, x) \wedge X_3(y)] \vee$$
$$(\neg X_1(x) \wedge \neg X_2(x))) \to X_3(x)]\}. \tag{3.69}$$

We have two strata: $\{(3.67), (3.68)\}, \{(3.69)\}$. Using Theorem 3.22, we first simultaneously eliminate $\exists X_1 \exists X_2$ and then $\exists X_3$.

The elimination of $\exists X_1 \exists X_2$ provides us with the following definition of X_1 and X_2:

$$\text{SLFP}[X_1(x) \equiv \exists y [R(y, x) \wedge X_2(y)],$$
$$X_2(x) \equiv (\forall y [R(y, x) \to X_1(y)] \vee Q_2(x) \vee x = z)]. \tag{3.70}$$

The elimination of $\exists X_3$ provides us with the definition of X_3 given by (3.61). The result of elimination is then

$$\neg \exists z \{\neg Q_2(z) \wedge \forall x [\neg X_1(x) \vee \neg X_2(x)] \wedge$$
$$\forall x [\neg X_1(x) \vee \neg X_3(x)] \wedge \forall x [\neg X_2(x) \vee \neg X_3(x)]\} \tag{3.71}$$

with X_1, X_2, X_3 respectively substituted by definitions provided by formulas (3.61) and (3.70).

Formula (3.71) can then be presented in a more readable form as:

$$\forall z \left\{ \left(\begin{array}{c} \forall x [\neg X_1(x) \vee \neg X_2(x)] \wedge \\ \forall x [\neg X_1(x) \vee \neg X_3(x)] \wedge \forall x [\neg X_2(x) \vee \neg X_3(x)] \end{array} \right) \to Q_2(z) \right\}.$$

Note that the quantifier $\forall z$ cannot be moved to $Q_2(z)$, since z appears in the antecedent as a part of definition of X_2 given by (3.70).

Since the resulting formula is a fixpoint formula, we have the following corollary.

Corollary 3.33. *Model checking problem for the theory $\mathcal{A}(R, Q_1, Q_2, Q_3)$ with circumscriptive policy expressed by* $Circ^\uparrow(\mathcal{A}(R, Q_1, Q_2, Q_3); Q_2; Q_1, Q_3)$ *is in* PTime *in the size of the model.* □

Minimization on Q_3

Consider the circumscription formula $Circ_\downarrow(\mathcal{A}; Q_3; Q_1, Q_2)$:

$$\mathcal{A}(R, Q_1, Q_2, Q_3) \wedge \tag{3.72}$$

$$\forall X_1 \forall X_2 \forall X_3[(\mathcal{A}(R, X_1, X_2, X_3) \wedge \forall x[X_3(x) \to Q_3(x)]) \to$$
$$\forall x[Q_3(x) \to X_3(x)]]. \tag{3.73}$$

We focus on (3.73), which is equivalent to

$$\neg \exists X_1 \exists X_2 \exists X_3 \{\forall x[X_1(x) \vee X_2(x) \vee X_3(x)] \wedge$$
$$\forall x[\neg X_1(x) \vee \neg X_2(x)] \wedge \forall x[\neg X_1(x) \vee \neg X_3(x)] \wedge \forall x[\neg X_2(x) \vee \neg X_3(x)] \wedge$$
$$\forall x[\forall y[R(y, x) \to X_1(y)] \to X_2(x)] \wedge$$
$$\forall x[\exists y[R(y, x) \wedge X_2(y)] \to X_1(x)] \wedge$$
$$\forall x[(\forall y[R(y, x) \to (X_1(y) \vee X_3(y))] \wedge \exists y[R(y, x) \wedge X_3(y)]) \to X_3(x)] \wedge$$
$$\forall x[X_3(x) \to Q_3(x)] \wedge \exists z[Q_3(z) \wedge \neg X_3(z)]]\}.$$

As in Section 3.6.3, we obtain

$$\neg \exists X_1 \exists X_2 \exists X_3 \{\forall x[X_3(x) \to Q_3(x)] \wedge \exists z[Q_3(z) \wedge \neg X_3(z)] \wedge \tag{3.74}$$

$$\forall x[\neg X_1(x) \vee \neg X_2(x)] \wedge \forall x[\neg X_1(x) \vee \neg X_3(x)] \wedge \forall x[\neg X_2(x) \vee \neg X_3(x)] \wedge \tag{3.75}$$

$$\forall x[\forall y[R(y, x) \to X_1(y)] \to X_2(x)] \wedge \tag{3.76}$$

$$\forall x[\exists y[R(y, x) \wedge X_2(y)] \to X_1(x)] \wedge \tag{3.77}$$

$$\forall x[(\forall y[R(y, x) \to (X_1(y) \vee X_3(y))] \wedge \exists y[R(y, x) \wedge X_3(y)] \vee$$
$$(\neg X_1(x) \wedge \neg X_2(x))) \to X_3(x)]\}. \tag{3.78}$$

We have two strata: $\{(3.76), (3.77)\}$ and $\{(3.78)\}$, which are the same as in Section 3.6.3. The result of elimination is then

$$\neg \{\forall x[X_3(x) \to Q_3(x)] \wedge \exists z[Q_3(z) \wedge \neg X_3(z)] \wedge$$
$$\forall x[\neg X_1(x) \vee \neg X_2(x)] \wedge \forall x[\neg X_1(x) \vee \neg X_3(x)] \wedge \forall x[\neg X_2(x) \vee \neg X_3(x)]\} \tag{3.79}$$

with X_1, X_2, X_3 respectively substituted by definitions given by formulas (3.60) and (3.61). Formula (3.79) can be presented as:

$$\left.\begin{array}{l} \forall x[X_3(x) \to Q_3(x)] \wedge \forall x[\neg X_1(x) \vee \neg X_2(x)] \wedge \\ \forall x[\neg X_1(x) \vee \neg X_3(x)] \wedge \forall x[\neg X_2(x) \vee \neg X_3(x)] \end{array}\right\} \to \forall z[Q_3(z) \to X_3(z)].$$

Since the result is a fixpoint formula, we have the following corollary.

Corollary 3.34. *Model checking problem for the theory $\mathcal{A}(R, Q_1, Q_2, Q_3)$ with circumscriptive policy expressed by $Circ_\downarrow(\mathcal{A}(R, Q_1, Q_2, Q_3); Q_3; Q_1, Q_2)$ is in* PTIME *in the size of the structure.* □

3.6.4 Reducing circumscription in the annotation theory related to approximate reasoning

Let us now illustrate techniques introduced in Section 3.6.2.

Consider the circumscription formula $Circ_\downarrow(\mathcal{R}; Q_1, Q_3; Q_2)$:

$$\mathcal{R}(R, Q_1, Q_2, Q_3)\wedge \tag{3.80}$$

$$\forall X_1 \forall X_2 \forall X_3[(\mathcal{R}(R, X_1, X_2, X_3)\wedge$$
$$\forall x[X_1(x) \rightarrow Q_1(x)] \wedge \forall x[X_3(x) \rightarrow Q_3(x)]) \rightarrow \tag{3.81}$$
$$\forall x[Q_1(x) \rightarrow X_1(x)] \wedge \forall x[Q_3(x) \rightarrow X_3(x)]].$$

We focus on (3.81).

We first apply Proposition 3.29, so attempt to eliminate second-order quantifiers from

$$\exists X_2\{\forall x[X_1(x) \vee X_2(x) \vee X_3(x)]\wedge$$
$$\forall x[\neg X_1(x)\vee\neg X_2(x)]\wedge \forall x[\neg X_1(x)\vee\neg X_3(x)]\wedge \forall x[\neg X_2(x)\vee\neg X_3(x)]\wedge$$
$$\forall x\forall y[(R(x, y) \wedge X_1(x)) \rightarrow X_1(y)]\wedge$$
$$\forall x\forall y[(R(x, y) \wedge X_2(x)) \rightarrow X_2(y)]\wedge$$
$$\forall x\forall y[(R(x, y) \wedge X_3(x)) \rightarrow X_3(y)]\}.$$

The above formula is equivalent to

$$\exists X_2\{\forall x[\neg X_1(x)\vee\neg X_2(x)]\wedge \forall x[\neg X_1(x)\vee\neg X_3(x)]\wedge \forall x[\neg X_2(x)\vee\neg X_3(x)]\wedge$$
$$\forall x\forall y[(R(x, y) \wedge X_1(x)) \rightarrow X_1(y)]\wedge$$
$$\forall y[((\neg X_1(y) \wedge \neg X_3(y)) \vee \exists x[R(x, y) \wedge X_2(x)]) \rightarrow X_2(y)]\wedge$$
$$\forall x\forall y[(R(x, y) \wedge X_3(x)) \rightarrow X_3(y)]\}.$$

Using Theorem 3.14, we obtain the following definition of X_2:

$$\text{L}_{\text{FP}}[X_2(y) \equiv ((\neg X_1(y) \wedge \neg X_3(y)) \vee \exists x[R(x, y) \wedge X_2(x)])].$$

Now we can use Lemma 3.28, which allows us to remove formulas where X_1 and X_3 appear only negatively, and we obtain that (3.81) is equivalent to

$$\neg\exists X_1\exists X_3[\mathcal{R}'(R, X_1, X_2, X_3)\wedge$$
$$\forall x[X_1(x) \rightarrow Q_1(x)] \wedge \forall x[X_3(x) \rightarrow Q_3(x)]\wedge \tag{3.82}$$
$$(\exists x[Q_1(x) \wedge \neg X_1(x)] \vee \exists x[Q_3(x) \wedge \neg X_3(x)])],$$

where $\mathcal{R}'$ is the conjunction

$$\forall x\forall y[(R(x, y) \wedge X_1(x)) \rightarrow X_1(y)]\wedge$$
$$\forall x\forall y[(R(x, y) \wedge X_3(x)) \rightarrow X_3(y)]. \tag{3.83}$$

Formula (3.82) is then equivalent to

$$\neg\{\exists X_1\exists X_3[\mathcal{R}'(R, X_1, X_2, X_3)\wedge \tag{3.84}$$
$$\forall x[X_1(x) \rightarrow Q_1(x)] \wedge \forall x[X_3(x) \rightarrow Q_3(x)]\wedge \tag{3.85}$$
$$\exists x[Q_1(x) \wedge \neg X_1(x)]]\vee \tag{3.86}$$
$$\exists X_1\exists X_2\exists X_3[\mathcal{R}'(R, X_1, X_2, X_3)\wedge \tag{3.87}$$
$$\forall x[X_1(x) \rightarrow Q_1(x)] \wedge \forall x[X_3(x) \rightarrow Q_3(x)]\wedge \tag{3.88}$$
$$\exists x[Q_3(x) \wedge \neg X_3(x)])]\}, \tag{3.89}$$

so in the scope of the outermost negation we have a disjunction of two second-order formulas, $((3.84)–(3.86)) \vee ((3.87)–(3.89))$. We start with the first disjunct:

$$\exists x \exists X_1 \exists X_3 [\forall y [\exists x [R(x, y) \wedge X_1(x)] \rightarrow X_1(y)] \wedge$$
$$\forall y [\exists x [R(x, y) \wedge X_3(x)] \rightarrow X_3(y)] \wedge$$
$$\forall x [X_1(x) \rightarrow Q_1(x)] \wedge \forall x [X_3(x) \rightarrow Q_3(x)] \wedge$$
$$Q_1(x) \wedge \neg X_1(x)].$$

Observe that for both X_1 and X_3 satisfy assumptions of Proposition 3.31, so the formula reduces to $\exists x [Q_1(x)]$. Similarly, the second disjunct reduces to $\exists x [Q_3(x)]$. When we move negation inside, disjunction is switched to conjunction, so the final result is

$$\forall x [\neg Q_1(x)] \wedge \forall x [\neg Q_3(x)]. \tag{3.90}$$

Since (3.90) is a classical first-order formula, we have the following corollary.

Corollary 3.35. *Model checking problem for the theory* $\mathcal{R}(R, Q_1, Q_2, Q_3)$ *with circumscriptive policy expressed by* $Circ_\downarrow(\mathcal{R}; Q_1, Q_3; Q_2)$ *is in* LOGSPACE *(so also in* PTIME*) in the size of the structure.* □

We have Corollary 3.35 because we did not place any positive facts as to Q_1 and Q_3. Such facts would contribute to the result. In such a case Theorem 3.13 would be applicable, so we have the following proposition.

Proposition 3.36. *Model checking problem for the theory* $\mathcal{R}(R, Q_1, Q_2, Q_3)$ *with circumscriptive policy expressed by* $Circ_\downarrow(\mathcal{R}; Q_1, Q_3; Q_2)$ *and additional positive facts concerning* Q_1 *and/or* Q_2 *is in* PTIME *in the size of the model.* □

3.6.5 Reducing circumscription in the formalization of semantics of logic programs

Consider the circumscription formula $Circ_\downarrow(\mathcal{L}; Q_2; Q_1, Q_3)$:

$$\mathcal{L}(R_+, R_-, Q_1, Q_2, Q_3) \wedge \tag{3.91}$$
$$\forall X_1 \forall X_2 \forall X_3 [(\mathcal{L}(R_+, R_-, X_1, X_2, X_3) \wedge \forall x [X_2(x) \rightarrow Q_2(x)]) \rightarrow$$
$$\forall z [Q_2(z) \rightarrow X_2(z)]]. \tag{3.92}$$

We focus on (3.92), which is equivalent to

$$\neg \exists z \exists X_1 \exists X_2 \exists X_3 \{\forall x [X_2(x) \rightarrow Q_2(x)] \wedge Q_2(z) \wedge \neg X_2(z) \wedge \tag{3.93}$$
$$\forall x [\neg X_1(x) \vee \neg X_2(x)] \wedge \forall x [\neg X_1(x) \vee \neg X_3(x)] \wedge \forall x [\neg X_2(x) \vee \neg X_3(x)] \wedge \tag{3.94}$$
$$\forall x [(\exists y [R_+(y, x) \wedge X_1(y)] \vee \exists y [R_-(y, x) \wedge X_2(y)]) \rightarrow X_1(x)] \wedge \tag{3.95}$$
$$\forall x [(\forall y [R_+(y, x) \rightarrow X_2(y)] \wedge \forall y [R_-(y, x) \rightarrow X_1(y)]) \rightarrow X_2(x)] \wedge \tag{3.96}$$
$$\forall x [(\forall y [R_+(y, x) \rightarrow (X_2(y) \vee X_3(y))] \wedge$$
$$\forall y [R_-(y, x) \rightarrow (X_1(y) \vee X_3(y))] \wedge \tag{3.97}$$
$$\exists y [((R_-(y, x) \vee R_+(y, x)) \wedge X_3(y))] \vee (\neg X_1(x) \wedge \neg X_2(x))) \rightarrow X_3(x)].$$

We have two strata: $\{(3.95), (3.96)\}$ and $\{(3.97)\}$. Using Theorem 3.22, we first simultaneously eliminate $\exists X_1 \exists X_2$ and then $\exists X_3$.

The elimination of $\exists X_1 \exists X_2$ provides us with the following definition of X_1 and X_2:

$$\text{SLFP}[X_1(x) \equiv (\exists y [R_+(y, x) \wedge X_1(y)] \vee \exists y [R_-(y, x) \wedge X_2(y)]),$$
$$X_2(x) \equiv (\forall y [R_+(y, x) \rightarrow X_2(y)] \wedge \forall y [R_-(y, x) \rightarrow X_1(y)])]. \tag{3.98}$$

The elimination of $\exists X_3$ provides us with the following definition of X_3:

$$\text{L}_\text{FP}[X_3(x) \equiv (\forall y[R_+(y,x) \rightarrow (X_2(y) \vee X_3(y))] \wedge$$
$$\forall y[R_-(y,x) \rightarrow (X_1(y) \vee X_3(y))] \wedge \qquad (3.99)$$
$$\exists y[((R_-(y,x) \vee R_+(y,x)) \wedge X_3(y))] \vee (\neg X_1(x) \wedge \neg X_2(x)))].$$

The result of elimination is then

$$\neg \exists z[\forall x[X_2(x) \rightarrow Q_2(x)] \wedge Q_2(z) \wedge \neg X_2(z) \wedge \forall x[\neg X_1(x) \vee \neg X_2(x)] \wedge$$
$$\forall x[\neg X_1(x) \vee \neg X_3(x)] \wedge \forall x[\neg X_2(x) \vee \neg X_3(x)]], \qquad (3.100)$$

with X_1, X_2, X_3 respectively substituted by definitions given by formulas (3.98) and (3.99). Formula (3.100) can be presented in a more readable form:

$$\left. \begin{array}{l} \forall x\{X_2(x) \rightarrow Q_2(x)] \wedge \forall x[\neg X_1(x) \vee \neg X_2(x)] \wedge \\ \forall x[\neg X_1(x) \vee \neg X_3(x)] \wedge \forall x[\neg X_2(x) \vee \neg X_3(x)] \end{array} \right\} \rightarrow \forall z[Q_2(z) \rightarrow X_2(z)].$$

Since the result is a fixpoint formula, we have the following corollary.

Corollary 3.37. *Model checking problem for theory* $\mathcal{L}(R_+, R_-, Q_1, Q_2, Q_3)$ *with circumscriptive policy expressed by* $\text{Circ}_\downarrow(\mathcal{L}; Q_2; Q_1, Q_3)$ *is in* PTIME *in the size of the structure.*   □

3.7 Conclusions

In this chapter we introduced the concept of annotation theories and showed that such theories, together with minimization/maximization policies expressed by means of circumscription, are rich enough to capture important phenomena appearing in many applications, including specific theories of argumentation, approximate reasoning as well as semantics of logic programs with negation.

Even if circumscription is substantially a second-order formalism, we provided a number of results allowing to eliminate second-order quantifiers. Even simpler methods, based on results from [112, 272] appear quite powerful and applicable to a wide class of circumscribed formulas (see also [110, 155]). The problem of quantifier elimination in annotation theories appears, in general, as difficult as the question whether PTIME =NPTIME, so considering particular annotation theories is an interesting research area, far from being completed.

Annotation theories deserve further investigations. In particular, an interesting problem is to search for algorithms for finding annotations, especially ones that construct the model incrementally on the graph. In general this problem is as difficult as PTIME =NPTIME, as shown in Section 3.4. However, in the case of stratified theories we can apply Theorem 3.22, allowing us to reduce complexity to PTIME.[11]

Also, using theory approximation [302, 303, 72, 73, 106] is worth investigating in the context of second-order quantifier elimination from circumscribed annotation theories. Namely, when such elimination is not possible using Theorem 3.22, one can approximate considered theories by theories admitting quantifier elimination. We leave these subjects for future research.

---

[11] This method is applicable to annotation theories considered in Sections 3.3.2, 3.3.3 and 3.3.4.

4

Logical Modes of Attack in Argumentation Networks

4.1 Introduction

An abstract argumentation network has the form (S, R), where S is a nonempty set of arguments and $R \subseteq S \times S$ is an attack relation. When $(x, y) \in R$, we say x attacks y.

The elements of S are atomic arguments and the model does not give any information on what structure they have and how they manage to attack each other.

The abstract theory is concerned with extracting information from the network in the form of a set of arguments which are winning (or 'in'), a set of arguments which are defeated (or are 'out') and the rest are undecided. There are several possibilities for such sets and they are systematically studied and classified. See Figure 4.1 for a typical situation. $x \to y$ in the figure represents $(x, y) \in R$.

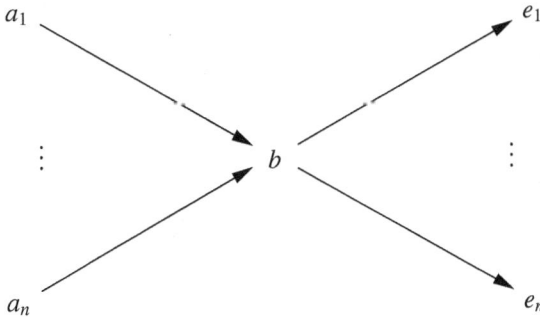

Fig. 4.1.

A good way to see what is going on is to consider a Caminada labelling. This is a function λ on S distributing values $\lambda(x), x \in S$ in the set {in, out, ?} satisfying the following conditions.

1. If x is not attacked by any y then $\lambda(x) = 1$
2. If $(y, x) \in R$ and $\lambda(y) = 1$ then $\lambda(x) = 0$
3. If all y which attack x have $\lambda(y) = 0$ then $\lambda(x) = 1$.

4. If one y which attack x has $\lambda(y) =?$ and all other y have $\lambda(y) \in \{0, ?\}$ then $\lambda(x) =?$.

Such λ exist whenever S is finite and for any such λ, the set $S_\lambda^+ = \{x \mid \lambda(x) = 1\}$ is the set of winning arguments, $S_\lambda^- = \{x \mid \lambda(x) = 0\}$ is the set of defeated arguments and $S_\lambda^? = \{x \mid \lambda(x) =?\}$ is the set of undecided arguments.

The features of this abstract model are as follows:

1. Arguments are atomic, have no structure.
2. Attacks are stipulated by the relation R; we have no information on how and why they occur.
3. Arguments are either 'in' in which case all their attacks are active or are 'out' in which case all their attacks are inactive. There is no in between state (partially active, can do some attacks, etc.). Arguments can be undecided.
4. Attacks have a single strength, no degrees of strength or degree of transmission of attack along the arrow, etc.
5. There are no counter attacks, no defensive actions allowed or any other responses or counter measures.
6. The attacks from x are uniform on all y such that $(x, y) \in R$. There are no directional attacks or coordinated attacks. In Figure 4.1, $a_1, \ldots, a_n$ attack b individually and not in coordination. For example, a_1 does not attack b with a view of stopping b from attaching e_1 but without regard to $e_2, \ldots, e_n$.
7. The view of the network is static. We have a graph here and a relation R on it. So Figure 4.1 is static. We seek a λ labelling on it and we may find several. In the case of Figure 4.1 there is only one such λ. $\lambda(a_i) = 1, \lambda(b) = 0, \lambda(e_j) = 1, i = 1, \ldots, , j = 1, \ldots, n$.

   We do not have a dynamic view, like first a_i attack b and b then (if it is not out dead) tries to attack e_i. Or better still, at the same time each node launches an attack on whoever it can. So a_i attack b and b attacks e_i and the result is that a_i are alive (not being attacked) while b and e_j are all dead.

   We use the words 'there is no progression in the network' to indicate this. The network is static.

We will address point 4 above in Chapter 9, but points 1–3, 5–7 remain only partially untreated by us. It is our aim in this chapter to give theoretical answers to these questions.

There are several authors who have already addressed some of these questions. See [47, 79]. We shall build upon their work, especially [79].

Obviously, to answer the above questions we must give contents to the nodes. We can do this in two ways. We can do this in the metalevel, by putting predicates and labels on the nodes and by writing axioms about them or we can do it in the object level, giving internal structure to the atomic arguments and/or saying what they are and defining the other concepts, e.g. the notion of attack in terms of the contents.

Example 4.1 (Metalevel connects to nodes). Figure 4.2 is an example of a metalevel extension.

The node a is labelled by α. It attacks the node b with transmission factor ε. Node b is labelled by β. The attack arrow itself constitutes an attack on the attack arrow from b to c. This attack is itself attacked by node b. Each attack has its own transmission factor. We denoted attacks on arrows by double arrows.

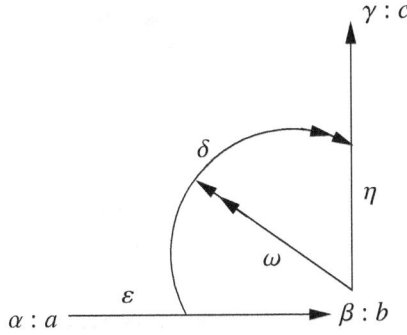

Fig. 4.2.

Formally we have a set S of nodes, here

$$S = \{a, b, c\}.$$

the relation R is more complex. It has the usual arrows $\{(a, b), (b, c)\} \subseteq R$ and also the double arrows, namely, $\{((a, b), (b, c)), (b, ((a, b), (b, c)))\} \subseteq R$. We have a labelling function $\mathbf{l}$, giving values

$$\mathbf{l}(a) = \alpha, \mathbf{l}(b) = \beta, \mathbf{l}(c) = \gamma,$$
$$\mathbf{l}((a, b)) = \varepsilon, \mathbf{l}((b, c)) = \eta,$$
$$\mathbf{l}(((a, b), (b, c))) = \delta$$
$$\mathbf{l}((a, ((a, b), (b, c)))) = \omega.$$

We can generalise the Caminada labelling as a function from $S \cup R$ to some values which satisfy some conditions involving the labels. We can write axioms about the labels in some logical language and these axioms will give more meaning to the argumentation network. See Chapter 9 for some details along these lines. The appropriate language and logic to do this is Labelled Deductive Systems (LDS) [136].

We shall not pursue the metalevel extensions approach in this chapter except for one well known construction which will prove useful to us later.

Example 4.2 (The logic program associated with an ordinary abstract network $N =$ (S, R)). Consider S as a set of literals. Let $\Rightarrow$ be the logic programming arrow and let $\wedge, \neg$ be conjunction and negation as failure. Consider the logic program $P(N)$ containing the following clauses $C_x, x \in S$

$$C_x : \bigwedge_{i=1}^{m} \neg y_i \Rightarrow x$$

where $y_1, \ldots, y_m$ are all the nodes in S which attack x (i.e. $(\bigwedge_{(y,c) \in R} y) \Rightarrow x$).

If no node attacks x then $C_x = x$.

C_x simply says in logic programming language x is in if all y which attack it are out (i.e. $\neg y$).

Compare with Example 4.10.

We are now ready for our second approach, namely giving logical content to nodes.

Assume we are using a certain logic **L**. **L** can be monotonic, nonmonotonic, algorithmic, etc. At this stage anything will do. This logic has the notion of formulas A of the logic, theories Δ of the logic and the notion of $\Delta \vdash A$ and possibly also the notion of Δ is not consistent.

The simplest approach is to assume the nodes $x \in S$ are theories Δ_x supporting logically a formula A_x (i.e. $\Delta_x \vdash A_x$ in the logic). The exact nature of the nodes will determine our options for defining attacks of one node on another.

We list the important parameters.

1. The nature of the logic at node x and how it is presented. The logic can be classical logic, intuitionistic logic, substructural logic, nonmonotonic logic, etc. It can be presented proof theoretically, or semantically or as a consequence relation, or just as an algorithm.
2. What is Δ_x? A set of wffs? A proof? A network (e.g. a Bayesian network) with algorithms to extract information from it? etc.
3. The nature of the support Δ_x gives A_x. We can have $\Delta_x \vdash A_x$, or we can have that A_x is extracted from Δ_x by some algorithm $\mathcal{A}_x$ (e.g. abduction algorithms, etc.).
4. How does the node x attack other nodes? Does it have a stock of attack formulas $\{\alpha_1, \alpha_2, \ldots\}$ that it uses? Does it use A_x? etc.
5. What does the node x do when it is attacked? How does it respond? Does it counter attack? Does it transform itself? Does it die (become inconsistent)?
6. To define the notion of an attack one must give precise formal definitions of all the parameters involved.

We give several examples of network and attack options.

Example 4.3 (Networks based on monotonic logic). Let **L** be any monotonic logic, with a notion of inconsistency. Let the nodes have the form $x = (\Delta_x, A_x)$ where Δ_x is a set of formulas such that $\Delta_x \vdash A_x$ and Δ_x is a minimal such set (i.e. no $\Theta \subsetneq \Delta_x$ can prove A_x).

Δ_x attacks Δ_y by forcing itself onto Δ_y (i.e. forming $\Delta_x \cup \Delta_y$). If the result is inconsistent then a revision process starts working and a maximal $\Theta_y \subseteq \Delta_y$ is chosen such that $\Delta_x \cup \Theta_y$ is consistent. The result of the attack on y is Θ_y. Of course if $\Delta_x \cup \Delta_y$ is consistent then the attack fails, as $\Theta_y = \Delta_y \vdash A_y$, otherwise $\Theta_y \nvdash A_y$ and the attack succeeds. However, the node y transforms itself into the logically weaker node $(\Theta_y, \bigwedge \Theta_y)$.

Note that unless Θ_y is empty, the new transformed Θ_y is still capable of attack.

To give a specific example, consider the two nodes:

$$x = (\neg A, \neg A) \text{ and } y = ((A, A \to B), B)$$

x attacks y and the result of the attack is a new $\Delta'_y = \{A \to B\}$. Δ'_y can still attack its targets though with less force.

Consider the following sequence

$$z = (\neg E, \neg E), x' = ((E, \neg A), \neg A \wedge E), y = ((A, A \to B), B)$$

Here z attacks x', x' as a result of the attack regroups itself into x and proceeds to attack y.

Note that we view progression from left to right along R.
Consider now the following Figure 4.3, as a third example:

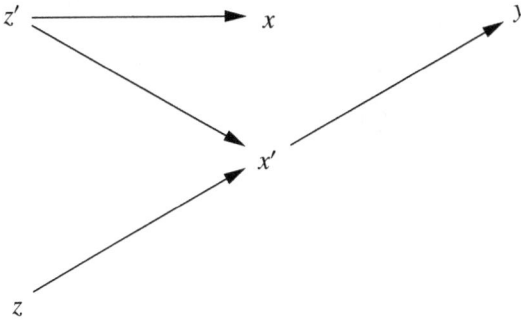

Fig. 4.3.

where $z' = (A, A)$ and z, x', x and y as before. Because of the attack of z', x' cannot regroup itself into x because x is also being attacked.
Consider now a fourth example, Figure 4.4

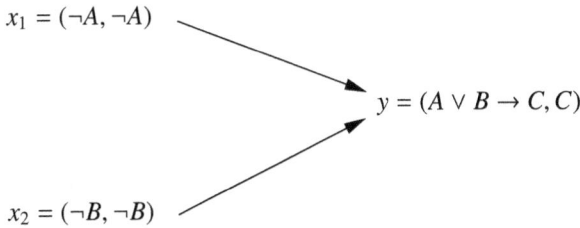

$x_1 = (\neg A, \neg A)$

$y = (A \lor B \rightarrow C, C)$

$x_2 = (\neg B, \neg B)$

Fig. 4.4.

Here neither x_1 nor x_2 can cripple y but a joint attack can.

Remark 4.4 (Summary of options for the monotonic example).

1. Attacks are done by hurling oneself at the target. This can be refined further by allowing sending different formulas at different targets.
2. Attacks can be combined.
3. The target may be crippled but can still 'regroup' and attack some of its own targets.
4. The nature of any attack is based on inconsistency and revision.
5. We can sequence the attacks as a progression along the relation R.
6. Attacks are not symmetrical since we use revision. So if A attacks $\sim A$, AGM revision for example, will give preference to A. So for $\sim A$ to attack A it has to do so explicitly, and the winner is determined by the progression of the attack sequence.

Example 4.5 (Networks based on nonmonotonic logic). This example allows for nodes of the form $x = (\Delta_x, A_x)$ where the underlying logic is a nonmonotonic consequence $\vdash$. In nonmonotonic logic we know that we may have $\Delta_y \vdash A_y$ but $\Delta_y + B \not\vdash A_y$.[1]

So if node $x = (\Delta_x, A_x)$ attacks node y, it simply adds Δ_x to Δ_y and we get $y' = \Delta_x \cup \Delta_u \vdash ?A_y$.

Here the attack is based on giving more information and not on inconsistency and revision.

to show the difference, let Δ_y be

1. Bird $(a) \mapsto$ Fly (a)
2. Penguin $(a) \wedge$ Bird $(a) \mapsto \sim$ Fly (a)
3. Bird (a)

where $\mapsto$ is defeasible implication.

Let A_y be Fly (a).

Let Δ_x and A_x be Penguin (a). x can attack by sending the extra information that Penguin (a). Another attack from another point x' can be by sending $\sim$Bird (a) i.e. let x' be $\Delta_{x'} = A_{x'} =\sim$ Bird (a).

Example 4.6 (Prolog programs). The theories here are Prolog programs and the argu-

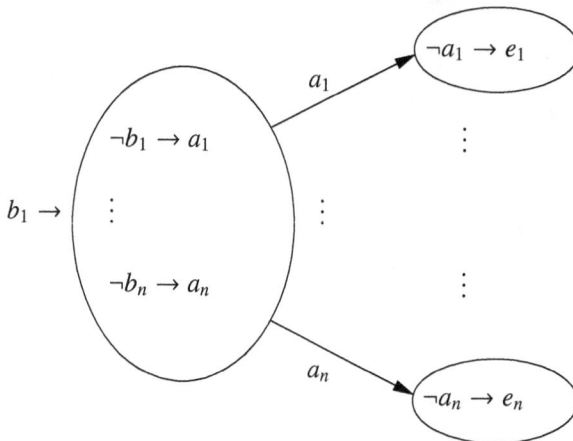

Fig. 4.5.

ments are the literals they prove.

An attack is executed by sending a literal from the attacking theory to the to target theory. See Figure 4.5

---

[1] A nonmonotonic consequence on the wffs of the logic satisfies three minimal properties

1. Reflexivity: $\Delta \vdash A$ if $A \in \Delta$.
2. Restricted monotonicity: $\Delta \vdash A$ and $\Delta \vdash B$ imply $\Delta, A \vdash B$.
3. Cut Rule: $\Delta, A \vdash B$ and $\Delta \vdash A$ imply $\Delta \vdash B$.

Note that $\vdash$ can be presented in many ways, semantically, proof theoretically or algorithmically.

Example 4.7 (Counter-attack). The Dung framework does not allow for counter-attacks being effective only when attacked but not before. The model is static. The attacks do not 'progress' along the network like a flow going through the nodes activating them as it goes along. However, if we perceive such progression, we can define the concept of counter-attack.

Consider Figure 4.6

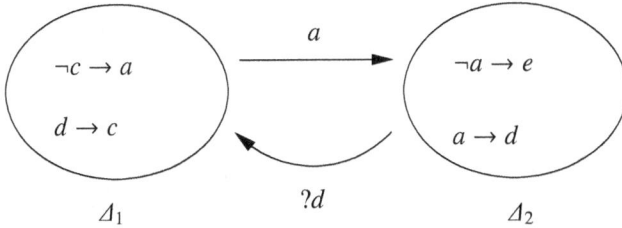

Fig. 4.6.

$\Delta_1 \vdash a$ and can attack Δ_2 by passing a along the attacking arrow. The $?d$ is a counter-attack. As long as Δ_2 is not attacked by Δ_1, d is not provable and so cannot be sent to Δ_1. Once Δ_2 is attacked then d becomes provable and can counter-attack Δ_1 and render a unprovable.

Example 4.8 (Directional attacks). The following is a more enriched logical model where more options are naturally available. We can let a node a be a nonmonotonic theory Δ_a such that $\Delta_a \vdash a$. We can understand an attack of a nonmonotonic node a on, say, node e_1 as the transmission of an item of data say α_1 (such that $\Delta_a \vdash \alpha_1$) to Δ_{e_1} with the effect that $\Delta_{e_1} + \alpha_1 \not\vdash e_1$. Since Δ_{e_1} is nonmonotonic, the insertion of α_1 into it may change what it can prove. See Figure 4.7.

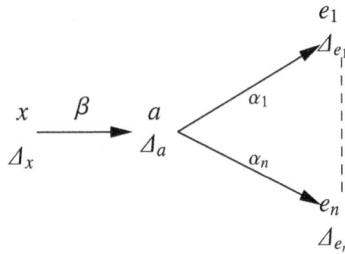

Fig. 4.7.

We have $\Delta_x \vdash \beta, \Delta_x \vdash x, \Delta_a \vdash a, \Delta_a \vdash \alpha_i, i = 1, \ldots, n$.
We may have

$$\Delta_a + \beta \not\vdash \alpha_1$$

and therefore the attack on e_1 fails, but we may still have that $\Delta_a + \beta \vdash \alpha_n$, hence the attack on e_n still succeeds. The attack by β is not a specific attack on the arrow from a

to e_1. It tansforms a to something else which does not attack e_1. So Figure 4.7 is not a good representation of it. It shows the result but not the meaning.

By the way, to attack the attack from x to a in Figure 4.7, we might add a formula β' to β, and so the attack changes from β to (β and β').

Example 4.9 (Abduction). Another example can be abduction.

The node y contains an argument of the following form. It says, we know of Δ_y and the fact that a formula E_y should be provable, but Δ_y cannot prove E_y. So we abduce A_y as the most reasonable additional hypothesis. So the node y is (Δ_y, A_y), where $A_y = Abduce(\Delta_y, E_y)$.

x can attack by sending additional information Δ_x. It may be that $\Delta_y \cup \Delta_x \nvdash E_y$, but $Abduce(\Delta_y \cup A_x, E_y)$ is some A'_y and not A_y.

An example that I like is from Euclid. Euclid proved that if we have a segment of length l we can construct a triangle ABC whose sides are all equal to length l. The construction is as in the diagram:

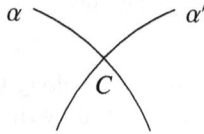

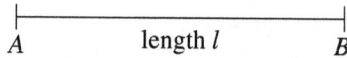

We construct the two arcs α and α' of radius l around A and B and they intersect at point C.

The gap in the proof is that the two arcs may slip through gaps in each other. In other words the point C may be a hole in the plane. The principle of minimal hypothesis for abduction would add the least hypothesis needed namely that all rational points in the field with $\sqrt{2}$ are allowed but not more. So the lines can still have gaps in them. This argument can be attacked by the additional information that the Greeks thought in terms of continuous lines and not in terms of the field generated by the rationals and $\sqrt{2}$. So we must abduce the hypothesis that lines have no gaps.

Example 4.10 (Replacement networks). Let $\vdash$ be a consequence relation. Consider a network $N = (S, R)$, where the set S contains atoms of the language of $\vdash$ and the nodes x have the theories (Δ_x, A_x) associated with them, where $\Delta_x = \{x\}$ and $A_x = x$.

The network (S, R) can now be viewed in two ways. One as an abstract network and one as a logical network with (Δ_x, A_x).

To have the two points of view completely identical we must assume about $\vdash$ that the following holds:

(*) whenever $y_1, \ldots, y_k$ are all the nodes that attack x (i.e. $y_i R x$ hold) then we have $\{y_i, x\} \nvdash x$, for each $i, i = 1, \ldots, k$.

When we have (*) the logical attack coincides with the abstract network attack. By the properties of consequence relation, we also have $y_i \not\vdash x$. Note that we do not know much about $\vdash$ beyond property (*) and so any nonmonotonic consequence relation satisfying (*) will do the job of being equivalent to the abstract network. So let us take a Prolog consequence ⊩ for a language with atoms, $\wedge$, $\Rightarrow$ and $\neg$ (negation as failure). Let

$$\Delta_x = (\bigwedge \neg y_i) \Rightarrow x$$
$$A_x = x$$

This ⊩ satisfies condition (*) and so can represent or replace $\vdash$ on the networks.

Compare with Example 4.2.

Remark 4.11. Example 4.10 leaves us with several general questions

1. Given a nonmonotonic $\vdash$ under what conditions can it be represented by a Prolog program?
2. What can we do with extensions of Prolog, say *N*-Prolog, etc. How much more can we get?
3. Given the above network, what do we get if we describe it in the meta-level as we did in Example 4.2
4. Given a reasonable $\vdash$, can we cook up a reasonable extension of Prolog to match it?

   Can we be systematic about it and have the same construction for any reasonable $\vdash$?

4.2 Methodological considerations

In order to present a methodologically robust view of logical modes of attack in argumentation networks, as intutiively described in the last section, we need to clarify some concepts. There is logical tension between two possibly incompatible themes.

Theme 1

Start with a general logical consequence $\vdash$, use databases of this logic as nodes in a network and define the notion of attack and then emerge with one or more admissible extensions.

Questions

These extensions are sets of nodes (the 'winning' nodes or the network 'output' nodes). They contain logic in them, being themselves theories of our base background logic. What are we going to expect from them? Consistency? Are we going to define a new logic from the process?

Theme 2

We start with some notion of proof (argument). We can prove opposing formulas or databases of some language **L**. We create a network of all the proofs we are interested in and define the notion of one proof (argument) attacking another. We emerge with several admissible or winning sets of proofs.

Questions

What are we to do with these proofs? Do we define a logical consequence relation using them? For example, let Δ be a set of formulas and rules. Let S be all possible proofs we can build up using Δ. Note that these proofs can prove opposing results, e.g. q and $\sim q$, etc. So we do not yet have a consequence relation for getting results out of Δ. Let R be a notion of attack we define on S. Let E be a winning extension chosen in some agreed manner. Then we define a new consequence by declaring $\Delta \vdash E$.

What connection do we require between this new consequence $\vdash$ and some other possibly reasonable consequence relation we can define directly using proofs (without the intermediary of networks)? We need rationality postulates on the notion of defeat (attack).

To make the above questions precise and gain some intuitions towards their solutions we need to examine some examples in rigorous detail. We begin with some puzzles critically examined in [79].

Example 4.12. This is example 4 in [136, p. 292]. The language allows for atoms, negation $\sim$, strict rules (implication) $\rightarrow$ and defeasible rules (implication) $\Rightarrow$. The theory Δ contains

1. *wr* (strict fact)
   Reading: John wears something that looks like a wedding ring.
2. *go* (strict proof)
   Reading: John often goes out until late with his friends.
3. $wr \Rightarrow m$
   m reads: John is married
4. $go \Rightarrow b$
   b reads: John is a bachelor
5. $m \rightarrow hw$
   hw reads: John has a wife
6. $b \rightarrow \sim hw$

If modus ponens (detachment) is the only rule we can use, we can construct the following arguments from Δ:

$$A_1 : wr$$
$$A_2 : go$$
$$A_3 : wr, wr \Rightarrow m$$
$$A_4 : go, go \Rightarrow b$$
$$A_5 : wr, wr \Rightarrow m, m \rightarrow hw$$
$$A_6 : go, go \Rightarrow b, b \rightarrow \sim hw.$$

Caminada and Amgoud write A_3 as $[A_1 \Rightarrow m]$, A_4 as $[A_2 \Rightarrow b]$, A_5 as $[A_3 \rightarrow hw]$ and A_6 as $[A_4 \rightarrow \neg hw]$. We find it a less visually transparent notation.

The following is implicit in the Caminada and Amgoud understanding of the situation.

Il $\Delta = \{1, 2, 3, 4, 5, 6\}$

I2 The argument network is *all* possible arguments that can be constructed from elements of Δ

I3 An argument is a sequence (chain) or elements from Δ that respect modus ponens (detachment).

I4 Given $x \rightarrow y$, we take it literally and do not say that we also have $\sim y \rightarrow \sim x$. If we want the latter we need to include it explicitly. (This assumption is clear since later Caminada and Amgoud do include such additional rules explicitly as part of their proposed solution to some anomalies.)

I5 One argument attacks another if the last head of the last implication is the negation of the last head of the last implication of the other.

Caminada and Amgoud point out an anomaly in this example. They point out that $A_1, \ldots, A_4$ do not have any defeaters. So they win. What $\{A_1, \ldots, A_4\}$ prove (their 'output' as they define it), is the set $\{wr, go, m, b\}$. Thus if the output is supposed to mean what is justified then both m and b are to be considered justified. Yet, and here is the anomaly, the strict rules closure of the output is inconsistent since it contains $\{hw, \sim hw\}$.

We now discuss this example. We take up this example again in Chapter 15, using tripolar argumentation networks.

First let us try to sort out some confusion. Are we working in Theme 1, where there is a background logic or in Theme 2 where we want to use an argumentation framework to define a logic?

If there is a background logic then does it include closure under strict rules? If yes, then A_3 and A_4 already attack each other. If no, then don't worry about the inconsistency of the output. We simply have defined an inconsistent theory using the tool of argumentation networks.

Caminada and Amgoud are aware that if we allow closure under strict rules at every stage then the anomaly is resolved. They attribute this solution to Prakken and Sartor [290]. They offer another example, which has anomaly, example 6, page 293, and where this trick does not work. We shall address this example later. Let us first consider Caminada and Amgoud's own solution to Example 4. They add two more contraposition rules to the database.

7. $hw \rightarrow \sim b$
8. $\sim hw \rightarrow \sim m$

With two more rules in the database, two more arguments can be constructed from the database:

A7 $wr, wr \Rightarrow m, m \rightarrow hw, hw \rightarrow \sim b$
A8 $go, go \rightarrow b, b \Rightarrow \sim hw, \sim hw \rightarrow \sim m$.

Now that our stock of arguments has A7 and A8, we have that A8 defeats A3 and A7 defeats A4. The set of winning arguments changes and the only justified arguments are $\{wr, go\}$, without the anomalies $\{b, m\}$.

We do not consider this as a solution to the anomaly. Caminada and Amgoud changed the problem (i.e. took a different, bigger database) and changed the underlying logic. Not always do we have that if $x \rightarrow y$ is a rule so is $\sim y \rightarrow \sim x$. We need to give a rigorous definition of the defeasible logic we are using, and then examine the problem of anomalies. We shall do this in Section 3. See Example 4.29 and Remark 4.32. The anomalies arise because the Dung framework does not allow for joint attacks. By the way, Caminada and Amgoud have done an excellent analysis of the anomalies. We are simply continuing their initial work.

Let us now address Example 6 of [79].

Example 4.13. The database has the following facts and rules

1. a, strict fact
2. d, strict fact
3. g, strict fact
4. $b \wedge c \wedge e \wedge f \rightarrow \sim g$
5. $a \Rightarrow b$
6. $b \Rightarrow c$
7. $d \Rightarrow e$
8. $e \Rightarrow f$.

Caminada and Amgoud consider the following arguments

A: $a, a \Rightarrow b$
B: $d, d \Rightarrow e$
C: $a, a \Rightarrow b, b \Rightarrow c$
D: $d, d \Rightarrow e, e \Rightarrow f$.

We also have the arguments

F1: a
F2: d
F3: g

The notion of one argument defeating another is the same as before, i.e. we need the two arguments to end their chains with opposite heads. Thus A, B, C, D do not have any defeaters. The justified literals are $\{b, c, e, f\}$ as well as the facts $\{a, g\}$.

Thus we get an anomaly: the closure of the winning facts under strict rules is not consistent.

We again ask the question, what exactly the underlying logic? We need a formal definition to assess the situation.

Is the following argument G also acceptable?

G: $A, C, B, D, 4$

In other words, G is

$$a, a \Rightarrow b; a, a \Rightarrow b, b \Rightarrow c; d, d \Rightarrow e; d, d \Rightarrow, e \Rightarrow f, b \wedge c \wedge e \wedge f \rightarrow \sim g$$

We first use A, C, B, D to prove the antecedent of 4 and then get $\sim g$.

If this argument is acceptable, then it must be included in the network, as the rules of the game is to include in the network all arguments which can be constructed from Δ, then G and $F3$ attack each other and so the winning set is only $\{b, c, e, f, a\}$ and we have no inconsistency.

If argument G is not acceptable because we cannot do modus ponens with more than one assumption, then the winning set is indeed $\{b, c, e, f, a, g\}$ but then we cannot get inconsistency because we cannot use modus ponens with $b \wedge c \wedge e \wedge f \rightarrow \sim g$.

So again we ask: we need a rigorous definition of the logic!

Depending on how the logic works, we may be able to deduce, for example, from A the rule $c \wedge e \wedge f \Rightarrow\sim g$ (since b defeasibly follows from a) and similarly from B we deduce $b \wedge c \wedge f \Rightarrow\sim g$ and from C we get $e \wedge f \Rightarrow\sim g$ and from D we get $b \wedge c \Rightarrow\sim g$.

If we are allowed to have that then we have that C and D defeat each other, and again we have no anomaly. So it all depends on the logic.

Let us now define one such a logic. We shall indicate what options we have.

Definition 4.14. *Let Q be a set of atoms. Let $\wedge$ be conjunction, $\sim$ a form of negation and $\rightarrow$ stand for strict (monotonic) implication and $\Rightarrow$ for defeasible implication.*

2. *A rule has the form*
   $\pm a_1 \wedge \ldots \wedge \pm a_n \rightarrow \pm b$ *(strict rule)*
   $\pm a_1 \wedge \ldots \wedge \pm a_n \Rightarrow \pm b$ *(defeasible rule)*
   where a_i, b are atoms, $+a$ means a and $-a$ means $\sim a$.
3. *A fact has the form $\pm a$ (we consider strict facts only).*
4. *A database Δ is a set of rules (strict or defeasible) and facts.*

Definition 4.15. *Let Δ be a database. We define the notion of the sequence π of formulas (actually a tree of formulas written as a sequence) is an argument for the literal $\pm a$ of length n and defeasible degree m, and specificity σ.*

1. *π is an argument of $\pm a$ from Δ of length 1 and degree 0 iff $\pm a \in \Delta$ and $\pi = (\pm a)$.
   Let $\sigma = \{\pm a\}$.*
2. *Assume $\pi_1, \ldots, \pi_k$ are all proofs of $\pm a$ from Δ of lengths n_i and degree m_i and specificity sets σ_i resp. for $i = 1, \ldots, k$. Assume $\bigwedge \pm a_i \rightarrow \pm b$ is a strict rule. Then $(\pi_1, \pi_2, \ldots, \pi_n, \bigwedge \pm a_i \rightarrow \pm b)$ is an argument for $\pm b$ of length $1 + \sum_i n_i$ and degree $\mathbf{f}_\rightarrow(m_1, \ldots, m_k)$, where $\mathbf{f}_\rightarrow$ is some agreed function representing the degree of 'defeasiblility' in the argument.
   Options for $\mathbf{f}_\rightarrow$ are*

   Option max
   $\mathbf{f}_\rightarrow = max(m_i)$
   Option sum
   $\mathbf{f}_\rightarrow = \sum m_i$
   $Let \ \sigma = \bigcup_{i=1}^{k} \sigma_i.$
3. *Assume π_i are arguments of $\pm a_i$. Let $\bigwedge \pm a_i \Rightarrow \pm b$ be a defeasible rule. Then $\pi_1, \ldots, \pi_k, \bigwedge \pm a_i \rightarrow \pm b$ is an argument of $\pm b$. The length of the argument is $1 + \sum n_i$ and the degree of the argument is $\mathbf{f}_\Rightarrow = 1 + \mathbf{f}_\rightarrow(m_1, \ldots, m_k)$, and $\sigma = \bigcup \sigma_i$.*

Remark 4.16.

1. The strict rules are not necessarily classical logic. So for example from $x \rightarrow\sim y$ and y we cannot deduce $\sim x$.
2. The definition of an argument watched for the complexity m measuring how many defeasible rules are used in the argument and the specificity σ recording the set of literals (i.e. the factual information) used in the argument. This measure is used later to define when one argument defeats another. We know from defeasible logic that the specificity of a rule is also important. So $a \wedge b \Rightarrow c$ is more specific that

$a \Rightarrow c$. The set σ is a rough measure of specificity. One can be more fine tuned. We can define a more complex measure say μ which reflects a finer balance between the number of defeasible rules used and their specificity.

Definition 4.17. *Let Δ be a database. An argument π is said to be based on Δ if all its elements are in Δ. We now define the notion of $\Delta \vdash \pm a$, a atomic, using Theme 1 point of view.*

We wish to do this in steps:

Step 1 $\Delta \vdash_1 \pm a$ *iff* $\pm a \in \Delta$

Step $m + 1$ $\Delta \vdash_{m+1} \pm a$ *iff there is a rule in Δ of the form $\bigwedge \pm a_i \Rightarrow \pm a$ such that $\Delta \vdash_{m_i} \pm a_i$, with $\sum m_i = m$, and for no rule in Δ of the form $\bigwedge \pm a_i' \Rightarrow \mp a$ (note $\mp a =\sim \pm a!$) do we have $\Delta \vdash_{m_i'} \pm b_j$ with $\sum m_i' < m$, or if $\sum m_i' = m$ then we do not have that $\bigcup \sigma_i \subsetneq \bigcup \sigma_i'$. (In words, $\pm a$ is proved using defeasible rule complexity m and specificity set σ and there is neither a less complex argument for $\mp a$ nor an argument for $\mp a$ for the same complexity but more specific, i.e. $\sigma \subsetneq \sigma'$.)*

We also agree that if $\Delta \vdash_m \pm a_i$ and $\bigwedge \pm a_i \to \pm a \in \Delta$ then $\Delta \vdash_m \pm a$.
We say $\Delta \vdash \pm a$ if $\Delta \vdash_m \pm a$ for some m.

Remark 4.18. The previous definition is one possibility of many. The important point to note is that any definition of $\vdash$ must say inside the induction step how one argument defeats another.

Let us give some examples.

Example 4.19.

1. Let Δ be $\{d, a, a \Rightarrow b, d \Rightarrow\sim c, d \wedge b \Rightarrow c\}$.
   We have that $\Delta \vdash_2 c$ because of the argument $a, a \Rightarrow b, d, d \wedge b \Rightarrow c$ is defeated by the argument $d, d \Rightarrow\sim c$, and thus $\Delta \vdash \sim c$.
   Some defeasible systems will say the argument for c defeats the argument for $\sim c$ because it is more specific. Our system says the argument for $\sim c$ defeats the argument for c because it uses less number of defeasible rules.
2. Our definition does say, for example, that for the database $\Delta' = \{a, d, a \Rightarrow c, a \wedge d \Rightarrow\sim c\}$. We have that $\sim c$ can be proved because it relies on a more specific information. See remark 4.16.
3. We could give a definition which measures not only how many defeasible rules are used but also give them weights according to how specific they are. Our aim here is not to develop the theory of defeasible systems and their options and merits but simply to show how to define the notion of defeasible consequence relation and to make a single most important point:

   > To define the notion of consequence relation for a defeasible system we must already have a clear notion of one argument defeating another.

Definition 4.20. *We now give a second definition of a consequence relation:*

1. *Let A, B be two arguments. Define the notion of A defeats B in some manner. Denote it by $A \mathcal{D} B$.*

2. *Let Δ be a theory, being a set of rules and literals. Let N be the set of all arguments based on Δ. Consider the network $(N, \mathcal{D})$ where $\mathcal{D}$ is the relation from (1) above. Let $\mathcal{A}$ be an algorithm for choosing a winning justified set of atoms from the net, e.g. let us $\mathcal{A}$ take the unique grounded extension which always exists. Then define $\Delta \vdash_{\mathcal{D}, \mathcal{A}} \pm a$ iff $\pm a$ is justified by the above process $\mathcal{A}$ in the $(N, \mathcal{D})$ network.*

We are now ready for some methodological comments.

Rationality postulates for defeat

We need rationality postulates on the notion $\mathcal{D}$ of one argument defeating another where the arguments are defined in the context of facts, strict rules and defeasible rules. Caminada and Amgoud give rationality postulates on the admissible sets derived from $\mathcal{D}$ but this is the wrong place for rationality. $\mathcal{D}$ is supposed to be such that it ensures we get a proper consequence relation relation $\vdash_{\mathcal{D}}$ out of $\mathcal{D}$, satisfying reflexivity, restricted monotonicity and cut.

Representation problem

1. Given a consequence relation $\vdash$ for defeasible logic (i.e. $\vdash$ contains defeasible and strict rules), can we extract from $\vdash$ a defeat notion $\mathcal{D} = \mathcal{D}_{\vdash}$ for arguments, and a network algorithm $\mathcal{A}$ such that the notion $\vdash_{\mathcal{D}, \mathcal{A}}$ is a subset of $\vdash$?
2. Given any consequence relation defined by any means (e.g. defined semantically), can we guess/invent a notion of argument and a notion of defeat $\mathcal{D}$ such that the associated $\vdash_{\mathcal{D}, \mathcal{A}}$ is a subset of $\vdash$?
3. If we don't have such a representation theorem in the case of (1) above, using a natural $\mathcal{D}_{\vdash}$, then we perceive this as an anomaly.

Any solution to the anomalies raised in the Caminada and Amgoud paper must, in my opinion, respect the above methodological observations. It must not be some technical *ad hoc* solution!

4.3 A rigorous case study — 1

This section shows in a rigorous way how Theme 2 works. We define a nonmonotonic consequence relation using Dung networks on arguments built up using rules.

Two comments

1. The strict rules need not be classical logic.
2. We use labelling to keep control of the proof process and possibly add strength to rules. However, the labels will not be used at first in our definitions and examples. Some strict logics require the labels in their formulation (e.g. resource logics) as well.

Definition 4.21.

1. *Let our language contain atomic statements $Q = \{p, q, r, \ldots\}$, the connective $\sim$ for negation, $\wedge$ for conjunction, $\rightarrow$ for strict rules and $\Rightarrow$ for defeasible rules.*
2. *A literal x is either an atom q or $\sim q$. We write $-x$ to mean $\sim q$ if $x = q$ and q if $x = \sim q$.*
   A rule has the form $(x_1, \ldots, x_n) \Rightarrow x$ (strict rule) or $(x_1, \ldots, x_n) \Rightarrow x$ (defeasible rule) where x_i, x are literals. We are writing $(x_1, \ldots, x_n) \rightarrow x$ instead of $\bigwedge x_i \rightarrow x$ to allow us to regard the antecedent of a rule as a sequence. This gives us a greater generality in interpreting the strict rules as not necessarily classical logic. We can also allow for $\varnothing \Rightarrow x$, where $\varnothing$ is the empty set.
3. *A rule of the form $(x_1, \ldots, x_n) \Rightarrow x$ is said to be more specific than a rule $(y_1, \ldots, y_m) \Rightarrow y$ iff $m < n$ for some $i_1, \ldots, i_m \leq n$ we have $x_{i_j} = y_j$. Of course, any rule $(x_1, \ldots, x_n) \Rightarrow x$ is more specific than $\varnothing \Rightarrow y$. Note that we are not requiring $y = \sim x$.*
4. *A labelled database is a set of literals, strict rules and defeasible rules. We assume each element of the database has a unique label from a set of labels Λ. Λ is a new set of symbols, not connected with Q or anything else.*
   So we present the database as

$$\Delta = \{\alpha_1 : A_1, \ldots, \alpha_k : A_k\}$$

   where α_i are different atomic labels from Λ and A_i are either literals or rules. The labels are just names at this stage, allowing us greater control of whatever we are going to do.
5. *Let Δ be a labelled database. We define by induction the strict closure of Δ denoted by Δ^S as follows:*
   a) Let $\Delta_0^S = \Delta$.
   b) Assume Δ_n^S has been defined. Let $\Delta_{n+1}^S = \Delta_n^S \cup \{\beta : x \mid \text{ for some } \alpha_i : x_i \in \Delta_n^S, \alpha : (x_1, \ldots, x_n) \rightarrow x \in \Delta \text{ and } \beta = (\alpha, \alpha_1, \ldots, \alpha_n)\}.$
   Let $\Delta^S = \bigcup_n \Delta_n^S$.
   Δ is consistent if for no literal x do we have $+x$ and $-x \in \Delta^S$.
6. *Note that we do not close under Boolean operations. The strict logic is not necessarily classical. We may have $\sim q \rightarrow r, \sim r \in \Delta$, this does not imply $q \in \Delta^S$.*
7. *Also note that only strict rules are used in the closure. So if Δ_0 is the set of defeasible rules in Δ, then $\Delta^S = \Delta_0 \cup (\Delta - \Delta_0)^S$.*

Definition 4.22 (Arguments). *We define the notion of an argument (or proof) π (based on a database Δ) its Δ-output $\theta_\Delta(\pi)$, its Head $H(\pi)$, its literal base $L(\pi)$, and its family of subarguments $A(\pi)$.*

1. *Any literal $t : x \in \Delta$ is an argument of level 1. Its head is $t : x$. Its Δ-output is the set of all literals in the strict closure of $\{t : x\}$ and its head rule $H(\Delta)$ is $t : x$. Its literal base is $\{t : x\}$ and its subarguments are $\varnothing$.*
2. *Let $\pi_1, \ldots, \pi_n$ be arguments in Δ of level m_i resp., and let $\rho : (x_1, \ldots, x_n) \Rightarrow x$ be a defeasible rule in Δ. Assume $\alpha_i : x_i$ can be proved using strict rules from the union of the outputs $\theta_\Delta(\pi_i)$. Then $(\pi_1, \ldots, \pi_n, \rho : (x_1, \ldots, x_n) \Rightarrow x)$ is a new argument π. Its output is all the literals in the strict closure of $\{x\} \cup \bigcup_i \theta_\Delta(\pi_i)$, $H(\pi) = \rho : (x_1, \ldots, x_n) \Rightarrow x$, $L(\pi) = \bigcup L(\pi_i)$, and $A(\pi) = \{\pi_1, \ldots, \pi_n\} \cup \bigcup_i A(\pi_i)$. The level of π is $1 + \max(m_i)$.*

3. *An argument is consistent if its output is consistent.*

4. *Note that there is no redundancy in the structure of an argument. If a, b are literals then (a, b) is not an argument. If π_1, π_2 are arguments then (π_1, π_2) is not an argument.*

Definition 4.23 (Notion of defeat for arguments of levels 1 and 2). *Let π_1, π_2 be two consistent argument. We define the notion of π_1 defeats $\pi_2, \pi_1 \mathcal{D}\pi_2$, as follows:*

1. *A literal $t : x \in \Delta$ considered as an argument of level 1 defeats any argument π of any level 1 with $-x$ in its output. Note that if our arguments come from a consistent theory Δ, then no level 1 argument can defeat another level 1 argument. They are all consistent together as elements of Δ^S.*

2. *Let*
$$\pi_1 = (t_1 : x_1', \ldots, t_n : x_n', r : (x_1, \ldots, x_n) \Rightarrow x)$$
$$\pi_2 = (s_1 : y_i', \ldots, s_m : y_m', s : (y_1, \ldots, y_m) \Rightarrow y)$$

be two arguments of level 2, then π_1 defeats π_2 if $r : (x_1, \ldots, x_n) \Rightarrow x$ is more specific than $s : (y_1, \ldots, y_m) \Rightarrow y$, and $\theta_\Delta(\pi_2)$ and $\theta_\Delta(\pi_1)$ are inconsistent together.[2]

3. *In (2) above, we defined how an argument of level 2 can defeat another argument of level 2. (It cannot defeat any argument of level 1). Note that it can defeat an argument of any level m if it defeats any of its subarguments of level 2.*

4. *Note that two arguments of level 2 cannot defeat each other.*

5. *We shall give later the general definition of defeat for levels m, n.*

6. *An argument π_1 attacks an argument π_2 if*
    a) *Their outputs are not consistent.*
    b) *The head rule of π_1 is more specific than the head rule of π_2, or π_1 is of level 1.*

7. *π_1 may attack π_2 but not defeat it. However a level 2 argument always defeats other arguments it attacks.*

Example 4.24.   1. Let $\Delta = \{a, a \Rightarrow x, a \Rightarrow y, x \wedge y \rightarrow \sim a\}$. Δ is consistent because $\Delta^S = \{a\}$.

The arguments
$$\pi_1 = (a, a \Rightarrow x)$$
$$\pi_2 = (a, a \Rightarrow y)$$

attack each other, but none can defeat the other because it has to be more specific. Compare with Example 4.29.

Example 4.25. This example does not use labels. We also write $\bigwedge x_i \Rightarrow x$, when we do not care about the order of x_i.

1. Consider the two arguments
$$\pi_1 = (d, a, d \wedge a \Rightarrow c)$$
$$\pi_2 = (d, a, a \Rightarrow b, a \wedge b \wedge d \Rightarrow \sim c).$$

π_1 is of level 2 and π_2 is of level 3. In this section, our definition of defeat will say that π_2 defeats π_1 because the head of π_2 is more specific than the head of π_1. We are not giving advantage to π_1 on account of it being shorter (contrary to Definition 4.17).

---

[2] Note that we do not require that $x = \sim y$, nor that $\{x, y\}$ is inconsistent. The requirement is that the outputs are inconsistent.

2. Consider now π_3

$$\pi_3 = (d, a, a \wedge d \Rightarrow \sim b, a \wedge d \Rightarrow c).$$

Does π_2 defeat π_3?

Its main head rule, $a \wedge b \wedge d \Rightarrow \sim c$ is more specific but its subproof $(d, a, a \Rightarrow b)$ is defeated by the π_3 subproof $(d, a, a \wedge d \Rightarrow \sim b)$.

So π_3 defeats π_2 according to this section (as opposed to Definition 4.17).

Example 4.26 (Cut rule). Again we do not use labels, and we do not care about order in the antecedents of rules.

Let Δ be

$$\Delta = \{b, d, d \wedge a \wedge b \Rightarrow c, d \Rightarrow c, a \wedge b \Rightarrow \sim c, c \Rightarrow a\}$$

we have

$$\Delta, a \vdash c$$

because of the proof

$$\pi_1 : (b, d, a, d \wedge a \wedge b \Rightarrow c\}$$

$\pi_2 = (a, b, a \wedge b \Rightarrow \sim c)$ is defeated by π_1.

We also have

$$\Delta \vdash a$$

This is because of π_3.

$$\pi_3 = (d, d \Rightarrow c, c \Rightarrow a).$$

We ask do we have $\Delta \vdash c$?

We can substitute the proof of a into the proof of c, that is we substitute π_3 into π_1. We get π_4.

$$\pi_4 = (b, d, d \Rightarrow c, c \Rightarrow a, d \wedge a \wedge b \Rightarrow c).$$

The question is, can we defeat π_4?

We can get a proof of $\sim c$ by substituting π_3 into π_2, to get π_5

$$\pi_5 = (d, d \Rightarrow c, c \Rightarrow a, b, a \wedge b \Rightarrow \sim c).$$

Example 4.27 (Mutual defeat). Let π_1 be $(a, b, a \wedge b \Rightarrow x)$. Let π_2 be $(a, b, c, a \Rightarrow \sim x, a \wedge b \wedge c \Rightarrow \sim x, \sim x \wedge \sim x \Rightarrow y)$. Then π_1 defeats a subargument of π_2, namely $(a, a \Rightarrow \sim x)$, and a subargument of π_2, namely $(a, b, c, a \wedge b \wedge c \rightarrow \sim x)$ defeats π_1. You may ask why does π_2 prove $\sim x$ twice in two different ways? Well, maybe the strict rules of the logic are not classical and so two copies of $\sim x$ are needed (in linear logic $\sim x \rightarrow (\sim x \rightarrow y)$ is not the same as $\sim x \rightarrow y$), or maybe that is the way π_2 is, however, a proof is a proof.

The output of π_1 is $\{a, b, x\}$ and the output of π_2 is $\{a, b, c, \sim x, y\}$. Each is consistent.

Definition 4.28 (Defeat for higher levels).

1. We already defined how any argument of level 1 can defeat any argument of level $m \geq 2$. No argument of level m can defeat an argument of level 1, (this is because all arguments are based on a consistent Δ).

2. We defined how an argument of level 2 can defeat another argument of level 2.
3. An argument π_1 of level 3 can defeat an argument π_2 of level 2 if
   a) one of its level 1 or level 2 subarguments defeats π_2
   or
   b) its head is more specific than the head of π_2 of level 2, its output is inconsistent with the output of π_2 and π_2 does not defeat any of its level 2 subarguments.
4. Assume by induction that we know how an argument π_2 of level 2 can defeat or be defeated by an argument π_1 of level $k \le m$. We show the same for level $m + 1$:
   - π_2 defeats π_1 if
     a) π_2 defeats some subargument of level $k \le m$ of π_1
     or
     b) the head of π_2 is more specific than the head of π_1, its output is inconsistent with that of π_1 and no subargument of π_1 of level $\le m$ defeats π_2.
   - The argument π_2 is defeated by π_1 if
     a) Some subargument of π_1 of level $\le m$ defeats π_2
     or
     b) the head of π_1 is more specific than the head of π_2 and its output is inconsistent with that of π_2 and π_2 does not defeat any subargument of level $\le m$ of π_1.
   We have thus defined how an argument of level 2 can defeat or be defeated by any argument of level m for any m.
5. Assume by induction on k that we defined for level k and any m how any argument of level k can defeat or be defeated by any argument of level m for any m. We define the same for level $k + 1$.
   We define this by induction on m. We know from item (4) how π_{k+1} can defeat or be defeated by an argument of level 2. Assume we have defined how π_{k+1} can defeat or be defeated by any argument π'_n of level $n \le m$. We define the same for level $n = m + 1$.
   a) π_{k+1} is defeated by an argument π'_{m+1} of level $m + 1$ if either π'_{m+1} defeats a subargument of π_{k+1} of level $\le k$ or if the head of π'_{m+1} is more specific than the head of π_{k+1}, its output is not consistent with the output of π_{k+1} and no subargument of π'_{m+1} of level $\le m$ is defeated by π_{k+1}.
   b) π_{k+1} defeats an argument π'_{m+1} if either it defeats a subargument of π'_{m+1} of level $\le m$ or its head is more specific than the head of π'_{m+1}, its output is not consistent with the output of π'_{m+1} and no subargument of π_{k+1} of level $\le k$ is defeated by π'_{m+1}.
6. We thus completed the induction step of (5) and we have defined for any k and m how an argument of level k can defeat or be defeated by an argument of level m for any m and k.
7. We need one more clause: π_1 defeats π_2 if some subargument π_3 of π_1 defeats π_2 according to clause (1)–(5) above.

Example 4.29 (Anomalies). Consider the following database Δ.

$$\Delta = \{a, b, c, a \Rightarrow d, b \Rightarrow e, c \Rightarrow f, a \wedge b \wedge c \wedge d \wedge e \rightarrow \sim f\}.$$
$$\Delta^S = \{a, b, c\}.$$

The arguments are besides the literals a, b, c the following:

$$\pi_1 : a, a \Rightarrow d$$
$$\pi_2 : b, b \Rightarrow e$$
$$\pi_3 : c, c \Rightarrow f$$

In our system all the arguments from an admissible winning set and we get an anomaly since the output is inconsistent. We have no more arguments since we use in our definition only defeasible rules. If we allow in arguments for strict rules, or turn the strict rule into a defeasible rule, $a \wedge b \wedge c \wedge d \wedge e \Rightarrow \sim f$, this might help. Δ itself becomes one big argument, and Δ defeats π_3 on account of it being more specific. But then Δ itself contains π_3 and so it is self defeating. Thus we are still left with $a, b, c, \pi_1, \pi_2, \pi_3$ as the winning arguments and the anomaly stands.

By the way, a well known rule of nonmonotonic logic is that if $a \vdash b$ monotonically then $a \mid\sim b$ nonmonotonically. So we can add/use the strict rules in our arguments.

We can add the axiom

$$\frac{(x_1, \ldots, x_n) \rightarrow x}{(x_1, \ldots, x_n) \Rightarrow x}$$

So why are we getting anomalies? The reason is not our particular definition of defeat or the way we write the rules or the like.

The reason is that we do not allow for *joint attacks*. You will notice that some of the devices of Caminada can help here, but they are not methodological. The problem is that we are getting anomalies because outputs of successful argument can join together in the strict reasoning part to get a contradiction but their sources (i.e. the defeasible arguments which output them) cannot be joint together in a joint attack. See Example 4.13 which is very similar to this example.

The difference now in comparison with Example 4.13 is that we have precise definitions for our notions of defeat etc. and so we can define joint attacks, change the underlying logic or take whatever methodological steps we need.

The simplest way to introduce joint attacks in our system without changing the definitions is to add the following rule axiom schema for any Δ.

$$(x_1, \ldots, x_n) \Rightarrow \top$$

for any $x_1, \ldots, x_n$, any n. Thus we would have the proofs

$$\eta_3 : (\pi_1, \pi_2, (d, e) \Rightarrow \top)$$
$$\eta_2 : (\pi_1, \pi_3, (d, f) \Rightarrow \top)$$
$$\eta_1 : (\pi_2, \pi_3, (e, f) \Rightarrow \top)$$

$$\eta_0 : (\pi_1, \pi_2, \pi_3, (d, e, f) \Rightarrow \top$$

The argument η_0 is inconsistent, and we ignore arguments like $(a, a \Rightarrow \top)$ or $(a, a \Rightarrow d, (a, d) \Rightarrow \top)$, which give nothing new.

Since attacks and defeats are done by the output of the arguments, we get that η_i attacks and is being attacked by π_i.

The resulting network will need a Caminada labelling and not all π_i, η_i will always be winning.

The outputs of the various arguments are as follows:

$$
\begin{aligned}
\text{output}(a) &= \{a\} \\
\text{output}(b) &= \{b\} \\
\text{output}(c) &= \{c\} \\
\text{output}(\pi_1) &= \{a, d\} \\
\text{output}(\pi_2) &= \{b, e\} \\
\text{output}(\pi_3) &= \{c, f\} \\
\text{output}(\eta_3) &= \{a, b, d, e\} \\
\text{output}(\eta_2) &= \{a, d, c, f\} \\
\text{output}(\eta_1) &= \{b, e, c, f\} \\
\text{output}(\eta_0) &= \{a, b, c, d, e, f, \sim f\}.
\end{aligned}
$$

Figure 4.8 shows the network (we ignore the arguments which give nothing new):

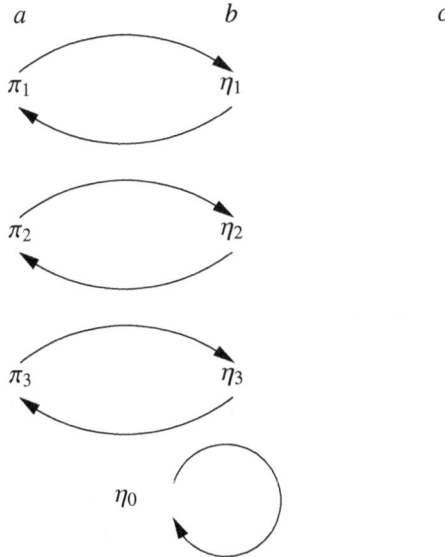

Fig. 4.8.

Clearly any Caminada labelling will choose one of the pairs $\{\eta_i, \pi_i\}$. The justified theory will be consistent!

Definition 4.30 (Consequence relation based on defeat). *We assume we allow joint attacks as suggested in Example 4.29. Let Δ be a consistent theory and let a be a literal. We define the notion of $\Delta \vdash a$ as follows:*

Let $\mathfrak{A}$ be the set of all consistent arguments based on Δ and let $\mathcal{D}$ be the defeat relation as defined above. Then $(\mathfrak{A}, \mathcal{D})$ is a Dung network. Let $\mathbb{T}$ be an admissible set of arguments (take some Caminada labelling or if you wish, take the unique grounded set) and let $\mathbb{A}_\Delta$ be the strict closure of the union of all outputs of the arguments in $\mathbb{T}$. Then we define

$$\Delta \vdash a \text{ iff } a \in \mathbb{Q}_\Delta.$$

Lemma 4.31. $\mathbb{Q}$ *is consistent.*

Proof. Otherwise we have several winning arguments. $\pi_i, i = 1, \ldots, n$ with $x_i \in \theta_\Delta(\pi_i)$ such that Δ^S and $\{x_i\}$ and the strict rules in Δ can prove y and $\sim y$. Assume n is minimal for giving a contradiction.

However the argument

$$\eta_i = (\pi_1, \ldots, \pi_{i-1}, \pi_{i+1}, \ldots, \pi_n, (x_1, \ldots, x_{i-1}, x_{i+1}, \ldots, x_n) \Rightarrow \top)$$

attacks and is being attacked by π_i.

So not all π_i can be winning!

Remark 4.32. The exact results for $\vdash$ depend on the admissible set winning but the important point is that now the system is aware of the anomaly (inconsistency) and so we have no anomaly!

To summarise, the devices we used are:

1. joint attacks through the axiom $\bigwedge x_i \Rightarrow \top$
2. arguments attack through their output and not just through the head of the last rule in the argument. (In other words, we always close under strict rules at every stage of the argument/proof.)

Remark 4.33 (Failure of cut — 1). This example shows that we cannot always chain proofs together.

Let $\Delta = \{u, u \Rightarrow \sim b, \sim b \Rightarrow a, a \Rightarrow b\}$.

Then $\Delta \vdash a$ because of $\pi_a = (u, u \Rightarrow \sim b, \sim b \Rightarrow a)$.

We also have $\Delta, a \vdash b$ because of $\pi_b = (a, a \Rightarrow b)$. However, we cannot string π_a and π_b together to get a proof for $\Delta \vdash b$ because $(u, u \Rightarrow \sim b, \sim b \Rightarrow a, a \Rightarrow b)$ is not consistent.

Thus cut fails for the consequence relation of Definition 4.30. The next example shows failure of cut even when the proofs π_a and π_b can consistently chain.

Example 4.34 (Failure of cut — 2). This is another example for the failure of cut for the consequence relation of Definition 4.30. Let $\Delta = \{u, u \Rightarrow a, a \Rightarrow v, v \Rightarrow b, x, x \Rightarrow v, x \wedge v \Rightarrow w, x \wedge u \rightarrow \sim w\}$.

Then $\Delta \vdash a$ because of π_a:

$$\pi_a = (u, u \Rightarrow a).$$

$\Delta, a \vdash b$ because of π_b:

$$\pi_b = (a, a \Rightarrow v, v \Rightarrow b).$$

The outputs of π_a and π_b together are $\{u, a, v, b\}$ and are consistent. So we can string the proofs together to π_b^a proving b from Δ.

$$\pi_b^a = (u, u \Rightarrow a, a \Rightarrow v, v \Rightarrow b).$$

This proof however is defetated by the proof η (which is consistent and undefeated).

$$\eta = (x, x \Rightarrow v, x \wedge v \Rightarrow w).$$

The output of η is $\{x, v, w\}$. The reason for the defeat is because

1. The head rule of η is more specific than that of π_b^a.
2. The union of the outputs of η and π_b^a is the set $\{u, a, v, b, x, w\}$ which is inconsistent because of the strict rule $x \wedge u \to \sim w$. η does not defeat π_b because we need u to get inconsistency.

Example 4.35 (Success of cut). Let $\varDelta$ be the following database

$$\varDelta = \{u, x, a \Rightarrow v, v \Rightarrow b, x \Rightarrow \sim a, \sim a \Rightarrow v, u \Rightarrow x \wedge v \Rightarrow \sim b\}.$$

The arguments we can construct from $\varDelta \cup \{a\}$ are as follows.

$$
\begin{aligned}
\pi_a &= (u, u \Rightarrow a) \\
\pi_b &= (a, a \Rightarrow v, v \Rightarrow b) \\
\eta &= (x, x \Rightarrow \sim a, \sim a \Rightarrow v, x \wedge v \Rightarrow \sim b) \\
A_1 &= (a, a \Rightarrow v) \\
A_2 &= (x, x \Rightarrow \sim a) \\
A_3 &= (x, x \Rightarrow \sim a, \sim a \Rightarrow v) \\
A_4 &= (u, u \Rightarrow a, a \Rightarrow v) \\
\pi_b^a &= (u, u \Rightarrow a, a \Rightarrow v, v \Rightarrow b) \\
B_1 &= (a, x, a \Rightarrow v, x \wedge v \Rightarrow \sim b) \\
B_2 &= (x, u, u \Rightarrow a, a \Rightarrow v, x \wedge v \Rightarrow \sim b).
\end{aligned}
$$

We also have the atomic arguments $(a), (u)$ and (x). We have $\varDelta \!\vdash\! a$ because of π_a and maybe $\varDelta, a \!\vdash\! b$ because of π_b, but this is attacked and defeated by B_1. We now look at π_b^a and ask whether it is undefeated and hence shows that $\varDelta \!\vdash\! b$. It is attacked by η and defeated.

4.4 Rigorous case Study- 2. Labelled Systems

The reader is referred to Chapter 21, where arguments are labelled and to attack an argument one can attack the label.

5

Modal Provability Foundations for Argumentation Networks

5.1 Introduction

Let **P** be an argumentation network. Figures 5.1, 5.2, 5.3 and 5.4 are typical examples.

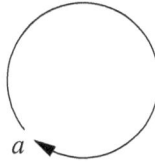

Fig. 5.1.

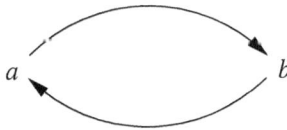

Fig. 5.2.

The network **P** has logical contents, saying which arguments are winning (denoted by a is 'on' or $a = 1$), which are losing (denoted by a is 'off' or $a = 0$) and which are undecided (denoted by a is 'undecided' or $a = ?$).

The content (or extensions) can be obtained from a Caminada labelling.

Definition 5.1 (Caminada labelling). *A function* $\mathbf{f} : \mathbf{P} \mapsto \{1, 0, ?\}$ *is a Caminada labelling iff the following holds.*

1. *If x is an initial point (no arrow leads into x) then $\mathbf{f}(x) = 1$.*
2. *If $y_1, \ldots, y_n$ are all nodes with arrows leading into y then*
   2.1. *if for some $i, \mathbf{f}(y_i) = 1$ then $\mathbf{f}(y) = 0$*
   2.2. *if for all $i, \mathbf{f}(y_i) = 0$ then $\mathbf{f}(y) = 1$*

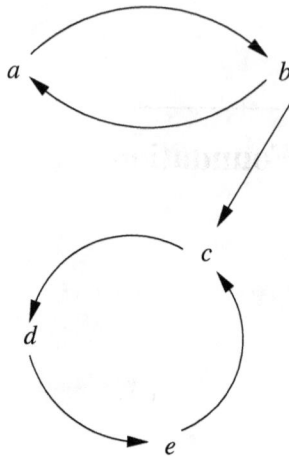

Fig. 5.3.

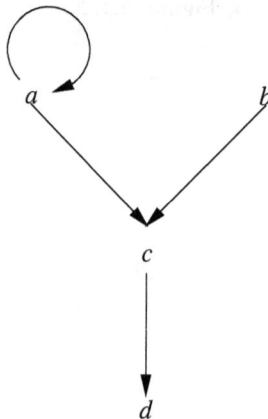

Fig. 5.4.

2.3. if for all i, $\mathbf{f}(y_i)$ is either 0 or ? and for at least one of y_i $\mathbf{f}(y_i)$ =? then $\mathbf{f}(y)$ =?.

The extension obtained from a Caminada labelling is $\{x|\mathbf{f}(x) = 1\}$. The family of extensions for any particular algorithm is obtained from conditions on the Caminada labelling. For example, let $E_{\mathbf{f}}^{y} = \{x|\mathbf{f}(x) = y\}$, where $y = 0, 1, ?$. Order $\{E_{\mathbf{f}}^{y}|\mathbf{f}$ a Caminada function}, for y fixed, by inclusion. We have the following:

Grounded extension: $E_{\mathbf{f}}^{1}$ and $E_{\mathbf{f}}^{-}$ are minimal. $E_{\mathbf{f}}^{?}$ is maximal.

Preferred extension: $E_{\mathbf{f}}^{1}$ and $E_{\mathbf{f}}^{0}$ are maximal.

Complete extension: Same as complete labelling.

Stable extension: $E_{\mathbf{f}}^{?} = \emptyset$, i.e. no undecided.

Semi-stable extension: $E_{\mathbf{f}}^{?}$ is minimal.

See [80] for a survey.

This kind of content depends on an algorithm yielding extensions and there are several types of algorithms, giving several types of extensions. Different networks may admit some extensions and possibly not others.

For example

1. Network $\mathbf{P}_1$ of Figure 5.1 has no extensions.
2. The network $\mathbf{P}_2$ of Figure 5.2 has the following types of extensions:
   - Grounded: ∅
   - Preferred: $\{a\}, \{b\}$
   - Complete: ∅, $\{a\}, \{b\}$.
3. The network $\mathbf{P}_3$ of Figure 5.3 has the following types of extensions:
   - Grounded: ∅
   - Preferred: $\{a\}, \{b, d\}$
   - Complete: ∅, $\{b, d\}$.
4. The network $\mathbf{P}_4$ in Figure 5.4 has the following types of extensions:
   - All types except stable: $\{b, d\}$

The different extensions can be obtained from a Caminada labelling of the nodes.

Our aim in this chapter is to associate with a network $\mathbf{P}$ a modal formula $\mathbf{m}(\mathbf{P})$ of a certain modal logic (which we call **LN1**) such that the following hold:

1. The modal formula $\mathbf{m}(\mathbf{P})$ contains the nodes of the network $\mathbf{P}$ as atomic propositions.
2. The possible worlds models of the modal formula correspond in one to one fashion to the Caminada labelling of the network.

To be specific, consider a three-point linear Kripke model of the form as in Figure 5.5.

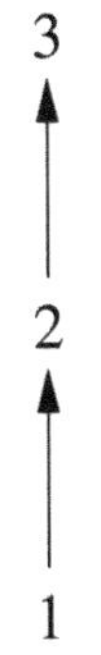

Fig. 5.5.

(i.e. if $<$ is the accessibility relation then we have $1 < 2 < 3$).

The modal logic **LN1** is such that for the Kripke model of Figure 5.5 above and for any modal formula $\mathbf{m}(\mathbf{P})$ for a network $\mathbf{P}$ such that $\mathbf{m}(\mathbf{P})$ holds in the model, only three types of assignment are possible, as in Figure 5.6

We named the types by 0, 1, ? to correspond to the Caminada labelling.

Any model of $\mathbf{m}(\mathbf{P})$ will give truth values to the atoms in each world and the atom will acquire a type. Thus we define a Caminada function $\mathbf{f}$ accordingly

Type 1 Type 0 Type ?

3	⊤	⊥	⊤
↑	↑	↑	↑
2	⊤	⊥	⊥
↑	↑	↑	↑
1	⊤	⊥	⊤

Fig. 5.6.

$$\mathbf{f}(x) = \begin{cases} 1 \text{ if } x \text{ is of type 1} \\ 0 \text{ if } x \text{ is of type 0} \\ ? \text{ if } x \text{ is of type ?} \end{cases}$$

Converseley, any Caminada labelling $\mathbf{f}$ of $\mathbf{P}$ can be turned into an assignment to the Kripke model of Figure 5.6 via the same correspondence and that will be a model of $\mathbf{m}(\mathbf{P})$. Thus the logic **LN1** and the formula $\mathbf{m}(\mathbf{P})$ contains in it the information of all Caminada labelling for $\mathbf{P}$ and therefore all types of extensions are retrievable from it by additional considerations on the models.

The modal logic formula is obtained from general methodological considerations which have nothing directly to do with argumentation. It is not something we 'cooked up' to correspond to the Caminada labelling. Rather, it is a result of general approach to self-referential loops in algorithmic logic, as can be seen from the next section.

5.2 Methodological motivation

The modal approach was first introduced by the author in 1986, in an Imperial College technical report [132] and later published in 1990 as [134].

The problem to be solved was to associate 'logical content' with any Prolog program π with negation as failure, especially for programs which loop.

Example 5.2. Consider the following four programs:

π_1: • a if $\neg a$

π_2: • a if $\neg b$
         • b if $\neg a$

π_3: • a if $\neg b$
         • b if $\neg a$
         • c if $\neg b \wedge \neg e$
         • d if $\neg c$
         • e if $\neg d$

π_4: • a if $\neg b$
         • b
         • c if $\neg a \wedge \neg b$
         • d if $\neg c$.

The problem with such programs is that they contain loops and it is not clear what they say.

The modal idea was to associate with each program π a formula $\mathbf{m}(\pi)$ of a modal logic **N1**, see [134][1] and identify $\mathbf{m}(\pi)$ via a fixed point solution of a modal equation in the logic **N1**. The machinery for doing so, including effective tractable algorithms for finding $\mathbf{m}(\pi)$ is presented, motivated and discussed at length in [134]. For our purposes it is enough to explain the idea through one simple example.

We illustrate the idea for the case of the program π_1 $(a$ if $\neg a)$, see [162, Example E4, p. 197].

Let X_1 be the logical content of the program π_1. If we understand the modality as 'provable' then

- q succeeds from π_1 iff

$$X_1 \vdash \Box(X_1 \to q)$$

- q fails from π_1 iff

$$X_1 \vdash \Diamond(X_1 \wedge \sim q)$$

- q loops from π_1 iff neither the above.

Therefore we need to solve something like the following modal fixed point equation for π_1

$$X \leftrightarrow \Box[a \leftrightarrow \Diamond(X \wedge \sim a)]$$

Considerations in the paper [134] lead us (for technical reasons) to the following exact equations

$$X \leftrightarrow G[\Box\bot \vee (a \leftrightarrow \Diamond(X \wedge \sim a)] \tag{$*$}$$

where $Gq = q \wedge \Box q$, and $\Box$ is the irreflexive and transitive modality of the logic **N1** yet to be introduced.

It is proved in [134] that to get a solution $X = Y$ it is enough to substitute $\top$ for X in the right hand side of the equation, i.e.

$$Y = G(\Box\bot \vee a \leftrightarrow \Diamond \sim a)$$

is a fixed point of equation $(*)$.

The connection with argumentation comes through the following translation of an argumentation network into a Prolog program

Definition 5.3. Translation τ.

Let $\mathbf{P}$ be a network and let $y \in \mathbf{P}$.

1. If no arrow goes into the node y then translate $\tau(y) = y$.

---

[1] In [134] we use the modal logic **N1** of finite trees, described in the next section. The case of argumentation networks is much simpler than general logic programs and it is easier to add an axiom of linearity to **N1** to obtain **LN1** and use the linear version for our purposes. All the fixed points machinery and theorems in [134] apply to **LN1** and so we save a lot of work in this chapter.

The logic **N1** is a slight variation of logic introduced by Löb, see Solovay [314] and Boolos [67], used to formulate the provability predicate of Peano arithmetic. So this modal logic has meaning and motivation totally independent of logic programming or argumentation.

2. *If $y_1, \ldots, y_n$ are all the nodes with arrows into y then translate*

$$\tau(y) = (y \text{ if } \bigwedge_i \neg y_i).$$

Let $\pi = \tau(\mathbf{P}) = \{\tau(y)|y \in P\}$. Then π is the associated Prolog program.

It is easy to see that for the networks $\mathbf{P}_1, \ldots, \mathbf{P}_4$ of Figures 5.1–5.4 the associated Prolog programs are $\pi_1, \ldots, \pi_4$.

Now using the machinery of [134] we obtain a modal formula $\mathbf{m}(\pi)$ for π giving 'logical content' to π in the logic $\mathbf{N1}$.

This same formula considered in the logic $\mathbf{LN1}$ will give logical content to $\mathbf{P}$.

Of course, in the next sections we will work on $\mathbf{P}$ directly and just help ourselves to technical lemmas from [134].

5.3 The modal logic LN1

This section introduces the modal logic $\mathbf{N1}$ of [134] which we use to give modal content to logic programs and consequently to argumentation networks, through its extension $\mathbf{LN1}$. The language contains atoms, the classical connectives and a modality $\square$ (and hence $\lozenge$). The models we envisage for this logic are Kripke models of the form $(S, <, a, h)$ where $(S, <, a)$ is a finite tree with root $a \in S$ and h is the assignment, giving each atomic q a subset $h(q) \subseteq S$.

The following must be satisfied.

1. If all non-endpoints are (resp. are not) in $h(q)$ then $h(q) = S$ (resp. $h(q) = \varnothing$).

Definition 5.4 (N1 semantics).

1. *A model has the form $(S, <, a, h)$, where S is a finite set, $<$ is a transitive and irreflexive relation on S, a is the actual world and the following holds:*
   a) *$(S, <, a)$ is a tree with root a.*
   b) *$h(q) \subseteq S$ for each atomic q. If $\bar{S}$ is the set of endpoints of S then $(S - \bar{S}) \subseteq h(q) \Rightarrow h(q) = S$.*
   $(S - \bar{S}) \cap h(q) = \varnothing \Rightarrow h(q) = \varnothing$.
2. *Satisfaction is defined in the usual way*

$$t \vDash \square A \text{ iff } \forall s(t < s \Rightarrow s \vDash A)$$

3. *The model satisfies A iff $a \vDash A$.*
4. *the model is an $\mathbf{LN1}$ model if the tree is a linear chain.*

Definition 5.5. *Axioms for $\mathbf{N1}$*

1. *All substitution instances of classical tautologies*
2. *All substitution instances of modal $\mathbf{K4}$ axioms*
   - $\square(A \to B) \to (\square A \to \square B)$
   - $\square A \to \square\square A$
   - $\vdash A \Rightarrow \vdash \square A$

3. *All substitution instances of Löb's axiom*

$$\Diamond A \rightarrow \Diamond(A \wedge \Box \sim A).$$

4. *Axiom for atomic q:*
   4.1. $q \rightarrow \Box(\sim q \rightarrow \Box q)$
   4.2. $\Box(\Box\bot \vee q) \leftrightarrow \Box q$
   4.3. $\Box(\Box\bot \vee \sim q) \leftrightarrow \Box \sim q.$
5. *Additional axiom of linearity for* **LN1**.
   - *All substitution instances of*

$$\Diamond A \wedge \Diamond B \rightarrow \Diamond(A \wedge B) \vee \Diamond(A \wedge \Diamond B) \vee \Diamond(B \wedge \Diamond A).$$

Theorem 5.6.   *1.* **N1** *is complete for the proposed semantics of finite trees.*
2. **LN1** *is complete for the proposed semantics of linear finite chains*

Proof. Well known. — see [127]

Definition 5.7. *Let* **P** *be an argumentation network. For any node y in* **P** *let* $y_1, \ldots, y_n$
be all the nodes attacking y (i.e. with arrows leading into y). Let $\tau(y) = y \leftrightarrow \bigwedge_i \Diamond \sim y_i.$
If y has no attackers let $\tau(y) = y.$
   Let GA be defined as $A \wedge \Box A.$
   Let **m(P)** *be*

$$G\left(\Box\bot \vee \bigwedge_{\substack{y \in \mathbf{P} \\ y \text{ has attackers } y_i}} \left(y \leftrightarrow \bigwedge_i \Diamond \sim y_i\right)\right) \wedge \bigwedge_{\substack{z \in \mathbf{P} \\ z \text{ without attackers}}} Gz$$

   Let **P** be a network. Let $y \in \mathbf{P}$ and let $y_i, i = 1, \ldots, n$ be all elements attacking y (with arrows leading into y). Let us understand X as the modal 'logical content' of **P**. Then for any $z \in P$, X proves z if z is 'on' and if z is not 'on' then $X \wedge \sim z$ is consistent. Thus for y to be 'on', we need to have

$$y \leftrightarrow \bigwedge_i \Diamond(X \wedge \sim y_i).$$

This holds for any $y \in \mathbf{P}$.
   If y is not attacked by anything then we want $X \vdash y$.
   Thus we expect X to satisfy more or less the equation

$$X \equiv \bigwedge_{\substack{y \in P \\ y \text{ not attacked}}} \wedge \bigwedge_{\substack{y \in \mathbf{P} \\ y \text{ attacked}}} \left(y \leftrightarrow \bigwedge_i \Diamond(X \wedge \sim y_i)\right)$$

An extensive technical discussion in [134] shows that the correct equation is the following:

$$X \equiv G \left(\Box\bot \lor \bigwedge_{\substack{y \in P \\ y \text{ is attacked by } y_i}} \left[y \leftrightarrow \bigwedge_i \Diamond(X \land \sim y_i) \right] \land \bigwedge_{\substack{y \in P \\ y \text{ not attacked}}} Gy \right)$$

A sequence of Lemmas in [134] shows (see Lemma L1, page 190), that the solution to the above equation is the formula **m(P)** of Definition 5.7.

Lemma 5.8. *Consider a 3 point* **LN1** *chain model of a formula* **m(P)** *of Definition 5.7. Then each atomic q has truth values of type 0 or of type 1 or of type ? of Figure 5.6.*

Proof. The axioms of **LN1** force the following for atomic q in the chain $1 < 2 < 3$:

- $1 \vDash q$ and $2 \vDash q \Rightarrow 3 \vDash q$
- $1 \vDash\sim q$ and $2 \vDash\sim q \Rightarrow 3 \vDash\sim q$
- $1 \vDash q$ and $2 \vDash\sim q \Rightarrow 3 \vDash q$.

We need to prove that the option

$$1 \vDash q \text{ and } 2 \vDash\sim q \text{ and } 3 \vDash\sim q$$

cannot arise.

Consider a point y such as y is being attacked. (If not then Gy is in **m(P)** and so y is of type 1.)

We have $G(\Box\bot \lor (y \leftrightarrow \bigwedge \Diamond \sim y_i))$ holds in the model.

Hence $y \leftrightarrow \bigwedge_i \Diamond \sim y_i$ holds at points 1 and 2.

We distinguish several cases:

1. If for any j, if $y_j = \bot$ at 3 then $\Diamond \sim y_j$ is true at 1 and 2 and it plays no role in the conjunction. If for all y_j, $3 \vDash\sim y_j$ then $y = \top$ at 1 and 2 and y is of type 1.
2. Some y_j are $\top$ at 3. Then $\Diamond \sim y_j$ is $\bot$ at 2 and $y = \bot$ at 2.
3. We now check whether y_j is true or false at 2. If for some j, $y_j = \top$ at 2 then $\Diamond \sim y_j = \bot$ at 1 and $y = \bot$ at 1 and hence $y = \bot$ everywhere, and hence y is of type 0.

If for all $j, y_j = \bot$ at 2 then $y = \top$ at 1 and we have that $y = \top$ at 1 and $y = \bot$ at 2 and we must have by the axiom 4.1 that $3 \vDash y$ and y is of type 3.

Example 5.9. We show that the axioms are needed. Consider the **P** of Figure 5.7

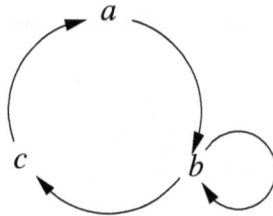

Fig. 5.7.

We have that $\mathbf{m(P)}$ for this figure is:

$$G(\Box\bot \vee ((a \leftrightarrow \Diamond \sim c) \wedge (b \leftrightarrow \Diamond \sim a \wedge \Diamond \sim b) \wedge (c \leftrightarrow \Diamond \sim b)))$$

We can have the assignment of Figure 5.8 and $\mathbf{m(P)}$ still holds.

$$
\begin{array}{ccc}
a & b & c \\
3 \;\; \bot & \top & \top \\
\uparrow & & \\
2 \;\; \bot & \bot & \bot \\
\uparrow & & \\
1 \;\; \top & \top & \top
\end{array}
$$

Fig. 5.8.

Theorem 5.10. *Let* **P** *be a network and let* $\mathbf{m(P)}$ *be its 'logical content' in the logic* **LN1**.

Then there exists a one-to-one correspondence between the three chain models of $\mathbf{m(P)}$ *and the possible Caminada labelling of* **P**. *The correspondence is given by the code of Figure 5.6.*

Proof. Direction 1. Let h be an assignment model satisfying $\mathbf{m(P)}$. By Lemma 5.8 every atom node x in P gets assigned values $h(x)$ of types 0, 1, or ?.

Define a Caminada candidate function $\mathbf{f}$ according to the types, namely $\mathbf{f}(x) = i$ iff $h(x)$ is of type $i, i \in \{0, 1, ?\}$. We now show that $\mathbf{f}$ satisfies the conditions of a Caminada function of Definition 5.1.

1   If x is an initial point then Gx is a conjunct of $\mathbf{m}(P)$, therefore $h(x)$ is of type 1 and hence $\mathbf{f}(x) = 1$.

2   If $y_1, \ldots, y_n$ are all nodes with arrows to y, then we have the conjunct $y \leftrightarrow \bigwedge_i \Diamond \sim y_i$ in $\mathbf{m(P)}$ in the clause

$$G(\Box\bot \vee \bigwedge_y (y \leftrightarrow \bigwedge_i \Diamond \sim y_i))$$

This means that at nodes 1 and 2 of the chain $y \leftrightarrow \bigwedge_i \Diamond \sim y_i$ must hold. Assume all y_i are of type 0, then $y = \top$ at 2 and 1. Hence by axiom 4.2, $y = \top$ also at node 3 and y is of type 1. This means that if all y_i are 'out' then y is 'in'.

Assume one of y_i is of type 1. Then for this y_i, $\Diamond \sim y_i$ is $\bot$ at nodes 1 and 2. Hence y is false at 1 and 2 and by axiom 4.3, $y = \bot$ at node 3. Hence y is of type 0.

Now assume all of y_i are either of type 0 or of type ?, and at least one of y_i say y_1 is of type ?. The y_i of type 0 have no influence on y because $\Diamond \sim y_i$ is $\top$ at nodes 1 and 2. The crucial nodes are the nodes like y_1 which are of type ?. This means that $1 \vDash y_1, 2 \vDash\sim y_1, 3 \vDash y_1$. Thus $\Diamond \sim y_1 = \bot$ at 2 and $\Diamond \sim y_1 = \top$ at 1. This holds for any type y_i of type ?. Thus $y = \top$ at 1 and $y = \bot$ at 2. Hence by axiom 4.1, $y = \top$ at 3 and hence y is of type ?.

Direction 2. Let **f** be a Caminada function Define a model by assigning values to the propositions according to the code of Figure 5.6. We claim **m(P)** holds in this model.

First all nodes y without arrows into them are assigned type 1 and hence Gy holds. For the other nodes we must check the formula

$$G(\Box\bot \lor \bigwedge_{y} \quad (y \leftrightarrow \bigwedge_i \Diamond \sim y_i))$$

y has arrows leading to it

and show it holds in the model.

We distinguish several cases. Assume all y_i are of type 0. This means $\mathbf{f}(y_i) = 0$ for all y_i. So $\Diamond \sim y_i$ is $\top$ at nodes 1 and 2. But since all $\mathbf{f}(y_i) = 0$ we get that $\mathbf{f}(y) = 1$ and hence $y \leftrightarrow \bigwedge_i \Diamond \sim y_i$ holds at 1 and 2 as required.

If one of y_i is of type 1, this means $\Diamond \sim y_i$ is $\bot$ at 1 and 2. Hence $\bigwedge_i \Diamond \sim y_i$ is $\bot$ at 1 and 2. But also since $\mathbf{f}(y_i) = 1$, we get $\mathbf{f}(y) = 0$ and so y is of type 0 and $y = \bot$ at 1 and 2.

Hence $y \leftrightarrow \bigwedge_i \Diamond \sim y_i$ holds at 1 and 2.

Now assume that all y_i are either of type 0 or type ?. This means $\mathbf{f}(y_i)$ is either 0 or ?. Assume that at least one y_i is of type ? (i.e. $\mathbf{f}(y_i) =$?). Then $\mathbf{f}(y) =$? and we have $y = \top$ at 1 and $y = \bot$ at 2. Let us check whether $\bigwedge_i \Diamond \sim y_i$ is $\bot$ at and $\top$ at 2. Since all y_i are of type ? or $\bot$ with at least one y_i of type ?, the type ? y_i will be $\top$ at node 2 and hence $\Diamond \sim y_i = \bot$ at node 1 and hence $y = \bot$ at node 1.

On the other hand all y_i are $\bot$ at node 3 and hence $\bigwedge_i \Diamond \sim y_i$ is $\top$ at node 2. This shows that $y \leftrightarrow \bigwedge_i \Diamond \sim y_i$ is $\top$ at nodes 1 and 2.

Thus the above argument shows that

$$G(\Box\bot \lor \bigwedge_{u} (y \leftrightarrow \bigwedge_i \Diamond \sim y_i))$$

holds at all nodes 1, 2, and 3.

This completes the proof of the theorem.

Remark 5.11 (Expressive power of the modal approach). We showed correspondence between the Caminada labelling and the 3 chain models of **m(P)**, for the logic **LN1**. Our modal logic approach is more powerful than the Caminada labelling. We can tell in more detail *why* a node gets value ?.

The key is axiom 4.1. We adopted this axiom

$$q \to \Box(\sim q \to \Box q)$$

in order to get correspondence between models an Caminada labelling. We are however better off without it. If we abandon this axiom we can take any model of **m(P)** and using it to give a better labelling.

Consider Figure 5.9

We have added a new type for 'unknown', type (?∗). This new type is allowed because we abandoned axiom 4.1

	Type 1	Type 0	Type ?	Type?*
3	T	⊥	T	⊥
↑				
2	T	⊥	⊥	⊥
↑				
1	T	⊥	T	T

Fig. 5.9.

To explain, consider and compare Figure 5.7 and Figure 5.10. Figure 5.10 has the **m(P)** as

$$G(\Box\bot \vee ((a \leftrightarrow \Diamond \sim c) \wedge (b \leftrightarrow \Diamond \sim a) \wedge (c \leftrightarrow \Diamond \sim b)))$$

This formula does not allow for any assignment of type ?*. While the formula for Figure 5.7, as given in Example 5.9, does allow type ?* for b. So by the models of our **m(P)**, we can classify better our networks, because we can distinguish between Figure 5.7 and 5.10.

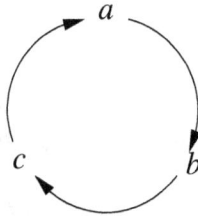

Fig. 5.10.

Caminada labelling will give a, b, c of Figure 5.10 all ? (unknown). Same as to a, b, c of Figure 5.7. There is only one Caminada labelling to these figures so we cannot make distinctions between the figures using other Caminada labellings. Therefore Caminada labelling has less expressive power with regards to why nodes are unknown. We can also use chains of length $n > 3$ as models and have many more types for unknown which can be more sensitive to how the network is.

The reader might wish to know what is the intuitive meaning of a type ?* label. The answer is this: If y is of type ?* then y never attacks alone, i.e. $\forall x [y$ attacks $x \rightarrow \exists z \neq y(z$ attacks $x)]$.

Also if x is attacked only by type ?* elements, then x is 'on'.

The new Caminada–Gabbay rule should be (for this case) as follows:

- If $y_1, \ldots, y_n$ attack y and each y_i is either of type 0 or type ? or type?*, and at least one of y_i is of type ? then y is of type ?.

Remark 5.12 (Further advantages of the modal approach).

1. The modal approach associates a modal formula **m(P)** with every argumentation network **P**. Thus we can define the concept of one network P_2 being a logical extension of another network P_1. We simply say that for all x **m(P_1)** ⊢ x implies **m(P_2)** ⊢ x or equivalently **m(P_2)** ⊢ **m(P_1)**.
   There is no simple way of defining this conecpt directly in network terms.

2. Another option available to us is the logical characterisation of extensions. We saw that following Definition 5.1 that the different labelling options **f** for **P**, corresponds to all the models for **m(P)** and that the various extensions can be characterised by properties of **f** and the sets E_f^1, E_f^0 and $E_f^?$. The grounded extension for example, corresponds to minimal (among all **f**) E_f^1 and E_f^0 and is unique. Therefore it can be characterised using the modal logic **LN1** as the set of all literals $z = x$ and $z = \sim x$ such that $\mathbf{m(P)} \vdash \Box z$.

   The other extensions such as the preferred extensions or the semi-stable extensions are not unique and need some additional language capability extending **LN1** to be expressed.

We shall study these points in detail in the next section.

5.4 Advantages of the modal approach

We saw in Remark 5.12 that associating a modal formula **m(P)** with an argumentation network **P**, allows us to treat **P** as a logical theory. In the realm of logic one can perform certain operations and one can ask certain questions.

The following are three examples

1. Given $\mathbf{P}_1$ and $\mathbf{P}_2$, give meaning to and check whether $\mathbf{P}_1 \vdash \mathbf{P}_2$.
2. Given $\mathbf{P}_1(x)$ and $\mathbf{P}_2$, form the substitution $\mathbf{P}_3 = \mathbf{P}_1(x/\mathbf{P}_2)$, and give it meaning.
3. If $\mathbf{P}_1, \mathbf{P}_2$ are constructs in a logic then there are standard ways of adding a modality ☒ to the system and considering expressions like ☒**P**. Give this construct meaning and check its properties.

We ask what would the above look like directly in terms of argumentation networks? Our strategy is as follows:

• To understand what certain logical operations would mean in terms of networks **P**, go to **m(P)**, perform the operations in the modal logic, see the results obtained and then guess how to do the corresponding operation directly on **P**.

The notion of $\mathbf{P}_1 \vdash \mathbf{P}_2$ is easy. Take it as $\mathbf{m(P}_1) \vdash \mathbf{m(P}_2)$ in the modal logic. The latter means that every 3 point chain Kripke model of $\mathbf{m(P}_1)$ is also a Kripke model of $\mathbf{m(P}_2)$. But we know that any Krpke model of $\mathbf{m(P}_1)$ corresponds to a Caminada labelling of $\mathbf{P}_2$. So we expect that this labelling also determines a Caminada labelling of $\mathbf{P}_2$.

So we have that for example the following

• network of Figure 5.10 $\vdash$ network of Figure 5.7.

But

• Network of Figure 5.7 $\nvdash$ network of Figure 5.10.

because in the network of Figure 5.7 node b can get the label $(?*)$.

Dealing with substitution is more tricky. The notion of substituting one network in another has been studied by us in several papers and appropriate definitions were given for the cases of Bayesian networks and for the case of neural networks. It seems that each type of network has its own natural way of doing substitution. See Chapter 10 of our book [85]. The problem in general can be illustrated by Figures 5.11 and 5.12 below. We have a network $\mathbf{P}_1(b, a_i)$ with a node $b \in \mathbf{P}_1$, and a_i are the rest of the nodes. We substitute for the node b another network $\mathbf{P}_2(e_j)$, whose nodes are e_j. We may allow equalities of some $a_i = e_j$.

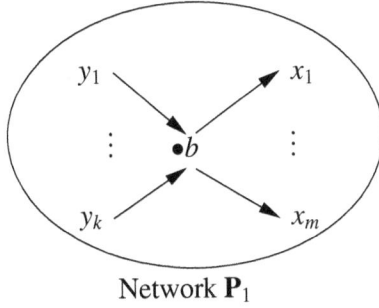

Network $\mathbf{P}_1$

Fig. 5.11.

In Figure 5.11 b is attacked by $y_1, \ldots, y_k$ and b attacks $x_1, \ldots, x_m$. Now consider Figure 5.12, obtained by substituting network $\mathbf{P}_2$ for b in $\mathbf{P}_1$.

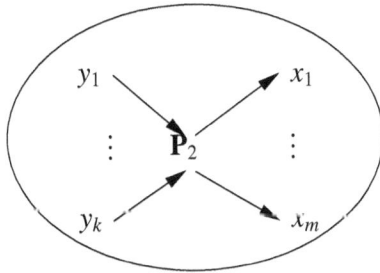

Fig. 5.12.

We ask what is the meaning of Figure 5.12? We have several options:

1. Allow by definition the attacks from $y_1, \ldots, y_k$ to go into $\mathbf{P}_2$ and attack some points in $\mathbf{P}_2$. For example, we can attack all points in $\mathbf{P}_2$ or only the initial points of $\mathbf{P}_2$.
2. Allow by definition for the attacks emanating from b to emanate from all ? or some ? points of $\mathbf{P}_2$?
3. What to do if $\mathbf{P}_2$ contains points of $\mathbf{P}_1$? (e.g. $a_1 = e_2$)?

We need to agree on some reasonable definition of substitution. We can try to figure out the appropriate definition by looking at $\mathbf{m}(\mathbf{P}_2)$ and $\mathbf{m}(\mathbf{P}_1)$.
We reason as follows:

- If $\mathbf{m}(\mathbf{P}_1)(b, a_i)$ represents the logical contents of $\mathbf{P}_1(b, a_i)$ and $\mathbf{m}(\mathbf{P}_2)(e_j)$ represents the logical contents of $\mathbf{P}_2(e_j)$, then surely $\mathbf{n} = \mathbf{m}(\mathbf{P}_1)(b/\mathbf{m}(\mathbf{P}_2), a_i)$ must represent the logical content of $\mathbf{P}_1(b/\mathbf{P}_2, a_i)$, i.e. the **n**, which is meaningful in modal logic may represent the logical contents of the network of Figure 5.12

We need to be careful here. The logic **LN1** we are using has special axioms not allowing certain assignment to atomic formulas. We are not allowing atoms to have assignments of the form $(\top, \top, \bot)$ or of the form $(\bot, \bot, \top)$. Thus if b is atomic in $\mathbf{m}(\mathbf{P}_1)$, we cannot substitute for it $(\mathbf{mP}_2)$ unless we are sure that it satisfies the restriction for atoms.

To achieve that we can adopt the axioms of item 4 of Definition 5.5. So we can append a conjunct to formula **n** which requires the axioms in item 4 of 5.5 to hold for $\mathbf{m}(\mathbf{P}_2)$. Call this $\mathbf{n}^*$. Thus $\mathbf{n}^*$ is the logical content we propose to associate with the network $\mathbf{P}_1(b/\mathbf{P}_2)$.

We now have two problems

1. Understand in logical terms what $\mathbf{n}^*$ says.
2. Derive from this understanding some algorithm for propagating the arrows of $\mathbf{P}_1$ into $\mathbf{P}_2$ and propagating arrows from $\mathbf{P}_2$ out to $\mathbf{P}_1$, (i.e. implement geometrically what $\mathbf{n}^*$ says).

This will have to be done in a separate publication. It is too involved.

However, our next chapter, Chapter 6, is devoted to adding modalities to argumentation networks.

6

Modal and Temporal Argumentation Networks

6.1 Introduction and orientation

This section will look at modal and temporal logics in a way which is compatible with argumentation networks. This will allow us to understand our options in introducing modal and temporal argumentation networks.

We adopt a context view of modal logic. Assume we are talking about various contexts, which we denote by $w_1, w_2, \ldots$, and we discuss a finite number of atomic facts, $Q = \{q_1, \ldots, q_n\}$. We write $w \models q$ to mean hat q holds in the context w. Modal logic allows us to talk from within a context w about other related contexts. Let R be an accessibility relation on the set W of contexts, so we write wRw' to mean w' is accessible to w.

If we add the modal connective $\Diamond$ to our language, we can write

* $w \models \Diamond q$ iff for some w', such that wRw', we have $w \models q$.

This gives us basic modal logic.

The first question we ask is where does the relation $w \models q$ come from? In traditional modal logic this is arbitrary.

A traditional modal model (for the logic **K**) comes as a triple $(W, R, \models)$ where W is the set of worlds (contexts), $R \subseteq W \times W$ is the accessibility relation and $\models$ is the relation which tells us which atoms $q \in Q$ hold at which world (i.e. when $w \models q$ holds). $\models$ is arbitrarily given. One can modify abstract modal logic and add more details about the origins of $\models$. For example, we can imagine that each context w has a database Δ_w associated with it and we write

$$w \models q \text{ iff } \Delta_w \vdash q$$

If we do this we no longer have the semantics arbitrary modal logic **K** and we may get some special versions/extensions of **K**.

In the case of argumentation we may associate with each context w an argumentation network $N_w = (Q, \rho_w)$, where Q is the set of atoms and $\rho_2 \subseteq Q \times Q$ is the attack relation. We need to choose an extension $E_w \subseteq Q$ calculated properly using traditional Dung rules from the attack relation ρ_w. We thus have

$$w \vdash q \text{ iff } q \in E_w.$$

Thus in this approach the argumentation networks are vehicles for defining $\vDash$. This has intuitive sense. In the world w, there is a local perception of attack ρ_w about the facts of w and a debate resulting in an extension E_w, and this extension tells us what facts hold in the context w. So really we should write

$$N_w = (Q, \rho_w, E_w)$$

If we adopt this approach, we will have a problem. Modal logic allows us to use expressions like $\Diamond q$ at any world w. So we need to allow $\Diamond q$ to enter the argumentation considerations taking place at world w.

Let $Q_\Diamond = Q \cup \{\Diamond q | q \in Q\}$. We have

$$N_w = (Q_\Diamond, \rho_w . E_w)$$

with

$$\rho_w \subseteq Q_\Diamond \times Q_\Diamond$$

and E_w an extension,

$$E_w \subseteq Q_\Diamond.$$

We ask how is this extension calculated?

Suppose we have in the world w, the case that $\Diamond q$ attacks p (i.e. $\Diamond q \rho_w p$) and $\Diamond q$ is not attacked by any other argument. We would, therefore, expect that $\Diamond q \in E_w$, i.e. that $w \vDash \Diamond q$ holds.

On the other hand, we also expect that for some w' such that wRw' we have $w' \vDash q$, i.e. that $q \in E_{w'}$. However, such a w' may not exist!

We get an internal incompatibility in the system. Obviously we need an internal device to deal with this. Either an additional compatibility postulate, or by something else. We solve this problem by introducing the concept of usability. We have a usability function h_w and we have in this case $h_w(\Diamond q) = 0$, i.e. we cannot use it in the considerations of finding an extension E_w.

Of course, there is independent motivation to the concept of usability. Its introduction is not just technical.

In the above considerations, the dominant view was that of modal logic, and the argumentation network view was auxiliary, it provided a means of defining $\vDash$. Can we look at the entire possible world system as one big argumentation network? Put differently, can we view $(W, R, N_w), w \in W$ as one big argumentation network?

This is possible to do using auxiliary arguments which eliminate R and W. We first explain this by example. Consider Figure 6.1. There are three worlds w, w_1, w_2. We have wRw_1 and wRw_2. We assume the grounded extension in each world. $\Diamond q$ is not attacked in w_1. It says that it is possible to have an accessible world in which q holds. So we must have either q in at the grounded extension of world w_1 or q in at the grounded extension of world w_2.

If this holds, we say $\Diamond q$ is usable in w, otherwise not. This usability is implemented in the object level by adding the point $h_w(q)$ as a meta-point outside the possible worlds, which point attacks $\Diamond q$ of the world w, and letting all the qs of the accessible worlds attack it. By the traditional rules of Dung argumentation, if at least one of the qs in w_1 or in w_2 is in, then $h_w(q)$ is out and so $\Diamond q$ is in. Otherwise $\Diamond q$ is out (not usable) in w. We do the same trick for $\Diamond p$. If we want to eliminate the big circles

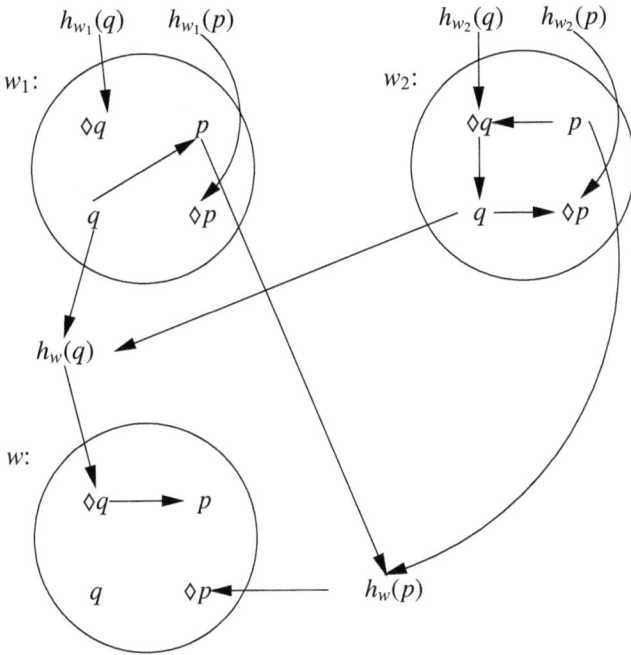

Fig. 6.1.

indicating the worlds, we have to annotate the atoms by the world name. If we do that we get Figure 6.2, which is one big uniform argumentation network where we seek the grounded extension.

The above discussion shows what options we have in principle in formulating modal and temporal argumentation.

1. Let modal logic dominate by keeping the possible worlds separate and allow for an argumentation network to say what holds in each world.
2. Incorporate the modal aspects inside one big argumentation network by using auxiliary meta-arguments. This way we are compiling modal logic into argumentation. This is technically possible to do, even possible to do nicely, in view of the results in Chapter 13 below, showing that argumentation is equivalent to classical logic in a nice way.
3. In the temporal logic case, there is a simple more immediate way of defining a temporal logic network. We simply time stamp each argument with a moment of time or an interval of time, these being the times when this argument is considered usable. Thus a network has the form (T, ρ, τ), where T is the set of arguments, ρ is the attack relation and τ is a function,

$$\tau : T \cup \rho \mapsto \text{ real numbers}$$

where $\tau(x)$ = moments of time where x is usable.
There is a very simple model, used in 2005 in [37], see also Chapter 9 below, in connection with temporally varying numerical strength of arguments.

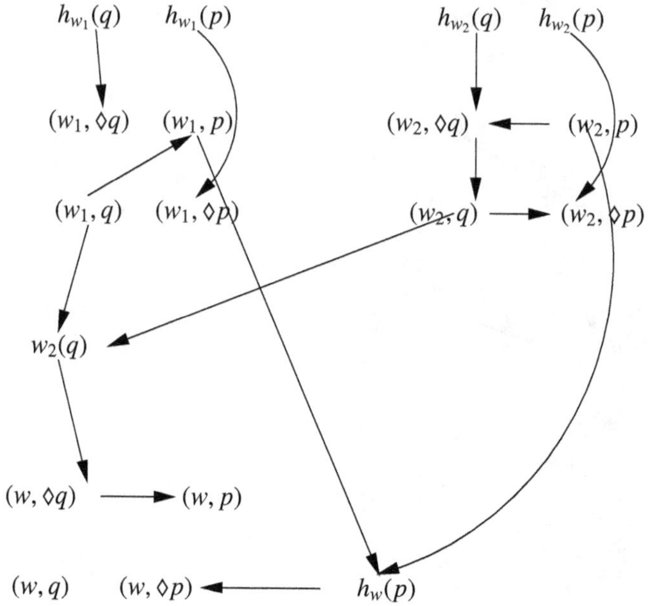

Fig. 6.2.

In itself the model is too simple and does not offer much, but one can easily add natural structure to it to indicate evolution over time, as done in Chapters 9 and 15. In Chapter 9, the change in time was of strength of argument and this can influence the argument's attack capabilities and in Chapter 19 the time stamping was used to resolve loops.

6.2 Introducing the global metalevel approach to temporal networks: Concept of usability

There are several good reasons why we should consider modal and temporal argumentation networks

1. **Temporal facts as arguments.** Past facts or future scheduled events can also be used as arguments for the present. An argument against the trustworthiness of a person may be the facts of past betrayals. An argument in favour of a higher mortgage loan may be a scheduled increment in salary next year. Unfortunately, this does not work with UK banks. An argument against a higher mortgage loan may be the possibility of redundancy in the future. Figure 6.3 is an example of how a scheduled redundancy exercise in the near future can be used as an argument against a high mortgage loan now, where d means the event of redundancy and m the general argument in favour of a mortgage. We shall discuss later why it is not reasonable to encapsulate 'Future d' as a single argument c attacking m now. We lose structure this way (compare with how we lose structure in propositional logic in case $\forall x A(x) \vdash A(\alpha)$. We need predicate logic to go into the structure of the sentences).

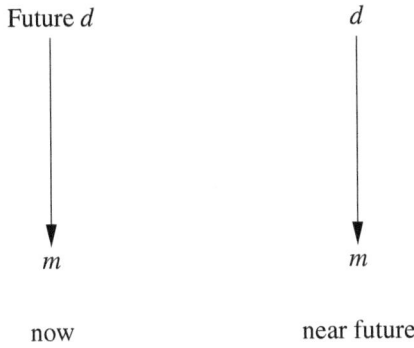

Future d d

m m

now   near future

Fig. 6.3.

2. **Fibring arguments.** Arguments from one domain may be brought into another domain. For example, expert medical arguments may be brought in as a package into a legal argument. This may be best treated in the context of modal logic, (bringing information from the medical world into the legal world), where $\Diamond x$ means bring in information x from another world, i.e. domain. See Figure 6.4. Let b be the legal argument to commit the accused to a one year prison sentence for tax evasion. Let c be the medical argument that the accused has cancer. This medical argument attacks b in the legal world. A hefty fine is more appropriate. c is part of a medical network of arguments and c emerges among the winning arguments of that network. Figure 6.4 illustrates the situation. Of course, in the legal world $\Diamond c$ might be attacked as unacceptable evidence on the basis of some procedural errors in putting in forward (not shown in diagram).

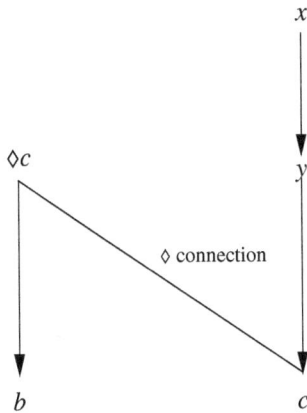

x

$\Diamond c$ y

$\Diamond$ connection

b c

Fig. 6.4.

3. **Future arguments.** The possibility that an argument a may be able to defeat another argument b. We denote this by $\Diamond a$. Such possibilities are central to threats and negotiations arguments where various future scenarios are discussed. For example, don't ask for more than a 10% salary settlement as it will never

be approved by the executives — there may be strong fairness arguments for claiming 10% but pragmatically it will not be affordable and thus will not get approved.

4. **Past arguments.** We can use the fact that an argument c was potent in the past (denoted by Pc) to attack another current argument. Figure 6.5 is an example of such a configuration where $\Diamond a$ indicates that argument a is possible and Pc indicates that argument c was considered in the past, but maybe is no longer taken seriously now, yet the fact that it was a serious argument in the past is sufficient to defeat b. For example, a mother might say "I have always cared for you, you cannot abandon me now".

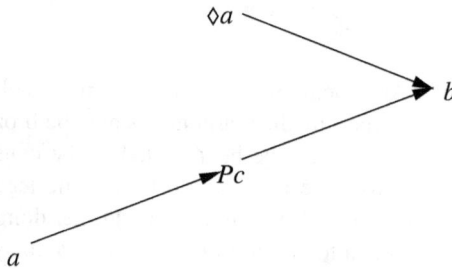

Fig. 6.5.

For example a female employee may threaten with a possible argument claiming harassment. It may be that one cannot argue harassment now but it is not clear what the circumstances would look like when reviewed in the future. So $\Diamond$harassment may have some force. We have had many such arguments when UK law was expected to be overruled by EU law. Many $\Diamond$(EUlaw) arguments were already defeating local UK arguments even before the EU law came into force in the UK.

Negotiations always involve evaluation of future scenarios of possible arguments and counter arguments and the possibilities of certain scenarios may be a strong argument at present.

Arguments from the point of view of tradition have always been successful in the past, e.g. we've always accepted the A'level standard as an appropriate university entrance qualification, so we continue to do so, even though many will argue that the level has dropped. We can have our doubts about the value of tradition now but yet an argument of the form "but it has always been the case that x" may still win out.

Any use of precedent is also akin to this form.

The basic notion is that of admissible extension. This is a subset $E \subseteq S$ such that

1. E is conflict free, i.e. for no $x, y \in E$ can we have $(x, y) \in R$.
2. E is self defending, i.e. for all $x, y \in S$, if $y \in E$ and $(y, x) \in R$ then there exists a $z \in S$ such that $(z, y) \in R$.

The above discussion suggests we introduce the concept of usability of arguments. We may have at a certain time or at a certain context some arguments that are talked about and are available in some real sense but these arguments cannot be used, for a variety of reasons. The formal presentation of such arguments can be to introduce them into the network but label them as unusable through a usability function h. If x is an argument then $h(x) = 1$ means it is usable and $h(x) = 0$ means it is not. The reader may ask why do we want to introduce them at all if they are not usable? Well, in the context of modal and temporal logic, it makes sense to talk about them. Maybe they were usable, maybe they will be usable or are possibly but not necessarily usable, or should have been usable, etc, etc. We give several examples.

Example 6.1 (The catholic super administrator). A UK university, operating an equal opportunities policy, advertises for a faculty administrator. There is a shortlist of three candidates and, because of a special request from one candidate, the interview date is moved.[1]

The top two candidates are: a woman aged 42, who knows 15 languages and 10 computer languages and has a PhD in economics and business administration from Harvard. She has lots of experience working for government administration; the other candidate is a man of similar age, but not with as strong a background as the lady.

There is an argument for hiring the lady candidate: she is the best!

There is an argument against hiring the lady candidate: she is Catholic, aged 42, recently married, and will probably waste no time in starting a family.

This latter argument is a strong subjective argument, which, following the proper procedures, cannot be used. Indeed, one cannot mention it, let alone even think it! h (this argument) $= 0$.[2]

Example 6.2 (The rape.). A girl complained she was raped by a man late at night in the street. The man claimed that she gave him reason to take the view that she was willing and available. The entire incident was filmed, video and audio, by a CCTV camera.

However, this camera was installed without a licence and hence, because of a legal technicality, any evidence from the CCTV is not admissible. The evidence from the CCTV clearly and unambiguously defeated the claim of the man, but because of its inadmissability, the jury was instructed to accept the man's claim.

In both cases we present the network in the traditional form (S, R), where S is the set of arguments and R is the attack relation, but with both usable and unusable arguments included, where those that are usable are marked via a function h. $h(a) = 0$ means we cannot use argument a. $h(a) = 1$ means we can use the argument.

It is important to note that unusability is temporary and can change. Circumstances can change, the law can change, new arguments can be brought forward an what was unusable can become usable.

---

[1] As an argument for wishing a new interview date, the candidate has declared that she is getting married in the local catholic church, and these dates coincide. As a result of this correspondence, the interviewers know she is Catholic and a new bride.

[2] There is a known case where a preferred candidate did not score as well as another candidate in an interview for a position in local government. In exceptional circumstances, the interview panel was reconvened and the outcome was that the preferred candidate's score actually fell!

1. **Unusability due to defeat.** Figure 6.4 can be an example of unusable argument. The notation $\Diamond c$, wants to bring the cancer argument from another network, the medical network into the present network, the legal one. In the figure c is a winning argument in the medical network. c is attacked by y in the figure but is defended by x. However, it is quite possible in a complex cross network situation, that we have a $\Diamond z$ such that in the appropriate network for z, z is defeated. In this case we can view $\Diamond z$ as unusable. Again, this is not permanent and may change.
2. **Unusability due to secrecy.** It is quite possible that an argument a is defeated by an argument $a*$ which cannot be recorded explicitly in the system. In this case it may be convenient not to mention $a*$ and to simply mark a as unusable.

6.3 Temporal networks: Formal considerations

We need to define the formal machinery and distinctions allowing us to put in context our approach to modal and temporal argumentation networks. So we define some basic notions in this section and move on to the Kripke models in the next section.

Definition 6.3 (General labelled networks).

1. A general labelled network has the form

$$\mathbb{N} = (T, \rho, \mathbf{l}, \mathbf{f})$$

where T is a set of nodes and $\rho \subseteq T \times T$ is a binary relation on T. $\mathbf{l}$ is a labelling function on $T \cup \rho$ giving labels from a set of labels $\mathbb{L}$ (usually $\mathbb{L} = \{0, 1\}$ or $[0, 1]$). The label $\mathbf{l}(t)$, for $t \in T$ can be thought of as the strength of the node. The label $\mathbf{l}(t, s)$ for $(t, s) \in \rho$ can be thought of as the transmission label from t to s.
The functional $\mathbf{f}$ is an update functional, it updates the labelling function $\mathbf{l}$ to a new one $\mathbf{f}(\mathbf{l})$. $\mathbf{f}$ is a pair of functions, $\mathbf{f}_1, \mathbf{f}_2$, which operate on multisets of elements to give a new element. For example, the function 'maximum' or the function 'take the sum of' is such a function. We write the value $\mathbf{f}_i(x, y, z, \ldots)$ where $(x, y, z, \ldots)$ is a sequence or a multiset). Given a node t, let $\rho(t)$ be $\{s \mid s\rho t \text{ holds}\}$. Then for any t, let

$$\mathbf{f}(\mathbf{l})(t) = \mathbf{f}_1(\mathbf{l}(t), \mathbf{l}(s), \mathbf{l}(s, t), s \in \rho(t))$$
$$\mathbf{f}(\mathbf{l})(s, t) = \mathbf{f}_2(\mathbf{l}(t), \mathbf{l}(s), \mathbf{l}(s, t))$$

be new labels at t, and at(s, t) given by the functional $\mathbf{f}$. $\mathbf{f}$ depends on ρ and $\mathbf{l}$, and on the labels and transmission labels, as depicted in Figure 6.6.

The way $\mathbf{f}$ is calculated is not described here. The reader can compare later in the Section below, where we give some examples of algorithms for $\mathbf{f}$ in terms of ρ and $\mathbf{l}$. $\mathbf{f}$ can be used for successive updating of the labelling of our network.
We define $\mathbf{l}_m$ for $m \geq 0$ by induction on m.
Step 0 : $\mathbf{l}_0 = \mathbf{l}$.
Step $m + 1$: $\mathbf{l}_{m+1} = \mathbf{f}(\mathbf{l}_m)$

Let $?$ be a fixed label in $\mathbb{L}$. We can define $\mathbf{l}_\infty$ using $?$ by letting $\mathbf{l}_\infty(x) = y$, if for some k we have $\mathbf{l}_m(x) = y$ for all $m \geq k$, and otherwise let $\mathbf{l}_\infty(x) = ?$. x is either a node $t \in T$ or a connection $(s, t) \in \rho$.

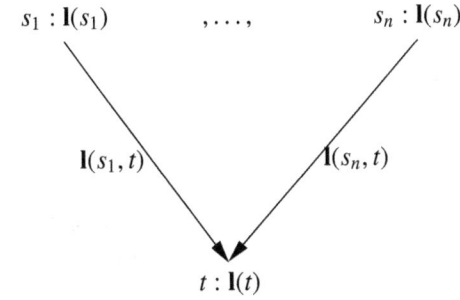

$$s_1 : \mathbf{l}(s_1) \qquad , \ldots , \qquad s_n : \mathbf{l}(s_n)$$

$$\mathbf{l}(s_1, t) \qquad\qquad \mathbf{l}(s_n, t)$$

$$t : \mathbf{l}(t)$$

We have $\rho(t) = \{s_1, \ldots, s_n\}$

The new label is $\mathbf{f}(\mathbf{l})(t) = \mathbf{f}(\mathbf{l}(s_i), \mathbf{l}(s_i, t)), i = 1, \ldots, n)$

Fig. 6.6.

We now have the machinery to look at argumentation networks and we use the Caminada labelling for them [81, 82].

Definition 6.4 (Argumentation model).

1. *Let $\mathbb{C}$ be the language of the classical propositional calculus with atoms Q and connectives $\neg, \wedge, \vee, \rightarrow, \top, \bot$. Q is the set of atomic arguments.*
2. *An argumentation model has the form $\mathbb{N} = (\mathbb{F}, \rho, h)$ where $\mathbb{F}$ is a set of formulas, ρ is a binary relation on $\mathbb{F}$ and h an assignment of truth values to the atoms Q. We can view h as a subset $h \subseteq Q$, and think of it as the set of usable arguments.*
3. *Given h, we can assign usability values to the formulas of $\mathbb{F}$ using the traditional truth table. We write $h(A)$ as the value of a wff A under h. h can now be regarded as a subset of $\mathbb{F}$.*
4. *A network is atomic iff $\mathbb{F} \subseteq Q$.*
5. *Note that h gives initial usability values which are not necessarily permanent and may change in the course of the recursive evaluation, see Definition 6.5.*

Definition 6.5. *Let $\mathbb{N} = (\mathbb{F}, \rho, h)$ be an argumentation model. We define an algorithm for extracting winning arguments out of $\mathbb{N}$ as follows. The definition is by levels. We define $h_m(A)$ for $A \in \mathbb{F}$ by induction on m, (compare with Definition 6.3).*

Level 0
$h_0(A) = h(A)$

Level $m + 1$
Let $A \in \mathbb{F}$ and let $\rho(A) = \{B_1, \ldots, B_k\}$ be all formulas B of $\mathbb{F}$ such that $B\rho A$ holds. These are the formulas which attack A according to ρ. There are several possibilities

1. *$h_m(B) = 0$ for all $B \in \rho(A)$. In this case let $h_{m+1}(A) = 1$.*
2. *For some $B \in \rho(A), h_m(B) = 1$. In this case let $h_{m+1}(A) = 0$.*
3. *$\rho(A) = \emptyset$, in which case let $h_{m+1}(A) = 1$*

Let π be the operation which defines h_{m+1} out of h_m, i.e. $h_{m+1} = \pi h_m$. Of course π depends on ρ. To be more explicit about the role of π, assume H is a function giving $\{0, 1\}$ values to all elements of $\mathbb{F}$. Using ρ and rules (1), (2), (3) above we can transform H

into H'. We write $H' = \pi H$.[3]

Level ∞

Let $h_\infty(A) = y \in \{0, 1\}$ iff for some k, $h_m(A) = y$ for all $m \geq k$.
Let $h_\infty(A) =?$ (undefined) otherwise.

h_∞ *is called the BG labelling of* $\mathbb{F}$.

Definition 6.6 (Caminada labelling). *Let $(\mathbb{F}, \rho)$ be an atomic network. A Caminada labelling on $\mathbb{F}$ is a function λ giving values in $\{0, 1, ?\}$ satisfying the following:*

1. *if $\rho(x) = \varnothing$ then $\lambda(x) = 1$*
2. *If for some $y \in \rho(x)$, $\lambda(y) = 1$, then $\lambda(x) = 0$.*
3. *If for all $y \in \rho(x)$, $\lambda(y) = 0$ then $\lambda(x) = 1$.*
4. *If for some $y \in \rho(x)$, $\lambda(y) =?$ and for no $y \in \rho(x)$ do we have $\lambda(y) = 1$, then $\lambda(x) =?$.*

Lemma 6.7. *Let $\mathbb{N} = (\mathbb{F}, \rho, \lambda)$ be an atomic network with the Caminada labelling λ. Then there exists an assignment h_0 such that $h_\infty = \lambda$.*

Proof. Let $h^+(x) = 1$ if $\lambda(x) = 1$ or $\lambda(x) =?$.
Let $h^+(x) = 0$ if $\lambda(x) = 0$.
Let $h^-(x) = 1$ if $\lambda(x) = 1$
Let $h^-(x) = 0$ if $\lambda(x) = 0$ or $\lambda(x) =?$
Let $h_0 = h^+$.
    We now prove

1. If $h_m = h^+$ then $h_{m+1} = h^-$
2. If $h_m = h^-$ then $h_{m+1} = h^+$.

Assume $h_m = h^\pm$. We calculate $h_{m+1}(x)$, $x \in \mathbb{F}$, and show $h_{m+1} = h^\mp$.
    If $\lambda(x) = 1$ then either $\rho(x) = \varnothing$ or for all $y \in \rho(x)$, $\lambda(y) = 0$. In this case $h_m(y) = 0$ and so $h_{m+1}(x) = 1$.
    If $\lambda(x) = 0$ then for some $y \in \rho(x)$, $\lambda(y) = 1$. In this case $h_m(y) = 1$ and so $h_{m+1}(x) = 0$.
    If $\lambda(x) =?$ then $\rho(x) \neq \varnothing$ and for some $y \in \rho(x)$, $\lambda(y) =?$ and for none of the other $y \in \rho(x)$ do we have $\lambda(y) = 1$. So let $\{y_1, \ldots, y_k, y_{k+1}, \ldots, y_r\} = \rho(x)$, with $k \geq 1$, $\lambda(y_j) =?$, $1 \leq j \leq k$ and $\lambda(y_{k+1}) = \ldots \lambda(y_r) = 0$.
    Clearly if $h_m = h^\pm$, then $h_m(y_{k+1}) = \ldots = h_m(y_r) = 0$.
    If $h_m = h^+$ then $h_m(x) = 1$ and $h_m(y_1) = \ldots h_m(y_k) = 1$ and hence $h_{m+1}(x) = 0$.
    This shows that $h_{m+1} = h^-$, since x was arbitrary.
    If $h_m = h^-$ then $h_m(x) = 0$ and $h_m(y_1) = \ldots = h_m(y_k) = 0$ and so for all $y \in \rho(x)$, $h_m(y) = 0$ hence $h_{m+1}(x) = 1$.
    Again since x was arbitrary we get $h_{m+1} = h^+$.
    So if we start with $h_0 = h^+$, we get $h_{2m} = h^+$, $h_{2m+1} = h^-$ and so $h_\infty = \lambda$.

Lemma 6.8. *The converse of the previous lemma does not hold. Not every h_∞ is a Caminada labelling.*

Proof. Consider the network in Figure 6.7

---

[3] In terms of Definition 6.3, H is a labelling **l** (no transmission labels) and π is a functional **f** whose algorithm uses clauses (1), (2), (3).

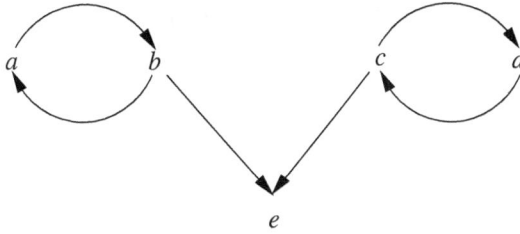

Fig. 6.7.

Start with $h_0(a) = h_0(b) = 1, h_0(c) = h_0(d) = 0, h_o(e) = 0$. We have $h_1(a) = h_1(b) = 0, h_1(c) = h_1(d) = 1, h_1(e) = 0$.

We also have

$$h_{2m} = h_0, h_{2m+1} = h_1.$$

Thus $h_\infty(a) = h_\infty(b) = h_\infty(c) = h_\infty(d) = ?$ and $h_\infty(e) = 0$.

The Caminada labelling rules do not allow for $\lambda = h_\infty$.

Remark 6.9.

1. The reason we could provide the example in Figure 6.7 is that h gave value 1 to
   the loop (a, b) and value 0 to the loop (c, d). So as the values in the loop oscillated, there was always one loop which attacked e. If all loops were to oscillate
   synchronously, $h(e)$ would have oscillated as well.
   How can we overcome this? We can use ultrafilters to get an exact value out of
   the oscillation. We need some concepts
   Let *Nat* be the set of natural numbers. A family of subsets $\mathbb{U}$ of numbers is called
   an ultrafilter if the following holds
   a) *Nat* $\in \mathbb{U}, \emptyset \notin \mathbb{U}$
   b) If $X, Y \in \mathbb{U}$ then $X \cap Y \in \mathbb{U}$
   c) either X or *Nat* $- X$ is in $\mathbb{U}$.
   $\mathbb{U}$ says which sets are 'big'.
   We also note that there exists an ultrafilter $\mathbb{U}$ such that all co-finite sets are in $\mathbb{U}$.
   We now give an alternative definition of h_∞. Call it h_ω.

   $$h_\omega(x) = 1 \text{ iff } U_x = \{m | h_m(x) = 1\} \in \mathbb{U}.$$

   Let us see what happens with the example of Figure 6.7 if we use h_ω instead of
   h_∞. We have
   $U_a = U_b = $ all even numbers
   $U_c = U_d = $ all odd numbers.
   One of two sets {odd numbers, even numbers} is in $\mathbb{U}$. From symmetry we can
   assume without loss of generality that it is the even numbers. We get

   $$h_\omega(a) = h_\omega(b) = 1, h_\omega(c) = h_\omega(d) = 0.$$

So h_ω is the same as h_0 and we have nothing.
Let us try another angle.

2. The discrepancy with the Caminada labelling and hence with the Dung network rules seem to arise in the case where a winning argument x according to Dung gets a value $h(x) = 0$.

Figure 6.8 gives two typical examples.

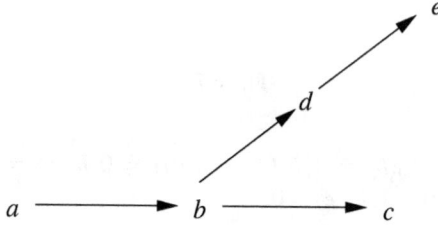

Fig. 6.8.

According to the Dung rules a, d, c are winning arguments. If h is such $h(a) = 0$ then b and e will be the winning arguments.

The question we ask is can we use a device which makes the two approaches compatible?

Suppose we say a is not usable (i.e. $h(a) = 0$) because there is an attack on a. Say $\mathbf{h}(a)$ is the argument which attacks a. $\mathbf{h}(a)$ is not in the network but the fact that a is attacked is recorded by $h(a) = 0$. There may be good reason why we don't want $\mathbf{h}(a)$ to be explicit in the network. Maybe $\mathbf{h}(a)$ is a different type of argument. Maybe it is a secret argument. Whatever the reason is, the real network is Figure 6.9. The above trick works for this network. Does it work in general?

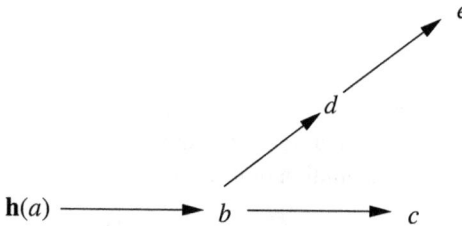

Fig. 6.9.

Given an atomic network $(\mathbb{F}, \rho, h)$ can we get the correct result by adding nodes $\mathbb{F}_1 = \{\mathbf{h}(x) | x \text{ such that } h(x) = 0\}$ and letting $\mathbb{F}' = \mathbb{F} \cup \mathbb{F}_1$ and $\rho' = \rho \cup \{(\mathbf{h}(x), x) \mid \mathbf{h}(x) \in \mathbb{F}_1\}$?

Do we have a general theorem that we can pair the winning subsets? I.e. can we have:

a) For any winning set $T' \subseteq \mathbb{F}'$, $T' \cap \mathbb{F}$ is a winning set of $(\mathbb{F}, \rho, h)$.

b) For any winning set $T \subseteq \mathbb{F}$, there exists a winning set $T' \subseteq \mathbb{F}'$ such that $T = T' \cap \mathbb{F}$.

We can hope for such results only if whenever h says an argument x is out then it is out permanently, because when we insert $\mathbf{h}(x)$ to attack x and force it out, it is out permanently.

Our algorithm in Definition 6.5 and later on in the section dealing with modal and temporal logic, does not keep unusable arguments out permanently, it does bring them in depending on the attack cycles.

Remark 6.10 (Discussion of the Dung network rules).

1. The discrepancy with the Caminada labelling is a serious one. The Caminada labelling is faithful to the Dung argumentation network rules, namely (see Definition 6.5).

   a) If all attacks on a node x are defeated (are *out*) then x is *in*.

   b) If some attacks on a node x are *in* then x is *out*.

   c) If there are no nodes attacking x then x is *in*.

   In the Dung framework these rules are *not defeasible rules*, they are *absolute*.

   So consider, for example, an argumentation network with one node and one argument x. Since nothing attacks x, x is a winning argument. If we look at a classical model for x with $x = 0$, namely x is unusable for whatever reason, then the Dung rule overrides the unusability of x and x is still a winning argument.

   Compare this with default logic. The default rule $\frac{x}{x}$ says that if x is consistent to add then add it by default. This will not override any given data about x.

   So rules (a), (b), (c) are too strong when we give a model interpretation to the arguments. An argument can win even though it is unusable, simply because it is not attacked.

2. We call for a new critical evaluation of rules (a)–(c). We would abandon rule (c) and modify rule (a).

   The proposed BG rules for a Dung network relative to an assignment or other evidence, are (a*)–(c*) below. We call the associated update functional π^* (compare with π of Definition 6.5).

   a*. If all attacks on x are defeated and there is no evidence that x is unusable then x is *in*

   b*. If some attacks on x are in the x is *out*.

   c*. If there are no nodes attacking x then x is *in* only if there is no evidence that it is unusable.

   Jakobovits and Vermeir [223] have already proposed that an argument which has all of its defeaters out need not necessarily be in. This view is criticised in [75]. Our (a*) and (c*) are in line with [223].

3. Let us call an assignment h a Caminada assignment to the atoms of Q if for some Caminada labelling λ we have
   - $h(x) = 1$ if $\lambda(x) = 1$.
   - $h(x) = 0$ if $\lambda(x) = 0$
   - $\forall x(\lambda(x) = ?$ implies $h(x) = 1)$.

   From Lemma 6.7 we know that $h_\infty = \lambda$ and hence Caminada assignments are compatible with Dung networks. If we restrict our Kripke model to Caminada assignments then maybe we will have no technical discrepancies.

Remark 6.11. The difference between BG and Caminada labelling can be appreciated by looking at the logic programming translation of a Dung network. Consider the network $(\{a, b\}, \{(ab)\})$, (that is a network with two arguments a and b with a attacking b). Its translation is ($\neg$ is negation as failure)

1. b if $\neg a$
2. a

In our model we also give usability assignments $h(x)$ to x. So we translate as follows into a logic programming program (we regard $\mathbf{h}(x)$ as a new literal dependent on x attacking x by virtue of an argument showing that x is unusable):

1*. b if $\neg a \wedge \neg \mathbf{h}(b)$
2*. a if $\neg \mathbf{h}(a)$
3*. $\mathbf{h}(x)$ if $\neg \mathbf{zero}(x)$, for all nodes x
4*. $\mathbf{zero}(x)$, for all x such that x = usable, under the assignment h.

$\mathbf{h}(x)$ and $\mathbf{zero}(x)$ are new literals, for each node x.

Note that we use $\neg \mathbf{h}(x)$ rather than $\mathbf{h}(x)$ in the program so that the program will have a corresponding Dung network. To make $\neg \mathbf{h}(x)$ fail we need to add $\mathbf{h}(x)$, for x = usable, under h.

The corresponding Dung network for the above network is Figure 6.10 below (see footnote 1):

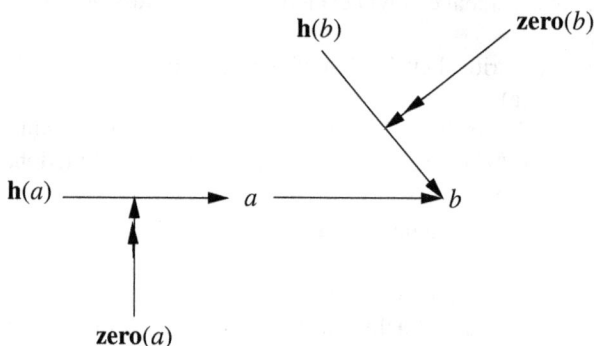

Fig. 6.10.

So the BG program is defined to contain the following clauses

1* If $x_1, \ldots, x_n$ attack y in the network we include the clause

$$y \text{ if } \neg x_1 \wedge \ldots \wedge \neg x_n \wedge \neg \mathbf{h}(y)$$

where $\mathbf{h}(y)$ is a new atom.
2* If y is not attacked by any node we include the clause
  y if $\neg \mathbf{h}(y)$
3* Add $\mathbf{h}(y)$ if $\neg \mathbf{zero}(y)$

4* For every y such that $h(y) = 1$ we include the literal **zero**(y).[4]

6.4 Kripke models for argumentation networks 1

We begin this section with some methodological remarks and general examples which will bring us to a point of view best suited for the presentation of modal and temporal argumentation networks.

We begin with a simple example. Consider the sentence

- John read le livre with interest.

This sentence contains some French words. To understand the sentence we need to go to a French dictionary and come back with values. We come back with the words 'the book'.

Now consider

- John is dishonest because he did not pay yesterday.

To check the value of the above, we must go to yesterday and verify that John did not pay.

Let t label the location of the main sentence and let s label the location of where we need to go. In each case we have the following situation.

In the process of evaluating the algorithm $\mathcal{A}_t$ at t, we hit upon a unit of the form

Take x to location s, find a value $y = V_s^t(x)$ for x at s and come back and plug it into our local algorithm $\mathcal{A}_t$ and carry on.

The notation $V_s^t(x)$, is the value you get for x at location s intended to be understood and used at t (so V_s^t is a French to English 'function' in the first example, and a function reading from a payment ledger for the second example).

Let us now take another example. Consider the three argumentation networks in Figure 6.11. In network t the node x is not an argument but an instruction to look for an accessible network in which there is a winning argument which can defeat b. The two accessible networks are s and r. In s, a is not a winning argument, but it is in r. Suppose a is capable of defeating b; this knowledge is not recorded in the network t but is known to us either extra-logically or intrinsically (for example, b is logically inconsistent with a). Then x will be argument a coming from network s. x can be as specific as needed for a successful search. Note that we could have written Figure 6.12. This is a fibring of network r at position x at network t. The attack on b comes from inside the network r from node a onto node b (in network t).

We can turn the situation into modal logic by using $\Diamond x$ (or even $\Diamond a$ if we know that $x = a$ will do the job) and we get a kind of Kripke model, see for example Figure 6.13. The zigzag arrow $\rightsquigarrow$ is accessibility and the ordinary arrow $\rightarrow$ is attack.

---

[4] If you recall the idea of attacks on attacks as developed in Section 2 of [41] and originally in [37], (coming in Chapter 9 below) you will realise we can present a network relative to additional nodes attacking connections. The **zero**(y) nodes are added according to assignment to attack the connection **zero**$(y) \rightarrow$ **h**(y). In fact there is no need to do it this way. We do not need **zero**(y). We can simply augment the original network with additional nodes **h**(y) attacking the node y for all y such that $h(y) = 0$. The problem with that is that h keeps on changing and so the **h**(y) will keep on being in and out.

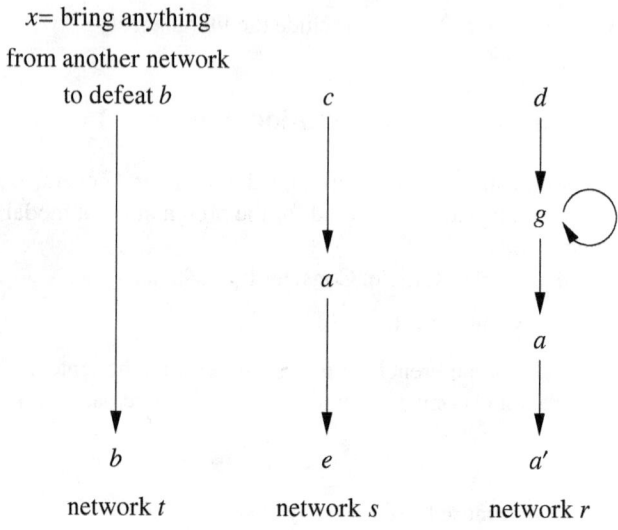

Fig. 6.11.

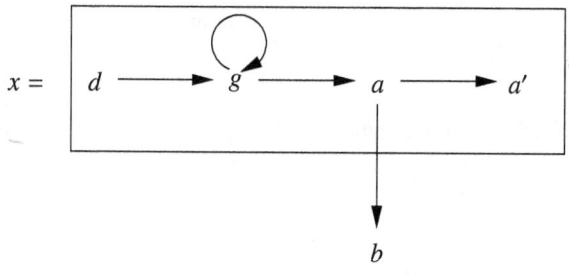

Fig. 6.12.

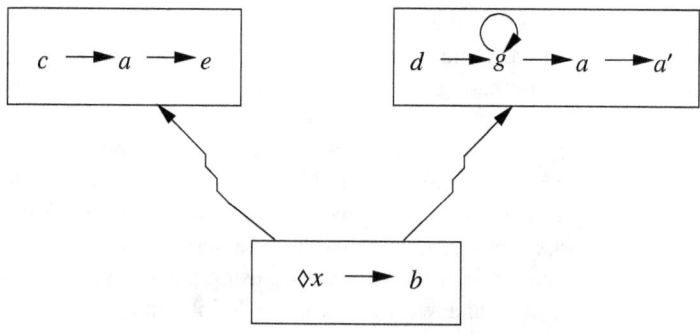

Fig. 6.13.

◊x (or ◊a) means find a winning argument a which can attack b. Here the meaning of ◊a is adminstrative. ◊ is a metalevel administrative connective. ◊a does not mean that a is a possible argument; and ◊ is not in the argument language.

 Consider now the following network

◊ storm → b

in which ◊ storm represents "it is possible there will be a storm" and b represents "setting sail now". The model envisages several possible futures, if in at least one of them there is a storm then that possibility defeats b. Here the possible ◊ is not an administrative but a temporal event. The language of ◊a is object level, and ◊ must be in the language of arguments.

Technically the mathematics of both cases, the administrative metalevel ◊ and the temporal event object level ◊, is very similar. In both cases we can put ◊x in the nodes of an argumentation network and seek winning arguments in accessible networks.

In either case of ◊x we need to search other networks for an appropriate winning value. So it is not clear until after the calculation and search, whether we have a usable argument here or not, especially if the family of networks is complex. Hence the need and the technical usefulness and value of a usability assignment. It simplifies matters during the calculations.

We now explore our options for Kripke models for argumentation networks. We begin with a simple first attempt which will turn out to need improvement. However, it is helpful to go through this first attempt for us to appreciate what is to be added. Consider the network of Figure 6.5 and let Figure 6.14 describe a Kripke model. The

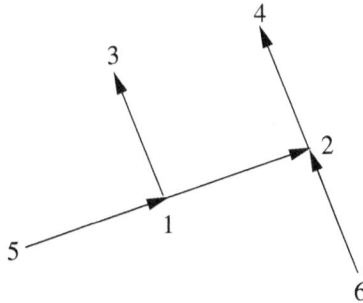

Fig. 6.14.

reader should bear this in mind when reading the next formal definition.

Definition 6.12 (Languages). *We recall here the languages of modal and temporal logic.*

1. *The classical connectives we use are ~ (negation), ∧, ∨, →. We reserve ¬ for negation as failure.*

2. *In temporal logic we use the connective PA for A was true in the past and FA for A will be true in the future. A temporal model is usually presented through a non-empty set T of moments of time and an earlier-later relation < on T. < is usually taken as irreflexive and transitive. The classical truth conditions for P and F are*
   - $t \vDash PA$ *iff for some s such that $s < t$, we have $s \vDash A$.*
   - $t \vDash FA$ *iff for some s such that $t < s$, we have $s \vDash A$*

   Temporal logic defines
   $HA =\sim P \sim A = A$ *has always been true*
   $GA =\sim F \sim A = A$ *will always be true.*

3. *Modal logic uses ◊A reading A holds in another accessible world. The set of worlds is denoted by S and has an accessibility relation R ⊆ S × S. We have*
   - *t ⊨ ◊A iff for some s such that tRs we have s ⊨ A.*
   □A usually means ∼ ◊ ∼ A.
4. *The usual temporal or modal logics have formulas evaluated at worlds. If we want to define the notion of modal and temporal networks we will need to deal with networks of formulas evaluated at worlds.*
5. *We can have in the language both the temporal connectives P, F and the modal connective ◊. In which case the semantics will need to have both R and <. We may allow for the future to be also a possibility for ◊, in which case we have:*

$$t < s \rightarrow tRs.$$

   Sometimes only P and ◊ are used in which case we can use only R and go backwards in it for evaluating P.
6. *The examples below use 'usability' instead of truth. So we have for example that "a is usable at world t" is treated mathematically the same way as we treat "a is true (holds) at t".*

Let us start with some examples in the simple language which contain arguments of the form x, (atomic), $◊x$, possibility of an argument and Px, past arguments. Think of $◊x$ as future possibility. So in the Kripke model, $◊$ goes up in the arrow direction and P goes down in the arrow direction.

In the usual Kripke evaluation procedures for classical logic, with a set Q of atomic arguments, we have an assignment h giving a truth value at each world t for each atomic proposition q. We write $h(t) \subseteq Q$ for the set of true propositions at t. In the argumentation case we do not have atomic propositions, but we do have argumentation networks which themselves contain atomic propositions. For example the network of Figure 6.5 contains the atoms a, b, c.

So we give the following definition. For each node n in the Kripke model we assign a set $h(n) \subseteq$ set of atoms appearing in the network. For an atom q, we assign a usability value 1 if $q \in h(n)$ and 0 otherwise. We also use the notation $h(n, q) = 1$ or $h(n) ⊨ +q$ to indicate that $q \in h(n)$, and $h(n, q) = 0$ or $h(n) ⊨ -q$ to indicate that $q \notin h(n)$.

Figure 6.15 is an example of such an assignment, where we write $\pm q$ to indicate the value of q.

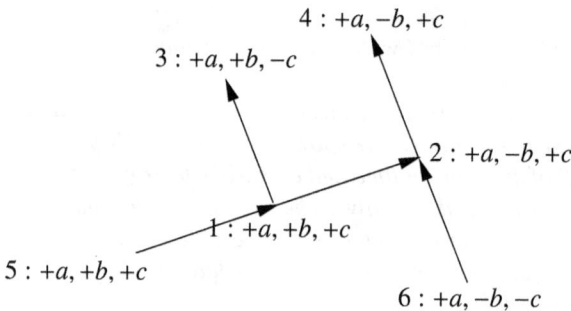

Fig. 6.15.

Suppose we want to know the value of the network of Figure 6.5 at the model at node 1. How do we evaluate Pc? We follow the traditional steps of evaluation in a Kripke model; we go down the accessibility relation and look for a world where c is usable.

At node 5 we have $+c$, so maybe we say $+Pc$ at node 1, but we notice that a attacks Pc in the network (see Figure 6.5). So does $1 \vDash Pc$ or not? Furthermore, Pc attacks b and so at node 2 does Pb hold or not?

We have $+b$ at node 1, so we would like to say $2 \vDash Pb$, however b may be successfully attacked at node 1. So we may not have Pb after all, so what do we have?

We need an agreed recursive definition.

Let us see some examples where we might have a loop, and try and get a clue by working the example out.

Example 6.13. Consider the network and Kripke model as described in Figure 6.16

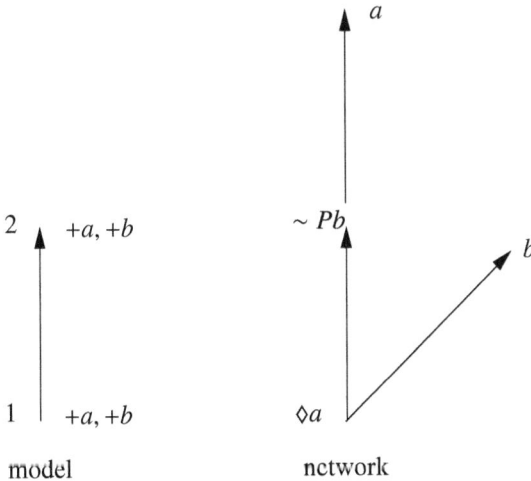

Fig. 6.16.

To evaluate the network at 1, we know that $\sim Pb$ = usable. But $\sim Pb$ is attacked by $\Diamond a$ and so we need the value of $\Diamond a$ at 1. For this we need the value of a at point 2. We have $+a$ at 2, but this is attacked by $\sim Pb$, which is not attacked by $\Diamond a$ at 2 since it is not usable at 2.

So we need to know the value of b at 1. We do have $+b$ at 1 but this is attacked by $\Diamond a$ so we need to know the value of a at 2 and we have a loop.

We need some process of evaluation which will give us a better chance to resolve the loops.

We do this by levels of recursive evaluation. We use two bits of notation.

1. $h_m(t, A)$ is the assignment of value at level m to the argument A. A may be atomic or a formula.
2. $t \vDash_m A$ is the network value of A at world t at level m. We use the BG algorithm for the functional π^* namely clauses $(1*), (2*)$ and $(3*)$ of Remark 6.10.

So to make the meaning of h_m and $\vDash_m$ crystal clear: h_m says which arguments in the network are usable at level m, while $\vDash_m$ says which arguments are winning (not defeated) at level m. An argument defeated at $\vDash_m$ is considered unusable by h_{m+1}.

Let us now apply this to Figure 6.16's example.

Level 0

h_0 is the assignment h to the atoms as indicated in the Figure, i.e. $h_0(1, a) = h_0(1, b) = h_0(2, a) = h_0(2, b) = 1$. $\vDash_0$ is defined for a b as the same value as h_0.

For $\Diamond a, Pb, h_0, \vDash_0$ is not necessarily defined.

It is convenient to use the notation

$h_m(t) \vDash +A$ to say $h_m(t, A) = 1$

$h_m(t) \vDash -A$ to say $h_m(t, A) = 0$.

The idea of h_m and $\vDash_m$ is as follows: h_m evaluates the temporal modalities using the Kripke model without regard to the argumentation network. Then $\vDash_m$ records the result of the attacks (i.e. the winning arguments) of the argumentation network at each world. The $\vDash_m$ may record the defeat of some atoms in the network, thus rendering them unusable at level $m + 1$ (by virtue of being defeated at level m) thus giving rise to a new assignment which can now be used to calculate h_{m+1}, and so on.

Level 1

$$h_1(1) \vDash +\Diamond a, +b, + \sim Pb, +a$$
$$h_1(2) \vDash -\Diamond a, +b, - \sim Pb, +a.$$

Now we have networks with nodes which have values and so we calculate the winning arguments. These are the ones holding at the worlds of the Kripke model at level one. We write:

$$1 \vDash_1 \Diamond a, a$$
$$2 \vDash_2 b, a$$

Level 2

The assignment we use to calculate h_2 is the atomic part of $\vDash_1$, namely $1 \vDash_1 a$ and $2 \vDash_1 a, b$.

$$h_2(1) \vDash +\Diamond a, -b, +a, + \sim Pb$$
$$h_2(2) : -\Diamond a, +b, + \sim Pb, +a$$

After evaluation of the networks we get

$$1 \vDash_2 \Diamond a, a$$
$$2 :\vDash_2 \quad \sim Pb, b$$

Level 3

$\vDash_2$ gives us a new assignment to the atoms, namely $1 \vDash_2 a$ and $2 \vDash_2 b$.

$$h_3(1) \vDash -\Diamond a, -b, +a, + \sim Pb$$
$$h_3(2) \vDash -\Diamond a, +b, + \sim Pb, -a$$

calculating the surviving arguments at each world we get

$$1 \vDash_3 \sim Pb, b$$
$$2 \vDash_3 \sim Pb, b$$

Level 4
We now get the assignment from $\vDash_3$ for atoms as

$$1 \vDash_3 b, 2 \vDash_3 b$$

We calculate h_4

$$h_4(1) \vDash -\Diamond a, - \sim Pb, +b, -a$$
$$h_4(2) \vDash -\Diamond a, - \sim Pb, +b, -a$$

We calculate $\vDash_4$ at each node using network rules

$$1 \vDash_4 \sim Pb, b,$$
$$2 \vDash_4 \sim Pb, b.$$

$\vDash_3$ and $\vDash_4$ are the same. So the answer is now stable.

Remark 6.14. The reader may wonder what has happened to the loop we observed before and what kind of interpretation (loop resolution) we are getting. To explain that, let us look at the traditional loop in the network of Figure 6.17, in which a and b

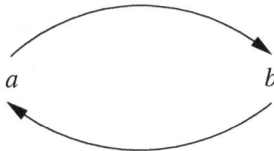

Fig. 6.17.

attack each other. We have three complete extensions $\{a\}, \{b\}, \emptyset.^5$ Let us do the level calculation intuitively.

Level 0
Start with $+a, +b$.
Level 1
Attack as suggested by level 0 we get $-a, -b$.
Level 2
Attack as suggested by level 2. We get $+a, +b$.

So we are infinitely looping and we can put a question mark on a and on b, leading to $\emptyset$.

Of course everything depends on the assignment at level 0. Other possible assignments are $\{+a, -b\}, \{-a, +b\}, \{-a, -b\}$, yielding $\{a\}, \{b\}$ and $\emptyset$ respectively.

5

1. A subset E of arguments E is conflict free if no member of it is attacked by another member.
2. A subset of arguments E is admissible, if whenever a member x in E is attacked by any other argument z, then there is a y in E which attacks z.
3. E is complete extension if it is a subset satisfying 1. and 2. and is maximal with respect to set inclusion.

Example 6.15. We try and evaluate the network of Figure 6.5 at the model of Figure 6.16. We use the *BG* π^* evaluation algorithm.

We do this by levels. Let h_0 be the assignment to the atoms as indicated in Figure 6.15. Thus h_0 satisfies

Level 0

$$h_0(5) \models +a, +b, +c$$
$$h_0(6) \models +a, -b, -c$$
$$h_0(1) \models +a, +b, +c$$
$$h_0(2) \models +a, -b, -c$$
$$h_0(3) \models +a, +b, -c$$
$$h_0(4) \models +a, -b, +c.$$

Level 1

$$h_1(5) \models +a, +b, +c, -Pc, +\Diamond a$$
$$h_1(6) \models +a, -b, -c, -Pc, +\Diamond a$$
$$h_1(1) \models +a, +b, +c, +Pc, +\Diamond a$$
$$h_1(2) \models +a, -b, +c, -Pc, +\Diamond a$$
$$h_1(3) \models +a, +b, -c, +Pc, -\Diamond a$$
$$h_1(4) \models +a, -b, +c, +Pc, -\Diamond a$$

Now the network of Figure 6.5 has values for each node at each world. We can compute the winning argument at each world. Note that c does not appear as a node in the network so its value we inherit from h.

$$5 \models_1 +a, -b, +c, -Pc, +\Diamond a$$
$$6 \models_1 +a, -b, -c, -Pc, +\Diamond a$$
$$1 \models_1 +a, -b, +c, +Pc, +\Diamond a$$
$$2 \models_1 +a, -b, +c, -Pc, +\Diamond a$$
$$3 \models_1 +a, +b, -c, +Pc, -\Diamond a$$
$$4 \models_1 +a, +b, +c, +Pc, -\Diamond a$$

Level 2

We evaluate h_2 using the assignment to the atoms suggested by $\models_1$.

$$h_2(5) \models +a, -b, +c, -Pc, +\Diamond a$$
$$h_2(6) \models +a, -b, +-, -Pc, +\Diamond a$$
$$h_2(1) \models +a, -b, +c, +Pc, +\Diamond a$$
$$h_2(2) \models +a, -b, +c, -Pc, +\Diamond a$$
$$h_2(3) \models +a, +b, -c, +Pc, -\Diamond a$$
$$h_2(4) \models +a, +b, +c, +Pc, -\Diamond a$$

We now calculate $\models_2$

$$5 \models_2 +a, -b, +c, -Pc, +\Diamond a$$
$$6 \models_2 +a, -b, -c, -Pc, +\Diamond a$$
$$1 \models_2 +a, -b, +c, +Pc, +\Diamond a$$
$$2 \models_2 +a, -b, +c, -Pc, +\Diamond a$$
$$3 \models_2 +a, +b, -c, +Pc, -\Diamond a$$
$$4 \models_2 +a, +b, +c, +Pc, -\Diamond a$$

We have stability since $\models_1$ equals $\models_2$.

Definition 6.16 (Temporal languages). *Let Q be a set of atoms. The basic temporal language based on Q uses the unary connectives $\mathbb{C} = \{\neg, \Diamond, P\}$. A temporal formula has the form $\alpha_1 \alpha_2 \ldots \alpha_m q$, where $q \in Q$ and $\alpha_i \in \mathbb{C}, i = 1, \ldots, m$.*

The language is said to be very basic if we allow $\Diamond x$ and Px and x only.

The full temporal language also allows the use of the classical connectives $\wedge, \vee \rightarrow$ and unrestricted use of $\Diamond$ and P.

Definition 6.17 (Temporal Kripke models). *Let (S, R, a) be a Kripke model and let $\mathbb{N} = (\mathbb{F}, \rho)$ be a network in the basic temporal language with $\neg, \Diamond, P$ and Q or in the full temporal language. Let h be an assignment giving for each world $t \in S$ and an atom $q \in Q$ a usability value $h(t, q) \in \{0, 1\}$.*[6]

We define by induction a sequence of new assignments $h_1, h_2, \ldots$ and semantic consequences $\vDash_1, \vDash_2 \ldots$ as follows:

1. *Let h_1 be defined as follows*
   $h_1(t, q) = h(t, q)$, *for q atomic*
   $h_1(t, \Diamond A) = 1$ *if for some $s \in S$ such that tRs we have $h_1((s, A) = 1$.*
   $h_1(t, PA) = 1$ *iff for some s, such that sRt we have $h_1(s, A) = 1$.*
   The definition for the classical connectives is the usual one.

2. *Let $\vDash_1$ be defined as follows:*
   First consider $h_1(t)$ as an assignment on $(\mathbb{F}, \rho)$ with t as a fixed parameter. We can consider $h_1(t)$ as a subset of $\mathbb{F}$. Consider the operator π of Definition 6.5.
   We define $\vDash_1$ by

   $$t \vDash_1 x \text{ iff } x \in \pi h_1(t).$$

   The reader can compare with the construction in Example 6.13.

Note that if we want to use defeasible rules as discussed in Remark 6.10 then we use π^* of Remark 6.10 instead of π of Definition 6.5.

   We now define $h_{m+1} \vDash_{m+1}$ for $m \geq 1$.

   h_{m+1} is obtained from $\vDash_m$ in the same way that h_1 was obtained from h, by regarding $t \vDash_m q, q \in Q$ as an assignment to the atoms. Note that all we need is the values of $\vDash_m$ on the atoms of Q.

   $\vDash_{m+1}$ is obtained from h_{m+1} in the same way that h_2 was obtained from h_1, i.e. for each t, we have $t \vDash_{m+1} A$ iff $A \in \pi h_{m+1}(t)$.

   This defines $h_m \vDash_m$ for all $n \geq 1$.

   We now define $t \vDash_\infty A$ for $t \in S$ and $A \in \mathbb{F}$ as follows:

   $t \vDash_\infty A$ holds (does not hold) if for some k, $t \vDash_m A$ holds (resp. does not hold) for all $n \geq k$.

   Otherwise if $t \vDash_m A$ oscillates, we say that $t \vDash_\infty A$ is undecided.

---

[6] This definition is parallel to the traditional one. $\Diamond$ goes up the accessibility relation and P goes down it. The evaluation is more complicated because the formulas are part of a network.

   As noted before if we insist that at each t the assignment $h(t) \subseteq Q$ is a Caminada assignment (need not be the same one for all t), then we will have less technical discrepancies with the Dung interpretation. We need to check, however, since our language has temporal operators, whether $h_m(t)$ remains a Caminada assignment. We will examine this point later in the section. Our guess is that further adjustment will be needed.

   We shall give better definitions later in the next section.

6.5 Kripke models for argumentation networks 2

Let us assess the situation we are in. In the previous section, we offered a simple model. This model can be improved. The problem is not so much the discrepancy with the Dung approach (see Remark 6.11) as this is not a unique possible world problem, but the difficulty is that we need to sharpen the intuitive meaning to Definition 6.17. Definition 6.17 is mainly technically motivated by a natural formal analogue of the semantical movements in a traditional modal and temporal Kripke model. We need to clarify more sharply the meaning of usability of arguments and its connection to truth and falsity and possibility of argument and facts.

There seems to be a fundamental difference between the modal operator $\Diamond$ and the temporal operator P. P goes into the past while $\Diamond$ goes to an alternative world of different reasoning/argumentation framework. In ordinary Kripke semantics for traditional modal and temporal logics, there is no technical distinction between the two. In argumentation context we need to give different technical treatment to these two connectives. The $\Diamond$ we treat as a fibring operator (go to another context, do something there and come back with the result, see [138]), while the operator P is still treated as purely temporal.

We illustrate with an example.

We want to argue against a political candidate c. We want to bring in the past facts that he double-crossed his partners, showing lack of loyalty and trustworthiness (call this Pd, d for "dirt"). However, the situation today is such that digging up the past on a candidate is counterproductive (call this $\sim p$). It is suggested therefore to wait 6 months for the facts to emerge naturally (i.e. $\Diamond d$ where $\Diamond$ here reads future possibility).

A counterargument against waiting is that by that time criteria for judging candidates will change and the argument will be defeated, say d will be attacked by e (e can mean who cares? ; it was a long time ago!).

Figure 6.18 shows the situation:

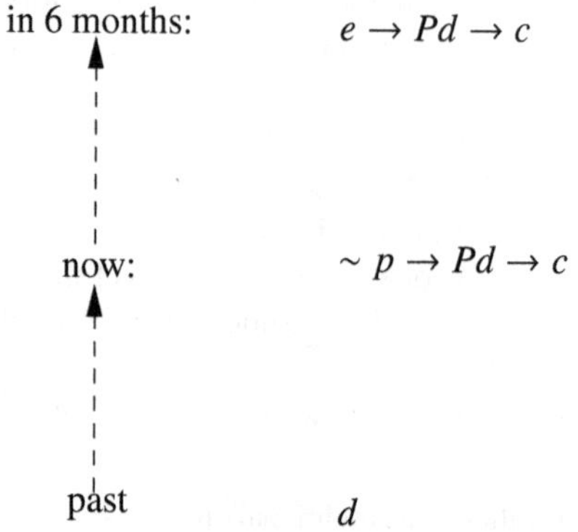

Fig. 6.18.

We notice the following discrepancies between the formal situation of Figure 6.18 and the formal definition we gave to modal and temporal argumentation networks, in Definition 6.17.

1. We have different networks at different possible worlds.
2. $\Diamond$ behaves like a fibring operator
3. P is purely temporal for facts.
4. With P the assignment h indicates $\pm$ usability by virtue of truth or falsity, while with $\Diamond$ the assignment h might indicate $\pm$ usability for other reasons.
5. We cannot have a proper temporal and modal treatment of arguments without looking into the details of what is the internal structure of arguments and how exactly do they attack one another in terms of such structure.

   Suppose we adopt the view that arguments are proofs and attacks disrupt such proofs. Let us examine how time T gets into the picture. Consider the following example:

   I park my car near Russell Square at 11am in the morning, do my business of the day and come back to it at 5 pm. I expect to find it there. If I don't find the car then nonmonotonic deduction allows me to conclude that the car was stolen. We can have a little logical system (nonmonotonic or Bayesian net) which compares the conclusions of "car stolen" against "car towed away by local council parking department". We can assume the latter conclusion can be defeated. We can also ignore some well known difficulties of persistence of arguments where one gets paradoxically that the car was stolen only a few seconds before my return (the stolen car paradox).

   There is one sure way to attack this argument and its conclusion. This is to prove the simple fact that I don't have a car.

   Figure 6.19 illustrates the situation

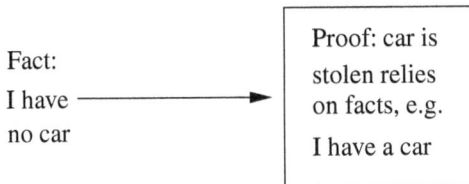

Fig. 6.19.

Facts can attack arguments most effectively. Also if the facts are undecided or are not available, then we claim the argument is not usable. See Section 6 for further discussion.

So when designing a new modal and temporal logic for argumentation, we need a pure $\{P, F\}$ temporal logic just for the facts.

6. Our next question is whether an argument itself can be time dependent. This is a bit tricky. In monotonic logic the answer is no. Euclid geometric proofs are as valid and good today as they were in ancient times. But in the nonmonotonic case the answer is yes. Nonmonotonic reasoning depends on context. Today a girl in a mini-skirt will not be considered immodest but go back 200 years and everyone at that time will nonmonotonically deduce she is 'fast'. So as we can see from this example the deduction mechanism itself can change in time. Thus we may argue for example along the lines 'you had better get yourself a decent long dress

now because soon people's perception will change and you will no longer be respectable wearing a mini-skirt'.

Definition 6.18 (Temporal Kripke models - udpated).

1. *Consider a language with the classical connectives $\{\sim, \wedge, \vee, \rightarrow\}$ and the temporal and modal connectives $\{P, F, \Diamond\}$. We also assume a set Q of atoms, and we use the atoms to construct formulas using the connectives in the traditional manner.*
2. *A Kripke model for the above language has the form $(S, R, \leq, a, h)$, where S is a nonempty set of worlds and $R \subseteq S \times S$ is a binary relation for $\Diamond$ and $<$ is an irreflexive and transitive relation on S for F and P. h is an assignment giving for each $t \in S$, a subset $h(t) \subseteq Q$. We consider h as a usability function on the atomic arguments of Q, saying which elements of Q are usable at world t.*
3. *For each t, let $\mathbb{N}(t) = (\mathbb{F}(t), \rho(t))$ be an argumentation system.*
4. *We define by induction a sequence of new assignments $h_1, h_2, \ldots$ and semantic consequences $\vDash_1, \vDash_2 \ldots$ as follows:*
    a) *Let h_1 be defined as follows*
       $h_1(t, q) = h(t, q)$, *for q atomic*
       $h_1(t, \Diamond A) = 1$ *if for some $s \in S$ such that tRs we have $h_1((s, A)) = 1$.*
       $h_1(t, PA) = 1$ *iff for some s, such that $s < t$ we have $h_1(s, A) = 1$.*
       $h_1(t, FA)$ *iff for some s such that $t < s$ we have $h_1(s, A) = 1$.*
       The definition for the classical connectives is the usual one.
    b) *Let $\vDash_1$ be defined as follows:*
       First consider $h_1(t)$ as an assignment on $(\mathbb{F}, (t), \rho(t))$ with t as a fixed parameter. We can consider $h_1(t)$ as a subset of $\mathbb{F}(t)$. Consider the operator π_* of Remark 6.10.
       We define $\vDash_1$ by

$$t \vDash_1 x \text{ iff } x \in \pi^* h_1(t).$$

We now define $h_{m+1} \vDash_{m+1}$ for $m \geq 1$.
h_{m+1} is obtained from $\vDash_m$ in the same way that h_1 was obtained from h, by regarding $t \vDash_m q, q \in Q$ as an assignment to the atoms. Note that all we need is the values of $\vDash_m$ on the atoms of Q.
$\vDash_{m+1}$ is obtained from h_{m+1} in the same way that h_2 was obtained from h_1, i.e. for each t, we have $t \vDash_{m+1} A$ iff $A \in \pi^* h_{m+1}(t)$.
This defines $h_m \vDash_m$ for all $n \geq 1$.
We now define $t \vDash_\infty A$ for $t \in S$ and $A \in \mathbb{F}$ as follows:
$t \vDash_\infty A$ holds (does not hold) if for some k, $t \vDash_m A$ holds (resp. does not hold) for all $n \geq k$.
Otherwise if $t \vDash_m A$ oscillates, we say that $t \vDash_\infty A$ is undecided.

6.6 Conclusion and discussion

We begin with a comparison with some key papers.

1. M. L. Cobo, D. C. Martines and G. Simari, a 2010 paper on admissibility in timed abstract argumentation networks, [94].

This paper offers a time stamping model, where each argument is time stamped when it is available. The model is similar but not the same as the ones used in Chapter 19 and Chapter 9 below (based on [37, 41]). In all cases the change is in the meta-level. There are no temporal connectives involved.

2. N. D. Rotstein, M. O Moguillansky, A. J. Garcia and G. R. Simari, a 2010 paper on dynamic argumentation framework [298].

   The paper offers a model where arguments are based on evidence structures. Thus the dynamics of change come from updates and changes in the evidence structure There are no modal connectives in the object level to connect between different evidence structures.

3. G. Boella, S. Kaci and L. van der Torre, a 2009 paper on dynamics in argumentation with single extensions, [56].

   This paper does not offer dynamics of argumentation networks but rather principles in the meta-level which can govern change in networks. In particular they offer properties for refinement of the attack relation. Thus if for example we have an interrelated family of argumentation networks (different worlds or different times) we can check whether this family satisfies the meta-level principles offered in this paper.

Let us conclude with our plans for future research. What is interesting is to look at argumentation network and partition the arguments into different subsets and regard them as subnetworks (i.e. different modal worlds) and put requirements on extensions as seen from the point of view of the different subnetworks. This is a very neat way of introducing modality into argumentation.

Fibring Argumentation Frames

7.1 Overview of our approach to fibring

Given a network (S, R) with a node $x \in S$, we want to view it as a variable for which we can substitute values. This question is a logical question and can be asked of any system of any sort provided it allows for atomic elements.

The notion of fibring of networks arises when we substitute for x another network, see Figure 7.1. There are two options here

- General fibring: we substitute any other network in any network, e.g. substitute a neural network for a node x in argumentation network.
- Self fibring: substitute networks of the same type. So in our case substitute an argumentation netwrok for a node in another argumentation network.

We are faced with two immediate problems

- Give meaning to the substitution
- Generalise the notion of the network so that it is closed under substitution.

We discuss the meaning of the substitution and its properties in Sections 2.1 and 2.2. In section 2.3 we generalise the notion of argumentation network so it is closed under substitution and we study its properties. We call these networks *higher-level networks*.

These are networks with conjunctive and disjunctive attacks. The results in section 2 give rise to methodological considerations and these are studied in Section 3. Our aim in Section 3 is to show the existence of labellings on our new networks. We do this by reducing the new networks to ordinary networks.

Section 4 compares our results with the literature and discusses further research possibilities.

7.2 Fibring argumentation networks

7.2.1 The fibring problem

On comparing general logics with networks, we find that certain logical operations, which are basic and widespread in logic, do not occur in networks. One such operation is substitution. Given a logical formula $A(q_1, \ldots, q_k)$ built up from the atoms

$q_1, \ldots, q_k$ we can take q_1 and substitute for it another formula $B(p_j, q_i)$ to obtain $A(B(p_j, q_i), q_2, \ldots, q_k)$. The analogous operation in networks of any kind is to take a node q_1 in a network $\mathbf{N}$ and substitute for the node another network $\mathbf{N'}$. This kind of operation is not naturally done in networks. If we want to integrate logics with networks, it might be helpful if we try to make sense of this operation for the cases of the various networks available such as neural nets, Bayesian nets, argumentation nets, etc.

Figure 7.1 shows the basic situation

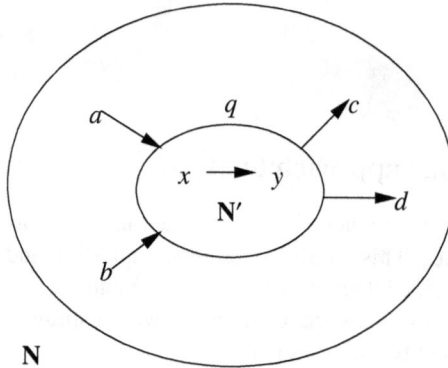

Fig. 7.1.

We substitute $\mathbf{N'}$ for the node q in the network $\mathbf{N}$, to obtain $\mathbf{N}(q/B\mathbf{N'})$. The node q is connected to the nodes c and d in $\mathbf{N}$ and nodes a and b are connected to it. The exact nature of the connection is not relevant and it depends on the nature of the network. In $\mathbf{N'}$ we also have internal nodes such as x and y and there is a connection from x to y. These nodes now become nodes in the new network $\mathbf{N}(q/\mathbf{N'})$.

Our problem here is to make reasonable sense of this situation. We need to address the following questions:

Question 1

How do we understand connections from a, b into $\mathbf{N'}$? Do they connect in any way to the internal nodes x and/or y?

Question 2

How do we understand connections emanating from $\mathbf{N'}$ to other points in $\mathbf{N}$? Do they emanate from some nodes in $\mathbf{N'}$?

Question 3

What do we do with nodes occurring in both $\mathbf{N}$ and $\mathbf{N'}$?

These connections were originally going to and from q of **N**, but now we have substituted **N′** for q, and so we need to answer question 1, 2, and 3. We also have to decide what to do with q itself. Do we leave it in **N** or is it out being replaced by **N′**?

Question 4

Some networks have several options internally for what can happen to q. Do we treat the substitution of **N′** differently by case analysis depending on what happens to q? (This is how we do it in Bayesian nets.)

We have already analysed this situation for the case of Bayesian nets and for the case of neural nets. We found natural solutions to the above questions but the solutions vary from network to network. The solution is completely different for neutral nets from the case of Bayesian nets. See [85] for a summary.

Question 5

How do we view **N′**?. **N′** is a network. In the case of Figure 7.1, **N′** is a two point network $x \rightarrow y$. The question we ask is do we 'process' network **N′** first (option 1) and then substitute the result, or do we substitute **N′**, as is (option 2), first and then process the result. In the case of **N′**, processing means taking an extension, so in this case there is only the ground extension, with $x =$ in, and $y =$ out. So if we process **N′** first, we ignore y since it is out and substitute only x. Otherwise, we substitute **N′** in its entirety. If **N′** is the net of Figure 7.5 then we substitute the set $\{a, c\}$ and we still have to address questions 1–4.

If we choose option 1 and process **N′** first we must also address the case where **N′** has more than one extension, see for example Figure 7.13. Do we do case analysis and different substitutions in each case?

We shall address question 5 in the first half of Section 2.3.

We now try to figure out a solution for argumentation nets and see what we can get.

7.2.2 A fresh look at argumentation networks

We quickly recall some basic definitions in order to present a new point of view on argumentation networks.

Definition 7.1.

1. An argumentation network has the form $\mathbf{P} = (S, R)$ where $S \neq \varnothing$ is the set of arguments and $R \subseteq S^2$ is the attack relation.
2. A Caminada labelling on $\mathbf{P}$ is comprised of three subsets of S, $Q_0, Q_1, Q_2 \subseteq S$ satisfying the following axioms Δ:
   a) $\forall x[Q_0(x) \lor Q_1(x) \lor Q_2(x)]$
   b) $\sim \exists x[Q_i(x) \land Q_j(x)]$ for $i \neq j, i, j = 0, 1, 2$.
      Q_0 is the set of out arguments.
      Q_1 is the set of in arguments
      Q_2 is the set of undecided arguments.

c) $\forall y[\forall x(xRy \to Q_0(x)) \to Q_1(y)]$
d) $\forall y[\exists x(xRy \wedge Q_1(x)) \to Q_0(y)]$
e) $\forall y[(\forall x(xRy \to (Q_0(x) \vee Q_2(x)))) \wedge \exists x(xRy \wedge Q_2(x)) \to Q_2(y)]$

3. *A network may have more than one Caminada labelling to its nodes, which satisfies Δ, see Figure 7.13. Each such option is called an Extension.*
4. *$Q_1(x)$ says that x is labelled* in *(or x = in), $Q_0(x)$ says x is labelled* out *(or x = out) and $Q_2(x)$ says that x is undecided (or x =?).*

The basic idea of the Caminada labelling can be expressed in the following general terms. Assume x_i are nodes, which we call *units*. (Since these nodes might end up, after substitution, as networks, we prefer to call them units.) These units execute among other things, an attack in the direction of the unit y.

The labelling must satisfy the following conditions:

We use the same numbering as in Definition 7.1, namely: items 2c, 2d and 2e of Definition 7.1.

For the unit y to be *in*, all attacking units x in the direction of y must be *out* as far as the direction of y is concerned.

For the unit y to be *out* it is sufficient that one of its attacking units x is *in* as far as the direction of y is concerned.

2.1.2e For a node to be *undecided* we must have that all the attackers in its direction are *out* or *undecided* with at least one of them being *undecided*.

Argumentation networks (called Argumentation Frames) were introduced by Dung in 1995 through a logic programming point of view. This point of view persists until this day and is adopted in the majority of papers on the subject. Given a network (S, R) as in Definition 7.1, one seeks subsets of S called extensions, satisfying certain fixed point properties. This is parallel to the various extensions of logic programming. Caminada was the first, as far as I know, to present the labelling point of view, as given here in Definition 7.1. However, the Caminada labelling point of view is still tied in with the logic programming extensions point of view.

We need to break away from this point of view and think in terms of labels as functions, giving values to the nodes in some algebraic or numerical range (usually the complex or real numbers). This is the point of view of [40] and [37], see Chapters 9 and 14, and this is what we need for the results of this chapter. We are not rejecting or even criticising the logic programming point of view, we simply need the functional point of view to be able to prove some theorems and be able to compare argumentation networks with other networks, following our agenda of unifying logic and networks.

Our point of view is best explained via some examples.

Consider Figure 7.13. The labelling/extensions point of view will say that this network has three extensions or three Caminada labellings.

1. $a = in, b = out, c = out$
2. $a = out, b = in, c = out$
3. $a = ?, b = ?, c = ?$.

The functional approach will say that we are looking for a numerical or algebraic labelling function $\lambda(q)$ of nodes $q \in S$, giving values in a field of values (the complex numbers will do) satisfying the conditions of Definition 7.1, (written appropriately for λ) as a set of equations.

The conditions to satisfy are:

(*1)   if $x_1, \ldots, x_n$ are all the attackers of y then $\lambda(y) = \prod_i(1 - \lambda(x_i))$
(*2)   if y has no attackers then $\lambda(y) = 1$.

To present λ for Figure 7.13 we use values in $\{0, 1, \frac{1}{2}\}$ or values over the complex numbers.

We get the following system of equations, using (*1) and (*2) to present the equations:

1. $\lambda(a) = 1 - \lambda(b)$
2. $\lambda(c) = (1 - \lambda(a))(1 - \lambda(b))$

To solve the equations, let $\lambda(a) = t$
   We get

1. $\lambda(b) = 1 - t$
2. $\lambda(c) = (1 - t)t$.

If we let t range over the values $\{0, 1, \frac{1}{2}\}$ we get the Caminada extensions (with $\frac{1}{2} =$ undecided).

We can also use values in an algebra. We define the algebra *Cam* as follows:
Let $\{in, out, ?\}$ be 3 values and define the operations of inverse and multiplication in *Cam* as follows:

$$x \mapsto \bar{x} \text{ (we also write } \bar{c} \equiv (1 - x))$$
$$x, y \mapsto x \cdot y$$

$\bar{in}$ $= out$
$\bar{out}$ $= in$
$\bar{?}$ $= ?$
$in \cdot x = x \cdot in = x$
$out \cdot x = x \cdot out = out$
$? \cdot ? = ?$

Note that in *Cam*, $x^2 = x$, for all x.
Let us now look at Figure 7.14. Here, we get the equation

$$\lambda(a) = 1 - \lambda(a)$$
$$\lambda(a) = \tfrac{1}{2}.$$

The difference between the two points of view can be clearly separated by the following three examples, as depicted in Figures 7.2, 7.3 and 7.4.
   For Figure 7.2 we get the equations

1. $\lambda(a) = 1 - \lambda(c)$
2. $\lambda(c) = 1 - \lambda(b)$
3. $\lambda(b) = 1 - \lambda(a)$

The only solution is $\lambda(a) = \lambda(b) = \lambda(c) = \frac{1}{2}$ in the complex numbers and $?$ in *Cam*.
   For Figure 7.3 we get the equations

1. $\lambda(a) = 1 - \lambda(c)$

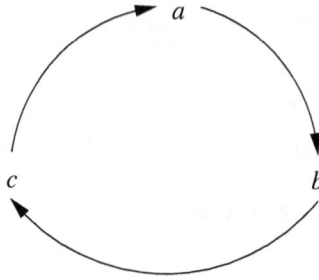

Fig. 7.2.

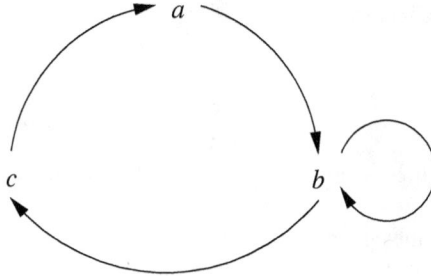

Fig. 7.3.

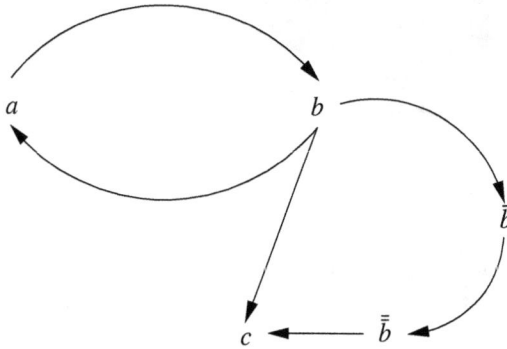

Fig. 7.4.

2. $\lambda(c) = 1 - \lambda(b)$
3. $\lambda(b) = (1 - \lambda(a))(1 - \lambda(b))$

To solve, we get

4. $\lambda(b) = (1 - \lambda(b))^2$

If our range of values is the algebra *Cam* then we can solve the equations by giving the unique solution

$$\lambda(a) = \lambda(b) = \lambda(c) = ?$$

If our range of values is the complex numbers, we can continue to solve for λ in the complex numbers. We get the equation

$$\lambda(b)^2 - 3\lambda(b) + 1 = 0$$

and solve

$$\lambda(b) = 1.5 \pm \sqrt{1.25}$$

The other values for $\lambda(a)$ and $\lambda(c)$ can be calculated.

The important point is that we can tell the difference between Figure 7.2 and 7.3, by choosing the right range for λ.

The Caminada labelling will give the nodes a, b, c value $? = undecided$ in both cases and will not be able to tell the difference. The argumentation theorist may not care about the difference between the two figures. Both are incoherent. He may concede that the functional point of view may be of interest in comparing with other networks but as far as the area of argumentation itself is concerned, he may believe that we do not need this new point of view.

In fact, however, the functional point of view does make a difference for argumentation theory itself. Figure 7.4 is the example for that.

Let us write the equations:

1. $\lambda(b) = 1 - \lambda(a)$
2. $\lambda(\bar{b}) = 1 - \lambda(b)$
3. $\lambda(\bar{\bar{b}}) = 1 - \lambda(\bar{b}) = \lambda(b)$.
4. $\lambda(c) = (1 - \lambda(b))(1 - \lambda(\bar{\bar{b}})) = (1 - \lambda(b))(1 - \lambda(b))$.

Note that from (2) and (3) we get that

5. $\lambda(c) = (1 - \lambda(b))^2$

Working in the complex numbers, let $\lambda(a) = t$ and use it as a parameter. Then

1. $\lambda(b) = 1 - t$
2. $\lambda(\bar{b} = t$
3. $\lambda(\bar{\bar{b}}) = 1 - t$
4. $\lambda(c) - t^2$

Working in *Cam* we observe the following when comparing Figure 7.4 with Figure 7.13.

Ignoring the nodes $\bar{b}$ and $\bar{\bar{b}}$, we get that Figure 7.13 is a subnetwork of Figure 7.4. In fact the equations we get for Figure 7.4 for the nodes a, b, c would be the same equations as those of Figure 7.4 provided that $(1 - \lambda(b)) \cdot (1 - \lambda(b)) = (1 - \lambda(b))$ holds. Indeed, this equation does hold for *Cam*. So as far as *Cam* is concerned Figure 7.4 is a conservative extension of Figure 7.13. Any λ on $\{a, b, c\}$ can be uniquely extended to a λ^* on Figure 7.4 in a consistent manner. The equations for Figure 7.4 give the same solutions to the nodes of Figure 7.13 as the solutions to the equations of Figure 7.13.

Let us consider another example, that of Figure 7.46.

We get the following equations:

1. $\lambda(\bar{x}) = 1 - \lambda(x)$
2. $\lambda(\bar{a}) = 1 - \lambda(a)$
3. $\lambda(\bar{c}) = 1 - \lambda(c)$
4. $\lambda(e(x, a)) = (1 - \lambda(\bar{a}))(1 - \lambda(\bar{x})) = \lambda(a)\lambda(x)$
5. $\lambda(e(x, c)) = (1 - \lambda(\bar{x}))(1 - \lambda(\bar{c})) = \lambda(x)\lambda(c)$

6. $\lambda(\bar{\bar{a}}) = 1 - \lambda(\bar{a})$
7. $\lambda(\bar{\bar{c}}) = 1 - \lambda(\bar{c})$
8. $\lambda(e(x)) = (1 - \lambda(\bar{\bar{a}}))(1 - \lambda(\bar{\bar{c}}))(1 - \lambda(\bar{x}))$
9. $\lambda(a) = (1 - \lambda(e(x)))(1 - \lambda(e(x,c)))$
10. $\lambda(c) = (1 - \lambda(e(x)))(1 - \lambda(e(x,a)))$.

Simplifying, we get

4. $\lambda(e(x,a)) = \lambda(a) \cdot \lambda(x)$
5. $\lambda(e(x,c)) = \lambda(c) \cdot \lambda(x)$
6. $\lambda(\bar{\bar{a}}) = \lambda(a)$
7. $\lambda(\bar{\bar{c}}) = \lambda(c)$
8. $\lambda(e(x)) = (1 - \lambda(a))(1 - \lambda(c))\lambda(x)$
9. $\lambda(a) = [1 - [\lambda(x)(1 - \lambda(a))(1 - \lambda(c))]] \cdot [1 - \lambda(x)\lambda(c)]$
10. $\lambda(c) = [1 - [\lambda(x)(1 - \lambda(a))(1 - \lambda(c))]] \cdot [1 - \lambda(x)\lambda(a)]$

Let us, for example, see what we get if we substitute in the equations the values $x = ?$ $a = in$ and $c = ?$.

First note that from equations (1)–(8) we get that the values for $\bar{x}, \bar{a}, \bar{c}, e(x,a), e(x,c), \bar{\bar{a}}, \bar{\bar{c}}$ and $e(x)$ are uniquely determined. The problem we face is whether our choice of the above values for $\{x, a, c\}$ is fortunate so that equations (9) and (10) also hold.

Let us check. We get

1. $\bar{x} = ?$
2. $\bar{a} = out$
3. $\bar{c} = ?$
4. $e(x,a) = (1 - out)(1-) = in\cdot? = ?$
5. $e(x,c) = (1-?)(1-?) = ?\cdot? = ?$
6. $\bar{\bar{a}} = in$
7. $\bar{\bar{c}} = ?$
8. $e(x) = (1 - in)(1-?)(1-?) = out \cdot? = out$
9. $a = (1 - out)(1-?) = in\cdot? = ?$
10. $c = (1 - out)(1-?) = in\cdot? = ?$

We get that equation 9 is contradictory. So we guessed wrong!

The equational approach is developed in Chapter 10.

7.2.3 Joint and disjunctive attacks

We are faced with the need to make sense of situation illustrated in Figure 7.5 as an example.

We got Figure 7.5 by substituting the network of Figure 7.6 for the position q in the network of Figure 7.7.

We need to answer Questions 1, 2, 3, 4 and 5 for this case. It is a simple case since the substituted network of Figure 7.6 has no loops and so it has a clear message, only one extension: a is in, c is in and b is out. So effectively b is out and plays no active 'in' role in the network.

We must clarify our concepts at this junction: Figure 7.5 shows a substitution of the network of Figure 7.6 at position q of Figure 7.7.

The first choice of options we have to do is to ask Question 5.

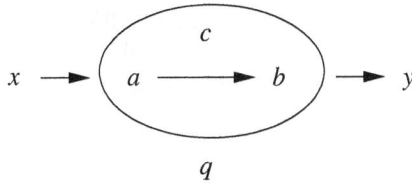

Fig. 7.5.

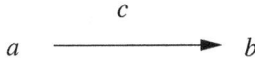

Fig. 7.6.

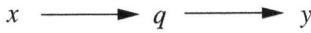

Fig. 7.7.

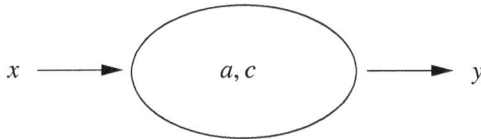

Fig. 7.8.

(*1) Do we process the network of Figure 7.6 first, i.e. choose an extension for it and then and only then substitute the result for q in Figure 7.7? (Call this Option 1.),

or

(*2) We substitute Figure 7.6 as is, as a network, and define whatever is supposed to happen (call this Option 2).

If we follow option 1 for the network of Figure 7.6, we get the network of Figure 7.8. It is as if we substitute the set $\{a, c\}$ for the node q in Figure 7.7.

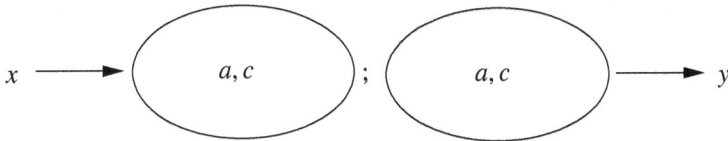

Fig. 7.9.

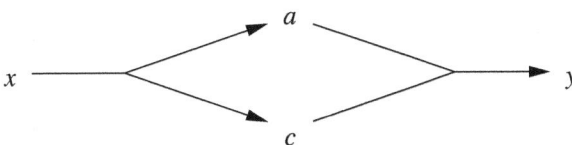

Fig. 7.10.

We begin by examining Option 1 and study the formal network of Figure 7.8. We need to decide how to define the concepts of attacks (see Figure 7.9). Figure 7.8 has two problematic parts, $\{a, c\}$ attacking y and x attacking $\{a, c\}$, as shown in Figure 7.9.

(*3) What does it mean for a unit (argument) x to attack a set $\{a, c\}$?
(*4) What does it mean for a set $\{a, c\}$ to attack a unit (argument) y?

The case (*4) is easy. For $\{a, c\}$ to be *in* we must have both a and c *in*. In this case y is *out*. See Figures 7.9 and 7.10.

For the case of (*3), when the unit $\{a, c\}$ is attacked by x, we have two possibilities:

Possibility 1 for (*3)
x must attack one of the elements of the set i.e. either x attacks a or x attacks b.

Possibility 2 for (*3)
x attacks the "*in*" status of $\{a, b\}$, that is we need to have either a *out* or b *out*.

Possibility 2 is the symmetrical counterpart to (*4). For $\{a, c\}$ to be *in* we must have "a is *in* and b is *in*", therefore for $\{a, c\}$ to be *out* we must have "a is *out* or c is *out*".

The two possibilities are not equivalent.

This will be discussed in detail at the beginning of Section 3 and in Section 4.2. We shall see later that Nielsen and Parsons [270, 271] proposed joint attacks, which is coming from a completely different point of view, follows possibility 1. We shall compare our approaches in Section 4.2.

Definition 7.2 (Joint and disjunctive attacks). *We adopt possibility 2 for (*3). So the definition is as follows (see Figure 7.11).*

- *x attacks $\{e_1, \ldots, e_m\}$ (disjunctively) means:*
  $x =$ in *implies* $\bigvee_{i=1}^{m} e_i =$ out
  (especially this can mean that if x is in *then several or more e_i are* out*).*
- *$\{e_1, \ldots, e_m\}$ (jointly) attack y means:*
  $\bigwedge_{i=1}^{m} e_i =$ in *implies y is* out*.*
- *$\{e_1, \ldots, e_m\}$ is in iff $\bigwedge_{i=1}^{m} e_i -$ in*
- *$\{e_1, \ldots, e_m\}$ is out iff $\bigvee_{i=1}^{m} e_i =$ out*

Example 7.3 (Caminada–Gabbay labelling for joint and disjunctive attacks). We illustrate/define our labelling using the typical case of Figure 7.11.

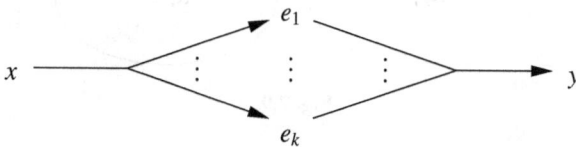

Fig. 7.11.

1. If x is labelled *in* and $e_j, j = 1, \ldots, k$ are the nodes being disjunctively attacked from x then at least one of e_j must be *out*.

2. If y is jointly attacked from e_j, $j = 1, \ldots, k$ and all of e_j are *in* then y must be *out*.

3. Suppose q is targeted by a disjunctive attack from $x_1, \ldots, x_n$. How can q be *undecided*? To see what must be done consider Figure 7.12 showing a typical situation, where q is attacked from two different directions.

In Figure 7.12, q is being attacked from two different directions. One attack is emanating from x and one attack emanating from y. The attack emanating from x involves also v_1 and v_2 which are also being disjunctively attacked and the attack emanating from y involves also w_1 and w_2. We imagine Figure 7.12 as a subnetwork of a larger network **N**. So it is quite possible that x or v_i or w_j are being attacked from other parts of the network. We assume that q is being attacked only from x and y.

Let us consider an arbitrary attack on q, emanating from a node z and involving nodes $u_1, \ldots, u_k$ which are also attacked. Thus to be explicit, z disjunctively attacks $\{u_1, \ldots, u_k, q\}$.

Let us use the terminology that q is subject to an attack emanating from z. We now want to define three concepts describing this attack.

(♯1)   the attack on q emanating from z is not a threat to q.

(♯2)   the attack from q emanating from z forces q *out*

(♯3)   the attack on q emanating from z makes q *undecided*.

Let us define these three concepts:

(♯1)   This attack is not going to be a threat to q if one of $u_1, \ldots, u_k$ is *out*. Say u_i is out because of an attack on it from the rest of the network **N**, in which Figure 7.12 is embedded. If one of the u_i is *out* then the attack succeeds without 'hitting' q. So this attack is no threat to q and q can have any value, *in*, *out* or *undecided*.

(♯2)   the attack on q from z forces q out if $z = in$ and all of $u_1, \ldots, u_k$ are *in*.

(♯3)   when do we say that the attack on q from the direction of z makes q undecided? There are two cases:

(a) z itself is undecided. So we do not know whether z is *in* or *out*. If some u_i is *out* then it does not matter that z is undecided, q is not threatened. If all of u_i are not *out*, but are either *in* or *undecided*, then there is a real possibility of attack on q. If z and all of u_i turns out to be *in* then q must be *out* and if one of $\{z, u_1, \ldots, u_k\}$ is *out* then q can be *in*. Thus we must make q *undecided*.

Now we ask can u_1 be *in*? Well, it cannot be *in* because from the point of view of u_1, we see the same situation as that which we saw from q. We see $z = ?$ and $u_1, \ldots, u_k$ and q are not *out*. So u_1 must also be *undecided*. So to maintain coherence of our rules we must declare (*) below:

(*) If a disjunctive attack emanates from z in the direction of $u_1, \ldots, u_k, q$ and z is *undecided* and none of $u_1, \ldots, u_k, q$ is *out* ($u_1, \ldots, u_k, q$ can be *out* owing to attacks from the rest of the network) then $u_1, \ldots, u_k, q$ are all *undecided*. Let us refer to this situation as u_i, q become *undecided* because of a disjunctive attack on it emanating from an undecided z.

(b)   Let us now check what happens if $z = in$. Can we still get $q = undecided$? The answer is yes. Suppose $u_1, \ldots, u_k$ are all attacked from some respective $z_1, \ldots, z_k$ which are all undecided. This will make $u_1, \ldots, u_k$ *undecided* according to (*) above.

So if $u_1 \ldots$ or $\ldots u_k$ ends up *out*, then q is safe, and if all of $u_1, \ldots, u_k$ end up *in*, then q has to be *out*. Thus we must declare q undecided. So we get rule (**) below:

(**)If a disjunctive attack emanates from z in the direction of $u_1, \ldots, u_k, q$ and at least one of u_i is undecided (because for example of a disjunctive attack on u_i emanating from an undecided z_i) then all of $u_1, \ldots, u_k, q$ are all undecided.

We can now define what it means for q to be undecided, by using (*) and (**):

(***)q is undecided if

(1) q is not forced out by any attack on it emanating from any z.

(2) q can be shown undecided by repeated applications of (*) and (**) above.

Let us now check when v_1 can be undecided in Figure 7.12 v_1 can be undecided in Figure 7.12 if for example x is undecided and there are no attacks that force any of v_1, v_2 or q *out*.

v_1 can be undecided if $x = in$ but y is undecided and there are no attacks forcing any of v_1, v_2, q, w_1, w_2 *out*.

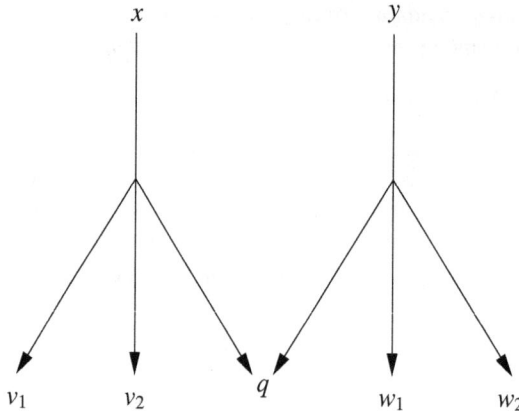

Fig. 7.12.

Now having adopted our notion of disjunctive and joint attacks in Definition 7.2, we are ready to discuss further whether to adopt Option 1 for network substitution, namely whether to choose an extension for the network first before we substitute and then substitute only the nodes that are *in* in that extension. This discussion will be somewhat lengthy (terminating just before Definition 7.4) and will result in rejecting option 1 in favour of option 2. The weakness of option 1 is that it gets complex in the presence of loops.

The above considerations explained what happens with option 1 when the substituted network has no loops so it has no undecided nodes and therefore it has only one extension. What do we do when the substituted network has loops? There are two typical cases to consider, as shown in Figures 7.13, 7.14.

Figure 7.16 is obtained by substituting Figure 7.13 into Figure 7.15 at node q and Figure 7.17 is obtained by substituting Figure 7.14 into Figure 7.15, at node q.

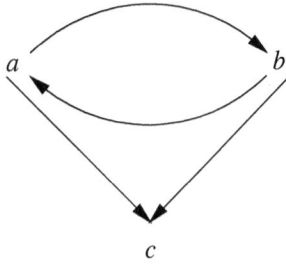

Fig. 7.13.

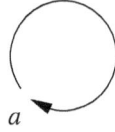

Fig. 7.14.

Fig. 7.15.

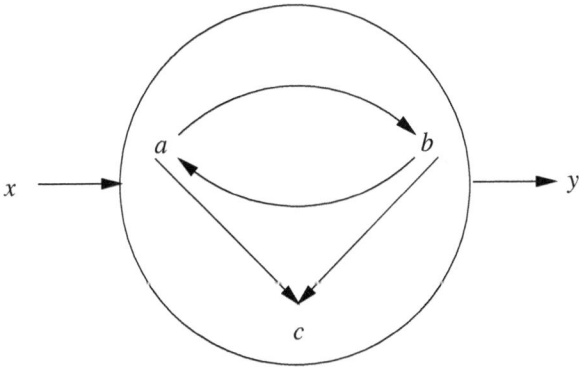

Fig. 7.16.

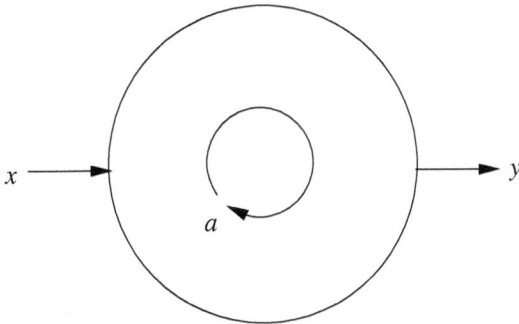

Fig. 7.17.

In the first case we have two options for the network, the extensions $\{a, c\}$ and $\{b, c\}$. Which one do we take? This is not really a problem because we can push the problem onto the labelling. So depending on the labelling we get either Figure 7.15

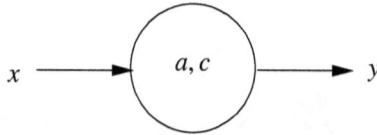

Fig. 7.18.

or Figure 7.19 we still have the problem of how do we represent the original net-

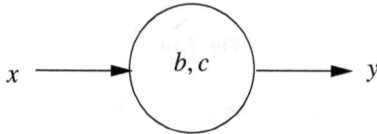

Fig. 7.19.

work?

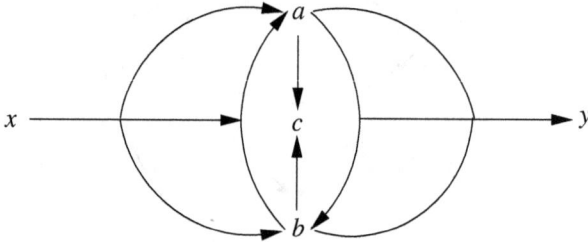

Fig. 7.20.

Let us try the representation as in Figure 7.20.

If a is *in* and c is *in*, we have a case of $\{a, c\}$ and so since x must be *in*, its attack on $\{a, c\}$ must take one of $\{a, c\}$ out. Similarly if b is *in* and c is *in*, we have a case of $\{b, c\}$ and so since x must be *in*, its attack on $\{b, c\}$ must take one of $\{b, c\}$ out.

We immediately see that we have a problem with the proposed representation of Figure 7.20. In the figure we put in the entire network of Figure 7.13 into Figure 7.15. Thus when b is *in*, a is *out* and when a is *in* we have that b is *out*. So formally, for the case let us say, that $\{a, c\}$ are *in* and b is *out*, we set that the attack from x is formally successful, because in our diagrams for this case b is *out* so $\{a, c\}$ can stay *in*, contrary to our intentions! Similarly, since always either a or b is *out* then the substituted network can never attack anything as represented in the diagram of Figure 7.20 because formally the joint attack from $\{a, b, c\}$ always has one node out (either *a* or *b*). Obviously, we need first to calculate the result for the substituted network of Figure 7.13 and get say that {*a, c*} are *in* and *b* is *out* and only afterwards address the attack from *x* and the attack on *y*.

But then in this case we must make it clear that *b*, being *out* does not play a role in the considerations of these attacks.

'Making things clear' means additional labels. Maybe several kinds of annotated '*out*' with information when each '*out*' is to be taken into account.

The labelling and the substitutions become hierarchical.

This is not good, not only because of complexity considerations but also conceptually. When we substitute we allow ourselves to use any arguments, so this *b* in question could be *x*. Now what do we do? In the hierarchical evaluation the *b* = *x* inside Figure 7.13 as substituted is *out* while the *b* = *x* originally outside the figure is *in*. But they are the same *b* = *x*! So what do we do?

Let us look at Figure 7.17 and see if we can get a clue as to what to do. Figure 7.14, as substituted into Figure 7.15 is both attacking (Figure 7.21) and is being attacked (Figure 7.22). So what do we do with Figures 7.21 and 7.22?

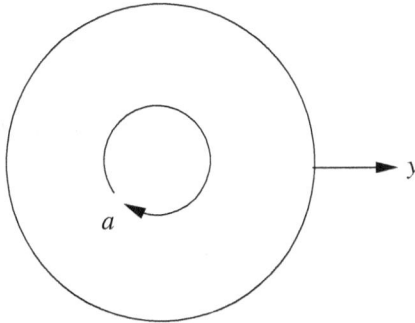

Fig. 7.21.

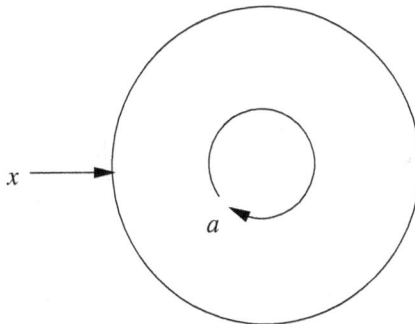

Fig. 7.22.

The latter, Figure 7.22 is clear, *x* can take *a out*.

On the other hand, we look at Figure 7.21, then since *a* is *undecided*, we must give *y* *undecided*. Now we can see that the proposed hierarchical evaluation, when applied to Figure 7.17 is no good. Evaluating hierarchically will make *a* *undecided*, *y* *undecided* and *x* *in* which is not a good solution for Figure 7.17. However, evaluating directly without any hierarchical considerations gives us that *x* is *in*, since it is not attacked, *a* is *out* and *y* is *in*. All nice and clear.

The above discussion shows that we had better reject Option 1 and view any sub-stituted network **N'** into another network **N** at a point $q \in$ **N**, in which q attacks another node $y \in$ **N**, as a joint attack from *all* the arguments of **N'**. If some of them are *out* in **N'** then we regard **N'** as not conflict-free. Thus the network of Figure 7.23 is not conflict-free.

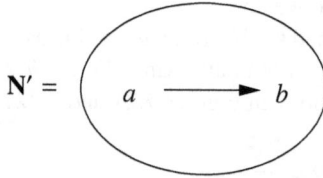

$$\mathbf{N'} = \qquad a \longrightarrow b$$

Fig. 7.23.

The reason we take this view is because when we substitute **N'** of Figure 7.23 into another network **N**, *b* and *a* may be in **N** in other places and we do not know what can happen, *a* may end up *out* or *b in*.

The above is our Option 2 which we shall adopt.

We now have an agreed sequence of definitions.

Definition 7.4 (Higher level networks).

1. *Let $S \neq \emptyset$ be a set of nodes. Let S^0 be the family of all finite non-empty subsets of S, identify the singleton $\{x\}$ with x for simplicity of notation.*
2. *A higher level argumentation network has the form (S, S^0, R) where S and S^0 are as above and $R \subseteq S^0 \times S^0$ is the attack relation.*
    XRY represents a joint attack from the set X disjunctively attacking the set Y. When $X = \{x\}$ and $Y = \{y\}$ we get an ordinary point to point attack and so when $R \subseteq S \times S$ we get the usual network definition.
    We represent the situation as in Figure 7.24

    $$X = \{x_1, \ldots, x_n\}.$$
    $$Y = \{y_1, \ldots, y_n\}.$$

    or perhaps Figure 7.25 is more clear.
3. *We understand Figure 7.24 as saying that the set $\{x_1, \ldots, x_n\}$ is jointly mounting a disjunctive attack on the set $\{y_1, \ldots, y_m\}$. So only if all the x_i are in can the attack go forward and in which case we expect at least one of the y_j to be* out.

We now want to define substitution of one such higher level network into another. The result will again be a higher level network.

We will understand better how to define the substitution after we do some examples.

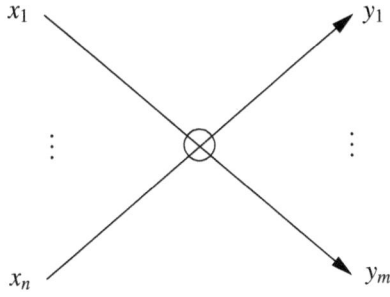

Fig. 7.24.

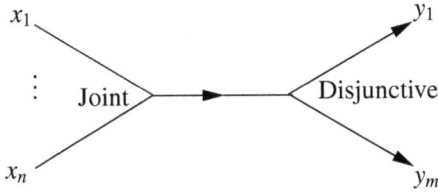

Fig. 7.25.

Example 7.5 (Network substitution). Start with a simple network of Figure 7.26

Fig. 7.26.

Now substitute Figure 7.27 for y and substitute Figure 7.28 for x and get Figure 7.29.

Fig. 7.27.

Fig. 7.28.

Figure 7.29 should be written as Figure 7.30

Now substitute Figure 7.31 for u in Figure 7.30 and get Figure 7.32.

Note that in Figure 7.32 we have a joint attack of a and '$a \to x$' on some target. According to our Option 2, we regard this as a joint attack from $\{a\} \cup \{a, x\}$ on the target, i.e. a joint attack from $\{a, x\}$.

Fig. 7.29.

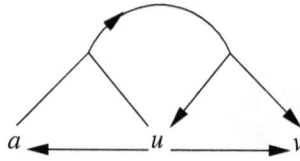

Fig. 7.30.

$$a \longrightarrow x$$

Fig. 7.31.

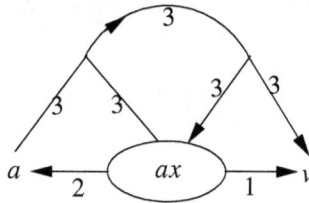

Fig. 7.32.

Of course Figure 7.32 needs to be simplified since a is written twice. We get Figure 7.33.

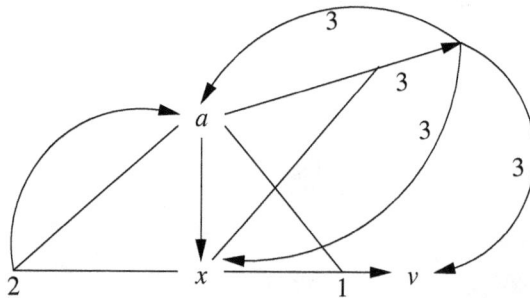

Fig. 7.33.

We numbered the attacks in Figures 7.32 and 7.33 so it will be clear what attack in Figure 7.32 became what attack in Figure 7.33.

We try as an exercise to substitute for u in Figure 7.30, a new figure, Figure 7.34. So instead of substituting Figure 7.27 as we did above, we substitute Figure 7.34.

In this case we get Figure 7.35.

Fig. 7.34.

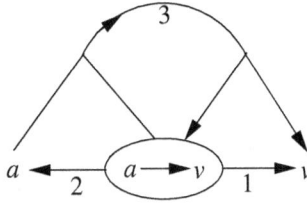

Fig. 7.35.

Now this figure needs to be simplified to Figure 7.36. In Figure 7.36 we have:
1 is a joint attack of $\{a, v\}$ on v
2 is a joint attack of $\{a, v\}$ on a
3 is a joint attack of $\{a, v\}$ which disjunctively targets a and v.

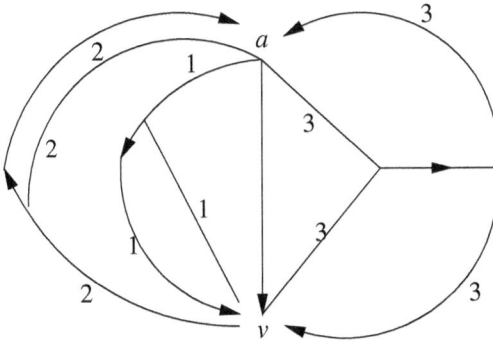

Fig. 7.36.

Remark 7.6. We can see that the graphs can get very complex. We note, however, that we cannot get everything, just by repeated substitution. For example, we believe we cannot get Figure 7.37 by mere substitutions:

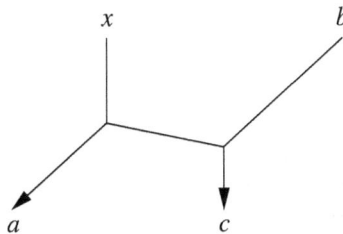

Fig. 7.37.

We cannot get that x sends a disjunctive attack in the direction of a and in the direction of c but the attack disjunct going towards c joins forces and becomes conjunctive with the attack from b on c.

The only possibilities we can generate are Figures 7.38 and 7.39

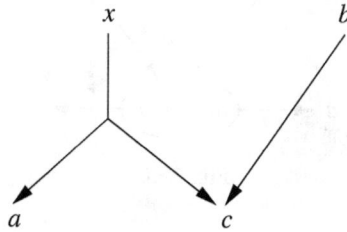

Fig. 7.38.

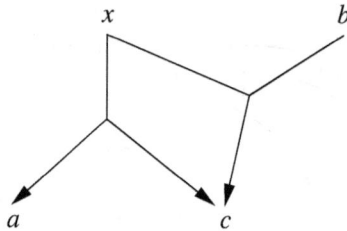

Fig. 7.39.

Figure 7.37 is perfectly OK, but I don't think we can get it by repeated substitution of ordinary networks. Figure 7.37 has three extensions:

1. $x = in$, $a = out$, $b = in$, $c = out$
2. $x = in$, $a = in$, $b = in$, $c = out$
3. $x = in$, $a = out$, $b = in$, $c = in$.

It is debatable whether extensions (2) and (3) are acceptable. This depends on how we understand Figure 7.37. See section 4 for a discussion.

We must bear in mind that we are not just defining generalisations of argumentation networks from a mathematical point of view. We are following the methodological manifesto of Section 1 and there are good reasons for substituting one network in another. We asked ourselves in Section 2 what general notion of networks we need to allow for substitution. That is when we substitute such a network in another such network, we get a result of the same kind. We found that we need the notion of higher level network of Definition 7.4. So generalising the definition of higher level networks, to allow for Figure 7.37 requires independent conceptual justification. Substitution is justified intuitively. Joint attacks exists in common sense arguments. Attacking joint attacks also makes sense and this gives rise to disjunctive attacks. How can the mathematical situation of Figure 7.37 be justified or explained?

We can intuitively say that the disjunctive attack arises from an attack on a joint unit such as x attack on $\{a, c\}$ in Figure 7.9. To be successful, x wants either a or c to be *out*. So why not joint forces with some b to attack c and increase his chances? This makes sense. We get Figure 7.37. x will not insist on $c = out$ if his attack on a succeeds. In Figure 7.39, $\{x, b\}$ insist on $c = out$ even if $a = out$. In Figure 7.39, $\{x, b\}$ attack on c is not part of a joint attack.

Let us postpone this to the discussion to Section 4.

We can now define higher level substitution

Definition 7.7. *Let* $\mathbf{P}_i = (S_i, S_i^0, R_i)$ *for* $i = 1, 2$ *be two finite higher level networks. Let* $x \in S_1$ *be a node. We want to define the network*

$$\mathbf{P} = \mathbf{P}_1(x/\mathbf{P}_2)$$

being the result of the substitution of $\mathbf{P}_2$ *for* x *in* $\mathbf{P}_1$.

The set of points of $\mathbf{P}$ *is* $S = S_1 \cup S_2$. *We need to define* $R \subseteq S^0 \times S^0$. *We shall follow the traditional practice used in substitution in logic and assume that* x *itself is not present in* S_2.

We therefore have two types of available relations.

1. *Type 1 from* $\mathbf{P}_1$:

$$\{a_1, \ldots, a_m, x\} R_1 \{b_1, \ldots, b_k, x\}$$

where x *may not appear among the* a_i *or not appear among the* b_j. *In this case we take*

$$\{a_1, \ldots, a_m\} \cup S_2 R \{b_1, \ldots, b_k\} \cup S_2$$

where S_2 *will appear wherever* x *appears.*
2. *Type 2 from* $\mathbf{P}_2$

$$X R_2 Y$$

in which case we take XRY.

We have thus really defined R *to be*

$$R =_{def} R_2 \cup R_1(x/S_2)$$

with the understanding that $\{e_j, S_2\} = \{e_j\} \cup S_2$.

Example 7.8. Let us examine again the network in Figure 7.37. We claimed that such a network cannot arise in our definition of higher level networks.

Since clearly a and c are being attacked and x and b are the attacking nodes, then only the following higher level attacks are possible in the form XRY, as shown in Figure 7.41.

Figure 7.41 shows the connections of Figure 7.40 using points as nodes. It is clear that the attack pattern of Figure 7.37 is not present in Figure 7.41.

Definition 7.9 (Caminada–Gabbay labelling). *Let* $\mathbf{P} = (S, S^0, R)$ *be a higher order network.*

Let λ *be a function giving values in* $\{0, 1, ?\}$ *to each* $s \in S$.

We say this function is a proper labelling iff the following holds.

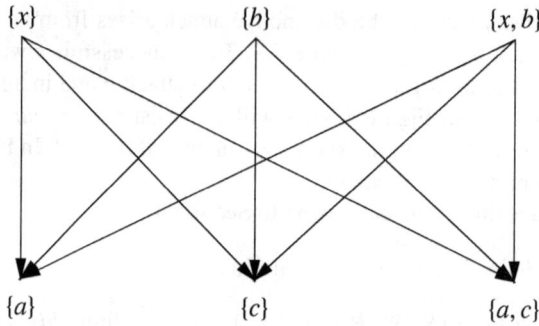

Fig. 7.40.

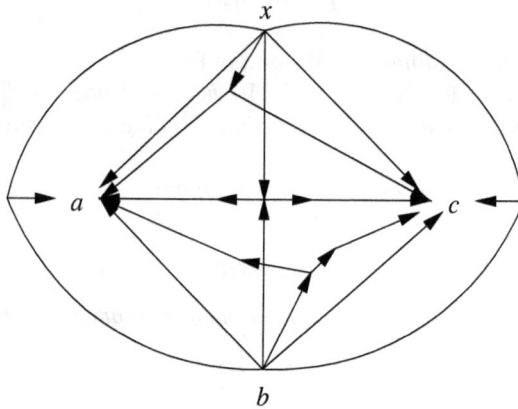

Fig. 7.41.

1. If XRY holds and $\forall x \in X(\lambda(x) = 1)$ then $\exists y \in Y\lambda(y) = 0)$.
2. If for every X, Y such that $y \in Y$ and XRY holds we have that $\exists x \in X$ s.t. $\lambda(x) = 0$ then $\lambda(y) = 1$.
3. If for all X, Y such that $y \notin X \cup Y$ and $XRY \cup \{y\}$, we have that $\forall z \in X \cup Y(\lambda(z) \in \{1, ?\})$ and $\exists z \in X \cup Y(\lambda(z) = ?)$. Then $\lambda(y) = ?$.

Remark 7.10. Figure 7.42. If $\lambda(x) = 1$ and $\lambda(a) = 0$ then $\lambda(c)$ can be either 1 or 0. Both cases are acceptable. If Figure 7.42 is part of a larger network and because of attacks from the larger network either $\lambda(x) = ?$ then we cannot have $\lambda(a) = 1$ and $\lambda(c) = ?$.

7.3 Methodological results

We are now in a position where we have a definition of a new kind of argumentation network, (S, S^0, R) as defined in Definition 7.4, where sets of arguments can attack sets of arguments. Never mind how this definition was motivated. We got such networks because we wanted our networks to be closed under fibring/substitution (Definition 7.7). The question now is, what are the extensions of this new kind of network? We need to do Dung style analysis of extensions, or develop some other means of answering this question.

Given $E \subseteq S$, is E an extension? Let us see what we need to do Dung style; something like the following:

The attacks are done by sets of arguments on sets of arguments. So we need to define the notion of a conflict free family $\mathbb{E}$ of sets of arguments. Then we define the notion of a set X of arguments being acceptable to $\mathbb{E}$. Then we define a function φ

$$\varphi(\mathbb{E}) = \{X \mid X \text{ is acceptable to } \mathbb{E}\}$$

Then we say that $\mathbb{E}$ is admissible if $\mathbb{E} \subseteq \varphi(\mathbb{E})$ and now we can define extensions, e.g. $\mathbb{E}$ is complete iff $\mathbb{E} = \varphi(\mathbb{E})$. $\mathbb{E}$ is a family of subsets of the set of arguments S. From $\mathbb{E}$ maybe we get the extensions say

$$E = \bigcup_{X \in \mathbb{E}} X$$

At this point we do not know if this can work or not. We need to try it and develop the correct definitions by trial and error. We are not going to do that. We are going to take another route. We are going to start with (S, S^0, R) and add new points S^* to S and get a new network $S \cup S^*$. We define an ordinary R^* on $S \cup S^*$ and then embed our network (S, S^0, R) into the network $(S \cup S^*, R^*)$ by implementing the idea of joint and disjunctive attacks using the additional points of S^*.

The new network $(S \cup S^*, R^*)$ is an ordinary Dung network, so it has extensions. Let E^* be an extension. We will define any $E = E^* \cap S$, as an extension of (S, S^0, R). For this to work properly we need the following critical property:

Critical property of S in $S \cup S^*$

Let λ_1^* and λ_2^* be two Caminada labelling on $S \cup S^*$. Assume that $\lambda_1^* \upharpoonright S = \lambda_2 \upharpoonright S$, then $\lambda_1 = \lambda_2$.

The critical property of S in $S \cup S^*$ ensures the embedding is faithful!

So now that we know what we are doing, let's roll!

We now reduce higher level networks to ordinary networks. We do this in several stages.

1. Reduce the disjunctive attacks to joint attacks.
2. Reduce the joint attacks to single attacks.
3. Derive the existence of labellings and extensions from (1) and (2).

7.3.1 Conceptual analysis of disjunctive attacks

Before we embark on any reductions, we must fully clarify the properties of disjunctive attacks.

Consider the disjunctive attack part of Figure 7.9. We have the following situation (see Figure 7.42):

- $x = in$ implies $a = out$ or $c = out$

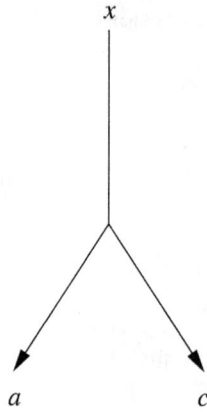

Fig. 7.42.

The following are the three possibilities

1. $x = in, a = in, c = out$
2. $x = in, a = out, c = in$
3. $x = in, a = out, c = out$

Figure 7.42 cannot be reduced to Figure 7.43 which is properly written as Figure 7.44.

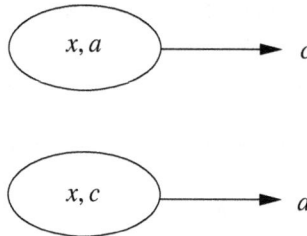

Fig. 7.43.

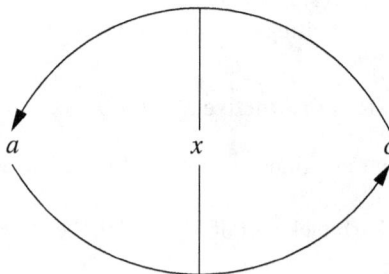

Fig. 7.44.

Figure 7.44 does not allow for the extension

4. $x = in$, $a = out$, $c = out$

The meaning of Figure 7.42 is a disjunctive attack on $\{a, c\}$ and not any specific attack on either a or c. This is also reflected in the problems we had in interpreting Figure 7.37.

We have three reading of Figure 7.42.

(r1) x attacks the coalition (joint) of $\{a, c\}$. x would be happy to see either $a = out$ or $c = out$ but is not mounting any specific attacks on a or on c. According to this interpretation Figure 7.37 does not have meaning. b cannot join any attack from x onto c. There is no such attack.

(r2) x sends two attacks, one in the direction of a and one in the direction of c. x is happy if at least one of them is successful, but would also be happy with both successful.

(r3) x sends two attacks as in (r2) above but does not want both to succeed, only one.

(r1) and (r2) above say:

• $x = in$ implies $a = out$ or $c = out$

while (r3) says:

• $x = in$ implies $(a = out \wedge c = in)$ or $(a = in$ and $c = out)$.

Figure 7.44 corresponds to (r3).

Note that we need not make a choice between (r1) and (r2) for the purpose of this section. We shall see in Section 4.2 that for the purpose of explaining Figure 7.37 we need to adopt (r2).

How can we represent (1) or (2)? What is the corresponding figure?

We need auxiliary points. Consider Figure 7.45:

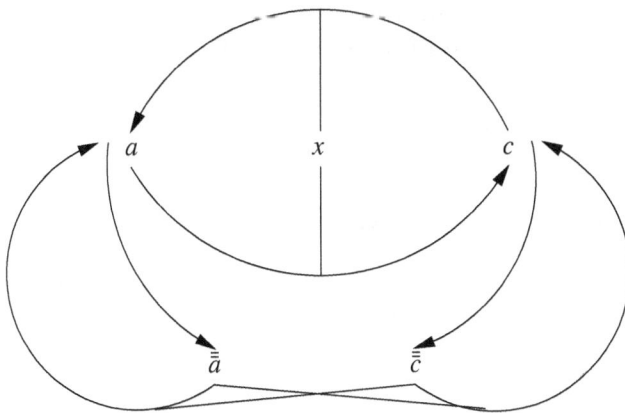

Fig. 7.45.

Figure 7.45 extends Figure 7.44 by adding two joint attacks from $\{a = out, c = out\}$ one on a and one on c. To do this properly we use the help of two new intermediary points $\bar{a}$ and $\bar{c}$. When $a = out$ and $c = out$, we get $\bar{a} = in$ and $\bar{c} = in$ and $\{\bar{a}, \bar{c}\}$ mount the joint attacks on a and c.

Let us see what happens in Figure 7.45. Since x is not attacked, $x = in$. We now consider our possibilities for a and c. If at least one of $\{a, c\}$ is in, the joint attacks of $\{\bar{a}, \bar{c}\}$ fail and we are back to Figure 7.44. Figure 7.44 behaves as we want except in the case of $a = out$ and $c = out$. But in this case, $\bar{a} = \bar{b} = in$ and so the joint attacks of $\{\bar{a}, \bar{c}\}$ on a and c must succeed, thus confirming the assumption that $a = b = out$.

We can display the situation without joint attacks in Figure 7.46.

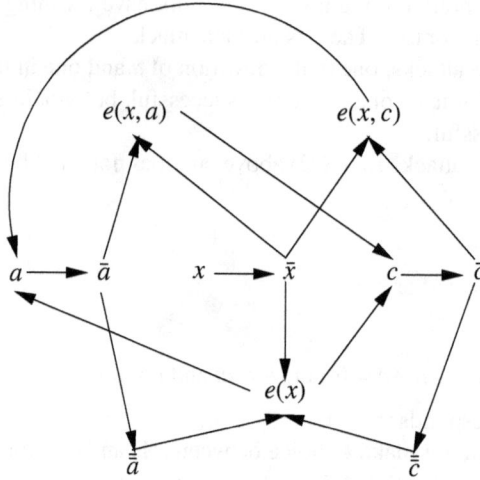

Fig. 7.46.

We used auxiliary points as follows:

1. With each node involved, x, a, c we added new nodes $\bar{x}, \bar{a}, \bar{c}$ and $\bar{\bar{a}}, \bar{\bar{c}}$.
2. We added the intermediaries $e(x), e(x, c), e(x, a)$.

Nodes x, a, c attack only via $\bar{x}, \bar{a}, \bar{c}$ respectively.

$e(x, a)$ is an intermediary doing the job of (x, a) attack on c. Similarly $e(x, c)$ is an intermediary for representing the attack of (x, c) on a.

These two represent the situation of Figure 7.44. To get also the option of $x = in$, $a = out$, $c = out$, we use $\bar{\bar{a}}, \bar{\bar{c}}$ which attack $e(x)$. $\bar{a}$ attacks $\bar{\bar{a}}$ and $\bar{c}$ attacks $\bar{\bar{c}}$.

Example 7.11 (Analysis of Figure 7.46). Figure 7.46 can be stand-alone or can be embedded inside a larger Figure. If Figure 7.46 is stand-alone, then x is the only node which is not attacked by anything. Hence $x = in$. Therefore $\bar{x} = out$. We now examine four options for a and c:

1. $a = out, c = out$
2. $a = in, c = out$
3. $a = out, c = in$
4. $a = in, c = in$

We check consistency of the *in–out* labelling, for the case $x = 1$.

Case 1

If $a = c = out$, we get $\bar{a} = \bar{c} = in$. Therefore $e(x, a) = e(x, c) = \bar{\bar{a}} = \bar{\bar{c}} = out$. Also $e(x) = in$.

$e(x)$ attacks a and c confirming they should be out.

So (1) is a consistent labelling.

Case 2

If $a = in$ and $c = out$ then $\bar{a} = out$ and $\bar{c} = in$. Hence $e(x, a) = in$ (it has not attackers, both $\bar{x}$ and $\bar{a}$ are *out*) and $e(x, c)$ is *out*.

Since $e(, a) = in$ and $e(x, a)$ attacks c it confirms c is *out*.

Since $\bar{a}$ is *out* we get $\bar{\bar{a}}$ is *in*, therefore $e(x)$ is *out*. Hence a is not attacked by anything (both $e(x)$ and $e(x, c)$ are *out*), confirming that a is *in*.

We also get $\bar{\bar{c}}$ is *out* in $\bar{c}$ is *in*. This does not affect $e(x)$, because $\bar{\bar{a}}$ is in and so $e(x)$ is *out*.

Case 3

This is the case of $a = out$ and $c = in$. It is the mirror image of Case 2 and follows from the symmetry of the Figure with a and c swapped.

Case 4

This is where $a = in$ and $c = in$. This case should come out inconsistent. Indeed, in this case both $\bar{a}, \bar{b}$ are *out*.

Hence since $\bar{x} = out$ we get $e(x, a)$ and $e(x, c)$ are *in* and since they attack c and a respectively, c and a cannot be *in*. An inconsistency. We also have that $\bar{\bar{a}}, \bar{\bar{c}}$ are *in* and $d(x)$ is *out*.

We now need to check what happens when Figure 7.46 is embedded inside a larger network. Figure 7.46 arose from the attempt to eliminate the disjunctive attack of Figure 7.42. So if Figure 7.42 is part of a larger network then Figure 7.46 will also be in the larger network. However, the larger network interacts (attacks or is being attacked) only by $\{x, a, c\}$ and not the new points. So it is crucial for us to show that if the labels of $\{x, a, c\}$ are fixed by the larger network then the labels of the new points are unique and they are consistent with the labels of $\{x, a, c\}$.

Lemma 7.12. *In Figure 7.46, if we fix the value {in, ?, out} of $\{a, c, x\}$ then the value of the other points $\{\bar{a}, \bar{c}, \bar{x}, \bar{\bar{a}}, \bar{\bar{c}}, e(x), e(x, a), e(x, c)\}$ are also uniquely fixed, and are consistent with the values of $\{a, c, x\}$.*

Proof. We need to show that if the values of $\{a, c, x\}$ are fixed then the values of the new points are uniquely determined.

Let **N** be a general network in which Figure 7.42 is a subnetwork. This means that there may be attacks on $\{a, c, x\}$ from other nodes of the network **N**, say from $\{d_1, \ldots, d_k\}$ and also that there may be attacks emanating from $\{a, c, x\}$ individually or jointly with others onto nodes $\{d'_1, \ldots, d'_{k'}\}$ in **N**.

We are now replacing in **N** Figure 7.42 by Figure 7.46. The word 'replacing' is not accurate. Figure 7.42 (i.e. the subnetwork of **N** comprising of nodes $\{a, c, x\}$ and the connections between them) remains in **N**, we are adding new points and connections to Figure 7.42 to form Figure 7.46 a part of **N**.

We want to show that any acceptable labelling λ on **N** which gives values to $\{a, c, x\}$ can be uniquely extended to the new points of Figure 7.46 in a manner consistent with λ. Call the new extension λ^*.

We need to make a case analysis on the values of $\lambda(a), \lambda(c), \lambda(x)$.

We have to check all cases. We follow the cases in three groups $x = 1, x = 0$ and $x =?$. For each group we check all cases of a and c as in Table 7.1.

Group x = in
Table 7.1 suggests 9 cases for a and c. However not all are possible in **N**. Since λ is an acceptable labelling and $x = in$, only cases in which either $a = out$ or $c = out$ are possible.

Table 7.1.

case	c	a
1	out	out
2	out	in
3	in	out
4	in	in
5	?	out
6	?	in
7	in	?
8	out	?
9	?	?

Thus we need examine only cases 1, 2, 3, 5, 8. Cases 1, 2, 3 have already been examined in Example 7.11. They are OK and are internally consistent. Cases 5 and 8 are completely symmetrical (since Figure 7.46 is symmetrical in a and c). So we need examine only Case 5.

Case 5
$a = out, c =?$
First we see which values for the new points are forced by this case.

$$a = out \Rightarrow \bar{a} = in$$
$$\bar{a} = in \Rightarrow \bar{\bar{a}} = out$$
$$\bar{a} = in \Rightarrow e(x, a) = out.$$

Claim
$e(x)$ cannot be *in*!

Otherwise, since $\bar{\bar{a}} = out$ then if $e(x)$ were *in* then $\bar{\bar{c}} = out$, hence $\bar{c} = in$ hence c is *out*. But we are given that $c =?$. So $e(x)$ is *not in*.

Therefore, since $e(x, a) = out$ and $e(x)$ is either *out* or ? we have a situation which is consistent with $c =?$. We now use the fact that $c =?$ to get $\bar{c} =?$ and hence $\bar{\bar{c}} =?$, and hence $e(x) =?$. This establishes a unique value for $e(x)$. We continue, since $\bar{c} = out$ and $\bar{c} =?$, we get $e(x, c) =?$.

We have that $e(x, c)$ attacks a. Does this contradict $a = out$? The answer is no. a is *out* because of some attack from **N**. So we got for Case 5 unique consistent values for the new points.

Group x = out
Since $x = out$, there are no attacks from x on a and c, therefore all values of $\{a, c\}$ of Table 7.1 are possible. We have to examine all cases.

We begin by checking what values are forced on the new points by the fact that $x = out$. We get

$$x = out \Rightarrow \bar{x} = in$$
$$\bar{x} = in \Rightarrow e(x, a) = e(x, c) = e(x) = out$$

Since there are no attacks from the new points of Figure 7.46 on the points $\{a, c\}$, their value consistently stand in the expanded network.

If $\lambda = 1, 0, ?$ use the notation $1 - \lambda$ to mean $-, 1, ?$ respectively. Thus the value of $\bar{a}$ is determined as $1 - \lambda(a)$ and the value of $\bar{c}$ is determined as $1 - \lambda(c)$. Thus the value of $\bar{\bar{a}}$ is the same as the value of a and the value of $\bar{\bar{c}}$ is the same as the value of c.

Hence for all cases of Table 7.1 we see that the values of the new points are unique and are consistent with the values of $\{a, c\}$ for our group case of $x = out$.

Group x =?
If $x =?$ and x disjunctively attacks a and c, we cannot have the case of $a = c = in$, because maybe x can be *in*. Similarly we cannot have the cases of $a = in, c =?$ nor $a =?$ and $c = in$.

So let us see first what we can get which holds for all cases

$$x =? \Rightarrow \bar{x} =?$$

This is all we can get.
We have to check cases 1–3, 5, 8, 9.

Case 1
$c = out$ and $a = out$.

$$a = out \Rightarrow \bar{a} = in \text{ and } e(x, a) = out$$
$$c = out \Rightarrow \bar{c} = in \text{ and } e(x, c) = out$$
$$\bar{a} = in \Rightarrow \bar{\bar{a}} = out$$
$$\bar{c} = in \Rightarrow \bar{\bar{c}} = out$$
$$\bar{x} =? \text{ and } \bar{\bar{a}} = \bar{\bar{c}} = out \Rightarrow e(x) =?$$

Now a and c are out, they are attacked by $e(x) =?$. This is still consistent.
So this case is OK.

Case 2
$c = out$ and $a = in$.

$$\bar{\bar{c}} = out \Rightarrow \bar{c} = in$$
$$\bar{c} = in \Rightarrow \bar{\bar{c}} = out$$
$$\bar{c} = in \rightarrow e(x, c) = out$$
$$a = in \Rightarrow \bar{a} = out$$
$$\bar{a} = out \Rightarrow \bar{\bar{a}} = in \text{ and } e(x, a) = out$$
$$\bar{\bar{a}} = in \Rightarrow e(x) = out.$$

We now get unique values for the new points. We have consistency because there are no attacks on $\{a, c\}$ from the new points.

Case 3
$c = in$ and $a = out$.
This is the symmetrical case of Case 2.

Case 5
$c =?$ and $a = out$.
We have $x =?$ and $\bar{x} =?$

$$
\begin{aligned}
a &= \text{ out} \Rightarrow \bar{a} = \text{ in} \\
\bar{a} &= \text{ in} \Rightarrow e(x, a) = \text{ out} \\
c &=? \Rightarrow \bar{c} =? \\
\bar{x} &=? \text{ and } \bar{c} =? \Rightarrow e(x, c) =? \\
\bar{a} &= \text{ in} \Rightarrow \bar{\bar{a}} = \text{ out} \\
\bar{c} &=? \Rightarrow \bar{\bar{c}} =? \\
\bar{\bar{a}} &= \text{ out} \text{ and } \bar{\bar{c}} =? \Rightarrow e(x) =?
\end{aligned}
$$

We now got unique values for all new points. Do we have consistency?
    $a = out$ is attacked by

$$e(x) = e(x, c) =?$$

Hence a remains *out*. We have consistency here.
    $c =?$ is attacked by $e(x, a) = out$, so we maintain consistency.

Case 8
$c = out$ and $a =?$
This case is symmetrical with Case 5, with a and c interchanged. By symmetry it is OK.

Case 9
$a =?$ and $c =?$
In this case we get everything has value ? and we have no problem.

The situation of moving from Figure 7.42 to Figure 7.46 is typical and the fact that we dealt with an attack from x on $\{a, c\}$ and we did not deal with attacks from a general set $X = \{x_1, \ldots, x_n\}$ on a general set $Y = \{y_1, \ldots, y_m\}$ does not incur any loss of generality. In the general case we proceed similarly. We need to add the auxiliary points $\bar{\bar{y}}_1(X), \ldots, \bar{\bar{y}}_m(X)$ add attacks of y_i on $\bar{\bar{y}}_i(X)$ and add joint attacks as in Figure 7.47

7.3.2 Eliminating disjunctive and joint attacks

We are now ready, following our conceptual discussion in Section 3.1, to give a series of definitions and Lemmas showing how networks with joint and disjunctive attacks can be reduced to ordinary Dung networks with only point-to-point attacks. The reduction is done with the help of auxiliary points, using the intuition described in Figure 7.48.[1]

---

[1] This translation shows implicitly that Brewka and Woltran's Abstract Dialectical Frameworks are implicit in this chapter , as they indeed admit in their paper [69].

$$X \cup (Y - \{y_1\}) \qquad ,\ldots, \qquad X \cup (Y - \{y_m\})$$

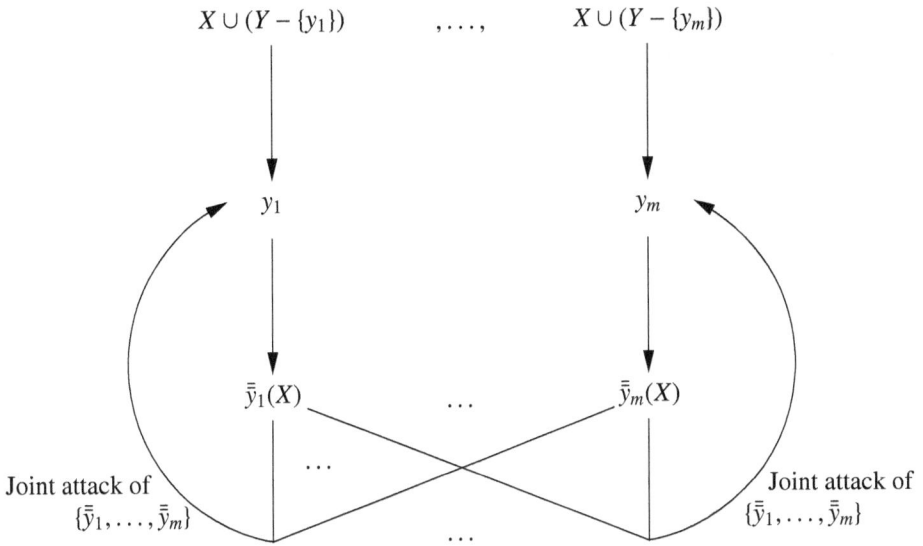

$$y_1 \qquad\qquad\qquad y_m$$

$$\bar{\bar{y}}_1(X) \qquad \ldots \qquad \bar{\bar{y}}_m(X)$$

Joint attack of $\{\bar{\bar{y}}_1,\ldots,\bar{\bar{y}}_m\}$

Joint attack of $\{\bar{\bar{y}}_1,\ldots,\bar{\bar{y}}_m\}$

Fig. 7.47.

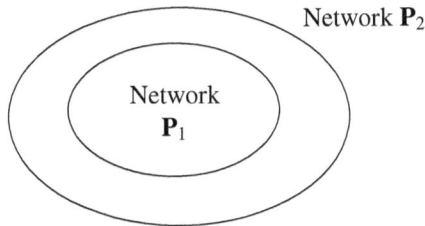

Network $\mathbf{P}_2$

Network $\mathbf{P}_1$

Fig. 7.48.

Definition 7.13 (Critical subsets). *Let* $\mathbf{P}_i = (S_i, R_i)$ *be two networks. Suppose all points of network* $\mathbf{P}_2$ *are embedded inside network* $\mathbf{P}_1$. *So* S_2 *is a subset of* S_1. *We say that* S_2 *is a* critical *subset of* S_1 *iff every Caminada labelling on* S_2 *can be extended uniquely to a labelling on* S_1. *This means that the additional nodes of* S_1 *only help clarify what is going on in* S_2 *and do not add any additional information. Any Caminada labellings of* S_1 *which agree on* S_2 *must be equal.*

This sort of characterisation is known from model theory. A set of classical models $\mathcal{M}_2$ in a language $\mathbb{L}_2$ is said to be EC_Δ iff it can be characterised as the set of all models of a first-order theory Δ_2 of $\mathbb{L}_2$. The set of models is said to be PC_Δ iff we can extend the language $\mathbb{L}_2$ such that the set there exists a theory Δ_1 in the language $\mathbb{L}_1$ such that any model $\mathbf{m}_2$ of $\mathcal{M}_2$ can be obtained as the restriction $\mathbf{m}_2 = \mathbf{m}_1 \upharpoonright \mathbb{L}_2$ to $\mathbb{L}_2$ of a unique model $\mathbf{m}_1$ of Δ_1. Furthermore, all restrictions to $\mathbb{L}_2$ of models of Δ_1 are models of Δ_2. 0 We are now ready to push on with our reduction.

Definition 7.14 (Eliminating disjunctive attacks). *Let* $\mathbf{P} = (S, S^0, R)$. *Let* $S_1 = S \cup \{\bar{\bar{s}}(X) | s \in S, X \subseteq S, X \neq \varnothing\}$.
Let $\mathbf{P}_1 = (S_1, S_1^0, R_1)$ *be defined as follows, see Figure 7.47:*

1. *Let XR_1a if XRa holds.*
2. *If XRY holds then let $X \cup (Y - \{y\})R_1y$ hold for every $y \in Y$.*
3. *Let $sR_1\bar{\bar{s}}(X)$ hold for every $s \in S, X \subseteq S, X \neq \emptyset$*
4. *If XRY holds let $\{\bar{\bar{y}}(X)|y \in Y\}R_1y$ hold of each $y \in Y$.*

Lemma 7.15. *For any $\lambda : S \mapsto \{0, 1, ?\}$ we have that λ is proper labelling on $\mathbf{P}$ iff λ is proper on $\mathbf{P}_1$.*

Proof. One can give a direct proof of this theorem by working directly of Figure 7.47, converting it to the analog of Figure 7.46. However a simpler rout is to use Lemma 7.12 for the case of disjunctive attacks involving only two elements. All we need to do is to convert a disjunctive attack on $n \geq 3$ elements, as in Figure 7.49 to a disjunctive attack on $n - 1$ elements as in Figure 7.50. We use auxiliary points as indicated.

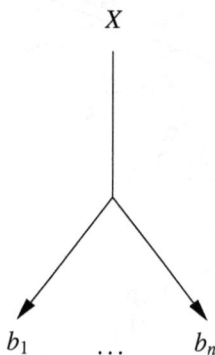

$$X$$

$$b_1 \quad \ldots \quad b_n$$

Fig. 7.49.

We can therefore assume that we are dealing only with disjunctive attacks on two points, since all other disjunctive attacks can be reduced to two points by repeated applications of the above procedure.

The case of disjunctive attacks of two points was dealt with by Figure 7.46 and Lemma 7.12.

Definition 7.16 (Eliminating joint attacks). *Let $\mathbf{P} = (S, S^0R)$ be a joint attack network. Define an ordinary network $\mathbf{P}^* = (S^*, R^*)$, with $S \subseteq S^*$ and S^*, R^* as follows. We add to S the following additional groups of points.*

Group G_1
For every $s \in S$ a new node $\bar{s}$
Group G_2
For every $X \subseteq S, X$ finite with two points or more add the node $e(X)$.

Let $S^* = S \cup G_1 \cup G_2$.
Define R^* on S^* as follows.
Assume $\{a_1, \ldots, a_n\}Rb_j$ hold for $n \geq 2, j = 1, \ldots, k$, where $b_1, \ldots, b_k$ are all the nodes jointly attacked by $\{a_1, \ldots, a_n\}$. Figure 7.51 shows the situation.
Note that we cannot represent this situation by writing Figure 7.52 because that would be a disjunctive attack.

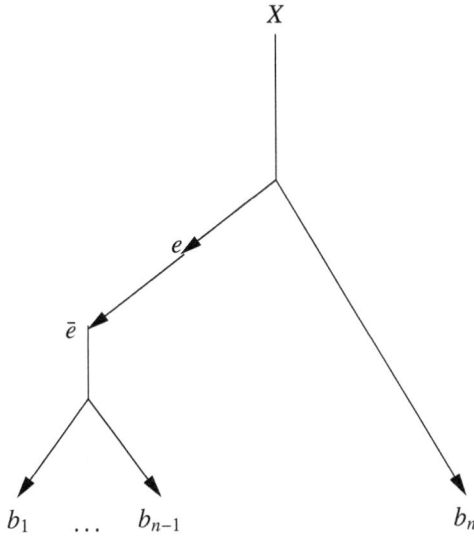

Fig. 7.50.

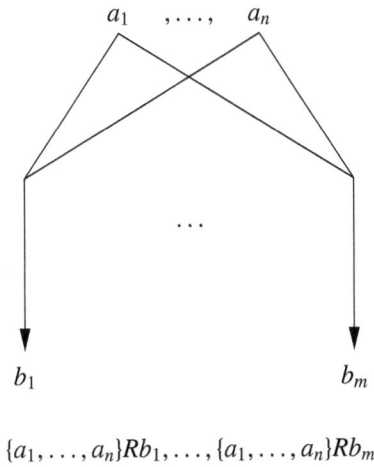

$$\{a_1, \ldots, a_n\}Rb_1, \ldots, \{a_1, \ldots, a_n\}Rb_m$$

Fig. 7.51.

We define R^* according to Figure 7.53. Thus Figure 7.51 with R becomes Figure 7.53 with R^*.

We have that the following hold in R^*:

1. *$sR^*\bar{s}$ for any $s \in S$.*
2. *tR^*s whenever $\{t\}Rs$*
3. *If $\{a_1, \ldots, a_n\}Rb_j$, holds with $n \geq 2$, and $j = 1, \ldots, m$ then let $\bar{a}_iR^*e(a_1, \ldots, a_n)$ hold for $i = 1, \ldots, n$ and let $e(a_1, \ldots, a_n)R^*b_j$ hold, for $j = 1, \ldots, m$.*

Lemma 7.17. *Let $\mathbf{P}$ and $\mathbf{P}^*$ be as in Definition 7.16. Observe the following*

1. *Since s is the only attacker of $\bar{s}$, we have for any Caminada labelling that*

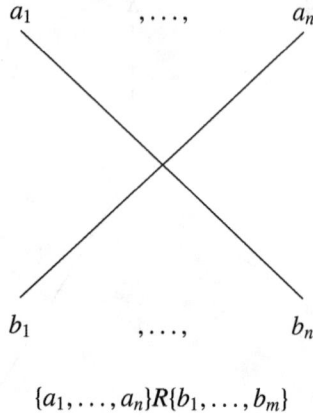

$$\{a_1, \ldots, a_n\} R \{b_1, \ldots, b_m\}$$

Fig. 7.52.

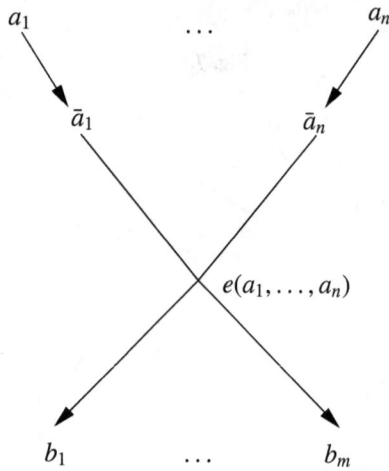

Fig. 7.53.

$$\lambda(s) = 1 \text{ iff } \lambda(\bar{s}) = 0$$
$$\lambda(x) =? \text{ iff } \lambda(\bar{s}) =?$$

2. In Figure 7.53

$$\lambda(e(a_1, \ldots, a_n)) = 1 \text{ iff}$$
$$\lambda(a_1) = \lambda(a_2) = \ldots = \lambda(a_n) = 1$$

3. *If $\lambda(a_i) = 1$ for $i = 1, \ldots, n$, then $\lambda(b_j) = 0$, $j = 1, \ldots, m$.*
4. *If for some i, $\lambda(a_i) = 0$ then $\lambda(e(a_1, \ldots, a_n)) = 0$ and then there is no attack from $e(a_1, \ldots, a_m)$ on any b_j, $j = 1, \ldots, m$.*
5. *Any Caminada–Gabbay labelling function on λ^* on S^* induces a labelling $\lambda = \lambda^* \upharpoonright S$ on S.*

To explain the relationship between **P** and **P*** we need some general concepts.

Definition 7.18. *Let* **P** = (*S*, *R*) *be ordinary Dung argumentation networks. Let* *E* ⊆ *S be a set of nodes. We say that E is a* critical *set in* **P** *iff the following holds:*

(*) *For any two Caminada labelling* λ_1 *and* λ_2 *on* **P***, if* λ_1 *and* λ_2 *agree on E then*
$\lambda_1 = \lambda_2$

Lemma 7.19. *In Definition 7.16, S is critical in* (*S**, *R**).

Proof. Follows from Lemma 7.17.

Remark 7.20 (Labelling of higher order networks). To show the existence of labelling for higher order networks **P** = (*S*, S^0, *R*), we reduce the network to an ordinary network **P*** = (*S**, *R**) by eliminating first the disjunctive attacks and then the joint attacks.

From our sequence of Lemmas we know that *S* ⊆ *S** is critical in *P**. Therefore any labelling λ* on *S** induces a labelling $\lambda = \lambda$* ↾ *S* on *S* which is acceptable in **P**. Furthermore, λ* can be uniquely retrieved from λ and thus any λ on *S* can be expanded to a unique λ* on *S**.

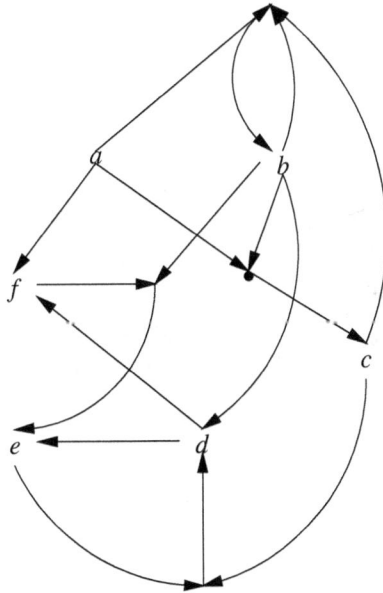

Fig. 7.54.

Example 7.21. Figure 7.54 displays the following network from [271].

$$S = \{a, b, c, d, e, f\}$$

R is defined as follows

$$\{a, c, d\}Rb$$
$$\{ab\}Rc$$
$$\{b\}Rd$$
$$\{c, e\}Rd$$
$$\{d\}Re$$
$$\{b, f\}Re$$
$$\{a\}Rf$$
$$\{d\}Rf$$

The grounded extension is $\{a\}$. The preferred extensions are $\{a, b, e\}$ and $\{a, c, d\}$.

We now construct our reduct ordinary network. We add points $\bar{a}, \bar{b}, \bar{c}, \bar{d}, \bar{e}\bar{f}$ and $e(a, b, d), e(a, b), e(c, e)$ and $e(b, f)$ and get Figure 7.55.

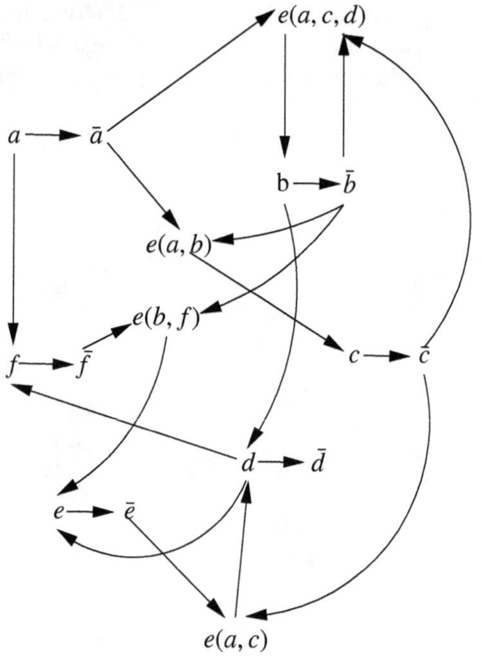

Fig. 7.55.

We have the following attacks:

$$a \to \bar{a}$$
$$b \to \bar{b}$$
$$c \to \bar{c}$$
$$d \to \bar{d}$$
$$e \to \bar{e}$$
$$f \to \bar{f}$$
$$\bar{a} \to e(a, c, d)$$
$$\bar{c} \to e(a, c, d)$$
$$\bar{d} \to e(a, c, d)$$
$$e(a, c, d) \to b$$
$$\bar{a} \to e(a, b)$$
$$\bar{b} \to e(a, b)$$
$$e(a, b) \to c$$
$$b \to d$$
$$\bar{c} \to e(c, e)$$

$$\bar{e} \to e(c, e)$$
$$e(c, e) \to d$$
$$d \to e$$
$$\bar{b} \to e(b, f)$$
$$\bar{f} \to e(b, f)$$
$$e(b, f) \to e$$
$$a \to f$$
$$d \to f$$

Graphically it is very easy to convert any joint network to an ordinary network as follows.

1. First replace any node x by $x \to \bar{x}$ and let any arrows emanating from x to emanate now from $\bar{x}$. Any arrows targeting x remain as is.
2. Any joint attack from $a_1, \ldots, a_n$ on b have arrows emanating from each a_i (now emanating from $\bar{a}_i$) to an unnamed junction where they all meet and then a single arrow go to b as in Figure 7.56.
   Name the junction $e(a_1, \ldots, a_n)$ and let the arrows be now attacking arrows of an ordinary network. We get Figure 7.57.

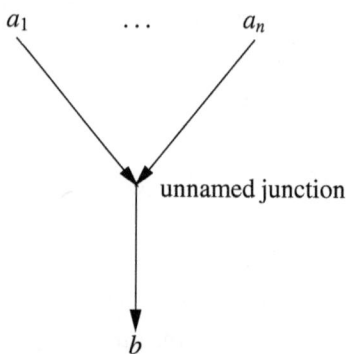

Fig. 7.56.

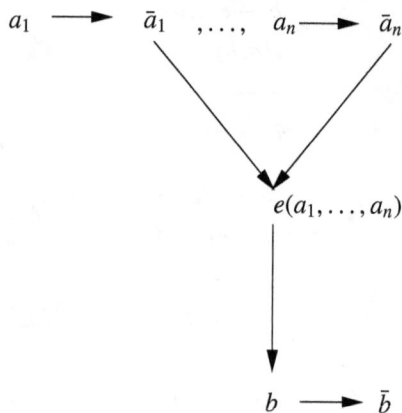

Fig. 7.57.

7.4 Comparison and discussion

7.4.1 Comparison with Nielsen and Parsons

We compare this work with the paper of Nielsen and Parsons [270, 271]. Nielsen and Parsons put forward a system where joint attacks are possible. They introduce attacks of the form XRy where X is a non-empty set of arguments and y is an argument.

The attack XRY, where Y is a set of arguments is reduced to

$$XRY \text{ iff } XRy \text{ for some } y \in Y.$$

This is the same as our approach for the case in which the networks we are dealing with have no arrows in them and in which there are no disjunctive attacks. The point of view of Nielsen and Parsons arises from a qualitative argument in favour of joint attacks. They argue their case convincingly and proceed to develop the theory beautifully in their first paper [270]. In the second paper [271] they continue with an algorithm for computing extensions.

We arrived at these attacks from a different point of view. We are substituting one network in another. We thus get a network attacking another network. Since the networks may have points in common they can internally influence one another.

To obtain the Nielsen and Parsons case our networks should have no internal arrows, and no disjunctive attacks. So we interpret an attack on a network as in Figure 7.42 reading (r2). Another difference between the two approaches is the reduction we make of the joint attack networks to ordinary Dung networks. The original network becomes a critical subset in an ordinary Dung network and so we can compute the extensions of our original network by computing the extensions of the target network.

We would like to quote and criticise a statement of Nielsen–Parsons in their paper [270] they say:

> We claim that it is never necessary to specify a non-singleton set of arguments as attacked, as in $\{A_1, \ldots, A_n\} \triangleright \{B_1, \ldots, B_m\}$: If collective defeat is taken to heart, the attack can be reformulates as a series of attacks
>
> $$\{A_1, \ldots, A_n\} \triangleright B_1$$
> $$\vdots$$
> $$\{A_1, \ldots, A_n\} \triangleright B_m$$
>
> It is easily seen that the above attacks would imply the attack, which is intended, as the validity of the A-arguments would ensure that none of the B-arguments are valid.
> If instead indeterministic defeat is required, the attack can be reformulated as
>
> $$\{A_1, \ldots, B_n, B_2, \ldots, B_m, \} \triangleright B_1,$$
>
> which ensures that in case the A arguments are valid, then B_1 cannot be a valid argument if the remaining B-arguments are also true, thus preventing the entire set of B-arguments from being valid at once, if the A-arguments are true.

Nielsen and Parsons use reading (r3) of Figure 7.42 (to use our terminology), namely they use Figure 7.44. Their notion of 'collective defeat' requires reading (r2) and not (r3) as Nielsen and Parsons claim.

7.4.2 Connection with attacks on attacks

In Chapters 8 and 9 we shall introduce the notion of attack on attack. Figure 7.58 is an example.

The network contains the arguments $\{a, b, c, d\}$. a attacks b. Denote this attack by '$a \rightarrow b$'. The attack '$a \rightarrow b$' is being attacked by c and it itself attacks the node d.

Our question is how does this relate to our joint and disjunctive attack networks? Consider Figures 7.59 and 7.60

Can we view Figure 7.58 as the result of substituting Figure 7.60 for x in Figure 7.59? The answer is that if we do that we get the wrong meaning. The subject is studied extensively in Chapters 8 and 9. The actual meaning of Figure 7.58 is that c attacks the argument (denoted by '$a \rightarrow b$') motivating and describing why a attacks b, and this very argument actually attacks d.

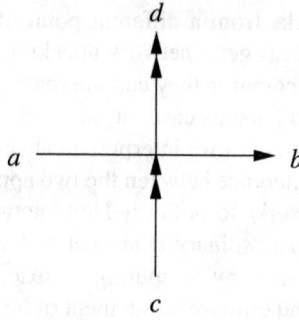

Fig. 7.58.

Fig. 7.59.

$$a \longrightarrow b$$

Fig. 7.60.

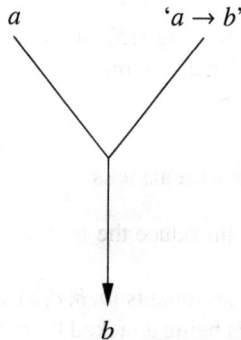

Fig. 7.61.

So the correct interpretation of Figure 7.60 is Figure 7.61.

a and '$a \to b$' mount6 a joint attack on b. So Figure 7.58 can be prestned as Figure 7.62

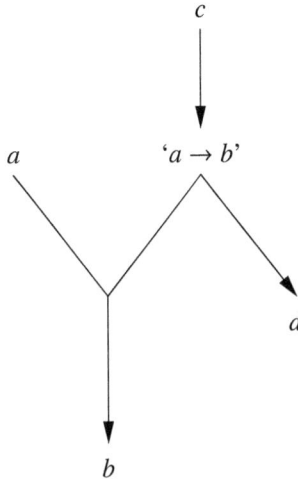

Fig. 7.62.

Of course if we want to be very pedantic we can say that c, which attacks the '$a \to b$' component of this joint attack, is attacking '$a \to b$' jointly with '$(c \to (\text{'}a \to b\text{'}))$' argument. Also '$a \to b$' which is attacking d is attacking it jointly with "$((\text{'}a \to b\text{'}) \to d)$' argument. So the correct figure corresponding to Figure 7.58 is Figure 7.63.

However, since the attacks of c on '$a \to b$' and of '$a \to b$' on d are not themselves being attacked, we do not need to unfold them in Figure 7.63 and Figure 7.62 is sufficient.

Note that if we systematically rewrite attacks on attacks on attacks, etc., as joint attacks and then use the semantics of Section 3 to define extensions, we end up with semantics for all these higher level attacks. This is done directly in Chapter 8.

7.4.3 Flow argumentation networks

This subsection deals with Figure 7.37. To explain Figure 7.37, we need a consistent point of view to explain the case of disjunctive attacks as in Figure 7.42. For Figure 7.42 we know that we require that

(*) $x = in$ implies $a = out$ or $c = out$

The questions relevant to our understanding of Figure 7.37 are the following

Question 1
Does x attack each of a and c and expects at least one of them to succeed (reading (r2) of Figure 7.37 in Section 3.1) or does x attack only the set $\{a, c\}$ and expects (*) above to hold? (Reading (r1) of Figure 7.37 in Section 3.1).

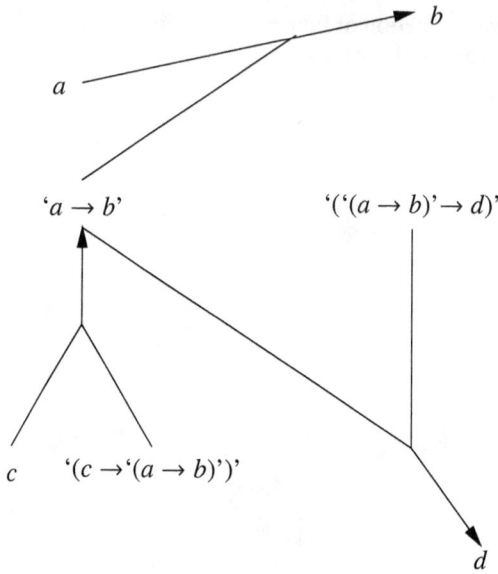

Fig. 7.63.

Obviously to be able to give Figure 7.37 any meaning we must adopt the view that x attacks each of a and c (reading (r2)).

We now ask our second question.

Question 2

When the attack of x on one of $\{a, c\}$ succeeds, e.g. $a = out$, is x still attacking c? In other words x is still attacking even though x does not care if the attack succeeds? If the answer is no, and there is no attack on c, then this means that in Figure 7.37, the labelling

1. $x = in, b = in, a = out, c = in$

is acceptable, because the success of the attack on a ($a = out$) entails that there is no attack from x onto c and hence the joint attack with b on c fails because x is not attacking and hence $c = in.$[2] If the answer is yes, and indeed there is a attack on c then (1) is not acceptable and (2) is acceptable.

2. $x = in, b = in, a = out$ and $c = out$

Of course, both views will accept (3) below.

3. $x = in, b = in, a = in$ and $c = out$

We adopt the view that x sends attacks in the direction of a and c and expects at least one of them to succeed (this is the (r2) reading of Figure 7.42 in Section 3.1).[3]

---

[2] In this case, $a = out$ makes x attack (with b) on c a voluntary attack. We investigate voluntary attacks in [65].

[3] Imagine mother wants to buy a special toy for her little girl for Christmas. There is great deamnd for the toy which is very popular and people queue up early in the morning at toy

Notice the words we use. We talk about 'x sending attacks in the direction of y'. We speak of 'joint attacks' and 'disjunctive attacks'. We need for formalise our intuitive model. We found that the best way to represent what is going on is in terms of 'flow', the attacks flow along the edges of the network. So we call these 'flow argumentation networks'.

It would help if we consider an example. Consider Figure 7.64, which is an expansion of Figure 7.37, with a new twist to it. The arguments of Figure 7.64 are $\{x, a, b, c, d, e\}$ and we also gave names to the flow (flow of attacks) junctions $\{\alpha, \beta, \gamma, \delta\}$. These are not arguments, just input output nodes in the graph, to help us follow the attacks flows.

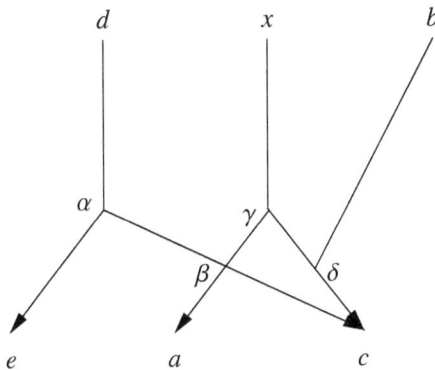

Fig. 7.64.

In Figure 7.64, d sends a disjunctive attack through junction α. The attack splits into an attack on e and an attack in the direction of a and c. x at the same time sends a disjunctive attack through junction γ which splits one in the direction of a and one in the direction of c. The x attack going to c joins forces with the attack on c emanating from b. At junction δ they join and become a joint attack on c.

The attack of x in the direction of a meets the attack of d in the direction of a and c at junction β and decide to mount a joint attack emanating from β on a and c. The flow from γ to δ is an attack of x intended for a, at β it joined the flow $d \to \alpha \to \beta$ and became a joint attack from β on a and on c. We can say x did not mind forming this coalition with d and adding c to the attack because x is attacking c anyway through the direction $x \to \gamma \to \delta \to c$.

The above description is in literary prose, giving intentions to the arrows in terms of flows. How do we do this formally? And how do we calculate labelling?

Suppose we label $c = out$, $a = in$, $e = in$. Is this OK with $d = x = b = in$? Or maybe we need also $e = out$ not $e = in$?

---

shops to make sure that they can get the product. So mother sends her husband and her brother to two different shops to queue up early in the morning in the hope that at least one of them will get to buy the toy before stocks run out for the day. Mother would be happy if at least one of them succeeds. Under this view each one of the two men is there buying a toy.

The other option is for mother to ask them to call the other mobile phone the minute one of them secures the toy. In that case the other may not buy the toy. This is the other option.

We need a formal model which follows the flow using $\alpha, \beta, \gamma, \delta$.
We leave this as the subject of another paper.

8

Semantics for Higher Level Attacks in Extended Argumentation Frames

8.1 Historical background

In our 2005 paper [37] (see Chapter 9) we introduced networks with higher level attacks of any kind. Figure 8.1 is a typical situation.

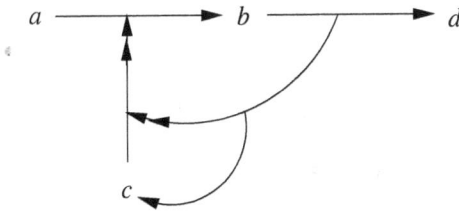

Fig. 8.1.

Figure 8.1 can represent any kind of network, not necessarily an argumentation network. It can be part of a Kripke model, an electrical network, a biological ecological network, etc. In each case the arrows have their own meaning. The paper [37] also allowed for value annotations to the nodes and arrows and gave algorithms for the propagation of these values.

In the case of argumentation networks the nodes are arguments and the arrows mean attacks. The values (annotations) can correspond to the Caminada labellings and the propagation of values are governed by the properties of the Caminada labelling (see Definition 8.2 below for Caminada labellng).

The aim of this chapter is to interpret (give semantics) to higher level attacks, in the case of argumentation networks.

Section 1 gives the background, Section 2 gives the conceptual ideas behind our approach, Section 3 gives the formal machinery of the model, Section 4 discusses other papers, Section 5 gives more formal results and Section 6 is the conclusion and discussion.

Let us look again at Figure 8.1, this time reading it as an argumentation frame.

In Figure 8.1 the argument c attacks the attack from a to b, (we use the notation $c \twoheadrightarrow (a \rightarrow b)$), while the attack from b to d attacks the attack emanating from c

(notation $(b \to d) \twoheadrightarrow (c \twoheadrightarrow (a \to b))$. This later attack attacks c (notation $((b \to d) \twoheadrightarrow (c \to (a \to b))) \twoheadrightarrow c$).

The question we ask is how to define the possible acceptable extensions for $\{a, b, c, d\}$ for the network of Figure 8.1. The reader should note that whatever approach we give for defining extensions it must come from reasonable general principles which are meaningful for general networks. It should not rely on very specific features of argumentation networks. The general principles can have a specialised meaning in the argumentation case, but then equally the principles can have their own meanings in the case of other networks. We shall see later that we shall use general network fibring principles.

We now describe some recent background developments in the context of argumentation networks. Let us denote by $\mathbf{BGW_0}$ the argumentation networks where nodes are allowed to attack attacks. So let S be a set of nodes. Denote by $x \to y, x, y \in S$, the attack of x on y. Call this attack of level $(0, 0)$. Assume z is a node and α is an attack of level $(0, n)$ then $x \twoheadrightarrow \alpha$ is an attack of level $(0, n+1)$. The index 0 in $(0, n)$ indicates that attacks emanate from points $t \in S$.

Recently, in 2009, S. Modgil [260] used preferences to 'attack' or 'nullify' attacks from argument a to argument b on the grounds that b is *preferable* to a.

Formally this gives rise to attacks on attacks of the form

$$x \twoheadrightarrow (y \to z)$$

Indeed, S. Modgil presented his system using our notation (arrows for attacks between arguments and double arrows for attacks on attacks) and presented ways of getting extensions for such networks.

His networks were required to satisfy additional conditions. This condition was motivated by the fact that the attacks on attacks come from preferences.

The Modgil condition is the following:

If x attacks $a \to b$ and y attacks $b \to a$ then x attacks y and y attacks x. $(x \twoheadrightarrow (a \to b))$ and $(y \twoheadrightarrow (b \to a))$ imply $(x \to y)$ and $(y \to x)$.

Let us denote the Modgil system by **M**. D. D. Hahn, P. M. Dung and P. M. Thang in [215] disagreed with the way S. Modgil was deriving his extensions, and presented an alternative way to derive extensions. Denote their system by **HDT**. At the same time, Dung called upon the authors of [37] to present a general semantics (for deriving extensions) for the general higher level case, especially their system **BGW**.

Quote from the end of Section 2 of [215].[1]

Though the Modgil's extended argumentation could be viewed as a special case of BGW framework [37], its semantics is based on the underlining intuition that attacks against attacks represent preferences between conflicting arguments. Hence the condition 4 in Definition 2.4 is introduced. This constraint plays a fundamental role in the definition conflict-freeness and hence in Modgil's semantics. This insight suggests that different intuitions and applications could lead to different classes of extended argumentation and different semantics for general BGW extended argumentation.

---

[1] We changed the original bibliographical reference numbers to those of the present book.

Quote from Section 5 of [215]

We proposed a solution to an intriguing problem concerning the nonmono-
tonicity of the characteristic function of Modgil's acceptability. We believe
that further study is needed to gain a better understanding of the semantics of
extended argumentation frameworks, especially the general BGW extension
of abstract argumentation. A key question in a semantics for BGW general
extended argumentation is how the notion of conflict-free should be generalized and what does it mean for an argument to be acceptable? It would be
interesting to see how works on logical modes of attacks [161] as well as
interpretations in [37] could be applied to provide a formal framework here.

Such higher level semantics is already implicitly available in our methodology,
and this chapter responds to Dung's call and presents our semantics explicitly.

Modgil and Dung disagree on the extensions of the networks of Figure 8.2 and
Figure 8.3.

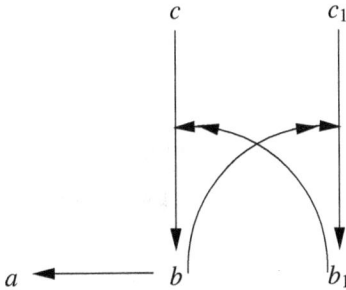

Fig. 8.2.

For the network of Figure 8.2, the acceptable M extensions are $\{c, c_1, a\}$ (which is
not acceptable to **HDT**) and $\{c, c_1, b, b_1\}$.

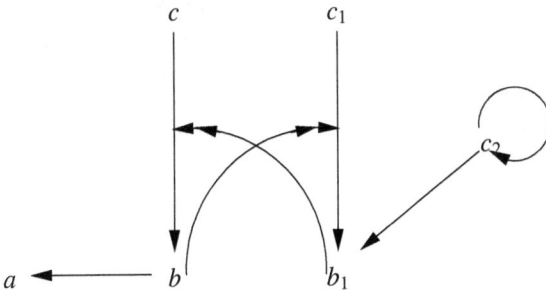

Fig. 8.3.

For the network of Figure 8.3, **M** allows for the extension $\{c, c_1, a\}$ while **HDT** allows for the extension $\{c, c_1\}$. Our general approach will settle this disagreement from general principles.

To be able to present our case we need some definitions.

Definition 8.1 (Higher level argumentation frames).

1. *An ordinary argumentation frame has the form* $\mathbf{A} = (S, R)$, *where* S *is the set of arguments and* $R \subseteq S \times S$ *is the attack relation.*[2]
2. *A level* $(0, 1)$ *argumentation frame has the form* $\mathbf{A}^{0,1} = (S, \mathbb{R})$ *where* S *is the set of arguments and* $\mathbb{R}$ *is a set of pairs of the form* (x, y), *or* $(z, (x, y))$, *where* $x, y, z \in S$. $(x, y) \in \mathbb{R}$ *means that* x *attacks* y *and* $(z, (x, y)) \in \mathbb{R}$ *means that* z *attacks the attack* (x, y).
3. *Level* $(0, n)$ *argumentation frames are defined as follows*
    a) *A pair* $(x, y) \in S \times S$ *is called a level* $(0, 0)$ *attack.*
    b) *If* $z \in S$ *and* α *is level* $(0, n)$ *attack then* (z, α) *is a level* $(0, n + 1)$ *attack.*
    c) *A level* $(0, n)$ *argumentation frame has the form* $\mathbf{A}^{0,n} = (S, \mathbb{R})$ *where* $\mathbb{R}$ *contains attacks* β *of level* $(1, m)$ *for* $m \leq n$, *including some elements of level* $(0, n)$.

Definition 8.2 (Camindada labelling). *Let* $\mathbf{A} = (S, R)$ *be ordinary argumentation frame. A function* λ

$$\lambda : S \mapsto \{0, 1, ?\}$$

is called a complete Caminada labelling if it satisfies the following conditions:

1. $\lambda(x) = 1$ *if for no* y *do we have* yRx
2. *If for some* y, yRx *and* $\lambda(y) = 1$. *Then* $\lambda(x) = 0$
3. *If for all* y *such that* yRx *we have* $\lambda(y) = 0$ *then* $\lambda(x) = 1$.
4. *If for all* y *such that* yRx *we have* $\lambda(y) \neq 1$ *and for some* y *such that* yRx *we have* $\lambda(y) = ?$ *then* $\lambda(x) = ?$.
5. *If* $\lambda(x) = 1$ *we say* x *is 'live' or 'on'. If* $\lambda(x) = 0$, *we say* x *is 'dead' or 'out'. If* $\lambda(x) = ?$ *we say* x *is 'undecided'.*

Remark 8.3 (Correspondece between Caminada labelling and extensions). There is a correspondence between Caminada labellings on (S, R) and extensions on S. This is extensively studied in Chapter 2. We refer the reader to Table 2.1.

From now on we work with Caminada labellings. They are more convenient mathematically and more general conceptually. We can have labelling in general algebraic structures and not just into the set $\{1, 0, ?\}$. This, however, is a story for another paper.

8.2 Preliminary conceptual discussion

We now discuss our options in giving semantics for attacks on attacks. We aim to analyse the basic situation of Figure 8.4 which is an acceptable network to **M**.

However, we need to begin with a simpler figure first.

---

[2] The usual notation is Ar =(Arg, Att), but this notation becomes awkward when we move to higher levels.

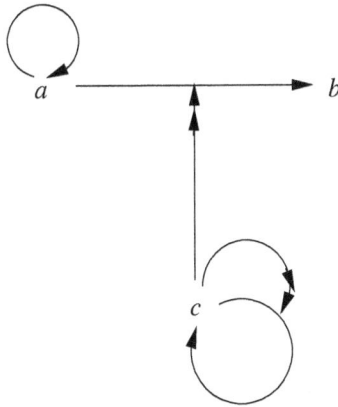

Fig. 8.4.

8.2.1 Conceptual discussion of attack on attack

Consider Figure 8.5

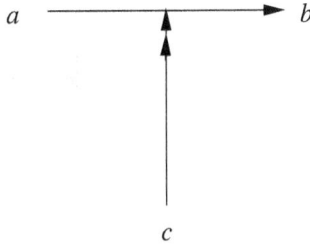

Fig. 8.5.

We ask how many 'arguments' or 'units' are in this figure? We have $S = \{a, b\}$. Do we have more? We believe that '$a \to b$' is a unit statement. It says that 'a attacks b'. The argument c attacks this statement. It may be attacking it because b is preferred to a, as in the S. Modgil approach **M**, or it may be attacking it because a is irrelevant to b. So our set of 'arguments' has not just the three elements $\{a, b, c\}$ but five elements $S^* = \{a, b, c, (a, b), (c, (a, b))\}$. Each of the elements of S^* can be 'live' or 'dead'.

To have a better appreciation of our problems let us follow a familiar story. We ask the reader to read again Example 1.2 from Chapter 1.

From this example we see two points

1. Attacks on attacks need not necessarily be due to preferences.
2. The attack on attack can be valid even though the actual attack is not executed.
   Put differently, d for example, attacks the arrow from a_2 to b in Figure 8.6, and this arrow is independent of whether a_2 is 'live' or is being attacked and is 'dead'.

Consider Figure 8.6.

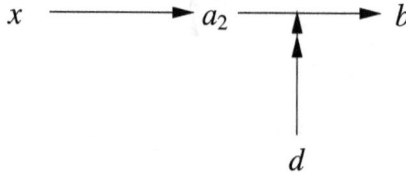

Fig. 8.6.

In this figure, for the purpose of any extension involving $\{x, a_2, b, d\}$ the attack of d on $a_2 \to b$ is not relevant, because a_2 is 'dead', defeated by x. But if we allow ourselves to give the attack $a_2 \to b$ an independent existence, d plays a role in defeating this attack, independently of the status of a_2.

To make the point even more clear and in focus, consider Figure 8.7

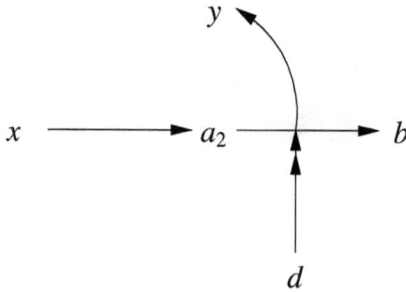

Fig. 8.7.

In Figure 8.7, the arrow from a_2 to b attacks some argument y. Clearly we want to know the acceptance of $a_2 \to b$ *as an attack* independently of the acceptance of a_2.[3]

Now consider Figure 8.4. What are our options for extensions?

We have one extension $\{c, a, b\}$, but we are not ready yet to explain why. We shall see later.

Let us now list the approaches we are going to take in providing semantics for higher level extended argumentation networks as defined in Definition 8.1. Let $(S, \mathbb{R})$ be a level $(0, n)$ argumentation network. We can provide semantics for it by one of the following methods.

1. *Translation option*
   We faithfully embed the network $(S, \mathbb{R})$ of level $(0, n)$ into an ordinary argumentation frame $(S^*, R), R \subseteq S^* \times S^*$ such that $S \subseteq S^*$. The new ordinary network has

---

[3] Of course **M** and **HDT** are less concerned about this because they do not allow for attacks to emanate from arrows, however, we shall argue that there are cases of this nature in argumentation practice.

more nodes. It satisfies certain conditions which enable us to extract semantics for $(S, \mathbb{R})$ from this embedding and from the traditional known semantics of (S^*, R).

2. *Labelling option*

   We extend the Caminada labelling concept to $(S, \mathbb{R})$, defining the notion of BGW labelling and using this BGW-labelling we define our extensions and give semantics for $(S, \mathbb{R})$. Alternatively one can consider an equivalent traditional Dung approach using the notions of conflict free sets, acceptability etc etc to get the extensions.

3. *Logic programming option*

   We translate $(S, \mathbb{R})$ faithfully into a logic program π. The literals are the nodes $S \cup \mathbb{R}$. We use $\mathbb{R}$ to write the clauses of π. The semantics we get for $(S, \mathbb{R})$ comes from the semantics of logic programs.

8.2.2 Discussion of the translation option

We now give the reader a quick preview of our translation option, which will be discussed and motivated from general principles in the next section.

Consider Figure 8.5 again. Consider the expanded frame shown in Figure 8.8.

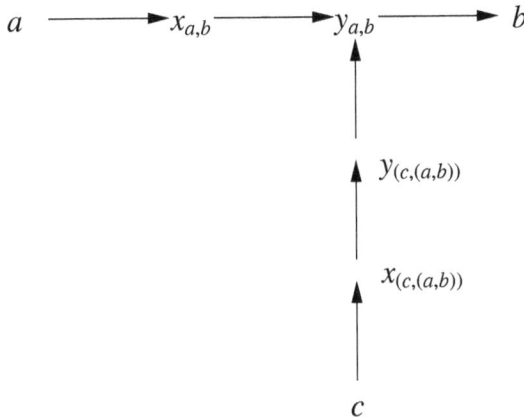

Fig. 8.8.

In Figure 8.8 we added for each attack arrow of the form $\alpha \to \beta$, two new points $x_{(\alpha,\beta)}$ and $y_{(\alpha,\beta)}$ and expanded the 'unit' $\alpha \to \beta$ into $\alpha \to x_{\alpha,\beta} \to y_{\alpha,\beta} \to \beta$.

Note that if α is 'live' then $x_{\alpha,\beta}$ is 'dead' and so $y_{\alpha,\beta}$ is 'live' and so β is 'dead'. The pair $(x_{\alpha,\beta}, y_{\alpha,\beta})$ represents the attacking arrow from α to β. So to attack the arrow unit '$\alpha \to \beta$' we attack $y_{\alpha,\beta}$. This is why the attacking c, which in Figure 8.5 attacks the arrow $a \to b$ in Figure 8.8 it attacks the point $y_{a,b}$.

In fact it is sufficient to do this trick only for arrows which are under attack. So for our purposes here (but not if we give a general mathematical definition), we can work with Figure 8.9, the simpler version of Figure 8.8.

We can similarly turn Figures 8.2 and 8.3 into Figure 8.10 and 8.11, resp.

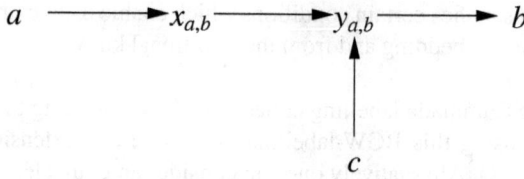

Fig. 8.9.

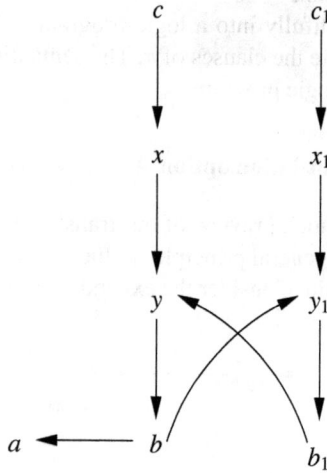

Fig. 8.10.

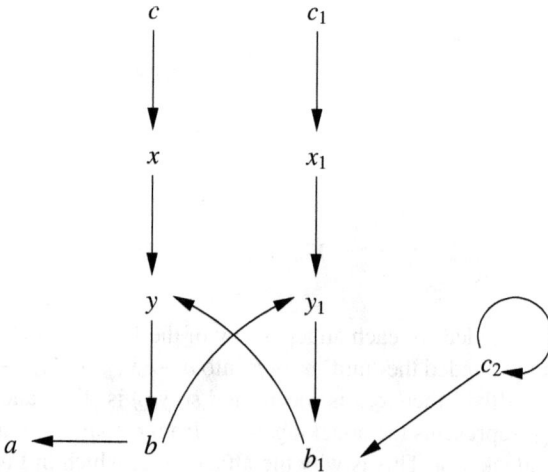

Fig. 8.11.

Figures 8.10 and 8.11 are now ordinary frames. Let us check their extensions. We get the extensions for Figure 8.10.

E_1. $\{c, c_1, y, y_1, a\}$
E_2. $\{c, c_1\}$
E_3. $\{c, c_1, b, b_1\}$

For the original language E_1 suggests $\{c, c_1, a\}$.

For Figure 8.11 we get $E^+ = \{c, c_1, y, y_1, a\}$.

E^+ suggest the extension $\{c, c_1, a\}$ for the original language, i.e. we are supporting the Modgil semantics **M**. We also have the extension $\{c, c_1\}$ supporting **HDT**.

Our strategy is therefore to start with an extended higher level frame based on the set S. We 'simplify' it by adding points X of the form $x_{a,\alpha}, y_{a,\alpha}$ for any $(a, \alpha) \in \mathbb{R}$, we adjust the attacks as discussed and illustrated in Figures 8.8 and 8.9 and get an ordinary frame based on the points $S \cup X$.

Let $E_1^+, E_2^+, \ldots$ be the extensions on $S \cup X$. Then $E_i^+ \cap S, i = 1, 2, \ldots$ are the extensions according to our translation option, on S.

The proper way to do this is in terms of Caminada labelling on $S \cup X$. The detailed machinery we develop in the next section.

The reader may now ask is this our solution? Are we happy now with the proposed translation option? The answer is no. We still need to continue with our conceptual analysis. We argued that we want to view '$a \to b$' as an independent unit, the attack of a on b, which can in itself be attacked by another argument c or itself attack another argument d. The 'trick' proposal of inserting the two intermediaries $x_{a,b}, y_{a,b}$ with $a \to x_{a,b} \to y_{a,b} \to b$ may work for attacks from c on $a \to b$, but it is not satisfactory for attacks from $a \to b$ onto other points d. We ask, where is this attack to emanate from? The obvious node is $y_{a,b}$. But if a is not 'live', then $x_{a,b}$ is 'live' and $y_{a,b}$ is 'dead'. So it cannot attack other nodes. Our approach requires that '$a \to b$' as a unit be kept 'live' unless attacked itself.

We therefore offer another solution. In $a \to b$ the node b is attacked by two arguments. The node a and the attack arrow unit '$a \to b$'. This is a joint attack on b, both of these participants must be 'live'.

So how do we represent that? Figure 8.12 shows how it is done.

In Figure 8.12, '$a \to b$' is just a node. We are saving on notation and not writing $z_{a,b}$. In Figure 8.12, only when both a and '$a \to b$' are 'live' will we have $y_{a,b}$ 'live' and hence b is 'dead'. So Figure 8.12 is an implementation of the joint attack of 'a' and '$a \to b$' on 'b'. See Figure 8.13.

So to attack the arrow in $a \to b$ we attack '$a \to b$'. For the arrow of $a \to b$ to attack another point we emanate the attack from '$a \to b$'.

Let us now translate the frame of Figure 8.3 into our new set up, we get Figure 8.14. We need only use a simplified version.

Figure 8.14 is an ordinary argumentation frame. Let us calculate its extensions. In the original frame of Figure 8.3, we have the following extensions.

Modgil accepts $\{c, c_1, a\}$ only
Dung accepts $\{c, c_1\}$ only

1. Let us check Modgil option in Figure 8.14. If a is 'live' then b is 'dead', so '$c_1 \to b_1$' must be 'live'. Since c_1 is also 'live', then both x_1 and x_{c_1,b_1} are 'dead' and so

$$a \qquad\qquad 'a \to b'$$

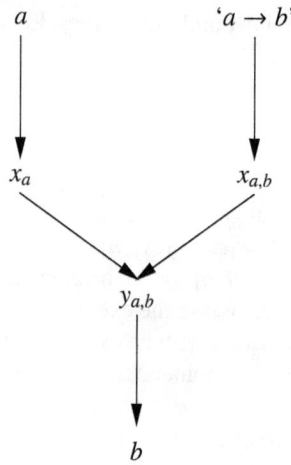

Fig. 8.12. Representation of $a \to b$

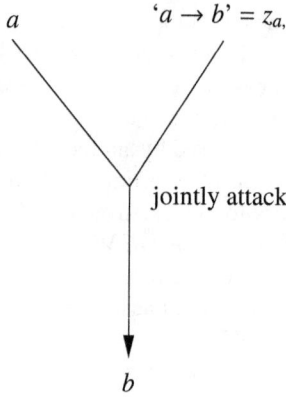

$$a \qquad\qquad 'a \to b' = z_{a,b}$$

jointly attack

$$b$$

Fig. 8.13.

y_{c_1,b_1} is 'live' and b_1 is 'dead'. So '$c \to b$' is 'live'. So since c is also 'live' we get that $x, x_{c,b}$ are 'dead'. Hence $y_{c,b}$ is live and b is dead. The loop is complete and this is consistent.

2. Can we have an extension with only $\{c, c_1\}$ live? as Dung proposes for Figure 8.3? The answer is yes, we can.

Summary of our translation policy

Let $(S, \mathbb{R})$ be an extended higher level network. Transform it in some algorithmic way to an ordinary network (S^*, R), such that the following holds

1. $S \subseteq S^*$
2. For every Caminada labelling λ_1, λ_2 on S^* we have

$$\lambda_1 \upharpoonright S = \lambda_2 \upharpoonright S \implies \lambda_1 = \lambda_2$$

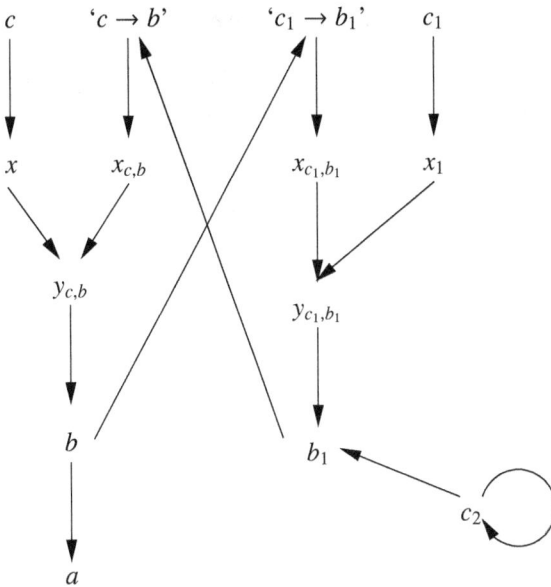

Fig. 8.14.

(i.e. if λ_i agree on all values on S then they agree on S^* as well.)[4]

3. Then the extensions of $(S, \mathbb{R})$ are all subsets of the form $E_\lambda = \{x \in S \mid \lambda(x) = 1$ where λ is a complete Caminada labelling on $(S^*, R)\}$.

8.2.3 The labelling option

Consider the situation in Figure 8.5. We asked ourselves how many 'units' are participating in this network? The answer is five units.

$$\{a, b, c, a \to b, c \twoheadrightarrow (a \to b)\}$$

In the BGW labelling they all get labels in $\{0, 1, ?\}$. So a attacks b through the channel '$\to$' in '$a \to b$'.

For the attack to come through, a must be 'live', and the channel must be 'live' and only then will b be attacked.

In fact, annotating nodes and arrows is exactly what we were doing in [37] in 2005. We gave them real numbers and the numbers on the arrows were transmission and capacity indicators. This material is discussed in detail in Chapter 9, section 9.2.

So we do the same here, the numbers are 'live' or 'dead', 'on', or 'off' indicators.

Definition 8.4 (BGW labelling). *Let $(S, \mathbb{R})$ be a higher level argumentation frame. A function $\eta : S \cup \mathbb{R} \mapsto \{0, 1, ?\}$ is a complete BGW labelling if the following holds.*

---

[4] This condition says that S is a *critical subframe* of S^*. This notation was introduced in Chapter 7 and is studied extensively there. It ensures faithful embedding of the source in the target!

We use Figure 8.15 for guidance. In Figure 8.15, $a_1, \ldots, a_k$ are all the nodes attacking β and for each arrow $a_i \to \beta$, the nodes $e_1, \ldots, e_{k_i}$ are all the nodes attacking it.

β is either a node or an arrow of any level.

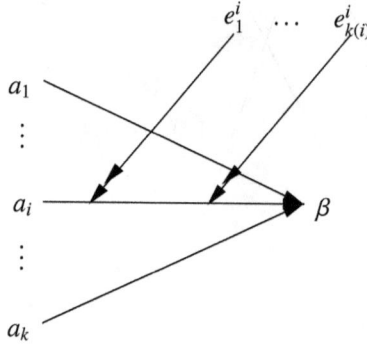

Fig. 8.15.

1. *$\eta(\beta) = 1$ if there is no y such that $(y, \beta) \in \mathbb{R}$.*
2. *$\eta(\beta) = 0$ if for some a_i we have $(a_i, \beta) \in \mathbb{R}$ and $\eta(a_i) = 1$ and $\eta((a_i, \beta)) = 1$ (i.e. both a_i and the attack arrow are 'live').*
3. *$\eta(\beta) = 1$ if for all a_i such that $(a_i, \beta) \in \mathbb{R}$ we have that either $\eta(a_i) = 0$ or $\eta((a_i, \beta)) = 0$.*
4. *$\eta(\beta) = ?$ if for all a such that $(a, \beta) \in \mathbb{R}$ we have that either $[\eta(a) = 0$ or $\eta(a, \beta) = 0]$ or $[\eta(a) = 1$ and $\eta(a, \beta) = ?]$ or $[\eta(a) = ?$ and $\eta(a, \beta) = 1]$ or $[\eta(a) = \eta(a, \beta) = ?]$. We also require that for some a such that $(a, \beta) \in \mathbb{R}$ either $\eta(a) = ?$ or $\eta(a, \beta) = ?$*

We now stand at a crossroads of how to continue. Using the Caminada like BGW labelling we can define extensions as done in Table 2.1 and then use this to give semantics for $(S, \mathbb{R})$ or we can follow a more traditional route and talk about admissibility of sets $\mathbb{E}$, acceptability of elements relative to $\mathbb{E}$, extensions, etc., etc.

From the point of view of more generality and connections with other networks, the BGW labelling is better. From the point of view of simplicty in the particular cases of argumentation networks, the tranditional way is better.

Luckily, there is a quick way to give semantics, using the connection we established in our discussion with joint attacks and the work already done by Nielsen and Parsons in [270].

Definition 8.5 (Frames with joint attacks).

1. *A joint attack frame has the form $\mathbf{J} = (S, \mathcal{R})$ where S is the set of arguments and $\mathcal{R} \subseteq S \times S \times S$ is a ternary relation. We understand $(x, y, z) \in \mathcal{R}$ as saying that the two nodes (x, y) are mounting a joint attack on z. Note that the notation $\mathcal{R}$ allows us to make a distinction, say that x is the main attacker and y is the assistant attacker. However, for our purpose here we do not need this distinction.*

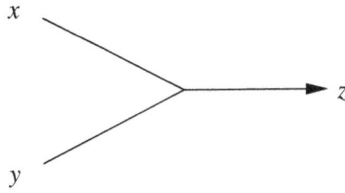

Fig. 8.16.

2. We represent a joint attack diagrammatically as in Figure 8.16
   Figure 8.17 is an example of a network with joint attacks. Note that according to
   our definition, we can still have single attacks, if $(x, x, y) \in \mathcal{R}$. So we generalise
   the standard networks.

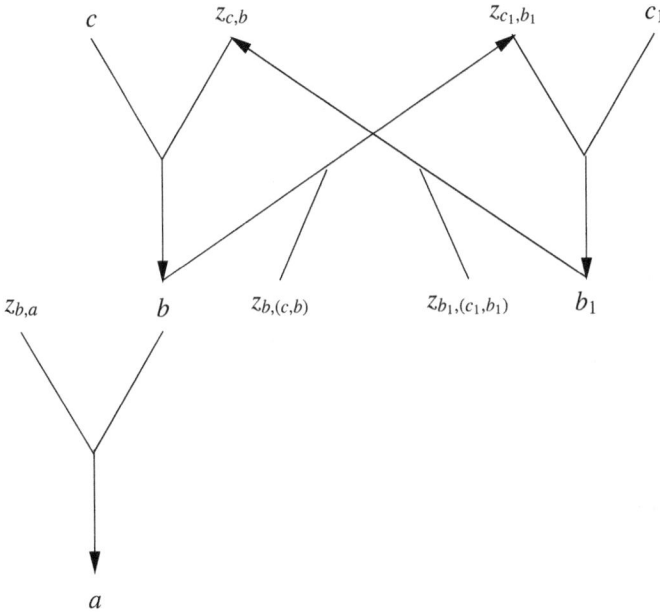

Fig. 8.17.

The network of Figure 8.13 is another example. If we read the code in Figure 8.17
that $z_{\alpha,\beta}$ stands for '$\alpha \to \beta$', then Figure 8.17 is a realisation of Figure 8.2 in
terms of joint attacks.

Definition 8.6 (Translation of an ordinary network into a joint attack network).
Let (S, R) be an ordinary network with $R \subseteq S \times S$. Define $\mathcal{R} \subseteq S \times S \times S$ as follows

$$\mathcal{R} = \{(x, y, y) \mid (x, y) \in R\}.$$

Definition 8.7 (Reduction of higher order networks to joint attack networks). Let
$(S, \mathbb{R})$ be a higher level network as defined in Definition 17.1. Define a corresponding
joint argumenation frame $(S^*, \mathcal{R})$ as follows:

$$S^* = S \cup \mathbb{R}$$
$$\mathcal{R} = \{(\alpha, \beta, \gamma) \mid \beta = (\alpha, \gamma) \text{ and } \beta \in \mathbb{R}\}.$$

If we look at Figures 8.2 and 8.17, we see that we used z_β to stand for β.

It is now clear from the definitions that if we have the machinery for defining extensions for joint attack networks of Definition 8.5, then we can give semantics for higher level networks. Such machinery exists in [270], done beautifully and in detail by Nielsen and Parsons for general theory of sets of arguments attacking jointly other arguments. See their definitions and lemmas in pages 59–65. The next Definition D4 displays the necessary concepts for our case.

Definition 8.8.   *1. A subset $\mathbb{E} \subseteq S \cup \mathbb{R}$ is conflict free if no $x \in \mathbb{E} \cap S$ and $(x, \beta) \in \mathbb{E}$ attacks any $\beta \in \mathbb{E}$.*
  2. β is acceptable for $\mathbb{E}$ if for any y such that $(y, \beta) \in \mathbb{R}$, there is a $z \in \mathbb{E}$ such that $(z, y) \in \mathbb{E}$ and $z \in \mathbb{E}$.
  3. Let $\varphi(\mathbb{E}) = \{\beta | \beta \text{ is acceptable by } \mathbb{E}\}$.
  4. $\mathbb{E}$ is admissible if $\mathbb{E} \subseteq \varphi(\mathbb{E})$.
  5. $\mathbb{E}$ is a preferred extension if $\mathbb{E}$ is maximal admissible.
  6. $\mathbb{E}$ is a complete extension if $\mathbb{E} = \varphi(\mathbb{E})$.
  7. $\mathbb{E}$ is stable if $\mathbb{E}$ attacks all $\beta \notin \mathbb{E}$.

Note that we reduced our problem of giving semantics to higher level networks of the form $(S, \mathbb{R})$ to the problem of giving semantics to the joint attack network $(S \cup \mathbb{R}, \mathcal{R})$ of Definition 8.7.

Remark 8.9 (Answering Dung's question). We can now give a very short answer to Dung's question from [215].
  HDT say (I quote):

"A key question in a semantics for BGW general extended argumentation is how the notion of conflict-free should be generalised and what does it mean for an argument to be acceptable"

My quick answer: Start with higher level $(S, \mathbb{R})$ of Definition 17.1. Translate into a joint attack network of Definition 8.5, using the translation of Definition 8.6. Apply the machinery of [81] as suggested in Definition 8.8 and now you are done.

We comment in passing that networks with joint attacks have been studied in Chapter 7. We have in Chapter 7 joint and disjunctive attacks as in Figure 8.18.

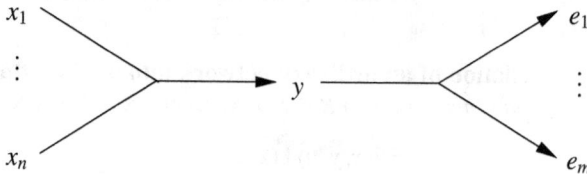

Fig. 8.18.

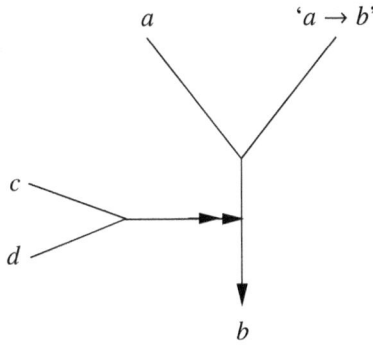

Fig. 8.19.

Remark 8.10 (Higher level joint attack networks). Of course we can allow attacks on joint attacks as well, as in Figure 8.19.

This can be iterated and can get quite involved, but this is the subject of another paper.

8.2.4 The logic programming approach

Let $(S, \mathbb{R})$ be a higher level extended program and let $S^* = S \cup \mathbb{R}$. We write a translation logic program π for $(S, \mathbb{R})$ as follows.

1. The literals of π are all the elements of S^*.
2. For each $\beta \in S^*$, if β is not attacked then take the clause β.
3. If β is attacked then Figure 8.15 is the typical situation.
   Take the clause

$$\beta \text{ if } \bigwedge (\neg a_i \vee \bigvee_{j=1}^{k(i)} (e_j^i, (a_i, \beta)) \wedge e_j^i))$$

We have
   a) $s \in S$ is in the ground extension iff $\pi \vdash s$
   b) $\mathbb{E}$ is an extension of S^* iff $\mathbb{E}$ is an answer set solution in π.
   The sets $\mathbb{E} \cap S$ gives for all extensions $\mathbb{E}$ give the extensions for S.

To be continued. What else are we going to do?

- We shall develop the mathematics underlying the three approaches (translation approach, labelling and extensions fixed points approach and the logic programming approach) in Section 3.
- In Section 4 we compare the three approaches and hope to show that all three approaches are equivalent.
  We also show that we favour the Dung approach for the case of level $(0, n)$ (Dung's level).
- Section 5 will deal for higher level networks where arrows can attack other arrows.
- Section 6 will place this chapter within the landscape of our general network methodology.

9

Temporal, Numerical and Metalevel Dynamics in Argumentation Networks

9.1 Introduction and Orientation

This chapter is an expansion of our 2005 Festschrift paper [37]. The argumentation community was not aware, until recently, of [37] and some of the ideas there were rediscovered in a related form by various people at a later date, see for example papers [263, 32, 33, 117, 118, 264, 252, 210, 237, 86, 87, 254, 94, 298, 56, 176]. However, [37] remains the most general approach, containing ideas still new to the community. See also [163], and Section 8.1 of the previous chapter.

In the first section we provide a landscape orientation and an introduction to give the reader a perspective, and in the last section we offer a comparison with the literature.

Our starting point is a network (S, R) where S is a set of arguments and $R \subseteq S \times S$ is an attack relation between arguments.

Figure 9.1 gives examples of two such networks.

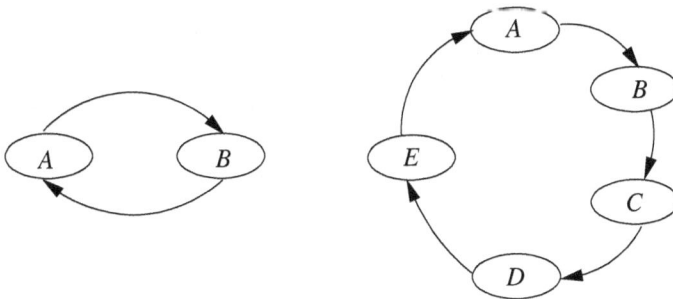

Fig. 9.1. Sample argumentation networks

The networks in Figure 9.1 are in fact loops involving even and odd number of arguments. An arrow $x \rightarrow y$ indicates an attack from node x to node y, i.e., that $(x, y) \in R$. There are three ways of looking at such networks.

1. Dung's original approach [114], where properties of subsets $E \subseteq S$ are considered and a set-theoretical definition of *extension* is put forward.

The basic notion is that of admissible extension. This is a subset $E \subseteq S$ such that
   a) E is conflict-free, i.e. for no $x, y \in E$ can we have $(x, y) \in R$.
   b) E is self-defending, i.e. for all $x, y \in S$, if $y \in E$ and $(y, x) \in R$ then there exists a $z \in S$ such that $(z, y) \in R$.
In the network on the left of Figure 9.1 there are the two extensions $E_1 = \{A\}$ and $E_2 = \{B\}$ as well as the empty extension $E_0 = \emptyset$. The network on the right of the figure has only the empty extension.

2. Recently, Caminada (see [75] and Chapter 2 for a survey) came forward with the brilliant idea of a Caminada labelling function $\lambda : S \longrightarrow \{\text{in, out, undecided}\}$ satisfying certain conditions that ensure a complete correspondence with Dung's extensions, see Chapter 2 and [82]. For example, for network on the left of Figure 9.1 we have the three functions λ_1, λ_2 and λ_0 below.

E_1	E_2	E_0
$\lambda_1(A) = \text{in}$	$\lambda_2(A) = \text{out}$	$\lambda_0(A) = \text{undecided}$
$\lambda_1(B) = \text{out}$	$\lambda_2(B) = \text{in}$	$\lambda_0(B) = \text{undecided}$

For network on the right of Figure 9.1, we have only the function λ such that $\lambda(A) = \lambda(B) = \lambda(C) = \text{undecided}$.

3. The third approach is Gabbay's equational approach [166, 167] discussed in Chapter 10 below. This approach views (S, R) as a mathematical graph generating equations for functions in the unit interval $U = [0, 1]$. Any solution f to these equations conceptually corresponds to an extension. Of course, the end result depends on how the equations are generated and we can get different solutions for different equations. Once the equations are fixed the totality of the solutions are viewed as the the totality of the equational extensions. Given the situation described in Figure 9.2, where $A \in S$ is any node and $X_1, \ldots, X_n$ are all of the nodes attacking A (i.e., $(X_i, A) \in R$), one equation we can possibly generate is Eq_{max}:

$$f(A) = 1 - \max_i\{f(X_i)\} \tag{9.1}$$

Another possibility is Eq_{inv}:

$$f(A) = \prod_{i=1}^{n}(1 - f(X_i)) \tag{9.2}$$

Gabbay has shown that in the case of Eq_{max} the totality of solutions corresponds to the totality of extensions in Dung's sense. The correspondence is best explained in terms of the Caminada labelling. We get the Caminada labelling through the correspondence below.

$f(A) = 1$ :: $\lambda(A) = \text{in}$
$f(A) = 0$ :: $\lambda(A) = \text{out}$
$0 < f(A) < 1$ :: $\lambda(A) = \text{undecided}$

The above discussion relates to abstract argumentation networks, where (S, R) is a pure graph. The nodes of the graph in this case have no internal structure and the relation $(x, y) \in R$ is given without any further structural explanation.

Caminada and other colleagues hold the view that such networks are too abstract and further structure is needed for practical applications. Caminada recommends a

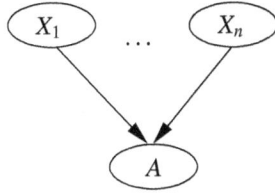

Fig. 9.2. Direct attacks on node A.

base logic L on some language and a base theory Δ. The arguments $A \in S$ are derived from proofs in L from Δ. Thus, A attacks B happens now for a reason: the contents of A attack (logically in L) the contents of B.

Another enrichment to purely abstract argumentation networks is to add a preference relation on S to tell us which abstract arguments are preferred to others.

A third alternative is obtained by adding a valuation $V : S \longrightarrow \text{VALUES}$, where VALUES is a set of abstract values, to give us a little more information about the elements of S. Mathematically one can get a preference relation out of a valuation and a valuation out of a preference relation. So let us view our position mathematically as being given a triple (V, S, R) where (S, R) is an abstract argumentation network and V a valuation function as before.

Given (V, S, R) we now have two main options.

1. to involve V in the definition of extensions
2. to define the extensions without using V, but involve it in the choice of extensions once these are generated. For example, if the valuation give us that A has the highest value in network R of Figure 9.1, we may adopt the new extension $\{A\}$.

Option 1. was widely used by Bench-Capon and colleagues, while option 2. is favoured by Caminada and his colleagues, as well as by Talmudic Logic, see our book [8] and also see what is discussed in this book in Chapter 19.

In the equational approach, it is reasonable to assume that V is a function $V : S \cup R \longrightarrow U$ and thus option 1. means that we use V in formulating the equations whereas option 2. means we use V to eliminate or modify some solutions f. For example, going back to Figure 9.2, we now also have the values $V(A), V(X_1), \ldots, V(X_n)$ and so we can either modify the equations and write, for instance, Eq_{inv} as

$$f(A) = V(A) \cdot \prod_{i=1}^{n}(1 - f(X_i))$$

or stick to the original equations but accept only solution functions f such that, for example, for all $A \in S$, $f(A) \geq V(A)$.

The approach we adopt in this chapter is numerical, that is we look at argumentation networks (S, R, V) where $V : S \cup R \mapsto [0, 1]$. In 2005, in our paper [37], this was a pretty innovative approach. In this framework we also put forward new concepts, like attack on attacks to any level, see Section 2 below, an idea which applies to any network (not necessarily numerical) and was later rediscovered by S. Modgil [260] and by P. Baroni [32].

A full discussion of the subject matter of higher level attacks, including priority can be found in Chapter 8, section 8.1.[1]

The study of this topic, in its present generality, has emerged from our previous research into argumentation frameworks (Gabbay and Woods [183, 181, 177, 179, 182] and Woods [339]), where we observed that circular loops in argumentation networks, no matter whether they are even or odd, can be viewed as, and possibly resolved as, local predator-prey networks. Our starting point is therefore a generalisation of abstract argumentation networks.

Let us now consider some examples.

The numerical view of networks allows us to connect argumentation networks with other major networks.

Example 9.1. Figures 9.3 to 9.5 are three examples of such networks. We assume that the arguments denoted by the nodes all have equal, and unit, strength, and similarly all attacks have unit strength. The result of a source node attacking a target (all of unit strength) is the refutation of the target. Evaluating the effects of such a network will determine which arguments survive, i.e. are *active*, and which ones are refuted, i.e. are *inactive*.

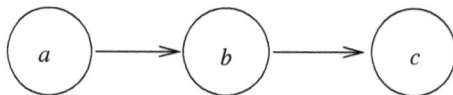

Fig. 9.3.

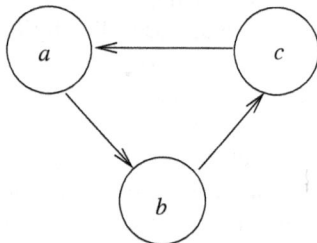

Fig. 9.4.

(a) The situation in Figure 9.3 is straightforward. The argument *a* is not attacked by anything, so it is evaluated as an active argument. Since *a* attacks *b*, *b* is a refuted argument and is evaluated as inactive, and so, as *c* is not attacked by an active argument, *c* is evaluated as active. We can write the net, evaluated, result of Figure 9.3 as {+*a*, −*b*, +*c*}.

---

[1] Note that the concept of attack on attack is geometrical and applies to any network and so priority really lies with [37].

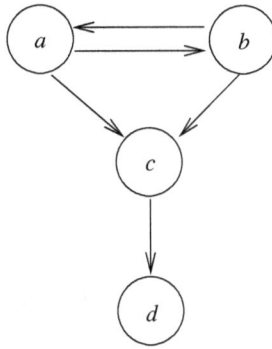

Fig. 9.5.

(b) The situation in Figure 9.4 is a complete loop. No argument can definitely be said to be active or inactive, and hence all three arguments denoted by a, b and c are evaluated as *unknown*. We write this as $\{?a, ?b, ?c\}$.

(c) The situation in Figure 9.5 is more interesting. Here, as a and b attack each other, we have those arguments evaluated as unknown, i.e. $\{?a, ?b\}$. Because of that, we also evaluate arguments c and d as unknown. If a or b were active, c would be refuted and evaluated as inactive and thus d would be evaluated as active. On the other hand, if neither a nor b were active, then c would evaluate as active, which in turn would cause d's evaluation to be inactive, (note that this does not give us an extension in the Dung-Caminada sense). However we can observe that both a and b attack c; so no matter which of a or b are active (i.e. whether we have $\{+a, -b\}$ or $\{-a, +b\}$), we always have $-c$, and so the net result could be taken to be $\{?a, ?b, -c, +d\}C$, (again note that this does not give us an extension in the Dung-Caminada sense). On the other hand, we might adopt the view that a, b cancel each other, in which case the net result would be $\{-a, -b, +c, -d\}$ (again note that this does not give us an extension in the Dung-Caminada sense).

Since circularity, loops and mutual attacks of arguments are very common in real life, it is obvious that much attention is required to resolving loops in argumentation networks. Abstract argumentation networks were generalised by Bench-Capon [43], where a colouring (representing the type of argument) was added to the network. The colours are linearly ordered by strength. A weaker coloured node cannot successfully attack a stronger coloured node. So a network with colours has the form (S, R, V) where, as before, $R \subseteq S \times S$ and V is a function giving, say, numbers to nodes: $V : S \mapsto Numbers$, and the numbers represent strength.

Thus in Figure 9.4 suppose $V(b) = r$ and $V(a) = V(c) = s$. Clearly if $r < s$, then the attack of b on c cannot take place and the net outcome of the network is $\{+b, +c, -a\}$. If $r > s$, then the attack of a on b cannot take place and the result is $\{+b, +a, -c\}$. If $r = s$, we get, as before, $\{?a, ?b, ?c\}$.

Note that technically the colouring function V is an instrument for cancelling attacks from some nodes to others. However, it is an instrument that requires restrictions. Not every proposed list of attacks to be cancelled can be implemented by a function V. Consider Figure 9.4. Suppose that we want to cancel all attacks. To cancel the attacks of a on b and of b on c we must have $V(a) < V(b) < V(c)$. By transitivity $V(a) < V(c)$, so the attack by c on a cannot be cancelled by V.

The main rationale behind the introduction of V is not necessarily the resolution of loops or cancellation of attacks, but the modelling of the intuition that arguments can be divided into kinds, and that some kinds of arguments are more important than others, some kinds of arguments are irrelevant[2], etc..

Remark 9.2. As we mentioned earlier, this chapter, following its earlier 2005 version [37], goes in the numerical direction. This enables us to connect argumentation networks with a diverse landscape of other very well known networks and the communities studying them. We note that following the idea of Caminada labelling, the traditional Dung approach to argumentation networks can also be considered as numerical. For a node x, we give the values $V(x) = 1$ for x being in, $V(x) = 0$ for x being out and $V(x) = 1/2$ for x being undecided. If $x_1, \ldots, x_n$ are all the attackers of y, then we let $V(y) = 1 - \max V(x_1), \ldots, V(x_n)$.

It is shown in [166, 167], see Chapter 10, that this particular three valued numerical approach is equivalent to the traditional Dung approach. In the sequel we shall always point out how the new ideas and definitions of this chapter manifest themselves in traditional Dung argumentation, by using the above special three valued numerical case.

The numerical view of networks allows us to connect argumentation networks with other major networks.

Let us list a few major ones.

A: For Argumentation theory, node b may represent an argument or statement and the supporting or attacking nodes a_i would be arguments in favour or against the argument represented by node b. The associated weights on the nodes and arrows would indicate the strength of each argument, the strength of each support or attack and the force of their presentation. The main problem for a given network of this type is to determine how to evaluate the effects of the various supports and attacks in an un-timed context with fixed weights as well as a timed context in which weights are treated as being time dependent. See Chapters 2–4.

B: The well-known Bayesian networks fall within our abstraction. Dependent nodes, i.e. target nodes, can be viewed as being evaluated through joint conditional probability on their parent nodes, i.e. source nodes. There are no individual weights on arrows unless there is independence of the conditional probabilities of the target on its sources.

F: In Flow networks, where directed edges may be labelled by various parameters, for example flow and capacity, typical problems are optimising other dependent parameters associated with the network, e.g. total cost, maximal flow, minimal circulation, etc.. In this context topological and graph-theoretic properties of the network can play a more prominent role.

---

[2] Attacking the character of a witness presenting an argument may or may not be relevant to the argument. For example, attacks on the private life of an expert witness are unlikely to be relevant to the credibility of her expertise and hence her arguments, whereas attacks on the credentials of the expert witness's expertise are more likely to be relevant.

N: In Neural networks, nodes represent neurons with capacity to fire on to their targets with adjustable weights on the edges. The main emphasis is on training the network, i.e. determining appropriate weights, through a variety of inputs for tasks in a given application. See [186].

E: In an Ecological setting, the network represents an ecology, where the nodes represent species and the arrows represent dependencies between species. An attack arrow signifies the source species being detrimental to the target species, e.g. source as predator, and the support arrow represents the source being beneficial to its targets, e.g. food. The parameters on nodes can represent relativized population numbers. The parameters on the edges denote interaction parameters between the species relevant to appropriate governing dynamical equations, e.g. the Logistic equation [255] or the Lotka-Volterra equation for predator-prey modelling [244, 332]. One of the major problems in this area is to identify solutions, e.g. steady-state, oscillatory, chaotic, etc., dependent on initial conditions. Clearly in this model there is already temporal dependence, even when the parameters do not change over time.

This chapter generalises argumentation networks in several directions.

1. It allows for nodes in argumentation networks not only to attack other nodes but also for support of other nodes. Moreover, we allow for varying strengths of attack and support. We further generalise the model by allowing that strengths of attacks or support themselves to be subject to attack or support.

2. It allows for the strengths of attack or support to be time dependent.
   This enables us to model the phenomenon of 'Let's lie low and wait for the argument to blow away'.

3. This chapter and several later chapters in this book also examine loop-resolution in argumentation networks, and explores similarities between such loops and predator–prey models in mathematical biology / ecology. One important outcome of such similarity is the possibility of getting extensions by choosing elements in a loop as starting points, assuming they are in, and propagating their attack recursively. In traditional Dung argumentation, this is how we can get the grounded extension. We start with the elements that are not attacked at all, these have to be in, and then we propagate their attacks recursively. See Section 2.

The plan of the chapter is as follows:

Section 9.2 will discuss "attack only" networks. There are three problems to be addressed in such networks. *Note that we study these problems in the numerical context only.*

1. The formal definition and motivation of a variety of attack networks.
2. The modes of attack, a discussion of various options as to how to evaluate the result of attacks.
3. The resolution of attack loops, such as Figures 9.4 and 9.5.

Section 9.3 is devoted to various methods for the resolution of attack loops. In the course of deciding how to handle loops, we explore formal connections between evaluating local loops occurring in numerical argumentation networks and determining steady-state solutions to network models occurring in mathematical biology / ecology.

Section 9.4 deals with numerical networks that allow for both attack and support arrows. We quickly ascertain the need to redefine the way in which attack and support are (numerically) carried out, and our considerations lead us to a surprising connection with the Dempster–Shafer rule and with the Cross-Ratio and projective metrics in geometry.

Section 9.5 deals with time-dependent attack and support of arguments. Here a connection with artificial intelligence time–action models is established, as well as a connection with dynamical systems and general temporal logics.

Section 9.6 discusses the results and indicates future directions of this work.

9.2 Attack-only networks with strength

In the 2005 version of this chapter, [37], this section was innovative in introducing the systematic study of numerical attacks. As we said before, the argumentation community overlooked paper [37] for some time and several authors introduced numerical features independently. We shall offer a comparison and discussion of these papers in the conclusion section. However, there are still features in this section which are new and apply equally to numerical and traditional Dung networks. We shall highlight and expand the discussion of these features. They have to do with the dynamic way in which we calculate extensions.

To explain this in principle, let our starting point be Figure 9.3. Viewed as a traditional Dung argumentation network we can calculate the grounded extension as follows:

Inductive algorithm

Step 1. a is in because it is not attacked.

Step 2. Let all x that are in execute their attack and take out their targets. In our case we get that b is out.

Step 3. Mark as in all nodes that are now not attacked. In this case c is in. *Step 4.* Repeat step 2 for the new nodes that are in.

We continue until there is no change. In our case we get a = in, b = out, c = in. In numerical form we get $a = 1, b = 0, c = 1$.

If we regard this network as numerical in our sense we need to give the nodes initial strength. Let us choose as an example $V(a) = 1, V(b) = 1, V(c) = 1$. We get Figure 9.6

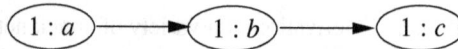

$$1 : a \longrightarrow 1 : b \longrightarrow 1 : c$$

Fig. 9.6.

To remain in the Dung argumentation world, we ignore the numbers suggested by V.

The innovative idea of this section is to use them.

Numerical Inductive Algorithm

Step 1. a and b and c are in because $V(a) = V(b) = V(c) = 1$.

Step 2. Let all x that are in execute their attack and take out their targets. We get b is out and c is out.

So the final extension we get is $a = $ in, $b = $ out, $c = $ out or numerically we get V' with $V'(a) = 1, V'(b) = V'(c) = 0$.

This is what we do in Example 9.5 for a much more complex network. Of course the very idea of these new networks like in Figures 9.8, 9.9, 9.3 and 9.13.

We now continue this section with an example motivating and explaining the idea of strength of a node and strength of attack on a node.

Example 9.3. Consider the election for Governor of California and the then candidate, actor Arnold Schwarzenegger. Let

$a = $ The candidate is alleged to have a certain attitude towards
    women, and to have behaved towards them accordingly.
$b = $ The candidate will run California very well.

These arguments may have different strengths based on evidence for case a and training and experience for case b. There is also another argument concerning the question of to what extent can a attack b. Is a relevant at all to b and to what degree? We represent this situation by the network in Figure 9.7

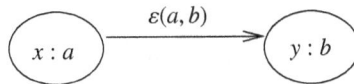

Fig. 9.7.

The value $\varepsilon = \varepsilon(ab)$, where (ab) is the attacking arc from a to b, represents the strength of the argument that a is relevant to b. It therefore can also be attacked, since one can argue against any connection between a and b.[3]

The perceptive reader might wonder how the two categories of weights (one on the nodes, one to the links) can be meaningfully instantiated for argumentation purposes.

The idea of associating some sort of strength with an argument (a weight on the nodes), has been proposed by some other approaches to argumentation, see for example [117, 118].

However, we may still ask, what do the weights on the links represent? If it is intended as some sort of contextual impact, then following Martin Caminada views, this factor is in fact grounded in interrelations between the content of arguments, which in

---

[3] The model also allows several attacks to emanate from the same argument, as in the figure below.

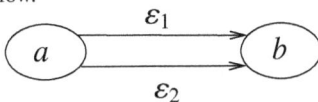

The idea here is that there are several different kinds of arguments as to why a is an attack upon b. This makes sense especially if a is a fact (see below). Such formal networks exist in the literature as *transition systems*, and the different arrows from a to b represent different actions, leading from state a to state b.

reality are structured and sometimes quite involved assertions, abstracted into atomic terms for argumentation frameworks purposes, see for example Chapter 4.

Can the weights attached to the links can adequately capture this?

our answer is no, numbers cannot adequately capture contextual relevance of attacks, (but see [252]). However, this chapter is about numerical networks and under this approach, this is the best that can be done.

Consider the situation described in Figure 9.8 where argument a has strength x. It attacks argument b, which initially has strength y.

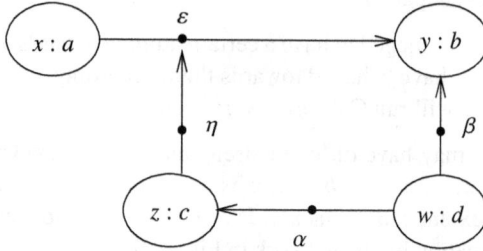

Fig. 9.8.

ε is the transmission factor, weakening b in a way that takes account of $x : a$.
b is also attacked by d with factor β.

However, factor ε is attacked by argument c, which is itself attacked by d, with transmission factor α.

This model has two innovations.

1. The strength of nodes and the transmission factor.
2. The idea that the transmission factor can itself be attacked. This is the so-called higher level attack[4], see [163] and Chapter 8 for discussion and further references. A restricted form of this idea was later independently discovered by Sanjay Modgil [260] and put to good use.

What kind of network does Figure 9.8 represent? First, note that the strength of nodes is actually a colouring of them. One might expect us to introduce a transmission factor between colours, then in Figure 9.7 ε could depend only on x and y. We choose to make ε depend on the nodes, taking into consideration that the transmission factor depends on the nature of the argument and not just on their strengths.

The option of attacking transmission factors enables us to delete attacks, one by one, by attacking (lowering) their transmission factor.

Example 9.4 (Modes of attack). Consider a simple numerical model. Assume all values are between 0 and 1. If a is an argument of strength x which is attacking an argument b of strength y, and the transmission rate is ε, then we get εx as the value

---

[4] We mention that the idea of higher level attacks is non other than the idea of reactive arrows in networks and Kripke models, originating in [147], and applied to argumentation networks in [37].

transmitted.[5] The question now is how does this value εx reduce the value y of b to a new value y'? We have two options. The first is that the attack reduces the value y of b in proportion, i.e. by εx. Thus the new value of b is $y(1 - \varepsilon x)$. The second option is that the new value of y is $y' = \varepsilon xy$. This second option makes sense if we view the attack of a on b as a pre-emptive protective measure, reducing a possible attack of b on a. If a is strong ($x = 1$) and $\varepsilon = 1$ then $1 - x\varepsilon = 0$ whereupon a destroys b. This is the previous option, being a genuine attack. However $\varepsilon xy = y$ when $\varepsilon = 1$ and $x = 1$; so b is not affected. But if x is small, then $y' = \varepsilon xy$ is small. So if b attacks a with transmission rate η, the value of this attack would be $1 - \eta y'$ and the attack would not be effective. Hence the second option can be used as a pre-emptive attack.

We now address the problem of combining attacks. In Figure 9.8, b is also attacked by d and this attack alone will reduce the value of b to $y(1 - \beta w)$. How do we combine them?

Here too there are two options:

1. Perform the operation of reduction consecutively (and commutatively), so that the new value of b after the joint attack is $y(1 - \beta w)(1 - \varepsilon x)$.
2. Add the two reductions, in which case the new value for b is the value $\max\{0, y - y\varepsilon x - y\beta w\} = \max\{0, y(1 - \varepsilon x - \beta w)\}$.

The advantages of option 1 are that it is simple and that the combination is independent of how the attack is calculated. Another major advantage is that it is compatible with ordinary Dung argumentation when the numerical values are restricted to $\{0, \frac{1}{2}, 1\}$. See Remark 9.2. For example, this can give as the new value of b the combination $\varepsilon x(1 - \beta w)$.

Example 9.4 above has put forward just one mode of attack. There are many other possible modes. Additional possibilities will be examined in Section 4, in conjunction of models with both attack and support. For non-numerical logical modes of attack, see Chapter 4.

In general, we have the situation shown in Figure 9.9. In this case, we require the following function: If b has value y and if $x_1 : a_1, \ldots, x_n : a_n$ attack $y : b$ with strengths $\varepsilon_1, \ldots, \varepsilon_n$ resp., then we need a function $\mathbf{f}$ such that the new value of node b is $y' = \mathbf{f}(y, x_i, \varepsilon_i)$. This situation is reminiscent of Bayesian networks, where $\mathbf{f}$ is the conditional probability of b on $a_1, \ldots, a_n$.[6]

We adopt option 1 as our mode of attack mainly because of its compatibility with ordinary Dung networks as discussed in Remark 9.2. So the new value $y' = V(b)$ in Figure 9.9 is

$$y' = y(1 - \varepsilon_1 x_1) \ldots (1 - \varepsilon_n x_n)$$
$$= y \prod_i (1 - \varepsilon_i x_i).$$

---

[5] In general the value transmitted should be a function $\tau(\varepsilon, x)$ of ε and x, monotonically increasing in ε and x, with $\tau(0, 0) = 0, \tau(1, 1) = 1$. We choose multiplication here by way of example.

[6] In Bayesian nets there are no $\varepsilon_1, \ldots, \varepsilon_n$. x_i are the probabilities associated with the nodes a_i and $\mathbf{f}$ is the conditional probability of node b relative to all the a_i. Thus the probability y of b can be calculated.

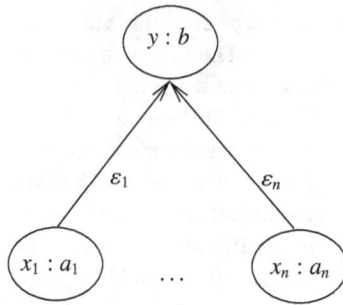

Fig. 9.9.

The magnitude $\Delta^- y$ which y decreases is

$$\Delta^- y = y - y' = y(1 - \prod(1 - \varepsilon_i x_i)).$$

Example 9.5. We calculate the transmission of values in Figure 9.8.

This is to be done in steps, by calculating a valuation function V on the nodes of Figure 9.8. At step $n, n \geq 1$ we define a partial valuation function V_n on the nodes, with some nodes being declared as having a final updated value. Our initial starting function is V_0, with $V_0(a) = x$, $V_0(b) = y$, $V_0(c) = z$, $V_0(d) = w$, $V_0(ab) = \varepsilon$, $V_0(db) = \beta$, $V_0(dc) = \alpha$ and $V_0(c(ab)) = \eta$. V_0 records the values given by the network graphs in Figure 9.8.

Step 1: The final updated value V_1 of node d is w, as it is not attacked by anything. Write $V_1(d) = w$. Similarly $V_1(a) = x$. We write V_1 because this is the value obtained as final at Step 1.

Step 2: The new value V_2 of nodes c and b are $V_2(c) = z(1 - \alpha w)$, $V_2(b) = y(1 - \beta w)$. Of course since nodes a and d have already obtained their final value, we can write: $V_2(a) = V_1(a)$, $V_2(d) = V_1(d)$. Node a cannot transmit, at this time (step 2), because we know from the figure that ε is being attacked, and so we need to wait for its value to change. Only when $V(ab)$ gets its final value will a be able to transmit. Thus node (ab) cannot get a final value at this step.

Step 3: The new value V_3 of the transmission connection (ab) is

$$V_3(ab) = \varepsilon(1 - \eta V_2(c))$$
$$= \varepsilon(1 - \eta z(1 - \alpha w)).$$

Of course, $V_3(a) = V_2(a)$, $V_3(d) = V_2(d)$, $V_3(c) = V_2(c)$, and $V_3(b) = V_2(b)$.

Step 4: Now node a can transmit to node b. This gives

$$V_4(b) = V_2(b)(1 - V_3(ab) \cdot x)$$
$$= y(1 - \beta w)(1 - \varepsilon x(1 - \eta z(1 - \alpha w))).$$

Of course, $V_4(a) = V_3(a)$, $V_4(d) = V_3(d)$, $V_4(c) = V_3(c)$ and $V_4(ab) = V_3(ab)$. Note that node b has had its value changed in bits and pieces. First, it was changed at Step 1 and then at Step 4. This is all right for the current way of changing values, because it is commutative and cumulative. However, the general definition will not allow for this!

This kind of model contains the traditional one as a special case, where all values are taken to be 1 and where there are no attacks on transmissions. Let us see what Figure 9.8 becomes in this case. Consider Figure 9.10 and note that it reduces to Figure 9.11.

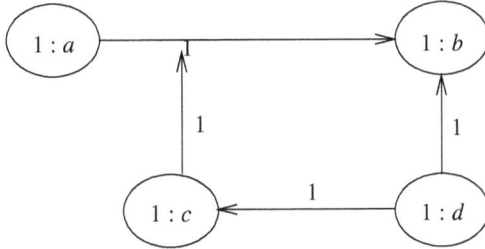

Fig. 9.10.

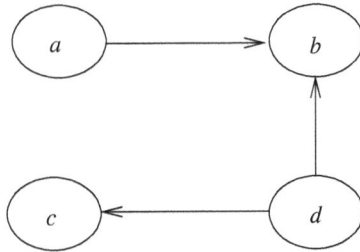

Fig. 9.11.

We can now give a definition of value propagation for acyclic networks. The general treatment of cycles, or loops, in networks will be addressed in section 9.3.

To give a definition we need to agree on the representation of the network. Let's do it for the case of Figure 9.8. We need a set of atomic nodes A. In the case of Figure 9.8, $A = \{a, b, c, d\}$.

To represent the attack of atomic x on y, i.e. the arrow from x to y, we write the expression $x \leftrightsquigarrow y$ (called *attacks*).[7] In Figure 9.8, we have attacks $a \leftrightsquigarrow b, d \leftrightsquigarrow c$ and $d \leftrightsquigarrow b$.

These arrows represent the attacks from a to b, d to c and d to b respectively. One of these attacks, namely $a \leftrightsquigarrow b$, is itself attacked by c. This is represented by the expression $c \leftrightsquigarrow (a \leftrightsquigarrow b)$.

Note that we *cannot* write an expression of the form $(x \leftrightsquigarrow y) \leftrightsquigarrow z$. This would mean that the fact that there is an attack from x to y is in itself an attack on z.[8] We are not saying that such reasoning does not exist. We deal with it in the context of fibring

---

[7] When the arrow is an attack we use the curly arrow instead of a straight one headed arrow $\rightarrow$. When it is a support (see Section 4) we use the double arrow $\twoheadrightarrow$.

[8] For example, if z is a supporter of a theory of love and peace, then any x attacking y is in itself an attack on z. Also in those applications where it makes sense to traverse the network

networks, see [164]. In other words, a whole network can be embedded as a node and attack another node.

Figure 9.8 can be represented by the set of nodes and attack arrows.

$$T = \{a, b, c, d, a \leftrightarrow b, d \leftrightarrow b, d \leftrightarrow c, c \leftrightarrow (a \leftrightarrow b)\}.$$

Note that this set T has the property that if $x \leftrightarrow y \in T$, then $x \in T$ and $y \in T$. What we still need are the numbers (valuations) in the figure. This we can represent by a function $V : T \to \mathbb{R}$, where $\mathbb{R}$ is the set of real numbers.

We are now ready for a formal definition.

Definition 9.6. [9] *Let A be a set of atomic nodes.*

1. *Define the notion of a attack arrow based on A as follows:*
   - *$a \leftrightarrow b$ is an attack arrow if $a, b \in A$.*
   - *$a \leftrightarrow x$ is an attack arrow if $a \in A$ and x is an attack arrow.*
2. *Let T be a set of attack arrows and atomic nodes. We say that T is an attack network if the following holds*
   - *$x \leftrightarrow y \in T$ implies $x \in T$ and $y \in T$.*
   We say that T is finitely branching (in the outgoing direction) if for every $t \in T$ $\{a|(a \leftrightarrow t) \in T\}$ is finite.
3. *A valuation function on T is a function $V : T \to \mathbb{R}$.*
4. *An attack network with a valuation is a triple $N = (A, T, V)$, where A is a set of atomic nodes, T is an attack network based on A and V is a valuation on T.*
5. *Let $\mathbf{f}$ be a functional giving for each string of real numbers of the form $(y, x_1, \ldots, x_n, \varepsilon_1, \ldots, \varepsilon_n)$ a new real number $y' = \mathbf{f}(y, \bar{x}_i, \bar{\varepsilon}_i)$ (where $\bar{z}_i$ abbreviates $z_1, \ldots, z_n$, for $z = x$ or $z = \varepsilon$). Note that n is arbitrary. We assume $\mathbf{f}$ to be continuous and symmetrical in the pairs of variables $(x_i, \varepsilon_i), i = 1, \ldots, n$ and generally nice.[10,11] For example, let $\mathbf{f}(y, \bar{x}_i, \bar{\varepsilon}_i) = y \prod_{i=1}^{n}(1 - \varepsilon_i x_i)$. See Section 4.2 for more options.*
6. *An argumentation attack model is a pair $(N, \mathbf{f})$, where N and $\mathbf{f}$ are as above.*

Examples 9.7 *Let us look at some examples. Consider Figure 9.12 in which a attacks b but also attacks its own attack. This is a case of a self defeating attack of a on b.*
   We have $T = \{a, b, a \leftrightarrow b, a \leftrightarrow (a \leftrightarrow b)\}$ and

---

along its edges, then traversing the edge from x to y may trigger an attack on z. The formal machinery of representing such attacks exists in Chapters 7–8.

[9] Compare with Definition 8.1, which deals with ordinary argumentation networks with $\{0, 1\}$ values.

[10] By restricting $\mathbf{f}$ to finite sequences, we are forced to impose the condition of finitely branching on T in Definition 9.11 below. However, $\mathbf{f}$ can be more general, for example, we can take

$$\mathbf{f}'(y, S) = \inf\{\mathbf{f}(y, \bar{x}, \bar{\varepsilon}) \mid (\bar{x}, \bar{\varepsilon}) \in S\},$$

where S can now be an infinite set. This will allow us more freedom in Definition 9.34 below.

[11] When we say $\mathbf{f}$ is symmetrical, we mean that for any permutation σ of $\{1, \ldots, n\}$ we have

$$\mathbf{f}(y, \bar{x}_i, \bar{\varepsilon}_i) = \mathbf{f}(y, \bar{x}_{\sigma(i)}, \bar{\varepsilon}_{\sigma(i)}).$$

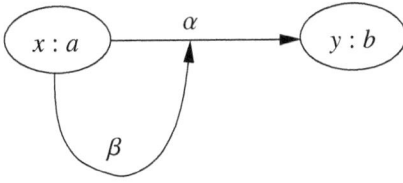

Fig. 9.12.

$$V(a) = x, V(b) = y,$$
$$V(a \rightsquigarrow b) = \alpha \text{ and}$$
$$V(a \rightsquigarrow (a \rightsquigarrow b)) = \beta$$

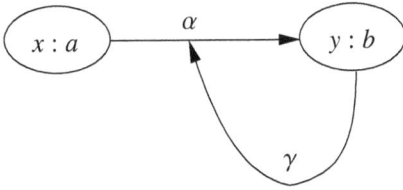

Fig. 9.13.

We can compare Figure 9.12 with Figure 9.13. In Figure 9.13 we can interpret γ as a feedback loop, attacking and reducing α. The weaker the argument b is the less we want to spend effort attacking it.

Definition 9.8 (Syntactic acyclicity[12]). *Let T be an attack network. Define $R_T \subseteq A^2$ as follows:*

$$aR_T b \text{ iff } a \rightsquigarrow b \in T \text{ or for some } x \in A, a \rightsquigarrow (x \rightsquigarrow b) \in T.$$

Let R_T^* be the transitive closure of T. We say T is syntactically acyclic iff there is no $x \in A$ such that $xR_T^* x$.
If $N = (A, T, V)$, we say N is syntactically acyclic if T is such.

Example 9.9. Figure 9.14 is cyclic while Figure 9.15 is acyclic and finitely branching.

Example 9.10. Figure 9.16 is cyclic syntactically, but is acyclic after evaluation using V.
Note that although the network is syntactically cyclic, since $V(\beta) = 0$, it is as if $b \rightsquigarrow a$ does not exist in T. We shall deal with this kind of *semantic* acyclicity later.

Definition 9.11 (Value propagation). *Let $(N, \mathbf{f})$ be a model, where N is syntactically acyclic and finitely branching. We shall propagate the values V through the model using $\mathbf{f}$. We do this in waves. Wave m will define values $V_m(x)$ for some $x \in T$ and x is then referred to as an updated element with the updated value $V_m(x)$.*

---

[12] The notion of semantic acyclicity is discussed at the end of the section.

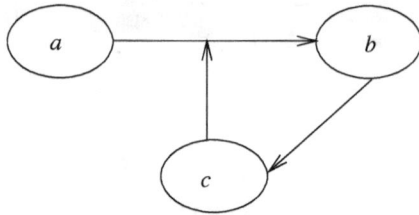

Fig. 9.14.

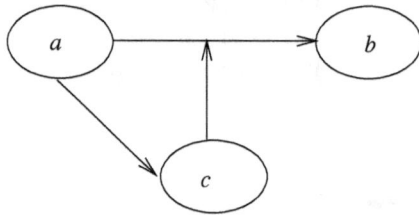

Fig. 9.15.

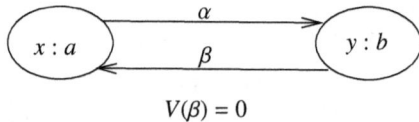

$$V(\beta) = 0$$

Fig. 9.16.

Wave 0

An element $a \in T$ is said to be syntactically free of attack if for every $e \in A$ we have $(e \hookrightarrow a) \notin T$. Let it be said that the updated elements of Wave 0 are the free of attack elements and let the updated value V_0 be $V_0(a) = V(a)$, for an updated a of wave 0.

Wave $n + 1$

Assume we have defined the updated elements of waves $k \leq n$ and their updated value V_k. Let b be any element and let $a_1, \ldots, a_m$ be all, if any, elements of T such that $(a_i \hookrightarrow b) \in T$. Assume for each i, that a_i, as well as $a_i \hookrightarrow b$, were updated at some earlier wave $k_i \leq n$ and $l_i \leq n$ respectively.

Define

$$V_{n+1}(b) = \mathbf{f}(V(b), \bar{V}_{k_i}(a_i), \bar{V}_{l_i}(a_i \hookrightarrow b)).$$

When the network is finite, the algorithm updates all the nodes and terminates with some $V_n = V'$ in quadratic time.[13]

Example 9.12 (Figure 9.8 revisited). Let us examine the network of Figure 9.8 again. We are listing the updated elements. Let us compare with Example 9.4.

Wave 0

---

[13] Consider a linear network of n nodes with the following connection structure. Label the nodes from 1 to n. Node i attacks all nodes numbered $> i$. Wave 0 will have to search n nodes. Wave $i < n$ will have to search $n - i$ nodes. The sum of all waves is $n(n-1)/2$.

$$w : d, x : a, \beta : d \twoheadrightarrow b,$$
$$\alpha : d \twoheadrightarrow c, \eta : c \twoheadrightarrow (a \twoheadrightarrow b).$$

Wave 1

$$z(1 - \alpha w) : c$$

Note that the only updated element in this wave is c. b is not updated because not all of its attackers (namely a) have been updated. In our earlier computation we did attack b at this stage, but we cannot do that under our current definition. We will not get a different result because our function **f** launches the attacks from separate nodes independently, cumulatively and commutatively.

Wave 2

$$\varepsilon(1 - \eta z(1 - \alpha w)) : a \twoheadrightarrow b.$$

Here $a \twoheadrightarrow b$ is being updated.

Wave 3
Now we can update b. We get

$$y(1 - \beta w)(1 - \varepsilon x(1 - \eta z(1 - \alpha w))) : b$$

Definition 9.13. *Let $(N, \mathbf{f})$ be a finite model. Propagate V using **f** in waves as defined above. Let the new valuation $V'(a), a \in T$, be the updated value of a. We call V' the result of the waves of attack in the network. Note that the propagation is executed only once.*

Remark 9.14 (Networks with values in $\{0, 1\}$). For such networks more can be said. Consider the situation in Figure 9.17

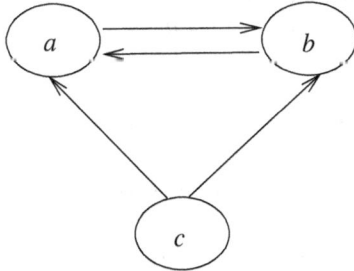

Fig. 9.17.

Although the network is not syntactically acyclic, it is what we can call semantically acyclic. Let us do waves.

Wave 0
The element c is updated to having value $+1$, since it is not attacked by anything.

Wave 1
The nodes a and b are attacked by c and therefore updated to value 0. We have a clear updated semantic solution V' in this case.

This works only when the functions $\mathbf{f}$ can give value to arguments which are not evaluated themselves. Let us do this formally.

We have $V_0(c) = 1$ and just let $V_0(a) = x$, $V_0(b) = y$ where x and y are any values.

Wave 0

$V_0(c) = 1$ because c is not attacked.

Wave 1

$$V_1(c) = V_0(c)$$
$$V_1(a) = \mathbf{f}(x, y, 1) = 0$$
$$V_1(b) = \mathbf{f}(y, x, 1) = 0$$

We accept $V_1(a)$, $V_1(b)$ as final updated values because for the value $V_0(c) = 1$, the function $\lambda x \lambda y \mathbf{f}(x, y, 1)$ is identically 0, no matter what x, y are.

Definition 9.15 (Semantic acyclicity Version 1). *Let $(N, \mathbf{f})$ be a model where N is finitely branching but not necessarily syntactically acyclic. We say that $(N, \mathbf{f})$ is semantically acyclic Version 1 if we can propagate V in waves as follows:*

Wave 0

An element b is said to be semantically (version 1) "free" of attacks if either (a) or (b) hold.

(a) *b is syntactically free of attack, i.e. for every $e \in A$ we have $e \leftrightarrow b \notin T$. In this case we set $V_0(b) = V(b)$ as the updated final value of b.*

(b) *Let $a_1, \ldots, a_m$ be all elements of A such that $a_i \leftrightarrow b \in T$. Assume that $y = \mathbf{f}(V(b), V(a_i), V(a_i \leftrightarrow b))$ depends on $V(b)$ only. Then let $V_0(b) = y$ be the updated value of b.*

Wave $n + 1$

Assume that we have defined the updated elements of waves k for all $k \leq n$. Let b be any element such that for some $a_1, \ldots, a_m$, $a_i \leftrightarrow b \in T$ for $i = 1, \ldots, m$. Consider the value $y = \mathbf{f}(V(b), x_1, \ldots, x_m, w_1, \ldots, w_m)$.

Assume that for any i, j such that the value x_i of a_i or w_j of $a_j \leftrightarrow b$ respectively has not been updated at some earlier stages k_i or $l_j \leq n$, respectively, we have that y does not depend on x_i, w_j respectively.

Then we say that b is updated to the value $V_{n+1}(b)$ at stage $n + 1$ and the value is y, where:

$$y = \mathbf{f}(V(b), V_{k_i}(a_i), V_{l_j}(a_j \leftrightarrow b))$$

where $V_{k_i}(a_i)$, $V_{l_j}(a_j \leftrightarrow b)$ are the updated values of a_i and respectively $a_j \leftrightarrow b$ if such values exist and otherwise we don't care. We can do this because we assumed that y does not depend on the nodes that were not updated).

If the algorithm terminates giving updated values to all nodes of T we say the network is semantically acyclic version 1.

Example 9.16. Let us check some of our previous figures for semantic acyclicity. Figure 9.4 is cyclic because for example

$$\mathbf{f}(V(b), V(a)) = V(b)(1 - V(a))$$

depends on $V(a)$ unless $V(b) = 0$. Similarly, Figure 9.5.

Example 9.17. Consider the situation of Figure 9.12. Assume that in this figure all initial values are 1, i.e. $x = y = \alpha = \gamma = 1$. So $V \equiv 1$.

Let us try to update by waves and see what happens.

Wave 0

The final updated value for a in this wave is $V_0(a) = 1$.

Wave 1

We cannot get a value for b, even semantically version 1, because the value of b is

$$V_1(b) = V(b)(1 - V_0(a)V_0(a \twoheadrightarrow b)).$$

This value depends on the value of $V_0(a \twoheadrightarrow b)$ and this value is not finally updated.

So we cannot continue. However, there is something we can do. The initial value $V(a \twoheadrightarrow b)$ is 1. So we can propagate the attack from a to b and get a new value $V_1(b) = 0$. With this value, b cannot attack $a \twoheadrightarrow b$ and so the initial value 1 of $a \twoheadrightarrow b$ remains unchanged. So we do get a stable situation and not an oscillating situation.

This example suggests another definition of semantic acyclicity, a new version 2.

Definition 9.18 (Semantic acyclicity Version 2). *Let $(N, \mathbf{f})$ be a model where N is finitely branching but not necessarily syntactically acyclic. We say that $(N, \mathbf{f})$ is semantically acyclic version 2 if we can propagate V in waves as follows:*

Wave 0

Let b be an element which is semantically free of attack according to version 1 as in wave 0 of Definition 2.13. Let b be updated to $V_0(b) = V(b)$.

For any other element c which is not free let $V_0(c)$ be not yet defined.

Wave $n + 1$

Assume V_n has been defined, giving values to some elements of T. Further assume that some of these values are declared final and the rest of these given values are declared temporary. For some elements of T V_n may not give a value at all.

Let b be any element such that for some $a_1, \ldots, u_m, a_i \twoheadrightarrow b \in T, i = 1, \ldots, m.$

Consider $w = V_m(b), x_i = V_n(a_i)$ and $y_i = V_n(a_i \twoheadrightarrow b), i = 1, \ldots, m.$

Some of these values are final, some are temporary and some are undefined because V_n does not give a value. We assume that b is such that for some a_i or some $a_i \twoheadrightarrow b$, V_n does give some value. If V_n does not give any value to any of a_i and to any of $a_i \twoheadrightarrow b$ we do not touch b at this wave. This condition we call the slow propagation condition and it is intended to avoid un-intuitive situations like the one in Example 2.17.

So we can assume that for some of w, x_i or y_i, $i = 1, \ldots, m$, $j = 1, \ldots, m$, V_n does give a value final or temporary. To be in a situation where all variables w, x_i, y have defined values, let the value for w, x_i and y_j be $V(b)$, $V(a_i)$ or $V(a_j \twoheadrightarrow b)$ in case V_n is not defined and does not give them values.

Thus we can now calculate

$$z = \mathbf{f}(w, x_1, \ldots, x_m, y_1, \ldots, y_m)$$

We distinguish three cases.

Case 1

V_n does not give a value for b: In this case declare $V_{n+1}(b) = z$ and declare this value as temporary.

Case 2

$V_n(b) = w$ and $z = w$: In this case reassert $V_{n+1}(b) = V_n(b)$ as final or temporary, giving b the same status as that which we have for V_n.

Case 3

$V_n(b) = w$ and $z \neq w$. In this case we stop and say that $(N, \mathbf{f})$ is not semantically acyclic version 2.

We say that $(N, \mathbf{f})$ is semantically acyclic version 2 if for some n, V_n is total and $V_n = V_{n+1}$.

Example 9.19. Consider again the situation in Figure 9.5. Let us check for semantic acyclicity version 2 without the slow propagation condition. Assume V gives values 1 and all transmission values are 1.

Wave 0

V_0 cannot be defined on any node.

Wave 1

We use the values of V, since V_0 gives no values and since b is attacked by a we get that its V_1 temporary value is 0. Similarly $V(a) = 0, V_1(c) = V_1(d) = 0$. All temporary values.

Wave 2

The values remain 0 and so the network stabilises at $V_2 \equiv 0$.

The above calculation was without slow propagation.

If we invoke the slow propagation condition, Wave 1 cannot be executed. So we get no values.

Example 9.20. Consider the situation of Figure 9.14. Let us test for semantic acyclicity version 2. Assume all transmission values and all nodes values are 1, i.e. $\lambda x V(x) \equiv 1$.

Wave 0

We have $V_0(a) = V(a) = 2$. We cannot have any more values.

Wave 1

We have $V_1(b) = 0$ because $V_0(a) = 1$ and we use the transmission value $V(a \leftrightarrow b) = 1$. We cannot get a temporary value for c because we do not have any value for b at this wave.

Remark 9.21. Let us assess what we have so far. Consider a network $(N, \mathbf{f})$ and consider maximal cycles with respect to the relation R_T of Definition 2.6. If $\mathbb{C}$ is such a cycle, then if there is an element out of the cycle with an arrow into it, then there is a chance for the waves of the evaluation of Definition 2.16 to semantically resolve the cycle. The problem arises when the cycle $\mathbb{C}$ has no attack arrows leading into any of its members. Definition 2.16 will not be able to do anything, as we see in Example 2.17 which deals with the situation in Figure 9.5. We therefore need a new idea or methodology for resolving complete cycles and giving some values to its nodes. This new idea and these new values come from the treatment of loops in ecological systems and we address this in Section 3.

Remark 9.22 (Equational Labelling). Consider the special case of a network with no higher level attacks. This is just a set S with a binary relation on it. We allow for nodes to have strength as real values between 0 and 1 inclusive, and allow for full transmission. In terms of the basic situation described in Figure 9.9, we have $\varepsilon_1,..., \varepsilon_n$ are all 1 and the value of y is given by the formula

$$y = e(x_1)...e(x_n) \qquad (9.3)$$

where the function e is defined as $e(x) = 1 - x$.

Note that $y = 1$ only if all the x_i are 0.

We can ask for a function V giving values to nodes and satisfying the equation above for any instance of Figure 9.9 in the network such that $a_1, ..., a_n$ are ALL the nodes attacking b. We can use linear programming methods to find such a V.

This approach was introduced in [164], see Chapter 7, and fully developed in [166, 167], see Chapters 10–12, and generalises the Caminada labelling approach, [82]. It has two advantages

1. We can let V take values in any commutative and associative multiplicative algebra. So for example the $\{0, 1, ?\}$ valued Caminada labelling can be viewed as equational labelling in the three valued commutative associative algebra satisfying the axioms
$$x1 = 1x = x$$
$$x? = ?x = ?$$
$$0x = x0 = 0$$
We let $e(1) = 0$, $e(0) = 1$ and $e(?) = ?$. See Chapters 2 and 7.

2. We can write general non local equations or constraints which V should satisfy and seek a solution. This allows for more general networks. Equation 9.3 is local, it involves a point and its immediate neighbours, it is not the most general type of equation.

Note that Figure 9.4 has the $a = b = c = ?$ solution in the 3-valued algebra and the $a = b = c = \frac{1}{2}$ in the real numbers.

9.3 Handling loops — ecologies of arguments

This section is about handling loops, both in numerical argumentation netwroks and in general numerical networks, especially ecological networks.

9.3.1 Loops in argumentation networks

In the case of argumentation the situation is quite simple. We make a distinction between even and odd loops. The even loops can give rise to non-trivial extensions, while the odd loops can only give rise to the trivial extension all undecided.

The situation can be frustrating when we have an odd loop at the top of a network (i.e. none of the members of the loops is being attacked from outside the loop). Figure 9.18 is a typical situation

The only extension we can have here is all undecided, because the top odd loop $\{a, b, c\}$ is a bottleneck.

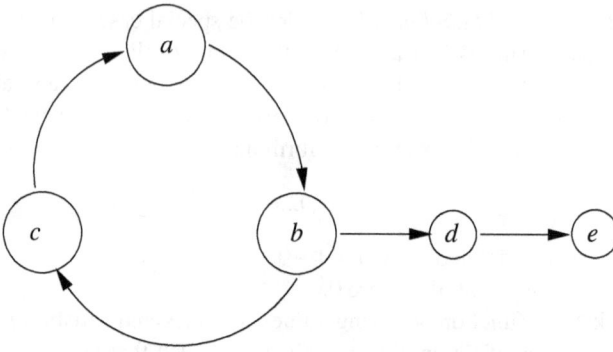

Fig. 9.18.

Recently authors looked for ways of breaking such bottlenecks by putting forward new semantics. See for example [7, 31, 53, 192, 30]. The CF2 semantics for example, will resolve the loop of Figure 9.18 by taking maximal conflict free subsets of the loop instad of admissible subsets. This would give us the following CF2 extensions:

$$\{b, e\}$$
$$\{a, d\}$$
$$\{c, d\}.$$

When we take a numerical point of view, we get equations to solve, and the solutions to the equations are the extensions. The equations in this case are (adopting option 1 discussed in Section 2 after Example 9.4) as follows:

1. $V(a) = 1 - V(c)$
2. $V(b) = 1 - V(a)$
3. $V(c) = 1 - V(b)$
4. $V(d) = 1 - V(b)$
5. $V(3) = 1 - V(d)$.

The only solution is

$$V(a) = V(b) = V(c) = V(d) = V(e) = \tfrac{1}{2},$$

which corresponds to all undecided. So the problem is not solved but is less frustrating, as we have more scope of writing modified equations, getting new solutions.

The traditional Dung approach gives a limited number of parameter concepts to play with. We have conflict-free, skeptical, credulous, attack, defence,etc., and can modify these to get ourselves out of odd loops. In the numerical approach we have the entire landscape of (what is known in fuzzy logic as) De Morgan norms to use. See Chapter 10 for a wider discussion. The De Morgan norms keep compatibility with traditional Dung approach to argumentation networks.

This numerical point of view certainly connects with other types of networks, such as ecological networks. The equations are different there, not compatible with argumentation, but the scope for resolving loops is much richer, as we discuss in this section.

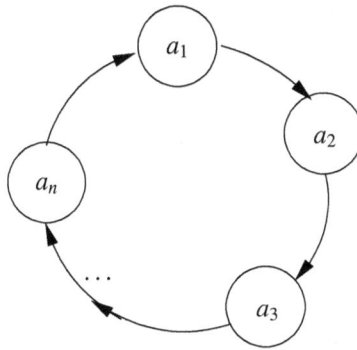

Fig. 9.19.

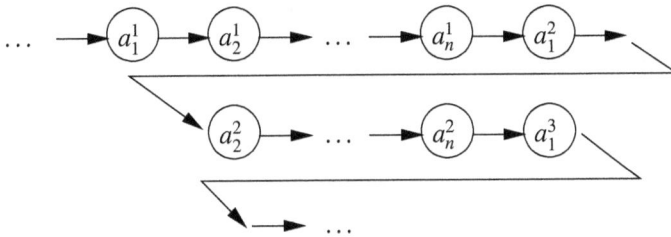

Fig. 9.20.

We add one more comment about unfolding loops. Consider the loop in Figure 9.19. In this figure n may be odd or even. When we unfold it, we get Figure 9.20.

There is no distinction in Figure 9.20 between odd or even, except that by writing copies a_k^i of the letter a_k, we indicate that we want either all a_k^i for all i to be in or all a_k^i for all i to be out.

With this restriction, if n is even, there is a solution and if n is odd, there is no solution.

Now that we have a general orientation between loops in traditional Dung argumentation and loops in general numerical networks, we can start the detailed discussion of this Section. We have encountered loops in Example 9.1 (b) and (c). In Figure 9.5 of (b), we need to resolve the loop $\{?a, ?b\}$ in order to propagate values to c. So technically all we need is some loop resolving compromise assignment of values to a and to b, and then the algorithm of Definition 9.11 can be invoked. How do we obtain such value?

The values we give to the loop depend on our interpretation of it. Hints for possible interpretations can be obtained from other possible interpretations of the entire network regarded as a mathematical entity. We shall therefore open this section by putting forward several points of view as to the meaning of labelled networks and their internal loops, which will then lead to ways of dealing with their loops.

To begin our discussion, consider the following Figure 9.21 which is the typical general numerical case of an even loop.

Let $\mathbf{f}_1(y, x, \varepsilon), \mathbf{f}_2(x, y, \eta)$ be the two transmission functions. We observe the following:

1. Figure 9.21 describes a syntactical loop.

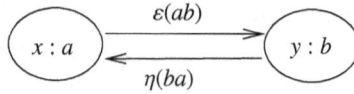

Fig. 9.21.

2. Depending on the values x, y, ε, η and depending on the functions $\mathbf{f}_1$ and $\mathbf{f}_2$, Figure 9.21 might not be a loop semantically. For example, if $x : a$ is much stronger than $y : b$ or if $\varepsilon = 0$ then this might not be a loop.

The general method is to find a solution for the pair of equations

$$y = \mathbf{f}_1(y, x, \varepsilon)$$
$$x = \mathbf{f}_2(x, y, \eta)$$

or, more generally, for the case of a maximally strongly-connected component of the network involving n nodes $x_1..x_n$, we solve the vector equation

$$\mathbf{X} = \mathbf{F}(\mathbf{X}, \overline{\overline{\Pi}})$$

We can then use a solution (if it exists) as the values of the nodes in the context of the original network. Thus for the loop of Figure 9.4 mentioned above, values for the nodes a and b can be determined using this method. We get the value for Figure 9.4 as $\frac{1}{2}$ for all nodes.

The reader should note that Brouwer fixed point theorem (see Wikipedia for reference) guarantees at least one solution to the above equation. There may be many solutions. It is up to us to decide whether we want to generate all solutions or seek a solution with certain properties/constraints (some constraints may not allow for a solution). If we have certain solutions in mind, then special numerical analysis techniques may have to be employed. See Gabbay's papers [166, 167] and Chapter 10 below. These options in seeking solutions is reminiscent of the discussion in Logic Programming (the so-called great logic programming schism).

Note that in Appendix B we turn to possible interpretations of loops in other kinds of networks. We examine these different interpretations in order to assess the relevance of their solutions to the situations arising in argumentation. We will also examine whether the argumentation network point of view may be applied to other networks.

9.3.2 Unfolding loops

There are various ways of treating loops.

(a) We can unfold them as done in, say, modal logic.
(b) We can let node a attack b, calculate the new value and then let b attack a, calculate the new value and then let a attack b and so on. This we call the parasite way of unfolding a loop.
(c) We can let a and b attack each other simultaneously, calculate the new values and then let them attack again and again. This is the predator-prey way of unfolding a loop.

(d) We can assume a steady state and solve the equations suggested by the geometry of the loop.

Let us now turn to Figure 9.21 and see what are our options for dealing with this loop.

Our first attempt at a solution is to regard (ab) and (ba) as the same channel and read the loop as feedback loops (see Appendix B). So a pushes εx towards b and b pushes ηy towards a. The net result is $(\varepsilon x - \eta y)$ in the direction of the positive value. So assuming $\varepsilon x \geq \eta y$ we get that Figure 9.21 is essentially reduced to Figure 9.22.

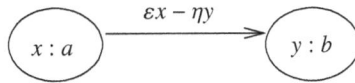

Fig. 9.22.

The solution is not satisfactory. It cannot deal with cases like Figure 9.4 unless we further commit the model to be a proper network flow model with various capacities, as studied in operational research. So let us try another approach. Assume in Figure 9.21 that we have $x = \eta = \varepsilon = y$.

Call the common value λ. We now get Figure 9.23

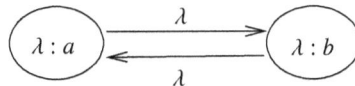

Fig. 9.23.

Let $V_0(a) = V_0(b) = \lambda$, the initial value, and let us transmit from a to b and back from b to a in cycles and see what presents itself. This is the modal logic approach.

We treat Figure 9.23 as equivalent to Figure 9.24:

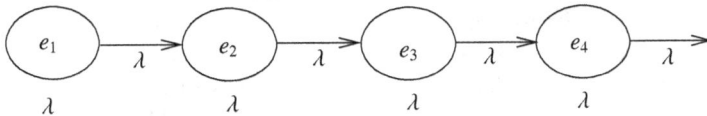

Fig. 9.24.

In Figure 9.24, nodes $e_1, e_3 \ldots$ represent node a of Figure 9.23 and needs $e_2, e_4 \ldots$ represent node b. So we start from $V_1(e_1) = \lambda$ and transmit to the right getting $V_n(e_n), n = 2, 3, \ldots$.

Step 1: Transmit λ^2 to e_2 to get $V_2(e_2) = \lambda(1 - \lambda^2) = \lambda - \lambda^3$.

Step 2: Transmit from e_2 to e_3 the value $\lambda V_2(e_2)$ and get $V_3(e_3) = \lambda(1 - \lambda V_2(e_2)) = \lambda - \lambda^3 + \lambda^5$.

We can continue by induction.

Lemma 9.23. *Suppose we have a node e with $V(e_n) = V_{\lambda,n} = \lambda - \lambda^3 + \lambda^5 - \ldots + (-1)^n \lambda^{2n+1}$ and suppose we are transmitting to a node $\lambda : e_{n+1}$ with value λ then we get $V(e_{n+1}) = V_{\lambda,n+1}$.*

Proof.

$$V(e_{n+1}) = \lambda(1 - \lambda V(e_n))$$
$$= \lambda - \lambda^3 + \lambda^5 \cdots - \lambda^2(-1)^n \lambda^{2n+1}$$
$$= \lambda - \lambda^3 + \lambda^5 \ldots + (-1)^{n+1} \lambda^{2(n+1)+1}$$

We now observe that when n goes to infinity, we get $V_{\lambda,\infty} = \dfrac{\lambda}{1 + \lambda^2}$.[14]

This means that Figure 9.23 stabilises into Figure 9.25. Note that the transmission rates in Figure 9.25 are all 0. This is because the values $V_{\lambda,\infty}$ obtained have already taken into account all recursive transmissions.

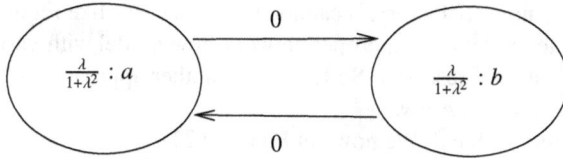

Fig. 9.25.

Note that Figure 9.24 represents one way of going through the cycle of Figure 9.23, i.e. the fuzzy modal logic approach. Another approach is what we called the parasite model, where we apply the transmission on Figure 9.23 directly, starting from node a to b with $V_1(a) = \lambda$, (corresponding to $V_1(e_2)$) we would get $V_2(b) = \lambda(1-\lambda^2)$, same as $V_2(e_2)$ and then transmit back to node a and get $V_3(a) = V_1(a)(1-\lambda V_2(b))$ (corresponding to $V_3(e_2)$). So far the values agree, but now there is a difference. Working directly on Figure 9.23 we transmit $1 - \lambda V_3(a)$ to node b whose *last* value is $V_2(b) = \lambda(1 - \lambda^2)$ and get $V_4(b) = \lambda(1 - \lambda^2)(1 - \lambda V_3(a))$. While in Figure 9.24, the value of node e_4 (which corresponds to b) is λ and so we get in Figure 9.24 $V_4(e_4) = \lambda(1 - \lambda V_3(e_3))$. So the question is, as we go through the cycle $a \to b \to a \to b \ldots$, do we use the new value or follow Figure 9.24 and keep the value at λ, the initial value!

Another possibility for dealing with Figure 9.23 is to adopt the predator-prey model and transmit simultaneously from node a to node b and from node b to node a, and then repeat the cycle. If $V_0(a) = V_0(b) = V_0 = \lambda$ is the initial value, then symmetry is maintained through the cycles and for step $n + 1$ we get

$$V_{n+1}(a) = V_{n+1}(b) = V_{n+1} = V_n(1 - \lambda V_n).$$

So we end up with a recursive equation

- $V_0 = \lambda, 0 \le \lambda \le 1$

---

[14] Note that if we solve the fixed point recursion equation $V_{\lambda,\infty} = \lambda(1 - \lambda V_{\lambda,\infty})$, we get $V_{\lambda,\infty} = \dfrac{\lambda}{1+\lambda^2}$.

- $V_{n+1} = V_n(1 - \lambda V_n)$

which for $0 \leq \lambda \leq 1$ gives $V_\infty = 0$, meaning that a and b cancel each other.[15]

The above considerations can be applied to other loops. The net result of Figure 9.4 will be similar to that of Figure 9.23.

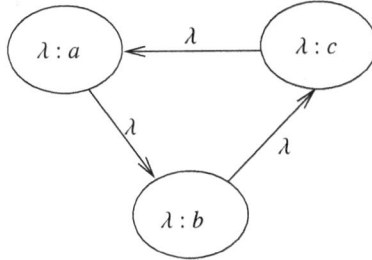

Fig. 9.26.

Consider Figure 9.26. Let us see what different options for resolving loops yield for the figure.

Option (a), the unfolding option would yield, by using Lemma 9.23 that Figure 9.26 stabilises as Figure 9.27

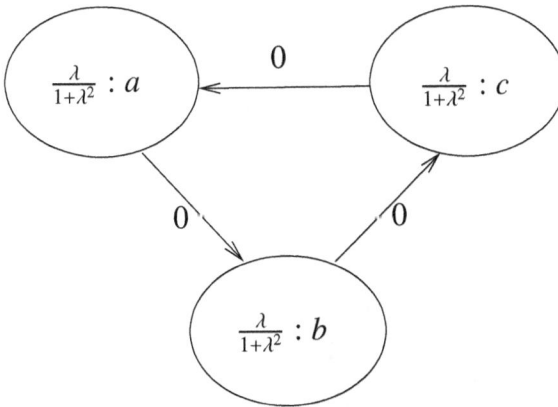

Fig. 9.27.

Option (b) yields the sequence of Figure 9.24 (which is not surprising because of symmetry) and therefore will give rise to Figure 9.27 again.

Option (c) again because of symmetry yields for each node the sequence of Figure 9.24. So again we get Figure 9.27 as the solution of the loop.

Option (d) is perhaps the simplest and most clear of all the options. We assume a steady state solution for nodes V_a, V_b, V_c. We get the equations

---

[15] The fixed point recursion equation for this case is $V = V(1 - \lambda V)$, yielding $V = 0$. We get the same equation if we follow option (d) for Figure 9.23.

1. $V_a = V_a(1 - \lambda V_c)$
2. $V_b = V_b(1 - \lambda V_a)$
3. $V_c = V_c(1 - \lambda V_b)$

Because of symmetry we can assume

$$V_a = V_b = V_c = V.$$

Thus we get $V = V(1 - \lambda V)$. The only solution is $V = 0$.

We can make one more move now. To resolve Figure 9.4, we consider Figures 9.26 and 9.27 and let λ approach 1. Thus we get the value $\frac{1}{2}$. Hence the net results of Figure 9.4 is Figure 9.28 below.

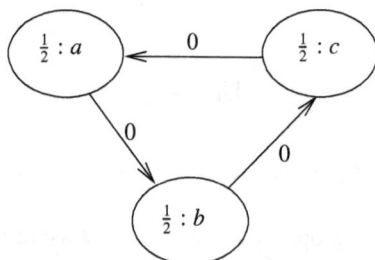

Fig. 9.28.

A similar net result obtains for Figure 9.29 below.

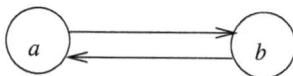

Fig. 9.29.

Note that now we can resolve the loop in Figure 9.5. We get $V(a) = V(b) = \frac{1}{2}$ and therefore $V(c) = \frac{3}{4}$ and hence $V(d) = \frac{1}{4}$.

We can also deal with an argument attacking itself. It will get $\frac{1}{2}$. We pause to remark that for the reader interested in argumentation networks only, the best method for resolving loops is option (d), namely assume a steady state solution and solve the resulting equation. See Gabbay's [166, 167], and Chapter 10 below. For other types of networks (see Appendix B) other options may be more natural.

There is still work to be done on resolving loops. We probably need a long paper dedicated just to loops in various networks and indeed we devote several chapters later on in this book to handling loops. In the context of our chapter here, we need to show the following:

1. How the results we get for the loop depend on the choice of numbers we assign to the nodes and for the transmission rates (we gave λ to all!).

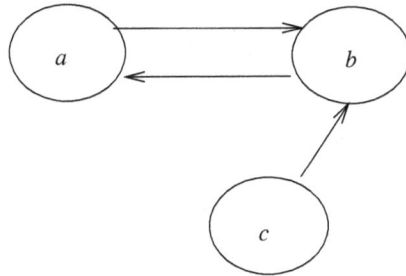

Fig. 9.30.

2. What happens when loops can be resolved but we use our method anyway, as in Figure 9.30.
   In Figure 9.30, the net result is

$$\{+c, -b, +a\}.$$

What do we get if we assign λ everywhere and get Figure 9.31?

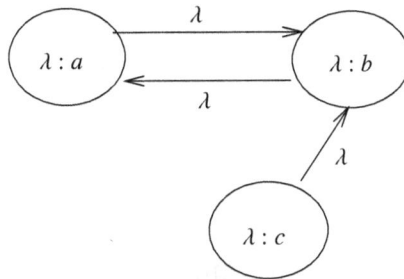

Fig. 9.31.

Here is the calculation: We start with $V_0(a) = V_0(b) = V_0(c) = \lambda$. Transmit from c to b and get $V_1(b) = \lambda - \lambda^3$. Transmit from b to a and get $V_1(a) = \lambda(1 - \lambda V_1(b)) = \lambda - \lambda^3 + \lambda^5$.

Obviously if we follow the loop we get as before $V_\infty(a) = V_\infty(b) = \frac{\lambda}{1+\lambda^2}$ and the net result is $\{1 : c, \frac{1}{2} : a, \frac{1}{2} : b\}$. This is not satisfactory.

It makes more sense to try to give c value 1 transmitting at rate 1, since c is not in a loop. This will give b value 0 and a value λ. When λ approaches 1 we get the right answer.

Perhaps we might follow the procedure of giving λ only to nodes in a loop?

3. Consider, however, the following loop in Figure 9.32.
   d is attacked twice and is attacking once, while a is attacking twice and is attacked once. Should we give them λ in the same way?

Example 9.24 (Resolving Figure 9.32). Let us try the fixed point approach on Figure 9.32. We begin with $V_0(a) = V_0(b) = V_0(c) = V_0(d) = y$ and with transmission λ.

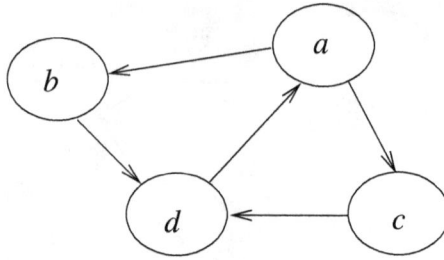

Fig. 9.32.

a) We start propagating from node a. We get

$$V_1(b) = V_1(c) = y(1 - \lambda y).$$
$$V_1(d) = y(1 - \lambda y(1 - \lambda y)^2)$$

and therefore

$$V_1(a) = y(1 - \lambda V_1(d)).$$

We need a fixed point solution to $V_1(a) = V_0(a)$. Hence

$$y(1 - \lambda V_1(d)) = y.$$

Excluding $y = 0$, we get

$$1 - \lambda V_1(d) = 1.$$

Hence

$$V_1(d) = 0.$$

This means

$$y(1 - \lambda y(1 - \lambda y)^2) = 0.$$

Hence

$$\lambda y(1 - \lambda y)^2 = 1.$$

Let $x = \lambda y$. We get $x(1 - x)^2 = 1$. This has a solution, x_0 of approximate value

$$x_0 \approx 1.755.$$

If we want $0 \le \lambda \le 1$ then there is no way $0 \le y \le 1$. Hence the only fixed point solution is $y = 0$.

b) Let us start at node d of Figure 9.32

$$V_0(d) = y$$
$$V_1(a) = y(1 - \lambda y)$$
$$V_1(b) = V_1(c) = y(1 - \lambda V_1(a))$$
$$V_1(d) = y(1 - \lambda V_1(b))^2$$

and try to solve the fixed point equation:

$$y = y(1 - \lambda V_1(b))^2.$$

Hence if we insist on $y \neq 0$,

$$1 = (1 - \lambda V_1(b))^2.$$

Hence

$$1 - \lambda V_2(b) = \pm 1.$$

So either
  i. $V_1(b) = 0$
     or
  ii. $V_1(b) = 2\lambda$.
For $V_1(b) = 0$ we get, if $y \neq 0$, that $V_1(a) = 1/\lambda$.
Hence $\lambda y(1 - \lambda y) = 1$. It is clear that this equation has no real solution.
Let us now try the case in which $V_1(b) = 2\lambda$.
Hence

$$y(1 - \lambda y(1 - \lambda y)) = 2\lambda$$
$$y - \lambda y^2(1 - \lambda y) = 2\lambda$$
$$y - \lambda y^2 + \lambda^2 y^3 - 2\lambda = 0$$

Does this have solutions? Remember $0 \leq \lambda \leq 1, 0 \leq y \leq 1$.
If we choose $\lambda = 0.133$ and $y = 0.275$, the value of the polynomial is 0.0006.
Since we are dealing with continuous functions, we can find proper solutions.
Let us now try another way of tackling Figure 9.32, which can be rewritten as Figure 9.33 below, where a_i represent a, b_i represent b, c_i represent c and d_i represent d.

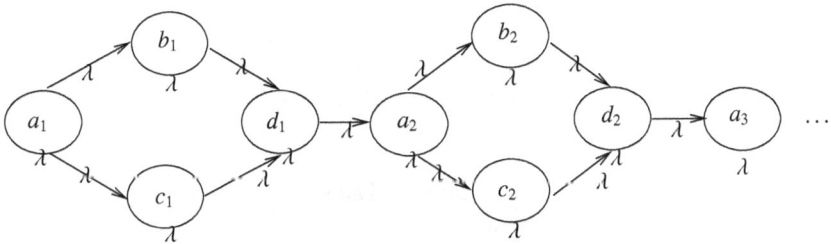

Fig. 9.33.

The neural net approach gives us an additional dimension. We can run the cycles in the loop but also transmit to the rest of the network, and possibly stop after so many cycles (say $n = 100$) and examine the values in all nodes of other network. If the time involved in the cycles has meaning in terms of the network itself changing in time (as modelled in Section 4 below), then we have added a new and interesting dimension to loops in these networks.
In other words, we are saying that attacks take time to be executed, a loop of the form "a attacks b and b attacks a" also takes time to unfold, and meanwhile the network can change.
To give an example of such a loop, think of contradicting witnesses and circumstantial evidence, one supporting a and one supporting $b = \neg a$. So the loop is as in Figure 9.34

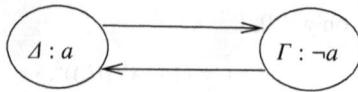

Fig. 9.34.

where Δ, Γ are themselves argument structures which are time dependent. This loop certainly takes time to unfold! There may be some facts in Δ or Λ that take time to verify or refute!

We conclude this section with a remark which may be of interest to a reader who wants to expand his horizon beyond he area of argumentation networks. We believe that the general treatment of loops should be done in the context of neural networks (see [195]), not because of a conceptual connection, but because these nets can technically reach equilibrium and resolve loops of the kind that arise there.

Note that every graph can be presented as an acyclic graph of nodes which are themselves maximally connected cycles. So when we are dealing with cycles we can make use of that. In fact Baroni and his colleagues took advantage of this idea in [30].

9.4 Attack and support networks

This section deals with attack and support. Since our approach is numerical, all we have is several numbers attacking or supporting another number. From the point of view of traditional argumentation networks this simplification may seem as really going too far. We lose the richness of discussions and modelling we have in a substantial number of papers on support which we recently have recently seen published in the argumentation community. However, outside argumentation, numerical attack and support make sense. In ecological networks, in flow networks, in electrical networks, etc.

So this section is concerned with numerical attack and support. The main problem is what numerical function to use in the case of nodes $x_i : a_i$ support the node $y : b$. What function $y' = \mathbf{g}(y, x_i)$ should give us the new supported value y'? We have this problem in argumentation too, but in our case it is all numerical.

Because we are dealing with numbers, we add the requirement that if $y : b$ is both supported and attacked by the same number, then the support and attack should cancel each other. This requirement is reasonable. Many valuations we know, including paper submissions to conferences and various impact factors used by universities in promotion considerations, involve numerical scores. It would be naive to assume that cancellation principles are not used, especially since many of the people involved may not be familiar with the subject matter.

So the beginning of this section, subsection 4.1 tries to find the right functions for attack and support. We do find a good way of doing it and we discover to our surprise that the solution connects with the Dempster-Shafer rule (arising in a completely different community of researchers) and even more surprising is the connection with the cross-ratio of projective geometry. (The cross-ratio is used in geometry to define metrics on spaces.) This is investigated in Section 4.2.

By the end of Section 4.2 we will have drifted away from argumentation quite a bit. So the argumentation reader might ask why this should be of interest to him? We would say please expand your horizons, after years of research in logic, our gut feeling is that this is important.

This section discusses the addition of support arrows to argumentation networks. We will see that in order to have equal attack and support cancel each other, we need to reconsider the way we calculate the values of attacks (and supports). We offer a new definition and establish a connection between the new definition, the Dempster–Shafer rule, and surprisingly, the Cross-Ratio and projective metric distance from geometry.

9.4.1 Discussion of support

Consider a connection from a to b in Figure 9.35.

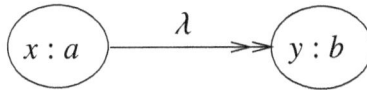

Fig. 9.35.

The double arrow indicates support. The simplest way to do it is to attack $(1 - y)$ which is the distance of b from 1.[16] Thus the new value of b is

$$1 - (1 - y)(1 - \lambda x) =$$
$$1 - [1 - \lambda x - y + \lambda xy]$$
$$= \lambda x + y - \lambda xy = y + (1 - y)\lambda x$$

If we have several supports, then $(1 - y)$ shrinks to

$$(1 - y)(1 - \lambda_1 x_1)(1 - \lambda_2 x_2) \ldots (1 - \lambda_k x_k)$$

and the new value y' becomes $1 - (1 - y) \prod_i (1 - \lambda_i x_i)$. The difference $y' - y$ becomes

$$\Delta^+ y = 1 - (1 - y) \prod_i (1 - \lambda_i x_i) - y$$
$$= (1 - y)(1 - \prod_i (1 - \lambda_i x_i))$$

and we have

$$y' = y + \Delta^+ y.$$

How do we deal with both attack and support? Consider Figure 9.36 In this figure $x : a$ attacks $y : b$ and $z : c$ supports it. So the new value for b is

$$y - \lambda xy + \mu z(1 - y).$$

It is not clear what to do with several simultaneous attacks and supports. The model must be commutative in the order of application.

Our solution is simple. b is at a distance y from 0 and distance $1 - y$ from 1. Let the attackers attack y to get it nearer to 0 and let the supporters attack $(1 - y)$ to get b

---

[16] This is Bernouli's rule of combination, see [306, pp. 75–76].

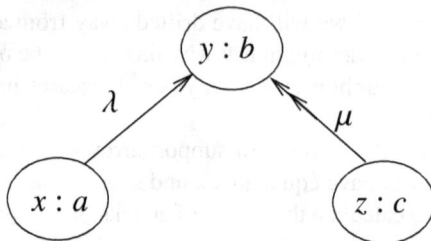

Fig. 9.36.

nearer to 1. Thus if $x_i : a_i$ attack $y : b$ with transmission λ_i and $z_i : c_i$ support $y : b$ with transmission μ_i we get y' as the new value at b, where

$$y' = y - \Delta^- y + \Delta^+ y$$
$$= y - y(1 - \prod_i(1 - \lambda_i x_i))$$
$$+ (1 - y)(1 - \prod_i(1 - \mu_i z_i))$$
$$= y \prod_i(1 - \lambda_i x_i) + (1 - y)(1 - \prod_i(1 - \mu_i z_i))$$

Note that there is something numerically wrong with our proposal. In Figure 9.35, if we let $z = x$ and $\mu = \lambda$, i.e. the attack and support have the same values, then, we would have expected that they cancel each other. However, this is not the case. The new value is $y' = y - 2\lambda xy + \lambda x$.

This should not surprise us. The closer y is to 1, the less is the numerical value of an attack on $1 - y$, and the more numerical value we get for an attack on y. So, for example, assume $y = 0.9$ in value. Then a support of 50% of y will half the distance of y from 1, i.e. will yield $\Delta^+ = 0.005$ in numerical value, while in comparison, a 50% attack on y will half the distance of y from 0 and will yield $\Delta^- = 0.45$. The net result of simultaneous attack and support will yield the new value $0.9 - 0.45 + 0.05 = 0.50$.

Can we remedy the situation? Perhaps we should attack by changing the ratio $r(y)$ of y to $1 - y$, (i.e. $r(y) = y/(1 - y)$), and then calculate the new y' which will give the new ratio. So suppose the transmitted value (of attack or support) is $0 \leq \theta \leq 1$.

If θ is an attack we want to reduce $r(y)$ and so we let $r'(y) = \theta r(y)$. If θ is a support, we want to increase y, so the new ratio is $r'(y) = r(y)/\theta$.

We now solve the equation

$$\frac{y'}{1 - y'} = r'(y)$$

and therefore we get

$$y' = \frac{r'(y)}{1 + r'(y)}.$$

We must now decide on what value θ to use. Let us use the same value we used before, as agreed in Example 9.4. In Figure 9.35, we have $x : a$ attacking $y : b$ with transmission rate λ and we therefore have $\theta = (1 - \lambda x)$.

Let us calculate the values of attack and support with θ.

Case of attack

$$r'(y) = \frac{y(1 - \lambda x)}{1 - y}$$

$$y' = \frac{y(1 - \lambda x)}{(1 - y)(1 + \dfrac{y(1 - \lambda x)}{1 - y})}$$

$$= \frac{y(1 - \lambda x)}{(1 - y + y - \lambda xy)}$$

$$= \frac{y(1 - \lambda x)}{(1 - \lambda xy)}$$

Case of Support

$$r'(y) = \frac{y}{(1 - y)(1 - \lambda x)}.$$

$$y' = \frac{y}{(1 - y)(1 - \lambda x)(1 + y/(1 - y)(1 - \lambda x))}$$

$$y' = \frac{y}{(1 - y)(1 - \lambda x) + y}$$

$$= \frac{y}{(1 - \lambda x + y\lambda x)}$$

$$= \frac{y}{1 - \lambda x + y\lambda x}$$

$$= \frac{y}{1 - \lambda x(1 - y)}$$

Let us now assume as before that the attack is 50%, e.g. $x = 0.5, \lambda = 1$.

We get $\theta = 0.5$. Assume as before $y = 0.9$. Hence $r'(y) = \frac{0.9}{0.1}.0.6 = 4.5$ and $y' = \frac{4.5}{1 + 4.5} = \frac{4.5}{5.5} = \frac{9}{11}$.

This should be compared with the previously attained value $0.45 = \frac{9}{20}$.

For the support we get

$$r'(y) = \frac{0.9}{0.1.0.5} = 18$$

So

$$y' = \frac{18}{1 + 18} = \frac{18}{19}$$

This should be compared with the value 0.05 we got previously.

How do we handle simultaneous attacks and supports? We follow the same principle as before. If $\theta_1, \ldots, \theta_n$ are attacking values and $\theta'_1, \ldots, \theta'_m$ are the supporting values then the new $r'(y)$ is

$$r'(y) = r(y)\frac{\prod_i \theta_i}{\prod_i \theta'_i}.$$

We shall argue below at the beginning of Section 4.2 that the choice of $\theta = (1 - \lambda x)$ is wrong. We can see that something is wrong qualitatively already. Consider Figure 9.36. We have that a attacks b and c supports b. So if a attacks b with $\theta =$

$(1 - \lambda x)$ and c supports b with $\theta' = (1 - \mu z)$, then we get according to our formula that $r'(y) = r(y)\dfrac{\theta}{\theta'}$.

Now let us ask: does a attack c? And, does c attack a? What are the qualitative implications of these equations? Let us calculate. Assume for simplicity that $\lambda = \mu = 1$. We get no clear answers to our questions! However, if we were to take (for the case $\lambda = \mu = 1$) the value $\theta = \dfrac{(1 - x)}{x}$, and $\theta' = \dfrac{(1 - z)}{z}$ then we get that the sequence, a attacks c and then the already attacked c supports b, gives the same result as in Figure 9.36, which is that a attacks c and b supports c in parallel. Notice that since $\theta = \dfrac{1}{r(z)}$ we get the following rules, as summarised in Remark 9.25 below:

Remark 9.25 (Rules for attack and support).
For x attacking y let $r'(y) = \dfrac{r(y)}{r(x)}$
For z supporting y let $r'(y) = r(y) \cdot r(z)$
For both we get $r'(y) = r(y) \cdot \dfrac{r(z)}{r(x)}$
These equations imply that x attacks z and also z attacks x. The following three scenarios give the same results:

> **x attacks z and the modified z supports y;**
> **z attacks x and the modified x attacks y;**
> **x attacks y and z supports y in parallel.**

Remark 9.26 (Loops involving support). Since we introduced support, we need to check what happens with loops involving support. For example, we need to check what happens with odd and even loops with support only.

In fact in this section we introduced a new algoirthm for numerical attack and support as summarised above in Remark 9.25.

So we need to check loops for this new type of numerical attack as well.

The first fact we observe is that by the definition of attack or support of $x : a$ on $y : b$, we use the ratios $r(x) = \frac{x}{1-x}$ and $r(y) = \frac{y}{1-y}$. For attack we divide by the ratio to get the new ratio $r'(y) = \frac{r(y)}{r(x)}$ and for support we multiply by the ratio to get the new ratio $r''(y) = r(y)r(x)$.

This means that
$(x : a)$ attacking $(y : b)$
is the same as
$((1 - x) : a)$ supporting $y : b$.

So when we have loops involving both attack and support, we can eliminate the supports by converting them into attacks.

Let us check Figure 9.37.

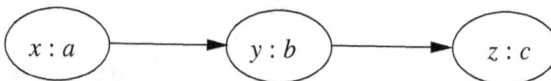

Fig. 9.37.

From $x : a$ attacking $y : b$ we get a new ratio for b namely $\frac{r(y)}{r(x)}$.
Now the new ratio for b attacks $z : c$ and so the new ratio for c is

$$r(z) \left| \frac{r(y)}{r(x)} = \frac{r(z)r(x)}{r(y)} \right.$$

The new ratio for $z : c$ is the same as the ratio we obtain if $y : b$ was attacking $z : c$ and $x : a$ was supporting $z : c$, as in Figure 9.38.

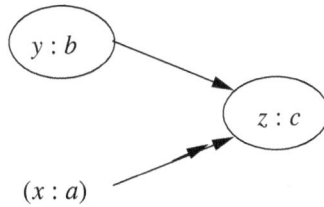

Fig. 9.38.

We get the new rate for c is $\frac{r(z) \cdot r(z)}{r(y)}$.
This is compatible with the intuition that if a attacks b which attacks c, then a actually supports c.

Let us now examine odd and even loops for the ratio type of attack. Consider Figure 9.39

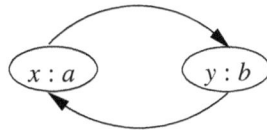

Fig. 9.39.

This is a typical even loop. Assume a steady state solution. This means the following equations hold:

1. Since b attacks a we must have

$$r(x) = \frac{r(x)}{(y)}$$

2. Since a attacks b we must have

$$r(y) = \frac{r(y)}{r(x)}$$

From the equations we get that $r(x) = r(y)$ and hence $r(x) = r(y) = 1$ and hence $x = y = \frac{1}{2}$.
This corresponds in Dung argumentation theory to $x = y = $ undecided.

In Dung argumentation we also have the solutions $x = $ in and $y = $ out and $x = $ out and $y = $ in.

These solutions correspond to the numerical solutions $(x = 0) \wedge (y = 1)$ and $(x = 1) \wedge (y = 0)$.

For $x = 0$ we have $r(x) = 1$ and for $x = 1$ we get $r(x) = \infty$.

Assume $f9x) = 1$ ad $r(y) = \infty$, and check the equations. For the first equation we get

$$1 = \frac{1}{\infty} = 0$$

which is impossible.

So we do not have these solutions in our case.

Let us now check the odd loop of Figure 9.40

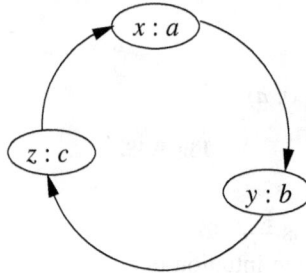

Fig. 9.40.

In a steady state solution we get the following equations:

1. $r(x) = \frac{r(x)}{r(z)}$
2. $r(y) = \frac{r(y)}{r(x)}$
3. $r(z) = \frac{r(z)}{r(y)}$

The only solution is $r(x) = r(y) = r(z) = 1$, namely $x = y = z = \frac{1}{2}$.

In Dung argumentation terms this corresponds to $a = b = c = $ undecided.

Now we ask what happens to these loops if some of the connections are supports? We saw that support of

$$(x : a) \twoheadrightarrow (y : b)$$

can be converted to attack

$$((1 - x) : a) \rightarrow (y : b).$$

The ratio $r(1 - x)$ equals the ratio of $\frac{1}{r(x)}$.

Now if we have a loop, odd or even, and some links are supports, we convert them into attacks with new ratio $= \frac{1}{\text{old ratio}}$. Since the only solutions we got was that the ratios of all nodes is 1, then $\frac{1}{\text{ratio}}$ is also 1 and this means we get all undecided as before.

We thus get that in our case the following:

- All loops, odd or even, no matter whether the links are attack or support, have only one solution: all undecided!

Remark 9.27 (Higher level attacks and support). The reader should note that our definition of numerical attacks and support as given in this section, really define how one number x can attack or support another number y. The number y' is calculated as summarised in Remark 9.25.

Therefore when we consider higher level attacks or support of nodes on arrows or arrows on other arrows, such as we have in Figure 9.8, we have no problem calculating them. We just look at the numbers. So in Figure 9.8, for example, the node $z : c$ attacks the number ε representing the transmission rate of $(x : a) \rightarrow (y : b)$ with the transmission rate η. So the real attacking number is $z\eta$, attacking the number ε. The new number ε' is terefore

$$\varepsilon' = \frac{r(\varepsilon)/r(z\eta)}{1+r(\varepsilon)/r(z\eta)}$$

$$= \frac{r(\varepsilon)}{r(\varepsilon)+r(z\eta)}.$$

Remark 9.28 (Comparison with biological networks). It is worthwhile comparing the recursion results we obtained with the kind of recursion one gets in mathematical biology. We use the table (Table 3.1) of [324, p. 53].[17]

1. Old attack formula

$$y_{n+1} = y_n(1 - \lambda x)$$

This can be compared with exponential population growth.
2. New attack formula

$$y_{n+1} = \frac{y_n(1 - \lambda x)}{1 - \lambda x y_n}$$

This can be compared with the Beverton–Hort formula in the table of [324, p. 53], see footnote 17.

Let us also examine what happens in case of loops. Consider Figures 9.23 and 9.24 again. We have $v_1(e_1) = \lambda$ and the recursion equation, according to Figure 9.24 is

$$V_{n+1}(e_{n+1}) = \frac{\lambda(1 - \lambda V_n(e_n))}{1 - \lambda^2 V_n(e_n)}.$$

The recursion fixed point equation for this case is

$$V = \frac{\lambda(1 - \lambda V)}{1 - \lambda^2 V}$$

or

---

[17] We quote: Some functional forms proposed for single-species discrete-time models of population growth ($\lambda_0 = \exp[r_0]$ is the intrinsic discrete or multiplicative rate of population increase, k is the carrying capacity, b some positive constant, and θ an exponent)

Label	Function
Exponential	$N_{t+1} = \lambda_0 N_t$
Quadratic map	$N_{t+1} = \lambda_0 N_t(1 - N_t/k)$
Ricker	$N_{t+1} = \lambda_0 N_t \exp[-bN_t]$
Gompertz	$N_{t+1} = \lambda_0 N_t^\theta$
Beverton-Holt	$N_{t+1} = \lambda_0 N_t/(1 + bN_t)$
Desensation	$N_{t+1} = \lambda_0 N_t^2/(1 + bN_t^2)$
Theta-Ricker	$N_{t+1} = \lambda_0 N_t \exp[-bN_t^\theta]$

$$V - \lambda^2 V^2 = \lambda - \lambda^2 V$$

$$\lambda^2 V^2 - \lambda^2 V - V + \lambda = 0$$

$$V^2 - \frac{(1 + \lambda^2)}{\lambda^2} V + \frac{1}{\lambda} = 0$$

Let λ approach 1, we get

$$V^2 - 2V + 1 = 0$$

and so $V_\infty = 1$.

If we do the recursion proper, as in Figure 9.23, we get

$$V_{n+1} = \frac{V_n(1 - \lambda V_n)}{1 - \lambda V_n^2}$$

The fix point equation becomes

$$V(1 - \lambda V^2) = V(1 - \lambda V).$$

If we discard the solution $V = 0$, we get

$$1 - \lambda V^2 = 1 - \lambda V$$

hence

$$V^2 = V$$

and hence $V = 1$.

For more on biological networks, see Appendix B.

9.4.2 Connection with Metric Projective Geometry and the Dempster–Shafer Rule

In the previous subsection, we agreed that in the situation of Figure 9.36, node a attacks node b by attacking the ratio:

$$r(y) = \frac{y}{1 - y}$$

We proposed that the attack value θ be $\theta = 1 - \lambda x$. We want in this subsection to re-examine our decision and see whether we want to use a different attack value. First to simplify our qualitative consideration, assume $\lambda = 1$ and $\mu = 1$. Second, let us focus on node c, which is supporting node b, with value z. Assume that z is very small, almost 0. One may feel that in many real applications, a very limited support is worse than nothing. It implies an attack on argument b, the hidden implication is that if b were any good why isn't c's support of it a bit stronger? This way of thinking would integrate the support and attack together. So if a node supports another node with value z then it simultaneously attacks it with value $1 - z$. If $z = 1$, then the support is complete. If $z \approx 0$ then the support is insulting and really accomplishes an attack to the value of $1 - z$

Let us look at Figure 9.36 again. There are two ways to look at this figure (with $\lambda = \mu = 1$). One way is that we have two nodes, $x : a$ and $z : c$, the first attacking the node $y : b$ and the second supporting it.

The other way is that there is a single node $z : c$ supporting the node $y : b$, but simultaneously attacking it to the value $1 - z$, as discussed above. Figure 9.36, with $x = 1 - z$ is a representation of this new point of view through the additional node $x = (1 - z) : a$.

Of course it is nicer to represent this new point of view directly, and indeed, Figure 9.42 represents this new point of view of support/attack mode by a double arrow.

Let us now calculate the new value y' of the attack and support configuration of Figure 9.36. We have:

$$r'(y) = \frac{y(1 - x)}{(1 - y)(1 - z)}$$

Hence

$$y = \frac{r'(y)}{1 + r'(y)}$$

$$= \frac{y(1 - x)}{(1 - y)(1 - z) + y(1 - x)}$$

$$= \frac{y(1 - x)}{1 - y - z + y(z + 1 - x)}$$

In order to compare with a later formula, let us rename the values. Let $z_2 = 1 - x$ and let $z_1 = z$. We get the equation (DS1) below:

$$(DS\,1) \qquad y' = \frac{yz_2}{1 - y - z_1 + y(z_1 + z_2)}$$

This equation means that a node $y : b$ is simultaneously supported by $z_1 : c$ and attacked by $(1 - z_2) : a$. Alternatively, we can say that the node is being [Support, Attacked] by the pair $[z_1, z_2]$. If $z_1 \leq z_2$ (i.e. $z + x \leq 1$), we can say we have a [Support, Attack] interval $[z_1, z_2], 0 \leq z_1 \leq z_2 \leq 1$.[18]

We adopt this terminology in preparation for the Dempster–Shafer point of view, yet to come. See item 3 of Example 9.29.

Let us now examine the case where $x = 1 - z$, i.e. $z = 1 - x$. We can view this as a [Support, Attack] pair $[z_1, z_2] = [z, z]$.[19]

We can view Figure 9.36 again and see that we are getting a situation of support value z from node c and attack value $1 - z$ from node a.

We have already calculated the new ratio $r'(y)$ for node b, it is

$$r'(y) = \frac{y(1 - (1 - z))}{(1 - y)(1 - z)} = \frac{yz}{(1 - y)(1 - z)}$$

---

[18] Actually the intervals involved are $[0, z_1], [1 - z_2, 1]$.

[19] Beware some possible confusion in notation. In Figure 9.36, the attack of a node is with value $x = 1 - z$ and the support is with value z. If we regard Figure 9.36 as representing the [Support, Attack] double arrow of Figure 9.42, we write it as $[z, 1 - x] = [z, z]$ and *not as* $[z, x]$. This is because $z_2 = (1 - x)$ appears in (DS1).

Let us write this equation as

(*)
$$r'(y) = \frac{y/(1-y)}{(1-z)/z}$$

We now calculate the new value y', it is

(**)
$$y' = \frac{r'(y)}{1 + r'(y)}$$

$$= \frac{yz}{(1-y)(1-z) + yz}$$

We thus get that a node $z : c$ supporting a node $y : b$ yields the new value $y' : b$, where:

(DS 2)
$$y' = \frac{yz}{1 - y - z + 2yz}$$
provided, of course, that $y + z - 2yz \neq 1$.

Let us say that (*) and (DS2) represent a combined [Support, Attack] result of a node to a value $[z, z]$, attacking a node with value y.

We now connect (DS2) to the Dempster–Shafer rule (see [306, 211]), and to the Cross-Ratio and projective metric from geometry (see [12, 124]).

Example 9.29 (Dempster–Shafer rule). The range of values we are dealing with is the set of all subintervals of the unit interval $[0,1]$. The Dempster–Shafer addition on these intervals is defined by

$$[a, b] \oplus [c, d] = [\frac{a \cdot d + b \cdot c - a \cdot c}{1 - k}, \ \frac{b \cdot d}{1 - k}]$$

where $k = a \cdot (1-d) + c \cdot (1-b)$, where '·', '+', '−' are the usual arithmetical operations. The compatibility condition required on a, b, c, d is

$$\varphi([a, b], [c, d]) \equiv k \neq 1.$$

The operation $\oplus$ is commutative and associative. Let $\mathbf{e} = [0, 1]$.
The following also holds:

- $[a, b] \oplus \mathbf{e} = [a, b]$
- For $[a, b] \neq [1, 1]$ we have $[a, b] \oplus [0, 0] = [0, 0]$
- For $[a, b] \neq [0, 0]$ we have $[a, b] \oplus [1, 1] = [1, 1]$
- $[a, b] \oplus [c, d] = \varnothing$ iff either $[a, b] = [0, 0]$ and $[c, d] = [1, 1]$ or

$[a, b] = [1, 1]$ and $[c, d] = [0, 0]$.

In this algebra, we understand the transmission value $[a, b]$ as saying that the actual transmission value lies in the interval $[a, b]$.
Let us make three comments:

1. Let x denote $[x, x]$. We get for $0 \le a \le 1$ and $0 \le c \le 1$ the following

$$a \oplus c = \left[\frac{ac + ac - ac}{1 - a(1 - c) - c(1 - a}, \frac{ac}{1 - a(1 - c) - c(1 - a)} \right]$$

$$= \left[\frac{ac}{1 - a - c + 2ac}, \frac{ac}{1 - a - c + 2ac} \right]$$

$$= \frac{ac}{1 - a - c + 2ac}$$

provided $(a + c - 2ac) \ne 1$.

We note immediately that (DS2) is $y \oplus z$. This is also the propagation method used by the MYCIN expert system. See [212].

2. Let us check for what values of a, c can we have equality, i.e. when can we have $a + c - 1 = 2ac$?

Assume $a \le c$.

We claim the only solution to the equation $a + c - 2ac = 1$ is $a = 0, c = 1$ for $a \le c$ and $a = 1, c = 0$ for the case $c \le a$. There is no solution for $c = a$.

To show this, let $c = a + \varepsilon, 0 \le \varepsilon \le c - a$.

Then assume

$$a + a + \varepsilon = 1 + 2a(a + \varepsilon)$$
$$2a + \varepsilon = 1 + 2a^2 + 2\varepsilon a$$
$$\varepsilon - 2\varepsilon a = 1 + 2a^2 = 2a$$
$$\varepsilon(\tfrac{1}{2} - a) = a^2 - a + \tfrac{1}{2}$$
$$= (a - \tfrac{1}{2})^2 - (\tfrac{1}{2})^2 + \tfrac{1}{2}$$
$$= (a - \tfrac{1}{2})^2 + (\tfrac{1}{2})^2$$

Hence

$$(a - \tfrac{1}{2})^2 + \varepsilon(a - \tfrac{1}{2}) + (\tfrac{1}{2})^2 = 0$$
$$[(a - \tfrac{1}{2}) + \tfrac{\varepsilon}{2}]^2 - (\tfrac{\varepsilon}{2})^2 + (\tfrac{1}{2})^2 - 0$$
$$(a - \tfrac{1}{2} + \tfrac{\varepsilon}{2})^2 = (\tfrac{\varepsilon}{2})^2 - (\tfrac{1}{2})^2$$
$$= (\tfrac{\varepsilon}{2} - \tfrac{1}{2})(\tfrac{\varepsilon}{2} + \tfrac{1}{2})$$

Hence $\varepsilon = 1$ and since $0 \le c = a + \varepsilon \le 1$ we must have $a = 0$ and $c = 1$.

In particular, we get that for $a = c = x$, $x \oplus x$ is always defined and we have

$$x \oplus x = \frac{x^2}{1 - 2x + 2x^2}$$

For example, we have

$$0 \oplus 0 = 0$$
$$1 \oplus 1 = 1$$
$$\tfrac{1}{2} \oplus \tfrac{1}{2} = \tfrac{1}{2}$$

3. Let us check what happens when $c = d$.

We get

$$[a, b] \oplus c = \frac{bc}{1 - a(1 - c) - c(1 - b)}$$

$$= \frac{bc}{1 - a + ac - c + bc}$$

$$= \frac{bc}{1 - a - c + c(a + b)}$$

The reader should compare this equation with the formula (DS1) obtained before.

Example 9.30 (Cross-Ratio). Consider the interval $[0, 1]$ and two points y and $1 - z$ in this interval. Let $A = 0, B = 1, C = y$ and $D = 1 - z$. Taking AC, CB, AD, DB as directed intervals, we have it that $AC = y, CB = 1 - y, AD = 1 - z$ and $DB = z$.

The projective Cross-Ratio between these points, denoted traditionally by $(A, B; C, D)$ is calculated as the ratio of ratios of the directed intervals.

$$(A, B; C, D) = \frac{AC/CB}{AD/DB} = \frac{y/(1 - y)}{(1 - z)/z} = \frac{yz}{(1 - y)(1 - z)}$$

Note that this is formula (*).

Note that this measures distance. In the Cayley–Klein metric, $\log(A, B; C, D)$ is used to describe the distance between points C and D. Figure 9.41 shows how it is done.

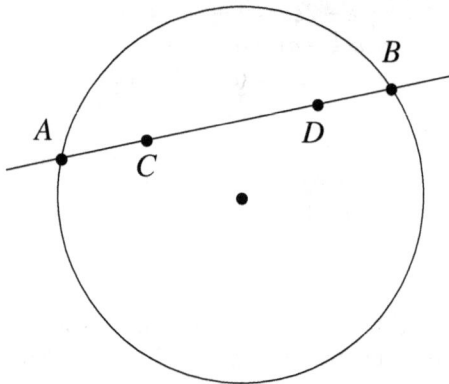

Fig. 9.41.

C and D are inside the unit circle. The chord connecting them meets the circle at A and B. See [12, Sections 4.10 and 11.7] and [124, Chapter 6].

Returning to Figure 9.36, we have

(*) $r'(y) = (0, 1; y, 1 - z)$

We can now define a new kind of support/attack arrow (with value $z/1 - z$) in a network, as displayed in Figure 9.42 by double arrow

We have for $0 \leq y, z \leq 1$

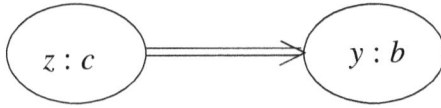

provided $y + z - 2yz \neq 1$.

Fig. 9.42.

($\sharp$1) $r(y) = \dfrac{y}{1 - y}$

($\sharp$2) $r'(y) = (0, 1; y, 1 - z)$

($\sharp$3) $y' = \dfrac{yz}{1 - y - z + 2yz} = y \oplus z$

($\sharp$4)   Furthermore, a formula (DS1) for a combined support to value z_1 and attack to value z_2, as in Figure 9.36 gives the result $y' = y \oplus [z_1, z_2]$.

Equations ($\sharp$2) and ($\sharp$3) and ($\sharp$4) open new opportunities for us.

1. Allow for values to be intervals because of the connection with Dempster–Shafer.
2. Allow for a connection with a more general non-Euclidean metric, using complex numbers.
3. Attack and support values need not be in $[0, 1]$.

For further investigations, see [257].

Example 9.31 (Cross-Ratio for intervals). This example will try to extend the notion of Cross-Ratio for intervals, i.e. we look for Cross-Ratio for $(0, 1; [y_1, y_2], 1 - z), 0 \leq y_1 \leq y_2; 0 \leq z \leq 1$.

We saw that the situation in Figure 9.43 can be described as follows:

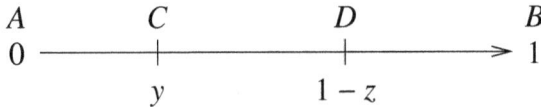

Fig. 9.43.

1.

$$r(y, z) = (0, 1; y, 1 - z)$$

$$= \frac{yz}{(1 - y)(1 - z)}$$

2. We also know that the Dempster–Shafer rule for the case of $y \oplus z = [y, y] \oplus [z, z]$ gives the value

$$y \oplus z = \frac{r}{1 + r} = \frac{yz}{1 - y - z + 2yz}$$

3. Our aim is to define Cross-Ratio $(0, 1; [y_1, y_2], 1 - z)$. We use (2): Consider

$$[y_1, y_2] \oplus z = \frac{y_2 z}{1 - y_1 - z + z(y_1 + y_2)}$$

4. Define by analogy with (2):

(∗1)
$$[y_1, y_2] \oplus z = \frac{r([y_1, y_2], z)}{1 + r([y_1, y_2], z)}$$

we do not know what $r^* = r([y_1, y_2], z)$ means. However, using (∗1) and solving for r^* we get:

$$r^* = \frac{[y_1, y_2] \oplus z}{1 - [y_1, y_2] \oplus z}$$

Fortunately, the expressions in the right-hand side are all numbers: Hence we get

$$r^* = \frac{\dfrac{y_2 z}{1 - y_1 - z + z(y_1 + y_2)}}{1 - y_1 - z + z(y_1 + y_2)(1 - \dfrac{y_2 z}{1 - y_1 - z + z(y_1 + y_2)})}$$

$$= \frac{y_2 z}{1 - y_1 - z + z y_1 + z y_2 - y_2 z}$$

$$= \frac{y_2 z}{1 - y_1 - z + z y_1}$$

$$= \frac{y_2 z}{(1 - y_1)(1 - z)}$$

$$= \frac{y_2}{y_1} \cdot \frac{y_1 z}{(1 - y_1)(1 - z)} = \frac{y_2}{y_1} r(y_1, z)$$

We therefore have

(∗2)
$$r([y_1, y_2], z) = \frac{y_2}{y_1} r(y_1, z).$$

We can therefore define

$$(*3) \qquad (0, 1; [y_1, y_2], 1 - z) =_{\text{def}} \frac{y_2}{y_1} (0, 1; y_1, 1 - z)$$

or more generally:

$$(\sharp) \qquad (A, B; [C_1, C_2], D) =_{\text{def}} \frac{AC_2}{AC_1} (A, B; C_1, D).$$

Let us check whether $(\sharp)$ is invariant under some projective transformations. Let us consider $\frac{y_2}{y_1}$. Think of it as a cross ratio as in the figure below

$$\frac{y_2 - 0}{y_1 - 0} \Big/ \frac{(y_2 - \frac{y_1 + y_2}{2})}{(y_1 - \frac{y_1 + y_2}{2})} = \frac{y_1}{y_1} \frac{y_2 - y_1}{y_1 - y_2} = \frac{-y_2}{y_1}$$

Thus

$$\frac{y_2}{y_1} = -(0, \frac{y_1 + y_2}{2}, y_1, y_2).$$

This Cross Ratio uses the midpoint between y_1 and y_2. Midpoints E between points A and B can be characterised as the Harmonic conjugate of the point at infinity relative to A and B.
So any transformation of the line leaving the point at infinity fixed will also preserve midpoints, i.e. if A goes to A', B to B' and E to E' and ∞ to ∞, then if E is the midpoint of AB then E' is the midpoint of $A'B'$.

5. Since $r(y, z)$ is commutative it stands to reason to define

$$r^{**} = r([y_1, y_2], [z_1, z_2]) = \text{def} \, \frac{y_2}{y_1} \cdot \frac{z_2}{z_1} \, r(y_1, z_1).$$

We now have a candidate definition for a Cross-Ratio for intervals.

$$r^{**} = \frac{y_2}{y_1} \frac{z_2}{z_1} \frac{y_1 z_1}{(1 - y_1)(1 - z_1)}$$

Hence

$$(*3) \qquad r^{**} = \frac{y_2 z_2}{(1 - y_1)(1 - z_1)}$$

Let $\bar{y} = [y_1, y_2], \bar{z} = [z_1, z_2]$. Therefore we can define a new $\boxplus$ using a similar connection as $(*2)$:

$$\bar{y} \boxplus \bar{z} = \frac{r^{**}}{1 + r^{**}}$$

$$= \frac{y_2 z_2}{(1 - y_1)(1 - z_1) + y_2 z_2}$$

Hence we summarise:

$$(*4) \qquad \bar{y} \boxplus \bar{z} = \frac{y_2 z_2}{1 - y_1 - z_1 + z_1 y_1 + z_2 y_2}$$

Let us compare $\boxplus$ with $\oplus$

We have

$$\bar{y} \oplus \bar{z} = \left[\frac{y_1 z_2 + y_2 z_1 - y_1 z_1}{1 - y_1 + y_1 z_2 - z_1 + y_2 z_1} , \frac{y_2 z_2}{1 - y_1 + y_1 z_2 - z_1 + y_2 z_1} \right]$$

$$= \left[\frac{y_1 z_2 + y_2 z_1 - y_1 z_1}{1 - y_1 - z_1 + y_1 z_2 + y_2 z_1} , \frac{y_2 z_2}{1 - y_1 - z_1 + y_1 z_2 + y_2 z_1} \right]$$

They are not the same, unless $z_1 = z_2$ or $y_1 = y_2$.

To see this let us ask when do we get a point interval? We equate the numerators of the interval endpoint and we get

$$y_1 z_2 + y_2 z_1 - y_1 z_1 = y_2 z_2$$

hence

$$z_1(y_2 - y_1) = z_2(y_2 - y_1)$$

i.e. either $y_1 = y_2$ or $z_1 = z_2$ i.e. one has to be a point

Summary

We have extended the Cross Ratio to a case of one interval, and it agrees with the Dempster–Shafer $\oplus$. We can also extend the Cross-Ratio to the case with two intervals, giving it the value

$$r^{**}(\bar{y}, \bar{z}) = \frac{y_2 z_2}{(1 - y_1)(1 - z_1)}$$

but it does not agree with the Dempster–Shafer $\bar{y} \oplus \bar{z}$.

We note, however, that since

$$r^{**}(\bar{y}, \bar{z}) = \frac{y_1}{y_1} , \frac{z_1}{z_1} r(y_1, z_1),$$

if we assume $y_1 = 1 - y_2, z_1 = 1 - z_2$ we get

$$r^{*}(\bar{y}, \bar{z}) = r(y_2, z_2) r(y_1, z_1)$$

We need to check what benefit this gives us!

Example 9.32 (Using Dempster–Shafer for attack and support). Consider again the basic situations depicted in Figures 9.36 and 9.42, or perhaps consider the more fundamental situation of Figure 9.7. Let us focus on the following Figure 9.44.

The new kind of arrow can stand in for attack, support or any combination transmitted from node c to node b. Our aim in this example is to review our options for the kind of values $\alpha, \beta, \varepsilon$ can take and the options available for the mathematical formulas for their combination and transmission.

Our previous discussion allows for the following Dempster–Shafer option

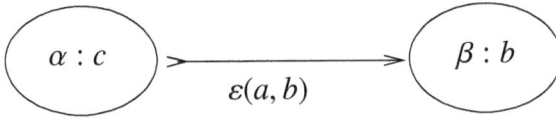

Fig. 9.44.

1. $\varepsilon = 1, \alpha = [z_1, z_2], \beta = y, 0 \le y \le 1, 0 \le z_1 \le z_2 \le 1$ and $y' = y \oplus [z_1, z_2]$ and the arrow is interpreted as [Support, Attack] connection as in formula (DS1). We saw the connection with the Cross Ratio as well.
2. To maintain symmetry, we must also allow β to be of the form $[y_1, y_2], 0 \le y_1 \le y_2 \le 1$ and we must write a formula for the [support, attack] on $\beta : b$. The obvious answer is to let

$$\beta' = \alpha \oplus \beta = [z_1, z_2] \oplus [y_1, y_2].$$

3. Another possibility is to take $\boxplus$, i.e. let $\beta'' = \alpha \boxplus \beta$ (as in (*4) of the previous example) but then β'' is a number not a proper interval.
4. Next let us ask what values can we give to ε? Again the simplest and most general value can be $\varepsilon = [u_1, u_2], 0 \le u_1 \le u_2 \le 1$. We need to say how to combine it with α to get a value transmitted? Again in analogy with expert systems in AI we can let the transmitted value to be $\alpha \oplus \varepsilon$. Thus the new value β' would be

$$\beta' = \alpha \oplus \varepsilon \oplus \beta.$$

9.5 Numerical temporal dynamics overview

The discussion so far was static. The network is fixed and does not change with time. We discussed support and attack options and discussed loops but we did not discuss change. When we have change we use the term *temporal dynamics* of networks. Let us begin.

9.5.1 Introduction

We devote this subsection to briefly outline some intuitive motivation for temporal dynamics. The reader should be aware that the temporal dynamic aspect of networks is central to the subject and will receive extensive study in future papers, see Chapter 7. Since our chapter is on numerical networks, it is worth while to give a brief discussion of the special aspects of numerical change. We can use calculus in this case and involve rate of change in attacks and support.

We begin our discussion with general considerations of how to introduce time into a system. There are two ways time can be introduced into the system. The (local) object level approach and the (global) metalevel approach. If the system is denoted by S and its components by $s_1, s_2, s_3, \ldots$ then the object level approach is to make each s_i time dependent, i.e. $s_i = s_i(t)$. The global meta approach is to take temporal snapshots of the whole system at different times.

To illustrate the two approaches, consider a system of particles in mechanics. The local object level approach is to give a trajectory function $s(t)$ for each particle s. The meta approach is to take snapshots of the whole system at different times. Since the particles may be interacting, the meta approach is to give general differential equations governing the behaviour of the system while the object level approach gives the solutions to these equations, being the trajectory for each particle.

In the case of argumentation networks, both approaches are meaningful. The metalevel approach takes snapshots of the argumentation system at different times. Thus we need an additional network of time points and we need to evaluate the argumentation network at the time network. We can use special connectives to express time behaviour and create new arguments involving time, out of the atomic arguments of the argument network. This is done at Sections 6–9 below, under the name of Modal and Temporal Argumentation Networks.

For example, if a represents a proof for the existence of God, then Ha can represent the fact that (new argument based on a) this proof has always been accepted as valid and thus one can use Ha as a new argument representing a stronger version of a.

The object level approach is to make the strength of a time dependent. So a may represent a visual argument (by way of television footage) against involvement in some foreign war. The strength of this footage goes down as time moves on.

The two approaches may be combined. For example, one can argue that the strength $a(t)$ of acceptance of the proof of the existence of God has been going down over the years (as measured by percents of the population who accept it), i.e. $\frac{da}{dt} << 0$, and therefore we should no longer take such proofs into consideration!

So this argument has the form:

$$H(\frac{da}{dt} << 0)\xrightarrow[\text{attack}]{} a.$$

There is a third temporal aspect to networks and this is the time it takes to traverse (evaluate the arguments of) the network. This aspect is dominant in biological networks. We now elaborate more on this aspect:

Given an attack and support network there are two interpretations for it which are intuitive and work well:

1. biological interpretation;
2. argumentation interpretation.

The biological network models population growth and dynamics varying in time. The network is fixed and time/population generations manifest themselves in propagating values in the network. Each complete cycle represents a generation in time.

In the case of argumentation networks, such cycles do not represent time but the calculation of the strength of the various participating arguments. A stable solution of the cycling/propagation of the argumentation network gives the final value of the strength of the arguments.

In contrast, a stable solution of the propagation of attack and support values in the biological network represent as a steady state population equilibrium.

An oscillatory solution in the biological case means population oscillation of various species, while in the argumentation case such oscillatory behaviour is a problem because we want a steady answer to the values of the various nodes. Further machinery is needed in the argumentation case to rescue the situation and get a value for each node.

What would be a temporal aspect in the argumentation case?
Consider the simple network of Figure 9.45.

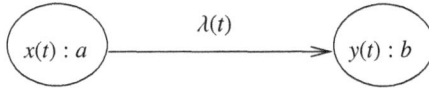

Fig. 9.45.

In this figure t is a time parameter. So the strength and transmission parameters of the net from a to b depend on time t.

The value of b is $y'(t) = y(t)(1 - \lambda(t)x(t))$.

We assume there that the transmission from a to b at any time is instantaneous. Thus what we have is a parameterised family of networks, with parameter t.

There are several ways in which such a system can be made more interesting and more applicable.

1. The variation of the network in time is continuous and has nice properties.
2. Transmission takes time as opposed to transmission being instantaneous. By transmission being instantaneous we mean that the process of Definition 9.11 takes no time at all.
3. The variation is in some sense evolutionary, i.e. the value of the net at time $t + \Delta t$ is somehow dependent on the value at t. This dependence is governed by some evolutionary equations.

Let us examine one such simple case. Consider Figure 9.16 again.

Assume that at time $t = 0$ we have $x(0) = 1$, $\lambda(0) = 1$ and $y(0) = 1$. In this case $y'(0) = 0$.

So really argument b is totally defeated. However if we know that the strength of $a(x(t))$ and its transmission rate $\lambda(t))$ decrease quickly, while the strength of $b, y(t)$ changes very slowly, then it is worth our while to wait a bit hoping the crisis (argument attack from a to b) will blow over.

Let us take for example

$$y(t) = \frac{1}{1 + t}$$

$$x(t) = \frac{1}{1 + te^t}$$

$$\lambda(t) = \frac{1}{1 + te^t}$$

Hence for a small $t = \varepsilon$

$$y'(\varepsilon) = \frac{1 - \dfrac{1}{(1 + \varepsilon e^\varepsilon)^2}}{1 + \varepsilon}$$

For $\varepsilon = 1$ we get $y(1) = \frac{1}{2}$ and in comparison, we have

$$y'(1) = \frac{1 - \dfrac{1}{(1+e)^2}}{2}$$

$$= \frac{1}{2} - \frac{1}{2(1+e)^2}$$

$$= \frac{1}{2} - 0.036$$

So there is a loss of about 7 percent as a result of the attack.

So if we are anxious to keep argument b, we might choose to wait a little (wait ε) for argument a and its transmission to weaken considerably.

Consider that we have

a = sex scandal

b = Governor to resign.

The chances are that public opinion will change quickly.

These time changes should be studied in the context of a time-action model. Suppose we have action $\mathbf{e}$ with precondition b and postcondition c. We want to take action $\mathbf{e}$ but if b is successfully attacked, we cannot do so. So we wait a bit. Conversely, suppose that we have d attacks a. Since d attacks a, b is available as $+b$ and so action $\mathbf{e}$ can be taken. But if d is weakening with time, we may choose to take action $\mathbf{e}$ immediately, while d is still 'saving' b by attacking a.

So a more sophisticated time–action–argument model will look at the speed of changes and will give values for actions to be taken.

We need to say more about what actions do in the model. We need to define the notion of a fact. We agree that syntactical facts e (as opposed to arguments), can be identified in our model by two properties:

1. $V(e) = 1$
2. e is not attacked by anything.

Of course there may be some arguments that have properties 1 and 2 above, but then for all practical purposes they are like facts.

There may be examples where it looks like some facts can be attacked by other facts. The fact that data is available on the computer may be attacked by the fact that a password was irretrievably lost. However, we can also look at the attack as focussing on the transmission rate of the fact and not the fact itself. We further accept that a node e is considered a semantical fact if $V(e) = 1$ and no attack arrows with positive transmission rate go into e. In a temporal dynamics model, these properties must hold at all times. If they hold only at some of the time, then e is not a fact but a commonly accepted truth which may sometime be attacked or doubted.

What do actions do? Actions create or destroy facts (see Gabbay and Woods [178]). So if at time t an action $\mathbf{e}$ is fired then the result is that some facts get deleted from the network and some new facts are added. We can also assume that all values V change as the result of the action.

For simplicity, let us assume that an action adds only one fact or deletes only one fact. Since we can formally delete by attacking we will only allow adding facts. By adding a fact we mean either a new fact or turning an existing argument into a fact. So an action has the form **e** = (preconditions, post conditions), where the precondition is a sequence of arguments $((x_i : a_i))$ and the postcondition is a sequence $((a \rightarrow y_i \rightarrow b_i))$. This means that we add the fact a and let it attack b_i with transmission rate $y_i, i = 1, \ldots, n$. In a given network if a is not a node then we add it as a node with value 1 and let it attack any node b_i which is in the network.

If a is already in the network, then "disconnect" all attacks on a by giving them value 0. Give a the value 1 and let a attack all existing b_i in the network. If b_i is already attacked by a with transmission rate u_i, then let the new transmission rate be $\max(u_i, y_i)$.

Note that **e** is stated independently of the network. To be activated we need the net final value of a_i to be at least x_i and then the postconditions act on the available b_i.

Example 9.33. Consider the network of Figure 9.8. Consider the action **e** with precondition $((x : a), (w : d))$ and postcondition $((b \rightarrow u \rightarrow c), (b \rightarrow y \rightarrow g))$. This action can be applied to the network of Figure 9.8.

The result is Figure 9.46 below. Note that since there is no node g in the network, $b \rightarrow y \rightarrow g$ is not implemented. This is equivalent to Figure 9.47.

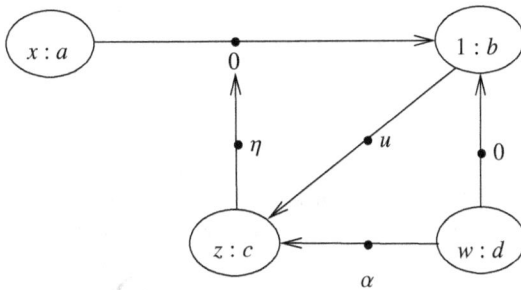

Fig. 9.46.

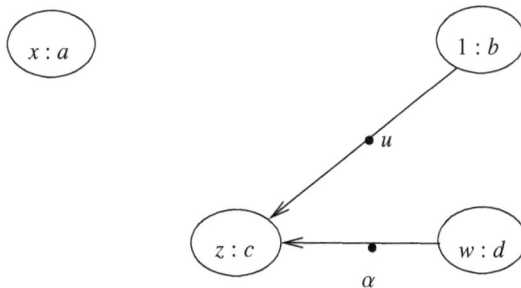

Fig. 9.47.

The next question for us to answer in a temporal network is the following. If action **e** is activated at time t, when do we see the result? If the network operates in discrete time, then the result is at time $t + 1$. Otherwise we have to treat the action like an *impulse* in a physical system, as when a ball hits another ball and gets it moving, and assume the result of the action **e** at t manifests itself immediately at all times s such that $t < s$. We have to give a reasonable definition of how the result of the action manifests itself. A good example for initial consideration is that if a new argument e is created by an action at time t then it shows up at all times $s > t$ and its strength at time $s > t$ decays slowly as s increases, say it has the form $V_s(e) = \dfrac{k}{1 + s - t}, k$ a constant ≤ 1. Similarly we can ask for a decay of the transmission rates.

If we bring into the argumentation system the idea that traversing the network also takes time (as we have in the biological case) we get two time movements: the traversing time and the network change time.

Such a combination exists in legal arguments. Court cases take time to argue and 'evaluate' and thresh out the evidence and in parallel the laws and public perception of justice get changed. In many cases they influence one another.

Let's go back to the biological case. Here the network does not change and the time components are the cycles (generations) through the network. So what can network change mean? This can be genetic mutations or genetic engineering which change the parameters in the system. The predator–prey relationship can change because of mutations in the prey, etc, etc. Major disasters can affect the ecology. Deterioration of the habitat and environment can affect a slow continuous change in the parameters.

Some arguments lose their potency with the passage of time. This is well known in political context. Politicians sometimes wait for the 'storm' to blow away, especially in matters of corruption and public protest. For example, some members of the UK Parliament (MPs) were recently exposed as claiming unjustifiably large expenses. There was a strong public protest to these findings, resulting in the resignation of some MPs. Many, however, have kept a low profile, awaiting for the public to forget. Let a represent the public protest over excessive expense claims, and b denote the standing of MPs, symbolically we may have that now a attacks b, see Figure 9.48, but not

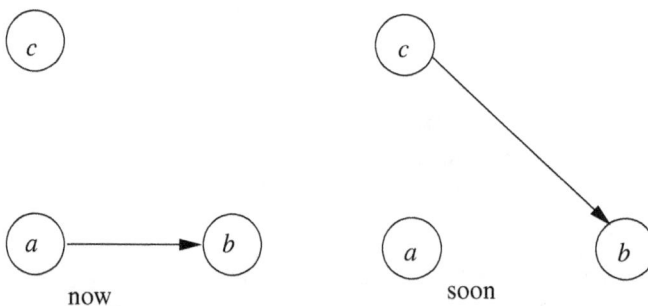

now   soon

Fig. 9.48.

for long, soon a will no longer attack b but a new attack on b, from c, say political in-fighting, may occur.

In such cases we can represent the arguments and the attacks as time dependent, $a(t), b(t)$ where t represents time. In contexts where arguments have strength (i.e. $a(t)$ is a number between 0 and 1) we can even consider the rate of change, $\frac{da}{dt}, \frac{db}{dt}$ and include it in our considerations. It may be convenient to represent the situation as in Figure 9.49. Where we do indicate attack arrows as on and off. Better still is to put

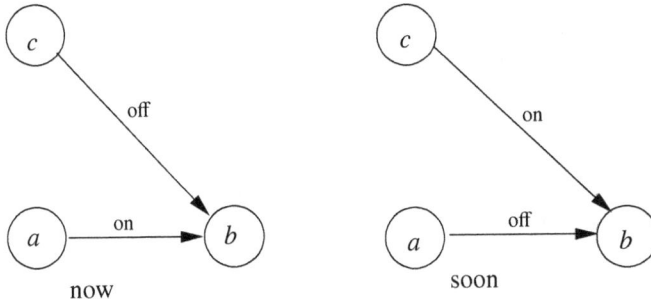

Fig. 9.49.

labels on the attacks, e.g. $l(c, b), l(a, b)$, which can be on or off, and consider these labels as time dependent.

9.6 Conclusion and discussion

We now conclude this chapter. We embarked from traditional argumentation networks in the numerical direction. This allowed us to compare and see the place of traditional argumentation in the landscape of general networks, (which are mostly numerical). We concentrated our discussion on the handling of numerical loops and on the concepts of numerical attack and support. We discovered interesting connections and new ideas. What is left to be done in this section is a comparison with some related papers published since 2005 and a discussion of further research. This we do now.

9.6.1 Comparison with related papers 2005-2011

Our current chapter is an expansion of our 2005 original paper [332]. This paper was not always noticed by the argumentation community and so some related results have been published since 2005. We discuss some of these papers here.

1. The paper by Rolf Haenni 2009, on probabilistic argumentation, [210].
   This paper relates to our paper in the following way. In our numerical argumentation networks we have nodes of the form $(x : a)$ and attacks of the form $(x : a) \to (y : b)$ or support of the form $(x : a) \twoheadrightarrow (y : b)$, x and y are numbers in $[0, 1]$. We take these numbers as given, and do not ask where they come from.So suppose a is a fact and it is established because several witnesses with varying reliabilities testified to a. If we have a system of belief functions or probabilities on the sources of information that supports a, we can calculate a final number x to annotate a. Paper [124] offers such a system. It is interesting to node that the Dempster-Shafer Rule does appear in this paper in the context of combining numbers and it also appears in our chapter as well as we have seen in Section 4.2

2. The paper by Leite and Martins 2011, on Social Abstract Argumentation, [237]. At the beginning of May 2011, J. Leite sent D. Gabbay a copy of this paper, which was submitted to IJCAI. He said he saw an abstract of Gabbay's paper [167] and thought it was related. Gabbay sent him papers [166, 167]. Paper [237] is indeed related to paper [166], which deals thoroughly with the equational approach to argumentation. In their paper [176], see also Chapter 20 below, Gabbay and Rodrigues addressed the ideas in paper [237]. Paper [237] has in it implicitly as a side-effect a valuable principle relating to the computation of steady state solutions to equations arising from numerical argumentation networks . It is a numerical analysis problem and is too involved to discuss here. We shall address it in the next version of [176].

3–4. The paper by Cayrol and Legasquie-Schiex 2005 on Graduality in Argumentation [86], and the paper by Matt and Toni 2008 on A Game-Theoretic Measure of Argument Strength for Abstract Argumentation, paper [254].

These two papers are concerned with the following problem: ordinary Dung argumentation and the Caminada labelling classify arguments into three classes only: *in*, *out* and *undecided*. These authors share the view that a finer classification is needed. Thus if we look at Figure 9.3, we see that a is not attacked by anything, while c is attacked by b and defended by a. There is a unique extension, grounded extension, $a = $ in, $b = $ out, $c = $ in.

The view of these authors is that a is much more valued in, while c is in, but not so much valued as a.

So the more attacked a node x is, the less value it has, even if it is in.

We therefore seek to define a numerical valuation function $\lambda x V(x)$ on arguments to reflect this situation. Paper [86] uses the geometry of the network to define such a function, $V(x)$, by looking at chains of attack and defence leading to the element x. Paper [254] also uses the geometry of the network, looking at the attackers of x and at the arguments which x attacks and by means of game theoretical methods gives a numerical value $V(x)$.

Our analysis of these papers is as follows:

a) First they offer a numerical valuation of points $x \in S$ in an argumentation network (S, R) which reflect their geometric-topological position in the network.

b) Second they claim that this numerical value can be used to give some meaningful value as to how "in" or "out" the element x is.

Giving values to points in (S, R) is purely mathematical and is a well established practice. This is how metrization theorems in topology are proved and this is how metrics are introduced in projective geometry, by means of the cross-ratio. However, connecting these values obtained with arguments and saying that one argument is better than another based on these numerical values is another matter. One may not subscribe to the basic idea that an argument not attacked at all is more "in" than an argument which is attacked and defended.

There are also technical reasons against this view. If (S_1, R_1) is a subsystem fo (S_2, R_2) the relative valuations of any two points $s, y \in S_1$ may change, depending on whether we view them as part of S_1 or part of S_2, giving the feeling that arguments have no individual merit in themselves, but depend only on the context in which they are used.

Now let us compare the systems of [86] and [254] with our system in this chapter. When we write $(x : a)$, giving value x to argument a, the value x is external, calculated by method outside the network, like how reliable a is, or x is a value obtained by probability on some evidence as in paper [210], etc. Since these values are obtained or given externally, we need to say how to calculate transmitted values following attacks and support. So if $(x : a)$ attacks $(y : b)$, we need to say what is the new value y' of b, following the execution of the attack or support.

The values obtained by the geometrical methods of paper [86] and paper [254], already take into consideration the geometry of the network. They are therefore final values. We cannot apply our machinery to them.

On the other hand, we can apply geometric machinery to networks with numerical values, considering for geometrical purposes the items in the network as slightly more complex, namely of the form $(x : a)$. So the values obtained by papers [86] and [254], have a different standing altogether, than what we do.

Consider for example, a node a in the network which is being attacked by other nodes. In our system we can give it value $x = 1$ (i.e. we allow $(1 : a)$). In the systems of papers [86] and [254], the value 1 is reserved (we think) only to points in the network with no geometrical attackers.

Papers [86, 254] can be compared, however, with the equational approach of Gabbay papers [166, 167], see also Chapter 12 below. The equations are written based on the geometry of the network and the solution to the equations give numerical values to nodes. This can be seen as a third method of making distinctions between nodes which also makes use of the geometry of the network.

5. The paper by Cayrol, Devred and Lasasquie-Schiex, 2008, on acceptability semantics accounting for strength of attacks in argumentation, paper [86].

This paper considers a finite number of types of attacks, organised in order of strength. We can present the network as say $(S, R_1, \ldots, R_n)$ with $R_i \subseteq S^2$ being pairwise disjoint, and with R_i being considered stronger than $R_{i=1}, 1 \leq i < n$. Consider for example Figure 9.50:

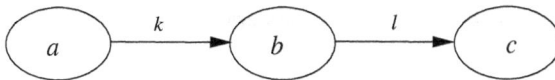

Fig. 9.50.

In this figure c is attacked by b but is defended by a. We would expect the extension $\{a, c\}$. However, we get this extension only if $k \geq l$.

Comparing this paper with our chapter, there are two ways to look at it:

a) Since the number of types of attacks is finite we can look at the paper as a contribution in the direction of classical logic argumentation either in the spirit of Chapter 3, or in the spirit of Chapter 13 below. The framework is classical logic, there are several types of attacks $R_1, \ldots, R_n$ and they are used in one way or another to define extension. The extra requirement that R_i are disjoint and that there is a meta-level ordering on $\{R_i\}$, is just one possible axiomatic restriction on the system (though these are exactly the axioms which allow us to talk about strength of attack).

b) The second point of view is to say the system of paper [86] is indeed a case of strength of attacks which is a special case of our system. We can simulate the system of [86] in our system as a special case. It is not our purpose here to write a mini paper about [86], but simply to compare and give the reader an orientation. To this end, consider Remark 9.26 and Figure 9.37 and compare it with Figure 9.50. They are very similar. Now consider Figure 9.38. This figure was shown equivalent in our system (as discussed in Remark 9.26) to Figure 9.37. Therefore we can look now at the "corresponding" Figure 9.51. In this figure b attacks c with strength l while a supports c with strength k. Clearly in this picture c will survive only if $k \geq l$.

The above is just one possibility of how we could simulate paper [86] as part of our system.

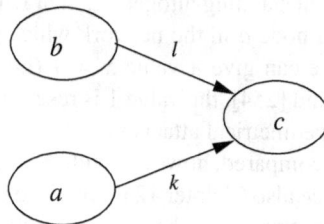

Fig. 9.51.

9.6.2 Further research

Let us summarise in this concluding section the new ideas emerging in this chapter, which show great potential and require further research.

1. NEURO-FUZZY ARGUMENTATION NETWORKS
   This is a natural generalisation of the numerical strength idea of sections 1 and 2. It has not been picked up in the argumentation community and not discussed in this chapter and book. We are going to pursue it extensively in a planned forthcoming book.
2. HIGHER LEVEL ATTACKS
   These are the reactive arrows emanating from arrows to arrows. The community have rediscovered this both as a concept and as a tool. In [163] we show how to reduce this concept to object level by adding more arguments. In fact this book deals with metalevel approach to argumentation.
3. CONNECTION WITH CROSS-RATIO OF PROJECTIVE GEOMETRY
   We believe this is very promising as well as mysterious. We still have to fully understand and exploit this connection. The key figure here is the situation described in Figure 9.41. C is attacking D and the result is calculated using the cross ratio with A and B. This is pure geometry on the line AB. So C and D can be vectors of length n, each containing multiple context incomparable arguments attacking the corresponding componentwise arguments. What would be the result? If we regard C and D as vectors in n-dimensional space and pass the line CD

through the n-dimensional unit sphere we get the points A and B and we can find the result. It is the point E on the line AB at a distance e from the point A where

$$\frac{e}{(1-e)} = \text{The Cross Ratio}(A, B; C, D)$$

This locates the point E which we can take as the result of C attacking D in n-dimensional space.

4. TEMPORAL DYNAMICS

   Much more needs to be done on the temporal aspects of arguments. Chapter 6 was just the basics. See also how we use temporal stamping to resolve loops in Chapter 19.

5. GEOMETRICAL NUMERICAL ARGUMENTATION NETWORKS

   We were impressed with papers [86] and [254] which were discussed in Section 6.1. We would like to offer our own view in the matter which includes both attack and support. arguments.

9.7 Appendices

9.7.1 Appendix A: Further examples of numerical networks

This appendix shows more familiar examples reviewed as numerical networks.

Modal Networks

We can read the nodes as possible worlds in a Kripke model and read the values as fuzzy truth values. ε is the fuzzy value of the accessibility of a to b (i.e. a arrow b means a is a possible world for b (i.e. bRa holds), while x is the fuzzy value of a being a possible world in the first place. So if $V_e(\varphi)$ gives a fuzzy value to the wff φ at world e, then $V_b(\Box\varphi) = \mathbf{f}(V_b(\varphi), \bar{V}_{a_i}(\varphi), \varepsilon_i)$, where a_i are all the nodes leading with an arrow into b.

It is worth giving a formal definition. See Chapter 14 for full details.

Definition 9.34.

1. *Let $\mathbb{L}$ be a propositional language with atoms $\{q_1, q_2, \ldots\}$, a modality $\Box$ and possibly other connectives $\mathbb{C}$. To fix our thoughts, say $\mathbb{C} = \{\Rightarrow, \neg\}$, where $\Rightarrow$ can be thought of as the Łukasiewicz many-valued implication (with truth values in $[0,1]$ and $0 = $ true and truth table [Value $(A \Rightarrow B) = Max(0, Value(B) - Value(A))$] and $\neg$ is a negation (with truth table [Value $(\neg A) = 1 - Value(A)$]).*

2. *A modal network model $\mathbf{m}$ is a family of models $\mathbf{m}_q = (A, T, V_q, \mathbf{f})$, q an atom of $\mathbb{L}$, such that each $\mathbf{m}_q$ is a finitely branching attack network model in the sense of Definition 9.6. Thus in $\mathbf{m}$ A, T and $\mathbf{f}$ are fixed and V_q varies with q. We assume that $\mathbf{f}$, V_q give values in $[0,1]$. We take $\mathbf{f}(y, x_i, \varepsilon_i) = \underset{i}{Sup}(\varepsilon_i \Rightarrow x_i) = \underset{i}{Sup} \, Max(0, x_i - \varepsilon_i)$.*

3. *For each $t \in T$ and each wff φ we define the value $V_\varphi^n(t)$, (for $n = 0, 1, 2, \ldots$) as follows:*

   a) *$V_q^0(t) = V_q(t)$, for atomic q, and $t \in T$.*

    b) $V_q^{n+1}(t) = \mathbf{f}(V_q^n(t), \bar{V}_q^n(a_i), \bar{V}_q^n(a_i \leftrightarrow t))$, where $a_1, \ldots, a_n$ are all the nodes such that $a_i \leftrightarrow t \in T$.

    c) $V_{A \Rightarrow B}^n(t) = Max(0, V_B^n(t) - V_A^n(t))$.

    d) $V_{\neg A}^n(t) = 1 - V_A^n(t)$.

    e) $V_{\Box A}^n(t) = V_A^{n+1}(t)$.[20]

4. We say $\mathbf{m}$ is stable iff for any wff A and any $t \in T$ there exists an n such that for all $m \geq n$ we have $V_A^m(t) = V_A^n(t)$. For stable models we can let $V_A^\infty(t) = \text{Lim}_i V_A^n(t)$.

5. We call a stable model $(A, T, V_A^\infty, \mathbf{f})$, a fuzzy modal model for $\mathbb{L}$.

Example 9.35 (Ordinary modal logic).

1. Let (S, R, h) be a traditional Kripke model for the language with $\{\rightarrow, \neg, \Box\}$, with S the set of possible worlds, R the accessibility relation and h the assignment to the atoms, (i.e. for each atomic $q, h(q) \subseteq S$). We assume that (S, R) is finitely branching, i.e. for each t the set $S_t = \{s | tRs\}$ is finite. Note that many modal logics are complete for a class of finitely branching models.

2. Let $A = S, T = S \cup \{a \leftrightarrow b | bRa\}$.

3. Let $V_q(a \leftrightarrow b) = 1$ for all atomic q and let $V_q(t) = 1$ iff $t \in h(q)$, for $t \in S$.

4. Let $\mathbf{f}(V(t), \bar{V}(a_i), \bar{V}(a_i \leftrightarrow t)) = 1$, where $a_1, \ldots, a_n$ are all nodes such that tRa_i holds, iff $\bar{V}(a_i) = 1$ for all $1 \leq i \leq n$.

5. We claim this model is stable. This can be proved by induction on the wff φ.

6. Note that we can get a new variety of modal logics by changing $\mathbf{f}$ from point to point, or by making $V_{\Box A}^n(t)$ dependent on $\{V_A^{n+m}(t) \mid m = 0, 1, 2 \ldots\}$ in a variety of ways.

Physics - special relativity

This has to do with composition of masses and velocities in special relativity. Consider the situation in Figure 9.52

Fig. 9.52.

    Particle 1 has rest mass m_1 and velocity u_1. Particle 2 has rest mass m_2 and velocity u_2. They are both moving in the same direction and they collide in an inelastic way, and form particle 3 with rest mass m_3 and velocity u_3. We view this as an attack of

---

[20] The reader should carefully note that we have huge scope here for defining a multitude of different modalities by choosing the dependence of $V_{\Box A}^n(t)$ on the set $\{V_A^{m+n}(t), m = 0, 1, \ldots\}$. What we here define is a $\mathbf{K}$-type modality. We can also define the hypermodality of [142] by letting:

$$V_{\Box A}^n(t) = \begin{cases} V_A^{n+1}(t), & \text{for } n \text{ odd} \\ Max(V_A^{n+1}(t), V_A^n(t)), & \text{for } n \text{ even} \end{cases}$$

node (m_1, u_1) upon node (m_2, u_2) with relativistic transmission 1 (full transmission) to form a new value $(m_2', u_2') = (m_3, u_3)$. The new value is the following, see [109, p. 51, exercise 4.7].

$$m_3^2 = m_1^2 + m_2^2 + 2m_2 m_2 \gamma_1 \gamma_2 (1 - u_1 u_2/c^2),$$
$$u_3 = \frac{m_1 \gamma_1 u_1 + m_2 \gamma_2 u_2}{m_1 \gamma_1 + m_2 \gamma_2},$$

with

$$\gamma_1 = (1 - u_1^2/c^2)^{-\frac{1}{2}}, \gamma_2 = (1 - u_2^2/c^2)^{-\frac{1}{2}}$$

where, as usual, c is the speed of light.

9.7.2 Appendix B: Interpretation of Loops

There are various interpretations for the situation depicted in Figure 9.21 besides our argumentation network interpretation.

The Ecology Interpretation
The network can be interpreted as an ecology, where nodes represent species. Species a feeds on species b and species b feeds on species a. The functions $\mathbf{f}_1$ and $\mathbf{f}_2$ give the success rates. This is a predator-prey situation.

Let V_n be the population of some species at generation n. We assume population growth is a discrete process taking place in cycles. Such biological examples are provided by many temperate zone arthropod populations, with one short-lived adult generation each cycle. One possible recurrence equation is the following
$V_{n+1} = V_n(1 + r(1 - \frac{V_n}{K}))$, where r and K are constants.

K is the maximum size for the population and r is a factor measuring dependence on the density of the population. The reader should compare this equation with the equation $V_b = \mathbf{f}_1(V_a, x, \varepsilon)$ arising from Figure 9.21. See [238, p. 324].

This equation is called *the non-linear logistic equation* which has the standard form

$$U_{n+1} = rU_n(1 - U_n), r > 0$$

This equation can exhibit chaotic behaviour depending on the value r, see [267].

A slightly different pair of equations has to do with parasitic life forms. Here we have, besides the population N_n, a parasitic population V_n. The recursive equations look like the following:

- $V_{n+1} = N_n - N_{n+1}/F$
- $N_{n+1} = FN_n f(N_n, V_n)$.

F is a factor indicating the proportion of those who escape the parasite. The difference between this equation for V_{n+1} and a direct recursion for V_{n+1} is that it is more complex. We get

- $V_{n+1} = N_n(1 - f(N_n, V_n))$
- $N_{n+1} = FN_n f(N_n, V_n)$

See [238, pp. 338].

Let us look at another example from biology. This is a model by M. P. Hasssell (1978) of two parasitoids (P and Q) and one host (N) model. The equations are (see [20, p. 295])

$$N_{t+1} = \lambda N_t f_1(P_t) f_2(Q_t)$$
$$P_{t+1} = N_t[1 - f_1(P_t)]$$
$$Q_{t+1} = N_t f_1(P_t)[1 - f_2(Q_t)]$$

where N, P and Q denote the host and two parasitoid species in generations t and $t+1$, λ is the finite host rate of increase and the functions f_1 and f_2 are the probabilities of a host not being found by P_t or Q_t parasitoids, respectively. This model applies to two quite distinct types of interaction that are frequently found in real systems. It applies to cases where P acts first, to be followed by Q acting only on the survivors. Such is the case where a host population with discrete generations is parasitized at different developmental stages. In addition, it applies to cases where both P and Q act together on the same host stage, but the larvae of P always out-compete those of Q, should multi-parasitism occur.

The functions f_1 and f_2 are:

$$f_1(P_t) = \left[1 + \frac{a_1 P_t}{k_1}\right]^{-k_1}$$
$$f_2(Q_t) = \left[1 + \frac{a_2 Q_t}{k_2}\right]^{-k_2}$$

where k_1 and k_2, a_1 and a_2 are constants.

To compare the biological model with the argumentation model, we put $a_1 = a_2 = 1$, $\lambda = 1$ and $k_1 = k_2 = -1$.
This gives

$$f_1(P_t) = 1 - P_t$$
$$f_2(Q_t) = 1 - Q_t$$

and therefore

$$N_{t+1} = N_t(1 - P_t)(1 - Q_t)$$
$$P_{t+1} = P_t N_t$$
$$P_{t+1} = Q_t N_t(1 - P_t)$$

giving us the appropriate functions for attack and counterattack for the situation in Figure 9.53:

In Figure 9.53, a and b attack c. c counterattacks a and b and a attacks b. The transmission rates are 1. Since c is attacked by a and b, the new value for c is $N(1 - P)(1 - Q)$. Since a is counterattacked by c, the new value for a is PN.[21] Since b is counterattacked by c and attacked by a the new value for b is $QN(1 - P)$.

To give Figure 9.53 some meaning, think of a, b, c as follows:

c = The US President has a strong case for re-election.
a = A deteriorating situation in Iraq (US soldiers killed) (attacks his chances).
b = Lack of success in combatting Al-Qaeda.

---

[21] See Example 9.4 as to why the counterattack value is PN and not $(1 - P)N$.

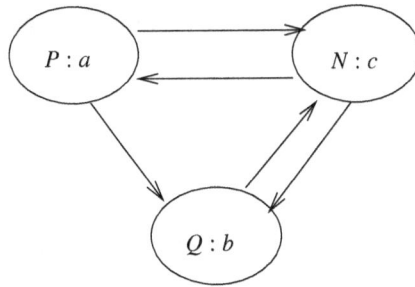

Fig. 9.53.

Clearly, if the value of c is low, less effort is required in attacking it from a and b. This explains the counterattack loop.

a attacks b by the argument that the situation in Iraq has diverted Al Qaeda away from attacking US territory proper.

To sum up, we have shown a connection with biological models. In view of this connection we would like to refer to loops as *ecologies* (of arguments).

Feedback Interpretation
We can consider the figures as a control-feedback situation. Say node b is a feedback for node a.

We consider a simple model of an electrical generator. We have three nodes H, P and I. H represents a hydraulic power source, moving a core in the power plant P producing electricity. The core rotates inside a magnetic field generated by a current coming from industry I. This produces a current which goes to I (Industry). I then sends part of the current it receives back to P and thus maintaining the magnetic field. Figure 9.54 describes the situation in its steady state.

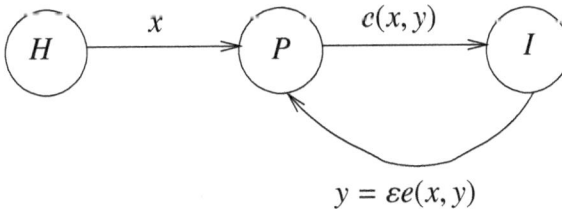

Fig. 9.54.

To keep the story simple, let us assume that there is full transmission from H to P, from P to I and from I to P. Let x be the transmission of hydraulic energy coming from H and let y be the current coming from I. Let $e(x,y)$ be the resulting current transmitted to I and let ε be the proportion of it coming back from I to P. Thus e is monotonic increasing in x and y and we have in a steady state that

- $y = \varepsilon e(x,y)$
- $e(x,0) = e(0,y) = 0$ (i.e. if no hydraulic power turns the core or no current is coming to maintain the electromagnetic field then we have no output e.)

- ε is such that $(1 - \varepsilon)e$ is maximal (i.e. we feedback current in such a way that what is left at I is maximal, given a fixed x).

10

An Equational Approach to Argumentation Networks

10.1 Introduction

This chapter expands on our equational ideas introduced in Chapter 7. A short introduction to the results of this chapter appears in *Proceedings of ECSQUARU'11*, Springer LNAI Series, see [136].

This section introduces the ideas involved in the chapter one by one through several subsections.

To achieve our goal, as outlined in Section 1.1 below, we adopt an equational approach, going back to the 19th century way of doing logic.

The Equational approach has its conceptual roots in the 19th century following the algebraic equational approach to logic by George Boole [64], Louis Couturat [103] and Ernst Schroeder [301].

The equational algebraic approach was historically followed, in the first half of the 20th century, by the Logical Truth (Tautologies) approach supported by giants such as G. Frege, D. Hilbert, B. Russell and L. Wittgenstein. In the second half of the twentieth Century the new current approach has emerged, which was to study logic through it consequence relations, as developed by A. Tarski, G. Gentzen, D. Scott and (for non-monotonic logic) D. Gabbay.

See Section 10.8.1 for further discussion.

10.1.1 Aims of this chapter

We have several good reasons for presenting this chapter.

1. To provide a general computational framework for Dung's argumentation networks; a framework in which the logical aspects, computational aspects and the conceptual aspects involved in Dung's original proposal can be isolated, highlighted and analysed, and thus paving the way for orderly responsible generalisations.

   The logical aspects involve the question of what is the logical content of an argumentation network and what inferences we can draw from it, as discussed in Chapter 5, and indeed, as will be further discussed in Chapters 11, 12, and 13. The computational aspects have to do with viewing the abstract argumentation networks as directed graphs or as finite models with binary relations on them and various algorithms for extracting subsets of such graphs or models. Recall Chapter 3 on annotation theories. The conceptual aspect is the reason behind the computation, involving concepts such as conflict free sets, admissibility and a variety of extensions. See Caminada's conceptual survey slides [83] and especially slide 19. At present Dung's networks are generalised in many ways by many capable researchers. Unfortunately, we have no general meta-level approach which the community can use for guidance and comparison.

2. To generalise Dung's argumentation networks in a natural way and connect and compare it with other networks communities, such as neural nets, Bayesian nets, biological–ecological nets, logical labelled deductive nets and more. See [30, 38, 37, 40] and Chapters 9 and 14.

   These networks have a different conceptual base but they look like abstract argumentation networks, i.e. they are directed graphs. We manipulate the graphs differently because they come from different applications. So the question to ask is whether we can we find common grounds (such as an equational approach to such graphs) which will bring the applications together at least on the formal mathematical side?

3. To introduce in a natural way various meta operations on networks such as distributed networks (modal logic), time dependence and fibring which exist in other types of networks and logics.

4. To connect with pure mathematics, numerical analysis and computational algebra. We will focus on Dung abstract argumentation semantics, although we also touch some other semantics such as CF2 semantics in Section 8.4.

The rest of this introductory section will explain through examples and discussion the ideas of this chapter and later sections will develop the mathematical machinery involved.

Dung's argumentation networks (see [114]) have the form (S, R_A) where S is a set of arguments, which for the current purposes we assume to be finite, and R_A is a binary attack relation on S. We are interested in subsets E of S of arguments which are admissible, that is self defending and conflict free, namely:

1. E is conflict free, namely for no x, y in E do we have that xR_Ay.
2. E defends each of its elements: Whenever for some x, we have xR_Ay and y is in E, there is some z in E defending y, i.e. we have zR_Ay. (E is self-defending.)
3. E is complete if E contains all the elements it defends.

The smallest such complete E is called the *grounded extension*, a maximal E (there may be several different such maximal sets) is called a *preferred extension*, and if we are lucky, we may also have a *stable extension* E, namely one which attacks anything not on it.

Such extensions are perceived as indicating coherent logical positions which can defend themselves against attacks.

Recall Chapters 2 and 3 for surveys. The above definitions is the original set theoretic way of introducing extensions. There is another very useful equivalent way of defining extensions, using the Caminada labelling, recall Chapter 2 for a survey. The Caminada labelling is the approach compatible with our Equational approach. It is described in the next subsection and it is the approach we are going to work with.

10.1.2 The conceptual vs. computational distinction

Consider the network of Figure 10.1

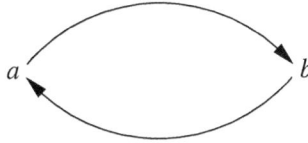

Fig. 10.1.

This network has two nodes $S = \{a, b\}$ and the attack relation R_A is described by the singel arrow "→". We have

$$R_A = \{(a, b), (b, a)\}.$$

We follow Caminada (see survey in [81]) and describe three extensions:

$$E_0 = \{a = b = \text{ undecided}\}$$
$$E_1 = \{a = \text{ in}, b = \text{ out}\}$$
$$E_2 = \{a = \text{ out}, b = \text{ in}\}$$

The rules governing the assignment of these values are the Caminada conditions:

Caminada conditions:

(C1)   If $a = $ in and a attacks b then $b = $ out.
(C2)   If all attackers x_i of b are out (or if there are no attackers) then $b = $ in.
(C3)   If all attackers x_i of b are either out or undecided with at least one such attacker is undecided then $b = $ undecided.

By looking at a Dung extension E we are saying two things:

1. *Computational statement:*
   Assign values from {in, out, undecided} to all nodes in such a way that the above connections are observed.
2. *Conceptual statement*
   Let
$$E_{\text{in}} = \{x | x = \text{ in}\}$$
   then we are saying that E_{in} is a set of arguments which is conflict free and represents a a coherent position one can take in adopting the arguments in E_{in}, a position which is able to defend itself.

Now consider Figure 10.2, where we add a third element and add the new relation of support R_S, indicated by the double arrow.
We have $S = \{N, a, b\}$. $R_A = \{(a, b), (b, a)\}$ and the support relation $R_S = \{(N, a)\}$. N is a special element for providing support.
We now have two problems in this new expanded network with support.

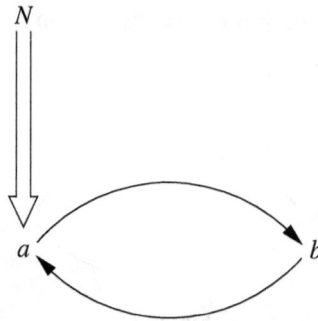

Fig. 10.2.

1. *Computational problem*

   How to assign values {in, out, undecided}, or possibly additional labels like weakly undecided, almost in, possibly out, etc.

2. *Conceptual problem*

   Extend conceptual meaning to "support" in the network. In other words, what do we mean by support. Of course our conceptual analysis will influence what labels we use in the computational approach.

The answer to (1) is to extend the algorithm governing the assignment of values {in, out, undecided, and other values} to cover cases of support.

The answer to (2) is to modify the network to a new network by some conceptual considerations, which take account of the support relation.

This can be done by modifying the notion of extension or the notion of attack or both.

Let us, by way of example, offer a minimalistic conceptual analysis of support, reading support as "endorse". If x supports y, then x does not give more strength to the argument y, but only shares its fate, having endorsed it. Let us see what we can offer as a corresponding computational approach.

The computational aspect can be solved, for example, by keeping within the labels {in, out, undecided}, and by adding new rules $(N1)$ and $(N2)$.

$(N1)$ If a supports b and $a = $ out then b is out.

$(N2)$ If a supports b and $b = $ in then $a = $ in.

Modify (C2) to say

$(NC2)$ If all attackers x_i of b are out (or if there are no attackers) and if b does not support any y then $b = $ in.

Another possibility of understanding support is as a licence for attack and defence. Here we are really adding strength to the supported node. We can declare for example one of the following:

$(Q1)$ A supported node cannot be attacked. (This reminds us of Bench-Capon value based networks. See [43].)

$(Q2)$ A node which is not supported cannot attack. (See, for example, [275]).

The effect of $(Q2)$ is to cancel the arrow $b \rightarrow a$.

We thus get the following:

i. If we adopt $(N1)$, $(NC2)$ and $(N2)$, we get the following extensions in Figure 10.2:

$$E_0 = \{N = \text{in}, a = b = \text{undecided}\}$$
$$E_1 = \{N = \text{in}, a = \text{in}, b = \text{out}\}$$
$$E_2 = \{N = \text{out}, a = \text{out}, b = \text{in}\}$$

We can, of course, make all kinds of other distinctions about points which support and there is an extensive literature on this. See, for example, our paper [331], and this will be discussed in Chapter 15 below.

The above is just an example to explain the computational vs. conceptual aspects distinctions, see Section 3 for the analysis of Support.

ii. If we adopt $(Q1)$ then Figure 10.2 is reduced to Figure 10.3.

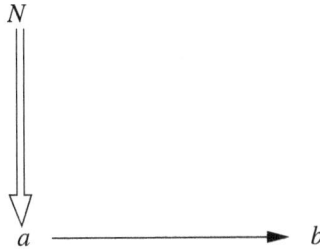

Fig. 10.3.

iii. If we adopt $(Q2)$ then Figure 10.2 is again reduced to Figure 10.3, but for a different reason. $(Q2)$ does not allow b to attack anything and $(Q1)$ does not allow b to attack a but it can attack other points.

We still need computational principles for Figure 10.3, though in this simple figure the only option is E_1.

10.1.3 Equational examples

This subsection is intended to motivate the formal equation section, Section 2. We give here several examples of the equational approach.

Let (S, R_A) be a Dung network. So $R_A \subseteq S^2$ is the attack relation. We are looking for a function $\mathbf{f} : S \mapsto [0, 1]$ assigning to each $a \in S$ a value of $0 \leq \mathbf{f}(a) \leq 1$ such that the following holds.

1. $(S, R_A, \mathbf{f})$ satisfies the following equations for some family of functions $\{\mathbf{h}_a\}, a \in S$:

   a) If a is not attacked (i.e. $\neg \exists x(x R_A a)$) then $\mathbf{f}(a) = 1$.

   b) If $x_1, \ldots, x_n$ are all the attackers of a (i.e. $\bigwedge_{i=1}^{n} x_i R_A a \wedge \forall y (y R_A a \rightarrow \bigvee_{i=1}^{n} y = x_i)$) then we have that $\mathbf{f}(a) = \mathbf{h}_a(\mathbf{f}(x_1), \ldots, \mathbf{f}(x_n))$.

Let us take, for example the same $\mathbf{h}_a = \mathbf{h}$ for all a and let

$$\mathbf{h}(\mathbf{f}(x_1), \ldots, \mathbf{f}(x_n)) = \prod_{i=1}^{n}(1 - \mathbf{f}(x_i))$$

The above equation we shall call Eq_{inverse}. We shall define other possible equations later on.

Thus we get

Eq_{inverse} for the function $\mathbf{f}$:

$$\mathbf{f}(a) = \prod_{i=1}^{n}(1 - \mathbf{f}(x_i)).$$

2. For any $(S, R_A, \mathbf{f})$, there exists a Caminada labelling of (S, R_A), such that

$$\lambda(a) = \begin{cases} \text{in, if } \mathbf{f}(a) = 1 \\ \text{out, if } \mathbf{f}(a) = 0 \\ \text{undecided, if } 0 < \mathbf{f}(a) < 1 \end{cases}$$

The equation

$$\mathbf{f}(a) = \prod_{i=1}^{n}(1 - \mathbf{f}(x_i))$$

ensures that:

If one of x_i (x_i are the attackers of a) is in then a is out.

If all the attackers are out then a is in.

If at least one of the attackers of a is undecided and none of the attackers of a are in then a is undecided.

The question remaining to be asked is the following:

If we have a Caminada labelling, can we find a function $\mathbf{f}$ solving the Eq_{inverse} equations which correspond to this Caminada labelling as in 2 above? The answer is negative, as Examples 10.7 and 10.11 show.

Let us now give a formal definition.

Definition 10.1 (Possible equational systems). *Let (S, R_A) be a networks and let a be a node and let $x_1, \ldots, x_n$ be all of its attackers.*

We list below several possible equational systems, we write $Eq(\mathbf{f})$ to mean the equational system Eq applied to $\mathbf{f}$:

1. $Eq_{\text{inverse}}(\mathbf{f})$

$$\mathbf{f}(a) = \prod_{i}(1 - \mathbf{f}(x_i))$$

2. $Eq_{\text{geometrical}}(\mathbf{f})$

$$\mathbf{f}(a) = [\prod_{i}(1 - \mathbf{f}(x_i))]/[\prod_{i}(1 - \mathbf{f}(x_i)) + \prod_{i}\mathbf{f}(x_i)].$$

We call this equation $Eq_{\text{geometrical}}$ because it is connected to the projective geometry Cross Ratio, see Section 9.4.

3. $Eq_{max}(\mathbf{f})$

$$\mathbf{f}(a) = 1 - \max(\mathbf{f}(x_i)).$$

We shall see the difference in the examples. In fact we shall see that this new function gives exactly the Caminada labelling.

4. $Eq_{suspect}(\mathbf{f})$

Let us further introduce a fourth system of equations which we call $Eq_{suspect}(\mathbf{f})$:

$$\mathbf{h}_a(\mathbf{f}(x_1), \dots, \mathbf{f}(x_n)) = \prod_i (1 - \mathbf{f}(x_i)), \ \ if \ \neg aR_A a \ holds$$

and

$$\mathbf{h}_a(\mathbf{f}(x_1), \dots, \mathbf{f}(x_n)) = \mathbf{f}(a) \prod_i (1 - \mathbf{f}(x_i)), \ \ if \ aR_A a \ holds.$$

Example 10.2. Consider Figure 10.1 again.
Let
$$\mathbf{f}(a) = \alpha$$
$$\mathbf{f}(b) = \beta$$

The equations are (using $Eq_{inverse}$, or indeed any other option, they are all the same on this example):

1. $\alpha = 1 - \beta$
2. $\beta = 1 - \alpha$

If we choose α, this fixes β.

For $\alpha = 1$ we get $\beta = 0$. For $\alpha = 0$, we get $\beta = 1$. For $0 < \alpha < 1$ we get $0 < \beta < 1$.

We thus get the three extensions, if we read any value strictly between 0 and 1 as undecided.

$$E_0 = \{a = b = \ undecided\}$$
$$E_1 = \{a = \ in, b = \ out\}$$
$$E_2 = \{a = \ out, b = \ in\}$$

Example 10.3. Consider Figure 10.4. Here the equation is (for $\mathbf{f}(a) = \alpha$, and all Equa-

Fig. 10.4.

tional options except $Eq_{suspect}$):
$$\alpha = 1 - \alpha$$

so $\alpha = \frac{1}{2}$.

Thus the extension is
$$\{a = \ undecided\}.$$

In the case of $Eq_{suspect}$, the equation is $\alpha = \alpha(1 - \alpha)$ and the solution is $\alpha = 0$.

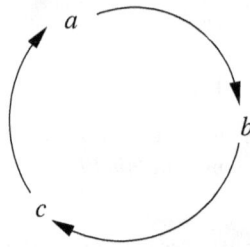

Fig. 10.5.

Example 10.4. Consider Figure 10.5
   Let $\mathbf{f}(a) = \alpha, \mathbf{f}(b) = \beta$ and $\mathbf{f}(c) = \gamma$.
   The equations are, under all options

1. $\alpha = 1 - \gamma$
2. $\beta = 1 - \alpha$
3. $\gamma = 1 - \beta$

From (1) and (3) we get $\alpha = \beta$ and from (1) and (2) we get $\alpha = \gamma$ and so $\alpha = \beta = \gamma = \frac{1}{2}$.
   The extension is $\{a = b = c = \text{ undecided}\}$.

Example 10.5. Consider Figure 10.6. Again we get, for $\mathbf{f}(a) = \alpha, \mathbf{f}(b) = \beta, \mathbf{f}(c) = \gamma$.

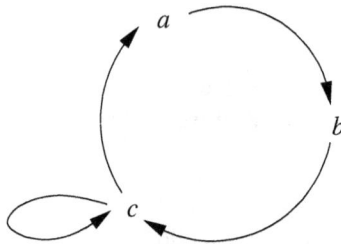

Fig. 10.6.

   Let us use the set $Eq_{\text{inverse}}(\mathbf{f})$:

1. $\alpha = 1 - \gamma$
2. $\beta = 1 - \alpha$
3. $\gamma = (1 - \beta)(1 - \gamma)$.

From (1) and (2) we get

4. $\gamma = \beta$

Therefore from (3)

5. $\gamma = \gamma(1 - \gamma)$

Therefore $\gamma = 0, \alpha = 1, \beta = 0$.
   This gives us the extension

$$\{a = \text{ in}, b = \text{ out}, c = \text{ out}\}.$$

Note that this corresponds to condition (2) on **f**. Since the only Dung extension says I don't know (undecided) on $\{a, b, c\}$, any value given by **f** is compatible.

Let us now consider Eq_{Suspect}. The equations are

1. $\alpha = 1 - \gamma$
2. $\beta = 1 - \alpha$
3. $\gamma = \gamma(1 - \beta)(1 - \gamma)$

From (1) and (2) we get

4. $\gamma = \beta$

therefore from (3) we get

$$\gamma = \gamma(1 - \gamma)(1 - \gamma)$$

The only solution is $\gamma = 0$, and therefore $\beta = 0, \alpha = 1$.

We now consider Eq_{max}. We get

1. $\alpha = 1 - \gamma$
2. $\beta = 1 - \alpha$
3. $\gamma = 1 - \max(\beta, \gamma)$.

From (1) and (2) we get

4. $\gamma = \beta$

From (3) we get

5. $\gamma = 1 - \gamma$

Therefore $\gamma = \frac{1}{2}$ and hence $\alpha = \beta = \frac{1}{2}$.

Let us now check $Eq_{\text{geometrical}}$. We get

1. $\alpha - 1 - \gamma$
2. $\beta = 1 - \alpha$
3. $\gamma = \frac{(1-\gamma)(1-\beta)}{(1-\gamma)(1-\beta)+\beta\gamma}$

From (1) and (2) we get

4. $\beta = \gamma$

and from (3) we get

5. $\gamma = \frac{(1-\gamma)^2}{(1-\gamma)^2+\gamma^2}$

Therefore

6. $\gamma(1 - \gamma)^2 + \gamma^3 = (1 - \gamma)^2$

and hence

$$\gamma^3 = (1 - \gamma)^3$$

The only solution is $\gamma = \frac{1}{2}$ and hence $\alpha = \beta = \frac{1}{2}$.

Remark 10.6. Actually there is a rationale to what is happening in the previous Example 10.5. Consider the network of Figure 10.6.

Start by assuming $a = $ in. Then b is unambiguously out. Now the question of whether c is in or out is not clear cut. We may wish to adopt a new policy in this case, different from the traditional one. c is not attacked by b but it does attack itself. So our new policy can consider whether to make c out or undecided. It cannot be in. That c is out is the best solution.

The other alternative, that a is out, entails b is in, therefore certainly c is out and so a must be in, a contradiction.

So the best solution is $a = $ in, $b = $ out and $c = $ out, as suggested by the equational network. Although the equational approach need not be compatible with Dung's approach, it being different and new, it is worth noting that in this case it does not contradict the Dung extension which says all undecided, but further refines it, as we have argued.

Let us look at the situation in terms of admissible extensions. The set $E = \{a\}$ is conflict free but not admissible because a is attacked by c and there is nothing in E which attacks c. This is the Dung concept of admissible.

Suppose we introduce and use a two new concepts:

- c is **Suspect** if c attacks itself, or possibly part of a loop of attacks which make it Suspect, (the details needs to be worked out).
- E is **Suspect-Admissible** if whenever a in E is attacked by c then either c is Suspect or E attacks c.

By exploiting the right concepts of **Suspect-Admissible** and of **Suspect** we might be able to find a qualitative set theoretic nonequational definition of extensions corresponding to our equational extensions.

Example 10.7. Consider Figure 10.7 and use the set $Eq_{\text{inverse}}(\mathbf{f})$

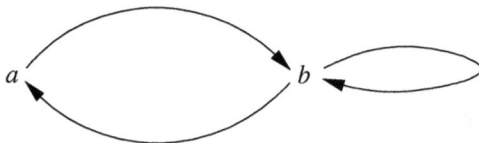

Fig. 10.7.

We have, under Eq_{inverse}

1. $\alpha = 1 - \beta$
2. $\beta = (1 - \alpha) \cdot (1 - \beta)$

Thus from (1) and (2) we get

3. $\beta = \beta(1 - \beta)$

So $\beta = 0, \alpha = 1$.

So we get the extension $\{a = 1, b = 0\}$.

It is clear from equation 3, that there is no way of letting $0 < \alpha < 1$ and $0 < \beta < 1$ (i.e. making a and b undecided) and satisfying the $Eq_{inverse}$ equations for this example. Again, see Remark 10.6.

Let us consider the other Eq options.

Under Eq_{max} we get

$$\alpha = \beta = \tfrac{1}{2}$$

Under $Eq_{suspect}$ we get the equations

1. $\alpha = 1 - \beta$
2. $\beta = \beta(1 - \alpha)(1 - \beta)$

From (1) and (2) we get

3. $\beta = \beta^2(1 - \beta)$

the only solution is $\beta = 0$.

Under $Eq_{geometrical}$ we get the equations

1. $\alpha = 1 - \beta$
2. $\beta = \frac{(1-\alpha)(1-\beta)}{(1-\alpha)(1-\beta)+\alpha\beta}$

From (1) and (2) we get

3. $\beta = \frac{\beta(1-\beta)}{\beta(1-\beta)+(1-\beta)\beta}$

We get $\beta = \tfrac{1}{2}$ and hence $\alpha = \tfrac{1}{2}$.

Example 10.8. Consider Figure 10.8 and use $Eq_{inverse}$.

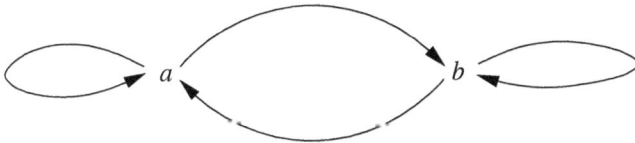

Fig. 10.8.

Here we have

1. $\alpha = (1 - \alpha)(1 - \beta)$
2. $\beta = (1 - \alpha)(1 - \beta)$

Therefore $\alpha = \beta$ and we have

$$\alpha = (1 - \alpha)^2$$
$$\alpha^2 - 3\alpha + 1 = 0$$
$$\alpha = 1.5 \pm \sqrt{1.25}$$

Only the $-\sqrt{1.25}$ makes sense as α must be in [01]. So $\alpha \approx 0.382, \beta = 0.382$. Note that while Dung says undecided, **f** is very specific about a and b.

Note that Eq_{max} and $Eq_{geometrical}$ give value $\tfrac{1}{2}$ to all nodes.

We now examine the case of $Eq_{suspect}$. The equations are

1. $\alpha = \alpha(1 - \alpha)(1 - \beta)$
2. $\beta = \beta(1 - \alpha)(1 - \beta)$

The only solution is $\alpha = \beta = 0$.

Example 10.9 (Caminada labelling and the Max function). We saw in Examples 10.5 and 10.7 and in Remark 10.6 that the functions of $Eq_{inverse}(\mathbf{f})$ do not give all possible Caminada labellings, but only some of them. To every function $\mathbf{f}$ there corresponds a Caminada labelling but some Caminada labelling may not have a corresponding $\mathbf{f}$. So we ask is there a system of equations which gives exactly the Caminada labelling? The answer is yes, it is the function $Eq_{max}(\mathbf{f})$. Let us check what we get for the network of Figure 10.6 under this labelling. The equations are:

1. $\alpha = 1 - \gamma$
2. $\beta = 1 - \alpha$
3. $\gamma = 1 - \max(\gamma, \beta)$.

The solution to these equations is $\alpha = \beta = \gamma = \frac{1}{2}$.

We get the same agreement for Example 10.7. We shall prove a general theorem in the next section.

Example 10.10. Let us do another example using all four options for equations, namely $Eq_{geometrical}$, $Eq_{inverse}$, Eq_{max} and $Eq_{suspect}$.

Consider Figure 10.9. We are looking for $\mathbf{f}$ solving the equations. Let $\mathbf{f}(a) = \alpha, \mathbf{f}(b) = \beta, \mathbf{f}(c) = \gamma$.

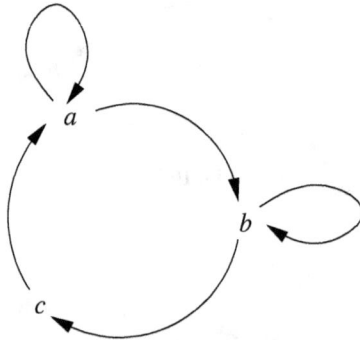

Fig. 10.9.

I We use $Eq_{inverse}$:

The equations are

1. $\alpha = (1 - \alpha)(1 - \gamma)$
2. $\beta = (1 - \beta)(1 - \alpha)$
3. $\gamma = 1 - \beta$

There are programs like Maple which can solve the equations of this sort and give all the solutions. We used one and got

$$\alpha = 1 - \frac{\sqrt{2}}{2}$$
$$\beta = \sqrt{2} - 1$$
$$\gamma = 2 - \sqrt{2}$$

Let's solve the equations by hand. We get from 3:

4. $\beta = 1 - \gamma$
5. From 4 and 1 we get that

$$\alpha = (1 - \alpha)\beta$$
$$\alpha = \beta - \alpha\beta$$
$$\alpha = \frac{\beta}{1+\beta}$$

Therefore

$$(1 - \alpha) = 1 - \frac{\beta}{1+\beta}$$
$$= \frac{1}{1+\beta}$$

6. So from 5 and 2 we get

$$\beta = \frac{1-\beta}{1+\beta}$$
$$\beta + \beta^2 = 1 - \beta$$
$$\beta^2 + 2\beta - 1 = 0$$
$$(\beta + 1)^2 = 2$$

So we get
7. $\beta = \sqrt{2} - 1$
   Hence
8. From 5 we get

$$\alpha = \frac{\sqrt{2} - 1}{\sqrt{2}} = 1 - \frac{1}{\sqrt{2}} = 1 - \frac{\sqrt{2}}{2}.$$

9. From (3) we get

$$\gamma = 2 - \sqrt{2}.$$

The interest in this case is that we are getting all kinds of values which shows that these equations are sensitive to the nature of the loops involved!

II. We use Eq_{max}:

The equations are

1. $\alpha = 1 - \max(\alpha, \gamma)$
2. $\beta = 1 - \max(\beta, \alpha)$
3. $\gamma = 1 - \beta$

We distinguish two cases:

Case 1: $\beta \geq \alpha$.
   Then from (2), $\beta = 1 - \beta$, i.e. $\beta = \frac{1}{2}$.
   So $\gamma = \frac{1}{2}$ and from (1)

$$\alpha = 1 - \max(\alpha, \tfrac{1}{2})$$
$$\alpha + \max(\alpha, \tfrac{1}{2}) = 1$$

The only way the last equation can be true is that $\alpha = \frac{1}{2}$.

So $\beta \geq \alpha$ implies $\alpha = \beta = \gamma = \frac{1}{2}$.

Case 2: $\beta < \alpha$.

From (2) we get

$$\beta = 1 - \alpha$$

and from (3) we get

$$\alpha = \gamma$$

So from (1) we get

$$\alpha = 1 - \alpha$$

So $\alpha = \frac{1}{2}$ and therefore $\beta = \gamma = \frac{1}{2}$.

So the final solution in this case is $\alpha = \beta = \gamma = \frac{1}{2}$.

III. We use Eq_{suspect}:

The equations are

1. $\alpha = \alpha(1 - \alpha) \cdot (1 - \gamma)$
2. $\beta = \beta(1 - \beta)(1 - \alpha)$
3. $\gamma = 1 - \beta$.

We check some cases.

Case 1: $\alpha = 0$. Then from (2) we get $\beta = \beta(1 - \beta)$ which forces $\beta = 0$ and from (3) $\gamma = 1$.

So the answer for case $\alpha = 0$ is that also $\beta = 0$ and $\gamma = 1$.

Case 2: $\alpha \neq 0$.

Can this case arise? From (1) we get

$$1 = (1 - \alpha)(1 - \gamma).$$

This cannot hold if $\alpha > 0$!

Thus the solution is $\alpha = 0, \beta = 0, \gamma = 1$.

This is compatible with Remark 10.6.

IV. We use $Eq_{\text{geometrical}}$.

The equations are:

1. $\alpha = \frac{(1-\alpha)(1-\gamma)}{(1-\alpha)(1-\gamma)+\alpha\gamma}$
2. $\beta = \frac{(1-\alpha)(1-\beta)}{(1-\alpha)(1-\beta)+\alpha\beta}$
3. $\gamma = 1 - \beta$.

We first note that the values $\alpha = 0$ or $\alpha = 1$ or $\beta = 0$ or $\beta = 1$ cannot be solutions and so in our calculations we can divide by α or β or $1 - \alpha$ or $1 - \beta$.

From equation (1) we get a contradiction if α is either 0 or 1, and from equation (2) we get a contradiction if β is either 0 or 1.

From (3) and (1) we get

4. $\alpha = \frac{(1-\alpha)\beta}{(1-\alpha)\beta+\alpha(1-\beta)}$

Solving (4) for β we get

$$\alpha(1-\alpha)\beta + \alpha^2(1-\beta) = (1-\alpha)\beta$$
$$\alpha(1-\alpha)\beta + \alpha^2 - \alpha^2\beta = (1-\alpha)\beta$$
$$\beta(1-\alpha - \alpha(1-\alpha) + \alpha^2) = \alpha^2$$

Therefore we get

5. $\beta = \frac{\alpha^2}{(1-\alpha)^2+\alpha^2}$

Therefore

$$1-\beta = \frac{(1-\alpha)^2}{(1-\alpha)^2 + \alpha^2}$$

Therefore

5a. $\frac{\beta}{1-\beta} = \left(\frac{\alpha}{1-\alpha}\right)^2$

We continue. Solving (2) for α we get

6. $\alpha = \frac{(1-\beta)^2}{(1-\beta)^2+\beta^2}$

Therefore $1-\alpha = \frac{\beta^2}{(1-\beta)^2+\beta^2}$

So

6a. $\frac{\alpha}{1-\alpha} = \left(\frac{1-\beta}{\beta}\right)^2$

From (5a) and (6a) we get

7. $\frac{\beta}{1-\beta} = \left(\frac{(1-\beta)}{\beta}\right)^4$

Let $x = \frac{\beta}{1-\beta}$.

Then x is s fifth root of unity, i.e. it solves the equation

8. $x^5 - 1 = 0$.

The fifth roots of unity can be solved by radicals, the only real solution is $x = 1$, which makes $\beta = \frac{1}{2}$ and so $\alpha = \gamma = \frac{1}{2}$.

The other solutions are complex numbers and are not relevant to us[1]

Example 10.11 (Comparing Eq_{max} and $Eq_{inverse}$). We shall show that these two equational systems may not yield the same extensions. The network is described in Figure 10.10.

Extensions according to Eq_{max}.
Let us compute the equations according to Eq_{max} and their possible solutions.
    The equations are (we write "x" instead of $\mathbf{f}(x)$):

1. $a = 1 - b$
2. $b = 1 - \max(a, b)$
3. $c = 1 - \max(b, e)$
4. $d = 1 - c$
5. $e = 1 - d$.

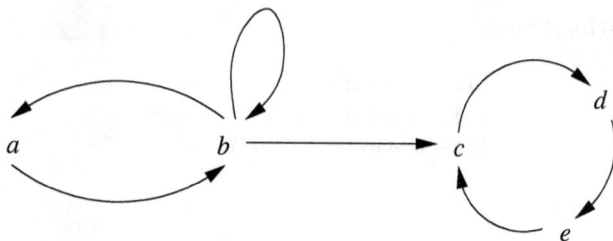

Fig. 10.10.

We get from (4) and (5)

6. $c = e$

From (1) and (2) we get

7. $b = 1 - \max(1 - b, b)$.

The only solutions to (7) are $b = 0$ and $b = \frac{1}{2}$.

Case $b = 0$
We get from (1) that

$$a = 1$$

From (6) and (3) we get

$$c = 1 - \max(0, c)$$

therefore

$$c = \tfrac{1}{2}.$$

Therefore also $d = e = \frac{1}{2}$.

Case $b = \frac{1}{2}$
Therefore from (1)

$$a = \tfrac{1}{2}.$$

From (3) and (6) we get

$$c = 1 - \max(\tfrac{1}{2}, e).$$

From (4) and (5) we get $c = e$. The only solution is therefore

$$c = \tfrac{1}{2}.$$

Therefore $e = \frac{1}{2}$ and $d = \frac{1}{2}$.

Summary for $Eq_{\max}$
We get two extensions

----
[1] The other solutions are are

$$[(1 - \sqrt{5}) \pm i\sqrt{10 + 2\sqrt{5}}]/4$$
$$[(1 - \sqrt{5}) \pm i\sqrt{10 - 2\sqrt{5}}]/4.$$

1. $\{a\}, (a = 1, b = 0, c = d = e = \frac{1}{2})$
2. $\phi, (a = \frac{1}{2}, b = \frac{1}{2}, c = d = e = \frac{1}{2})$

Compare this result with Theorem 10.21 below.

Extensions according to $Eq_{inverse}$
We now deal with Figure 10.10 using $Eq_{inverse}$. The equations are:

1. $a = 1 - b$
2. $b = (1 - b)(1 - a)$
3. $c = (1 - b)(1 - e)$
4. $d = 1 - c$
5. $e = 1 - d$.

We can have one solution with $a = 1$. (1) yields $b = 0$, which agrees with (2). From (3) and (5) we get $c = d$ and from (4) we get $c = \frac{1}{2}$ and so $d = e = \frac{1}{2}$.

We now ask is there a solution with $a < 1$? Equations (1) and (2) must agree. From (1)

$$b = 1 - a$$

Let us substitute in (2). We get

$$(1 - a) = a(1 - a).$$

Since $a \neq 1$ we can divide and get

$$a = 1$$

a contradiction.

Summary of extensions for $Eq_{inverse}$
We can have only one extension

$$\{a\}, (a = 1, b = 0, c = d = e = \frac{1}{2}).$$

Example 10.12. Let us now introduce support into our networks. Let us consider Figure 10.2 again.

The computational problem in our context is simply to say what equation to give for nodes like a in Figure 10.2 which is also additionally supported. The conceptual problem remains the same.

Let us propose an equation for support, just as an example. Suppose we have the situation as follows:

1. All of the attackers of node a are $x_1, \ldots, x_m$.
2. All the nodes supported by a are $y_1, \ldots, y_n$.

Then we require the following equation, obtained by modifying the $Eq_{inverse}$ equation:

$$\mathbf{f}(a) = \prod_{i=1}^{m}(1 - \mathbf{f}(x_i)) \prod_{j=1}^{n} \mathbf{f}(y_j).$$

If a node x is neither attacked nor supported and it does not support anything then $\mathbf{f}(x) = 1$.

Let us write the equations for Figure 10.2.
Let $\mathbf{f}(a) = \alpha, \mathbf{f}(b) = \beta, \mathbf{f}(N) = \nu$.

1. $v = \alpha$
2. $\alpha = v(1 - \beta)$
3. $\beta = 1 - \alpha$.

We get that α is arbitrary and $v = \alpha$ and $\beta = 1 - \alpha$.

Note that there is a principle involved here in how to obtain equations which include support from equations which include attacks. Take any equation for attacks of nodes x_i and y_j on a node a, which include the participation of the node y as an attacker. Then $\mathbf{f}(y)$ appears in the equation involved, say one of the equations in Definition 10.1.

Now suppose we change the role of y from attacker to supporter, then the equation changes by substituting "$1 - \mathbf{f}(y)$" for "$\mathbf{f}(y)$" in the equation.

So the principle we used can be summarised as follows:

The support by $\mathbf{f}(y)$ is equivalent to the attack by $1 - \mathbf{f}(y)$.

10.1.4 More on what equations can do

Example 10.13 (Boolean functions, Abstract dialectical framework of [69]). The function $\mathbf{f}$ in $(S, R_A, \mathbf{f})$ can be governed by a different equation for each node. So we have, for each $a \in S$ a different function $\mathbf{h}_a$. So if x_i for $i = 1, \ldots, n$ are all the nodes linking a (i.e. $x_i \to a$ is in R_A), we have

$$\mathbf{f}(a) = \mathbf{h}_a(\mathbf{f}(x_1), \ldots, \mathbf{f}(x_n))$$

For example $\mathbf{h}_a$ can be different Boolean functions in $x_1, \ldots, x_n$. Let $\mathcal{B}_a$ be a Boolean expression in $x_1, \ldots, x_n$ of the form

$$\bigvee_{i=1}^{k} \bigwedge_{j=1}^{n} \varepsilon_{i,j}$$

where $\varepsilon_{i,j} \in \{+x_j, -x_j\}$.

Let

$$\mathbf{e}(\varepsilon_{i,j}) = \begin{cases} \mathbf{f}(x_j) \text{ if } \varepsilon_{i,j} = +x_j \\ (1 - \mathbf{f}(x_j)) \text{ if } \varepsilon_{ij} = -x_j \end{cases}$$

Let

$$\mathbf{h}_a(\mathbf{f}(x_1), \ldots, \mathbf{f}(x_n)) = 1 - \prod_{i=1}^{k}(1 - \prod_{j=1}^{n} \mathbf{e}(\varepsilon_{ij})).$$

$\mathbf{h}_a$ implements the Boolean expression $\mathcal{B}_a$.

We chose basically to use Eq_{inverse} as our basis, but we could have used other functions from Definition 10.1. The functions we chose are most appropriate for dealing with logic programming, as we shall do below.

Note that since our graph (S, R_A) is a Dung network, it comes with the convention that points which are not attacked should get the value 1.

Thus the function $\mathbf{h}_a$, for any such point should be 1.

To render this example most general, we need to give up this convention, and regard (S, R_A) as a general directed graph.

We can of course still retain the Dung convention, in which case the Boolean equations need to respect that certain points have pre-determined value of 1.

Compare with [69, 70].

Note that this type of function already appears in [37]. Boolean functions are discussed in [331], where it is shown how to translate Brewka-Woltran Boolean conditions [69] into ordinary Dung networks using methods of [164]. See also Example 10.25 and Remark 10.31 below.

Remark 10.14 (Order among attackers). Note that in the Boolean functions Example 10.13, the local functions $\mathbf{h}_a$ are not symmetrical but uses some order on the attackers. By giving the attackers different names and writing a Boolean expression in them we cannot permute the names. This should be compared with ordinary Dung networks where there is no order among the attackers.

10.2 Formal theory of the equational approach to argumentation networks

In this section we formally develop our equational approach. We start with equational networks based on the unit real interval $[0, 1]$. We then consider Boolean equations and compare with work of Brewka and Woltran.

Conceptually the nodes and the Equations attached to them is the network and the solutions to the equations are the complete extensions, as we have seen in the examples of Section 1.

The meaning of the real numbers attached to nodes varies from one application to another and depends on the nature of the network. If the networks describes liquid flow then the numbers are relative capacities. If the network describe some ecology, (for example a biological predator prey network) then the numbers represent percents of populations in equilibrium. If the network is a logical network (see Example 10.58 and Section 8.3) then the numbers are truth values.

In the case of argumentation networks, numbers strictly between 0 and 1 generally mean that the node is undecided. However, the numbers can be connected to the geometry of the network and the type of loops there are in it. We saw such examples in the previous section where we got different solutions for different loops (compare for example the solutions to the networks in Figures 10.5, 10.6, and 10.9). We still need to investigate the connection between loops and solutions, we have no theorems in this area. See however our discussions and examples in Section 8.4.

10.2.1 Real numbers equational networks

Definition 10.15 (Real equational networks).

1. *An argumentation base is a pair (S, R_A) where $S \neq \varnothing$ is a finite or an infinite set and $R_A \subseteq S^2$ is a finitary relation, that is for all x in S the set $\{y | y R_A x\}$ is finite.*
2. *A real equation function in k variables $\{x_1, \ldots, x_k\}$ over the real interval $[0, 1]$ is a continuous function $\mathbf{h} : [0, 1]^k \mapsto [0, 1]$ such that*
   a) $\mathbf{h}(0, \ldots, 0) = 1$
   b) $\mathbf{h}(x_1, \ldots, 1, \ldots, x_k) = 0$
   Sometimes we also have condition (c) below, as in ordinary Dung networks, but not always.

c) $\mathbf{h}(x_1, \ldots, x_k) = \mathbf{h}(y_1, \ldots, y_k)$ *where* $\{y_j\} = \{x_i\}$ *are permutations of each other.*

3. *An equational argumentation network over* $[0, 1]$ *has the form* $(S, R_A, \mathbf{h}_a)$, $a \in S$
where

   a) (S, R_A) *is a base*

   b) *For each* $a \in S$, $\mathbf{h}_a$ *is a real equation function, with the suitable number of variables* k, k *being the number of nodes attacking* a.

   c) *If* $\neg \exists y(y R_A a)$ *then* $\mathbf{h}_a \equiv 1$.

4. *An extension is a function* $\mathbf{f}$ *from* S *into* $[0, 1]$ *such that the following holds:*

   - $\mathbf{f}(a) = 1$ *if* $\neg \exists y(y R_A a)$

   - *If* $\{x_1, \ldots, x_k\}$ *are all the elements in* S *such that* $x_i R_A a$, *then* $\mathbf{h}_a$ *is a* k *variable function and* $\mathbf{f}(a) = \mathbf{h}_a(\mathbf{f}(x_1), \ldots, \mathbf{f}(x_k))$.

Theorem 10.16 (Existence theorem, the finite case). *Let* $(S, R_A, \mathbf{h}_a)$, $a \in S$ *be a finite network as in Definition 10.15. Then there exists an extension function* $\mathbf{f}$ *satisfying (4) of Definition 10.15.*

Proof. Let n be the number of elements of S. For each $a \in S$ consider $\mathbf{h}_a$ as a continuous function from $[0, 1]^S \mapsto [0, 1]$. Note that in reality a may be attacked by only k nodes, with k possibly less than n. In this case $\mathbf{h}_a$ is a function of k variables, but we can consider it as a function of n variables. This is common practice in mathematics. Let $\mathbf{h}$ be the continuous function from $[0, 1]^S$ into $[0, 1]^S$ defined component wise by $\mathbf{h}(\alpha_1, \ldots, \alpha_n) = (\mathbf{h}_{a_1}(\alpha_1, \ldots, \alpha_n), \ldots, \mathbf{h}_{a_n}(\alpha_1, \ldots, \alpha_n))$.

This is a continuous function on a compact cube of n dimensional space and has therefore, by Brouwer's fixed point theorem, a fixed point $(x_1, \ldots, x_n) = \mathbf{h}(x_1, \ldots, x_n)$.

Let $\mathbf{f}$ be defined by $\mathbf{f}(a_i) = x_i$. Then we have that for each $a \in S$

$$\mathbf{f}(a) = \mathbf{h}_a(\mathbf{f}(a_1), \ldots, \mathbf{f}(a_k))$$

where a_i are all the points in S attacking a.

Remark 10.17. For Brouwer's fixed point theorem see Wikipedia.[2]

Remark 10.18. The perceptive reader should note that Theorem 10.16 ensures that some solution "extension" exists for any continuous function assigning to each argument a value in [01] on the basis of values assigned to all other arguments (the attack relation is irrelevant for the result).While this ensures that one has not to care for existence problems in defining equations, it also means that both meaningful and meaningless (in the sense that they have nothing to do with the notion of extension) equations provide some result. It is therefore an important issue to identify criteria for selecting meaningful equations for argumentation semantics. We provided some interesting examples in Definition 10.1, but in future work a generalization to families of equations will be considered. See also Remark 10.67 below about De Morgan Norms (the word "norm" is used here in the functional analysis sense). Such norms are used as fuzzy versions of the classical connectives $\neg$, $\wedge$ and $\vee$, and equationally they would be similar to Eq_{inverse}, Eq_{max} and $Eq_{\text{geometrical}}$, which are all derived from De Morgan norms. See [269, Chapter 6].

---

[2] See http://en.wikipedia.org/wiki/Brouwer_fixed_point_theorem and Sobolev, V. I., "Brouwer theorem", in Hazewinkel, Michiel, "Encyclopaedia of Mathematics, Springer, 2001.

Remark 10.19 (Existence theorem for the infinite finitary case). Let $(S, R_A, \mathbf{h}_a), a \in S$ be an infinite but finitary network, as in Definition 10.15. We ask, does there exists an extension $\mathbf{f}$ satisfying (4) of Definition 10.15? The answer is that the general case of this remark depends on the properties of the functions $\mathbf{h}_a$ and general existence theorems of analysis and will be investigated in a separate paper. It is also connected to the following question one can ask for the case of finite argumentation networks:

- Given some fixed values for some nodes, is there a solution to the equations of the network respecting these values?

Lemma 10.20. *Let (S, R_A) be a Dung argumentation network. Let $\lambda : S \mapsto \{in, out, undecided\}$ be a legitimate Caminada labelling, yielding an extension E_λ. Consider the functions $\mathbf{h}_a, a \in S$ as follows:*

1. $\mathbf{h}_a \equiv 0$ *if $\lambda(a) = out$*
2. $\mathbf{h}_a \equiv 1$ *if $\lambda(a) = in$.*
3. $\mathbf{h}_a$ *equals $\frac{1}{2}$, otherwise.*

Then there exists, by Theorem 10.16 an extension function $\mathbf{f}$ such that for all $a \in S$

$$\mathbf{f}(a) = \mathbf{h}_a(\mathbf{f}(x_1), \ldots, \mathbf{f}(x_k)),$$

where $\{x_i\}$ are all the nodes attacking a.

Note that what the above does is to find a system of equations characterising exactly a single extension.

The general problem is, given an argumentation network and a selection of some but not all of its extensions, can we write a system of equations which will yield exactly the selected extensions?

We do not know the answer to this question.

To get exactly all the extensions, i.e. all the Caminada labelling, we use the next theorem, Theorem 10.21.

Theorem 10.21 (Caminada complete labelling functions and Eq_{max}). *Consider the function*

$$\mathbf{h}_{max}(x_1, \ldots, x_n) = 1 - \max(x_1, \ldots, x_n).$$

This function is continuous in $[0, 1]^n \mapsto [0, 1]$ and therefore falls under Definition 10.15.

1. *Let $(S, R_A, \mathbf{h}_{max})$ be an equational network with $\mathbf{h}_{max}$ and let $\mathbf{f}$ be an extension, as in item 4 of Definition 10.15. Define a labelling $\lambda_\mathbf{f}$ dependent on $\mathbf{f}$ as follows*

$$\lambda_\mathbf{f}(a) = \begin{cases} in \text{ if } \mathbf{f}(a) = 1 \\ out \text{ if } \mathbf{f}(a) = 0 \\ undecided \text{ if } 0 < \mathbf{f}(a) < 1 \end{cases}$$

Then $\lambda_\mathbf{f}$ is a proper Caminada extension of (S, R_A).

2. *Let λ be a complete Caminada extension for (S, R_A). Let $\mathbf{f}_\lambda$ be the real number function defined as follows*

$$\mathbf{f}_\lambda(a) = \begin{cases} 1 \text{ if } \lambda(a) = in \\ 0 \text{ if } \lambda(a) = out \\ \frac{1}{2} \text{ if } \lambda(a) = undecided. \end{cases}$$

Then $\mathbf{f}_\lambda$ is a proper equational extension for $(S, R_A, \mathbf{h}_{max})$, i.e. $\mathbf{f}_\lambda$ solves the equations $\mathbf{f}_\lambda(a) = 1 - \max(\mathbf{f}_\lambda(x_1), \ldots, \mathbf{f}_\lambda(x_n))$ where x_i are all the attackers of a.

Proof.

1. We show that $\lambda_{\mathbf{f}}$ satisfies the Caminada conditions (C1)–(C3).
   Case C1 Assume x_1 attacks a and $\lambda_{\mathbf{f}}(x_1) = in$. This means that $\mathbf{f}(x_1) = 1$. Let $x_2, \ldots, x_n$ be the other attackers of a. Then $\mathbf{f}(a) = 1 - \max(\mathbf{f}(x_1), \ldots, \mathbf{f}(x_n))$ and hence $\mathbf{f}(a) = 0$ and hence $\lambda_{\mathbf{f}}(a) = out$.
   Case C2 Assume a has no attackers then $\mathbf{f}(a) = 1$ and $\lambda_{\mathbf{f}}(a) = in$.
   Otherwise let as before $x_1, \ldots, x_n$ be all the attackers of a, and assume $\lambda_{\mathbf{f}}(x_i) = out$, for all i. This means $\mathbf{f}(x_i) = 0$ for all i. Hence $\max(\mathbf{f}(x_i)) = 0$ and hence $\mathbf{f}(a) = 1$ and hence $\lambda_{\mathbf{f}}(a) = in$.
   Case C3 Assume $\lambda_{\mathbf{f}}(x_i) = out$ or undecided, with say $\lambda_{\mathbf{f}}(x_1)$ at least is undecided. This means that $\mathbf{f}(x_i) < 1$ for all i and for at least x_1 we have $\mathbf{f}(x_1) > 0$. This means that $0 < \max(\mathbf{f}(x_i)) < 1$.
   Hence $0 < 1 - \max(\mathbf{f}(x_i)) < 1$. Hence $0 < \mathbf{f}(a) < 1$. Hence $\lambda_{\mathbf{f}}(a) = undecided$.
2. Let λ be a proper Caminada extension. We show that $\mathbf{f}_\lambda$ solves the equations with **h**.
   a) If a has no attackers then $\lambda(a) = in$ and $\mathbf{f}_\lambda(a) = 1$.
   b) Let $x_1, \ldots, x_n$ be all attackers of a.
      i. If for some i, $x_i = in$ then $\mathbf{f}_\lambda(x_i) = 1$.
         Also in this case $\lambda(a) = 0$ and so $\mathbf{f}_\lambda(a) = 0$.
         But $\max(\mathbf{f}(x_i)) = 1$ and hence indeed $\mathbf{f}_\lambda(a) = 1 - \max(\mathbf{f}(x_i))$.
      ii. If all $\lambda(x_i) = out$ then $\lambda(a) = in$. So $\mathbf{f}_\lambda(a) = 1$ and $\mathbf{f}_\lambda(x_i) = 0$. Thus $\max(\mathbf{f}_\lambda(x_i)) = 0$. So indeed $\mathbf{f}_\lambda(a) = 1 - \max(\mathbf{f}_\lambda(x_i))$.
   c) If all $\lambda(x_i)$ are either out or undecided with at least $\lambda(x_1) = undecided$ then $\lambda(a) = undecided$ and so all $\mathbf{f}_\lambda(x_i)$ are either 0 or $\frac{1}{2}$ with at least $\mathbf{f}_\lambda(x_1) = \frac{1}{2}$, and $\mathbf{f}_\lambda(a) = \frac{1}{2}$.
      Hence $\max(\mathbf{f}(x_i)) = \frac{1}{2}$ and indeed $\mathbf{f}_\lambda(a) = 1 - \max(\mathbf{f}_\lambda(x_i))$.

Remark 10.22 (Caminada complete labelling and Eq$_{inverse}$*).* Theorem 10.21 does not hold for $Eq_{inverse}$. This follows from Example 10.11.

10.2.2 Critical translations of logic programs and Boolean networks

This subsection deals with reductions of one argumentation network to another. The key notion is that of a critical subset. The perceptive reader might wonder about the relevance of this subsection to the equational approach. Why are we having this subsection in this chapter? After all these reductions are purely logical and transfer the equational approach with them. The answer is that this is exactly the point. Any system which can be translated into argumentation theory can be endowed with the equational approach. The details of the translation yield the equations. This is why we show here how to translate logic programs. We can get answer set solutions to logic programs using equations. This is a new angle worth developing!

Definition 10.23 (Critical subsets). *Let* $(S, R_A, \mathbf{h}_a), a \in S$ *be an equational network. Let* $T \subseteq S$ *be a subset of some of the nodes of* S. *We say that* T *is a* critical subset *of nodes with respect to* $\{\mathbf{h}_a\}, a \in S$ *iff for any two solutions* $\mathbf{f}_1, \mathbf{f}_2$ *of the equations* $\{\mathbf{h}_a\}$ *(i.e.* $(S, R_A, \mathbf{f}_1, \mathbf{h}_a)$ *and* $(S, R_A, \mathbf{f}_2, \mathbf{h}_a)$), *we have that if* $\mathbf{f}_1 \upharpoonright T \equiv \mathbf{f}_2 \upharpoonright T$ *then* $\mathbf{f}_1 \equiv \mathbf{f}_2$. *I.e. if* $\mathbf{f}_1$ *and* $\mathbf{f}_2$ *agree on* T *then they agree on* S.

Remark 10.24. The notion of critical subset is important for meta-level considerations. Given a new type of network (T, R_A^T), if we can embed it into a Dung network (S, R_A^S) as a critical subset (with $R_A^S \upharpoonright T = R_A^T$) then conceptually we have reduced (or "implemented") the new features of T as attack features of S using additional nodes.

For example this was done extensively in Chapters 7 and 8. A full discussion of this notion and comparison with the recent paper [120] will be given in Section 10.8.2.

Example 10.25 (Critical translation for Logic Programs and Boolean nets). Consider the situation in Figure 10.11.

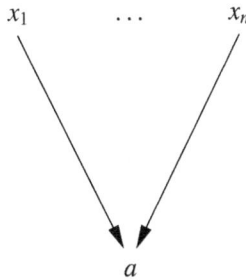

Fig. 10.11.

Assume a Boolean condition of the form

$$a = \text{ in iff } \bigvee_{i=1}^{k} \bigwedge_{j=1}^{n} \varepsilon_{i,j} \qquad (*)$$

where $\varepsilon_{i,j}$ can be either $x_j = $ in (can also be written as $+x_j$) or $x_j = $ out (can also be written as $-x_j$).

Consider the Boolean function $\mathbf{h}_a$ corresponding to the Boolean condition $(*)$, as presented in Example 10.13.

$$\mathbf{h}_a(\mathbf{f}(x_1), \ldots, \mathbf{f}(x_n)) = 1 - \prod_{i=1}^{k}\left(1 - \prod_{j=1}^{n} \mathbf{e}(\varepsilon_{i,j})\right)$$

where

$$\mathbf{e}(\varepsilon_{i,j}) = \begin{cases} \mathbf{f}(x_j) \text{ if } \varepsilon_{i,j} \text{ is } ``x_j = \text{in''} \\ 1 - \mathbf{f}(x_j) \text{ if } \varepsilon_{i,j} \text{ is } ``x_j = \text{out''} \end{cases}$$

We now implement Figure 10.11 as a critical subset of Figure 10.12.
We distinguish three cases

Case 1:

Some nodes x do attack the node a, as shown in Figure 10.11.

Case 2:

a is not attacked at all and $\mathbf{h}_a = 1$.

Case 3:

a is not attacked at all but $\mathbf{h}_a = 0$.

We now treat each case

Case 1:

We have new points as follows

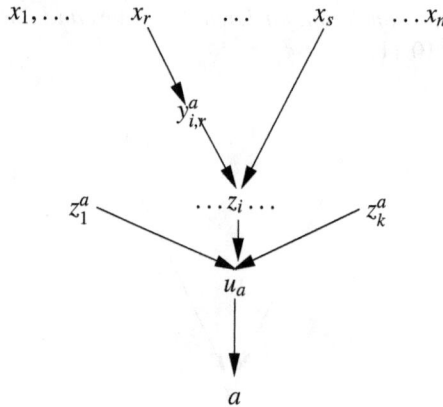

Fig. 10.12.

1. points $u_a, z_1^a, \ldots, z_k^a$ as in Figure 10.12
2. For every $\varepsilon_{i,r}$ which says $x_r = $ on, we add a new point $y_{i,r}^a$ for $i = 1, \ldots, k$ and $j = 1, \ldots, n$. The points $\{a, u_a, z_1^a, \ldots, z_k^a, y_{i,r}^a\}$ are connected as in the Figure 10.12.

the other points in Figure 10.12 are $x_1, \ldots, x_n$.

The point x_r is connected to $y_{i,r}$ if $\varepsilon_{i,r}$ is $x_r = $ in, for $i = 1, \ldots, k$ and $r = 1, \ldots, n$. The point x_s is connected directly to z_i^a if $\varepsilon_{i,s}$ says $x_s = $ out. (Note that in this case $y_{i,s}^a$ does not exist in Figure 10.12.)

This is done for $i = 1, \ldots, k$ and $s = 1, \ldots, n$.

Let us now calculate $\mathbf{f}(a)$ in terms of $\mathbf{f}(x_1), \ldots, \mathbf{f}(x_n)$.

We get

$$\mathbf{f}(y_{i,r}^a) = 1 - \mathbf{f}(x_r)$$

$$\mathbf{f}(z_i) = \pi(1 - \mathbf{f}(y_{i,r}^a)) \cdot \pi(1 - \mathbf{f}(x_s))$$

$$= \pi\mathbf{f}(x_r) \qquad \cdot \qquad \pi(1 - \mathbf{f}(x_s))$$

$$\varepsilon_{i,r} = \text{``}x_r \text{ is in''} \qquad \varepsilon_{i,s} = \text{``}x_s \text{ is out''}$$

$$= \textstyle\prod_{j=1}^{n} \mathbf{e}(\varepsilon_{i,j}).$$

We also have

$$\mathbf{f}(u_a) = \prod_{i=1}^{k}(1 - \mathbf{f}(z_i^a))$$

and

$$\mathbf{f}(a) = 1 - \mathbf{f}(u_a).$$

So in total we get

$$\mathbf{f}(a) = 1 - \mathbf{f}(u_a)$$
$$1 - \prod_{i=1}^{k}(1 - \mathbf{f}(z_i^a))$$
$$1 - \prod_{i=1}^{k}(1 - \prod_{j=1}^{n}\mathbf{e}(\varepsilon_{i,j}))$$

As you can see the old points do not depend on the new points.

Case 2

In this case do nothing, The graph for node a is node a as it is.

Case 3

In this case we want a graph that will give a value 0. So we take a two point graph, with the points a and the point u_a, with point u_a attacking a.

So to summarise, Figure 10.12 can be either of 3 figures according to which case occurs for the node a.

Construction 10.26 *We now show how to embed a general equational Boolean net* $(S, R_A, \mathbf{f}, \mathbf{h}_a), a \in S$ *in an ordinary Dung network* $(T, \rho), T \supset S$ *such that the S is a critical subset of T.*

Let $a \in S$ and let $x_1, \ldots, x_n$ be all the attackers of a. We are now in the situation of Figure 10.11, in either Case 1 or Case 2 or Case 3. Index this figure by a and let (T_a, ρ_a) be the network associated with the corresponding Figure 10.12. Note that all new points $u_a, z_i^a, y_{i,r}^a$ are already indexed by a.

Note also that u_a is a unique attacker of a (in Cases 1 and 3) and appears only in (T_a, ρ_a). In any other (T_b, ρ_b) in which a may appear, then a attacks a $y_{i,r}^b$ or a z_i^b but is never attacked by any of the T_b new points.

Let $T = \bigcup_{a \in S} T_a$ and $\rho = \bigcup_{a \in S} \rho_a$.

Then (T, ρ) is the required Dung networks. Note that for each a we are adding at most n^3 points.

Example 10.27. Let us do a simple example, see Figure 10.13

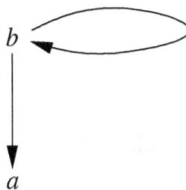

Fig. 10.13.

The conditions are:

$$a \text{ if } b$$
$$b \text{ if } \neg b$$

We can understand a if b as a sort of support, but it does not matter, it is a Boolean condition. We translate the figure into Figure 10.14

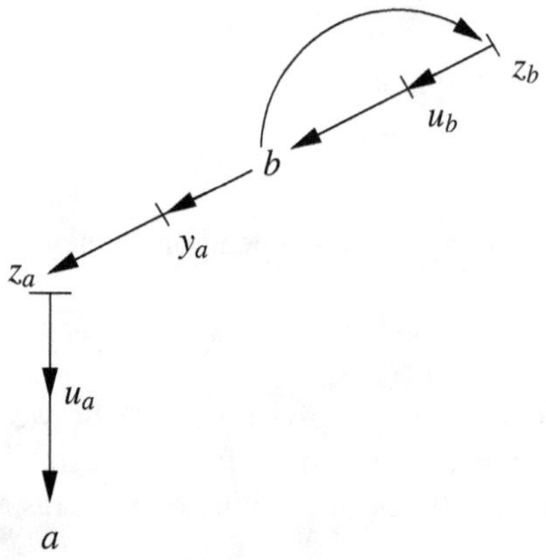

Fig. 10.14.

We have

1. $\mathbf{f}(z_b) = 1 - \mathbf{f}(b)$
2. $\mathbf{f}(u_b) = 1 - \mathbf{f}(z_b)$
3. $\mathbf{f}(b) = 1 - \mathbf{f}(u_b)$
4. $\mathbf{f}(y_a) = 1 - \mathbf{f}(b)$
5. $\mathbf{f}(z_a) = 1 - \mathbf{f}(y_a)$
6. $\mathbf{f}(u_a) = 1 - \mathbf{f}(z_a)$
7. $\mathbf{f}(a) = 1 - \mathbf{f}(u_a)$

From (1) and (2) we get

8. $\mathbf{f}(b) = \mathbf{f}(u_b)$ which together with (3) give $\mathbf{f}(b) = \frac{1}{2}$

Following (4)–(7) gives

9. $\mathbf{f}(a) = \frac{1}{2}$

Notice that y_a, z_a, u_a are just "transmitters" of values from b to a.

Remark 10.28 (Equations for logic programs). Note that what we called Boolean nets in Example 10.25 where we give up the Dung convention and regard the network as just a general directed graph, are actually logic programs.

A propositional logic program contains clauses of the form

$$a \text{ if } \bigwedge_i b_i \wedge \bigwedge_j \neg c_j$$

Note that we may have a clause of the form

$$a$$

in which case we write

$$a \text{ if } \top.$$

For a fixed literal a there may be several such clauses with a as the head. We can therefore write them all together as a disjunction of the form

$$x_r = \bigvee_i \bigwedge_j x_{r,i,j}^{\varepsilon_{r,i,j}} \qquad (**)$$

where x_r ranges over all heads, i ranges over all clauses with heads x_r and $x_{r,i,j}$ ranges over all literals in the ith clause for x_r.

We also have $\varepsilon_{r,i,j} \in \{0, 1\}$ and the convention that

$$x_{r,i,j}^1 = x_{r,i,j}$$

and

$$x_{r,i,j}^0 = \neg x_{r,i,j}.$$

Compare the above with equation (*) of Example 10.25.

Following the discussion in Example 10.25, the equations we get for the logic program (**) are the following

($\sharp$1) $\mathbf{f}(x_r) = 1 - \prod_i (1 - \prod_j \mathbf{f}^{\varepsilon_{r,i,j}}(x_{r,i,j}))$

where $\mathbf{f}^1(y) = \mathbf{f}(y)$ and $\mathbf{f}^0(y) = (1 - \mathbf{f}(y))$ for any y.

The equations are with the variables $\{x_r, x_{r,i,u}\}$.

($\sharp$2)   If x_r is a head of a clause without a body then $\mathbf{f}(x_r) = 1$.
($\sharp$3)   If $x_{r,i,j}$ is not a head of a clause then $\mathbf{f}(x_{r,i,j}) = 0$

Remark 10.29 (Embedding logic programs into argumentation networks). We wrote in Remark 10.28 equations ($\sharp$1)–($\sharp$3) directly for the logic program and did not say how to translate the logic program as a critical subset of some argumentation network. This translation we do now: We need first to transform the original logic program above to a new equivalent logic program in which every literal is a head of a clause.

We do this as follows:

There may be literals x in the old logic program which appear in body of clauses but are themselves not heads of any clause. In this case we provide a clause for x of the form

$$x \text{ if } \neg\top$$

our understanding that $\top$ is a literal which always succeeds (gets value 1). So the literal x always fails.

Thus we get a new program with one more literal, $\top$, and some additional clauses, and this new program is equivalent to the original, and in this program every literal (except $\top$) has a Boolean equation like (**) above associated with it.

Now we construct a Boolean network (S, R_A) as follows:

1. The set S of nodes are all the literals of the new logic program

2. Let x_r be any literal in S different from $\top$. Assume x_r is connected in the new logic program via (∗∗) to the literals $x_{r,i,j}$. Then let $x_{r,i,j} R_A x_r$ hold.
3. Let the Boolean equation for x_r be (∗∗).

We now have a Boolean network and we use Construction 10.26 to embed it as a critical subset of a Dung network.

The overall result is the embedding of the original logic program as a critical subset of a Dung network.

Remark 10.30. We noted in Chapter 4 that ordinary Dung networks can be translated to logic programs in a very direct manner. If $x_1, \ldots, x_m$ are all the nodes attacking x then we write the clause

$$x \text{ if } \bigwedge_i \neg x_i.$$

In [340] we discuss in detail the connection between argumentation networks and logic programs.

The equational approach allows us to give fuzzy values to logic programs. The much discussed loop problem resolves itself automatically. Consider Example 10.10. Figure 10.9 considered as a logic program yields

$$c \text{ if } \neg b$$
$$a \text{ if } \neg c \wedge \neg a$$
$$b \text{ if } \neg a \wedge \neg b.$$

We got fuzzy values as solutions.

It is worth while to investigate an equational approach to fuzzy logic programming. See Section 10.8.3.

Remark 10.31 (Discussion of [69]). Note that in [69] Brewka and Woltran introduce and study what they call abstract dialectical framework (what we call Boolean nets or Logic Programs). They need to define extensions for their networks. We got Boolean functions either as a side effect of fibring in [164] or as a natural equational example in the current chapter. We find our extensions by solving equations, or by critically translating into ordinary Dung networks. Brewka and Woltran need to do something similar. Their options are as follows:

1. Define the new notions of extensions for their networks and provide algorithms as Dung did for framework; or
2. Use the present chapter with Brouwer fixed point theorem and use MATLAB or MAPLE or NSolve to find the solution; these solutions might be more refined than the extensions they would get in [69], as we have seen in Remark 10.6; or
3. Another option is contained in Chapter 7 on fibring argumentation networks. In Chapter 7 I studied what happens when we substitute one Dung network inside another. In the course of this study I got networks with Boolean conditions of the form
   - a is out if $\bigwedge_i x_i = \text{in}$;
   and the form
   - $\bigvee_i x_i = \text{out}$ if $a = \text{in}$.

to find the extensions of such networks I translated in detail with examples the Boolean conditions into ordinary Dung networks and defined the notion of critical subsets, etc.

These notions are used in this chapter in Example 10.25.

Note that for equational networks we do not need the translation and the additional nodes. They disappear from the final equation, as we saw in the calculations of Example 10.25.

See also Example 10.41 below.

Remark 10.32. The reader is advised to consult Chapter 13 below, especially Section 2.4.entitled: Why are we not surprised by results of Gabbay, Brewka and Woltran?

The chapter shows the equivalence between Dung's argumentation networks and Classical logic with the Peirce-Quine dagger connective, and therefore Dung's argumentation is as expressive as classical logic, Dung's networks can represent any Boolean network or Logic Program, as shown in Chapter 13.

Remark 10.33 (Finding extensions). The question of how to use the equations for finding extensions will be discussed in a later section.

We can always find all solutions to the equations and this will give us all the extensions and then use extra secondary examination to classify them. However, this is not what we mean.

We want to modify the equations to give us the desired extensions as solutions.

We can do this now for finding stable extensions as follows:

The stable extensions give either 0 or 1 values to all nodes. This means that the new variable y defined below must solve to value 0:

$$y = (\sum_x [\mathbf{f}(x) \cdot (1 - \mathbf{f}(x))]).$$

So we add a variable y with equation as above and tell the math program such as MAPLE or MATLAB or NSolve to output solutions for $y = 0$, if possible.

Summary Remark 10.34 (Advantages of the equational approach) *We now list the advantages of our approach:*

First let us highlight the fact that given a traditional argumentation network with attacks only, we regard it as a graphical generator of equations, and we use equations as a conceptual framework. We no longer talk about concepts like defense, acceptability, admissible extensions and other extension, but talk instead about solutions to the equations.

Therefore conceptually we have

- *an extension is a solution to the equations and different extensions (grounded, preferred, stable, semi-stable, etc.) are characterised by further constraints/equations on these solutions functions using say Lagrange Multipliers see Section 6 below.*

Within this framework we note the following:

1. *To find all possible extensions we solve equations. We feed the equations into existing well known mathematical programs such as MAPLE or MATLAB or NSolve and get the solutions. See also Remark 10.33.*

   There are many papers which calculated computational complexity of finding various extensions, when we reduce the problem to that of solving equations in MAPLE or MATLAB or NSolve, complexity is not reduced, it can only increase. What do we gain then?

   - *A new uniform framework, not only for argumentation networks, but also for other types, Ecological, flow, etc., etc.*
   - *Possibility of finding different heuristics for equations which will work faster for most cases, giving an advantage over non-equational algorithms*
   - *Ordinary people such as lawyers etc., to the extent that they use argumentation at all and are not averse to formal logic, they may find that it is psychologically easier to plug the problem into the computer, go and make a cup of tea and then check the results.*

   Furthermore, if we insist on certain arguments being in or out, we can experimentally feed this into the equations and test the effect on other arguments. MATLAB itself does not generate all the solutions automatically but requires initial input, which is an advantage if we have a special set of arguments in mind.

   For example the question of whether a set of arguments belongs to some extension (being credulous) of a certain type or whether the set belongs to all extensions of a certain type (being sceptical) can very naturally be handled in the equational framework.

   To generate all extensions we need to keep plugging initial conditions into MATLAB, i.e., plug in all possible candidates for extensions (this is exponential in the number of nodes but we saw in Remark 10.29 and Construction 10.26 that any Boolean set of functions can be embedded in argumentation networks, and so the complexity is exponential anyway).

   Another possibility is to use NSolve which does generate solutions, see `http://reference.wolfram.com/mathematica/ref/NSolve.html`. *Another disadvantage of this is that we might get approximate solutions. So if we get $x = 0.999$ we ask is this for real or is the solution supposed to be $x = 1$?*

   On the other hand an advantage of using such programs is that it makes it easy to incorporate argumentations feature into other larger AI programs, as almost anything allows for solving equations.

2. *We have a framework for introducing support as we shall see in the next section*
3. *Note the word of caution and comments in Section 3.1 below.*

10.3 Analysis of support

The purpose of this section is to develop equations for networks of the form (S, R_A, R_S) where R_A is the attack relation and R_S is the support relation. Figure 10.2 is an example of such a system, but in this section we will have proper equational theory and many examples.

10.3.1 Caution about support

We begin with a word of caution. Our equational model for attack ends up with numbers between 0 and 1 labelling the nodes of the graph (S, R_A) with $x = 1$ meaning "x is in" and $x = 0$ meaning "x is out". The value $\mathbf{f}(a) = \mathbf{h}_a(\mathbf{f}(x_1), \ldots, \mathbf{f}(x_n))$ gives the impression that x_1 is to be considered as attacking a if by increasing $\mathbf{f}(x_1)$ we decrease $\mathbf{f}(a)$ and therefore we are tempted to say that x_2 supports a if by increasing $\mathbf{f}(x_2)$ we are increasing the value of $\mathbf{f}(a)$. This view is encouraged by situations as in Figure 10.15.

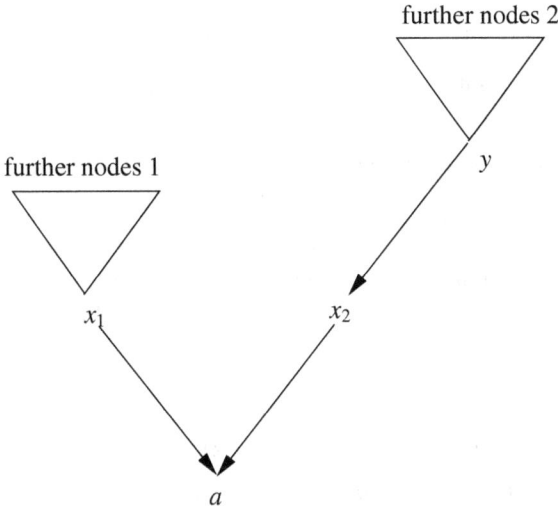

Fig. 10.15.

The equations are:

1. $\mathbf{f}(a) = (1 - \mathbf{f}(x_1))(1 - \mathbf{f}(x_2))$
2. $\mathbf{f}(x_2) = 1 - \mathbf{f}(y)$
3. $\mathbf{f}(x_1) = $ some function of its further nodes
4. $\mathbf{f}(y) = $ some function of its further nodes.

We therefore see that

$$\mathbf{f}(a) = (1 - \mathbf{f}(x_1)) \cdot \mathbf{f}(y).$$

So y is supportive, because when $\mathbf{f}(y)$ increases then $\mathbf{f}(a)$ increases. We can also see this immediately from Figure 10.15 because y attacks x_2 which itself attacks a.

So the idea we might consider is that of the clear support of any node y of a node a can be read from the functional equations. The reading may not be immediate, as can be seen from Figure 10.16, but we are able to define what support is!

In Figure 10.16 the equations are:

1. $\mathbf{f}(a) = (1 - \mathbf{f}(x))(1 - \mathbf{f}(u))$
2. $\mathbf{f}(x) = (1 - \mathbf{f}(y))(1 - \mathbf{f}(u))$
3. $\mathbf{f}(y) = 1 - \mathbf{f}(u)$

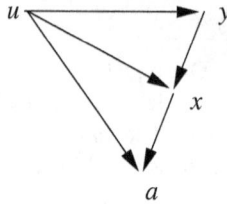

Fig. 10.16.

Hence

$$\mathbf{f}(a) = (1 - \mathbf{f}(u))(1 - \mathbf{f}(u)(1 - \mathbf{f}(u))).$$

Whether $\mathbf{f}(a)$ is increasing in $[0, 1]$ when $\mathbf{f}(u)$ is increasing needs to be checked by tracing the curve.

Our cautionary comment is that this is the wrong approach!

The numbers we get for the nodes as the result of the equation have no qualitative meaning beyond the fact that they indicate which points are in, out or undecided. They do not indicate strength of argument and therefore cannot be used to indicate support.

To make this point crystal clear, consider again Figure 10.9. We got solutions for this figure, namely:

$$\mathbf{f}(a) = 1 - \frac{\sqrt{2}}{2}$$
$$\mathbf{f}(b) = \sqrt{2} - 1$$
$$\mathbf{f}(c) = 2 - \sqrt{2}$$

These numbers say that the nodes a, b, c are all undecided and may indicate the types of loops we have in the original graph (though we have not developed such a theory yet, this is postponed to a subsequent paper. See however, Section 10.8.4 where we discuss the methodology of loops and the subsequent Chapters 18 and 19). They are not indicating strength of arguments!

For comparison, let us view Figure 10.9 as arising from some ecology on an island. a, b and c are species which prey on one another.

a can eat b but can also cannibalise themselves. b can eat c and also b and c can eat only a. Of course there should be some ecological balance where the number $0 \leq \mathbf{f}(x) \leq 1$ tells us the stable percents of population x which exists under equilibrium.

The equation

$$\mathbf{f}(x) = \mathbf{h}_a(\mathbf{f}(x_1), \ldots, \mathbf{f}(x_n))$$

tells us the size of species a facing its predators (attackers) $x_1, \ldots, x_n$ as a function of their size $\mathbf{f}(x_i)$. Thus the solutions $\mathbf{f}(a) = 1 - \frac{\sqrt{2}}{2}, \mathbf{f}(b) = \sqrt{2} - 1$ and $\mathbf{f}(c) = 2 - \sqrt{2}$ represent a stable ecology.

One can see immediately that the set of equations Eq_{max} is not suitable for the ecological model. Predators combine their effect, not politely allow the strongest to do the job. $Eq_{inverse}$ is more suitable and from these equations we get the above solution.

These kinds of ecological and other flow networks were studied extensively in Chapter 9 and contained the equational argumentation networks as a limiting case. In this context we can indeed say that y support x if an increase in $\mathbf{f}(y)$ generates an increase in $\mathbf{f}(x)$.

We recommend the reader to look in detail at Chapters 9 and 14, we believe they are of value to the argumentation community.[3]

10.3.2 Equational model of support

We saw in Section 3.1 that we need a conceptual motivation for support rather than use the equations produced by the attack. So let us offer a model and compare its results with the literature. Our starting point is a network of the form (S, R_A, R_S), where S is a set of arguments, $R_A \subseteq S^2$ is the attack relation and $R_S \subseteq S^2$ is the support relation. We need to decide what kind of equations $\mathbf{h}_a$ to offer for such a network. Our considerations must be qualitative in the context of argumentation (as opposed to, for example, ecology, fluid flow, etc.). Before we start we need a word of caution. Support is not a simple concept and in argumentation many serious authors tried to address it. In this serious context the reader should not expect much from the numerical equational approach. When we use numbers we are forced to simplify. We academics know very well that when we apply for promotion, our performance is reduced to impact factors and number of papers and other numbers, which in many cases does not reflect our real contributions. So the most we can expect from the numerical equational approach is common sense reasonable soundness. In our numerical paper [41] for example, recall Chapter 9, we even considered the possibility of an assumption that attack and support by the same strength should cancel each other. This makes sense as a soundness condition on the numerical functions.

Let us start with the simplest of diagrams, Figure 10.17

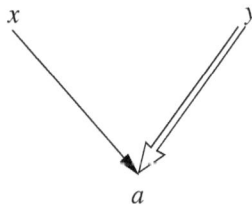

Fig. 10.17.

The argument a is attacked by x and is supported by y.

$$S = \{a, x, y\}. \quad R_A = \{(x, a)\}. \quad R_S = \{(y, a)\}.$$

The question is should a be in or out, or more generally, what value to give a as a function of the values of x and y. This of course depends on the intended meaning of

---

[3] We have in these papers not only weighted nodes (arguments) and weighted arrows (attacks) but also higher level attacks, temporal dependence and a lot more. These notions are being reproduced now by many authors. See, for example, [32, 114] for higher level attacks originating in [37], and our discussions and further development in Chapters 5, 7, and 8.

These chapters are part of a general methodological approach to applied logic and connect many areas together.

support, but is restricted by the parameters available to us in the numerical context. We cannot capture much meaning with numerical functions.

This is discussed extensively in Section 4.1 of [37], recall Chapter 9, with a view (in [37]) that equations should be sought which allow x and y to cancel each other if they have the same strength.

We shall not go into this in this chapter. See, however, Remark 10.50 in Section 4.

Our graph networks have no strength and so there is nothing we can say about Figure 10.17. We can adopt a risk averse approach and say if a is attacked by anything then a is out. This is certainly the attitude of government funding bodies; one negative review and your application is out! So the risk averse approach will allow for the extension a = out, $x = y$ = in.

Let us offer a definition. The notion of Threat is further extensively discussed in Chapter 15.

Definition 10.35 (Threats). *Let (S, R_A, R_S) be an argumentation network with attack and support. Define two auxiliary relations for the purpose of defining equations. The definition will be schematic, dependent on a parameter relation R, which will allow us to perform iterations. Let $R \subseteq S^2$ be any relation defined already using R_A and R_S. We define $\rho_1(R)$ and $\rho_2(R)$.*

Option 1

$$u\rho_1(R)x \text{ iff def. } uRx \vee [\exists z \exists n(xR_S^n z \wedge uRz)]$$

Option 2

$$u\rho_2(R)x \text{ iff def. } uRx \vee [(\exists z \exists n(zR_S^n x \wedge uRz)]$$

The meaning of $u\rho_i(R_A)x$ is that x is under threat from u.

Figure 10.18 explains the situation:

In figure 10.18, node u_0 attacks x directly, it is therefore both $\rho_1(R_A)$ and $\rho_2(R_A)$ threat to x.

u_1 attacks z which is at the end of a chain of support from x. According to $\rho_1(R_A)$, u_1 is a threat to x. It undermines x's credibility.

u_2 attacks a great grandparent supporter of x. Again can be a threat to x.

The process can be iterated. We can look at $\rho_1(\rho_1)$ and $\rho_2(\rho_1), \rho_1(\rho_2), \rho_2(\rho_2)$. See Example 10.37.

Let ρ_e be an arbitrary $\rho_{e_1}(\rho_{e_2}(\ldots \rho_{e_n}(R_A)\ldots)$ where $\mathbf{e} = (e_1, \ldots, e_n)$ and each $e_i \in \{1, 2\}$. So ρ_e is an arbitrary iteration of ρ_1, ρ_2.

We can use ρ_e to define the equations for the network. This means that we use n-level threats to define our equations. So the role of support in the network is to propagate different levels of threats on a node x through the networks by means of direct R_A attacks on various supporters or supported nodes related to x.

A policy of handling support will choose ρ_e or a combination and define the equations accordingly, as in the next definition 10.36. See also Remark 10.50 below.

Definition 10.36 (Equations for support). *Let (S, R_A, R_S) be a support network and let $\rho = \rho_e$ be as in Definition 10.35. Define $Eq_\rho(\mathbf{f})$ to be*

$$\mathbf{f}(a) = \prod_{i=1}^{n}(1 - \mathbf{f}(u_i))$$

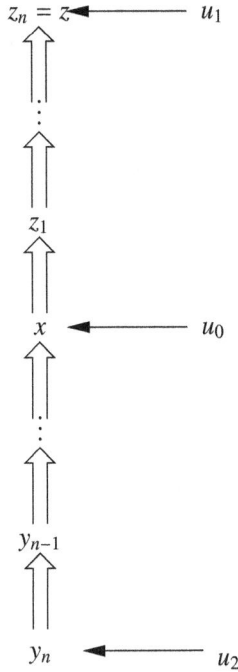

Fig. 10.18.

where,

$$\{u_1, \ldots, u_m\} = \{u | u \rho_e a\}.$$

So for the case of $\rho_i(R_A)$ above there are 2 definitions. One for each $\rho_i(R_A)$.
We understand the empty product to yield 1 (i.e. $\mathbf{f}(a) = 1$ if $\neg \exists u u \rho a$).

Note that our later discussions of higher level attacks in Section 4 below also suggest further equations for support. See Remarks 10.50 and 10.51.

We now examine examples from the literature. We use ρ.

Example 10.37 (Figure 4 from [331]). Consider Figure 10.19.

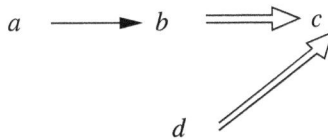

Fig. 10.19.

We have that $a\rho_2(R_A)c$ and $a\rho_1(\rho_2)(R_A))d$, but a is not a $\rho_1(R_A)$ threat to c. So depending on how far we want to look for threats, our extensions can be either $\{a, d, c\}$ if we look at $\rho_1(R_A)$ only, or $\{a\}$ if we look at all of $\rho_1(\rho_2(R_A))$, or we can look at both $\rho_1(R_A)$ and $\rho_2(R_A)$.

We see no need to explicitly write the equations here, as the results are clear.

For comparison with the literature, note that according to our paper [331] (which does not use equations but metalevel variables) we do get $\{a, d, c\}$. Cayrol and Lagasquie-Schiex [95, 96] gets $\{a\}$, while Oren *et al.* [275] also get $\{a, d, c\}$. A more detailed discussion will be done in Chapter 15 below.

We shall comment that the equational approach is very simple while [95, 96, 275] is very involved. See Remark 10.34. We further use the equational approach in Chapter 15 to solve these problems.

It is still open to characterise other approaches such as Cayrol, Oren or Brewka in terms of equations.

Example 10.38 (Figure 14 from [331]). Consider Figure 10.20

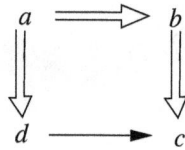

Fig. 10.20.

In Figure 10.20 node d $\rho_1(R_A)$ attacks node a. Node d also $\rho_1(R_A)$ attacks nodes b and c. Thus the only extension according to $\rho_1(R_A)$ equations is $\{d\}$. Both Cayrol and Lagasquie-Schiex and Oren *et al.* [95, 96, 275] get $\{a, d\}$. See Chapter 15 for a discussion.

Example 10.39 (Gabbay-Villata comparing example). Consider the following example for the purpose of comparison, see Figure 10.21.

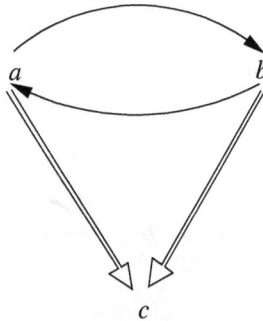

Fig. 10.21.

This is the fighting parents a and b example, who each support their child c.

a $\rho_2(R_A)$ attacks b and c

b $\rho_2(R_A)$ attacks a and c

However, if we do not use ρ_2, then ρ_1 does not affect c.

So

1. According to $\rho_1(R_A)$ equations in the extensions are
   a) $a = 1, b = 0, c = 1$
   b) $a = 0, b = 1, c = 1$
   c) $a = \frac{1}{2}, b = \frac{1}{2}, c = 1$
2. According to [331] c is always in and $\{a, b\}$ from a two elements loop.
3. According to Oren *et al.*, we have arguments as sets, so we have two possibilities. See Figures 10.22 and 10.23.

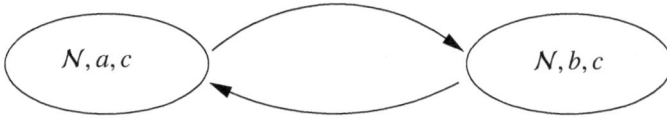

Fig. 10.22.

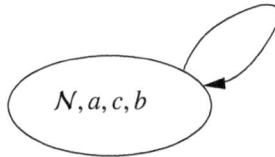

Fig. 10.23.

Note that Oren [275] always adds an almighty source of support $\mathcal{N}$.
4. Cayrol [95, 96] gets essentially the same as Oren.

Example 10.40 (Brewka and Woltran example in [69]). Brewka and Woltran example in Figure 10.24 is an interesting key example. It is also discussed in [331] as Figures 5 and 18 of our paper [331].
   Here it is:

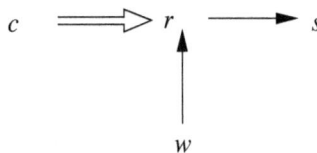

Fig. 10.24.

Brewka and Woltran put this forward as a counter example to Cayrol who supports the extension $\{w, s\}$. If we adopt $\rho_1(R_A)$ equation we get $\{w, c, s\}$. If we adopt $\rho_2(R_A)$ equations, then w is a threat to c, as well as to r. So we get $\{w, s\}$.
   Brewka and Woltran give a specific meaning to this example and argue that the attack of w on r is stronger than the support of c to r. As we discussed above in

Section 3.1, when we add strength to our networks we are playing a different game. In our paper [331] we interpret Brewka and Woltran's argument as a higher order attack as in Figure 10.25.

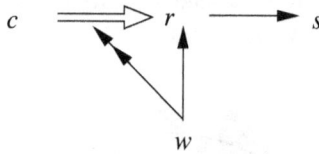

Fig. 10.25.

The double arrow attacks the support arrows.

This type of double arrow was extensively used in our 2005 paper [37] and in our paper [163], discussed in this book in Chapters 8 and 9, and was also independently used by Modgil [260] and treated by Baroni *et al.* [32].

It is what I call a *reactive arrow*, originally introduced by Gabbay in 2004, see [157] and widely applied in many areas, such as deontic logic, automata theory, multi-agent systems, formal grammars, and more.

It is a different game altogether and we shall analyse its equational semantics in a later section.

Example 10.41 (Brewka and Woltran [69]). In their paper [69] Brewka and Woltran gave the following Boolean example, see Figure 10.26. The connection with support is that the condition on c is that $c =$ in provided that the values of a and b are different. So if $b =$ out then the arrow $a \to c$ is support and if $b =$ in then the arrow $a \to c$ is attack. Following Example 10.25, Figure 10.26 can be represented as the ordinary Dung network of Figure 10.27.

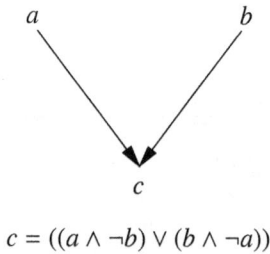

$$c = ((a \wedge \neg b) \vee (b \wedge \neg a))$$

Fig. 10.26.

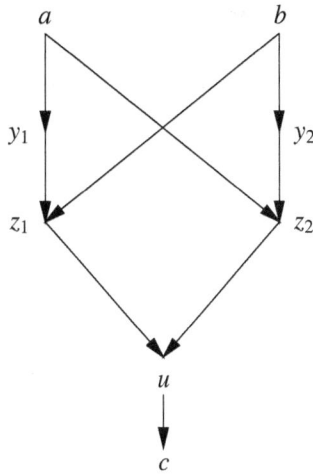

$$(a = \text{in}, b = \text{out}) \vee (a = \text{out}, b = \text{in})$$

Fig. 10.27.

10.4 Equations for higher level attacks

The case of higher level attacks is strong evidence for the success of the equational approach. The historical story runs as follows.

In Chapter 8 we introduced networks with higher level attacks of any kind. Figure 10.28 is a typical situation.

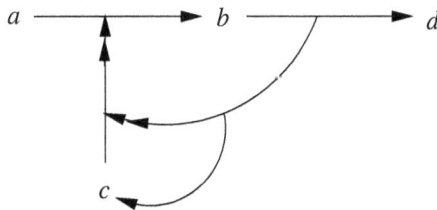

Fig. 10.28.

Figure 10.28 can represent any kind of network, not necessarily an argumentation network. It can be part of a Kripke model, an electrical network, a biological ecological network, etc. In each case the arrows have their own meaning. In Chapter 9 we also allowed for value annotations to the nodes and arrows and gave algorithms for the propagation of these values.

In the case of argumentation networks the nodes are arguments and the arrows mean attacks. The values (annotations) can correspond to the equational labellings and the propagation of values are governed by the properties of the equations.

The aim of this section is to interpret (give equational semantics) to higher level attacks, in the case of argumentation networks.

Let us look again at Figure 10.28, reading it as an argumentation frame.

In Figure 10.28 the argument c attacks the attack from a to b, (we use the notation $c \twoheadrightarrow (a \rightarrow b)$), while the attack from b to d attacks the attack emanating from c (notation $(b \rightarrow d) \twoheadrightarrow (c \twoheadrightarrow (a \rightarrow b))$. This later attack attacks c (notation $((b \rightarrow d) \twoheadrightarrow (c \rightarrow (a \rightarrow b))) \twoheadrightarrow c$).

The question we ask is how to define the possible acceptable extensions for $\{a, b, c, d\}$ for the network of Figure 10.28. The reader should note that whatever approach we give for defining extensions it must come from reasonable general principles which are meaningful for general networks. It should not rely on very specific features of argumentation networks. The general principles can have a specialised meaning in the argumentation case, but then equally the principles can have their own meanings in the case of other networks. We shall use equational principles.

In his 2007 paper [261] Modgil independently defined a higher level attack networks where arrows emanating from nodes can attack other arrows (like $c \twoheadrightarrow (a \rightarrow b)$ in Figure 10.28). Modgil defined certain restrictions on his networks and tried to define a suitable notion of extension in the traditional setting of admissible sets. His paper was studied by Hanh et al. [214] on account that his definition of extensions was not monotonic. Hanh et al. supplied their own discussion and definitions. The disagreement focussed around the illustrative example of Figure 10.29. Modgil allowed the extension $\{c, c_1, a\}$ while Hanh et al. insisted on $\{c, c_1\}$.

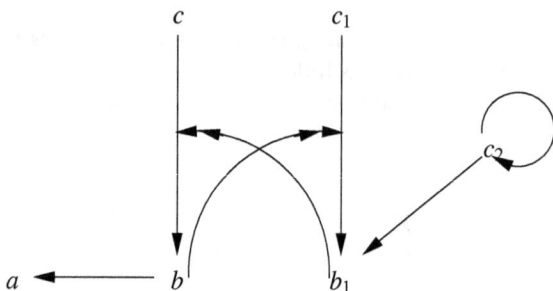

Fig. 10.29.

Meanwhile, two other papers addressed the semantics of higher level attacks in the Dung style. These were Baroni et al. [32] and Gabbay [163] (see our Chapter 8 in this book) both of 2009. For a survey and historical discussion see Chapter 8, section 8.1 entitled Historical Background.

We shall offer two equational systems for higher level attacks which will settle in two pages the long and tortuous discussions and definitions of the above papers.[4]

Note that since we are writing equations, we are guaranteed the existence of extensions!

---

[4] Hanh et al. point out that their approach is skeptical generalization of the standard argumentation framework while others like Sanjay's, Gabbay's or Baroni et al.'s are rather credulous. Hence in each extension of these approaches, there is a skeptical part that is one of Hanh et al. extensions. Hanh et al. also point out that Sanjay's generalization of grounded semantics is rather more liberal than theirs, and hence his characteristic function is not monotonic.

To be able to present our case we need some definitions.

Definition 10.42 (Higher level argumentation frames for attacks emanating from points).

1. *An ordinary argumentation frame has the form* $\mathbf{A} = (S, R)$, *where* S *is the set of arguments and* $R \subseteq S \times S$ *is the attack relation.*
2. *A level* $(0, 1)$ *argumentation frame has the form* $\mathbf{A}^{0,1} = (S, \mathbb{R})$ *where* S *is the set of arguments and* $\mathbb{R}$ *is a set of pairs of the form* (x, y), *or* $(z, (x, y))$, *where* $x, y, z \in S$. $(x, y) \in \mathbb{R}$ *means that* x *attacks* y *and* $(z, (x, y)) \in \mathbb{R}$ *means that* z *attacks the attack* (x, y).
3. *Level* $(0, n)$ *argumentation frames are defined as follows*
   a) *A pair* $(x, y) \in S \times S$ *is called a level* $(0, 0)$ *attack.*
   b) *If* $z \in S$ *and* α *is level* $(0, n)$ *attack then* (z, α) *is a level* $(0, n + 1)$ *attack.*
   c) *A level* $(0, n)$ *argumentation frame has the form* $\mathbf{A}^{0,n} = (S, \mathbb{R})$ *where* $\mathbb{R}$ *contains attacks* β *of level* $(1, m)$ *for* $m \le n$, *including some elements of level* $(0, n)$.

To explain our ideas and address Figure 10.29, let us first define equations for level $(0, 1)$ attacks only. A typical situation is described in Figure 10.30.

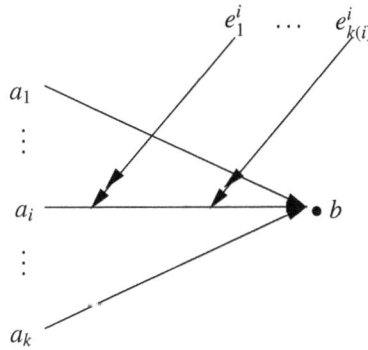

Fig. 10.30.

The attackers of b are $a_1, \ldots, a_k$ and the attacking arrow $a_i \to b$ is attacked by $e_1^i, \ldots, e_{k(i)}^i$. Since we restricted ourselves to level $(0, 1)$ there are no attacks on the double arrows $e_j^i \twoheadrightarrow (a_i \to b)$.

We now offer the following two options for equations for $\mathbf{f}(b)$, ("H" stands for "higher" and "1" stands for "(0, 1) attacks").

Option $H^1 Eq_{\text{inverse}}(\mathbf{f})$:

$$\mathbf{f}(b) = \prod_{i=1}^{k}(1 - \mathbf{f}(a_i) \cdot \prod_{j=1}^{k(i)}(1 - \mathbf{f}(e_j^i)))$$

Option $H^1Eq_{max}(\mathbf{f})$:

$$\mathbf{f}(b) = 1 - \max_i(\mathbf{f}(a_i) \cdot (1 - \max_j(\mathbf{f}(e_j^i))))$$

We now see what kind of extensions we get for Figure 10.29.

Example 10.43 (Modgil vs. Hanh et al.).

I We first use the $H^1Eq_{inverse}$ equations for Figure 10.29. We write x instead of $\mathbf{f}(x)$:
1. $a = 1 - b$
2. $c_2 = 1 - c_2$
3. $c = 1$
4. $c_1 = 1$
5. $b = 1 - c(1 - b_1)$
6. $b_1 = (1 - c_2)(1 - c_1(1 - b))$

From (2), (3), (4) we get $c = c_1 = 1$. $c_2 = \frac{1}{2}$. From (5) we get
7. $b = b_1$

and from (6) we get
8. $b_1 = \frac{1}{2}b_1$

The only solution is $b_1 = b = 0$.
From (1) we get $a = 1$. Thus the only extension is $a = c = c_1 = 1, b = b_1 = 0, c_1 = \frac{1}{2}$.
This corresponds to the Modgil extension.

II We now use H^1Eq_{max}. The equations are
1. $a = 1 - b$
2. $c_2 = 1 - c_2$
3. $c = 1$
4. $c_1 = 1$
5. $b_1 = 1 - \max(c_2, c_1(1 - b))$
6. $b = 1 - c(1 - b_1)$

We get $c_2 = \frac{1}{2}, c = c_1 = 1$. $b = b_1$. Thus from (6) we get
7. $b_1 = 1 - \max(\frac{1}{2}, 1 - b_1)$.

There are two solutions $b_1 = \frac{1}{2}$ and $b_1 = 0$.
We also get two corresponding solutions for a: $a = \frac{1}{2}$ and $a = 1$.
We thus have the second extension for the case of $b_1 = \frac{1}{2}$ is the Hanh *et al.* extension $c = c_1 = 1$ and $b = b_1 = a = c_2 = \frac{1}{2}$.

We now address the general case of arbitrarily high level of attacks on attacks. To do this right we need high level of induction and it can be complicated, as you can see from Figure 10.31.

This iteration can go on for many levels and we cannot practically handle it for each b and write very long iterated equations. We need an idea, a trick, to go around it. Fortunately we have it already in our 2005 paper [37], see Example 2.3 of our paper [37] in which we calculate the values for Figure 9.8 (our Figure 9.8 is Figure 5 in [37]). The idea is very natural and very simple.

In paper [37] we have arrows attacking arrows and both arrows and nodes were annotated with numbers. The numbers annotating nodes correspond to strength (when

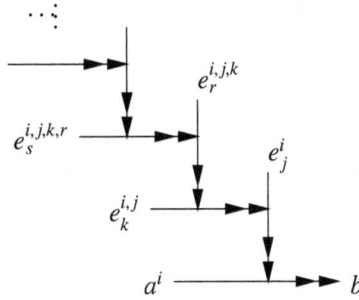

Fig. 10.31.

the node is an argument this is the strength of argument) and the numbers annotating arrows indicate strength of transmission. This was discussed in detail in Section 9.2.

So what do we do in our case of Figures 10.30 and 10.31 we regard the arrows of any kind for example $x \twoheadrightarrow \beta$ as transmission arrows and regard the attack of the form $y \twoheadrightarrow (x \twoheadrightarrow \beta)$ as an attack of y on the transmission factor $\varepsilon(x,\beta)$.

All we need to do now is to give the equation for an attack of the form of Figure 10.30.

But this situation and the equation for it is very obvious and intuitive. If an argument a of strength x attacks an argument b with transmission factor $\varepsilon(a,b)$, then the "flow" of the attack is the product $x \cdot \varepsilon(a,b)$, and since this is an attack we must have $y = 1 - x \cdot \varepsilon(a,b)$. If we have several such attackers, from say $a_1, \ldots, a_k$ then the formula is as follows. (We write b for $\mathbf{f}(b)$.)

$$b = \prod_{i=1}^{k}(1 - a_i \cdot \varepsilon(a_i,b))$$

Now if $\varepsilon(a_i, b)$ is attacked by say $e_1^i, \ldots, e_{k(i)}^i$ with transmission factor 1 then we get

$$\varepsilon(a_i, b) = \prod_{j=1}^{k(i)}(1 - e_j^i \cdot 1)$$

Putting the two equations together we get

(*) $$b = \prod_{i=1}^{k}(1 - a_i \cdot \prod_{j=1}^{k(i)}(1 - e_j^i)).$$

One can see that (*) is $H^1 Eq_{\text{inverse}}$ for Figure 10.30.

By analogy we use max as our other option.

So to summarise, here is the algorithm for writing equations for an arbitrary higher level argumentation network.

Definition 10.44 (Higher level equations). *Given a higher level network as in Definition 10.42, we define the equations for it in steps as follows:*

Step 1. *Insert a transmission factor $\varepsilon(x,\beta)$ for every arrow of the form $x \twoheadrightarrow \beta$ or of the form $a \to \beta$ in the network.*

Step 2. *View all higher order attacks on arrows as attacks on the transmission factor of the arrow. So the transmission factor becomes a sort of pseudo-node.*

Step 3. *For any pseudo-node β, whether it is a real node or a transmission factor, let Figure 10.32 represent all of attacks on it. The attack arrows have their own transmission factors as indicated in Figure 10.32. Note that for the purpose of writing the equations, transmission factors are not regarded as nodes. They are regarded as nodes only when they are attacked and play the role of β. Therefore in Figure 10.32 the transmission factors are to be seen as such and not as pseudo-nodes.*

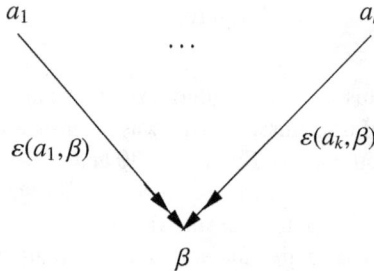

Fig. 10.32.

We now have two options for equations.
Option $HEq_{\text{inverse}}(\mathbf{f})$:

$$\mathbf{f}(\beta) = \prod_{i=1}^{k}(1 - \mathbf{f}(a_i) \cdot \mathbf{f}(\varepsilon(a_i, \beta)))$$

Option $HEq_{\text{max}}(\mathbf{f})$:

$$\mathbf{f}(\beta) = 1 - \max_i(\mathbf{f}(a_i)\mathbf{f}(\varepsilon(a_i, \beta)))$$

Step 4. *Solve the equations for the variables $\mathbf{f}(x)$, both for real nodes x and for transmission factors $\varepsilon(x, y)$. All your solutions $\mathbf{f}$ represent all the extensions for the network.*
If $\mathbf{f}(x) = 1$ then $x = in$.
If $\mathbf{f}(x) = 0$ then $x = out$.
If $0 < \mathbf{f}(x) < 1$ then x is undecided.

This is a typical case for solving equational problems for variables $\mathbf{f}(b), b$ a node with the help of additional variables $\mathbf{f}(\varepsilon(a, b)), \varepsilon(a, b)$ a transmission factor. This idea for solving equations using additional variables is not new, it is hundreds of years old.

Remark 10.45. The perceptive reader will note the simplicity of our equations as compared with the complexity of the definitions of admissibility and extensions of Baroni, Hanh *et al.*, Modgil and others!

Remark 10.46 (Baroni comment). The perceptive reader will further note that our current paper constitutes a positive response to a comment from Section 6 of the paper of Baroni *et al.* [32]:

The idea of encompassing attacks to attacks in abstract argumentation framework has been first considered in [37], in the context of an extended framework encompassing argument strengths and their propagation. In this quite different context, deserving further development, Dung style semantics issues have not been considered.

Example 10.47. Let us do Figure 10.28. Use HEq_{inverse}. First we add transmission. This can be seen in Figure 10.33.

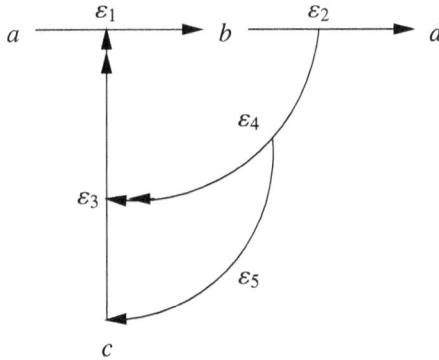

Fig. 10.33.

The equations are:

1. $a = 1$ (not attacked)
2. $b = 1 - \varepsilon_1 a$
3. $d = 1 - b\varepsilon_2$
4. $\varepsilon_2 = 1$ (not attacked)
5. $\varepsilon_3 = 1 - \varepsilon_4 \cdot \varepsilon_2$
6. $\varepsilon_4 = 1$ (not attacked)
7. $\varepsilon_1 = 1 - c\varepsilon_3$
8. $c = 1 - \varepsilon_4\varepsilon_5$
9. $\varepsilon_5 = 1$ (not attacked).

Simplifying the equations we get

10. $b = 1 - \varepsilon_1$
11. $d = 1 - b$
12. $\varepsilon_3 = 0$
13. $c = 0$
14. $\varepsilon_1 = 0$

The extension is
$$a = 1, b = 0, d = 0, c = 0.$$

Example 10.48 (Loop). Consider Figure 10.34

Let ε be the transmission $a \to a$.

The transmission $a \twoheadrightarrow (a \to a)$ is 1 since it is not attacked. We get again using HEq_{inverse}:

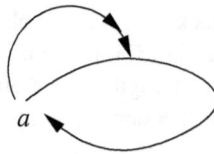

Fig. 10.34.

1. $a = 1 - \varepsilon a$
2. $\varepsilon = 1 - a$.

So

$$a = 1 - (1 - a)a$$
$$a = 1 - a + a^2$$
$$a^2 - 2a + 1 = 0$$
$$a = 1$$

Thus the extension is $\{a\}$.

Although a attacks itself it also attacks its own attack and so a is in.

Example 10.49 (Figure 1 from [214] entitled: A Bizzare framework).

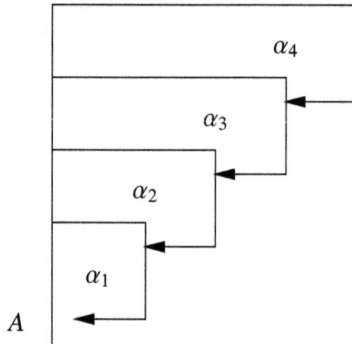

Fig. 10.35.

The previous example 10.48 can be generalised as in Figure 10.35. This figure was put forward by Hanh *et al.* [214] as an example showing how their approach to higher level attacks differs from Gabbay [163] and from [32].

We quote from [214], (we adjust the figure references to ours here).

For illustration consider a framework in Figure 10.35 consisting of attacks $\alpha_1 = (A, A)$ and $\alpha_{i+1} = (A, \alpha_i)$ for $i \geq 1$.

It is rather hard to imagine any practical interpretation of this framework. Hence, as a sceptical reasoner, one would not want to draw any conclusion, i.e. does nto accept A. An agent arguing for A has to rely on an infinite line of defence $\alpha_2, \alpha_4 \ldots$. The semantics of both Gabbay and Baroni *et al.* has a unique preferred extension $\{A, \alpha_2, \alpha_4 \ldots\}$, while our semantics has the empty set as the only extension. The example suggests that extended argumentation in a too literal form would allow counter-intuitive extensions.

Let us see what equations we get for our example.

Case $n = $ infinity
from Figure 10.35 we get

1. $A = (1 - A \cdot \alpha_1)$
2. $\alpha_n = (1 - A \cdot \alpha_{n+1})$

or equivalently

1. $\alpha_1 = (1 - A)/A$
2. $\alpha_2 = (2A - 1)/A^2$
3. $\alpha_3 = (A^2 - 2A + 1)/A^3$
   $\ldots$

We can see that $A = 1$ is a solution.

Let us now continue to examine more cases, let us suppose that we have in Figure 10.35 only a finite number of higher level attacks.

Case $n = 3$
The graph has the nodes $\{A, \alpha_1, \alpha_2, \alpha_3\}$.
What do we get?
Clearly $\alpha_3 = 1$, since it is not attacked.
Hence $\alpha_2 = 1 - A$
and

$$\alpha_1 = (1 - A(1 - A)).$$

Therefore we get

$$A = (1 - A \cdot \alpha_1) = (1 - A \cdot (1 - A(1 - A))).$$

We get the equation

$$A^3 - A^2 + 2A - 1 = 0.$$

There is a solution for A equals approximately around 0.5698402909980533, say $A = 0.57$, and so we get $\alpha_3 = 1, \alpha_2 = 0.43$ and $\alpha_1 = 0.755$.
This means that all nodes can be taken as undecided except α_3.

Case $n = 4$
The graph has the nodes $\{A, \alpha_1, \alpha_2, \alpha_3, \alpha_4\}$. Clearly $\alpha_4 = 1$ since it is not attacked.
We have
$$\alpha_3 = 1 - A$$
$$\alpha_2 = 1 - A + A^2$$
$$\alpha_1 = 1 - A + A^2 - A^3$$
$$A = 1 - A + A^2 - A^3 + A^4$$

1. We have one solution $A = 1, \alpha_4 = 1, \alpha_3 = 0, \alpha_2 = 1, \alpha_1 = 0$.

2. There is also another solution of the degree 4 equation,

$$A = 0.6823278038280193.$$

Again in this case we get all nodes except the top α_4 are undecided.

Case $n = m$
In this case the general equation for A is

$$A = (1 + \sum_{\text{from } k=1 \text{ to } k=m-1} (-A)^k)$$

Remark 10.50 (Support). The equations for higher level attacks allow us to interpret the notion of support in an equational way. Our interpretation of support in Section 3 was qualitative. We did not write equations for the support arrows but used them to define new attack relations (called threats) in Definition 10.35, and then wrote equations using the threat relation. So for example, we had nothing to say about Figure 10.17.
    We can now put forward a new idea:

• **Equational Support**
    a supports b if a attacks the transmission factor of any c attacking b

Thus we read Figure 10.17 as Figure 10.36.

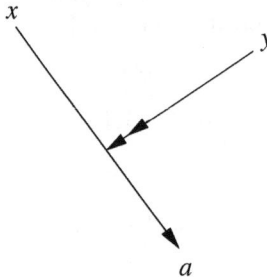

Fig. 10.36.

The equation for support is therefore

$$a = (1 - x(1 - y))$$

Note that in this way we are lead to the notion of *directed* support. See next remark.

Remark 10.51 (Directed support). Let E be a set of arguments. Let e, b be arguments. e is the E directed support of the node b iff e attacks the transmission factor any $a \in E$ which attacks b.
    Thus Figure 10.37 would reduce to Figure 10.38.
    We shall address directional equational support in the next section.

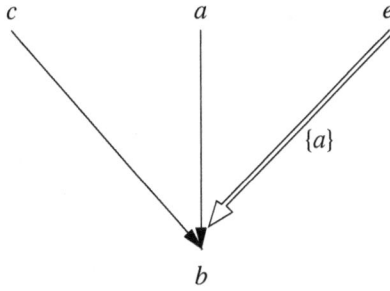

Fig. 10.37.

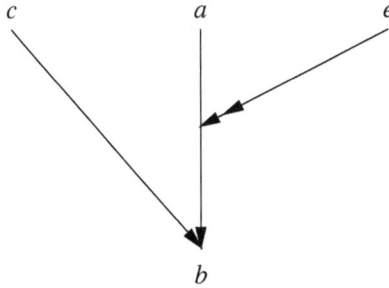

Fig. 10.38.

10.5 Approximately admissible extensions

A very natural next step in our equational approach is that of approximation. Whenever you have equations and their numerical solutions in mathematics, it is natural to examine approximations where certain aspects are neglected. In our case the argumentation network is given equational semantics, the complete extensions sets are the solutions to the equations, therefore it is natural to look at approximate solutions. What we get is a notion of approximate complete extensions. We can regard as being in those nodes getting a value near 1 (the notion of near needs to be defined) and to regard as being out those nodes getting a value near 0. Of course these "approximate extensions" may not satisfy the usual constraints, e.g. a node may be undecided with an attacker being in (actually, approximately in) but we have to accept that.

We shall see that this section naturally relates to an important recent paper of P. Dunne, A. Hunter, P. McBurney and M. Wooldridge [118].

Our starting point is Remark 10.6, where the notion of Admissible sets is changed to that of

- E is **Suspect-Admissible** if whenever a in E is attacked by c then either c is Suspect or E attacks c.

The above notion of Suspect was introduced in Remark 10.6 qualitatively using the loops in the attack relation, i.e. it is a geometrical notion.

If we are dealing with networks with transmission factors, as we have dealt extensively in Chapter 9, then we can use the transmission factor to define the notion of Negligible

- The attack of node a on node b is **Negligible** if its transmission factor is very small
- E is **Approximately-admissible** if whenever an element b in E is attacked by c then either the attack of c on b is **Negligible** or E attacks c via an attack which is not **Negligible**.

It is easy to see that the effect of using this notion is equivalent to disregarding all attacks with negligible transmission.

This immediately connects with paper [118], see their Definition 6.

In fact from our point of view where extensions are solutions of equations, the notion of Approximate Admissibility is automatically taken care of by the equations, because for a very low transmission factor ε of x attacking y, the expression $\varepsilon. x$ which appears in the equations, is very near zero and has almost no effect on the equations.

Thus the correct notion for us is to take as in not just the points x for which $\mathbf{f}(x) = 1$, but also points where the value of $\mathbf{f}$ is very near 1.

Frankly there is not much more to say about this, except to compare with [118].

P. Dunne *et al.* have not used in their paper [118] any of the machinery of Chapter 9. In Chapter 9 networks had transmission values (corresponding to weights) and calculations and propagations of values were performed, with extensive study of the influence of loops.

They say in their paper and I quote:

So, whilst there is gathering momentum for representing and reasoning with the strength of arguments or their attacks, there is not a consensus on the exact notion of argument strength or how it should be used. Furthermore, for the explicit representation of extra information pertaining to argument strength, we see that the use of explicit numerical weights is under-developed. So, for these reasons, we would like to present weighted argument systems as a valuable new proposal that should further extend and clarify aspects of this trend towards considering strength, in particular the explicit consideration of strength of attack between arguments.

I saw their paper just before it was going into print. I contacted the authors and I am grateful that they managed some last minute adjustments of their text.

We include one example comparing our approximation approach with that of [118].

Example 10.52 (Figure 2 of [118]). We show the connection by treating Figure 2 of [118] using our system.

Consider Figure 10.39

In Figure 10.39 the weights are from the numbers x from the set $\{1, 2, 3, 4, 5\}$. They become in our system transmission values, which are numbers y from $[0, 1]$. So the simple transformation $y = x/5$ will change the weights to fractions of the maximal weight 5.

Now looking at Figure 10.39 with weights adjusted, clearly the problem is the loop

$a4 \rightarrow a5$ with strength $1/5$

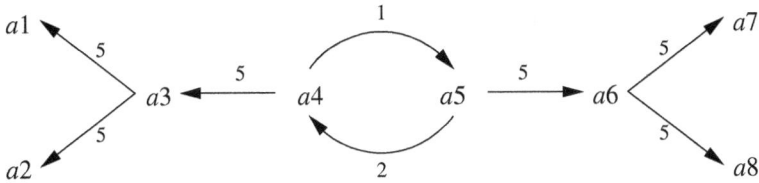

Fig. 10.39.

and

$a5 \rightarrow a4$ with strength $2/5$

The equations just for this loop are therefore

$$a5 = (1 - 0.2a4)$$
$$a4 = (1 - 0.4a5)$$

The solution is

$$a4 = 30/46$$
$$a5 = 40/46$$

the other values are easily calculated to be

$$a6 = (1 - a5) = 6/46$$
$$a7 = a8 = (1 - a6) = 40/46$$
$$a3 = (1 - a4) = 16/46$$
$$a1 = a2 = (1 - a3) = 30/46$$

If we take the approximate extension E of all nodes of values from $40/46$, we get
$E = \{a5, a7, a8\}$

If we take the approximate extension E' of all nodes of values from $30/46$, we get
$E' = \{a1, a2, a4, a5, a7, a8\}$

E corresponds to disconnecting the attack of $a4$ on $a5$ and E' corresponds to also disconnecting the attack of $a5$ on $a4$.

This is in full agreement with [118].

Remark 10.53 (Advantages of our method over [118]). We now list the advantages of our method over [118]. First advantage we can approximate without weights. Consider the loop of Example 10.10. Here we have no weights and no transmissions. The numbers come from the equations, yet because we have numbers we can approximate.

Second advantage is that there are approximations which are not generated by weights and cancelled links, but only from numbers. Some numbers arise from multiple attacks. Some numbers come from the equations.

Most importantly, when we accept nodes with numbers near 1 as "in" we are not changing the values of other nodes!

So for example in Example 10.10 and Figure 10.9 we can accept node c as "in" on account of it being the only one with value more than 0.5, but this does not make the value of node a 0.

Note by the way that $\{c\}$ is he only Suspect Admissible extension when we consider nodes a and b as Suspect on account of their attacking themselves.

A much more important application of these approximations is the handling of loops. This will be discussed in Section 10.8.4.

10.6 The equational view of semantics (preferred, stable, semi-stable, grounded)

This section deals with numerical and computational aspects of our equational models, and how they can calculate semantic extensions.

We begin with options for calculating extensions in ordinary Dung networks and their comparison with Caminada labelling. Our embarkation point is a Table from Chapter 2, Table 2.1.

We now write equations whose solutions give the correct extensions. We assume a set of equations Eq which is sound for Dung semantics, such as offered in Definition 10.1.

Our network is (S, R_A). Assume, in case we want to refer to the points of S explicitly, that $S = \{x_1, \ldots, x_n\}$.

We begin with a word of explanation and orientation addressed to the perceptive reader. In the set theoretic context of Dung extensions and semantics, there is uniformity and beauty in defining the semantics and the extensions, in terms of maximality and minimality of complete admissible sets. One can also define new types of semantics such as CF2 or stage or ideal or skeptical semantics, again using set theoretic concepts such as conflict freeness and intersections of extensions.

So for example a preferred extension is defined as a maximal complete extension.

When we move to the world of equations, we can get the basic complete extensions as functions $\mathbf{f}$ which are solutions to equations. We can still talk about maximal or minimal functions relative to pointwise numerical ordering. Thus we can define a preferred solution $\mathbf{f}$ as a maximal solution to the equations.

However in the context of equations we are tempted to use further equations and constraints and get the semantics directly as solutions of equations with constraints. This can get messy and we lose uniformity.

What we gain is that we can use solvers to find the maximal solutions.

What we lose for example is that we cannot form infinite products of functions, which correspond to infinite intersections of extensions.

Note also that if we use equations which are different from Eq_{max}, we obtain something not corresponding to Dung complete semantics, but something analogous to it for the type of equations used. Let us call it pseudo-complete semantics. Then it may be interesting to investigate notions like pseudo-stable or pseudo-grounded semantics in a purely numerical context. We leave this to future work.

Case complete extensions
Solve the Eq_{max} equations. Any solution $\mathbf{f}$ is an extension.

Case stable extensions
   Add a new variable y such that $y \notin S$, and write the additional equation

$$\mathbf{f}(y) = \mathbf{h}_y = \sum_{x \in S} \mathbf{f}(x)(1 - \mathbf{f}(x)).$$

If the solution $\mathbf{f}$ to the new expanded set of equations is a stable extension, then $\mathbf{f}(x)(1 - \mathbf{f}(x))$ is 0 for all $x \in S$ and hence $\mathbf{f}(y) = 0$. Conversely, if $\mathbf{f}(y) = 0$ then $\mathbf{f}$ is stable. Thus to check for stable extensions, we check $\mathbf{f}(y)$.[5]

Case of semi-stable extensions
This case minimises the undecided. We do the following.

Consider the quantity

$$\mu = \sum_{\substack{a \in S \\ x_1, \ldots, x_n \in S \\ \text{are all} \\ \text{attackers of } a}} [a - \mathbf{h}_a(x_1, \ldots, x_n)]^2.$$

In μ we regard all elements of S as variables. The equation $\mu = 0$ has a solution. We regard $\mu = 0$ as a constraint and minimise the expression

$$\nu = \sum_{x \in S} x(1 - x)$$

subject to the constraint $\mu = 0$.

This can be done using the method of Lagrange multipliers (see Wikipedia).

Case of grounded extensions
This is like the semi-stable case except that we minimise the expression $1 - \nu$.

Case of preferred extension
Let (S, R_A) be given. We want to identify all preferred extensions. For each $E \subseteq S$, let $S_E = \{(x, E) \mid x \in S\}$. Let $R_A^E = \{((x, E), (y, E)) \mid (x, y) \in R_A\}$. Thus (S_E, R_A^E) is an exact replica of (S, R_A), and they are all pairwise disjoint!

Let $S^* = \bigcup_E S_E$ and let $R_A^* = \bigcup_E R_A^E$. Thus (S^*, R_A^*) contains as many copies of (S, R_A) as there are subsets of S. For each $(x, E) \in S^*$, let (x', E) be a new point. Let $y_E^+, y_E^-, E \subseteq S$ be two more new test points. Let

$$S' = S^* \cup \{(x', E), y_E^+, y_E^- \mid x \in S, E \subseteq S\}$$

$$R_A' = R_A^* \cup \{((x, E), (x', E)), ((x', E), (x', E)) \mid x \in S, E \subseteq S\}$$

$$\cup \{((x', E), y_E^+) \mid x \in E\} \cup \{((x, E), y_E^-) \mid x \in S - E\}.$$

The situation is described in Figure 10.40, for a single copy S_E. The attacks are defined for each E.

Let us use Eq_{max} on this network and let μ be as before, namely

---

[5] This is a neater condition than the one mentioned in Remark 10.33, which was more explanatory than efficient.

Set S_E

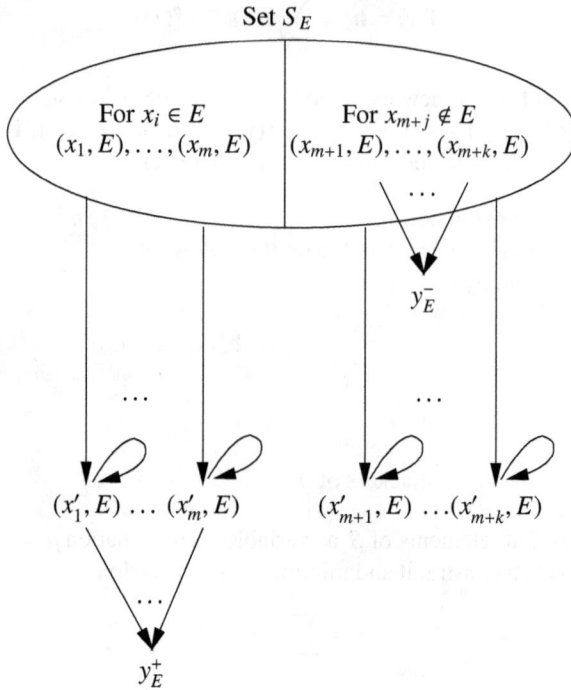

Fig. 10.40.

$$\mu = \sum_{\substack{a \in S \\ x_1, \ldots, x_n \text{ are} \\ \text{all attacker of } a}} [\mathbf{f}(a) - h_a(x_1, \ldots, x_n)]^2$$

Consider the expression

$$\gamma = \sum_E [\sum_{x \in E} K\mathbf{f}((x', E)) + \sum_{x \notin E} \mathbf{f}((x', E))]$$

where K is a large constant.[6]

To minimise $\sum_{x \in E} K\mathbf{f}((x', E))$ we need to have all of $\mathbf{f}((x, E)) = 1$. Any sacrifice of $\mathbf{f}((x, E)) \neq 1$ for $x \in E$ and compensating by $\mathbf{f}((x, E)) = 1$ for $x \notin E$ will not really minimise this sum because K is large. Thus minimising γ requires absolutely to try for each $E \subseteq S$ to have all of $\mathbf{f}((x, E))$ for $x \in E$ to be 1. Thus for each E, we get that if possible all of $\mathbf{f}((x, E)) = 1$ and hence all of $\mathbf{f}((x', E)) = 0$ and hence $\mathbf{f}(y_E^+) = 1$.

Let us now minimise γ subject to the constatn $\mu = 0$.

1. If $\mathbf{f}(y_E^+) = 1$, and $\mathbf{f}(y_E^-) = 0$, then there exists a preferred extension $E' \supsetneq E$.
2. If $\mathbf{f}(y_E^+) = 1$ and $\mathbf{f}(y_E^-) = \frac{1}{2}$, then E is a preferred extension.

If $\mathbf{f}(y_E^+) \neq 1$ then there does not exist a preferred extension $E' \supseteq E$.

---

[6] We need to see the network first, then choose a large K depending on the network.

The reader would note that we added an exponential number of points but this is unavoidable, and anyway it is all uniformly fed into a program like MATLAB, MAPLE or NSolve.

Remark 10.54 (Connection with the intertranslatability of argumentation semantics). In a brilliant recent paper [120], Dvorak and Woltran investigate translations of one type of extension into another. This has bearing to this section and will be discussed together with our notion of Critical Translations (Definition 10.23 in subsection 2.2) later in Section 10.8.2.

10.7 Time dependent networks

There are two ways to make a system time dependent.

1. Make each atomic component depend on time.
2. Take snapshots of the system at different times.

If we follow the first method for the case of argumentation networks, we will make each argument strength time dependent and the transmission factors also time dependent. So, for example, if we look again at Figure 9.7 as a typical figure, and make it time dependent, then the coefficients $x : a$ and $y : b$ become time dependent, as does the transmission factor $\varepsilon : (ab)$. This is how we already presented time dependence in our 2005 paper [37], recall Chapter 9, section 9.5. Thus we have $x(t, a), y(t, b), \varepsilon(t, a, b)$, where t is a time variable.

If we follow the second approach we again use a time variable t but our networks have the form $(S_t, R_{A,t})$, where $R_{A,t} \subseteq S_t$.

The first approach is more general. We can take a time dependent model of the first approach and get from it a model of the second approach. If we allow $x(t)$ and $\varepsilon(t)$ to take only $\{0, 1\}$ values, we can let

$$S_t = \{a | x(t, a) = 1\}$$
$$R_{A,t} = \{(a, b) | x(t, a) = y(t, b) = \varepsilon(t, a, b) = 1\}.$$

Thus we get a model $(S_t, R_{A,t})$ according to the second approach.

Note that if we do not involve the time element in the equations of the attack all we would have is lots of non-interacting networks sitting around. So to get some interaction, let us review the basic temporal configurations and the options they afford us.

Figure 10.41 is a typical situation, we allow for three nodes for simplicity.

Let us assume that $0 \leq x_i, \varepsilon_j \leq 1$ and also that $0 \leq t \leq 1$, so that we shall always have solutions to our equations. The value $y(t)$ depends on $x_i(t), \varepsilon_i(t)$ as in Definition 10.44.

(*) $$y(t) = (1 - x_1(t)\varepsilon_1(t))(1 - x_2(t)\varepsilon_2(t)(1 - x_3(t)\varepsilon_3(t))).$$

Let

$$\alpha_i(t) = (1 - x_i(t)\varepsilon_i(t))$$

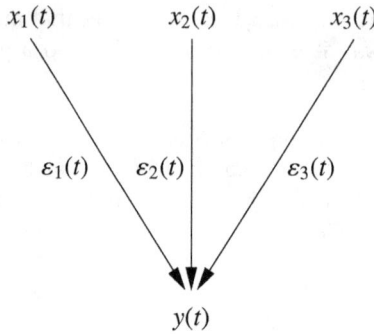

Fig. 10.41.

$\alpha_i(t)$ is the effective attack of node $x_i(t)$. The total attack is

$$y(t) = \prod_i \alpha_i(t).$$

We want to modify $\alpha_i(t)$ in view of the time dependence.

Equation (*) does not express any interaction in time. Suppose we want to add some temporal interactions, what option can we offer?

Option

We can make the attract at time t depend also on the rate of change at t.

Surely if the strength of the attack is on the decline it is less effective. So we can look at the derivative $\dfrac{dx_i(t)}{dt}$ and $\dfrac{d\varepsilon_i(t)}{dt}$ and look at their value at the point t.

If the rate is increasing, we make the attack of $x_i(t)$ on y a bit stronger, otherwise a bit weaker.

Think you are trying to buy a house and the price of the house is attacking the argument that you should own your house (rather than rent). If prices are going up, the argument for buying now is stronger. If they are going down, it is weaker, all compared with a temporal situation where prices are not changing.

Let us use the same notation that is used in physics. $\dot{x}_i(t)$ is the value of the definative at t. Similarly $\dot{\varepsilon}_j(t)$.

Define $r_i(t)$ and $s_i(t)$ as follows.

1. If $\dot{x}_i(t) = 0$, let $r_i(t) = 1$.
   If $\dot{x}_i(t) > 0$, let $r_i(t) = \frac{1}{1+\frac{1}{\dot{x}_i(t)}}$
   If $\dot{x}_i(t) < 0$, let $r_i(t) = \frac{1}{1-\frac{1}{\dot{x}_i(t)}}$.
   So $r_i(t) = \frac{1}{1+\frac{1}{|\dot{x}_i(t)|}}$.
   Define $s_i(t)$ similarly using $\dot{\varepsilon}_i(t)$.
   Now consider the attack $x_i(t)$ on $y(t)$. We want to modify the original expression

$$\alpha_i(t) = [1 - x_i(t)\varepsilon_i(t)]$$

into a new expression $\beta_i(t)$ and use that to define the attack.

Clearly if we think the force of the attack is going to decline, then the attack should be less effective. Thus for example if $\dot{x}_i(t)$ is very large and negative this means that the attack of x_i is going down very quickly and $r_i(t)$ is almost 1. We can almost ignore the attack. So we multiply $x_i(t)$ by $1 - r_i(t)$.

Thus we get

$$\beta_i(t) = 1 - x_i(t)\varepsilon_i(t)(1 - r_i(t))$$

Since $1 - r_i(t)$ is almost 0, we get that $\beta_i(t)$ is almost 1, thus we are almost ignoring this attack.

If $\dot{x}_i(t) > 0$ and is very large, again $r_i(t)$ is amost 1 but the attack of x_i is strengthened. So we multiply α_i by $(1 - r_i(t))$, and the attack becomes stronger. $\beta_i(t) = \alpha_i(t)(1 - r_i(t))$.

Similarly, if $\varepsilon_i(t) < 0$, this means transmission is going down so we multiply $\varepsilon_i(t)$ by $(1 - s_i(t))$.

If $\dot{\varepsilon}_i(t) > 0$, then $\varepsilon_i(t)$ is going up, then we multiply α_i by $(1 - s_i(t))$.

So this defines $\beta_i(t)$ and we let

$$y_{\text{new}}(t) = \prod_i \beta_i(t).$$

To take an example, assume for simplicity that the transmission factor is constant in time but that $x(t)$ declines exponentially in $[0, 1]$.

Let

$$x_i(t) = e^{-t}$$

Therefore

$$\alpha_i(t) = (1 - \varepsilon e^{-t})$$

and

$$r_i(t) = \frac{1}{1 + e^{-t}},$$

and so

$$1 - r_i(t) = \frac{1}{1 + e^t}.$$

We therefore have

$$\beta_i(t) = (1 - \frac{\varepsilon e^{-t}}{1 + e^t}).$$

So the attack is less effective!

Final comment

The equational approach allows us to take into consideration the rate of change of the strength of the attacks. This is not possible or even expressible in the traditional Dung networks.

Let us now see what options we have in the snapshot approach. To simplify, let us assume three points of time, and simple networks, as in Figure 10.42

Note the following:

1. If there is no connection between the times, then this is just a flat network without regard to time.

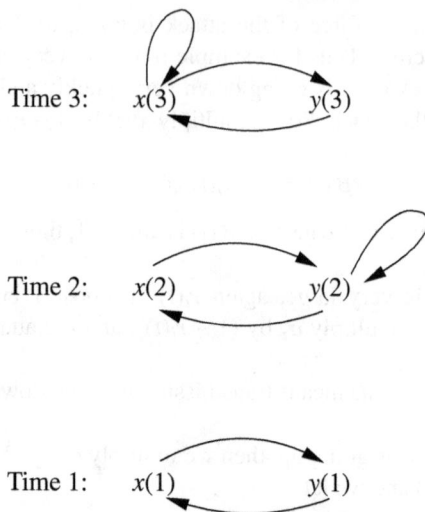

Time 3: $x(3)$ $y(3)$

Time 2: $x(2)$ $y(2)$

Time 1: $x(1)$ $y(1)$

Fig. 10.42.

2. Even if we add attacks from future to past, e.g. $x(2) \rightarrow y(1)$ then it is still a flat network.

   The only way is to say that for example

$$y(1) = (\exists t > 1)y(t).$$

So we have an existential quantifier here and $y(1)$ is say "in" if some $y(t)$ for $t > 1$ is "in" or, if we are writing equations then we might have, for example

$$y(1) = (1 - x(1))(1 - \max_{(t>1)}\{y(t)\}).$$

The reader can see that this approach is different from the previous "rate of change" approach.

   This approach is investigated in my paper with S. Modgil [159]. See also Chapter 6.

10.8 Further topics

This section expands on some further topics connected with this chapter. The issues mentioned in each subsection will be fully developed in subsequent papers. The intention here is to give the reader an idea of what is coming.

10.8.1 Equational approach to logic

We explain the general idea via some examples, and this would give the reader a better perspective on our equational approach to argumentation networks. The reader should also consult Chapter 13 below, in which we show that Dung's networks are essentially classical logic with the Peirce-Quine dagger connective.

Example 10.55 (Disjunctive inference). Consider a simple inference:

1. $(p \vee q)$
2. $p \rightarrow r$
3. $q \rightarrow r$

From (1)–(3) we want to infer

4. r

We proceed as follows, assuming our logic satisfies the Deduction theorem:

$$E \text{ and } x \text{ proves } y \text{ iff } E \text{ proves } x \rightarrow y.$$

5.   1. Assume p
     2. Get r from (5.1) and (2) using modus ponens.
6.   1. Assume q
     2. Get r from 6.1 and (3), using modus ponens
7. Get r from (1), (5-1–5.2) and (6.1–6.2) and the rule for disjunction elimination.

We now compare this with an equational approach.

Note that the above proof theoretic inference is valid in many logics, such as classical logic, intuitionistic logic and Łukasiewicz infinite valued logic.

When we write equations for the above inference, we have to choose in which logic we are operating. There will be different equations for different logics.

Definition 10.56 (Boolean negation disjunction network). *Let* $(S, R_\neg, R_+)$ *be a network with two binary relations and the following properties:*
Let $R = R_\neg \cup R_+$. *Then the following holds*

1. $xR_\neg y \wedge xR_\neg y' \rightarrow y = y'$.
2. $\neg \exists y_1 y_2 (xR_\neg y_1 \wedge xR_+ y_2)$
3. $xR_+ y \rightarrow \exists! z (z \neq y \wedge xR_+ z)$.
4. $\neg xRx$.

We associate the following functions wth R_+ *and* $R_\neg$.

1. *If* $xR_\neg y$ *then* $x = 1 - y$.
2. *If* $xR_+ y_1 \wedge xR_+ y_2$ *then* $x = [1 - (1 - y_1)(1 - y_2)]$.

Example 10.57 (Equational approach to disjunctive inference in classical logic). Consider Figure 10.43. This is a construction tree for the wffs involved in Example 10.55 from the point of view of classical logic. In classical logic R_+ indicates disjunction.

Let us apply our equational definition of Definition 10.56 to Figure 10.43. We get in terms of p, q, r the following equations

1. $u = [1 - (1 - p)(1 - q)]$
2. $z = [1 - p(1 - r)]$
3. $y = [1 - q(1 - r)]$

The disjunctive inference problem of Example 10.55 becomes the following equational problem

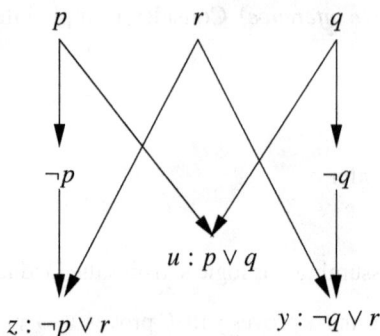

Fig. 10.43.

- Given that $u = z = y = 1$, solve for r. $r =$?

Let us see how to do it. We get

(*1) $(1 - p)(1 - q) = 0$
(*2) $p(1 - r) = 0$
(*3) $q(1 - r) = 0$

The way the proof procedure of Example 10.55 proceeds is to do case analysis. From (*1) either $p = 1$ or $q = 1$ and in each case from (*2) (resp. *3) we get $r = 1$.
   This is not equational solving but reasoning about the equations to prove that $r = 1$.
   We want to be more direct. Let us expand the equations.

(*1) $1 - p - q + pq = 0$
(*2) $p - pr = 0$
(*3) $q - qr = 0$

Add up all three equations and get

(*4) $1 + pq - pr - qr = 0$

or

(*5) $1 + pq = (p + q)r.$

Let us now add (*2) and (*3), we get

(*6) $p + q - pr - qr = 0$

or

(*7) $(p + q) = (p + q)r.$

We need to show that $p + q$ is not 0 so that we can divide by it.
   From (*7) and (*5) we get

(*8) $1 + pq = p + q$

   We can also deduce from (*8) that $p + q \neq 0$ and so we divide by $p + q$.
   So from (*7) by dividing by $p + q$ we get

(*9) $r = 1.$

Example 10.58 (Equational approach to disjunctive inference in Łukasiewicz many valued logic). Łukasiewicz logic is formulated using $\rightarrow$ and $\neg$, with the following truth table:

1. atoms get values in $[0, 1]$
2. $x \rightarrow y = \min(1, 1 - x + y)$
3. $\neg x = 1 - x$
4. A wff is a tautology iff its truth value is always 1
5. Define $x \oplus y = \neg x \rightarrow y$ and so we have

$$\neg x \rightarrow y = \min(1, x + y)$$

We can define therefore

$$x_1 \oplus \ldots \oplus x_{n+1}$$

to be

$$\neg x_1 \rightarrow (\neg x_2 \rightarrow \ldots \rightarrow (\neg x_n \rightarrow x_{n+1}) \ldots)$$

and its table is

$$\min(1, x_1 + \ldots x_{n+1}).$$

Consider now the network of Definition 10.56. We use new functions for the case of $x R_+ y_1$ and $x R_+ y_2$, we let

$$x = \min(y_1 + y_2).$$

We note that in Łukasiewicz logic the disjunction $x \vee y$ has the table

$$x \vee y = \max(x, y)$$

and can be defined as

$$(x \rightarrow y) \rightarrow y.$$

We can define conjunction $x \wedge y$ but

$$x \wedge y = \neg(\neg x \vee \neg y)$$

we have:

$$x \wedge y = \min(x, y).$$

6. The consequence relation for Łukasiewicz logic can be defined in several ways. We use the options which allows for the Deduction theorem, because the disjunctive proof in Example 10.55 uses it.
   - $A_1, \ldots, A_n \vdash B$ iff $(A_1 \oplus \ldots \oplus A_n) \rightarrow B$ is a tautology.
   The above means that $\sum_i \text{value}(A_i) \leq \text{value}(B)$.
   Consider now the network of Figure 10.44.
   Note that R_+ in the figure indicates the connective $x \oplus y = \neg x \rightarrow y$, therefore we have

$$x \rightarrow y = (\neg x) \oplus y.$$

and disjunction $x \vee y$ is defined as $(x \rightarrow y) \rightarrow y$, therefore

$$x \vee y = [\neg(\neg x) \oplus y)] \oplus y.$$

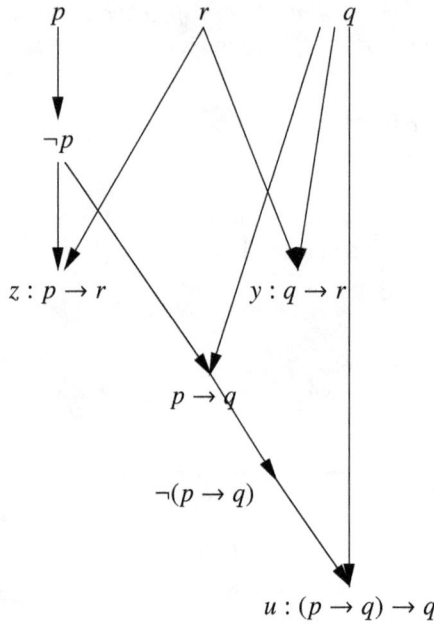

Fig. 10.44.

We ask the following question:

- Given that $\min(1, u + z + y) \le r$, what values can r have?

This means that given

$$\min(1, \max(p, q) + \min(1, 1 - p + r) + \min(1, 1 - q + r)) \le r$$

what can r be?

Assume without loss of generality that $p \ge q$.[7]

We ask can r be less than 1?

Assume $r < 1$ and get a contradiction.

Case 1

$r < q \le 1$.

In this case we get

$$(p + 1 - p + r + 1 - q + r) \le r$$

$$2 + 2r - q \le r$$
$$2 + 2r \le r + q$$

Since $r < q$ we get

---

[7] There is no loss of generality because the expression

$$\min(1, \max(p, q) \min(1, 1 - p + r) + \min(1, 1 - q + r)) \le r$$

is symmetrical in p and q, even though our starting Figure 10.44 is not.

$$2 + r + q \leq r + q$$
$$2 \leq 0$$

not possible.

Case 2

$1 > r \geq q$

We get

$$p + 1 - p + r + 1 \leq r$$
$$1 + 1 \leq 0$$

not possible.

Therefore $r = 1$ and so the left hand side is

$$[\min(1, p + 1 + 1 = 1] \leq [r = 1]$$

Remark 10.59 (Equational reasoning). Let us put on our meta-level hat and analyse what is happening here.

1. The logical database $\Delta = \{p \vee q, p \rightarrow r, q \rightarrow r\}$ became a set of nodes E_Δ in the network of Figure 10.43..
2. The inference problem
   - $\Delta \vdash ?r$
   becomes the following equational question
   - If the points in E_Δ all have value 1 does r have to get value 1 too?
   or more generally, let **e** be a function assigning values to points in E and let $y \notin E$. We can ask
   - If E gets the values indicated by the function **e**, what values are forced on y?
3. The equational question in (2) is meaningful for any network. Take for example an argumentation network and take any set of nodes E_0 and $y \notin E_0$. We can ask
   - Let $E \supseteq E_0$ be any extension of a certain type (say E a stable extension) are we forced to have $y \in E$?
   In which case we can write
   - $E \Vdash y_{\text{stable}}$
4. What is the analogous feature in the case of logic to the notion of extension in argumentation networks?
   We know that any set of nodes corresponds to a database. So the algorithms generating extensions correspond to a way of generating databases.
5. The notion of "consistency" in logic corresponds to "having a solution" in an equational network.
   Let **e** be a function associating values to the points in E. Then (e, E) is equationally consistent, iff there exists a solution **f** to the equations such that $\mathbf{f} \upharpoonright E = \mathbf{e}$.

Remark 10.60 (Comparison with Caminada view). For absolute clarity we need to compare the way we presented our equational reasoning with Caminada [83]. We reproduce here, as Figure 10.45, the Caminada slide 19 from [83].

We start with an inconsistent database Δ. Construct an argumentation network out of proofs $\{\pi\}$ from assumptions taken from the database. This gives us a set S of arguments. We have π_1 attacks π_2 if π_1 logically contradicts π_2.

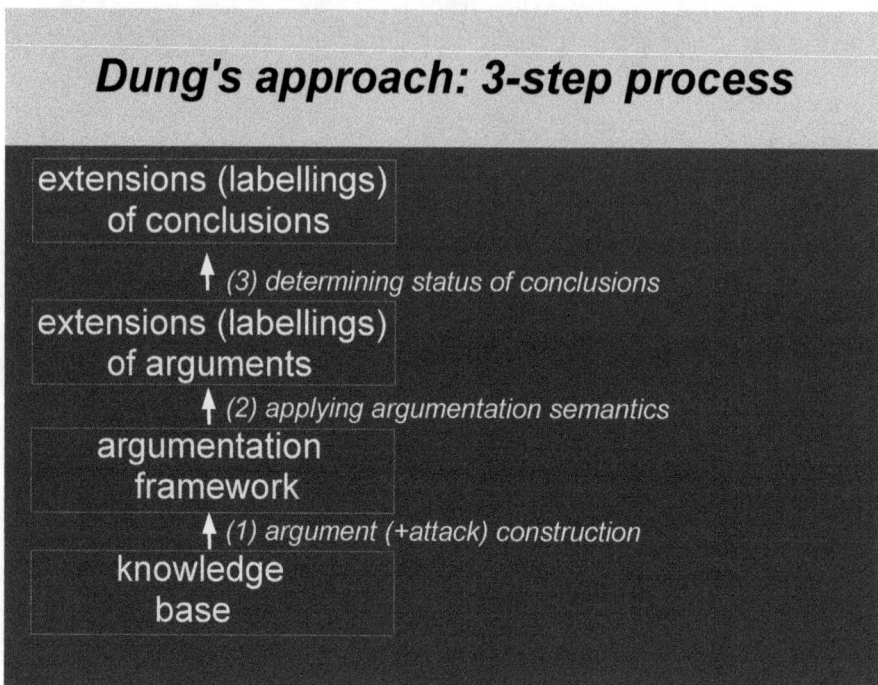

Fig. 10.45.

We now apply the Dung network theory and get an extension E of some sort. This extension defines a subset $\Delta_E \subseteq \Delta$. We expect Δ_E to be consistent and its consequence closure to be consistent.

As you can see, the above is a different concept than our Example 10.57.

See also the excellent detailed paper of Nikos Gorogiannis, Anthony Hunter [205]. We do not need such detailed analysis.

10.8.2 Critical sets and translations

We best explain our methodological ideas about critical sets through a series of examples.

Example 10.61 (Networks with stability tester). Consider networks of the form (S, R_A, y), where (S, R_A) is an argumentation network and $y \notin S$ is an additional point.

Let us use $Eq_{\max}$ for nodes of the (S, R_A) network and in addition we use for point y the equation

- $\mathbf{f}(y) = 1 - \max_{(x \in S)}\{1 - \max\{\mathbf{f}(x), (1 - \mathbf{f}(x))\}\}$

clearly any solution $\mathbf{f}$ of these equations in which $\mathbf{f}(y) = 1$ forces $\mathbf{f}$ to be a stable extension on (S, R_A).

(S, R_A, y) is, however, not an argumentation network.

Let us embed it in an argumentation network (S_1, R_1).

Let S_1 be:

$$S_1 = S \cup \{y\} \cup \{e_x | x \in S\} \cup \{\bar{x} | x \in S\}$$

Let R_1 be

$$R_1 = R_A \cup \{(x, e_x), (e_x, \bar{x}), (\bar{x}, y) | x \in S\}.$$

Figure 10.46 illustrates the role that the additional points play and the attack relation of the new points relative to any x in S.

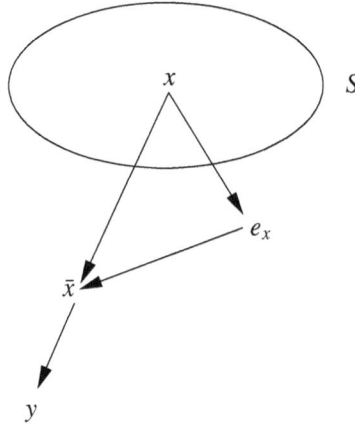

Fig. 10.46.

The equations for points within S remain the same in this new network (S_1, R_1). The equations for the additional points are as follows.

(*1) $\mathbf{f}(e_x) = 1 - \mathbf{f}(x)$
$\quad \mathbf{f}(\bar{x}) = 1 = \max(\mathbf{f}(x), \mathbf{f}(e_x))$
$\quad \quad = 1 - \max(\mathbf{f}(x), 1 - \mathbf{f}(x))$
(*2)
$\quad \mathbf{f}(y) = 1 - \max_{(x \in S)}(\mathbf{f}(\bar{x}))$
$\quad \quad = 1 - \max_{(x \in S)}\{1 - \max[f(x), (1 - f(x))]\}$

Note the following

(C1) The equations for the points of $S \cup \{y\}$ in (S_1, R_1) do not use the new points $\{e_x, \bar{x}\}$.

(C2) For any solution $\mathbf{f}$, the values of $\mathbf{f}$ on the new points are uniquely determined (see (*1) and (*2)) by the values of $\mathbf{f}$ on S.

Conditions (C1) and (C2) determine that the embedding of (S, R_A) into (S_1, R_1) is a faithful critical embedding.

Note that condition (C1) implies the following condition (C1*):

(C1*) Any extension $\mathbf{f}$ on (s, R_A, y) can be uniquely expanded to an extension $\mathbf{f}_1$ on (S_1, R_1).

Definition 10.62 (Critical embedding of networks). *Let* (S_1, R_1, Eq_1) *and* (S_2, R_2, Eq_2) *be two networks with nodes* S_i, *relation* $R_i \subseteq S_i^2$ *and respective equational systems* Eq_i.

Let $\mathbf{e} : S_1 \mapsto S_2$ *be a* $1 - 1$ *embedding of* S_1 *into* S_2.
We say this embedding is a critical embedding iff the following holds:

1. *For every solution* $\mathbf{f}_1$ *of* Eq_1 *in* (S_1, R_1), *there exists a solution* $\mathbf{f}_2$ *of* Eq_2 *in* (S_2, R_2) *such that for all* $x \in S_1$ *we have*
   - $\mathbf{f}_1(x) = \mathbf{f}_2(\mathbf{e}(x))$.
2. *Let* $S_1^{\mathbf{e}} \subseteq S_2$ *be the image of* S_1 *under* $\mathbf{e}$, *as a subset of* S_2. *Let* $\mathbf{f}_2$ *and* $\mathbf{f}_2'$ *be two solutions of* Eq_2 *in* (S_2, R_2), *then the following holds*
   - $\mathbf{f}_1 \upharpoonright S_1^{\mathbf{e}} = \mathbf{f}_2' \upharpoonright S_1^{\mathbf{e}} \Rightarrow \mathbf{f}_2 = \mathbf{f}_2'$
   In other words, the values of any Eq_2 *solution on* $S_1^{\mathbf{e}}$ *determines the rest of the* S_2 *values uniquely.*

Remark 10.63 (Different extensions on different parts of the network). Note that the device used in Figure 10.46 can be used not necessarily to the entire set S of arguments but only to a subset of it. In fact by using different additional points y we can use the device for different subsets. Thus we can force that our extension solution function $\mathbf{f}$, will be stable on certain subsets of our choice (if possible, of course).

The ideas of Section 10.6 where we use equations to force the type of extensions, when combined with what we are doing here, allow us to easily force different types of extensions on different subsets.

So to make it clear we can ask for a function $\mathbf{f}$, which is a preferred extension on one subset and a stable extension on another subset and a grounded extension on a third subset and the subsets need not be disjoint. It all depends on what equations we write.

Of course we might ask but not get a solution!

Example 10.64 (A non-critical embedding). Consider the network of Figure 10.47 and compare it with Figure 10.1.

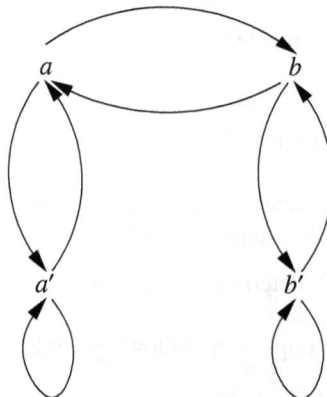

Fig. 10.47.

The equations for the network of Figure 10.1, (using Eq_{max}) are

($\sharp$1) $a = 1 - b$
($\sharp$2) $b = 1 - a$

This means a can take any value $0 \le \alpha \le 1$ and b would then be $b = 1 - \alpha$.
  We now consider the equations of Figure 10.47. There are (using Eq_{max}):

(*1) $a = 1 - \max(b, a')$
(*2) $b = 1 - \max(a, b')$
(*3) $a' = 1 - \max(a', a)$
(*4) $b' = 1 - \max(b', b)$

We now do a case analysis, to determine what are the possible solutions. Obviously we can have
$$\{a = 1, b = 0, a' = 0, b' = \tfrac{1}{2}\}$$
and
$$a = 0, b = 1, a' = \tfrac{1}{2}, b' = 0\}.$$
We shall prove below that the only other third solution to equation (*1)–(*4) is
$$a = b = a' = b' = \tfrac{1}{2}.$$

This means that the embedding of Figure 10.1 into Figure 10.47 is not critical, because it restricts the solutions for $\{a, b\}$ of Figure 10.1 only to $\{a = b = \tfrac{1}{2}\}, \{a = 1, b = 0\}$ and $\{a = 0, b = 1\}$.

Lemma 10.65. *For the network of Figure 10.47, the only possible solutions for Eq_{max} are*

1. $a = 1, b = 0, a' = 0, b' = \tfrac{1}{2}$
2. $a = 0, b = 1, a' = \tfrac{1}{2}, b' = 0$
3. $a = b = a' = b' = \tfrac{1}{2}$

Proof. 1. Clearly if $a = 1$, then $b = 0$ and $a' = 0$ and $b' = \tfrac{1}{2}$.
  2. if $a = 0$, then from (*3), $a' = \tfrac{1}{2}$.
    From (*1) we get
$$0 = 1 - \max(b, \tfrac{1}{2})$$
  so $b = 1$ and so $b' = 0$.
  3. By symmetry we also have if $b = 1$ then we have case 2 of the lemma and if $b = 0$, we have case 1 of the lemma.
  4. So assume $a \ne 0, a \ne 1, b \ne 0, b \ne 1$. By symmetry we must have $a = b = x$, and similarly $a' = b' = y$.
    We need to find x and y
  5. Let us substitute $a = b = x$ in the equations. We get
    (**1) $x = 1 - \max(x, y)$
    (**2) $x = 1 - \max(x, y)$
    (**3) $y = 1 - \max(y, x)$
    (**4) $y = 1 - \max(y, x)$

We immediately get that $x = y$.

Therefore from (**1) we get $x = 1 - x$

Therefore $x = y = \frac{1}{2}$ which is case 3 of the lemma.

Remark 10.66 (Paper of Dvorak and Woltran). In their paper [120] Dvorak and Woltran give some brilliant embeddings of various extensions in one another.

We refer to their translation 1 as given in Figure 3 of their paper, which we reproduce here as our Figure 10.48

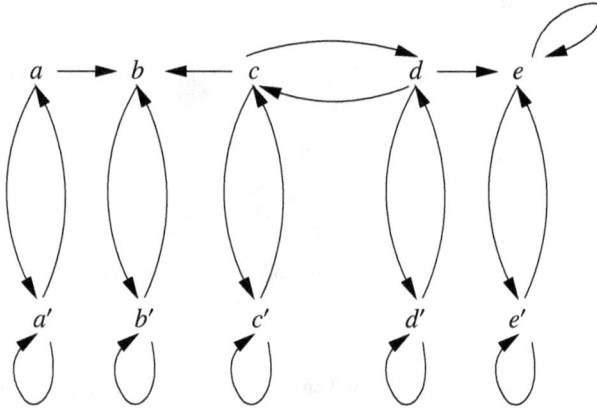

Fig. 10.48.

The network with $\{a, b, c, d\}$, i.e. Figure 10.49,

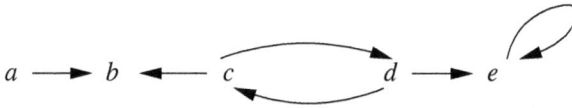

Fig. 10.49.

is presented as a faithful embedding in Figure 10.48.

This is true for the Caminada labelling, where the values are

$x =$ "in" corresponding to $x = 1$

$x =$ "out" corresponding to $x = 0$

$x =$ undecided corresponding to $0 < x < 1$

Compare with our Theorem 10.21.

As we saw in our Example 10.64 and Lemma 10.65, this embedding is not faithful if we consider the full range of values the nodes can take. We also know that the full range of values can tell differences among loops, as we saw in Example 10.5 (Figure 10.6) and Example 10.10 (Figure 10.9).

We suspect similar results hold for Dvorak and Woltran other embeddings.

We therefore invite Dvorak and Woltran to apply their considerable skills to finding new critical embeddings for their paper. These can be useful in the computational context of Section 6.

10.8.3 Connection with fuzziness

This subsection studies methodological ideas on

- How to make your argumentation network fuzzy.

This is really a subject for another paper, so here we only give the basic ideas.

The Problem with making a system fuzzy
The problem with making any system fuzzy is to make it in a meaningful methodological way and not in an ad hoc partisan way. One can always take the basic components of the system and make each one of them fuzzy. So for example in the case of argumentation network (S, R_A), we can make S fuzzy, R_A fuzzy, the Caminada labelling fuzzy etc., etc. The difficulty with that is that if we have similar networks formulated differently, with different components, we get different results. This was the case in fuzzy logics. Equivalent formulations of the same logic, gave rise to different fuzzy versions. The situation was really chaotic.

In my papers [137, 139], I used the Fibring Methodology of [138] and showed that, to make a logic fuzzy, you fibre it with Łukasiewicz infinite valued logic. The different methods of fibring (which are conceptually independent of the notion of fuzziness) yield all the options existing in the literature for making a logic fuzzy. So we got order out of chaos.

What are our options of how to make an argumentation network fuzzy?
Option 1. Fibring
Fibre the network with Łukasiewicz infinite valued logic, as discussed above, and see what you get.

Option 2. Equational approach
We saw in Section 10.1 that the equational approach gives us fuzzy values to nodes in argumentation networks , especially if the network contains loops. See Example 10.10.

Note however that in Fuzzy Logic we can assign any values in $[0, 1]$ to the variables, while in the equational approach the values emerge from the necessity of solving the equations, and so not all possible assignments may be obtained.

Option 3. Compare with fuzzy logic programming
We also saw the connection between argumentation networks and logic programs, see Remark 10.30.

In view of the above, we can look at some fuzzy logic programming papers , e.g. [120], and translate into argumentation networks and see what we get.

Option 4. Check existing fuzzy argumentation proposals
The cultural climate in argumentation theory is to offer good motivating discussions. So it is worthwhile to check existing proposal for fuzzy argumentation networks and compare. See for example [222].

330   10 An Equational Approach to Argumentation Networks

Remark 10.67 (Fuzzy norms). There is a connection between our Equations of Definition 10.1 and what is known as fuzzy t-norms and t-co-norms, see for example [230].

A t-norm is a function $T : [0, 1] \times [0, 1] \rightarrow [0, 1]$ which satisfies the following properties:

Commutativity: $T(a, b) = T(b, a)$
Monotonicity: $T(a, b) \leq T(c, d)$ if $a \leq c$ and $b \leq d$
Associativity: $T(a, T(b, c)) = T(T(a, b), c)$

The number 1 acts as identity element: $T(a, 1) = a$

Note that owing to associativity, we can write $T(x_1, \ldots, x_n)$ without ambiguity.

Given a situation as in Figure 10.11, where $x_1, \ldots, x_n$ all attack node a, we can use the Equation$_T$ approach and write

$$a = T(1 - x_1, \ldots, 1 - x_n)$$

In fact, we take any function v from $[0, 1]$ to $[0.1]$ which is anti-automorphism of order 2 and use it instead of the function $1 - x$.

We thus get

$$a = T(v(x_1), \ldots, v(x_n))$$

Any system (T, v) is known as a De Morgan System, see [269, Chapter 6].

10.8.4 Equations and loops

Our starting point is the much appreciated scholarly survey of H. Prakken ad G. A. Vreeswijk [291] in our *Handbook of Philosophical Logic*. It concerns Figures 10.50 and 10.51 (reproduced from pages 248 and 272 of [291]).

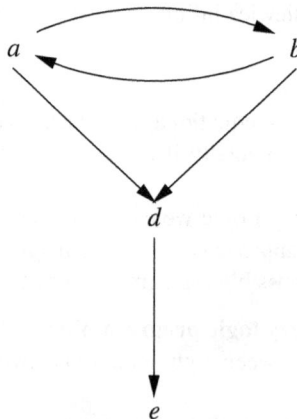

Fig. 10.50.

In Figure 10.50 the ground extension gives all undecided and two stable extensions give d = out and e = in.

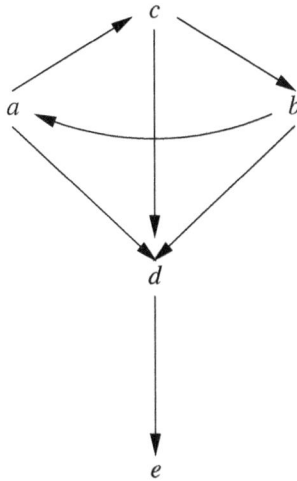

Fig. 10.51.

Under this situation, it might be argued that we really should consider d as out and e as in.

The situation is different in Figure 10.51. Here there are no other extensions than the ground extension which gives all undecided, and yet, one might feel that we still may intuitively conclude that e should be in and d should be out.

We quote Praakken and Vreeswijk (page 272 of [291]), we adapted the letters to our notation:

> "The difference between [Figure 10.50] and [Figure 10.51] is that the even defeat loop between two arguments is replaced by an odd defeat loop between three arguments. One view on [Figure 10.51] is that this difference is inessential and that, for the same reasons as why in [Figure 10.50] the argument e is justified, here, in [Figure 10.51], the argument e is ultimately undefeated: although e is strictly defeated by d, it is reinstated by all of a, b and c, since all these arguments strictly defeat d.

> skip ... skip

> However, an alternative view is that odd defeat loops are of essentially different kind than even defeat loops, so that our analysis of [Figure 10.50] does not apply here [i.e. to Figure 10.51]..."

Later on, on page 276 in their evaluation section, Prakken and Vreeswijk say, (I quote):

> "In fact this seems one of the main unsolved problems in argument-based semantics."

Let us now see what the equational view can give us. Let us use Eq_{inverse} and apply it to Figure 10.51.

We get the following equations:

$$a = 1 - b$$
$$b = 1 - c$$
$$c = 1 - a$$
$$d = (1 - a)(1 - b)(1 - c)$$
$$e = 1 - d$$

Solving the equations we get

$$a = b = c = \tfrac{1}{2}$$
$$d = \tfrac{1}{8}$$

$$e = \tfrac{7}{8}$$

Consider now Figure 10.52

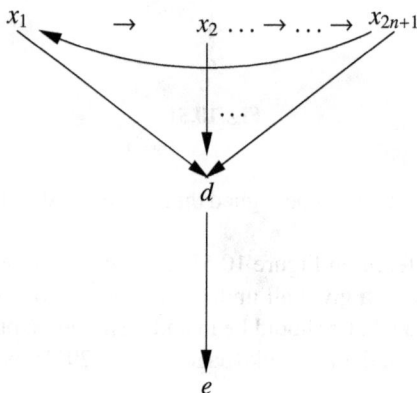

Fig. 10.52.

The equations here are

$$x_1 = 1 - x_{2n+1}$$
$$x_2 = 1 - x_1$$
$$\vdots$$
$$x_{2n+1} = 1 - x_{2n}$$
$$d = \prod_{i=1}^{2n+1}(1 - x_i)$$
$$e = 1 - d$$

The solution is

$$x_1 = x_2 = \ldots = x_{2n+1} = \tfrac{1}{2}$$
$$d = \tfrac{1}{2^{2n+1}}$$
$$e = 1 - \tfrac{1}{2^{2n+1}}$$

Thus the greater is n, the greater is e.

If we apply our concept of approximation from Section 5, we can confidently accept e as in.

Let us now look at Figure 10.53

Figure 10.53 is essentially Figure 10.9 of Example 10.10 attacking node d which attacks node e. Since Figure 10.9 is a closed loop, its solution is the same as that obtained in Example 10.10, namely:

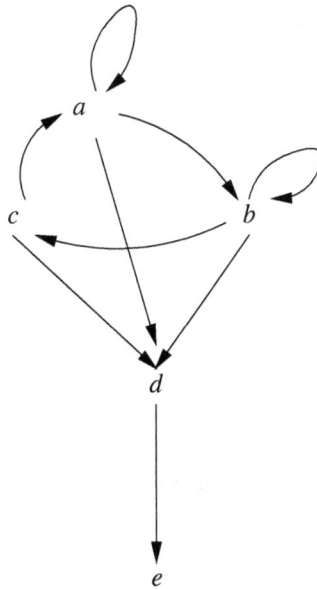

Fig. 10.53.

$$a = 1 - \frac{\sqrt{2}}{2}$$
$$b = \sqrt{2} - 1$$
$$c = 2 - \sqrt{2}$$

and therefore we get

$$d = \frac{\sqrt{2}}{2} \cdot (2 - \sqrt{2})(\sqrt{2} - 1)$$
$$\approx (0.707)(0.586)(0.414)$$
$$\approx 0.1715$$
$$e \approx 1 - 0.1715$$

d is relatively small and e is close enough to 1.

The general loop of this type has the form

$$L = \{x_1, \ldots, x_{2n+1}\}$$

where each $x_{i+1}, 1 \le i \le 2n$ is being attacked by $\{x_i, y_1^i, \ldots, y_{k(i)}^i)$ and x_1 is being attacked by $x_{2n+1}, y_1^{2n+1}, \ldots, y_{k(2n+1)}^{2n+1}\}$ with $y_j^i \in L$.

The equations are

$$x_{i+1} = (1 - x_i)(\prod_{j=1}^{k(i)}(1 - y_j^i))$$
$$x_1 = (1 - x_{2n+1})(\prod_{j=1}^{k(2n+1)}(1 - y_j^{2n+1}))$$
$$d = \prod_i(1 - x_i)$$

Clearly the more attacks among the x_i the smaller d is and therefore the nearer to 1 is the value of e.

We shall not develop these ideas further in this chapter.

We shall address them in a subsequent paper. We shall use as our starting base the fundamental paper of Baroni *et al.* [30].

Before we continue to the next subsection, let us do some more comparative examples of loops taken from [31, 53, 192].

Example 10.68. This is example 8 from [31]. Consider Figure 10.54, (being Figure AF$_8$ from [31]).

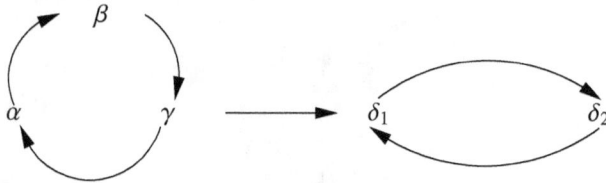

Fig. 10.54.

Baroni and Giacomin complain as follows, I quote:

In the example above, the absence of non-empty extensions for the three-length cycle prevents the existence of extensions including δ_1, leaving δ_2 as the only accepted argument, while this would not happen with an even-length cycle.

The CF2 semantics solves this problem. The extensions obtained by the CF2 semantics of Baroni and Giacomin [31] are as follows (see Example 8 of [31]):

$$E_1 = \{\alpha, \delta_1\}$$
$$E_2 = \{\alpha, \delta_2\}$$
$$E_3 = \{\beta, \delta_1\}$$
$$E_4 = \{\beta, \delta_2\}$$
$$E_5 = \{\gamma, \delta_2\}$$

The problem they have is the SCC set $\{\alpha, \beta, \gamma\}$ which is an odd loop, which can only be undecided and which makes the rest of the nodes all undecided, and so they take maximal conflict free subsets of $\{\alpha, \beta, \gamma\}$ which are $\{\alpha\}$, $\{\beta\}$ and $\{\gamma\}$ and add to them what can be added from $\{\delta_1, \delta_2\}$. For α and β we can add each of δ_1 and δ_2 and for γ we can add only δ_2. This way they get the above list of CF2 extensions.

If we adopt an equational point of view, we do not have to worry about loops. The equations will take care of that. So we know from Example 10.4 that the cycle $\{\alpha, \beta, \gamma\}$ solves to $\alpha = \beta = \gamma = \frac{1}{2}$, both under Eq_{max} and $Eq_{inverse}$. Under Eq_{max} the rest of the nodes also get value $\frac{1}{2}$, and we know this is the all undecided solution, but $Eq_{inverse}$, does solve the Baroni and Giacomin complaint. The solution is immediate and straightforward. Let us check the rest of the equations under $Eq_{inverse}$.

We get

1. $\delta_1 = \frac{1}{2}(1 - \delta_2)$
2. $\delta_2 = 1 - \delta_1$

The solution is as expected by Baroni and Giacomin, namely

$$\delta_1 = 0, \delta_2 = 1.$$

Note that we still have in the equational solution the value $\frac{1}{2}$ for α, β and γ, thus they are still considered undecided.

Example 10.69. This example is from [53]. [53] presents a new semantics to handle loops, called tolerant semantics, and example 27 of [53], shown here in Figure 10.55, is intended to show that tolerant semantics can give different extensions from CF2 semantics.

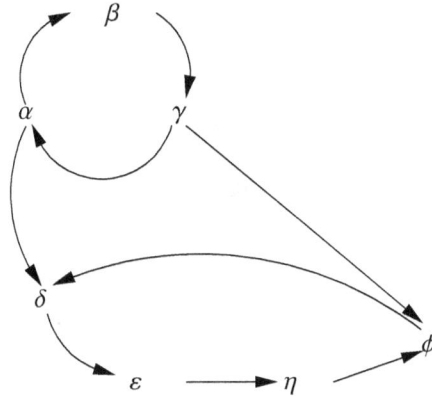

Fig. 10.55.

They show that $\{\alpha, \eta\}$ is a tolerant extension and not a CF2 extension. Let us see what our equations say:

As before, since $\{\alpha, \beta, \gamma\}$ is an independent non-attacked cycle, and so we get (using Eq_{inverse})

$$\alpha = \beta = \gamma = \tfrac{1}{2}$$

and

1. $\delta = \frac{1}{2}(1 - \phi)$
2. $\varepsilon = 1 - \delta$
3. $\eta = 1 - \varepsilon$
4. $\phi = \frac{1}{2}(1 - \eta)$

We get $\eta = \delta$.
    So

5. $\phi = \frac{1}{2}(1 - \delta)$
    $1 - \phi = \frac{2-1+\delta}{2} = \frac{1+\delta}{2}$

    Therefore

6. $\delta = \frac{1}{4}(1 + \delta)$

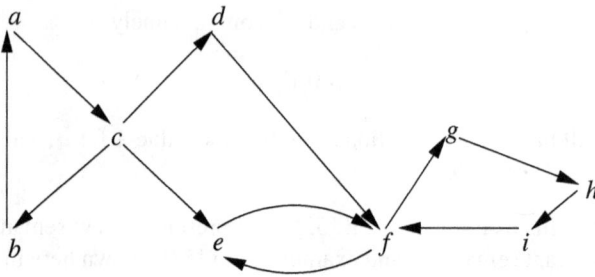

Fig. 10.56.

Hence $\delta = \frac{1}{3}, \varepsilon = \frac{2}{3}, \eta = \frac{1}{3}, \phi = \frac{1}{3}$.

Example 10.70. This example is from [191]. Consider Figure 10.56.

[191] is a brilliant paper dealing with CF2 semantics of [53] and [30]. The CF2 extensions of Figure 10.56 are (calculated by a program they have):

$$E_1 = \{c, f, h\}$$
$$E_2 = \{c, g, i\}$$
$$E_3 = \{b, d, e, g, i\}$$
$$E_4 = \{a, d, e, g, i\}.$$

Let us now apply $Eq_{inverse}$ to our example of Figure 10.56.
We get

1. $a = b = c = \frac{1}{2}$
2. $d = 1 - c = \frac{1}{2}$
3. $e = (1 - c)(1 - f) = \frac{1}{2}(1 - f)$
4. $f = (1 - c)(1 - e)(1 - i) = \frac{1}{2}(1 - e)(1 - i)$
5. $g = 1 - f$
6. $h = 1 - g$
7. $i = 1 - h$

From (5) and (6) we get

7. $g = h = 1 - i$

Therefore from (4), (3) and (7)

8. $f = \frac{1}{2}\frac{(1+f)}{2}f$

The only solution in $[0, 1]$ is $f = 0$. Therefore we get

$$e = \frac{1}{2}, g = 1, h = 0, i = 1$$

As the reader can see, that this is a different solution.

Remark 10.71. At this point the perceptive reader might say: "OK, so the equational approach is different, it can handle loops as a matter of course, without the need of special treatment. Is this all that the equational approach can offer?"

Baroni, Giacomin, Gaggl, Wolfram, Bodanza and Tohmé and their colleagues would like to remain basically in the {in, out} world and thus need to find special ways to handle loops. Can you (Dov Gabbay) say more in favour of the equational approach for loops?

My answer to that is the {in,out} approach is an open invitation for difficulties. We show this by means of a very simple example. Consider Figure 10.4 and compare it with Figure 10.5. These two figures are essentially equationally equivalent, see Remark 10.72 below. In Figure 10.4 a attacks itself and its equation is

$$a = 1 - a$$

In Figure 10.5, a attacks itself via the intermediaries b and c. The equations are

$$a = 1 - c$$
$$c = 1 - b$$
$$b = 1 - a$$

solving for a, we get

$$a = 1 - a.$$

We can regard b and c as "transmitters" of the attack. The idea of using intermediary "transmitters" has been gaining recognition in the literature in recent years. Whenever we have

$$a \rightarrow b$$

we can add intermediary transmitters $x_{a,b}, y_{a,b}$ and have

$$a \rightarrow x_{a,b} \rightarrow y_{a,b} \rightarrow b.$$

These metavariables may have meaning in the application area in question or they may be introduced purely formally to facilitate some technical point. From the equational point of view they have no effect on the logical content of the network containing the attack

$$a \rightarrow b.$$

We have used this device in Section 2.2 above. See also Example 10.27.

Now that we have argued for Figures 10.4 and 10.5 to be equationally equivalent, let us see what the CF2 approach does, when we replace the cycle $\{a, b, c\}$ in Figure 10.56 by the cycle $c \rightarrow c$. We get Figure 10.57.

$CF2$ will get a single $CF2$ extension for Figure 10.57, it is $\{d, e, g, i\}$. This is calculated as follows: For the first cycle $SCC, \{c\}$, it selects the empty set, then for the $SCC\{d\}$ the naive extension is the single set $\{d\}$. Argument f is attacked from outside the component so it is out. then, the next SCC is separated in the single SCC's $\{e\}, \{g\}, \{h\}$ and $\{i\}$, where e and g are not attacked from outside there component, so they are in, then h is attacked by g and is out, finally i is also in.

We thus see that as far as the set of nodes $\{c, d, e, f, g, h, i\}$ we get different subsets participating in $CF2$ extensions depending on whether they are part of Figure 10.56 or Figure 10.57. The equational approach, in comparison, will be undisturbed and give the same results for $\{c, d, e, f, g, h, i\}$ in Figure 10.57 as in the case of Figure 10.56.

To summarise, we show a difference in the way the equational approach deals with odd cycles (they are all essentially equationally equivalent) and the way traditional

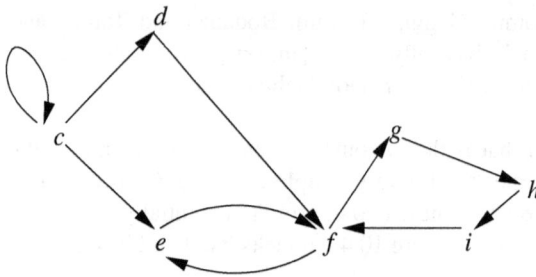

Fig. 10.57.

approaches deal with odd cycles (they are not all equivalent) by comparing with the CF2 semantics.

Remark 10.72 (Comparison with Emilia Oikarinen, and Stefan Woltran's notion of Strong Equivalence). In their paper [273] Emilia Oikarinen and Stefan Woltran introduce the notion of strong equivalence between argumentation networks. This needs to be compared with our equational approach in general and in particular our statement above about Figure 10.4 being essentially equivalent to Figure 10.5, on account that Figure 10.5 is obtained from Figure 10.4 by adding intermediary transmitters which do not affect the equations governing the original Figure 10.4.

Let us quote what [273] says about strong equivalence

"As a central notion we want to study in this paper, we define strong equivalence as the relation between argumentation frameworks F and G which holds, if F and G are equivalent under any extensions of the two *AF*s, i.e., if for each further *AF* H, $F \cup H$ and $G \cup H$ are equivalent. The study of such an equivalence notion is motivated by the following observations:

- Implicit vs. explicit information: as we have seen in the example above, the two frameworks store different information. However, the semantics do not make this difference visible, unless the AFs are suitably extended, i.e. some new information is added.

Strong equivalence can thus be understood as a property which decides whether two argumentation frameworks provide the same implicit information."

This is not the place to go into details about Oikarinen and Woltran's brilliant paper. Let us just mention that to reconcile our statement with their results we should replace the phrase

F and G are equivalent under any extensions of the two AFs, i.e., if for each further AF H, $F \cup H$ and $G \cup H$ are equivalent

by the phrase

F and G are equivalent under any non interfering extensions of the two AFs, i.e., if for each further AF H, which does not interfere with the intermediate transmitters of F nor with those of G, we have that $F \cup H$ and $G \cup H$ are equivalent.

The study of loops is continued in Chapters 18 and 19 and especially the section of Chapter 18 which deals with the Equational approach to CF2 semantics.

10.8.5 Equational modal consistency networks

Modal argumentation networks have already been studied in [38], in [40], in [162], Chapters 5, 9 and 14, and in [159].

The paper [40], see our Chapter 14, on network modalities, describes a general equational approach. A general discussion can also be found in [85]. So we've got the area pretty well covered. What we need in this section is some definitions for the equational modality where $\Diamond A$ means "A is consistent", which in the context of argumentation means that A is consistent with the extension in which $\Diamond A$ resides. We need this notion in order to deal with default.

Let our language contain atomic arguments of the form a and $\neg a$ as well as modal arguments of the form $\Diamond a$ and $\Diamond \neg a$.

Definition 10.73. *A modal consistency argumentation network has the form* (S, R), *where S contains elements of the form* $a, \neg a, \Diamond a, \Diamond \neg a$ *and R satisfies the following constraints.*

1. *If both $\neg a$ and a appear in the network then $(a, \neg a)$ and $(\neg a, a)$ are both in R.*
2. *$\Diamond q$ is attacked only by $\neg q$, if $\neg q \in S$. Otherwise it is not attacked.*
3. *$\Diamond \neg q$ is attacked only by q if $q \in S$, otherwise it is not attacked.*

We need the above type of network to deal with defaults. It is a modal network in the sense of Gabbay and Modgil [159], for the modal logic with exactly one posible world which is reflexive, i.e. for the frame $(\{a\}, \{(a, a)\})$.

Thus we have

$$a \vDash \Diamond q \text{ iff } a \vDash q.$$

Hence $\Diamond q$ is attacked by $\neg q$ and $\Diamond \neg q$ is attacked by q.

The requirement that the above are only attacks on $\Diamond q$ and $\Diamond \neg q$ corresponds to the meaning of $\Diamond$ as a consistency operator.

Example 10.74. Consider the network of Figure 10.58. The equations are as follows (we use Eq_{inverse}).

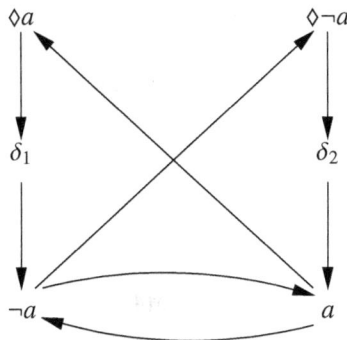

Fig. 10.58.

1. $\mathbf{f}(\Diamond a) = 1 - \mathbf{f}(\neg a)$
2. $\mathbf{f}(\delta_1) = 1 - \mathbf{f}(\Diamond a)$
3. $\mathbf{f}(a) = (1 - \mathbf{f}(\delta_1))(1 - \mathbf{f}(\neg a))$
4. $\mathbf{f}(\neg a) = (1 - \mathbf{f}(a))(1 - \mathbf{f}(\delta_2))$
5. $\mathbf{f}(\delta_2) = 1 - \mathbf{f}(\Diamond \neg a)$
6. $\mathbf{f}(\Diamond \neg a) = 1 - \mathbf{f}(a)$.

Solutions are as follows: From (1) and (2)

$$\mathbf{f}(\delta_1) = \mathbf{f}(\neg a).$$

Hence from (3)

7. $\mathbf{f}(a) = [1 - \mathbf{f}(\neg a)]^2 = \mathbf{f}(\Diamond a)^2$

Similarly

$$\mathbf{f}(\delta_2) = \mathbf{f}(a)$$

and so from (4)

8. $\mathbf{f}(\neg a) = [1 - \mathbf{f}(a)]^2 = \mathbf{f}(\Diamond \neg a)^2$.

We can guess two solutions:

Solution 1.
$$\mathbf{f}(\neg a) = \mathbf{f}(\Diamond \sim a) = 0$$
$$\mathbf{f}(a) = \mathbf{f}(\Diamond a) = 1.$$

Solution 2.
$$\mathbf{f}(a) = \mathbf{f}(\Diamond a) = 0$$
$$\mathbf{f}(\neg a) = \mathbf{f}(\Diamond \neg a) = 1.$$

For more solutions we need to solve the equations.
Let
$$x = \mathbf{f}(a)$$
$$y = \mathbf{f}(\neg a).$$

Then we have
$$x = (1 - y)^2$$
$$y = (1 - x)^2.$$

So
$$1 - y = 1^2 - (1 - x)^2 = (1 - (1 - x)(1 + 1 - x)$$
$$= x(2 - x)$$

Therefore
$$x = x^2(2 - x)^2$$

If $x = 0$ we get Solution 2. Assume $x \neq 0$ and get

$$1 = x(2 - x)^2 = x(4 - 4x + x^2)$$

Therefore we need to solve

$$x^3 - 4x^2 + 4x - 1 = 0$$

The solution is

$$x = 1 \text{ (this is Solution 1)}$$
$$x \equiv 0.382$$
$$x \equiv 2.618$$

Only $x \equiv 0.382$ is acceptable, and this gives Solution 3.

If we use Eq_{max} the third solution is $x = \frac{1}{2}$, i.e. all undecided.

10.8.6 Equations and defaults

This section will quickly indicate how the equational approach can deal with defaults. We reserve the full treatment to a separate paper. We rely on the survey of G. Antoniou [21], whose treatment of default is more compatible with our equational approach. Dung also addressed defaults in his seminal paper [114], but he uses the machinery of extensions as Reiter does. The same applies to [63].

The basic situation is that we have a language $\mathbb{L}$ and a notion of consistency for $\mathbb{L}$. A default system has the form $(\mathbb{W}, \mathbb{D})$ where $\mathbb{W}$ is a consistent set of wffs of L and $\mathbb{D}$ is a set of defaults of the form

$$\delta_i = \frac{x_i : y_i}{z_i}$$

We seek an extension $\mathbb{E} \supseteq \mathbb{W}$ satisfying for each δ_i the following:

(*)   If $x_i \in \mathbb{E}$ and y_i is consistent with $\mathbb{E}$ then also $z_i \in \mathbb{E}$.

We seek the set of all such extensions.

Our method is to embed $(\mathbb{W}, \mathbb{D})$ into an equational network and find all extensions as the solutions to network equations.

Given a default system $(\mathbb{W}, \mathbb{D})$ we now show how to get our network. We assume some simplifying assumptions on $(\mathbb{W}, \mathbb{D})$, not because we cannot handle the more complex cases (by methods of subsection 8.2) but because we need a full paper to deal with it. See our paper [168].

Our simplified assumptions are:

1. We deal only with atoms or their negations (i.e. with literals)
2. Each literal appears at the z-position of at most one rule.

We now translate a default rule

$$\delta = \frac{\pm x : \pm y}{\pm z}$$

into a fragment of a network. The networks we use are the modal consistency networks of Section 8.5.

Translate

$$\delta = \frac{\pm x : \pm y}{z}$$

into the fragment of Figure 10.59

The letters δ^1, δ^2 are called auxiliary letters.

Assume we are given a default system $(\mathbb{W}, \mathbb{D})$. For each $\delta_i \in \mathbb{D}$, let $\mathbf{F}_i$ be the corresponding figure with the letters $\pm x_i, \pm y_i, \pm z_i$ and the auxiliary letters δ_i^1, δ_i^2 as needed.

let (S_i, R_i) be the network suggested by Figure $\mathbf{F}_i$.

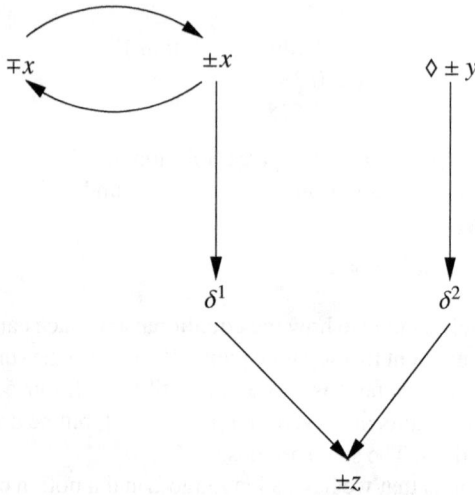

Fig. 10.59.

Let $\mathbf{F}_{\mathbb{D}}$ be the figure obtained by taking the union of all figures $\mathbf{F}_i$. Note that the same letter $\pm x$ may appear in several $\mathbf{F}_i$. We *do not* replicate these letters. In other words, we are letting $S_{\mathbb{D}}^0 = \bigcup_i S_i, R_{\mathbb{D}}^0 = \bigcup_i R_i$. We add the following additional letters and connections (arrows) in $\mathbf{F}_{\mathbb{D}}$ as follows:

1. For any $\Diamond \neg x$ add the arrows

$$x \to \Diamond \neg x$$

2. For any $\Diamond x$ add the arrows

$$\neg x \to \Diamond x$$

3. If both $\neg x$ and x are around add the arrows $x \to \neg x, \neg x \to x$.

We now have a network $\mathbb{F}_{\mathbb{D}}$ with nodes of the form $x, \neg x, \Diamond x, \Diamond \neg x, \delta_i^1, \delta_i^2$.
    All these nodes are considered *atomic*!
    Let therefore

$$S_{\mathbb{D}} = S_{\mathbb{D}}^0$$
$$R_{\mathbb{D}} = R_{\mathbb{D}}^0 \cup \{(\neg x, \Diamond x) \mid \neg x, \Diamond x \in S_{\mathbb{D}}^0\}$$
$$\cup \{(x, \Diamond \neg x) \mid x, \Diamond \neg x \in S_{\mathbb{D}}^0\} \cup$$
$$\cup \{(x, \neg x), (\neg x, x) \mid x, \neg x \in S_{\mathbb{D}}^0\}.$$

Given a default system $(\mathbb{W}, \mathbb{D})$, construct the network $\mathbb{F}_D$ (i.e. $(S_{\mathbb{D}}, R_{\mathbb{D}})$) as indicated.

How do we find the default extensions?

We seek all equational solutions $\mathbf{f}$ to $(S_{\mathbb{D}}, R_{\mathbb{D}})$ using Eq_{max} subject to the following constraints

1. $x \in \mathbb{W} \to \mathbf{f}(x) = 1$
2. $\neg x \in \mathbb{W} \to \mathbf{f}(\neg x) = 1$.

We now look at examples from [21].

Example 10.75. $\mathbb{W} = \{a\}$. $\mathbb{D} = \{\delta_1, \delta_2\}$, where

$$\delta_1 = \frac{a : \neg b}{d}$$

$$\delta_2 = \frac{True : c}{b}$$

The only default extension is $E = \{a, b\}$.
   Let us do the equational approach. $\mathbf{f}_{\mathbb{D}}$ is Figure 10.60.

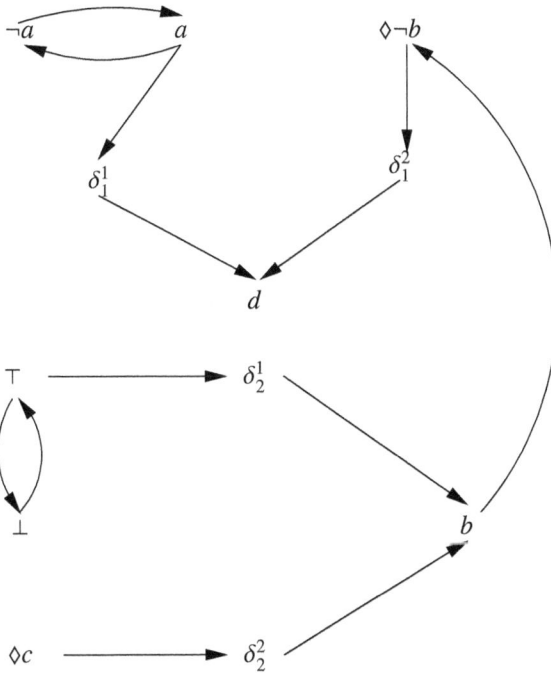

Fig. 10.60.

We have for any solution $\mathbf{f}$ satisfying the constraints $\mathbf{f}(\top) = 1$ and $\mathbf{f}(a) = 1$, that:

1. $\mathbf{f}(\Diamond c) = 1$, since not attacked.
2. $\mathbf{f}(\delta_2^2) = \mathbf{f}(\delta_2^1) = 0$
3. $\mathbf{f}(b) = 1$
4. $\mathbf{f}(\Diamond \neg b) = 0$
5. $\mathbf{f}(\delta_1^2) = 1$, $\mathbf{f}(\delta_1^1) = 0$
6. $\mathbf{f}(d) = 0$.

There is therefore only one extension $\{a, b\}$.

344   10 An Equational Approach to Argumentation Networks

Example 10.76. Let $\mathbb{W} = \emptyset$. $\mathbb{D} = \{\delta_1, \delta_2\}$, where

$$\delta_1 = \frac{\text{True} : p}{\neg q}$$

$$\delta_2 = \frac{\text{True} : q}{r}$$

This theory has exactly one default extension $E = \{\neg q\}$. Let us use the equational method. We draw $\mathbf{F_D}$ in Figure 10.61. We look for solution $\mathbf{f}$ with no constraints (of course $\mathbf{f}(\top) = 1$).

1. $\mathbf{f}(\Diamond p) = 1$
2. $\mathbf{f}(\delta_1^1) = \mathbf{f}(\delta_1^2) = 0$
3. $\mathbf{f}(\neg q) = 1$
4. $\mathbf{f}(\Diamond q) = 0$
5. $\mathbf{f}(\delta_2^2) = 1$
6. $\mathbf{f}(r) = 0$

The extension is therefore $\{\neg q\}$.

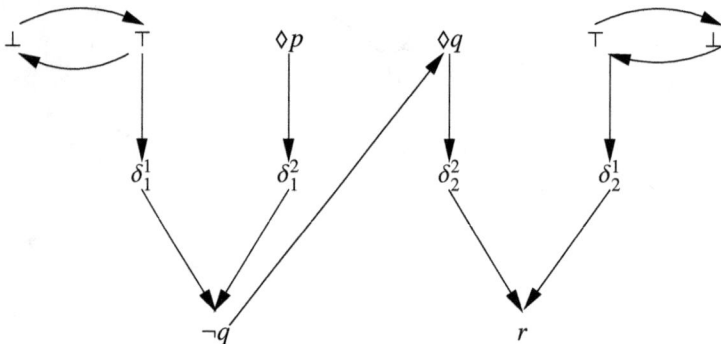

Fig. 10.61.

Example 10.77. $\mathbb{W} = \{g, a\}$.
$\mathbb{D} = \{\delta_1, \delta_2\}$ where:

$$\delta_1 = \frac{g : \neg l}{\neg l}$$

$$\delta_2 = a : \frac{l}{l}$$

There are two default extensions: $\{g, a, l\}$ and $\{g, a, \neg l\}$.
Let us check the equation in Figure 10.62.
The constraints are

$$\mathbf{f}(a) = \mathbf{f}(g) = 1.$$

This figure is full of cycles. Let us do case analysis on the cycle $\{l, \neg l\}$.

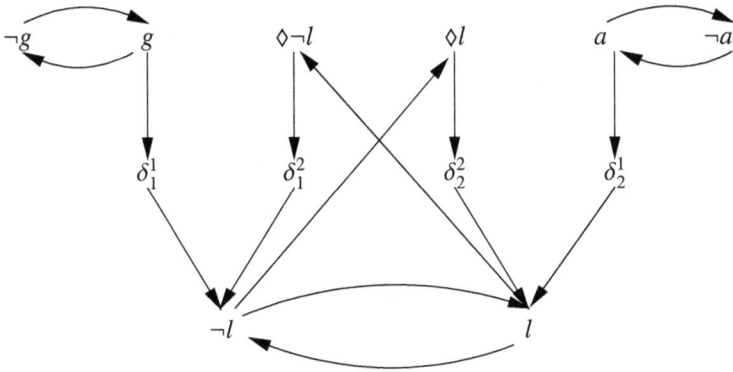

Fig. 10.62.

$$\mathbf{f}(l) = 1 \Rightarrow \mathbf{f}(\neg l) = 0 \Rightarrow \mathbf{f}(\delta_1^2) = 1 \Rightarrow \mathbf{f}(\Diamond \neg l) = 0.$$

Also

$$\mathbf{f}(l) = 1 \Rightarrow \mathbf{f}(\delta_2^2) = 0 \Rightarrow \mathbf{f}(\Diamond l) = 0.$$

A consistent cycle.

Similarly by symmetry $\mathbf{f}(\neg l) = 1$ gives a consistent cycle.

Moreover one can also consider the solution with all values equal to 0.5.

10.8.7 Argumentation networks and predator–prey networks

We have said much about the unifying value of the equational approach and how it allows us to have a common point of view for argumentation networks and other networks such as flow networks, neural networks, ecological networks, etc. This subsection will give details on the connection with predator–prey ecological networks.

Consider Figure 10.63. Consider especially the nodes under the names $\{N, P, Q\}$. This is a simple argumentation network.

Let us write the equations for this figure. We use Equation$_{\text{inverse}}$.

1. $a = 1 - P$
2. $c = 1 - N$
3. $b = 1 - Q$
4. $P = (1 - a)(1 - c) = PN$
5. $N = (1 - P)(1 - c)(1 - Q) = N(1 - P)(1 - Q)$
6. $Q = (1 - b)(1 - c)(1 - P) = QN(1 - P)$

If we look at he subnetwork with $\{N, P, Q\}$ only, we get a network with equations (4)–(6) only.

We are now ready for comparison with a predator–prey network. Our starting point is a model by M. P. Hasssell (1978) of two parasitoids (P and Q) and one host (N) model. The equations are (see [20, p. 295])

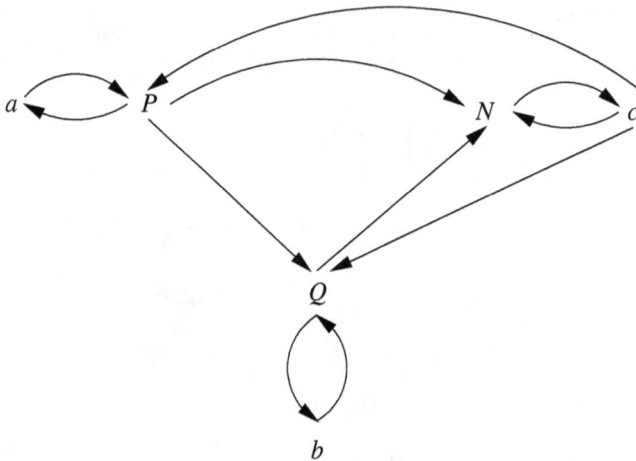

Fig. 10.63.

$$N_{t+1} = \lambda N_t f_1(P_t) f_2(Q_t)$$
$$P_{t+1} = N_t[1 - f_1(P_t)]$$
$$Q_{t+1} = N_t f_1(P_t)[1 - f_2(Q_t)]$$

where N, P and Q denote the host and two parasitoid species in generations t and $t+1$, λ is the finite host rate of increase and the functions f_1 and f_2 are the probabilities of a host not being found by P_t or Q_t parasitoids, respectively. This model applies to two quite distinct types of interaction that are frequently found in real systems. It applies to cases where P acts first, to be followed by Q acting only on the survivors. Such is the case where a host population with discrete generations is parasitized at different developmental stages. In addition, it applies to cases where both P and Q act together on the same host stage, but the larvae of P always out-compete those of Q, should multi-parasitism occur.

The functions f_1 and f_2 are:

$$f_1(P_t) = \left[1 + \frac{a_1 P_t}{k_1}\right]^{-k_1}$$
$$f_2(Q_t) = \left[1 + \frac{a_2 Q_t}{k_2}\right]^{-k_2}$$

where k_1 and k_2, a_1 and a_2 are constants.

To compare the biological model with the argumentation model, we put $a_1 = a_2 = 1$, $\lambda = 1$ and $k_1 = k_2 = -1$.
This gives

$$f_1(P_t) = 1 - P_t$$
$$f_2(Q_t) = 1 - Q_t$$

and therefore we get the equations

$$N_{t+1} = N_t(1 - P_t)(1 - Q_t)$$
$$P_{t+1} = P_t N_t$$
$$Q_{t+1} = Q_t N_t(1 - P_t)$$

giving us the appropriate functions for attack and counterattack for the situation in Figure 10.64:

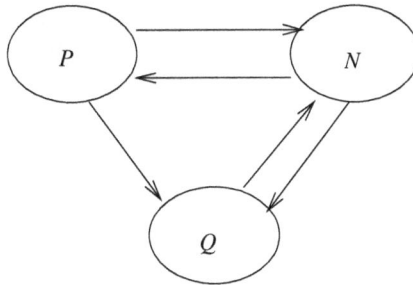

Fig. 10.64.

In Figure 10.64, P and Q attack N. N counterattacks P and Q and P attacks Q. Since N is attacked by P and Q, the new value for N is $N(1 - P)(1 - Q)$. Since P is counterattacked by Q, the new value for P is PN. Since Q is counterattacked by N and attacked by P the new value for Q is $QN(1 - P)$.

If we seek a steady state solution for these predator–prey equations, we need to solve equations (4)–(6) above.

10.9 Conclusion and future work

10.9.1 Discussion

Let us put forward how we see the results of this chapter. This is my personal view and it explains the methodology of this chapter.

There are several areas in logic where major monotonic and fixed point operations take place. For example

1. Logic programming, where the content of the logic program is determined by fixed points of suitable monotonic operations on the Herbrand universe.
2. Pure logic (e.g. classical or Łukasiewicz logics) where the deductive closure of theories is determined by fixed points of the consequence operations.
3. Default logics, autoepistemic logics and the like where the extensions are defined as fixed points of certain operations.

In each of the above areas, let us identify roughly two major "components".

a. The unit operational concepts.
b. The fixed point extensions as defined using (a).

In logic programming (a) is the Horn clause and negation as failure.
In logic (a) is the notion of proof rules (or immediate consequence)
In default logic (a) is the default rule.
(b) is more or less defined using (a) in a predictable way.

In the year 1995, Dung [114] came with another area:

4. Argumentation frames.

His basic concept (a) was attack and admissibility, and using that he defined fixed points and extensions. His seminal paper contains more. He identified his own (a) notions in the (a) notions of default logic and in logic programming and thus was able to present default logic and logic programming in the framework of argumentation.

In my view all of these areas (1)–(4) belong to the same family of systems and share a set-theoretic fixed point approach.

Being in the same family manifests itself also in the predictability and ease of translation from one system (of (1)–(4)) to the other. This was already pointed out in Gabbay [162, 164, 161] and in Sections 10.2.2 and 10.8.2 of this chapter. It was also pointed out by other researchers, such as Brewka and Woltran [69] and Brewka *et al.* [70].

In fact in the next Chapter 13, I show that Dung argumentation theory has the power of classical propositional logic, and this of course explains its expressive power.

The equational approach is a different family. It has its root in the early 19th centruy and it is not set theoretic. The fixed point feature does manifest itself in the equational approach through Brouwer's fixed point theorem, but otherwise the equational approach can go in different directions. We now elaborate on these new directions:

1. Connections with other purely numerical networks such as flow or ecology networks.
2. The extensions in the set-theoretic family are the solutions to equations in the equational approach. The equations can be solved in different ways.
   2.1. We can impose constraints.
   2.2. We can insist on the solutions to follow certain algorithmic restrictions.
   We already saw that we need this to implement defaults, but we also can get new results in Logic, as D'Agostino and Gabbay show in [105].
3. We can approximate solutions, as we mention in Section 5 of this chapter.

10.9.2 Comparison with other papers

Some initial results are put forward to C. Gratie and A. Florea in the preliminary submission [206]. In communication with them, it appears that they were not aware of our [164]. Also translations into argumentations networks, like the ones in Sections 2.2 and 10.8.2, were already introduced in Gabbay's 2009 paper [164] and independently in Brewka *et al.*'s 2011 paper [70].

Argumentation and Negation as Failure

11.1 Introduction

This chapter discusses soundness and completeness of Prolog's computation trees w.r.t. program completion. Since there is a connection between argumentation semantics and negation by failure, we get new (equational) calculus for argumentation.

Clark's completion was one of the first semantics proposed for programs containing negation as failure. The problem of soundness and completeness, for programs with negation, with respect to Clark's completion semantics has been studied from the early days of logic programming. The chapter shows how to resolve this problem using so-called equational logic.

The chapter starts with the semantics of negation as failure based on Clark's completion. Roughly speaking, it strengthens the program by replacing 'not' with classical negation and by interpreting implications as equivalences. Although natural it may seem the account has an obvious drawback, which Clark himself was aware of. Soundness of Prolog w.r.t. this kind of semantics can be shown to hold for both success and failure. But counter-examples to completeness can be given as long as negative literals are allowed in the bodies of the clauses. In the chapter three counter-examples are provided, Examples 2.7, 2.8 and 2.9. The peculiar thing about the last two is that they involve programs with negative recursion — viz programs of form 'not $c \rightarrow c$', in which c depends (directly or indirectly) on a negative occurrence of itself. For them the completion gives an inconsistency. This shows that Clark's completion fails to provide a proper semantics for this kind of program.

In the second part, we explore the consequences of using intuitionistic logic as the underlying logic. The notion of completion is redefined within this set-up, and soundness with respect to this alternative semantics established. But completeness fails for the same reason as in the original semantics. This observation motivates the attempt made in the third part of the chapter to use equational logic instead.

Completion, then, is defined as in the original Clarke set-up except that it is read in equational logic rather than classical logic. The main contribution is a soundness and completeness result with respect to the equational completion semantics.

We conclude the chapter with comparison with other proposal for semantics for negation as failure in the literature and a final discussion.

We now make some methodological comments.

Let P be a database in some language, for example a propositional logic programming database. Let $\mathcal{A}$ be an algorithm for extracting answers to queries from P. To give a concrete example, let P be the Horn clause

$$a \wedge b \to a$$

and let the algorithm $\mathcal{A}$ be the usual Prolog algorithm, regarding $a \wedge b$ as a sequence (a, b) and first asking for $?a$ and then for $?b$.

The algorithm $\mathcal{A}$ can succeed in its query $P?q$, or it can stop and say it failed or it can loop.

We would like to translate P into a logic $\mathbf{L}$, via a translation τ to obtain a database $\tau(P)$ of $\mathbf{L}$ such that

- If $P?q$ succeeds then

$$\tau(P) \vdash_{\mathbf{L}} \tau(q)$$

- If $P?q$ finitely fails then

$$\tau(P) \vdash_{\mathbf{L}} \neg\tau(q)$$

- If $P?$ loops then

$$\tau(P) \nvdash_{\mathbf{L}} \tau(1) \text{ and } \tau(P) \nvdash_{\mathbf{L}} \neg\tau(q).$$

A logic $\mathbf{L}$ and a translation τ satisfying the above conditions may give the impression that they give a declarative meaning to the algorithm $\mathcal{A}$. However, this may not be the case. The logic $\mathbf{L}$ may describe the algorithm $\mathcal{A}$ acting as a metalevel language and the expression

$$\tau(P) \vdash_{\mathbf{L}} \tau(q)$$

says nothing more than that $P?q$ succeeds, and the expression

$$\tau(P) \vdash_{\mathbf{L}} \neg\tau(q)$$

says nothing more than that $P?q$ finitely fails. All the above can be achieved by describing the steps $\mathcal{A}$ takes inside $\mathbf{L}$.

To take a well known example of such a strategy, think of the metalevel description of Turing machine execution in classical logic. This metalevel interpretation is good for proving the undecidability of classical logic, but it does not give any logical meaning to the operations of a Turing machine.

So we need to add the additional requirement to the translation τ that it is more than a simulation and that it gives a logical meaning to the algorithm $\mathcal{A}$. This requirement which we call "the logical meaning criterion" cannot be formally defined, but for each $\mathcal{A}$ and each τ it can be debated whether it is fulfilled. We could have some borderline cases were the answer is not clear cut.

Let us take the program

$$a \wedge b \to a$$

mentioned above and let us take an algorithm $\mathcal{A}$ which is Prolog goal directed with a loop checker. Thus we get either success or failure for each query. We consider looping as a failure.[1]

---

[1] In later Chapters we will consider other variants of the Prolog algorithm, see Section 12.1 and Section 18.8.

The algorithm would give that $\neg a \wedge \neg b$ succeed (i.e. both a and b finitely fail). The Clark completion for this program is

$$(a \wedge b) \leftrightarrow a$$
$$\neg b$$

This database in classical logic does prove $\neg a \wedge \neg b$.

The Clark completion does fulfil our logical meaning criterion. It is not a metalevel simulation of the algorithm.

To further illustrate our point, let us modify the algorithm a bit more, and turn it into a relevance algorithm.

Example 11.1 (Relevance Prolog algorithm). Let P be a Horn clause database where clauses of the form of a single q are written as $\top \to q$ for atomic q, and consider the computation rules as follows. We keep track during the computation of all the clauses that have been used. We write

$$P?q; E$$

where E is a set of clauses said to have been used in the computation. E is a global variable which is updated with each computational step and gets bigger and bigger during the computation. The computation rules are:

1. $P?q; E$ succeeds if for some $x_1 \wedge \ldots \wedge x_n \to q \in P$ we have that for all i $P?x_i$ succeeds. We let the new E be $E \cup \{x_1 \wedge \ldots \wedge x_n \to q\}$.
2. $P?q\top, E$ succeeds.
3. We say that $P?q$ succeeds if $P?q; P$ succeeds, i.e. during the computation all of P is used.

This is a relevance algorithm, wanting all clauses of P to be used in the computation. Thus from the database

$$a \to b$$
$$c$$
$$a$$

the query $?b$ does not succeed because there is no way in which clause c can be used.

If we look at the Clark completion, it is:

$$a \leftrightarrow b$$
$$b$$
$$c$$

we ask: in what logic does this completion prove $\neg a$?

The answer is that I don't know. We may succeed in formulating such a logic. As far as I know, nobody ever tried. Thus we do not understand at this moment the logical content of negation as fialure with relevance; as presented above.

We can always simulate the computation and get a translation τ but this τ wil not explain the logical meaning of failure in the relevance case.

For ordinary logic programming, if we allow nagation as failure in the body of clauses, for example

$$\neg a \to a$$

then the Clark completion does not work. It does not do the job. Many authors have put forward proposals for translations τ into suitable logics. We also offer in this chapter a translation of our own into equational logic. All these proposals will be discussed and compared and assessed with respect to the question of whether they fulfill the logical content criterion.

11.2 Introducing negation as failure[2]

We are now ready to investigate the logical properties of negation as failure. To do that, our first step is to present negation as failure formally and describe the well-known difficulties associated with it. We deal with the propositional case first. The logical problems with the notion of "failure" as "negation" arises in the propositional case already. Negation as failure in Prolog has difficulties in the quantificational case as well, but these have to do with the special way the quantifiers are treated and this can be changed.

Consider a propositional language with propositional atoms $\{q_1, q_2, q_3, \ldots\}$ and the connectives $\wedge, \vee, \rightarrow$ and $\neg$. $\neg$ is supposed to be negation as failure and $\rightarrow$ causal implication.

We may later use other connectives like $\equiv$ (if and only if) or $\leftrightarrow$ (which also stands for if and only if).

Definition 11.2 (Prolog goals and clauses).

1. *Any atomic q is both a goal and a clause.*
2. *If A and B are goals so are A $\wedge$ B and $\neg$A.*
3. *If A is a goal q atomic then A $\rightarrow$ q is a clause. q is said to be the* head *of the clause and A is the* body *of the clause.*

Definition 11.3 (Prolog computation trees). *Let P be a set of clauses and G be a goal. Let x be a number 0 or 1. We define the notion of a labelled tree of the form $(T, <)$, where T is the set of nodes of the tree and $<$ is the binary relation on the tree, being a computation tree for (P, G, x).*

if (P, G, x) has a finite computation tree (as defined below) we write $P?G = x$ succeeds. If (P, G, x) does not have a finite computation tree but an infinite one we write $P?G = x$ loops.

1. *The top node of the tree is labelled (P, G, x).*
2. *Let t be any node and let $s_1, \ldots, s_k$ be the nodes immediately below t. If $k = 0$, this means that t is a bottom node. Let the labels be $(P, A(t), x(t))$ and $P, A(s_i), x(s_i))$ respectively. Then the following holds:*
    a) *If $A(t) = B \wedge C$ and $x(t) = 1$ then $k = 2, x(s_1) = x(s_2) = 1$ and $A(s_1) = B$ and $A(s_2) = C$.*
    b) *If $A(t) = B \wedge C$ and $x(t) = 0$ then $k = 1$ and $A(s_1) \in \{B, C\}$ and $x(s_1) = 0$.*
    c) *If $A(t) = \neg B$, then $k = 1$ and $A(s_1) = B$ and $x(s_1) = 1 - x(t)$.*
    d) *If $k > 0$ and $A(t) = q$, q atomic and $x(t) = 1$, then $k = 1$ and $A(s_1) \rightarrow q \in P$ and $x(s_1) = 1$.*

---

[2] Here begins the 1985 paper, [130].

    e) If $k > 0$ and $A(t) = q$ and q atomic and $x(t) = 0$, then $A(s_1) \to q, \ldots, A(s_k) \to q$ are all the clauses with heads q in P and $x(s_1) = \ldots = x(s_k) = 0$.

    f) If $k = 0$, i.e. t is an endpoint, then $A(t) = q$, q atomic, and:

       i. If $x(t) = 1$, then $q \in P$

       ii. If $x(t) = 0$, then q is not the head of any clause in P, in particular also $q \notin P$.

Examples 11.4

1. $\{\neg a \to b\}?b = 1$
   $\{\neg a \to b\}?a = 0$
2. $\neg c \to c$?c loops (i.e. has an infinite computation tree)
3. $\left.\begin{array}{l}\neg c \to a \\ a \to c\end{array}\right\}$ $\begin{array}{l}?c \text{ loops} \\ ?a \text{ loops}\end{array}$

Definition 11.5 (Clark's completion of P). *The following defines the Clark completion* Com(P) *of any program P. See [243, 99].*

1. *If a appears in some clause P and is not the head of any clause in P then put $\neg a \in$ Com(P).*
2. *If a is the head of exactly the clauses $A_i \to a$, $i = 1, \ldots, k$ in P, then put $(\bigvee A_i) \leftrightarrow a \in$ Com(P).*

Theorem 11.6 (Clark, Lloyd et al.). *For P without negation and any A:*

$$P?A = x \text{ iff } \text{Com}(P) \vdash_C A^x$$

where $A^1 = A; A^- = \neg A$ and $\vdash_C$ is provability in classical logic.

Remark 11.7. $\vdash$ *in the above theorem can be taken as intuitionistic logic, for the propositional case. The theorem is true for the predicate cases also for $\vdash_C$.*

Example 11.8. The problem arises for the case when we have negations in the body of clauses in the program: Take for example the program[3]

$$\neg a \to b$$
$$a \to b$$
$$\neg b \to c$$
$$a \to a$$

The completion is

$$b \leftrightarrow a \vee \neg a$$
$$c \leftrightarrow \neg b$$
$$a \leftrightarrow a$$

Com(P) $\vdash \neg c$ but $P?c$ loops.

    Clark did, however, prove soundness.

---

[3] We do not need the item $a \to a$, but we include it because it is allowed as a clause. It makes no difference to the example, which aims to show that Clark's completion does not work if we allow negations in the body of clauses.

Theorem 11.9. *For any program P we have that for any A P?A = x implies* $\mathrm{Com}(P) \vdash A^x$.

The completion is not effective in many cases. Consider

Example 11.10.

1. $\neg c \rightarrow c$ is logically expected to be $c \lor c$ which is equivalent to c, but $\{\neg c \rightarrow c\}?c$ loops.
   What is worse, $\mathrm{Com}(\neg c \rightarrow c)$ is a contradiction. Even worse still, we have the following
2. $P = \{\neg a \rightarrow b\}$ is a perfectly nice program. b succeeds (i.e. $P?b = 1$) and a fails (i.e. $P?a = 0$). The Clark completion does its job, i.e.

$$\mathrm{Com}(P) = \neg a \land (\neg a \leftrightarrow b)$$
$$= b \land \neg a.$$

However, if we add to the program the additional clause $\neg c \rightarrow c$, i.e. have

$$P' = \neg a \rightarrow b$$
$$\neg c \rightarrow c$$

$(P'?a)$ and $(P'?b)$ as goals are not affected because they have nothing to do with c. Clark's completion gives us, however,

$$\mathrm{Com}(P') = \neg a \land (\neg a \leftrightarrow b) \land (\neg c \leftrightarrow c)$$
$$= \text{contradiction}$$

Example 11.11. Consider the program

$$a \rightarrow c$$
$$\neg c \rightarrow a$$

When viewed as sentences of classical logic we get the "logical content" of

$$\neg a \lor c$$
$$c \lor a$$

which is equivalent to c.

Computationally, c succeeds *only through* the clause $a \rightarrow c$ and hence a and c are "married". They succeed or fail together.

The Clark completion is again a contradiction.

The above presented the Prolog notion of negation as failure and some of its problems. Our task in this part of the chapter is the following.

1. Given a general program P with $\neg$ in its clauses, find a way to associate with it some "logical content". Most likely a new form of completion P^* such that for any q:
$$P?q = x \text{ iff } P^* \vdash q^x.$$
2. Check in view of our results in the previous section of this chapter, whether the logical meaning for $\neg$, which $\neg$ inherits in P^*, really makes $\neg$ a negation.

We have three strategies open to us:

1. To introduce some natural loop checking device into the Prolog computation in such a way that Clark's theorem goes through. The completion of a program P will have to take account also of the loop checking rules.
2. To find a proper new logic or a new completion, for which Clark's theorem holds.
3. Use a combination of (1) and (2).

Note that all our definitions and concepts *must be* natural and intuitive (whatever that means).

In the following sections we shall introduce and examine several ways of dealing with our problem. These are:

1. use equational calculus of failure
2. use intuitionistic logic
3. use temporal and relevance logic[5]
4. use a modal provability[6]
5. use special loop checking.[7]

11.3 The completion in intuitionistic logic

We can get a slightly better completion, if we use intuitionistic logic as the underlying logic. We use the fact that the Horn clause fragment is the same for intuitionisitic logic and for classical logic.

The following holds for any program P without negation and any atom q.

Theorem 11.12.

$$\mathrm{Com}(P) \vdash_I q \text{ iff } P?q = 1$$

where $\vdash_I$ denotes provability in intuitionistic logic. This theorem follows directly from the Clark, Lloyd et al. theorem for the classical logic case.

J. Shepherdson [308, 309] has already pointed out that intuitionistic logic may be more suitable for the analysis of negation as failure. Indeed in my papers on N-Prolog (see [128]), I add intuitionistic implication to Prolog.

We shall now introduce the $I - Com(P)$, the intuitionistic completion of P. It will not solve our problems but it will do slightly better than Clark's $Com(P)$.

First let us look at an example. The $I - Com(P)$ is formally defined later in Definition 11.18.

Example 11.13. Consider the clause

1. $a \wedge b \rightarrow c$
2. $d \wedge e \rightarrow c$

---

[5] This part was not written.
[6] This was done in [134].
[7] There is another unpublished 1985 manuscript doing this.

Clark's completion is

$$(a \wedge b) \vee (d \wedge e) \leftrightarrow c$$

and

$$\neg a \wedge \neg b \wedge \neg d \wedge \neg e.$$

In the part of the completion relations to c the idea is to say that c succeeds through clauses (1) and (2) and *only through them*, and hence to enable us to prove $\neg c$ if c fails. In intuitionistic logic to prove c from (1) and (2) *we have to* prove the antecedent of (1) or of (2) first. This is the subformula property. So we *already have the only if part*. However, we still have to deal with failure, and say when does c fail.

Thus we want to say:

c fails if in each clause body with head c at least one literal fails.

Thus we have to add to the completion the following:

3. $\neg a \wedge \neg d \rightarrow \neg c$
4. $\neg a \wedge \neg e \rightarrow \neg c$
5. $\neg b \wedge \neg d \rightarrow \neg c$
6. $\neg b \wedge \neg e \rightarrow \neg c$

Thus $I - Com$ of the above is:

1. $a \wedge b \rightarrow c$
2. $d \wedge e \rightarrow c$

together with (3), (4), (5), (6) and $\neg a \wedge \neg b \wedge \neg d \wedge \neg e$.

In classical logic we have that (1)–(6) can be written as

$$[(a \wedge b) \vee (d \wedge e)] \leftrightarrow c.$$

This is *not true in intuitionistic logic*. We need the law of excluded middle of classical logic to manipulate (1)–(6) into $[(a \wedge b) \vee (d \wedge e)] \leftrightarrow c$ and in intuitionistic logic this rule is not available.

Let us give more examples.

Example 11.14.

$$P = \neg c \rightarrow c.$$

Clark's completion is, $\neg c \leftrightarrow c$, a contradiction.

The intuitionistic completion is:

$$(\neg c \rightarrow c) \wedge (\neg\neg c \rightarrow \neg c)$$

which is a contradiction. No improvement over Clark's completion.

Example 11.15. Consider the program P:

$$\neg a \rightarrow b$$
$$a \rightarrow b$$
$$\neg b \rightarrow c$$
$$a \rightarrow a$$

Clark's completion $s \vdash_C \neg c \wedge b$, even though P loops.

Our $I - Com$ is:[8]

$$\neg a \to b \qquad \neg\neg a \wedge a \to \neg b$$
$$a \to b \qquad \neg\neg b \to \neg c$$
$$\neg b \to c \qquad \neg a \to \neg a$$
$$a \to a$$

which is equivalent in intuitionistic logic to:

1. $(a \vee \neg a) \to b$
2. $\neg b \to c$
3. $\neg\neg b \to \neg c$

This completion can prove $\neg c$ (in propositional intuitionistic logic, but not in intuitionistic predicate logic). It cannot prove b, however.

The reason it can prove $\neg c$ is because in intuitionistic logic

$$\frac{x \to y}{\neg\neg x \to \neg\neg y}$$

and

$$\vdash \neg\neg(a \vee \neg a)$$

are both valid. We thus get from the first assumption that

$$\vdash \neg\neg(a \vee \neg a) \to \neg\neg b$$

i.e. $\vdash \neg\neg b$.

Hence $\vdash \neg c$ from the third assumption. b, however, cannot be proved.

Compared with the ordinary Clark completion, we are slightly better off. Clark's completion can prove $\neg c \wedge b$. The intuitionistic completion can prove $\neg\neg b \wedge \neg c$ which is weaker. It cannot prove b. Thus we can say that because a loop, we cannot assert b but we certainly can say through a sort of *lazy evaluation* that $\neg\neg b$ (i.e. that b cannot finitely fail) and hence also $\neg c$.

We can attempt to take the above set of wff as completion but base it on some relevance logic, in which $\nvdash \neg\neg(a \vee \neg a)$, i.e. the notion of $\neg$ will be some relevance sort of $\neg$. I will not go into details here.[9]

Notice that $I - Com(P)$ is sound because classical logic is sound for the Clark completion.

---

[8] Since $\neg$ is read as negation as failure, then the meaning of $\neg\neg$ is read as failure of negation as failure, namely as success. Thus we have $\neg\neg x \leftrightarrow x$ holding in the logic program and we can restrict all our consideration to one level of nested $\neg$. However if we start with a logic program with only one level of negation as failure and move to the intuitionistic completion where we read $\neg$ as intuitionistic, we get nested intuitionistic $\neg\neg$ in the process of constructing the I completion.

[9] In the intuitionistic case, Clark's completion is consistent with the idea of failure as failure by loop, i.e. well-founded semantics.

The intuitionistic completion seems to produce the same conclusion as Kunen semantics [235], and consequently with Fitting's semantics [125]. We shall address these issues in Section 6.

To continue with our example 11.15, notice that if $a \rightarrow a$ is taken out of the database, then the I-completion will have $\neg a$ in it and it will be

$$\neg b \rightarrow c$$
$$a \rightarrow b$$
$$\neg a \rightarrow b$$
$$\neg a$$
$$\neg\neg b \rightarrow \neg c$$
$$\neg\neg a \wedge \neg a \rightarrow \neg b$$

which proves $b \wedge \neg c$.

We can now formally define the set $I - Com(P)$, for any program P. First we need a lemma.

Lemma 11.16. *Let P be a propositional program and let (1) and (2) be clauses in P. Where*

1. $A \wedge \neg\neg a \rightarrow b_1$
2. $A \wedge \neg(a \wedge d) \rightarrow b$.

Let P' be the program obtained from P by deleting clauses (1) and (2) and adding instead the following clauses (1*), (2*), (3*).

(1*) $A \wedge a \rightarrow b_1$
(2*) $A \wedge \neg a \rightarrow b$
(3*) $A \wedge \neg d \rightarrow b$

Then for any goal G

$$P?G = x \text{ iff } P'?G = x.$$

In other words the same goals succeed or fail.

Proof. By induction on the computation tree.

Lemma 11.17. *Any program P is equivalent to a unique program P' with no embedded negations in the body of clauses, i.e. where each clause has the form*

$$\bigwedge a_i \wedge \bigwedge \neg b_j \rightarrow c$$

where a_i, b_j, c are atomic and P' is equivalent to P'' means that the same goals succeed or fail from each.

Proof. Follows from the previous lemma.

Definition 11.18.

1. Assume P is a program with clauses with no embedded negations in the body of clauses. We define the I-completion of P, denoted by $I - Com(P)$, as follows. We add to P the following additional clauses:
   a) $\neg x$ for atom x not the head of any clause.

b) *If x is a head of some clauses, assume these are c_j, $j = 1, \ldots, m$ and write*

$$C_j = \bigwedge_i A(i, j) \to x$$

where $A(i, j)$ are either atomic or negations of atomic $a_{i,j}$.
Let f be any selection function such that for each j it selects one $f(j)$ such that $A(f(j), j)$ is a conjunct in the body of the clause C_j. We then add to the completion $I - \text{Com}(P)$ all clauses of the form

$$B(f) = \bigwedge_j B(f(j), j) \to \neg x$$

where $B(i, j)$ is defined as:

$$B(i, j) = \begin{cases} \neg a_{i,j} & \text{if } A(i, j) = a_{i,j} \\ a_{i,j} & \text{if } A(i, j) = \neg a_{i,j} \end{cases}$$

$I - \text{Com}(P)$ is considered a theory in intuitionistic logic.
2. *For a general program P, let P' be the unique program promised by Lemma 11.17 and let $I - \text{Com}(P)$ be defined as $I - \text{Com}(P')$ of (1) above.*

Theorem 11.19 (Soundness). *For any P and atomic q,*

$$P? = 1 \text{ implies } I - \text{Com}(P) \vdash_I q$$
$$P?q = 0 \text{ implies } I - \text{Com}(P) \vdash_I q.$$

Proof. Can be proven by induction on the computation trees.

There is no completeness for $I - \text{Com}(P)$. As we have seen

$$\neg c \to c \vdash_I c$$

but $\{\neg c \to c\}?c$ loops.

11.4 The equational calculus of failure

We introduce here a special reduction calculus for failure. We intend it to capture the "logical content" of negation as failure.

Definition 11.20. *Consider a propositional language with the connectives $\neg, \wedge, \vee, \equiv, \top, \bot$.*

  $\neg$ is negation (as failure);
  $\wedge, \vee$ are conjunction and disjunction;
  $\equiv$ is equality; $\top, \bot$ are truth and falsity.

1. *Define the notion of a term of this language as follows:*
   a) any variable x is a term;
   b) $\top$ and $\bot$ are terms;
   c) If A_i $i = 1, \ldots, m$ are terms so are $\bigvee A_i$, $\bigwedge A_i$ and $\neg A_1$.

2. *Let A be a term, then A has a labelled construction tree, defined as follows:*
   a) *If x is a variable or ⊤ or ⊥ its construction tree is as indicated*

$$\bullet x$$

   b) *If A is obtained from A_i $i = 1, \ldots, m$ by the $\bigwedge$ symbol (i.e. $A = \bigwedge A_i$) then its construction tree is*

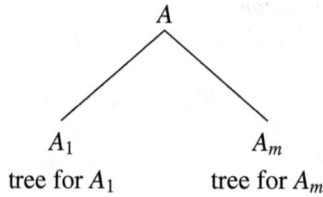

tree for A_1 tree for A_m

   c) *Similar to (2b) for $\bigvee$ symbol.*
   d) *If A is ¬B then its construction tree is*

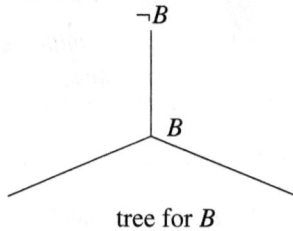

tree for B

3. *An equation is an expression of the form $A \equiv B$, were A and B are terms.*

Definition 11.21. *The axioms of the system are equations and rules:*

1. $\neg\neg x \equiv x$
   $x \wedge y \equiv y \wedge x$
   $x \vee y \equiv y \vee x$
2. $\neg(x \vee y) \equiv \neg x \wedge \neg y$
   $\neg(x \wedge y) \equiv \neg x \vee \neg y$
3. $\neg\bot \equiv \top$
   $\neg\top \equiv \bot$
   $x \wedge \top \equiv x$
   $x \wedge \bot \equiv \bot$
   $x \vee \top \equiv \top$
   $x \vee \bot \equiv x$

The rules of the system are reduction rules. *We have the rules*

$$\frac{x_i \equiv y_i}{A(x_i) \equiv A(y_i)}$$

$$x \equiv x$$

$$\frac{x \equiv y}{y \equiv x}$$

$$\frac{x \equiv y, y \equiv z}{x \equiv z}$$

$$\frac{x \wedge y \equiv \top}{x \equiv \top}$$

and

$$\frac{x \vee y \equiv \bot}{x \equiv \bot}$$

Definition 11.22.

1. *By an equational theory we mean a set of equations. An equational theory where all the equations are either $x \equiv \bot$ or $x \equiv \top$ is called an assignment theory.*
2. *Let $\mathbf{E}$ be an equational theory and let B be an equation. We write $\mathbf{E} \vdash B$ to mean that there exists a sequence of equations $B_1, B_2, \ldots, B_n = B$ such that each B_i in the sequence is either from $\mathbf{E}$ or is obtained from previous members of the sequence by an application of a reduction rule or an axiom.*
3. *Let A be a term, and let $\mathbf{E}$ be an assignment, i.e. a set of equations of the following special form:*

$$x_i \equiv \top \ or \ x_j \equiv \bot.$$

Let $A(x_i, y_j)$ be a term built up from x_i, y_j. We define by induction the notion of A is $\mathbf{E}$-directly reducible to A', notation $A \vdash_{\mathbf{E}} A'$, as follows:

   a) $\top \vdash_{\mathbf{E}} \top, \bot \vdash_{\mathbf{E}} \bot$
   b) $x \vdash_{\mathbf{E}} x$ if neither $x \equiv \top$ nor $x \equiv \bot \in \mathbf{E}$
      $x \vdash_{\mathbf{E}} \top$ if $x \equiv \top \in \mathbf{E}$
      $x \vdash_{\mathbf{E}} \bot$ if $x \equiv \bot \in \mathbf{E}$
   c) If A is obtained (by a construction tree) from A_i by $A = \bigwedge A_i$ and if $A_i \vdash_{\mathbf{E}} y_i$ then we say that
        $A \vdash_{\mathbf{E}} \bot$ if some $y_i = \bot$
        $A \vdash_{\mathbf{E}} \top$ if all $y_i = \top$
        $A \vdash_{\mathbf{E}} \bigwedge_{y_j \neq \top} y_j$ otherwise
   d) If A is obtained from A_i by $A = \bigvee A_i$ and $A_i \vdash_{\mathbf{E}} y_i$, then we say that:
        $A \vdash_{\mathbf{E}} \top$ if some $y_i = \top$
        $A \vdash_{\mathbf{E}} \bot$ if all $y_i = \top$
        $A \vdash_{\mathbf{E}} \bigvee_{y_i \neq \bot} y_i$ otherwise.
   e) If $A = \neg B$ and $B \vdash_{\mathbf{E}} y$ then we say that:
        $A \vdash_{\mathbf{E}} \top$ if $y = \bot$
        $A \vdash_{\mathbf{E}} \bot$ if $y = \top$
        $A \vdash_{\mathbf{E}} \neg y$ otherwise.

Lemma 11.23.

1. *Let* **E** *be an assignment theory. Assume that* $A \vdash_E B$, *then* $\mathbf{E} \vdash A \equiv B$.
2. $\mathbf{E} \subseteq \mathbf{E}'$ *implies* $\vdash_E \subseteq \vdash_{E'}$.

Proof. By induction on the reduction steps of A to B.

Definition 11.24. *We describe a* model *for our equational theory. Let V be an infinite set and let each atomic proposition x be interpreted in the model as a pair of sets $x = (x_1, x_2)$.*
   Let $\top = (V, \varnothing), \perp = (\varnothing, V)$.
   Define the operations as follows. For $x = (x_1, x_2), y = (y_1, y_2)$.

$$\neg(x_1, x_2) = \mathrm{def.}(x_2, x_1)$$
$$(x_1, x_2) \wedge (y_1, y_2) = (x_1 \cap y_1, x_2 \cup y_2)$$
$$(x_1, x_2) \vee (y_1, y_2) = (x_1 \cup y_1, x_2 \cap y_2)$$
$$x \equiv y = \mathrm{def.} x_1 = x_2 \text{ and } y_1 = y_2$$

Looping will be associated with $x_1 = x_2 = x$ *(i.e. (x, x) as we shall see later).*

Lemma 11.25. *The system is sound for the model.*

Proof. We check:

1. $\neg\neg(x_1, x_2) = \neg(x_2, x_1) = (x_1, x_2)$
2. $\neg(x \vee y) = \neg(x_1 \cup y_1, x_2 \cap y_2) = (x_2 \cap y_2, x_1 \cup y_1) = \neg x \wedge \neg y$
3. $\neg(x \wedge y) = \neg(x_1 \cap y_1, x_2 \cup y_2) = (x_2 \cup y_2, x_1 \cap y_1) = \neg x \cup \neg y$
4. $\neg\top = \neg(V, \varnothing) = (\varnothing, V) = \perp$
   $\neg\perp = \neg(\varnothing, V) = (V, \varnothing) = \top$
5. Clearly $x \wedge \top = x, x \wedge \perp = \perp, x \cup \top = \top, x \cup \perp = x$
6. Clearly if $x \equiv y$ then $A(x) \equiv A(y)$.

The proof of the other inference rules is straightforward, and is omitted.

Definition 11.26. *The equational completion $E - \mathrm{Com}(P)$ of a program P is the same syntactically as the Clark's completion, except that it is read in the equational calculus instead of classical logic. If $a \in P$ we put $a \equiv \top \in E - \mathrm{Com}(P)$ and if a is not the head of any clause in P we put $a \equiv \perp \in E - \mathrm{Com}(P)$.*

Example 11.27.

1. P_1 *is the program*

$$\neg a \rightarrow b$$

   The equational completion is $E - \mathrm{Com}(P)$

$$a \equiv \perp$$
$$\neg a \equiv b$$

   It certainly follows in the equational calculus that

$$E - \mathrm{Com}(P_1) \vdash \neg a \equiv \top \text{ and } b \equiv \top$$

2. Take P_2 to be:

$$\neg a \rightarrow b$$
$$\neg c \rightarrow c$$

The Clark completion of this program is

$$(b \equiv \neg a) \wedge \neg a \wedge (\neg c \equiv c) = \text{ falsity}$$

This is very bad for Clark because $\neg c \rightarrow c$ has nothing to do with $\neg a \rightarrow b$, and at least we should get $Com(P_2) \vdash b \wedge \neg a$.
How about our new $E - Com(P_2)$?
We get: $E - Com(P_2)$ is:

$$(b \equiv \neg a) \text{ and } (a \equiv \bot) \text{ and } (\neg c \equiv c).$$

$E - Com(P_2)$ is consistent, it proves $b \equiv \top, a \equiv \bot$ and also proves $\neg c \equiv c$, which is a consistent statement in the equational calculus.
Let us see what happens with these equations in our model:

$$a \equiv \bot \text{ means } (a_1, a_2) = (\emptyset, V)$$
$$b \equiv \neg a \text{ means } (b_1, b_2) = (V, \emptyset)$$
$$\neg c \equiv c \text{ means } c_2 = c_1$$

Example 11.28. Here is another example where we do better than Clark. P_3

$$\neg a \rightarrow b$$
$$a \rightarrow b$$
$$\neg b \rightarrow c$$
$$a \rightarrow a$$

Clark's completion is:

$$b \equiv \neg a \vee a$$
$$\neg b = c$$
$$a \equiv a$$

The problem here is that $Com(P_3) \vdash b \wedge \neg c$ without either b succeeding or c finitely failing.
Let us examine the equational completion directly in our model:
$E - Com(P_3)$ is

$$(b_1, b_2) \equiv (a_2, a_1) \vee (a_1, a_2)$$
$$(b_2, b_1) \equiv (c_1, c_2)$$

From this we get, by the definition of the operation for disjunction that:

$$b_1 = c_2 = a_1 \cup a_2$$
$$b_2 = c_1 = a_1 \cap a_2$$

The above is nonconclusive because (*) below cannot be proven:

$$* = \left\{ \begin{array}{l} b_1 = V, b_2 = \emptyset \\ c_1 = \emptyset, c_2 = V \end{array} \right\}$$

However, if the clause $a \rightarrow a$ is deleted, the Clark completion for the new program works, but also our completion, because we would have the clause $a \equiv \bot$, i.e.

$$a_1 = \emptyset, a_2 = V$$

which yields (*) above immediately.

Example 11.29. Consider the program

$$a \rightarrow c$$
$$\neg c \rightarrow a.$$

The completion is

$$a \equiv c$$
$$\neg c \equiv a$$

and the equational solution is

$$a_1 = a_2 = c_1 = c_2.$$

Lemma 11.30 (Soundness). *Let P be a program and let $E - \mathrm{Com}(P)$ be its equational completion. Then for any goal G we have:*
G succeeds from P implies $E - \mathrm{Com}(P) \vdash G \equiv \top$.

Proof. Assume x succeeds from P. Prove by induction on the length of the computation that $E = \mathrm{Com}(P) \vdash x \equiv \top$.

1. Length 1: Then either x is a positive atom and $x \in P$, in which case $\top \equiv x \in E - \mathrm{Com}(P)$ or x is not the head of any clause in P, in which case $\bot \equiv x \in E - \mathrm{Com}(P)$. In either case $P \vdash \neg x \equiv \top$.
2. Length m: If $x \wedge y$ succeed then x succeeds and y succeeds hence $E - \mathrm{Com}(P) \vdash x \equiv \top$ and $\vdash y \equiv \top$ and hence $\vdash x \wedge y = \top$. If $\neg(x \wedge y)$ succeeds then $x \wedge y$ finitely fails then at least x fails or y fails. Hence either $\vdash \neg x \equiv \top$ or $\vdash \neg y \equiv \top$. Since $\neg(x \wedge y) \equiv \neg x \vee \neg y$ and for any $Z\ Z \vee \top = \top$ and we get $\vdash \neg(x \wedge y) \equiv \top$.

Assume now that $\neg x$ succeeds and x is atomic. Then x finitely fails. Let $\{(A_{ij} \rightarrow x_i\}$ be all clauses with head x and A_{ij} is either an atom or negation. We have for each a, a $j(i)$ such that $A_{i,j(i)}$ finitely fails.

Hence we have

$$\vdash \bigvee_i \bigwedge_j A_{i,j} \equiv x$$

and

$$\vdash \neg A_{i,j(i)} \equiv \top$$

By the equational axioms:

$$\vdash \bigwedge_i \bigvee_j \neg A_{i,j} \equiv \neg x$$

Since each conjunct $\bigvee_j \neg A_{i,j}$ an $A_{i,j(i)}$ such that $\vdash \neg A_{i,j(i)} \equiv \top$ we get $\vdash \bigvee j \neg A_{i,j} \equiv \top$ for each i and hence

$$\vdash \bigwedge_i \bigvee_j \neg A_{ij} \equiv \top$$

hence

$$\vdash \neg x \equiv \top.$$

Lemma 11.31 (Completeness). *Let P be a program and let x be atomic or a negation of atomic. If $E - \mathrm{Com}(P) \vdash x \equiv \top$ then x succeeds from P.*

Proof. To prove completeness we need further lemmas and constructions.

Lemma 11.32. *Assume that $\mathbf{E} \vdash B$ in the equational calculus and let x be a variable in $\mathbf{E}$ and in B. Then $\mathbf{E}' \vdash B'$, where E', B' are obtained from $\mathbf{E}, B$ by substituting x' for x.*

Proof. The same sequence which proves B from E will prove B' from E'.

Definition 11.33. *Let $\mathbf{E}$ be a theory of the form $\{x_i \equiv A_i\}$, where x_i are atomic. Define by induction a sequence of assignment theories $\mathbf{E}_j, j = 0, 1, \ldots$ as follows*

$$E_0 = \{x_i \equiv \top \in \mathbf{E}\} \cup \{Y_j \equiv \bot \in \mathbf{E}\} \cup \{\top \equiv x_i \in \mathbf{E}\} \cup \{\bot \equiv y_j \in \mathbf{E}\}.$$

Suppose $\mathbf{E}_k$ was defined.
     Let $\mathbf{E}_{k+1} = \{y \equiv \top$ for some $A \equiv B \in \mathbf{E}, A \vdash_{E(k)} y$ and $B \vdash_{\mathbf{E}_k} \top\}$
     $\cup \{\top \equiv y \mid$ *same condition*$\}$
     $\cup \{y \equiv \bot \mid$ *for some $A \equiv B \in \mathbf{E}, A \vdash_{\mathbf{E}_k} y$ and $B \vdash_{\mathbf{E}_k} \bot\}$*
     $\cup \{\bot \equiv y \mid$ *same condition*$\}$,
     Let $\mathbf{E}^*$ be $\bigcup \mathbf{E}_k$.

Lemma 11.34.

1. *For any equation B, if $\mathbf{E} \vdash B$, then $\mathbf{E}^* \vdash B^*$ where B^* is obtained from B by substituting $\top$ or $\bot$ for the variables according to the assignment of $\mathbf{E}^*$.*
2. *If $\mathbf{E} \vdash x \equiv \top$ then $x \equiv \top \in \mathbf{E}^*$.*
   If $\mathbf{E} \vdash \neg x \equiv \top$ then $x \equiv \bot \in \mathbf{E}^*$

Proof of (1)
Using Lemma 11.32 and Lemma 11.23.

Proof of (2)
Define a model of sets of the equational theory.
     Let $y = (V, \emptyset)$ if $y \equiv \top \in \mathbf{E}^*$.
     Let $y = (\emptyset, V)$ if $y \equiv \bot \in \mathbf{E}^*$.
     let $y = (\emptyset, \emptyset)$ otherwise.
     We claim that all the equations of $\mathbf{E}$ are valid in this model. From this fact it follows that if $\mathbf{E} \vdash x \equiv \top$ or if $\mathbf{E} \vdash \neg x \equiv \top$ then $x \equiv \top$ or $\neg x \equiv \top$ (respectively) must also be valid in the model. By construction of the model this will not be the case unless $x \equiv \top$ or $x \equiv \bot$ (respectively) are in $\mathbf{E}^*$.

Lemma 11.35. *Let P be a program and let $\mathbf{E} = E - \mathrm{Com}(P)$. Then*

1. *if $\mathbf{E} \vdash x \equiv \top$ then $P?x = 1$*
2. *if $\mathbf{E} \vdash \neg x \equiv \top$ then $P?x = 0$.*

Proof. A previous lemma showed that the condition of (1) happens exactly when $x \equiv \top \in \mathbf{E}^*$ or $x \equiv \bot \in \mathbf{E}^*$ respectively. We therefore have to show that if $x \equiv \top \in \mathbf{E}^*$ then $P?x = 1$ and if $x \equiv \bot \in \mathbf{E}^*$ then $P?x = 0$.
     The proof is induction on k which defines $\mathbf{E}^* = \bigcup_k \mathbf{E}_k$.

Note that $\mathbf{E} = E - Com(P)$ has a very special form, i.e. the form

$$x = \bigvee_i \bigwedge_j A_{ij}.$$

If $x \in P$ we put $x \equiv \top \in \mathbf{E}$.

If x is not the head of any clause we put $x \equiv \bot \in \mathbf{E}$. Thus $\mathbf{E}_0$ is the set of $x \equiv \top$ for $x \in P$ and $x \equiv \bot$ for x not the head of any clause. $\mathbf{E}_1$ is obtained by substituting the value assignments from $\mathbf{E}_0$. If $x_1 \equiv \top$ or $x_1 \equiv \bot \in \mathbf{E}_1$ this means that an equation of the form $x_1 = \bigvee \bigwedge_j A_{i,j}$ was reduced to

$$x_1 \equiv \bot (\text{or } x_1 \equiv \top).$$

The reduction process is similar to the prolog computation process. $x_1 \equiv \top$ only if one of the disjuncts say the i-th disjunct) $\bigwedge_j A_{ij} \equiv \top$ and this means each $A_{ij} \equiv \top$. $x_1 \equiv \bot$ only if *each* disjunct is $\bot$ and this holds exactly when each disjunct has a conjunct which is $\bot$, i.e. x_1 finitely fails.

The induction proof follows these lines.

11.5 Negation as failure as a special case of parallel programming

The duality (or symmetry) we saw between a and $\neg a$ can be brought forward much more forcefully.

Example 11.36. We begin with an example. Consider the program

$$P_* : c \wedge \neg a \to b$$
$$c.$$

The intuitionistic completion of this program is obtained by writing as additional clauses the failure conditions. These are

$$\bar{P} : \neg\neg a \to \neg b$$
$$\neg c \to \neg b$$
$$\neg a$$

Call $\neg b$ by $\bar{b}$ and $\neg\neg b$ by b^* (this is just a notation, to bring out certain structures).

$$\text{Thus} \quad \neg x = \bar{x} \quad (\text{notation})$$
$$\neg\neg x = x^*$$

The program becomes:

$$P^* : c \wedge \bar{a} \to b$$
$$c$$

$\bar{P}$ = the addition to P^* to obtain the I-completion:

$$a^* \to \bar{b}$$
$$\bar{c} \to \bar{b}$$
$$\bar{a}$$

We now read a goal $\bar{b}$ to mean: compute b from the program $\bar{P}$ and read a goal g^* to mean: compute g from the program P^*. The idea is to present $\neg x$ to mean: compute x from the other program.

Let us ask when b fails from the original program, i.e. from $\dfrac{c \wedge \neg a \to b}{c}$

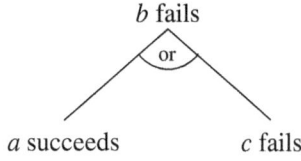

$$b \text{ fails}$$

$$\text{or}$$

$$a \text{ succeeds} \qquad\qquad c \text{ fails}$$

To compute $\bar{b}$ from $\bar{P}$ we have

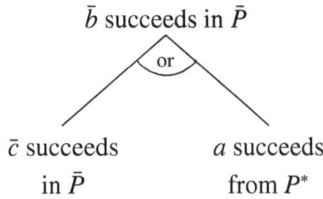

$$\bar{b} \text{ succeeds in } \bar{P}$$

$$\text{or}$$

$$\bar{c} \text{ succeeds} \qquad\qquad a \text{ succeeds}$$
$$\text{in } \bar{P} \qquad\qquad\qquad \text{from } P^*$$

The above shows the two computations are the same.

Definition 11.37. *We define the language for parallel processing from programs* $(P_1, P_2, \ldots)$.
The language contains atoms $\wedge, \to$ *and the operators* $n^\urcorner, n = 1, 2, 3, \ldots$ ($n^\urcorner A$ *can be read as "compute A from* P_n".)

1. *Any atom q is both a clause and a goal.*
2. *If A and B are goals so are $A \wedge B$ and $u^\urcorner A$.*
3. *If A is a goal and q atomic, then $A \to q$ is a clause.*

Definition 11.38. *Let* $P = (P_1, P_2, P_3, \ldots)$ *be a sequence of finite sets of clauses. We define the parallel computation of the goal A from P. We define two notions:*

$$P?A = success \text{ and } P?A = failure$$

1. *We start the computation of $P?A = success$ by asking $P_1?A = success$. We now continue to define the computation of $P_k?A = success$ by induction.*
2. *$P_k?A \wedge B = success$ if $P_k?A = success$ and $P_k?B = success$.*
3. *$P_k?n^\urcorner A = success$ if $P_n?A = success$.*
4. *$P_k?q = success$, for q atomic, if $q \in P_k$.*
5. *$P_k?q = success$, for q atomic, if for some $A \to q \in P_k$ and $P_k?A = success$. We now define the computation for $P?A = failure$.*
6. *For $P?A = failure$, ask for $P_1?A = failure$.*
7. *$P_k?A \wedge B = failure$ iff
   $P_k?A = failure$ or $P_k?B = failure$.*
8. *$P_k?n^\urcorner A = failure$ iff
   $P_n?A = failure$.*
9. *$P_k?q = failure$ if q is not the head of any clause in P_k.*
10. *$P_k?q = failure$ if for all clauses of the form $\bigwedge A_i \to q$, where A_i is either atomic or $n^\urcorner A_i'$ we have that some i_0 exists such that $P_k?A_{i_0} = failure$.*

Definition 11.39. *P_k, P_n are said to be* dual *if the following holds:*

$$P_k?A = \begin{cases} success \\ failure \end{cases} \text{ iff } P_n?A = \begin{cases} failure \\ success \end{cases}$$

In our notation the two programs $P^*, \bar{P}$ are dual.

In the predicate case, of ordinary negation as failure, $\neg A(x)$ can be understood as a special way of going to the dual.

E.g. "check ?$A(x)$ for success but do not *bind variables" or "don't execute unless grounded", etc. These are conditions for going from P_k to P_n.*

In this context, the condition of asking $\neg A(x)$ and not *binding x for the successful computation (i.e. the failure of $A(x)$) seems arbitrary.*

If there are difficulties with negation as failure these are difficulties of the particular management of the passage to the dual. Having recognised what it is, we can try and correct and improve it.

Remember : $\neg a$ negates a but $\neg(\neg a)$ negates $\neg a$ and is the same as a. By symmetry any objections you take for $\neg a$ as a negation are equally valid for a as an assertion.[10]

11.6 Discussion and comparison with post 1985 literature

This section will compare our results with the relevant literature, i.e. to papers offering semantics for negation as failure. The reader should bear in mind our "logical content criterion" introduced in Section 1

1. S. Cerrito. A linear axiomatization of negation as failure, [89]

   We simply start by quoting Cerrito's abstract:

   This paper is concerned with the axiomatization of success and failure in propositional logic programming. It deals with the actual implementation of SLDNF in PROLOG, as opposed to the general nondeterministic SLDNF evaluation method. Given any propositional program P, a linear theory LT_P, is defined (the linear translation of P) and the following results are proved for any literal A: soundness of PROLOG evaluation: If the goal A PROLOG-succeeds on P, then LT_P proves A, and if A PROLOG-fails on P, then LT_P, proves $\sim A$, and completeness of PROLOG evaluation: If LT_P proves A, then the goal A PROLOG-succeeds on P, and if LT_P proves $\sim A$, then A PROLOG-fails on P. Here "prove" means provability in linear logic, and $\sim A$ is the linear negation of A.

   Cerrito translation is a simulation of the procedure in linear logic. Cerrito is aware of this possible criticism and says in his concluding discussion , we quote:

   Now let us play the devil's advocate. A false impression that could arise from a quick reading of our work is that the ability of our translation LT_P, to reflect PROLOG success and failure does not depend on the specificity of linear logic: LT_P is just a paraphrase of how (standard) PROLOG searches through the clauses of P.

---

[10] Here ends the 1985 paper, [130].

Cerrito replies to this criticism. Our take of his reply is that he is saying that he uses the object level connectives and proof theory of linear logic in his translation and therefore this is not a simulation but a proper translation. With this I agree but I still think it is a borderline case with respect to the fulfillment of the "logical content criterion" because linear logic can simulate computational steps in the object level.

Cerrito further strengthens his answer by pointing out that other semantics, such as the one given by Mints, are obviously metalevel simulations,

> In [259] and [311] Mints proposes an axiomatization for PROLOG evaluation which applies to the general class of first-order logic programs; PROLOG is sound and complete with respect to such an axiomatization. As we already said, the main difference between Mints's approach and our approach is that Mints's axiomatization is a formal calculus which provides a paraphrase of PROLOG evaluation, rather than a logical theory which analyses such an evaluation by means of logical operators of negation, conjunction, disjunction, etc. The advantage of MintsÕs axiomatization is that it is not limited to propositional programs, while its weakness — in our opinion — lies in its rather ad hoc nature.

It would be instructive to take the program $\neg a \to a$ and see its Cerrito completion.

Remember that Cerrito translates the actual computation, so she translates the two sentences:

1. a succeeds if a fails, translated in linear logic into $\sim a \to a$ (this is equivalent to $a \vee a$) and
2. a fails if a succeeds, translated in linear logic into $a \to \sim a$ (this is equivalent to $\sim a \vee \sim a$)

Since linear logic has no contraction, neither a nor $\sim a$ can be proved, which is correct

To compare, our equational completion for this case is

$$a \equiv \neg a$$

which implies that $a = (x, x)$ for some x, and so a cannot be equivalent to neither $\top$ nor to $\bot$.

2. M. Fitting. Kripke–Kleene semantics for logic programs, [125].

Fitting's approach starts from the fixed point semantics for logic programs but uses Kleene three valued logic. He associates with each program a monotone operator on a space of three-valued logic interpretations, or better partial interpretations. This space is not a complete lattice, and the operators are not, in general, continuous. But least and other fixed points do exist.

These fixed points are shown to provide suitable three-valued program models. They relate closely to the least and greatest fixed points of the operators used in [22].

A program corresponds to a monotone operator in the space and its semantics is its fixed points.

This interpretation does satisfy the "logical content criterion" but s rather complex. Fitting admits this point, and I quote

Because of the extra machinery involved, our treatment allows for a natural consideration of negation, and indeed, of the other propositional connectives as well. And because of the elaborate structure of fixed points available, we are able to clearly differentiate between programs that "behave" the same but that we "feel" are different. Finally, we show the result is far too powerful. We can now write logic programs semantically characterizing the Π_1^1 relations, not just the recursively enumerable ones. Thus semantic behaviour is not generally machine realizable.

3. K. Kunen. Negation in logic programming, [235].

Kunen deals with predicate logic programming. His approach gives rise to a semantics which is a cross between the completed database obtained by Clark [99] and the 3-valued logic approach advanced by Fitting [125]. The Kunen semantics is more restrictive than either of these approaches, in the sense that any query which follows from the database under Kunen semantics also follows under both [99] and [125]; but not conversely.

4. D. Gabbay. Modal provability interpretation for negation by failure, [134].

This paper essentially continues Clark's attempts in finding a suitable "completion" by presenting "negation by failure as a modal provability notion. In fact, we use a variation of the modal logic of Solovay, originally introduced to study the properties of the Goedel provability predicate of Peano Arithmetic, and show that $\neg A$ can be read essentially as 'A is refutable from the (logical content of the) program'.

Given a program P, it contains lots of negation by failure symbols $\neg$ in it. We want to build a matrix E in the modal logic **L** which is associated with P and is obtained syntactically from P. The matrix will be of the form $E(x)$ with a free new propositional variable x. $E(x)$ i s obtained from P by replacing in the program each $\neg a$ by Not $(x \rightarrow a)$, and by possibly other additions. $E(x)$ is a theory in the modal logic **L**.

Intuitively one can understand $E(x)$ as the result of translating $\neg a$ in the program by Not$(x - a)$, where x is supposed to be the program itself.

We now solve, in the modal logic **L**, the equation $x \leftrightarrow E(x)$. If we choose a modal logic which always allows for a unique solution $x_0 = E(x_0)$, then this unique x_0 can be taken as the translation $P*$ of P into the modal logic **L**.

This modal provability semantics clearly satisfies the "logical content criterion".

For comparison, the translation of the program $\neg a \rightarrow a$ is $\Box(\Box\bot \vee (a \leftrightarrow \Diamond\neg a)) \wedge (\Box\bot \vee (a \leftrightarrow \Diamond\neg a))$.

An atomic q to succeeds from the program iff the modal completion can prove $\Box q$ and it finitely fails iff the modal completion can prove $\Box\neg q$.

5. J. Vauzeilles. Negation as Failure and Intuitionistic Three-Valued Logic, [327].

The author present a three-valued intuitionistic version of Clark's completion, denoted by Comp3I(P). He proves the soundness of SLDNF-resolution with respect to Comp3I(P), and the completeness both for success and failure, as far as allowed programs are concerned. He also compares his results to Kunen [235], Cerrito [89] and

Shepherdson's [308, 309, 310], which are based on classical three-valued logic, linear logic, and on a system of rules which are valid in both intuitionistic logic and three-valued logic.

Vauzeilles's work can compare with our intuitionistic discussion in Section 3. It seems that by introducing intuitionistic 3 valued logic the author was able to give semantics for negations as failure. We did not pursue such an option, it did not occur to us at the time (remember our paper [130] was written in 1985 and Vauzeilles's paper is from 1991)

6. R. Staerk. Cut property and negation as failure, [317].

This paper gives semantics for negation as failure by proof-theoretic methods. A rule based sequent calculus is used in which sequent is provable if, and only if, it is true in all three-valued models of the completion of a logic program. The main theorem is that proofs in the sequent calculus can be transformed into SLDNF-computations if, and only if, a program has the cut-property. A fragment of the sequent calculus leads to a sound and complete semantics for SLDNF-resolution with substitutions. It turns out that this version of SLDNF-resolution is sound and complete with respect to three-valued possible world models of the completion for arbitrary logic programs and arbitrary goals. Since we are dealing with possibly nonterminating computations and constructive proofs, three-valued possible world models seem to be an appropriate semantics. Staerk remarks that his sequent calculus is just a formalization of the three-valued approach to logic programming proposed by Fitting in [125].

We make a general comments that proof theoretical approaches are difficult to assess with respect to the "logical content criterion". Both the prolog algorithms and the proof theory involve syntactical manipulations and so it is easy to embed the algorithm inside the proof theory.

7. Other papers

[128, 201, 66, 274] discuss negation as failure in intuitionistic logic programming. Gabbay and Reyle [128] were the first, in 1984, to suggest that implications should be allowed in logic programming in the body of clauses. The system they offered, called N-Prolog, is really the implicational fragment of intuitionistic logic augmented with negation as failure. This was further studied by Gabbay in his 1985 paper [133]. The problem of the logical content of negation as failure with intuitionistic implication was indeed challenging and it was solved in [274].

We note that initially the Logic Programming community objected to N-Prolog but five years later the community came to realise that implications in the body of clauses are needed, and there were many papers published thereafter on intuitionistic logic programming. Although the community always referenced Gabbay and Reyle work, they never explicitly acknowledged them as pioneering the area of intuitionistic logic programming (five years earlier).

[27] is another modal logic approach.

[126, 265] use a four valued bilattice approach.

None of the above are directly related/comparable to our chapter, which uses the equational approach.

11.7 Conclusion

It seems from our discussion and comparison with the post 1985 literature that our equational approach is still unique. All the other proposed semantics for negation as failure are some sort of combinations of proof theoretical 3 valued or modal logic interpretations. The equational approach, introduced in 1985, is still different and furthermore it connects with different type of research community, the argumentation community. For the future , the connection with argumentation should be explored.

The connection with argumentation is as follows:

Let (S, R) be an argumentation network. S is the set of arguments and R is the attack binary relation on S. We say that x attacks y if xRy holds.

Argumentation theory in its formal abstract form is concerned with admissible subsets E of S satisfying some conditions among them are the following:

1. x is in E if it is not attacked
2. points in E do not attack one another
3. if y attacks a point x in E then there is a point z in E attacking y.

Recall Chapter 2 for an overview..

We translate the argumentation network into a logic program as follows:

Let the atoms of the program be S. For each x in S let $y_1, \ldots, y_n$ be all elements of S attacking x. Then put the following clause in the corresponding logic program.

$$\bigwedge \neg y_i \to x.$$

The equational semantics for the argumentation network comes from this translation. We assign $h(x) = (V, \varnothing)$ if x is a node in the network that is not attacked and require that h satisfies the following, where y_i are all the attackers of x:

$$h(x) = \bigwedge \neg h(y_i).$$

Note that through this translation, the various semantics for Logic programs with negation as failure offered in the literature by various researchers can be transferred to semantics for argumentation networks. For example, our modal provability interpretation for negation as failure, [134], was transferred to argumentation in [136], see Chapter 5. The semantics of Cerrito [89, 90] and the semantics of Vauzeilles [327] are especially promising, and I shall pursue the problem of their adaptation to argumentation networks.

11.8 Afterword

The material for this chapter was written in 1985, intended to develop as a joint paper with Marek Sergot, giving semantics to negation as failure. This was following our previous work, later published as paper [131]. Nothing happened in 1985 and the paper stayed in my drawer until a few months ago. I looked at the paper again in February 2011, when I was writing a paper on the equational approach to argumentation, papers [166, 167], see Chapter 10, and discovered that this old paper actually gave argumentation networks Boolean equational semantics through the equational approach to negation as failure. This made the paper relevant to this book.

An Equational Approach to Logic Programming

12.1 Equational approach to logic programs

This section will present the equational approach to logic programs. We begin with purely formal definitions. The examples will come afterwards.

Consider a propositional language with propositional atoms $\{q_1, q_2, q_3, \ldots\}$ and the connectives $\wedge, \vee, \rightarrow$ and $\neg$. $\neg$ is supposed to be negation as failure and $\rightarrow$ causal implication.

We may later use other connectives like $\equiv$ (if and only if) or $\leftrightarrow$ (which also stands for if and only if).

Definition 12.1 (Prolog goals, clauses and computation).

1. *Any atomic q is both a goal and a clause.*
2. *If A and B are goals so are $A \wedge B$ and $\neg A$.*
3. *If A is a goal q atomic then $A \rightarrow q$ is a clause. q is said to be the* head *of the clause and A is the* body *of the clause.*
4. *Given a program P and a goal G, we define the notion of success (failure) of G from P, notation $P?G = 0$ or 1.*
    a) *If $G = q$, q atomic, then q succeeds if q is in P or if for some clause $A \rightarrow q$ in P, A succeeds. q fails if it is not the head of any clause in P, or if for each clause $A \rightarrow q$ in P, A fails.*
    b) *If $G = A \wedge B$ then it succeeds if both conjuncts succeed and fails if at least one conjunct fails.*
    c) *$\neg G$ succeeds (fails) iff G fails (succeeds).*

Definition 12.2.

Let P be a set of clauses as defined in Definition 12.1. We write P in the following form

$$P = \bigcup_c P(c)$$

where c ranges over all atomic literals appearing in P, and $P(c)$ is the set of all clauses in P with head c.

Note that for the sake of uniformity of mathematical notation, we write

- *$P(c) = \emptyset$, if there are no clauses with head c*

- $\top \wedge \neg\bot \rightarrow c$ if the clause is just c
- $\top \wedge \bigwedge_k \neg b_k \rightarrow c$ if the clause if $\bigwedge_k \neg b_k \rightarrow c$, and
- $\bigwedge_i a_i \wedge \neg\bot \rightarrow c$, if the clause is $\bigwedge_i a_i \rightarrow c$.

Thus if c is the head of some clauses in P then $P(c)$ is the set of clauses of the form

$$\bigwedge_j a_{i,j} \wedge \bigwedge_k \neg b_{i,k} \rightarrow c$$
$$i = 1, \ldots, r(c)$$
$$j = 1, \ldots, J(i,c)$$
$$k = 1, \ldots, K(i,c).$$

1. We consider P as a syntactic object generating equations in the compact unit real interval $[0,1]$. The type of equations that P generates can be $Eq_{inverse}(P)$ or Eq_{max} or other types. The equations themselves, for any program P, are denoted by $Eq_{inverse}(P)$ or $Eq_{max}(P)$. The literals of P are considered the variables of the equations.
   We define

$$Eq(P) = \bigcup_c Eq(P(c))$$

and we define the following equations for the variables/literals in P:
- $\top = 1$
- $\bot = 0$
- $c = 0$ if $P(c) = \varnothing$
- If $P(c) \neq \varnothing$ then the equation for $P(c)$ is the following for the case of $Eq_{inverse}(P(c))$:

$$c = \mathbf{h}_{inverse}(c) = 1 - \prod_{i=1}^{r(c)}(1 - \prod_{j=1}^{J(i,c)} a_{i,j} \cdot \prod_{k=1}^{K(i,c)}(1 - b_{i,k}))$$

where, as we said, $c, a_{i,j}, b_{i,k}$ are considered variables ranging over $[0,1]$, and $\top$ is 1 and $\bot$ is 0.

2. We define $Eq_{max}(P)$ as follows. $Eq_{max}(P(c))$ is
- $c = 0$ if $P(c) = \varnothing$
   $c = \mathbf{h}_{max}(c) = \min(1 - \max_{j,k}(1 - a_{i,j}, b_{i,k}))$, otherwise

We use the following abbreviations when we talk about $Eq(P(c))$.

$$\prod_k \text{ refers to } \prod_{k=1}^{K(i,c)}(1 - b_{i,k})$$
$$\prod_j \text{ refers to } \prod_{j=1}^{J(i,c)} a_{i,j}$$
$$\prod_i \text{ refers to } \prod_{i=1}^{r(c)}(1 - \prod_j \prod_k).$$

Example 12.3. 1. Consider the program $P_1 = \{a \rightarrow b, \neg a \rightarrow b, \neg b \rightarrow c, a \rightarrow a\}$.
The $Eq_{inverse}$ and Eq_{max} equations are (both types are) the same for this program. They are as follows:
   a) $b = 1 - (1 - a)a$
   b) $c = 1 - b$
   c) $a = a$
From (a) and (b) we get

d) $c = (1 - a)a$

The solutions are $c = (1 - a)a, b = 1 - (1 - a)a$ and $a \in [0, 1]$ is arbitrary.

2. Consider the program $P_2 = \{\neg c \to c\}$.

The equation is

$$c = 1 - (1 - (1 - c)) = 1 - c$$

The solution is $c = \frac{1}{2}$.

3. Consider the program $P_3 = \{a \to c, \neg c \to a\}$.

The equations are

a) $c = 1 - (1 - a) = a$

b) $a = 1 - (1 - (1 - c)) = 1 - c$

The solution is $c = a = \frac{1}{2}$.

4. Consider the program $P_4 = \{d \wedge \neg p \to p, \neg d \to r, \neg r \to d\}$.

The equations are (using $Eq_{inverse}$):

a) $p = 1 - (1 - d(1 - p))$

b) $r = 1 - (1 - (1 - d)) = 1 - d$

c) $d = 1 - (1 - (1 - r)) = 1 - r$

The solution is $p = \frac{d}{1+d}, r = 1 - d, d$ arbitrary in $[0, 1]$.

If we use Eq_{max} the first equation becomes

(m1) $p = (1 - \max(p, 1 - d)) = \min(d, 1 - p)$

To solve the equations, we distinguish two cases.

Case 1. $1 - p \le d$.

Then $p = d$ and $r = 1 - d$ and d is arbitrary.

Case 2. $d \le 1 - p$.

then $p = \frac{1}{2}$ and $r = 1 - d$ and d equals any value $\le \frac{1}{2}$.

Theorem 12.4 (Existence). *Let P be a program and let $Eq_{inverse}(P)$ or $Eq_{max}(P)$ be the associated system of equations. then there exists at least one function $\mathbf{f}$: Literals of $P \mapsto [0, 1]$ which solves the equations.*

Proof. Consider the vector $(c_1, \ldots, c_n)$ listing all the literals in P. Consider the vector function $\mathbf{F}$ from $[0, 1]^n \mapsto [0, 1]^n$ defined as follows:

$$\mathbf{F}(c_1, \ldots, c_n) = (F_1(c_1, \ldots, c_n), \ldots, F_n(c_1, \ldots, c_n))$$

where

$$F_i = \begin{cases} 0, & \text{if } P(c_i) = 0 \\ \mathbf{h}_{inverse}(c_i) & \\ (\text{or resp. } \mathbf{h}_{max}(c_i)), & \text{if } P(c_i) \ne \varnothing \end{cases}$$

$i = 1, \ldots, n$.

This function is continuous on $[0, 1]^n$ and therefore by Brouwer fixed point theorem [71], has at least one fixed point solution $\mathbf{f}$.

$$(\mathbf{f}(c_1), \ldots, \mathbf{f}(c_n)) = F_1(\mathbf{f}(c_1), \ldots, \mathbf{f}(c_n)), \ldots, F_n(\mathbf{f}(c_1), \ldots, \mathbf{f}(c_n)).$$

Theorem 12.5 (Soundness). *Let P be a program and let $Eq_{inverse}(P)$ be its equational system. Let $\mathbf{f}$ be a solution to the equations. Then following the Prolog computation of Definition 12.1 we have for any literal c*

1. *If P?c = 1 then* $\mathbf{f}(c) = 1$.
2. *If P?c = 0 (i.e. P?¬c = 1) then* $\mathbf{f}(c) = 0$.

Proof. By induction on the depth of the computation of Definition 12.1 (i.e. the depth of the Prolog computation). We write c for $\mathbf{f}(c)$, to simplify our notation.

Case 1

If $P?c = 1$ in one step then the clause c is in P. The equation for Eq_{inverse} for c is

$$Eq(c) = c = 1 - \prod_i (1 - \prod_j a_{i,j} \cdot \prod_k (1 - b_{i,k})).$$

Since $c \in P$, we have the clause $\top \wedge \neg\bot \to c$ being used in constructing $\prod_i$. Say it contributes the ith clause. The factor for it in $\prod_i$ is

$$1 - 1 \cdot (1 - 1) = 0.$$

Therefore $\prod_i = 0$ and so $c = 1$.

If $P?c = 0$ in one stage, then c is not a head of any clause and the equation for c is
$c = 0$

Case $m + 1$

Assume $P?c = 1$ in $m + 1$ steps. Then for some clause (c, i):

$$\bigwedge_j a_{i,j} \wedge \bigwedge_k \neg b_{i,k} \to c$$

we have that

$$P?a_{i,j} = 1 \text{ for all } j$$

and

$$P?b_{i,k} = 0 \text{ for all } k$$

By the induction hypothesis, we have $\mathbf{f}(a_{i,j}) = 1$ for all j and $\mathbf{f}(b_{i,k}) = 0$ for all k.

Substituting these values in the equation $Eq(c)$ for the factor in $\prod_i$ corresponding to clause (c, i) we get

$$1 - \prod_j 1 \cdot \prod_k (1 - 0) = 0$$

Thus $c = 1 - 0 = 1$.

Assume now that $P?c = 0$. This means, according to the Prolog computation, that for each clause (c, i) for c, for $i = 1, \ldots, r(c)$, there exists either a $j(i)$ or a $k(i)$ such that either $P?a_{i,j(i)} = 0$ or $P?b_{i,k(i)} = 1$.

This means that by the induction hypothesis, for each factor of $\prod_i$ we have that it is equal to one of the two forms

$$1 - 0 \cdot \prod_k = 1$$

or

$$1 - \prod_j \cdot 0 = 1$$

thus $\prod_i = 1$ and so $c = 1 - \prod_i = 0$.

Example 12.6. The converse of Theorem 12.5 does not hold. If we look at program P_1 of Example 12.3, it has a solution $b = 1, c = 0, a = 0$ but $P?a$ is not 0.

The reader may ask what if $\mathbf{f}(c) = 1$ (resp. $\mathbf{f}(c) = 0$) in *every* solution $\mathbf{f}$ of the equations $Eq(P)$, does this imply that $P?c = 1$ (resp. $P?c = 0$)? The answer is negative as P_6 of Example 12.7 shows.

Example 12.7 (Comparing Eq_{inverse} and Eq_{max}). Consider the program $P_6 = \{\neg a \wedge \neg b \rightarrow b, \neg b \rightarrow a\}$.

The Eq_{inverse} equations are

1. $b = 1 - (1 - (1 - a)(1 - b)) = (1 - a)(1 - b)$
2. $a = 1 - b$

From (1) and (2) we get

3. $b = (1 - b)b$

The only solution is $a = 1, b = 0$.

The Eq_{max} equations are

1. $b = 1 - \max(a, b)$
2. $b = 1 - a$

From (1) and (2) we get

3. $b = 1 - \max(1 - b, b)$

Hence $a = b = \frac{1}{2}$.

Remark 12.8. Let P be a program with negation as failure in the body of clauses. Then we consider $Eq(P)$ as the Equational semantics/completion for P. We consider any solution $\mathbf{f}$ as a model for P.

12.2 Comparison with answer set semantics

We begin by introducing answer set programming, see [242, 315].

Definition 12.9. *Let P be a program as in Definition 12.1. Let X be a set of positive literals. Let P^X be the following program obtained from P using X.*

1. *Delete from P any rule of the form*
   (*) $\bigwedge_j a_{i,j} \wedge \bigwedge_k \neg b_{i,k} \rightarrow c$
   for which some $b_{i,k} \in X$.
2. *Replace any remaining rule (after the deletions in (1)) by its positive part, i.e. if (*) is such that $b_{i,k} \in X$ for all k then we replace (*) by the rule (*X)*
   (*X) $\bigwedge_j a_{i,j} \rightarrow c$
3. *P^X is obtained from P and X by executing (1) and (2) above.*

Definition 12.10.

1. *Let P be a program, and let X be a set of positive atoms appearing in P. Consider P^X as defined in Definition 12.9. P^X is a positive program. Define a set $Y(P^X)$ of atoms out of X and P^X as follows:*

a) Let $Y_0 = \{a | a \in P^X\}$

b) Assume Y_n has been defined. Let $Y_{n+1} = Y_n \cup \{c | \bigwedge_j a_{i,j} \to c$ is in P^X and $a_{i,j} \in Y_n$ for all $j\}$.

c) Let $Y(P^X) = \bigcup_n Y_n$.

2. We say that X is an answer set for the program P if $X = Y(P^X)$
   In words: $Y(P^X)$ is the minimal model of P^X and so for X being an answer set means that X is being the minimal model of P^X.

3. Note that if X is an answer set for P, then since X equals Y then the elements of x of X are ranked by the minimal n such that x is in Y_n. This comes in useful in proofs by induction involving X.

Example 12.11.   1. Let $P_5 = \{\neg b \to a, \neg a \to b\}$.
Then $P_5^{\{a\}} = a$ and $P_5^{\{b\}} = b$.

The $\{a\}$ and $\{b\}$ are two answer sets. $\{a, b\}$ is not an answer set because $P_5^{\{a,b\}} = \varnothing$ and its minimal model is $\varnothing$.

2. Let $P_6 = \{\neg a \wedge \neg b \to b, \neg b \to a\}$.
Then $P_6^{\{a\}} = \{a\}$ and $P_6^{\{b\}} = P_6^{\{a,b\}} = \varnothing$.
Thus only $X = \{a\}$ is an answer set.
Compare with Example 12.7. The only Eq_{inverse} solution there is $a = 1, b = 0$ agreeing with $\{a\}$ being the only answer set. The only Eq_{max} solution does not agree with $\{a\}$ being an answer set.

Theorem 12.12 (Soundness w.r.t. answer sets). *Let P be a program and let X be an answer set for P. Define a function $\mathbf{f}_X$ by $\mathbf{f}_X(c) = 1$ iff $c \in X$.*
   Then $\mathbf{f}_X$ is a solution to $Eq_{\text{inverse}}(P)$.

Proof.   1. If c is not a head of any clause in P then both $\mathbf{f}_X(c) = 0$ and the equation for c is $c = 0$.

2. If $c \in P$, then the equation for c is $c = 1$, but also since X is a minimal model for P^X and $c \in P^X$, then $\mathbf{f}_X(c) = 1$.

3. Consider the equation for c, using

$$c = 1 - \prod_i (1 - \prod_j a_{i,j} \cdot \prod_k (1 - b_{i,k}))$$

the clauses in $P(c)$ are

$$\text{clause } (c, i) = \bigwedge_j a_{i,j} \wedge \bigwedge_k \neg b_{i,k} \to c.$$

Case 1.
In clause (c, i) some $b_{i,k} \in X$. Then $\mathbf{f}_X(b_{i,k}) = 0$ and so under $\mathbf{f}_X$, the ith conjunct in $\prod_i$ is 1.
Let I_0 be the set of all such indices i.

Case 2.
In clause (c, i) all the $b_{i,k}$ are not in X. Thus under $\mathbf{f}_X$ the ith conjunct in $\prod_i$ is $1 - \prod_j a_{i,j}$.
Let I_1 be the set of all such indices i. Thus under $\mathbf{f}_X$ we have

$$\mathbf{f}_X(c) = 1 - \prod_{i \in I_1}(1 - \prod_j \mathbf{f}_X(a_{i,j})).$$

4. Assume $c \in X$. We now prove by induction on the rank of c (since we assumed $c \in X$, it has a rank as explained in item 3 of Definition 12.10), that $\mathbf{f}_X(c) = 1$. For rank 1, i.e. $c \in P^X$ we have $\mathbf{f}_X(c) = 1$ from item (1) above.
For rank $m + 1$, we have that for some i_0

$$\prod_j a_{i_0,j} \to c$$

is in P^X and we have that $a_{i_0,j} \in X$ and $a_{i_0,j}$ are of lower rank. Thus by the induction hypothesis $\mathbf{f}_X(a_{i_0,j}) = 1$.
From item (3) above we have that

$$\mathbf{f}_X(c) = 1 - \prod_{i \in I_1}(1 - \prod_j \mathbf{f}_X(a_{i,j})).$$

If $i_0 \in I_1$ then we are finished, because then we would have $\prod_{i \in I_1} = 0$.
Let us show that indeed $i_0 \in I_1$. We ask: How did $\bigwedge_j a_{i,j} \to c$ get into P^X?
This was because all $b_{i,k}$ were not in X. But this is the same condition for i_0 to get into I_1. Thus we get $\mathbf{f}_X(c) = 1$.
5. Assume $c \notin X$. then $\mathbf{f}_X(c) = 0$. We also have that c has no rank. This means that we never have that for some i

$$\bigwedge_j a_{i,j} \to c \in P^X$$

and for all $j, a_{i,j} \in X$.
Thus for each i, there is a $j(i)$ such that $a_{i,j(i)} \notin X$, i.e. $\mathbf{f}_X(a_{i,j(i)}) = 0$.
We now check whether the equation associated with $P(c)$ is satisfied by $\mathbf{f}_X$. We need to check whether

$$\mathbf{f}_X(c) =? 1 - \prod_i(1 - \prod_j \mathbf{f}_X(a_{i,j}) \cdot \prod_k(1 - \mathbf{f}_X(b_{i,k})))$$

Let us concentrate on the right hand side of the equation. Since for each i there is a $j(i)$ such that $\mathbf{f}_X(a_{i,j(i)}) = 0$, we get that $\prod_j = 0$ for each i.
Hence $\prod_i = 1$ and so the right hand side equals 0. So $\mathbf{f}_X(c)$ should be 0, which is indeed the case. We indeed get $0 = 0$.
So $\mathbf{f}_X$ solves this equation also for the case $c \notin X$.
From (4) and (5) we get the theorem.

Example 12.13. The converse of Theorem 12.12 is not true. Consider the program $P = \{\neg c \land a \to b, a \to x, x \to a\}$. The equations for P are

$$c = 0$$
$$b = 1 - (1 - a(1 - c))$$
$$a = x$$

Consider the function $\mathbf{f}$ such that

$$\mathbf{f}(c) = 0$$
$$\mathbf{f}(a) = \mathbf{f}(x) = \mathbf{f}(b) = 1$$

We get for b

$$1 = 1 - (1 - (1 - 0) \cdot 1$$
$$= 1 - (1 - 1)$$
$$= 1$$

Let $X_\mathbf{f} = \{x | \mathbf{f}(x) = 1\} = \{x, a, b\}$. Hence $P^{\{a,b,x\}} = \{a \to b, b \to x, x \to a\}$. The answer set for P is therefore $\varnothing$.

It is not $\{a, b, x\}$.

Thus, although every answer set X for P yields a solution function $\mathbf{f}_X$ with values $\{0, 1\}$, not every such solution function yields an answer set.

Example 12.14.

1. Consider the program P_4 of Example 12.3. The only answer set for P_4 is $X = \{r\}$. Indeed for the solution choice of $d = 0$ we get $r = 1$ and $p = 1$. This corresponds to the $\{r\}$ answer set.
2. The program $\neg c \to c$ has no answer sets, but its equation

$$c = 1 - (1 - (1 - c))$$
$$= 1 - c$$

does have solutions, e.g. the unique solution $c = \frac{1}{2}$.
3. The program $\{\neg p \to q, p \to p\}$ has $\{q\}$ as the only answer set. $\{p\}$ is not an answer set, even though the function $p = 1, q = 0$ is a minimal $[0, 1]$ function.

Theorem 12.15 (Completeness for programs containing only negated atoms in the bodies of its clauses). *Let P be a program containing only negated atoms in the bodies of its clauses and let $\mathbf{f}$ be a solution to the system $Eq_{\text{inverse}})(P)$, such that $\mathbf{f}(c) \in \{0, 1\}$ for each atom c appearing in P.*

Let $X_\mathbf{f}$ be $\{c | \mathbf{f}(c) = 1\}$. Then $X_\mathbf{f}$ is an answer set for the program P.

Proof. Let c be any atom appearing in P.

1. If c is not the head of any clause in P then certainly $\mathbf{f}(c) = 0$ and certainly c is not the head of any clause in P^X for any X and so c is not an element of any answer set of P.
2. Let us focus on c which is a head of some clauses in P. If $c \in P$ then $c \in P^X$ for any X and $\mathbf{f}(c) = 1$ always for any solution $\mathbf{f}$. So there is also agreement for this case.
3. Assume c is the head of some clauses and list them as all clauses of the form

$$\text{clause}(c, i) : \bigwedge_k \neg b_{i,k} \to c, \text{ for } i = 1, 2, \ldots$$

The equation for c is as follows, where we regard the atoms of P as variables in the equation

$$c = 1 - \prod_i (1 - \prod_k (1 - b_{i,k})) \qquad (*)$$

Let us analyse what can happen with this equation:

a) If $\mathbf{f}(c) = 1$ then $\prod_i$ must be 0. To be 0, we must have for some i_0 that

$$\prod_k (1 - \mathbf{f}(b_{i_0,k})) = 1.$$

So $\mathbf{f}(b_{i_0,k}) = 0$ for all k.

Thus $b_{i_0,k} \notin X_{\mathbf{f}}$ for all k.

This means that c is in $P^{X_{\mathbf{f}}}$.

b) If $\mathbf{f}(c) = 0$, then $\prod_i$ must be 1 and so for each i in equation (*) we must have that

$$\prod_k (1 - b_{i,k}) = 0.$$

So for each i, we must have a $k(i)$ such that

$$\mathbf{f}(b_{i,k(i)}) = 1$$

Hence c is not in $P_{\mathbf{f}}^X$, because all the clauses with head c are ignored.

4. We now summarise our situation from the points of view of atoms and clauses in $P^{X_{\mathbf{f}}}$

   a) If c is not a head of any clause in P, then c is not a head of any clause in $P^{X_{\mathbf{f}}}$.
   b) If $c \in P$, then $c \in P^{X_{\mathbf{f}}}$
   c) If c has some clauses in P for which it is head, then
      i. If $\mathbf{f}(c) = 1$, i.e. $c \in X_{\mathbf{f}}$ then because of some i_o, we have that c is in $P^{X_{\mathbf{f}}}$.
      ii. If $\mathbf{f}(c) = 0$ then c is not in $P^{X_{\mathbf{f}}}$.

5. It is now clear that $X_{\mathbf{f}}$ is an answer set for P, because we got that $P^{X_{\mathbf{f}}} = \{c | \mathbf{f}(c) = 1\}$.

Theorem 12.16. *Let P be a program containing only negated atoms in the bodies of its clauses. Then the answer sets X for P are exactly the characteristic sets $X_{\mathbf{f}}$ of functions $\mathbf{f}$ such that $\mathbf{f}$ is a $\{0, 1\}$ solution for $Eq_{\text{inverse}}(P)$.*

Proof. Follows from Theorems 12.15 and 12.12.

Remark 12.17. Theorem 12.16 cannot be stronger in view of item 3 of Example 12.14.

Theorem 12.18. *Let P be a program with clauses of the form*

$$\text{clause } (c, i) : \bigwedge_j a_{i,j} \wedge \bigwedge_k \neg b_{i,k} \to c.$$

Let the Clark completion of P be the set of all clauses

$$[\bigvee_i (\bigwedge_j a_{i,j} \wedge \bigwedge_k \neg b_{i,k})] \leftrightarrow c \qquad (*c)$$

if c is a head of some clauses in P and

$$\neg c \qquad (*\neg c)$$

if c appears in P, but is not the head of any clause in P.

Then the $\{0, 1\}$ solutions to $Eq_{\text{inverse}}(P)$ are exactly the $\{0, 1\}$ models of Comp(P).

Proof. Let **f** be a solution to the equations. Then we have

$$\mathbf{f}(c) = 1 - \prod_i (1 - \prod_j \mathbf{f}(a_{i,j}) \cdot \prod_k (1 - \mathbf{f}(b_{i,k}))) \tag{$\sharp c$}$$

when c is the head of some clauses in P and

$$\mathbf{f}(c) = 0 \tag{$\sharp \neg c$}$$

when c appears in P but is not the head of any clause in QP.

We write $\mathbf{f}(c) = 0$ or $\mathbf{f}(c) = 1$, when we regard **f** as a $\{0, 1\}$ solution to the equations $Eq_{\text{inverse}}(P)$ and we write $\mathbf{f} \vDash c$ or $\mathbf{f} \vDash \neg c$ when we regard **f** as a $\{0, 1\}$ model to $\text{Comp}(P)$.

We now prove the equivalence.

We have

$$\mathbf{f} \vDash c$$

iff, by definition

$$\mathbf{f}(c) = 1,$$

iff for some i

$$1 = \prod_j \mathbf{f}(a_{i,j}) \prod_k (1 - \mathbf{f}(b_{i,k}))$$

iff for some i and all j and all k

$$\mathbf{f}(a_{i,j}) = 1 \text{ and } \mathbf{f}(b_{i,k}) = 0$$

iff for some i and for all j and all k

$$\mathbf{f} \vDash a_{i,j} \text{ and } \mathbf{f} \vDash \neg b_{i,k}$$

iff for some i

$$\mathbf{f} \vDash \bigwedge_j a_{i,j} \wedge \bigwedge_k \neg b_{i,k}$$

iff

$$\mathbf{f} \vDash \bigvee_i (\bigwedge_j a_{i,j} \wedge \bigwedge_k \neg b_{i,k}).$$

So we have that **f** solves the equations iff for all c, ($\sharp c$) and ($\sharp \neg c$) holds for **f**. If ($\sharp c$) holds for **f**, i.e. c is the head of some clauses in P, then the above proof shows that

$$\mathbf{f} \vDash c \text{ iff } \mathbf{f}(c) = 1.$$

If c is not the head of any clause in P, then clearly by definition of $Eq_{\text{inverse}}(P)$ and of $\text{Comp}(P)$ we have $\mathbf{f}(c) = 0$ (i.e. ($\sharp \neg c$)) and ($*\neg c$) is in $\text{Comp}(P)$. So we do get for this case that

$$\mathbf{f} \vDash \neg c \text{ iff } \mathbf{f}(c) = 0.$$

We thus get

$$\mathbf{f} \text{ solves the equations iff } \mathbf{f} \vDash \text{Comp}(P)$$

12.3 Comparison with argumentation networks

An argumentation network has the form $\mathcal{A} = (S, R)$, where S is a non-empty set of points called arguments, and $R \subseteq S \times S$, called the attack relation. If $(x, y) \in R$, we say x attacks y.

The argumentation community is interested in subsets E of S called admissible extensions, satisfying the following:

1. If x is not attacked then $x \in E$
2. E is conflict free, i.e. if $x, y \in E$ then (x, y) is not in R, i.e. x does not attack y.
3. E protects itself, namely if z attacks x and $x \in E$ then for some $y \in E$, y attacks z.

The research in this area looks at different extensions for various networks and investigates their properties. See Chapter 2 and [271] for more information. For example, a stable extension E is an admissible set such that for all $x \notin E$ there is a $y \in E$ s.t. y attacks x.

The object level connection with logic programming we are interested in, was already introduced in Chapter 4, see also Chapter 7, is as follows: Given $\mathcal{A} = (S, R)$ regard any $x \in S$ as an atom in a logic program $P = P(\mathcal{A})$. For each $x \in S$, let the following clause C_x be in P.

$$C_x : \bigwedge_{k=1}^{m(x)} \neg y_k \rightarrow x$$

in case x does have attackers and where $y_1, \ldots, y_{m(x)}$ are all the attackers of x in (S, R).

If x has no attackers then the clause is $\neg \bot \rightarrow x$.

A logic program P can be identified as (arising from) argumentation network based on the atoms of P if the following holds

1. $\bot$ is also considered an atom.
2. All the clauses in P have the form $\bigwedge \neg y_k \rightarrow x$.
3. Each atom x has exactly one clause C_x in P for which it is head.

We can define (S, R) from P as follows:

4. S is the set of atoms of P excluding $\bot$.
5. yRx iff $y \neq \bot$ and $\neg y$ appears in the body of the unique clause of C_x of x.

In Chapter 10 we introduced the equational approach to argumentation and used, among others, $Eq_{inverse}$ and Eq_{max}.

For $x \in S$ we write the equations for $Eq_{inverse}$:

- $x = \prod_{k=1}^{m(x)} (1 - y_k)$ where $y_1, \ldots, y_{m(x)}$ are the attackers of x
- $x = 1$, if x is not attacked.

If we compare the above equations with the equations we offered for a logic program, we see they are the same. The equation for the unique C_x in P, namely for

$$\bigwedge_k \neg y_k \rightarrow x$$

is

$$x = 1 - (1 - \prod_k (1 - y_k))$$
$$= \prod_k (1 - y_k)$$

For unattacked x the clause is $\neg\bot \to x$ and we get $x = 1 - (1 - (1 - 0)) = 1$.
For Eq_{max} we write

- $x = 1 - \max(y_1, \ldots, y_k)$
- $x = 1$ if it has no attackers.

Recall that in Chapter 10 we proved the following:

Theorem 12.19. *1. If we use $Eq_{max}(\mathcal{A})$ then the solutions $\mathbf{f}$ correspond exactly to the Dung extensions of A. Namely*

- $\mathbf{f}(x) = 1$ *corresponds to $x = in$*
- $\mathbf{f}(x) = 0$ *corresponds to $x = out$*
- $0 < \mathbf{f}(x) < 1$ *corresponds to $x = undecided$.*
  The actual value in $[0, 1]$ reflects the degree of odd looping involving x.

2. If we use $Eq_{inverse}$, we give more sensitivity to loops. For example the more undecided elements y attack x, the closer to 0 (out) its value gets.

Remark 12.20. Note for example that an answer set for this program $P(\mathcal{A})$ yields a stable extension for the original argumentation system $\mathcal{A}$. In fact the stable extensions of $\mathcal{A}$ are the same as the answer sets of $P(\mathcal{A})$. Further note the interesting Corollary 12.21 below.

Corollary 12.21. *Let $\mathcal{A}$ be any argumentation system. The following holds: A function $\mathbf{f}$ is a $\{0, 1\}$ solution of $Eq_{max}(\mathcal{A})$ iff it is a $\{0, 1\}$ solution of $Eq_{inverse}(\mathcal{A})$.*

Proof. Follows from Theorem 12.16 used in conjunction with Theorem 12.19.

A general program P can be translated into an argumentation network using auxiliary additional variables. We need to overcome two obstacles:

1. Deal with clauses of the form $\bigwedge_j a_j \wedge \bigwedge_k \neg b_k \to x$, i.e. where un-negated atoms appear in the body.
2. Deal with the case where a head x have more than one clause for which it is head.

This is dealt with in Chapter 10. Let us give a quick explanation.
For clauses like

$$a \wedge \neg b \to c$$

we add an auxiliary atom a' and write

$$\neg a \to a'$$
$$\neg a' \wedge \neg b \to c.$$

If c is the head of two clauses such as

$$A \to c$$
$$B \to c$$

we add the new auxiliary atoms c_A and c_B and c' and write

$$A \to c_A$$
$$B \to c_B$$
$$\neg c_A \to c'$$
$$\neg c_B \to c'$$
$$\neg c' \to c$$

For details of the actual translation, see Chapter 10.

Remark 12.22.   1. It is instructive to see what happens with the program of item 3 of
Example 12.14.

The program is P

$$\neg p \to q$$
$$p \to p$$

The answer set for this program is $\{q\}$.

The equations are

$$q = 1 - p$$
$$p = p$$

The Clark completion is

$$p \leftrightarrow \neg q$$
$$p \leftrightarrow p$$

The equations have the two solutions $\{p = 1, q = 0\}$ and $\{p = 0, q = 1\}$ in
agreement with the two models of the completion, as predicted in Theorem 12.18.
Let us now eliminate the clause $p \to p$, as indicated in this section. We add a new
auxiliary atom p' and write a new program P'.

$$\neg p \to q$$
$$\neg p' \to p$$
$$\neg p \to p'$$

The equations are now

$$q = 1 - p$$
$$p = 1 - p'$$
$$p' = 1 - p$$

The solutions remain the same on $\{p, q\}$. The solutions are

$$\{q = 0, p = 1, p' = 0\} \text{ and } \{q = 1, p = 0, p' = 1\}.$$

The answer sets for P' must correspond now to the $\{0, 1\}$ solutions to the equations
of $Eq_{inverse}(P')$, by Theorem 12.16.

They are

$$\{q, p'\} \text{ and } \{p\}$$

Indeed they correspond to the P solutions (if we ignore p'). Thus the additional
non-answer set $\{0, 1\}$ solutions for P became answer set retracts in P'.

2. It is worthwhile to have a general formulation of this phenomenon.

Let P be a logic program. Then there exists a logic program $P' \supseteq P$ such that the
following holds:

a) X is an answer set for P' iff X is a model of Comp(P').
b) Comp(P) is logically equivalent to Comp$(P') \restriction P$
c) If X is an answer set for P then X is a retract of a unique answer set X' of P'.
d) The move from P to P' is functorial, it translates the language of P into the
language of P', and does not make use of the logical content of P. So we have

$$(P + P_1)' = P' + P'_1$$

e) In fact let Q be the atoms to be used in P or any of its possible extensions. Let us add the atom $\top$, and for any program P and any atom c in the language of P, let us add the clause

$$\neg\top \to c.$$

Let us have the commitment that $\top$ always succeeds and it numerical value is always 1.

This gives us a new program P^*, in the augmented language with $\top$.

The program P^* has the same logical contents as P, but it does have the additional property that any literal in P^* is the head of a clause. This property will facilitate the translation from P^* into P'^*. Under the above notation and assumptions on P and P^*, we can translate the clauses of P^* as follows: Let Q' be the set of the new atoms $\{x'|x \in Q\}$, not that $\top' = \bot$. Let

$$C = \bigwedge x_i \wedge \bigwedge \neg y_j \to z$$

be a clause in the language of Q.

Then the following set of clauses is the translation C' of C

$$\bigwedge \neg x'_i \wedge \bigwedge \neg y_j \to z$$
$$\neg x_i \to x'_i$$

Any program P^* of the language of Q is translated into a program P'^* of the language $Q+Q'$ by translating its clauses as above. Note that in the translation no new clauses are added which have heads in Q, the language of P^*.

Theorem 12.23. *Let P^*, P'^* be two programs related as in item (2e) of Remark 12.22. then the following holds:*

1. *$Eq_{\text{inverse}}(P'^*)$ reduced to the language of P^* are the same as $Eq_{\text{inverse}}(P^*)$.*
2. *Let $\mathbf{f}$ be any solution of $Eq_{\text{inverse}}(P^*)$. Then $\mathbf{f}$ can be extended uniquely to a solution $\mathbf{f}'$ of $Eq_{\text{inverse}}(P'^*)$.*
3. *Let $\mathbf{f}$ be any $\{0, 1\}$ solution to $Eq_{\text{inverse}}(P^*)$. Let $\mathbf{f}'$ be its unique extension as in (2) above. Then $\mathbf{f}'$ is a $\{0, 1\}$ function and $X_{\mathbf{f}'} = \{y \text{ in } P'^*|\mathbf{f}(y) = 1\}$ is an answer set for P'^*.*

Proof. 1. Let c be a literal in P^* and assume it is the head of the clauses

$$\text{clause } (c, i) : \bigwedge_j a_{i,j} \wedge \bigwedge_k \neg b_{i,k} \to c.$$

The equations for c are in $Eq_{\text{inverse}}(P^*)$ are

$$c = 1 - \prod_i (1 - \prod_j a_{i,j} \cdot \prod_k (1 - b_{i,k} \to c))$$

Clause (c, i) is translated in P'^* into clauses'(c, i):

- $\bigwedge_j \neg a'_{i,j} \wedge \bigwedge_k \neg b_{i,k} \to c$
- $\neg a_{i,j} \to a'i, j$

The equations for clauses'(c, i) are in $Eq_{\text{inverse}}(P'^*)$ are:

- $c = 1 - \prod_i(1 - \prod_j(1 - a'_{i,j}) \cdot \prod_k(1 - b_{i,k}))$

- $a'_{i,j} = 1 - a_{i,j}$
- $c' = 1 - c$

It is clear that since $a'_{i,j} = 1 - a_{i,j}$ that we get the same equations.
2. Follows from the proof of (1)
3. Let $\mathbf{f}$ be a $\{0, 1\}$ solution of $Eq_{\text{inverse}}(P^*)$. Clearly its extension $\mathbf{f}'$ is a $\{0, 1\}$ solution of the equations of $Eq_{\text{inverse}}(P'^*)$.
We now show that

$$X_{\mathbf{f}'} = \{y \text{ im } Q \cup Q' | \mathbf{f}'(y) = 1\}$$

is an answer set for P'^*.
We show that

$$P'^{*X_{\mathbf{f}'}} = X_{\mathbf{f}'} \tag{*}$$

a) Let $c \in X_{\mathbf{f}'}$ and assume c is in the language of P^*. The only way that c is kept out of $P'^{*X_{\mathbf{f}'}}$ is that all clauses with head c are deleted from being candidates for inclusion in $P'^{*X_{\mathbf{f}'}}$. This means that for every i, there is a $j(i)$ or a $k(i)$ such that either $a'_{i,j(i)} \in X_{\mathbf{f}'}$ or $b_{i,k(i)} \in X_{\mathbf{f}'}$.
This implies that for each i, either $\mathbf{f}(a_{i,j(i)}) = 0$ or $\mathbf{f}(b_{i,k(i)}) = 1$.
From the equation for c we get that this implies $\mathbf{f}(c) = 0$. But since we are given that $\mathbf{f}'(c) = 1$, we must have that for some i_0, all $a'_{i_0,j}$ for all j and all $b_{i_0,k}$ for all k are not in $X_{\mathbf{f}'}$. In this case we have $c \in P'^{*X_{\mathbf{f}'}}$.
b) Assume $c' \in X_{\mathbf{f}'}$ and c' is in the language of P'^*. This means that $c \notin X_{\mathbf{f}'}$, therefore $c' \in P'^{*X_{\mathbf{f}'}}$, because the only clause with head c' is the clause $\neg c \to c'$ and this clause is not deleted if $c \notin X_{\mathbf{f}'}$, and its head c' is put in $P'^{*X_{\mathbf{f}'}}$.
c) Assume, for the other direction of the equality $(*)$, that $c \in P'^{*X_{\mathbf{f}'}}$. If c is in the language of P^* then the only way c can get into $P'^{*X_{\mathbf{f}'}}$ is that for some clause (c, i)

$$\bigwedge_j \neg a'_{i,j} \wedge \bigwedge_k \neg b_{ik} \to c$$

we have that $a'_{i,j} \notin X_{\mathbf{f}'}$, for all j and $b_{i,k} \notin X_{\mathbf{f}'}$ for all k. This means that $\mathbf{f}(a_{i,i}) = 1$ for all j an $\mathbf{f}(b_{i,k}) = 0$ for all k.
But in this case we get from the equations that $\mathbf{f}$ satisfies that $\mathbf{f}(c) = 1$, i.e. $c \in X_{\mathbf{f}'}$.
d) Assume $c' \in P'^{*X_{\mathbf{f}}}$ and that c' is in the language of P'^*. The only clause which can let c' into $P'^{*X_{\mathbf{f}}}$ is $\neg c \to c'$. This means that $c \notin X_{\mathbf{f}'}$, i.e. $\mathbf{f}(c) = 0$, so $\mathbf{f}'(c') = 1$ ad $c \in X_{\mathbf{f}'}$.

From $(*)$ it is clear that $X_{\mathbf{f}'}$ is an answer set for P'^*.

12.4 Conclusion

We introduced the equational approach to logic programs and showed its soundness. This is part of a general methodological drive of applying the equational approach to argumentation, modal logic, default logic, logic programming, liar paradoxes and more. In the case of logic programming, we can regard the equational approach as offering new semantics, each solution to the equation to be considered a model. Note that the problem of the inconsistency of Clark completion for programs with negation as failure in the body of clauses does not arise in the equational semantics! We also mention that there is need to research the correspondence between properties of the solutions of the equations $Eq(P)$ and properties/answer sets/models of the program P.

13

Abstract Argumentation and the Peirce–Quine Dagger

13.1 Classical logic as a network

The primary aim of this chapter is to integrate logic with Dung's abstract argumentation networks. We get as a by-product of this effort various generalisations of Dung's network as well as applications to some problems in the area of argumentation.

So the chapter is about argumentation but the motivation behind the particular steps taken is integration with logic. The local researcher, immersed in practical argumentation theory, may perceive some of our moves as possibly unnecessary pure mathematics. See, however, the discussion in Section 2.5.

In the past 40 years logic has undergone a serious evolutionary development. The meteoric rise of the applied areas of computer science and artificial intelligence put pressure on traditional logic to evolve. There was the urgent need to develop new logics in order to provide better models of human behaviour and actions. Such models are used to help design products which aid/replace the human in his daily activity. As a result, a rich variety of new logics have been developed and there was the need for a new unifying methodology for the chaotic landscape of the new logics.

Thus the question of

"what is a logical system"

is repeatedly being asked and answered, by myself as well as other colleagues.

An important step in this search is the problem of integration of general network reasoning (Bayesian nets, Neural nets, Argumentation nets, Inheritance nets , Ecological nets, etc., etc.) with discrete logical systems of reasoning (classical logics, defeasible logics, logic programming, temporal and modal logics, etc., etc.).

We are going to present classical logic in a certain way, to make it most compatible with argumentation networks. This is done in Section 1.1, where we formulate classical logic with a special unary connective which we call the Peirce-Quine-Dung dagger.

Section 1.2 gives semantic tableaux for our logic and in Section 1.3 we axiomatise the consequence relation of this logic.Later in Section 2 we shall present certain restricted version of Dung's networks which correspond exactly to our logic. We then map argumentation concepts to logical concepts. For example in section 2.3 we show that semantic tableaux correspond to admissible sets in argumentation.

13.1.1 Peirce's arrow, Quine's dagger and the Sheffer stroke

As a first step towards showing the equivalence of classical logic with abstract argumentation frames we present the above connectives for classical logic.

Classical propositional logic is traditionally formulated with a set of atomic propositions and some or all of the connectives below:

- $1 = \text{truth} = \top$
- $0 = \text{FALSE} = \bot$
- $x \wedge y = \min(x, y)$
- $x \vee y = \max(x, y)$
- $\neg x = 1 - x$
- $x \rightarrow y = \max(1 - x, y)$.

Other connectives were put forward for the classical propositional calculus, among them the Sheffer stroke $x \uparrow y$ introduced in 1913 [307], and Peirce's arrow $x \downarrow y$, discovered in 1880 and published thirty years later [277]. Peirce's arrow was also discovered by Quine under the name Quine's dagger, see report in [194]. These connectives were introduced for logical methodological reasons in the foundations of classical logic. They turned up to be also very useful as gates in circuit design, as the NAND and NOR gates. So, in fact, these connectives are as fundamental as the usual ones.

The table for these connectives is

$$x \uparrow y = \max(1 - x, 1 - y)$$
$$x \downarrow y = 1 - \max(x, y).$$

The following is the truth table for these connectives:

Table 13.1.

A	B	$A \uparrow B$	$A \downarrow B$
0	1	1	0
0	0	1	1
1	1	0	0
1	0	1	0

The Sheffer stroke is equivalent to

$$A \uparrow B \equiv \neg A \vee \neg B$$

and the Peirce-Quine dagger is equivalent to

$$A \downarrow B \equiv \neg A \wedge \neg B.$$

The ordinary Boolean connectives are definable from these unary connectives as follows.

- $\neg A \equiv A \uparrow A \equiv A \downarrow A$
- $A \wedge B \equiv \neg (A \uparrow B) \equiv (\neg A) \downarrow (\neg B)$
- $A \vee B \equiv (\neg A) \uparrow (\neg B) \equiv \neg (A \downarrow B)$.

There exist axiomatisations of classical propositional logic in terms of the Sheffer stroke as the only connective, see [292, 307], and I assume one can do the same for the case of the Peirce-Quine dagger. The two unary connectives are interdefinable:

- $A \downarrow B \equiv \neg((\neg A) \uparrow (\neg B))$
- $A \uparrow B \equiv \neg((\neg A) \downarrow (\neg B))$

We now formulate classical logic in a certain way, ready to be turned into a Dung network.

Let Δ be a finite multiset of wffs, $\Delta = \{A_1, \ldots, A_n\}$.[1] Define the connective $\Downarrow \Delta$ as follows:

- $\Downarrow \Delta = 1$ iff $\bigwedge_{i=1}^n A_i = 0$

Thus

- $\Downarrow \Delta = \bigwedge_{i=1}^n \neg A_i = \bigwedge_{A \in \Delta} \neg A$

Thus the truth function for $\Downarrow \{x_1, \ldots, x_n\}$ is[2]

- $\Downarrow (x_1, \ldots, x_n) = 1 - \max(x_i)$.

We have that the traditional connectives $\neg$ and $\wedge$ are definable

- $\neg A \equiv \Downarrow \{A\}$
- $\bigwedge_{i=1}^n A_i \equiv \Downarrow \{\Downarrow \{A_i\} | i = 1, \ldots, n\}$
- $\bigvee_{i=1}^n A_i \equiv \Downarrow\Downarrow \{A_1, \ldots, A_n\}$.

In fact it might be easier to preset an axiom system for $\Downarrow$ than for $\downarrow$.

Let us call this connective the Peirce-Quine-Dung Dagger. This name is appropriate because as we shall see in the next section, the Dung attack relation (done by a dagger) corresponds to this connective!

We shall refer to $\Downarrow$ by the short name "Dagger".

This connective is definable using $\downarrow$ because $\wedge$ and $\neg$ are definable from $\downarrow$, as we have seen. It generalises $\downarrow$ because $\Downarrow \{A, B\} = A \downarrow B$.

Definition 13.1 (Dwffs: wffs with the Dagger connective). *A Dwff is defined by induction:*

1. *Any atomic q is a Dwff of level 0.*
2. *If $\Delta = \{A_1, \ldots, A_n\}$ is a multiset of Dwffs then $A = \Downarrow \Delta$ is also a Dwff.*
   A_i are said to be immediate subformulas of A. The level of A is $1 + \max\{level of A_i\}$.

Definition 13.2 (Trees). *A system (S, R, t) is a finite tree iff the following holds:*

1. *$R \subseteq S \times S$. t is an element of S, and S is finite.*
2. *For any $x \in S, x \neq t$, there exists a unique y such that yRx. y is called the predecessor of x.*
3. *For every x there exists a unique sequence $(t, x_1, \ldots, x_n = x)$ such that $tRx_1 \wedge x_1Rx_2 \wedge \ldots \wedge x_{n-1}Rx_n$.*

---

[1] We use $\{\ldots\}$ to denote multisets.

[2] Note that this truth function is meaningful for three values as well, i.e. if x_i range over $\{0, 1/2, 1\}$.

For x = t we have (t) as the sequence.

Definition 13.3 (Construction tree for a D-formula). *Let A be a Dwff. Then a tuple (S, R, t, α) is a construction tree for A iff (S, R, t) is a tree and α is a function associating a Dwff $\alpha(x)$ with each $x \in S$ such that the following holds.*

1. $\alpha(x)$ is atomic if x is an endpoint (i.e. $\neg \exists y(xRy)$).
2. For all $x \in S$ which are predecessors, we have

$$\alpha(x) = \Downarrow \{\alpha(y) | xRy\}.$$

3. $\alpha(t) = A$.

Definition 13.4 (Acyclic decorated ordering). *A system (S, R, α) is said to be an acyclic decorated ordering iff the following holds:*

1. (S, R) is a finitary acyclic ordering. This means that $R \subseteq S \times S$ is a binary relation such that for all x, $\{y | xRy\}$ is a finite set and such that for no $x_1, \ldots, x_m$ do we have $x_1 R x_2 \wedge x_2 R x_3 \wedge, \ldots, \wedge x_m R x_1$.
2. (S, R) may have a root t. This means that for all $x \neq t$, there exists a (not necessarily unique) sequence $x_1, \ldots, x_m = x$ such that $tRx_1 \wedge x_1 R x_2 \wedge, \ldots, x_{m-1} R x$.
3. Note that we have not imposed that S is finite, only finitary!
4. α is a function associating with each $x \in S$, a Dwff $\alpha(x)$ such that
    a) $\alpha(x)$ is atomic if x is an endpoint (i.e. $\neg \exists y(xRy)$).
    b) If x is not an endpoint then $\alpha(x) = \Downarrow \{\alpha(y) | xRy\}$.
5. Note that for the definition of (2) to be alright we need that (S, R, t) is acyclic and finitary. In fact the relation R needs to be well founded for α to exist!

Example 13.5 (Canonical ordering of Dwffs).

1. Let $Q = \{q_1, \ldots, q_k\}$ be atoms. Let Θ_Q be the set of all Dwffs built up from $\{q_1, \ldots, q_k\}$. Define R on Θ_Q by
   * *ARB iff B is an immediate subformula of A.*[3]
   Then (Θ_Q, R) is acyclic and for each non-atomic A we have

$$A = \Downarrow \{B | ARB\}.$$

---

[3] We have a notational compatibility problem. In argumentation one writes xRy to say that x attacks y. In logic we write ARB to say that B is an immediate subformula of A. The problem is that later we will have that the immediate subformula B of A actually attacks A.

To avoid confusion, let $\check{R}$ be the converse relation of R So we have

$$xRy \text{ iff (definition) } y\check{R}x$$

or using the ordered pair notation

$$(x, y) \in R \text{ iff } (y, x) \in \check{R}$$

In the sequel we shall use either R or $\check{R}$ depending on context. The reader should always remember that when we write $y\check{R}x$ or xRy, then y attacks x.

In our figures we display xRy by $y \to x$.

2. For Q = set of all atoms, let $\Theta = \Theta_Q$. Then (Θ, R) is the canonical ordering of all Dwffs.

Definition 13.6 (Models). *Let (S, R, α) be a decorated acyclic ordering.*

1. *A function h assigning a value $h(q) \in \{0, 1\}$ to any atom q of the language is called a model. In the case of three valued logic, h assigns values in $\{0, \frac{1}{2}, 1\}$.*
2. *Given a model h we can use it to give a truth value in $\{0, 1\}$ to every formula and node in (S, R, α) as follows (call this function $\mathbf{f}(x)$ for $x \in S$).*
   a) For x an endpoint let $\mathbf{f}(x) = h(\alpha(x))$.
   b) for x not an endpoint let

$$\mathbf{f}(x) = 1 - \max_{(xRy)}(\mathbf{f}(y))$$

3. *Note that this definition follows exactly the truth table of $\Downarrow$. We have*

$$\alpha(x) = \Downarrow \{\alpha(y) | xRy\}.$$

So assuming $\mathbf{f}(y)$ is the truth value of $\alpha(y)$ under h (in the model h), then $\mathbf{f}(x)$ would be the truth value of $\alpha(x)$ in the model h.

13.1.2 Semantic tableaux for $\Downarrow$

We now give a tableaux formulation for our connective $\Downarrow$. We assume two valued logic. The tableaux would be slightly different for three valued logic. We first recall the usual tableaux formulation of classical logic with say, $\neg$ and $\wedge$.

The basic computational units are tables of the form

$$\tau = [A_1, \ldots, A_n \| B_1, \ldots, B_k]$$

The A_i are on the left and the B_j are on the right. We are seeking an assignment h to the atoms such that $h(A_i) = 1$ and $h(B_j) = 0$. So what is on the left is intended to be true and what is on the right is intended to be false.

In the middle of the computation we have a set $\mathbb{T}$ of such tableaux. We pick a tableau $\tau \in \mathbb{T}$ and replace it by new tableaux one or more, say $\tau_1, \ldots, \tau_m$ by performing certain tableaux operations. We get a new set of tableaux $\mathbb{T}'$. The new tableaux $\tau_1, \ldots, \tau_m$ are of less complexity than τ.

Depending on what we need the tableaux system for, we may or may not remember the connection between τ and $\tau_1, \ldots, \tau_m$.

The above description is not mathematically precise, it is only to remind the reader. Our definition of tableaux for $\Downarrow$ will be precise.

The following are the tableaux rules for $\neg$ and $\wedge$ of classical logic, just to remind the reader.

$(\neg l)$ Rule for $\neg$ on the left

Replace

$$\tau = [A_i, \neg A \| B_1, \ldots, B_k]$$

by

$$\tau' = [A_i \| B_1, \ldots, B_k, A]$$

($\neg r$) Rule for $\neg$ on the right

Replace

$$\tau = [A_1, \ldots, A_n \| B_1, \ldots, B_k, \neg B]$$

] by

$$\tau' = [A_1, \ldots, A_n, B \| B_1, \ldots, B_k]$$

($\wedge l$) Rule for $\wedge$ on the left

Replace

$$\tau = [A_1, \ldots, A_n, C \wedge D \| B_1, \ldots, B_k]$$

by

$$\tau_1 = [A_1, \ldots, A_n, C, D \| B_1, \ldots, B_k]$$

($\wedge r$) Rule for $\wedge$ on the right

Replace

$$\tau = [A_1, \ldots, A_n \| B_1, \ldots, B_k, C \wedge D]$$

by the two tableaux

$$\tau_1 = [A_1, \ldots, A_k \| B_1, \ldots, B_k, C]$$

and

$$\tau_2 = [A_1, \ldots, A_n \| B_1, \ldots, B_k, D]$$

Closure of a tableau

A tableau

$$[A_1, \ldots, A_n \| B_1, \ldots, B_k]$$

is closed cf. for some i, j

$$A_i = B_j.$$

Definition 13.7 (Tableaux for $\Downarrow$).

1. A unit tableau for $\Downarrow$ has the form

$$\tau = [A_1, \ldots, A_n \| B_1, \ldots, B_k]$$

where A_i, B_j are Dwffs.
The tableau is closed if for some $i, j, A_i = B_j$.

*2. Reduction rules for tableaux. Given a tableau τ, we perform possible applicable rules on τ to obtain new tableaux from it. At each step we apply exactly one rule. If the result of the application of the rule are tableaux $\tau_1, \ldots, \tau_n$, then we write $\tau \rho \tau_1, \ldots, \tau \rho \tau_n$. so We now just define the rules. The actual construction process is defined later. ($\Downarrow l$) **Rule for $\Downarrow$ on the left***

Replace

$$\tau = [A_1, \ldots, A_n, \Downarrow \{C_1, \ldots, C_m\} \| B_1, \ldots, B_k]$$

by

$$\tau' = [A_1, \ldots, A_n \| B_1, \ldots, B_k, C_1, \ldots, C_m]$$

write $\tau \rho \tau'$.

($\Downarrow r$) **Rule for $\Downarrow$ on the right**
Replace

$$\tau = [A_1, \ldots, A_n \| B_1, \ldots, B_k \Downarrow \{C_1, \ldots, C_m\}]$$

by $\tau_i, i = 1, \ldots, m$ where

$$\tau_i = [A_1, \ldots, A_n, C_i \| B_1, \ldots, B_k]$$

Write $\tau \rho \tau_i$ for $i = 1, \ldots, m$.
Note that if $\tau \rho \tau'$, the complexity of τ' is less than that of τ!
3. *Given an initial family of tableaux*

$$\mathbb{T}_0 = \{\tau, \tau', \tau'', \ldots\}$$

With relation $\rho_0 = \varnothing$, we build by induction a family of tableaux $(\mathbb{T}_n, \rho_n)$, with ρ_n a tree relation on $\mathbb{T}_n$, as follows.

Step $n + 1$
Assume that $(\mathbb{T}_n, \rho_n)$ has been defined. Assume that ρ_n is a tree relation such that the following holds:
(*) If $\tau \in \mathbb{T}$ is not an endpoint, then there exists a rule ($\Downarrow r$) (or ($\Downarrow l$)) and a formula $\Downarrow \{C_1, \ldots, C_m\}$ which is at the right of τ (resp. at the left of τ) such that the ρ_n immediate successors of τ are exactly the result of the application of the rule on the above formula.
We now define $(\mathbb{T}_{n+1}, \rho_{n+1})$ as follows:
3.1. *Choose any endpoint tableau τ in $\mathbb{T}_n$. If all the formulas in τ are atomic then look for another endpoint tableau.*
3.2. *Assume τ has a Dwff $\Downarrow \{C_1, \ldots, C_m\}$ on the left. Let τ' be the result of applying the ($\Downarrow l$) rule to this formula. Let $\mathbb{T}_{n+1} = \mathbb{T}_n \cup \{\tau'\}$ and let $\rho_{n+1} = \rho_n \cup \{(\tau, \tau')\}$.*
3.3. *Assume τ has a Dwff $\Downarrow \{C_1, \ldots, C_m\}$ on the right. Let $\tau'_1, \ldots, \tau'_m$ be the resulting tableaux from applying the ($\Downarrow r$) rule to τ.*
Let $\mathbb{T}_{n+1} = \mathbb{T}_n \cup \{\tau'_i\}$ and let ρ_{n+1} be $\rho_n \cup \{(\tau, \tau'_i) | i = 1, \ldots, m\}$.
3.4. *Items (3.1)–(3.3) above are the steps which define $(\mathbb{T}_{n+1}, \rho_{n+1})$ from $(\mathbb{T}_n, \rho_n)$ through the use of an arbitrary choice of an endpoint in $(\mathbb{T}_n, \rho_n)$. Clearly it satisfies (*).*
3.5. *If the original $\mathbb{T}_0$ was infinite, then we want a fair sequence of choices, which will not neglect any endpoint tableaux.*
4. *Let $(\mathbb{T}_\infty, \rho_\infty) = \bigcup_n (\mathbb{T}_n, \rho_n)$. We get a tree. Let Π be a maximal path in the tree. Define two sets Π_r and Π_l as follows:*

$$\Pi_l = \{A | Dwff\ A \text{ is the left side of some } \tau \text{ in } \Pi\}$$
$$\Pi_r = \{A | Dwff\ A \text{ is in the right hand side of some } \tau \text{ in } \Pi\}$$

Π is said to be admissible if $\Pi_r \cap \Pi_l = \varnothing$.
5. *Let Π be admissible. Then the following holds:*
($\natural$) If $\Downarrow \{C_1, \ldots, C_m\} \in \Pi_l$ then $C_1, \ldots, C_m \in \Pi_r$.
Furthermore, if C_i is not atomic, i.e. $C_i = \Downarrow \{D_1^i, \ldots, D_{k(i)}^i\}$ then for some j, $D_j^i \in \Pi_l$.

13.1.3 Axiom system for ⇓

The logic of Dagger is not difficult to axiomatise. When we write axioms for a consequence relation of the form $\Delta \vdash A$, the elements of Δ are together as a conjunction. We also have negation because $⇓ A$ is $\neg A$. So basically we have what we need.

Definition 13.8 (Axioms for ⇓). *Consider the following rules as a consequence relation for the language with ⇓. Δ is a finite set of Dwffs.*

1. **Reflexivity**
   $\Delta \vdash A$ *if* $A \in \Delta$
2. **Monotonicity**
   $$\frac{\Delta \vdash A}{\Delta, B \vdash A}$$
3. **Cut**
   $$\frac{\Delta, A \vdash B ; \Delta, ⇓ A \vdash B}{\Delta \vdash B}$$
4. $⇓ \Delta \vdash ⇓ A$, *if* $A \in \Delta$
5. $⇓ A_i, i = 1, \ldots, n \vdash ⇓ \{A_i \mid i = 1, \ldots, n\}$
6. $\Delta, A \vdash B$ *iff* $\Delta, ⇓ B \vdash ⇓ A$
7. $\Delta, ⇓⇓ A \vdash A$.

Definition 13.9. *We can define $\Delta \vdash A$ for Δ infinite as well.*
Let $\Delta \vdash A$ iff for some finite $\Delta_0 \subseteq \Delta$ we have $\Delta_0 \vdash A$.

Definition 13.10. *A set Θ of wffs is consistent if for no Dwff A and finite $\Delta \subseteq \Theta$ do we have $\Delta \vdash A$ and $\Delta \vdash ⇓ A$.*
A theory is complete iff for any A, either $A \in \Delta$ or $⇓ A \in \Delta$.

Lemma 13.11. *Every finite consistent theory can be extended to a complete theory.*

Proof. The proof follows the usual lines known for classical logic. We first show that if Δ_0 is consistent and B a wff then either $\Delta_0 \cup \{B\}$ or $\Delta_0 \cup \{⇓ B\}$ are consistent. For otherwise for some A_1 and A_2 and finite $\Delta_0^1, \Delta_0^2 \subseteq \Delta_0$, we have

1. $\Delta_0^1, B \vdash A_1$
2. $\Delta_0^1, B \vdash ⇓ A_1$
3. $\Delta_0^2, ⇓ B \vdash A_2$
4. $\Delta_0^2, ⇓ B \vdash ⇓ A_2$.
   From (1) and axiom (2) and axiom (6) we get for $\Delta_0' = \Delta_0^1 \cup \Delta_0^2$

   $$\Delta_0', ⇓ A_1 \vdash ⇓ B$$
   $$\Delta_0', A_1 \vdash ⇓ B$$

   And by cut rule we get
5. $\Delta_0' \vdash ⇓ B$
   From (3) and (4) and axiom (6), we get

   $$\Delta_0', ⇓ A_2 \vdash B$$
   $$\Delta_0', A_2 \vdash B$$

   And by cut we get

6. $\Delta_0' \vdash B$.

Thus Δ_0 is not consistent, a contradiction.

We can now follow the usual construction and extend any finite theory Δ_0 to a complete theory. Enumerate all wffs of the language $A_1, A_2, A_3, \ldots$.

Let

$$\Delta_{n+1} = \begin{cases} \Delta_n \cup \{A_n\} \text{ if consistent} \\ \Delta_n \cup \{\Downarrow A_n\}, \text{ otherwise} \end{cases}$$

By what we have just proved, Δ_{n+1} is consistent.

Let $\Delta_\infty = \bigcup_n \Delta_n$.

Δ_∞ is consistent for otherwise for some finite $\Delta' \subseteq \Delta_\infty$ and some X we have $\Delta' \vdash X$ and $\Delta \vdash \Downarrow X$.

Since Δ' is finite, then for some $n, \Delta' \subseteq \Delta_n$ contradicting the consistency of Δ_n.

Theorem 13.12 (Completeness theorem). *The system of Definition 13.8 is sound and complete for the classical two valued semantics for* $\Downarrow$.

Proof. 1. Soundness can be seen from interpreting

$$\Downarrow \Delta = \bigwedge_{A \in \Delta} \neg A$$

2. To show completeness, let Δ_0 be a finite consistent theory. Extend Δ_0 to a complete theory Δ. Define a model h by

(*) $h(q) = 1$ iff $q \in \Delta$, q atomic.

We now show by induction that for any $\Downarrow \{A_1, \ldots, A_n\}$ we have

(*) $h(\Downarrow \{A_1, \ldots, A_n\}) = 1$ iff $\Downarrow \{A_1, \ldots, A_n\})$ is in Δ.

Case 1. Assume $h(\Downarrow \{A_1, \ldots, A_n\}) = 1$. We show $\Downarrow \{A_1, \ldots, A_n\} \in \Delta$.
From the assumption we get that $h(A_i) = 0, i = 1, \ldots, n$. Hence by the induction hypothesis, $A_i \notin \Delta$ and hence $\Downarrow A_i \in \Delta, i = 1, \ldots, n$.
Therefore for some $\Delta_m, \Downarrow A_i \in \Delta_n, i = 1, \ldots, n$.
From axiom 5 we have $\Delta_n \vdash \Downarrow \{A_1, \ldots, A_n\}$. We claim $\Downarrow \{A_1, \ldots, A_n\} \in \Delta$. Otherwise $\Downarrow\Downarrow \{A_1, \ldots, A_n\} \in \Delta$ and so for some large $m' \Downarrow \{A_1, \ldots, A_n\} \in \Delta_{m'}$ and $\Downarrow\Downarrow \{A_1, \ldots, A_n\} \in \Delta_{m'}$, contradicting the consistency of $\Delta_{m'}$.

Case 2. Assume $\Downarrow \{A_1, \ldots, A_n\} \in \Delta$. Show $h(\Downarrow (A_1, \ldots, A_n)) = 1$. Otherwise, by the induction hypothesis for some $i, A_i \in \Delta$. Thus for some $m, A_i \in \Delta_m$. For some large enough $k \geq m$ we also have $\Downarrow \{A_1, \ldots, A_n\} \in \Delta_k$ and hence $\Downarrow A_i \in \Delta_k$ by axiom 4, contradicting the consistency of Δ_k.
Thus h is our required model because $h(\Delta_0) = 1$, since $\Delta_0 \subseteq \Delta$, and (*) holds.

This concludes the completeness theorem.

13.2 Argumentation frames

This section introduces the basic notions of argumentation frames and then proceeds to show equivalence with classical propositional logic.

We must make our methodology absolutely clear. We begin by comparing a very restricted form of argumentation frames, called nearly acyclic orderings (Definition

13.13), with the Dagger logic of Section 1. Such frames can be regarded as logic models or as argumentation models at the same time. Using that we will map some correspondences between logical concepts and argumentation concepts, in Sections 2.2 and 2.3.

Once we have some idea on what may correspond to what, we start the technical machinery in earnest in Section 3 and generalise in Sections 4 and 5.

13.2.1 Argumentation preliminaries

An *argumentation framework* [114] consists of a set of arguments and an attack relation on these arguments. We only consider finite argumentation frameworks.

We advise the reader to recall and read Sections 2.2 and 2.3 from Chapter 2.

13.2.2 Discussion of challenges

It is clear from the above that although both classical logic and argumentation frames use orderings as a basis, (classical logic uses the sub-formula relation on the set of formulas and argumentation uses the attack relation on the set of arguments), they deal with them in completely different ways. Let us see if we can get these two areas closer together. Consider the typical situation of Figure 13.1.

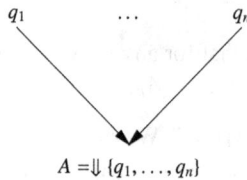

$$A = \Downarrow \{q_1, \ldots, q_n\}$$

Fig. 13.1.

Imagine that this is an ordering as part of the canonical ordering of Dwffs as in Example 13.5. We have $ARq_1 \wedge \ldots \wedge ARq_n$.

We can also regard this figure as an argument frame, by taking the attack relation to be the converse relation $\check{R}$ of R.

According to classical logic for $\Downarrow$ we have

- $\mathbf{f}(A) = 1$ iff $\bigwedge_{i=1}^{n} \mathbf{f}(q_i) = 0$

According to argumentation frames we have for any stable extension $\mathcal{A}rgs$:

- A is in iff $\bigwedge_{i=1}^{n} (q_i$ is out).

  or

- $\lambda_{\mathcal{A}rgs}(A) = 1$ iff $\bigwedge_{i=1}^{n}(\lambda_{\mathcal{A}rgs}(q_i) = 0)$.

The formula is the same. If we identify $\mathbf{f}(x) = 1$ as "x is in", then we can write with some abuse of notation that

- $\mathbf{f}(A) = 1 - \max(\mathbf{f}(q_i))$
- $\lambda_{\mathcal{A}rgs}(A) = 1 - \max(\lambda_{\mathcal{A}rgs}(q_i))$

Obviously we recognised that the recipe for being in any extension, given by Dung, corresponds to the truth table for $\Downarrow$.

There are, however, further differences. We now list them and then address and overcome them.

(D1) *From argumentation into logic.*
      In classical logic the ordering should be acyclic. In argumentation we can have any binary relation, e.g. as in Figure 2.1. We ask: how can we regard an arbitrary argumentation network as logic?

(D2) *From logic into argumentation.*
      In logic the endpoints (i.e. $\neg\exists y(xRy)$) x can be assigned any value. In argumentation the endpoints can be value 1 (they are in).
      If we consider Figure 13.1, the endpoints $q_1,\ldots,q_n$ can be assigned by h (the assignment) any value, for example

$$h(q_1) = h(q_2) = \ldots = h(q_n) = 0.$$

In this case A will get value 1.
Argumentation theory, when viewing the ordering of Figure 13.1, will allow only value 1.

$$h(q_1) = h(q_2) = \ldots = h(q_n) = 1.$$

(D3) In classical logic the values one gets are always 0 or 1. In argumentation we can get values 0,1 and undecided (in case of loops).

Let us address these points one by one.

(S1) Solution to D1
      Consider the network in Figure 2.1, how can we regard it as Logic? We now explain:
      Almost in every elementary textbook in logic, there are the following exercises:
      Exercise 1. The statement a says I am false (liar paradox). Can you give it a truth value?
      The answer is no, we cannot. If we give it $\top$ then it should be $\bot$ and if we give it $\bot$ then it should be $\top$.
      Exercise 2. You have two statements, a and b. a says b is false and b says a is false, can you give them a consistent truth value assignments?
      The answer is that there are two possibilities;

$$a = \top \text{ and } b = \bot$$

and

$$a = \bot \text{ and } b = \top$$

Exercise 3. We have n statements, $a_1,\ldots,a_n$.

$$a_1 \text{ says that } a_2 \text{ is false}$$

$$\vdots$$

$$a_{n-1} \text{ says that } a_n \text{ is false}$$
$$a_n \text{ says } a_1 \text{ is false.}$$

Can you consistently give these statements truth values?

The answer is that if n is odd we cannot but if n is even there are two solutions.

The above Exercises correspond to argumentation networks

Exercise 1 corresponds to the network containing a single node a attacking itself.

Exercise 2 corresponds to the network with two nodes a and b, where a attacks b and b attacks a.

Exercise 3 corresponds to the network which is a cycle of n nodes, as described below, where $\rightarrow$ is attack:

$$a_1 \rightarrow a_2 \rightarrow \ldots \rightarrow a_n \rightarrow a_1.$$

The logic exercise question:

what consistent truth value assignments can you give to the statements, including no assignment to some ?

Corresponds to the question:

what are the extensions to the network, including undecided ?

Now let us look at the network of Figure 2.1 and turn it into a logic exercise.

Exercise 4. Given the following statements, can you give them consistent truth value assignments (including no assignment to some) ?

A says B is false

B says A is false and C is false

C says D is false

D says E is false

E says C is false.

I think by now the reader has a fair idea of how we regard an argumentation network as an exercise in logic.

By the way, some logic textbooks have some hard exercises corresponding to some pretty complicated networks!

For general formal construction, see the subsection on Boolean Networks below, especially Theorem 13.21 and Remark 13.22

(S2)   Solution to D2

We still have the problem of the value assignments to endpoints. Argumentation theory requires us to give them value 1. This can be overcome by splitting every endpoint x into 2 points x and x^*, with $x^* R x$ and $x R x^*$.

So Figure 13.1 becomes Figure 13.2

This figure is not meaningful from the logic point of view. But from the argumentation point of view we can have 2^n stable extensions and 3^n complete extensions. For the case of two valued logic we use only stable extensions. Let E be any stable extension. Let $h_E(q_i) = 1$ iff $q_i \in E$. Now we have 2^n assignments h to the atoms.

For the case of three valued logic, we use complete extensions. Let E be a complete extension, then let $h_E(q_i) = 1$ if q_i is in E, let $h_E(q_i) = 0$ if q_i^* is in E and let $h_E(q_i) = \frac{1}{2}$ if neither is in E.

For the moment, this discussion is only intuitive. We need to set up the formal machinery to explain why and how we can pair Figure 13.1 with Figure 13.2.

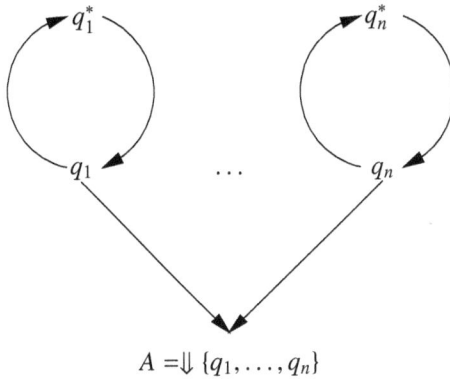

$$A =\Downarrow \{q_1,\ldots,q_n\}$$

Fig. 13.2.

The perceptive reader might ask: in what way does the discussion in (S2) address the difficulty (D2)?

Is it not just a trick?

We are saying:

"the graph of Figure 13.1 doesn't work, so let's switch to the supergraph of Figure 13.2".

The answer is that no, it is not a trick; it is common practice in mathematics to translate one system into another. See for example `http://en.wikipedia.org/wiki/Complex_number♯Matrix_representation_of_complex_numbers`, for the translation of complex numbers into matrices.

In our case we are translating Figure 13.1, which is a figure in the area of logic, into Figure 13.2, which is a figure in the area of argumentation networks.

Take the letter q_1 in Figure 13.1; here we are in the realm of logic, and q_1 is an atom, so we can talk about substituting another logical formula B for q_1.

Take now q_1 in Figure 13.2; here we are in the realm of argumentation networks, q_1 is an abstract argument and the question of substituting a formula B of logic for q_1 is meaningless.

By the way, we have investigated in [164] the possibility of substituting another argumentation network for q_1, but this is new research!

To make this point crystal clear, recall our answer (S1) above, as to how we translate from argumentation networks into logic. This is the other direction of showing that argumentation networks are equivalent to logic.

For a full discussion see Section 6.

(S3) Solution to D3

To solve the problem of undecided values in argumentation we can proceed in two ways

(a). Restrict argumentation extensions to stable extensions, or more specifically allow for orderings that give rise to stable extensions. Fortunately the orderings needed to solve (D2) are such orderings (e.g. Figure 13.2). This is not, however, really a solution to (D3) because we are making the problem disappear. Much better is:

(b). Allow and move to a 3-valued logic for the connective $\Downarrow$. This option is already incorporated into our definitions above.

We shall follow both possibilities. We shall address (b) in Section 3.

Definition 13.13 (Nearly acyclic ordering). *An ordering (S^*, R^*) is said to be nearly acyclic iff it is obtained from an acyclic ordering (S, R) as follows.*

1. Let T be all the endpoints of (S, R), i.e. $T = \{x | \neg \exists y \in S \, xRy\}$.
2. Let T^* be a set of new points of the form

$$T^* = \{x^* | x \in T\}.$$

Let

$$S^* = S \cup T^*$$

and

$$R^* = R \cup \{(x, x^*), (x^*, x)\}.$$

3. We say that (S, R) and (S^*, R^*) are paired.

13.2.3 Integrating argumentation frames with classical logic. An informal view

We are now ready to present in principle (technical definition will be given later) our method of integration. The idea is very simple. Present two networks (S, R) and (S^*, R^*) with $S \subseteq S^*, R \subseteq R^*$ such that they form a pair as in Definition 13.13. We can then view (S, R) either as classical logic formula construction tree as done in Section 1.1 or as an argumentation frame, a subframe of (S^*, R^*), and do what is natural to do from each point of view and compare what each movement means from the other point of view.

Definition 13.14 (The canonical argumentation frame based on Dagger logic).
Our starting point is the general canonical model (Θ, R) of Example 13.5.

The elements of Θ are all Dwffs of the language of classical logic based on $\Downarrow$ and the atoms Q. R is defined by

• *ARB iff B is an immediate subformula of A.*

Recall that this means that

• $\Downarrow (A_1, \ldots, A_n) R A_i, i = 1, \ldots, n.$

Regarding (Θ, R) as just an acyclic ordering, we can construct its pair $\Theta^* \supseteq \Theta, R^* \supseteq R$ as in Definition 13.13.

We add a set

$$Q^* = \{q^* | q \in Q\}$$

that is $\Theta^* = Q^* \cup \Theta$, and $R^* = R \cup \{(q, q^*), (q^*, q)\}.$

We now compare (Θ, R) from two points of view.

1. From the point of view of logic, when viewed as a canonical model for all Dwffs of the language of $\Downarrow$ and where we allow for arbitrary assignments h to the atoms of Q.

2. From the point of view of argumentation theory, where $(\Theta^*, \check{R}^*)$ is regarded as just an argumentation frame in which $\check{R}$ is the converse relation of R and where we examine the effects of argumentation extensions of $(\Theta^*, \check{R}^*)$ on the subnetwork $(\Theta, \check{R})$.

We now go through the concepts one by one.

(C1) **Concept of attack**

B attacks A in the argumentation network means that B is an immediate subformula of A in logic. Note that since

$$A = \Downarrow (B_1, \ldots, B_n) = \bigwedge_{i=1}^{n} \neg B_i$$

when B_i attacks A we have that $\vdash \neg(A \wedge B_i)$.

Thus this concept corresponds to the inconsistency attacks used by Besnard and Hunter in their book [47]. Note for example that $\neg\neg B_i = \Downarrow\Downarrow B_i$ does not attack A because it is not an immediate subformula of A.[4]

(C2) **The concept of an extension**

A complete extension in (Θ^*, R^*) can be generated by any function $\lambda_0 : Q \to \{0, \frac{1}{2}, 1\}$ generating a subset of Q of all q in Q which get value 1, a subset of Q^* of all q^* such that the value of q is 0 and the rest get value $\frac{1}{2}$.

The ordering (Θ^*, R^*) allows for stable extensions. Stable extensions give rise to $\{0, 1\}$ values on Q and Q^*. So let E^* be any stable extension. E^* must choose one from any mutually attacking pair q, q^*. So let $\lambda_0(q) = 1$ iff $q \in E^*$.

The values $\lambda_*(A), A \in \Theta$ are now determined uniquely by λ_0.

We have

$$\lambda_*(\Downarrow (A_1, \ldots, A_n\}) = 1 - \max_i \lambda_*(A_i).$$

Now let us look at λ_0 as a logical assignment $h = \lambda_0$ to the atoms.

For the two valued case we have, after propagating truth values, that (see Definition 13.6)

- $\Downarrow \{A_1, \ldots, A_n\}$ holds at the model $h = \lambda_0$ iff for all i, A_i does not hold in h.

The above means that we have the correspondence

- stable extension in argumentation corresponds to a complete theory (or equivalently a model) in classical and two valued logic.

For the three valued case it is slightly more complicated but we get a similar situation. We need to define the notion of what is a complete three valued theory. This can be defined (we will do this in Section 3) and we have:

- complete extensions correspond to complete three valued theories in three valued logic

There are other types of extensions in argumentation as shown in Figure 2.2, for orderings allowing for undecided elements. We shall deal with this in Section 3.

(C3) **The concept of a logical Dwff**

This corresponds to the notion of an argument node.

The notion of a logical theory therefore corresponds to the notion of a set of argument nodes.

---

[4] We believe we can draw conclusions from our chapter to the debate about the inferential argumentation frames , as championed by Martin Caminada. We need more time to assess our results. See, however, Section 3.3 and the discussion in Chapter 15 that follows.

(C4) **The concept of consequence $A \vdash B$ in logic**

This means that in any model of A is a model of B.

Since what corresponds to "model" is "stable extension" for two valued logic and a "complete extension" for three valued logic, the corresponding notion is as follows:

- Let (S, R) be an argumentation frame. Let $x, y \in S$. We say $x \vdash_{(S,R)} y$ iff in any stable (resp. complete) extension E, if $x \in E$ then $y \in E$.

Similarly we have a correspondence between theories (and subsets of arguments) proving another Dwff (another argument).

(C5) **What corresponds in logic to conflict free set?**

This concept has no logical counterpart, as a conflict free set of Dformulas may be inconsistent, as Figure 13.3 shows.

The nodes q and $\Downarrow\Downarrow\Downarrow\Downarrow q$ are conflict free but inconsistent in logic.

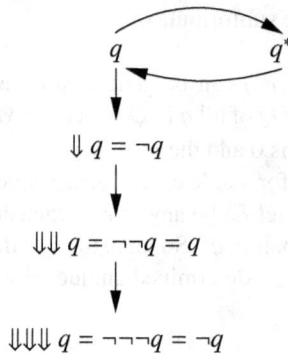

$$q \qquad q^{*}$$
$$\Downarrow q = \neg q$$
$$\Downarrow\Downarrow q = \neg\neg q = q$$
$$\Downarrow\Downarrow\Downarrow q = \neg\neg\neg q = \neg q$$

Fig. 13.3.

The reader should not despair, this is to be expected. If we look at the literature on how Dung's abstract argumentation framework is actually instantiated (by various scholars) we see that what is needed to produce consistency is not conflict-freeness but the stronger condition of admissibility. The really important concept is that of an admissible set. See Caminada, [83].

(C6) **The concept of admissible set**

This corresponds to the notion of an admissible path in the tableaux system for $\Downarrow$, as described in items (4) and (5) of Definition 13.7.

Let me do this systematically. Let E be an admissible set. It is

(1) Conflict free,

and

(2) If A attacks $B \in E$ then for some $C \in E$, C attacks A.

Let us see what (1) and (2) mean in terms of logic. (Note that condition (1) alone does not have a logical meaning, as we saw in (C5) above, but (1) and (2) together do have a logical meaning)

(1) means that we do not have any $\Downarrow \{C_1, \ldots, C_n\}$ and any C_i both in E.

(2) means that if $B = \Downarrow \{A_1, \ldots, A_n\} \in E$ then for some $X_i \in E$, we have that X_i attacks A_i, for each $i = 1, \ldots, n$. This must hold since A_i attacks B.

We remember that $E \subseteq \Theta^*$. So if A_i is an atom q_i, this means $q_i^* \in E$. If $A_i =\Downarrow$ $\{D_1^i, \ldots, D_{k(i)}^i\}$ then X_i is some $D_j^i \in E$.

In terms of tableaux think of the following set

$$E_l = \{X | X \in E \cap \Theta\}$$
$$E_r = \{X | \text{for some } Y \in E | XR^*Y, \text{ i.e } Y \text{ attacks } X\}.$$

We claim there is a path Π in the sense of Definition 13.7. We need to find the initial tableaux to start the process. We can take $[E_l || E_r]$ as our tableaux and observe that the rules (1) and (2) are really tableaux rules which ensure a non-closed path. The conflict free property ensures that for each atomic q we cannot have both $q, q^* \in E$. (Remember E is admissible in (Θ^*, R^*)!)

So an admissible set corresponds to an admissible path in the tableaux proof restricted to the members of the set. There is a nicer way to do this by taking minimal elements of E but we do this in the next section.

We shall also address and discuss what happens in the case of the empty set being an admissible set. Is it the empty tableau?

(C7) **The logical concept of proof**

We now examine how to represent in an argumentation frame the notion of logical proof. This is a dynamic concept and obviously we shall have to do something dynamic; like execute a logical walk over the nodes of the frame.

According to (C3), a Dwff corresponds to a node in the argumentation frame. According to (C4) the notion of

$$\text{consequence } A \vdash B,$$

corresponds to the notion of

B is an element of any extension (of the correct type) which contains A

Thus if we have a step-by-step proof of B from A, i.e.

$$D_1 = A$$
$$\vdots$$
$$D_n = B$$

Then $(D_1, \ldots, D_n)$ is a logic "walk" along nodes. We must define the rules of such a "walk" in terms of the geometry of the argumentation frame, and show that such a "walk" exists iff $A \vdash B$.

Our starting point is modus ponens. This has the form

$$A, A \rightarrow B \vdash B$$

The formula for $A \rightarrow B$ in the language of $\Downarrow$ is

$$A \rightarrow B \equiv \neg(\neg\neg A \land \neg B)$$
$$\equiv \Downarrow\Downarrow \{\Downarrow A, B\}$$

Figure 13.4 shows what is happening

According to (C3), a logical theory is a set of argument nodes, i.e. a subset $T \subseteq S$. So if $A, A \rightarrow B \in T$, this means that we can expand T and let also $B \in T$.

We can therefore offer the following definition for geometrical modus ponens (GMP):

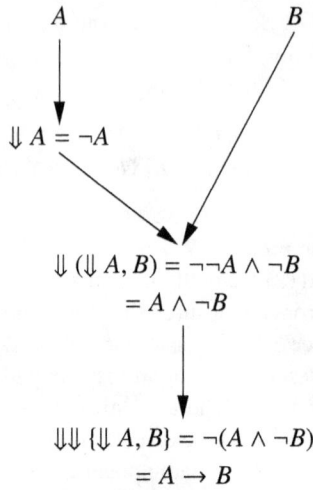

$$A \qquad\qquad\qquad B$$

$$\Downarrow A = \neg A$$

$$\Downarrow (\Downarrow A, B) = \neg\neg A \wedge \neg B$$
$$= A \wedge \neg B$$

$$\Downarrow\Downarrow \{\Downarrow A, B\} = \neg(A \wedge \neg B)$$
$$= A \rightarrow B$$

Fig. 13.4.

Definition 13.15 (Geometrical modus ponens).

a) Let (S, R) be an argumentation frame. The pattern subframe of Figure 13.5 is called modus ponens pattern. This pattern is directional. We can recognise the bottom z' and the left top x (distance 4 from the bottom) and the right top y, being the top at distance 3 from the bottom.

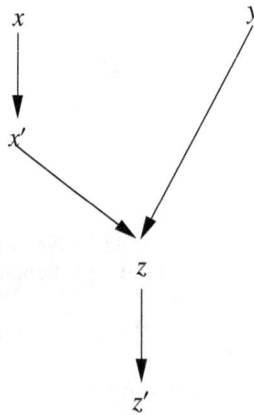

$$x \qquad\qquad\qquad y$$

$$x'$$

$$z$$

$$z'$$

Fig. 13.5.

b) The modus ponens closure rule on a set $T \subseteq S$ is according to this pattern the following:

- *if $x, z' \in T$ then let $y \in T$.*

c) We now look at the $\rightarrow$ elimination rule.

To show $A \rightarrow B$, assume A and prove BÖ

So geometrically to add the node z' of Figure 13.5 to T we need to identify z' as part of the pattern of Figure 13.5 and temporarily add the node x to T, show that the geometry allows us to add y to T ∪ {x}. Then we discharge x and end up with T ∪ {z'}.

d) *The concept of T ↦ x can be defined that starting from T we can follow a sequence of geometrical moves which allow us to expand T to T' containing x.*

An example will help.

Example 13.16. Consider the following derivation

a) $B \rightarrow (A \rightarrow C)$, assumption
b) A, assumption
c) We aim to prove $B \rightarrow C$
d) Show $B \rightarrow C$ from subproof
    i. B, assumption
    ii. $A \rightarrow C$, from (a) and (i)
    iii. C from (b) and (ii)
    iv. Exit subproof, discharge B.
Consider Figure 13.6.

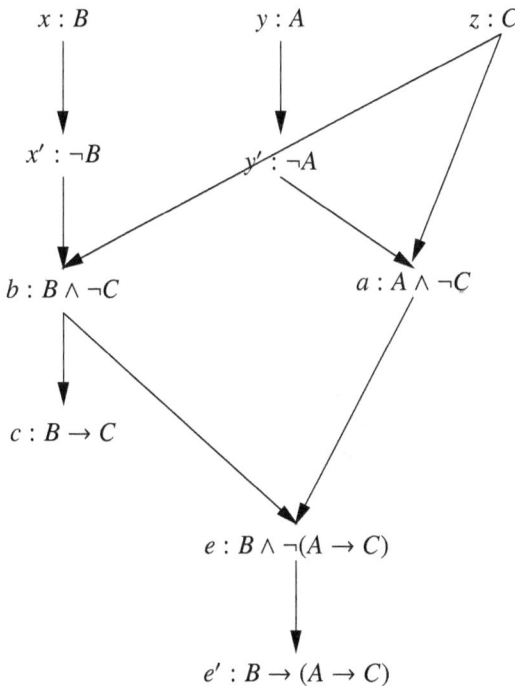

Fig. 13.6.

Look only at patterns. The wffs written is only to help read the pattern. We start with

$$T = \{y, e'\}$$

We want to show that $T \mapsto c$.

Step 0. Recognise that c is the bottom of a pattern $\{x, x', b, c, z\}$, with x left top and z right top.

Step 1. Add x as an assumption to T to form $T_1 = \{x, y, e'\}$.

Step 2. Using the GMP pattern $\{x, x', d, e, e'\}$ and the fact that $x, e' \in T_1$, we add d as an inferred item, to T_1 to form $T_2 = \{x, e', d, y\}$ (note that d stands for $A \to C$).

Step 3. Using the pattern $\{y, y', a, d, z\}$ we can add z to T_2 to obtain $T_3 = T_2 \cup \{z\}$.

Step 4. Having started with $T \cup \{x\}$ and ended up with $T \cup \{x, z\}$, we can now add c to T because c is the bottom node of a modus ponens pattern whose top left is x and top right is z and we have shown that $T \cup \{x\} \mapsto z$.

Thus the proof is a sequence of nodes each obtained from T and previous members of the sequence using geometrical (GMP rules) considerations.

Remark 13.17. The reader should note that our discussion here is only an intuitive introduction, to get the idea of what a proof is. The exact definition of a GMP pattern need yet to be given in the correct context. To get an idea of what we mean, consider Figure 13.7.

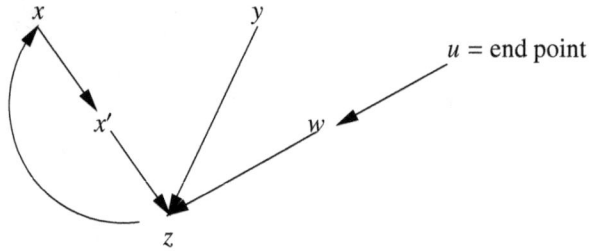

Fig. 13.7.

This figure may not conform to the pattern of Figure 13.5. First we have x itself acting as z'. We may ignore that. After all if $A \equiv (A \to B)$ and we have A we can still get B. A more interesting problem is the presence of node w attacking z. This really breaks the pattern. However, w is attacked by u which is an endpoint. So we do have the pattern if we take that into account.

We need a proper technical definition for our intuitions.

Remark 13.18. [Inference based argumentation and Geometrical Modus Ponens] We note that we can make a contribution to the debate about inference based argumentation. There is the view , championed by senior figures in the field among them Martin Caminada, Henry Prakken and others see [79] and [288], that there are three stages in constructing an argumentation network

1. start with base logic and a knowledge base, which can be used to construct proofs
2. Use the proofs from 1. to construct arguments, the attack relation and an argumentation frame

3. Use the network of 2. to construct extensions and expect the winning arguments in any extensions in 2. to give rise to a consistent theory of the base logic in 1.

See for example [288, 83]. The notion of geometrical Modus Ponens allows us to offer an abstraction of step 1 above in the sense that we can include it in the abstract within step 2. See Section 3.3 below.

The advantages of doing this are two fold

- We continue Dung's spirit of abstraction and not regress backwards into specialised systems. If in step 1 we already have the detailed notions of base logic, proof and inconsistency, why add step 2 to dress it up in some high level concepts which are obvious and available already in step 1?.
- The geometrical concepts of walking along a network to simulate the proof procedures are not just a local technical device cooked up especially to enable our local purpose — it is a commonly used apparatus of traversing graphs and networks used in many areas, including graph theory and Kripke semantics yielding a rich variety of results.

Summary of correspondence between logic and arguments

It would be helpful to summarise the correspondence between abstract argumentation and Dagger classical logic. See Table 13.2.

Table 13.2.

	Logic concept	Argumentation concept
1.	Classical logic formulated with $\Downarrow$	A specific canonical argumentation frame presented in Definition 13.14
2.	A is an immediate subformula of B	A attacks B
3.	Complete logical theory (or equivalently) a model	Complete extension for the case of 3-valued model, and a stable extension for the case of a two valued model.
4.	A Dwff	An argument (a node in the argumentation frame)
5.	Consequence $A \vdash B$	node b is present in any complete extension containing node a
6.	No direct correspondence	conflict-free set
7.	A maximal path in the tableaux (based on the elements of the admissible extension)	Admissible extension
8.	Modus ponens or other proof rules	Geometrical patterns on the abstract argumentation network
9.	A proof sequence from A to B using certain proof rules.	A "walk" in the network from point a to point b respecting and using certain geometrical patterns

13.2.4 Boolean networks

The main result of this section is to show that every argumentation network can be obtained in a systematic way from logic. In Section 2.2, under (S1), we discussed this. We intuitively shown in (S1) that every argumentation network can be obtained from a logic exercise. This however, is not mathematically systematic.

We show a stronger result, that every Boolean network can be obtained from logic.

To motivate Boolean networks, let us start with an example of Brewka and Woltran.

The correspondence between classical Dagger logic and argumentation makes the results of Chapter 7 and of Gabbay [168] and Brewka and Woltran [69, 70], about translating an arbitrary Boolean network into argumentation network, rather predictable.

We simply translate the relevant Boolean formulas into the language with Dagger and they will automatically be embedded into our canonical argumentation frame (Θ^*, R^*) of Definition 13.14.

Let us begin with an example of Brewka and Woltran.

Example 13.19 (Brewka and Woltran Boolean example). Brewka and Woltran [69] put forward the example in Figure 13.8, in which nodes a and b are neither attacking nor supporting node c. In their abstract dialectical framework one can write Boolean conditions on the nodes. In this case we want $c \equiv [(a \wedge \neg b) \vee (\neg a \wedge b)]$, i.e. we want $c = $ in when *exactly one* of $\{a, b\}$ is in.

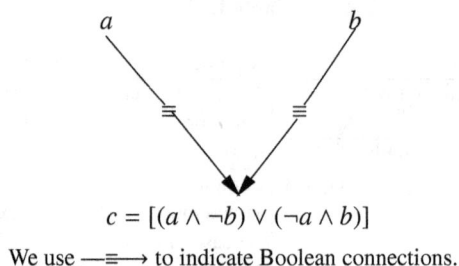

$$c = [(a \wedge \neg b) \vee (\neg a \wedge b)]$$

We use $\longrightarrow\!\!\!\!\!\!{\scriptstyle\equiv}\longrightarrow$ to indicate Boolean connections.

Fig. 13.8.

This example was addressed specifically in our paper [61, Example 14]. It was translated into Dung argumentation networks using additional nodes. Paper [61] then follows to indicate using [164] (see Chapter 7) how any abstract dialectical framework can be so translated into a Dung network.

We quote Figure 27 from [61] as our Figure 13.9.

If argument a is acceptable and argument b is acceptable then arguments u_a, u_b are not acceptable and argument y is acceptable and argument c is not acceptable.

If both arguments a and b are not acceptable then argument x is acceptable and so argument c is not acceptable.

If argument a is not acceptable and argument b is acceptable then argument x is not acceptable, argument u_a is acceptable and u_b is not acceptable. Because u_a is

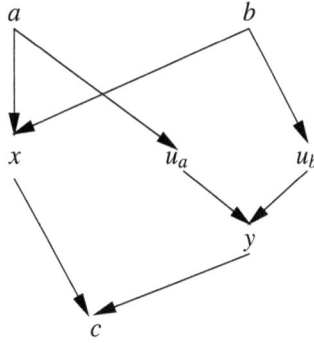

Fig. 13.9.

acceptable then we get y not acceptable. Since both x and x are not acceptable then we get argument c as acceptable, as desired.

If argument a is acceptable and argument b is not acceptable then we get argument x is not acceptable and argument u_b is acceptable and argument y is not acceptable. Thus argument c is acceptable, as desired.

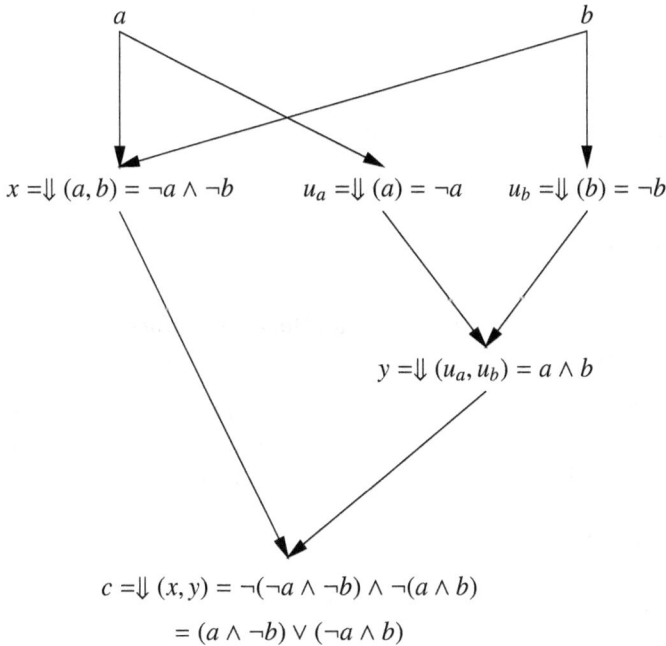

$x = \Downarrow (a, b) = \neg a \wedge \neg b \qquad u_a = \Downarrow (a) = \neg a \qquad u_b = \Downarrow (b) = \neg b$

$y = \Downarrow (u_a, u_b) = a \wedge b$

$c = \Downarrow (x, y) = \neg(\neg a \wedge \neg b) \wedge \neg(a \wedge b)$

$= (a \wedge \neg b) \vee (\neg a \wedge b)$

Fig. 13.10.

The general case requires the equational algebraic approach, we address it in Chapter 10.

Consider Figure 13.11

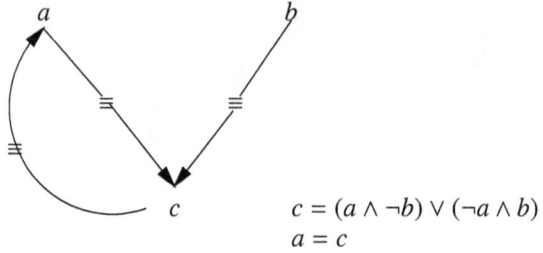

$$c = (a \wedge \neg b) \vee (\neg a \wedge b)$$
$$a = c$$

Fig. 13.11.

This presents a problem at the moment because it is cyclic.

Definition 13.20 (Boolean networks). *A Boolean network has the form* $(S, \check{R}, \Psi_t), t \in S$, *where* $\check{R} \subseteq S \times S$ *and for each* t, Ψ_t *is a Boolean wff in the set of atoms* $\{y\check{R}t\}$. *For* t *an endpoint,* Ψ_t *is either* $\top$ *or* $\bot$. *We write* $\Psi(y_1, \ldots, y_k)$, *where* $\{y_1, \ldots, y_k\} = \{y\check{R}t\}$.

A solution to the Boolean network is a function h *giving values in* $\{\top, \bot\}$ *to each node* $t \in S$, *such that the following holds*

(*) $h(t) = \Psi_t(h(y_1), \ldots, h(y_k))$

Note that we may not have a solution. Take for example the network

$$(\{t\}, t\check{R}t, \Psi_t = \neg t).$$

A solution for this network requires a function h *such that*

$$h(t) = \neg h(t).$$

This is not possible.

Theorem 13.21 (Representation theorem for Boolean networks). *Every finite acyclic Boolean network can be faithfully embedded into an argumentation network.*

Proof. Let $(S, \check{R}, \Psi_t)$ be an acyclic network. Write each $\Psi_t(y_1, \ldots, y_k)$ as a Dwff in the y's using the language with $\downarrow$. Since we are now using logic, we use the relation R instead of $\check{R}$. Remember that one is the converse of the other. So y attacks x can be written either as xRy or as $y\check{R}x$ or $y \to x$.

So our network has the form (S, R, Ψ_t^*) where Ψ_t^* is written with $\downarrow$. (We will have to accept $\top$ and $\bot$ as formulas as well.)

Now define new formulas $\varphi_t^*, t \in S$ as follows:

1. For t an endpoint (i.e. $\neg \exists y(tRy)$) let $\varphi_t^* = \Psi_t^*$.
2. Let $t, y_1, \ldots, y_k$ be such that $\{y|tRy\} = \{y_1, \ldots, y_k\}$.

Let Ψ_t^* be written as $\Psi_t^*(y_1, \ldots, y_k)$. Then let

$$\varphi_t^* = \Psi_t^*(y_1/\varphi_{y_1}^*, \ldots, y_k/\varphi_{y_k}^*)$$

It is easy to see that

1. tRs iff φ_s^* is a subformula of φ_t^*.
2. For any solution h to (*) we have $h(t) = h(\varphi_t^*)$.

From the above it is clear that the subset $\{\varphi_t^* | t \in S\}$ is a subnetwork of the canonical network (Θ^*, R^*) based on the atoms S of Definition 13.14.

This proves the theorem.

Remark 13.22 (Boolean network with loops). What do we do in the case of (S, R, Ψ_t) with loops where a solution h may or may not exist?

Again we assume Ψ_t is written with $\Downarrow$ only.

Define Dwffs $\Psi_t^n, n = 1, 2, \ldots$.

1. $\Psi_t^1 = \Psi_t$
2. $\Psi_t^{n+1} = \Psi_t(\{y / \Psi_y^n | tRy\})$.

We are getting an infinite number of wffs but all of these appear in the canonical model based on S as defined in Definition 13.14.

Note that for any solution h satisfying (*) we also have

$$h(\Psi_t^n) = h(\Psi_t).$$

Now consider the equivalence relation on the canonical model (Θ^*, R^*) as follows:

$x \approx_t y$ iff for some n, m we have that $x = \Psi_t^n \wedge y = \Psi_t^m$. Let us denote by $E(t)$ the set of all x such that $x \approx_t t$.

Let $x \approx y$ be defined by induction as follows

1. $x \approx y$ iff $x \approx_t y$ for some t
2. $\Downarrow \{x_i\} \approx \Downarrow \{y_i\}$, iff $x_i \approx y_i$, for $i = 1, 2, \ldots$.

$\approx$ is well defined. Take the factor $(\Theta^* / \approx, R^* / \approx)$ network with Θ^* being the equivalence classes and the attack relation $R^* / \approx$ is defined by

$$x / \approx R^* / \approx y / \approx \text{ iff for some } x' \approx x, y' \approx y \text{ we have } xR^*y.$$

We claim (S, R, Ψ_t) is faithfully embedded in $(\Theta^* / \approx, R^* / \approx)$.

Let t be mapped onto the equivalence class $\{\Psi_t^n\}$.

Assume tRs holds. Then $\Psi_t^{n+1} = \Psi_t(, \ldots \Psi_s^n \ldots)$ by definition. So for any h satisfying (*) we also have

$$h(\Psi_t / \approx) = \Psi_t(\{h(\Psi_s / \approx) | tRs\}.$$

Conversely if there is an h on the canonical model giving all elements of the same class the same value then there is an h satisfying (*) on the original Boolean network.

Let us give some examples illustrating the process of Remark 13.22.

Example 13.23. Consider the loop in Figure 13.12

In this Figure we have $\Psi_a = \neg b$ and $\Psi_b = \neg a$. We therefore have

$$\Psi_a^1 = \neg b \quad \Psi_b^1 = \neg a$$
$$\Psi_a^2 = a \quad \Psi_b^2 = b$$
$$\Psi_a^3 = \neg b \quad \Psi_b^3 = \neg a$$
$$\vdots \qquad \vdots$$

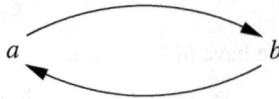

Fig. 13.12.

The initial equivalence classes are therefore

$$E(a) = \{a, \neg b\} \text{ and } E(b) = \{\neg a, b\}.$$

Consider the canonical model, based on the atoms $\{a, a^*, b, b^*\}$. Figure 13.13 describes part of it. Recall that $\neg A$ is $\Downarrow A$ in the Dagger language.

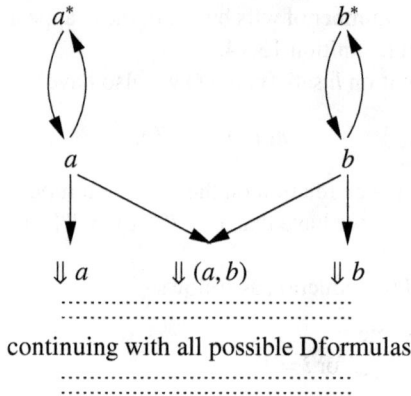

Fig. 13.13.

Taking into account the equivalence classes we get Figure 13.14.

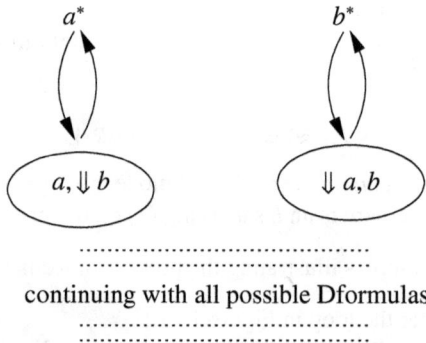

Fig. 13.14.

Example 13.24. Let us try an example where there is no solution in $\{0, 1\}$, as in Figure 13.15.

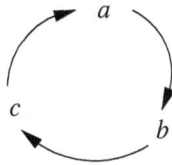

Fig. 13.15.

We get

$$
\begin{array}{lll}
\Psi_a^1 = \neg c & \Psi_b^1 = \neg a & \Psi_c^1 = \neg b \\
\Psi_a^2 = b & \Psi_b^2 = c & \Psi_c^2 = a \\
\Psi_a^3 = \neg a & \Psi_b^3 = \neg b & \Psi_c^3 = \neg c \\
\Psi_a^4 = c & \Psi_b^4 = a & \Psi_c^4 = b \\
\Psi_a^5 = \neg b & \Psi_b^5 = \neg c & \Psi_c^5 = \neg a \\
\Psi_a^6 = a & \Psi_b^6 = b & \Psi_c^6 = c \\
\Psi_a^7 = \neg c & \Psi_b^7 = \neg a & \Psi_c^7 = \neg b \\
\vdots & \vdots & \vdots
\end{array}
$$

Obviously there is one equivalence class

$$\{a, b, c, \neg a, \neg b, \neg c\}$$

The graph we get is in Figure 13.16

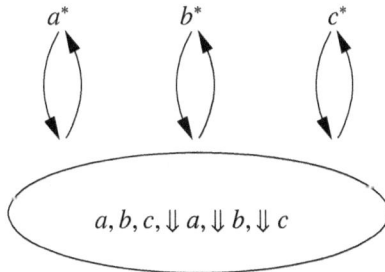

Fig. 13.16.

There is no solution in $\{0, 1\}$. The solution in $\{0, 1, \frac{1}{2}\}$ is $a = b = c = a^* = b^* = c^* = \frac{1}{2}$, so also $\Downarrow a = \Downarrow b = \Downarrow c = \frac{1}{2}$.

13.3 Technical results on integrating argumentation as logic

13.3.1 What is a logical system

Our first step is to clarify what we mean by logic. We give precise definitions of a view of "what is a logical system" which is most friendly to argumentation.

We now explain our ideas. Consider Figure 13.9. This is just a network. How can we turn it into a logic? One obvious way of doing it is to go through a semantical interpretation. This is an ordering, a set of worlds with a binary relation and so we can regard it as a Kripke model for some logic. It can thus define a modal logic, or an intuitionistic intermediate logic, or a provability logic, etc., etc.

In Chapter 5, we took this approach and interpreted argumentation networks into the modal logic of provability. We got surprising results about what is the logical content of argumentation network, but this does not present the network as a logic.

The construction in Section 1 of this chapter does turn the network into a logic. It says the nodes t are formulas and the attack of $x_1, ..., x_n$ on t means "being an immediate subformula of". So Figure 13.9 actually represents Figure 13.10.

So far so good, the problem is that argumentation networks can have loops. So how do we interpret Figure 13.7?

Our idea is to abstract from the actual syntax of a formula and regard it as an object and have the relations of "being an immediate subformula of" as the interpretation of the arrow. If we do that, all we have left is just an ordering? Where is the logic in it?

The answer is simple. All we need to add is the truth table functions which tell us for each node t and its attacking nodes $x_1, ..., x_n$, how to get the truth value of the formula t (whose structure we do not know) from the values of its immediate subformula $x_1, ..., x_n$, (whose structure we also do not know). Let us call such a truth propagating function

$$\mathbf{h}_t(x_1, ..., x_n).$$

We may not know the syntactic structure of the formulas but we do know how to calculate their values! When we know how to calculate values we do not mind cycles, they just give rise to equations to be solved.

Clearly $\mathbf{h}_t$ is also part of the logic.[5]

Now that we have the idea, all that remains is to figure out the coherent technical details. This we now do.

Definition 13.25 (Function spaces). *By a function space we mean the following*

1. A set V with a family of functions $\mathbb{F}$ of the form

$$\mathbf{h} : V^n \mapsto V$$

For $n = 1, 2, ...$.
The set V may have some operations on it by which means the functions $\mathbf{h}$ in $\mathbb{F}$ are defined. We give two examples:
a) $V = [0, 1]$, the set of all real numbers between 0 and 1 and $\mathbb{F}$ is the set of all continuous functions with any number of variables.

---

[5] The perceptive reader might be a bit puzzled here. He may reason as follows:

If I know the immediate subformulas of t and the function yielding the truth value of t from the truth value of its immediate subformulas, then this means that the logic is truth-functional (at least there) and therefore this function characterizes the main connective of t, so I know the structure of t after all (and, recursively, I would know the structure of the subformulas, applying the same principle).

Note, however, that h_t can vary with t and may not be Boolean. We may also have loops in the network.

b) *V is a complete partial order (i.e. it has a partial ordering on it and any subset $V_0 \subseteq V$ has a least upper bound and a greatest lower bound) and $\mathbb{F}$ is the set of all monotonic functions on V in any number of variables.*

2. *We assume $(V, \mathbb{F})$ satisfy the following. Let $\mathbf{h}_1, \ldots, \mathbf{h}_n$ be n functions in $x_1, \ldots, x_n$ (same n). Consider the vector function on V^n defined by*

$$\mathbf{y} = \mathbf{h}(\mathbf{x})$$

Where $\mathbf{x} = (x_1, \ldots, x_n)$, $\mathbf{y} = (y_1, \ldots, y_n)$ and $\mathbf{h} = (\mathbf{h}_1, \ldots, \mathbf{h}_n)$.
Then we assume that any such vector function for any n has at least one fixed point $\mathbf{x}_0$, i.e. we have

$$\mathbf{x}_0 = \mathbf{h}(\mathbf{x}_0).$$

Note that the fixed point condition holds for $[0, 1]$ because of Brouwer's fixed point theorem and for the complete partial orders because of Tarksi's fixed point theorem.

Definition 13.26 (Logics, models, theories and inconsistency). *Let $(V, \mathbb{F})$ be a function space.*

1. *By a $(V, \mathbb{F})$ logic we mean a system $(S, R, \mathbf{h}_t), t \in S$ where S is a finite set, $R \subseteq S^2$, and for each $t \in S$, the following holds:*
   Let $x_1, \ldots, x_n$ be all points in S such that tRx_i holds. Then $\mathbf{h}_t$ is an n-place function from $\mathbb{F}$.
   It is important to note that when we write $\mathbf{h}_t$, then the variables $x_1, \ldots, x_n$ are ordered inside $\mathbf{h}_t$. We are not assuming that $\mathbf{h}_t$ is a symmetric function in its variables. Where does this order come from? It comes from the geometry of (S, R).
   Consider for example Figure 13.7, the points x', y, w are distinguishable individually by the geometry of the graph and so we can order them as we see fit when we plug them into $\mathbf{h}_z$.
   It is expected that should the geometry not distinguish between any two points, say x_1 and x_2, then $\mathbf{h}_t$ must be symmetrical in these variables.
2. *By a model for the logic $(S, R, \mathbf{h}_t)$ we mean any function $\mathbf{f}$ from S into V such that for all $t \in S$ we have*
   (*1) $\mathbf{f}(t) = \mathbf{h}_t(\mathbf{f}(x_1), \ldots, \mathbf{f}(x_n))$.
   *Such $\mathbf{f}$s exist because of the fixed point theorem for $(V, \mathbb{F})$. $\mathbf{f}$ is not unique for solving (*1)*
3. *Let $(S, R, \mathbf{h}_t)$ be a logic and let $s, t \in S$. We say $s \vDash t$ iff for any model $\mathbf{f}$ we have*
   (*2) $\mathbf{f}(s) = 1 \Rightarrow \mathbf{f}(t) = 1$.
4. *Note that (S, R) corresponds to the set of wffs of the language and $\mathbf{f}$ to truth value assignments to the wffs. The partial ordering of (1b) of Definition 13.25 above corresponds to the Lindenbaum algebra of the logic. The functions $\mathbf{h}_t$ are the truth tables of the connectives. R corresponds to the inductive construction tree of the wffs, except that it needs not the tree but could be a general binary dependence relation!*
   This is a sort of "free style" notion of a wff.
   A wff is just an abstract point with a relation R saying xRy, which means y is an immediate subformula of x.

5. *A theory in traditional logic is a set of wffs required to be true. In our context, to require a set of nodes to be true means to impose constraints on the models, i.e. on the function* **f**, *i.e. on the solution of the equations of (*1).*

Since **f** *is a general function from S into V, there are many types of constraints we can impose, e.g. that* **f** *has a minimal number of 0 values or that it has a minimal number of* $\frac{1}{2}$ *values (this would yield circumscription minimal models in the right context and right formulation for the case of 0 values, or would yield semi-stable semantics for the case of argumentation and value* $\frac{1}{2}$*).*

So we define

(*3) A logical theory is a set of constraints on the solutions (models) **f***s. It is inconsistent if the constraints have no solution.*

See [168] for examples.

Example 13.27 (Geometrical proof rules, geometric inconsistency).

1. Let (S,R) be an argumentation frame. A unary geometrical proof rule **r** is a finite ordering (which can be represented visually by a figure) of the form **r** $=$ (P_r, R_r, x, y) where P_r is a set of nodes and R_r is a binary relation on P_r, and x, y are two distinct points in P_r. x is called the input point (the premise of the rule) and y is called the output point (the conclusion of the rule). We require that x and y be geometrically definable in (S,R) in terms of information in (P_r, R_r). Figure 13.17 shows two such rules corresponding to the rules

$$\mathbf{r}_1 : \frac{A}{\neg\neg A}$$

And

$$\mathbf{r}_2 : \frac{\neg\neg A}{A}$$

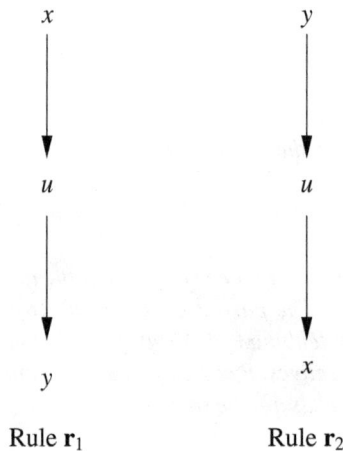

Fig. 13.17.

2. A binary geometrical proof rule for (S, R) has the form $\mathbf{r} = (P_{\mathbf{r}}, R_{\mathbf{r}}, x, z', y)$ where $(P_{\mathbf{r}}, R_{\mathbf{r}})$ is finite ordering and $x, z', y \in P_{\mathbf{r}}$. x is the input point, z' is the base point and y is the output point. We require that x, y, z' be geometrically identifiable in (S, R) in terms of information available in $(P_{\mathbf{r}}, R_{\mathbf{r}})$.
3. What do we mean by a point being geometrically identifiable in (S, R) in terms of information available in $(P_{\mathbf{r}}, R_{\mathbf{r}})$? Consider Figure 13.18.

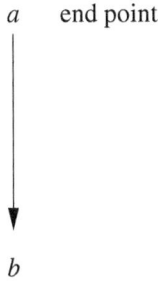

a end point

b

Fig. 13.18.

This pattern requires in (S, R) two points; the first being an endpoint attacking the second. The information $a = $ endpoint is for (S, R).
4. What do we mean by geometric inconsistency? We need to have a family of subsets of the network S which are marked unacceptable. Intuitively each such subset corresponds to a constraint of, say, wanting an equational solution which gives all members of the subset value 1. So the family is of all subsets for which there is no solution.

A formal definition is given in Definition 13.28.

Another example of a pattern for geometrical modus ponens is Figure 13.5. The points x, y, z' in this figure are definable using $R_{\mathbf{r}}$.

Definition 13.28 (Formal definition of geometric proof rules and inconsistency).

1. *Let (S, R) be an argumentation network with $R \subseteq S \times S$. Consider first order predicate logic with a binary relation R. Let $\Psi(x)$ be a formula in this language with a free variable x. Let $a \in S$. We can ask*

$$(S, R) \models ? \Psi(a)$$

For example

$$\Psi_1(x) = \neg \exists y (xRy)$$

Says that x is an endpoint in (S, R) or

$$\Psi_2(x) = \exists y (xRy \wedge yRy)$$

We call such formulas $\Psi(x)$ as additional information about node x formulated in the first order predicate logic of the binary relation R.

2. A geometric rule **r** *has the form*

$$\mathbf{r} = (P_{\mathbf{r}}, R_{\mathbf{r}}, \Psi_t(x), a_1, \ldots, a_k, c, b), \text{ for } t \text{ in } P_{\mathbf{r}}$$

Where $a, a_i, c, b \in P_{\mathbf{r}}$.
$a_1, \ldots, a_k$ *are inputs. c is the base and b is the output.*
Intuitively think of it as $a_1 \wedge \ldots \wedge a_k \Rightarrow_c b$.
$(P_{\mathbf{r}}, R_{\mathbf{r}})$ *is a finite ordering and for each* $t \in P_{\mathbf{r}}, \Psi_t(t)$ *is additional information about t in the sense of item (1) above.*

3. Let **r** *be a geometric rule, and let* (S, R) *be an argumentation network. Let* μ : $P_{\mathbf{r}} \mapsto S$ *be a one to one embedding of* $P_{\mathbf{r}}$ *into S. We say* μ *identifies the pattern of the rule* **r** *in* $\mu(P_{\mathbf{r}})$ *iff the following holds.*

  a) $x R_{\mathbf{r}} y$ *iff* $\mu(x) R \mu(y)$ *for all* $x, y \in P_{\mathbf{r}}$.
  b) For all $x \in P_{\mathbf{r}}, (S, R) \models \Psi_x(\mu(x))$.

4. We require also that the points $\mu(a_1), \ldots, \mu(a_k) \mu(c)$ *and* $\mu(b)$ *are uniquely identifiable in* (S, R). *More precisely there are properties* $\varphi_{a_1}(x), \ldots, \varphi_{a_k}(x), \varphi_c(x), \varphi_b(x)$ *such that for each* $y', y \in \{a_1, \ldots, a_k, c, b\}$ *we have*

$$(S, R) \models \varphi_y(\mu(y)) \wedge \bigwedge_{y' \neq y} \neg \varphi_{y'}(\mu(y)).$$

5. An inconsistency notion is just a family of subsets of S. For monotonic logic the family is closed under enlargement

Example 13.29. Consider the pattern of Figure 13.18.
   Here

$$\Psi_a(x) = \neg \exists y x R y$$
$$\Psi_b(x) = x = x.$$

Consider the following embeddings of this pattern into Figure 13.7. (I chose this figure at random, just for illustration).

   If we match (a, b) with (y, z), we get a good embedding because y satisfies Ψ_a, but if we embed (a, b) as (x', z), then the conditions are not satisfied.

   Note that from now on to simplify notation we regard μ as the identity and talk about $P_{\mathbf{r}} \subseteq S$ and $R_{\mathbf{r}} \subseteq R$.

Definition 13.30 (Geometrical proofs). *Let* (S, R) *be part of a logic as defined in Definition 13.26. Let* $T \subseteq S$ *be any subset. Let* $\mathbf{r}_1, \ldots, \mathbf{r}_k$ *be proof rules. We define the notion of*

- *the sequence* $(x_1, \ldots, x_n), n \geq 1, x_i \in S$ *is a proof of level* $m \geq 0$ *of* x_n *from T, using* $\mathbf{r}_1, \ldots, \mathbf{r}_k$.

The definition is by induction on m and n.
Case $n = 1, m = 0$
x_1 *is a proof from T if* $x_1 \in T$.
Case $n + 1, m = 0$
$(x_1, \ldots, x_{n+1})$ *is a proof from T iff one of the following holds:*

  1. $x_{n+1} \in T$

2. x_{n+1} is obtained from some $x_i, i \leq n$ using a unary geometrical rule $\mathbf{r} = (P_{\mathbf{r}}, R_{\mathbf{r}}, x, y)$ such that $P_{\mathbf{r}} \subseteq S, R_{\mathbf{r}} \subseteq R$ and x_i is the input $(x = x_i)$ and x_n is the output $(y = x_n)$.

3. For some $x_{j_1}, \ldots, x_{j_k}, x_j, i, j \leq n$ and some geometrical rule $\mathbf{r} = (P_{\mathbf{r}}, R_{\mathbf{r}}, x_1, \ldots, x_k, z', y)$ we have $P_{\mathbf{r}} \subseteq S, R_{\mathbf{r}} \subseteq R, x_{j_i} = x_i, z' = x_j$ and $y = x_{n+1}$.

Case level $m + 1$

Assume that for each T and any $m' \leq m$ and any n we have defined the notion of $x_1, \ldots, x_n$ is a proof of x_n of level $\leq m'$. We now define this notion for level $m + 1$ and $n \geq 1$. Let cases (1)–(3) be as above for level $\leq m$. We add more cases

4. **Case** $m + 1, n = 1$

For some rule $\mathbf{r} = (P_{\mathbf{r}}, R_{\mathbf{r}}, x'_j, z', y)$ we have that there exists $(y_1, \ldots, y_{n'}), y_{n'} = y$ which is a proof of level $\leq m$ of y from $T \cup \{x'_j\}$. We also have $z' = x_1$.

5. **Case** $m + 1, n > 1$

For some rule as in (4), we have that there exists $(y_1, \ldots, y_{n'}) \, y_{n'} = y$, which is a proof of level $\leq m$ of y from $T \cup \{x'_j\} \cup \{x_1, \ldots, x_{n-1}\}$. We also have $z' = x_n$.

Remark 13.31. Note that in logic based argumentation networks (see [79] or [220]) only level 0 proofs are used. The rules have the form $A_1 \wedge \ldots \wedge A_n \Rightarrow_c B$ and only $\Rightarrow_c$ eliminations are used.

Definition 13.32 (Soundness of rules). *Let $(S, R, \mathbf{h}_t)$ be a logic in the sense of Definition 13.26. Let $\mathbf{r}_1, \ldots, \mathbf{r}_k$ be rules in the sense of Definition 13.28. We say the rules are sound iff whenever b is proved from a in (S, R) as in Definition 13.30 for $a, b \in S$ then $a \vdash b$ holds as defined in Definition 13.26.*

We say the rules are complete iff we have

- *$a \vdash b$ iff b is provable from a using the rules.*

If a subset of S is marked inconsistent then the constraint arising from that set cannot be solved.

Example 13.33 (Defeasible rules). Ordinary implication (strict implication) we can write as $A_1 \wedge \ldots \wedge A_n \rightarrow B$ or as $A_1 \rightarrow (A_2 \rightarrow \ldots \rightarrow (A_n \rightarrow B) \ldots)$.

Let us do geometrically $A \rightarrow B$ and $A \Rightarrow B$. All we need are some markers in the figures representing these two implications, to distinguish one from another. See Figures 13.19, 13.20

Consider now (S, R) of Figure 13.21 and consider the rule $\mathbf{r}$ of Figure 13.20. Take the theory $T = \{a_1, \ldots, a_n\} \cup \{b_n\}$. What can it prove using $\mathbf{r}$?

The answer is that it can prove b_0 and all $b_{n-1}, \ldots, b_1$ along the way. The deduction is essentially a_j and $b_j = [a_j \Rightarrow (a_{j-1} \Rightarrow \ldots (a_1 \Rightarrow b_0) \ldots)]$ yields for $b_{j-1} = [a_{j-1} \Rightarrow \ldots \Rightarrow (a_1 \Rightarrow b_0)]$.

We can turn this ordering into a logic if we give the functions $\mathbf{h}_t$, for any t of Figure 13.21. Try $\mathbf{h}_{e_i} = \frac{1}{2} \cdot \mathbf{h}_{a_i} = $ arbitrary. $\mathbf{h}_{a'_i} = \mathbf{h}_{a_i} \cdot \mathbf{h}_{b_0} = $ arbitrary. $\mathbf{h}_{b_j} = \min(1, 1 - a'_j + b_{j-1})$ for $j \geq 1$.

We need to show the rule is sound in this semantics, but in this case it is clear because the rules are versions of modus ponens and the function $\mathbf{h}_x$ are from Łukasiewicz many valued logic.

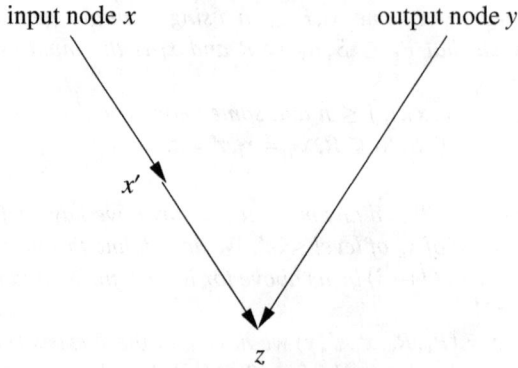

$z = x \rightarrow y$, base node x' is an auxiliary node so that we can tell the difference between input and output. When identifying this pattern in an argumentation network it is required that nodes z and x' bear exactly the attacks shown in the pattern.

Fig. 13.19.

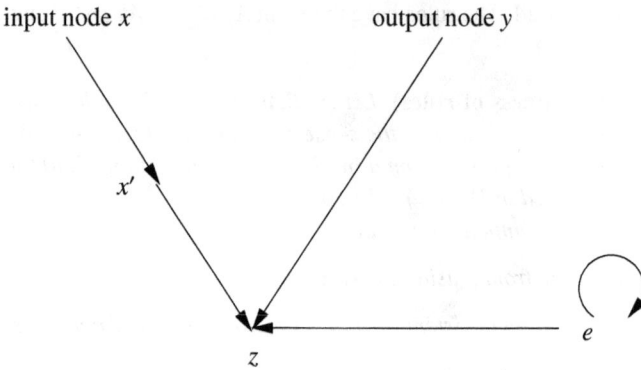

$z = x \Rightarrow y$, base node x' is an auxiliary node. When identifying this pattern in an argumentation network it is required that nodes z, e and x' bear exactly the attacks shown in the pattern.

Fig. 13.20.

13.3.2 Revisiting the canonical model

In Definition 13.14 we introduced a canonical ordering (Θ^*, R^*) made up of all Dagger wffs as well as an additional atoms Q^*.

Let us now treat it as a logic in the sense of Section 3.1.

Take $V = \{0, \frac{1}{2}, 1\}$ and

- $\mathbf{h}_t(x_1, \ldots, x_n) = 1 - \max(x_i)$

Let the relation $A \approx B$ be defined as saying

- $A \approx B$ for $A, B \in \Theta, A, B$ wff iff A and B have the same classical truth table.

This relation is an equivalence relation and is decidable.

Let $\Theta_0 = \Theta/ \approx$

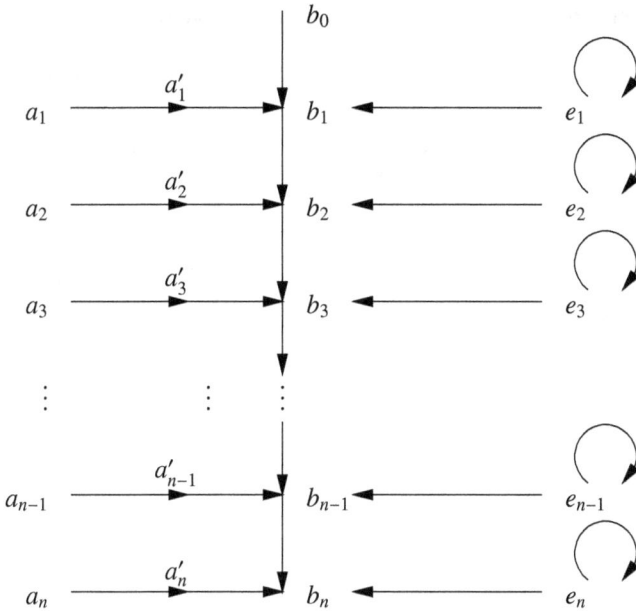

Fig. 13.21.

Let T be the set of all $\approx$ equivalence classes of Θ^*. Define an attack relation **Att** on T by $A/\approx$ **Att** $B/\approx$ iff for some $A' \approx A$ and $B' \approx B$ we have $B'RA$.
We claim[6]

- $A/\approx$ **Att** $B/\approx$ iff $A \vdash \neg B$ in classical logic.

What we need to show is that if $A \vdash \neg B$ then for some $A' \approx A$ and $B' \approx B$, A' is an immediate subformula of B'. This is not difficult to show. We can assume A and B are written in the language with $\Downarrow$ only.
Since $A \vdash \neg B$, we have $B \vdash \neg A$, hence $B \approx B \wedge \neg A$.
Consider

$$\Downarrow (\Downarrow B, A)$$

This is equivalent to B and indeed A is an immediate subformula of it.
We thus got a canonical model of $(T,$ **Att**$)$ of classical formulas (up to equivalence).
The above notion of Att is symmetric and is based on inconsistency in classical logic. In the literature, There are a number of frameworks for modelling argumentation in logic. A common assumption for logic-based argumentation is to start with a knowledge base **KB** in some logic (usually a version of defeasible logic based on **KB**) and construct arguments as pairs (Δ, A) where Δ is a minimal subset of the knowledge base such that Δ is consistent and $\Delta \vdash A$. Hunter [220] call the logic used for consistency and entailment, the base logic. Different base logics provide different definitions for consistency and entailment and hence give us different options for argumentation.

---

[6] This is a standard way of transferring any relation between points in a set to their equivalence classes.

13.3.3 Argumentation systems arising from knowledge bases

We are going to use our results to analyse problems and abnormalities existing in argumentation systems based on knowledge bases formulated in some base logic see [79, 220, 288, 83].

We begin with a specific paradoxical example, which according to Caminada and Amgoud [79] seems to defy solution.

Example 13.34. This is example 6 of [79, page 293].

Our base logic has both strict rules with $\rightarrow$ and defeasible rules with $\Rightarrow$.

Our knowledge base has the following data:

Strict data

1. a
2. d
3. g
4. $b \wedge c \wedge e \wedge f \rightarrow \neg g$

Defeasible data

7. $a \Rightarrow b$
8. $b \Rightarrow c$
9. $d \Rightarrow e$
10. $e \Rightarrow f$

From the above knowledge base we construct arguments. We follow [79] but use our notation to indicate the chain of reasoning.

Arguments

$$A = [a, a \Rightarrow b]$$
$$B = [d, d \Rightarrow e]$$
$$C = [a, a \Rightarrow b, b \Rightarrow c]$$
$$D = [d, d \rightarrow e, e \Rightarrow f]$$

The problem with this example is that A, B, C, D have no defeaters and so one has to accept them and their conclusions which is the set $\{b, c, e, f\}$ is justified which together with the data $\{a, g\}$ should be consistent but it is not consistent because of item 4 of the data.

Example 13.35 (Representing Example 6 of [79] in our system).

Figure 13.22 represents Example 6 of [79] as described above in our Example 13.35.   The   nodes $a', b', b'', c'', e', e'',$ f'', x', y', z', w' are auxiliary nodes. The original nodes mentioned are

$$a, b, c, d, e, f, g, \neg g$$

and

$$x = (a \Rightarrow b)$$
$$y = (b \Rightarrow c)$$
$$z = (d \Rightarrow e)$$
$$w = (e \Rightarrow f)$$
$$u = b \wedge c \wedge e \wedge f \rightarrow \neg g.$$

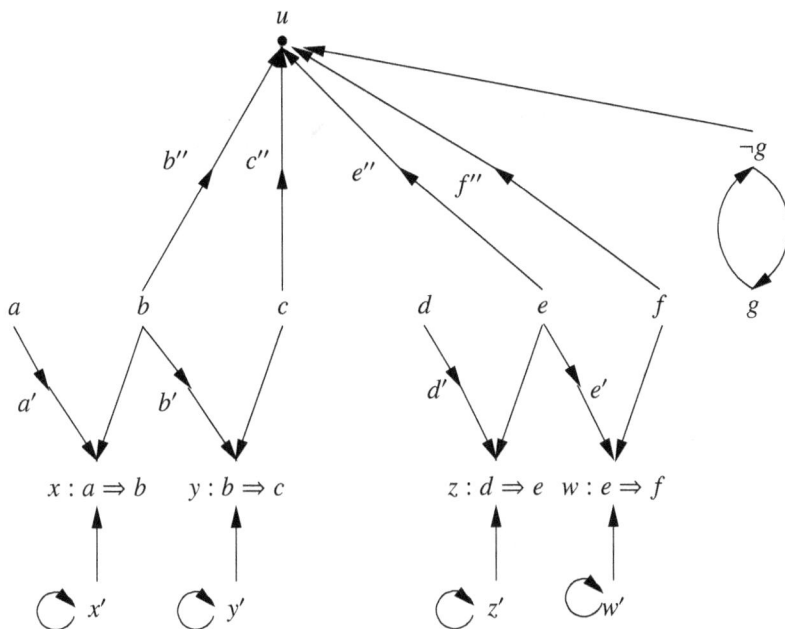

Fig. 13.22.

The rules of modus ponens are represented as in Figures 13.19 and 13.20. This is why we need the auxiliary points. The arguments represent walks along Figure 13.22, which respect the geometry of the rules.

They are:

$A = \text{walk } (a, b)$

$B = \text{walk } (d, e)$

$C = \text{walk}(a, b, c)$

$D = \text{walk}(d, e, f)$

We perceive the walk as a proof augmenting the knowledge base (which is a set of nodes), with additional nodes. So if **KB** is the original knowledge base, then we perceive the arguments as follows:

A as $\textbf{KB} \cup \{a, b\}$

B as $\textbf{KB} \cup \{d, e\}$

C as $\textbf{KB} \cup \{a, b, c\}$

D as $\textbf{KB} \cup \{d, e, f\}$.

The perceptive reader might say: OK, fine. So you represented the problem your way. How are you going to solve the anomaly? A new representation does not solve any original problem! So?

The answer is that the new representation may suggest more readily a new idea for a solution to the problem. Once we get the new idea we can apply it equally well to the old original representation.

OK, so what is the new idea?

I say that what is wrong (in my opinion) with the old way of deriving arguments from a knowledge base is in the way they define attacks between arguments.

The concept is flawed and this is the source of difficulty. Let us recall Chapter 4 Logical modes of attack in argumentation networks. We presented there a model of attack between nonmonotonic knowledge bases, say Δ_1 and Δ_2. Δ_1 can attack Δ_2 by sending one or more items of data from itself (i.e. Δ_1) into Δ_2. This may render Δ_2 inconsistent or may hinder Δ_2 in its nonmonotonic deductive tasks. See Chapter 4, Example 4.8.

How are we going to apply this definition in our case?

We saw that an argument becomes a walk along a graph. During the walk we collect points into our set. The more points we have collected, the more geometric patterns we can use to collect even more points. So imagine two people walking along a graph (say the streets of the old district in town). We first notice that the two fellows may not be using the same tourist guide (or map). So they may not have the same walks. In geometrical proof terms maybe the first fellow can use rule **r** while the second fellow does not use it. OK, now we ask: how can one obstruct the walk of the other?

1. One can be an immediate obstruction. Stop the other immediately in his tracks. This corresponds to a rebuttal (you should stop now or needed to stop before) or undercut (you got here by going through a "forbidden to walk" path!).
2. Give the guy a wrong direction so he will reach a dead end and will have to stop. How is this done in practice?

Remember each of these walkers is collecting nodes in the graph. If one of them gives the other some nodes then the enlarged set of the other can be inconsistent, or unacceptable (this is case (1) above), or he may be tempted to use the additional nodes to carry on walking and get to a pending inconsistency, (this is case (2) above).

Recall inconsistency was defined in Definition 13.28.

It is (2) above that is missing from the concept of attack. I hit upon a similar problem in the early 1990s when I was writing my book on Labelled Deductive Systems [136]. The notion was labelled revision theory and inconsistency. I could have a labelled theory which receives an input and really eventually becomes inconsistent but it was not seen immediately. You needed to carry on proving all kinds of things from it to eventually realise it is inconsistent. So how are we going to use this idea in our context?

Let $(S, \check{R})$ be an argumentation network with geometrical rules $\mathbf{r}_1, \ldots, \mathbf{r}_k$. Let $T \subseteq S$. This network is our base logic. We are using the fact that we can view a logic as network as we have shown in Section 3.1.

$(S, \check{R}, T, \mathbf{r}_1, \ldots, \mathbf{r}_k)$ is our knowledge base **KB**, presented as a network. An argument X is a walk along this network. Let $X = \{x_1, \ldots, x_n\}$ and $Y = \{y_1, \ldots, y_m\}$ be two arguments. We say that Y attacks X if $T \cup \{x_1, \ldots, x_n, y_1, \ldots, y_m\}$ is not acceptable or not consistent according to the rules of X.

We may not see this immediately so we can talk about levels of pending inconsistency, inconsistency revealed after so many steps.

Note that the relation of attack is not symmetrical. X may not be able to consistently continue according to his rules but Y's rules may allow him to continue.

In our example the situation is symmetrical and C and D attack each other.

13.4 Predicate argumentation

The discussions of previous sections allow us to put forward argumentation theory where the arguments are not atomic but are involved in predication, either by having parameters themselves, i.e. the arguments themselves are like predicates in predicate logic, or by being predicated upon , being themselves the elements of a meta-predicate.

Two immediate examples come to mind

1. The argument x itself involves a claim about certain domains, e.g.

$$x = \text{all men are mortal.}$$

   See [47] for examples and references.
2. There is a predicate Q, a value predicate, operating on the argument x to form $Q(x, e)$,

$$Q(x, e) = \text{The value of } x \text{ is } e,$$

   like that introduced by Bench Capon, [43, 44], and there is a preference relation on values.

In this case the argument itself is being predicated upon by the value predicate.

This section will address predication and arguments.

The basic situation is shown in Figure 13.23.

In this figure a is an element of the domain. $P(a)$ is a predicate about this element, say 'a is red'.

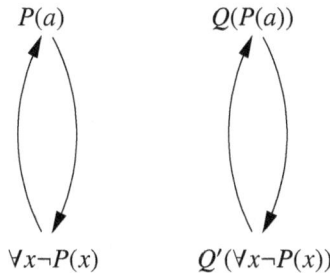

$$P(a) \qquad\qquad Q(P(a))$$

$$\forall x \neg P(x) \qquad\qquad Q'(\forall x \neg P(x))$$

Fig. 13.23.

The argument $\forall x \neg P(x)$ is a sweeping general statement saying that nothing is red. The metapredicates Q and Q' tell us what source the statements come from. So $Q(P(a))$ says $P(a)$ comes from source Q and $Q'(\forall x \neg P(x))$ tells us that $\forall x \neg P(x)$ comes from source Q'. We consider Q' less reliable than Q. We can write $Q' < Q$.

In fact, the source might be a chain of hearsay. Q'' says he heard from Q' that .... So we might be comparing

$$Q_1 Q_2 \ldots Q_k P(a)$$

with

$$Q_1' Q_2' \ldots Q_{k'}' \forall x \neg P(x).$$

So really we need to handle chains of predicates $\alpha = (Q_1, \ldots, Q_n)$. In real situations there may be several such chains as different witnesses tell stories which support the argument. See [91] for some examples and discussion.

Anyway, in our case the attack from $\forall x \neg P(x)$ on $P(a)$ cannot be accepted.

The figure has two features at the same time

1. Bench-Capon type value predicates Q, Q'.
2. The arguments themselves are predicate logic arguments.

We therefore need a language which allows us to express the following

1. Given a wff φ and a predicate $Q(x)$, we are allowed to write $Q(\varphi)$. (Think of Q as a provability predicate, for example.)
2. Try and make quantified $\forall x \varphi(x)$ behave like some instantitated $\varphi(x_0)$, in other words, eliminate quantification. Quantification can be eliminated if the domain is finite and known. Say $D = \{a_1, \ldots, a_n\}$ then we can write

$$\exists x \varphi(x) = \bigvee_i \varphi(a_i)$$
$$\forall x \varphi(x) = \bigwedge_i \varphi(a_i)$$

But if we do not make such an assumption we still need a solution to how to eliminate quantifiers.

If we succeed in the above two tasks we will have reduced the problem of predicate argumentation to ordinary Dung networks. How are we going to do it?

There is a language introduced by D. Gabbay in [136, 138], called **HFP**, Hereditarily Finite Predicates. It allows to write expressions like $P(\varphi)$.

It was studied in [138] in the chapter entitled "Self fibring of predicate logics". The language is powerful and has been applied to the logics of security among other applications.

To eliminate quantifiers we use Hilbert Epsilon symbol see [236].

We can write $\varepsilon x \varphi(x)$. This is a term which picks up in the model an element satisfying φ and if no such element exists (i.e. $\forall x \neg \varphi(x)$ holds) then it picks up an arbitrary element. Thus we have

$$\varphi(\varepsilon x \varphi(x)) \to \exists y \varphi(y)$$
$$\neg \varphi(\varepsilon x \varphi(x)) \to \forall y \neg \varphi(y)$$

So by using the Epsilon symbol, we don't need quantifiers and the language can be treated like it were propositional.

So Figure 13.23 becomes Figure 13.24

This is a propositional network. We need to somehow extract from it by geometric means the reason for the attack relation and why the Q' cannot attack the Q. Let us begin first by taking predicate logic formulated with the ε-symbols. We choose monadic logic for simplicity. We have a set $\mathbb{P}$ of unary predicates, variables V and the connective $\Downarrow$ and the ε-symbol.

Definition 13.36. *Formulas and terms are defined as follows*

1. *x is a term with x free, for $x \in V$.*
2. *If $\mathbf{t}(x_1, \ldots, x_n)$ is a term with x_i free then $P(\mathbf{t}(x_1, \ldots, x_n))$ is a fomula with x_i free.*
3. *If $\varphi(x, x_1, \ldots, x_n)$ is a formula with x, x_i free then $\varepsilon x \varphi(x)$ is a term with x_i free.*
4. *If A_i are wffs with $\{x_j^i\}$ for each $i = 1, 2, \ldots$ then $\Downarrow \{A_i\}$ is a wff with $\{x_j^i\}$ free.*

$$P(a) \qquad\qquad Q(P(a))$$

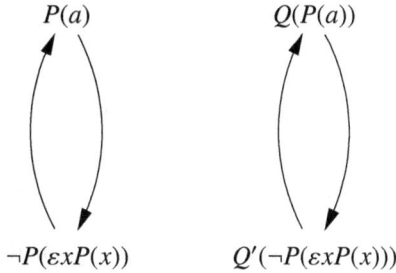

$$\neg P(\varepsilon x P(x)) \qquad Q'(\neg P(\varepsilon x P(x)))$$

Fig. 13.24.

Definition 13.37. *Now that we have the wffs of the language we can define a canonical network as follows.*

Let S^* be the set of all wffs together with the new additional predicates $P^*(x)$, for any predicate P and variable x.

Define R^* on S^* by

1. *$P(x)R^*P^*(x)$ and $P^*(x)R^*P(x)$.*
   This gives us an assignment
2. *AR^*B if B is an immediate subformula of A*
   This gives us the table for $\Downarrow$
3. *$\Downarrow \varphi(\varepsilon x\varphi(x))R^*\varphi(y)$, y any term*

   This corresponds to the quantifier rule $\forall x\neg\varphi(x) \rightarrow \neg\varphi(y)$.

Remark 13.38.

1. Note that we do not have the axiom:

   $$(\sharp) \qquad \forall x(P(x) \leftrightarrow Q(x)) \rightarrow (\varepsilon x P(x) = \varepsilon x Q(x)).$$

   This means that $\varepsilon x P(x)$ and $\varepsilon x(P(x) \wedge P(x))$ may choose different elements. So our semantics for the logic without $(\sharp)$ is syntactic. A model is a function **m**, giving values to all wffs such that

   $$(*) \qquad \mathbf{m}(P(y)) = 1 \Rightarrow \mathbf{m}(P(\varepsilon x P(x))) = 1$$

   If we do adopt $(\sharp)$ as an axiom, we can have set theoretic models, where we have a domain D and a selection function **s** giving selection from any subset $D_0 \subseteq D$ and element $\mathbf{s}(D_0) \in D_0$. If $D_0 = \varnothing$, then $\mathbf{s}(\varnothing) \in D$.
   We now have

   $$\varepsilon x P(x) = \mathbf{s}\{x \mid P(x) \text{ holds}\}.$$

2. The axiom $(\sharp)$ is written with $\rightarrow$, the universal quantifier and $=$. The language we use has $\Downarrow$, the Epsilon symbol and no equality. We can express $\rightarrow$ using $\Downarrow$ and express the universal quantifier using the Epsilon symbol. To overcome the lack of equality we can write the axiom as

   $$(\sharp 1) \qquad \forall x(P(x) \leftrightarrow Q(x)) \rightarrow (A(\varepsilon x P(x)) \rightarrow A(\varepsilon x Q(x)))$$

Where A is an arbitrary new predicate. This expression is OK because the axiom has implicit quantifier $\forall$, i.e., it is

$$\forall P \forall Q \forall A(\sharp 1).$$

Proposition 13.39. *The stable extensions of the canonical network* (S^*, R^*) *of Definition 13.37, yield exactly all predicate models with ε-symbols based on $\Downarrow$ and V.*

Proof. Let $\mathbf{m}$ be a model for the language. Following Remark 13.38, $\mathbf{m}$ is a syntactical model, giving values in $\{0, 1\}$ to all wffs, and satisfying

$$\mathbf{m}(\varphi(y)) = 1 \rightarrow \mathbf{m}(\varphi(\varepsilon x \varphi(x))) = 1.$$

Consider now the extension $\mathbb{E}_\mathbf{m}$ defined by

$$P(x) \in \mathbb{E}_\mathbf{m} \text{ if } \mathbf{m}((x)) = 1$$
$$P^*(x) \in \mathbb{E}_\mathbf{m} \text{ if } \mathbf{m}(_(x)) = 0.$$

We now check an arbitrary point $A \in S$, that A is in $\mathbb{E}_\mathbf{m}$ iff $\mathbf{m}(A) = 1$. This we do by structural induction which is also induction on the tree (S, R).

1. A is attacked by all its main subformuals Y. If $A =\Downarrow \{Y\}$, then really $A = \bigwedge_Y \neg Y$ and so if $\mathbf{m}(Y) = 1$, which by the induction hypothesis means that Y is in the extension $\mathbb{E}$, for any Y, then $\mathbf{m}(A) = 0$, and also A is not in the extension $\mathbb{E}$. If for all $Y, \mathbf{m}(Y) = 0$, then by the induction hypothesis, no such Y is in $\mathbb{E}_\mathbf{m}$. Therefore A is in $\mathbb{E}_\mathbf{m}$, and we also have that $\mathbf{m}(A) = 1$, because $A = \bigwedge \neg Y$. Thus we conclude for this case that A is in $\mathbb{E}_\mathbf{m}$ iff $\mathbf{m}(A) = 1$.
2. We now check the case of $A =\Downarrow B$ where $B = \varepsilon x \varphi(x)$. In this case we have that A is attacked by B and also by any $\varphi(y)$, for any term y.
   We must show that if any of the attackers Z gets value 1 (i.e. $\mathbf{m}(Z) = 1$ which means by the induction hypothesis that Z is in $\mathbb{E}_\mathbf{m}$) then $\mathbf{m}(A) = 0$, and A is not in the extension $\mathbb{E}_\mathbf{m}$. Otherwise $\mathbf{m}(A) = 1$.
   If $\mathbf{m}(B) = 1$, then clearly $\mathbf{m}(A) = 0$. If $\mathbf{m}(\varphi(y)) = 1$, then by (*) $\mathbf{m}(B) = 1$ and so $\mathbf{m}(A) = 0$. If $\mathbf{m}(\varphi(y)) = 0$ for all y, then $\mathbf{m}(\neg B(\varepsilon x B(x))) = 1$, but also from the argumentation network point of view, $A \in \mathbb{E}_\mathbf{m}$.
3. Assume we have a stable extension $\mathbb{E}$. Let $\mathbf{m}_\mathbb{E}$ be defined by

$$\mathbf{m}_\mathbb{E}(P(x)) = 1 \text{ iff } P(x) \in \mathbb{E}.$$

To show that this holds for any $A \in S$, we follow similar reasoning as in case (1).

Let us now also deal with predicates on predicates of the form $Q(P(x))$.
There are difficulties with the ε-symbol in this case, as we shall see.
Let us give the formal definitions first and then follow with a discussion.

Definition 13.40 (Syntax). *Consider a language with variables* $V = \{x_1, x_2, \ldots\}$, *the Dagger connective* $\Downarrow$ *and the ε-operator* (εx) *and a set of unary predicates*

$$\mathbb{P} = \{P_1, P_2, \ldots\}.$$

We define the notion of a term and a formula with free variables

1. *Any $x \in V$ is a term. x is free in the term x.*
2. *If P is an atomic predicate and $\mathbf{t}(x_1, \ldots, x_n)$ is a term with the free variables $x_1, \ldots, x_n$, then $P(\mathbf{t}(x_1, \ldots, x_n))$ is a formula with the free variables $x_1, \ldots, x_n$.*
3. *Let $\alpha = (P_1, \ldots, P_k)$ be a sequence of elements from $\mathbb{P}$. If $\varphi(x, x_1, \ldots, x_n)$ is a formula with the free variables $x, x_1, \ldots, x_n$, then $\varepsilon^\alpha x \varphi(x)$ is a term with the free variables $x_1, \ldots, x_n$.*
4. *If Q is a predicate and $\varphi(x_1, \ldots, x_n)$ is a formula with free variables $x_1, \ldots, x_n$ then $Q(\varphi)$ is a formula with free varaibles $x_1, \ldots, x_n$.*
5. *If $\varphi_i, i = 1, \ldots, k$ are formulas with free variables $x^i_j, i = 1, \ldots, k, j = 1, \ldots, n(i)$, respectively, then $\Downarrow \{\varphi_i\}$ is a formula with the free varaibles*

$$\{x^i_j \mid i = 1, \ldots, k, j = 1, \ldots, n(i)\}$$

6. *Let $\mathbb{L}(\mathbb{P})$ be the language defined by clauses (1), (2), (5). $\mathbb{L}(\mathbb{P}, \varepsilon)$ the language defined by (1), (2), (3), (5) and $\mathbb{L}(\mathbb{P}, \varepsilon, \text{Fib})$ be the language defined by (1)–(5).*

Definition 13.41 (Models). *Let $\mathbb{P}^*$ be the set of all finite sequences from $\mathbb{P}$. For each such sequence $\alpha \in \mathbb{P}^*$, let $\mathbf{m}_\alpha$ be a classical monadic model for the language $(\mathbb{P}, \Downarrow)$, i.e. $\mathbb{L}(\mathbb{P})$. We turn the system $\{\mathbf{m}_\alpha\}$ into a model of our syntax.*

We can assume all models $\mathbf{m}_\alpha$ have domain V.

First we convert each $\mathbf{m}_\alpha$ into a model of the language $\mathbb{L}(\mathbb{P}, \varepsilon)$. We need to deal with $\varepsilon^\alpha x \varphi(x)$ and assign it a value. We use induction.

1. *For a wff φ without ε, let $\varepsilon^\alpha x \varphi(x)$ be any x such that $\mathbf{m}_\alpha \vDash \varphi(x)$. Otherwise let $\varepsilon^\alpha x \varphi(x)$ be any element. Note that if $\mathbf{m}_\alpha \vDash \exists x \varphi(x)$ then $\varphi(\varepsilon^\alpha x \varphi x)$ holds, otherwise $\neg\varphi(\varepsilon^\alpha x \varphi(x))$ holds. We say that $\varepsilon^\alpha x \varphi(x)$ has been assigned a value.*

    Also note, following Remark 13.38 that we are not committed to assigning the same element to the Epsilon symbol applied to two logically equivalent formulas.
2. *Let $\Psi(x, \mathbf{t}_1, \ldots, \mathbf{t}_k)$ be a wff of classical logic with ε and assume all $\varepsilon^\beta x \varphi(x)$ in Ψ have been assigned values in $\mathbf{m}_\alpha$ for any α, β. This means that Ψ itself can be evaluated in $\mathbf{m}_\alpha$, because all expressions $\varepsilon^\beta x \varphi(x)$ in Ψ are assigned elements in $\mathbf{m}_\alpha$. So let $\varepsilon^\alpha x \Psi(x)$ be some arbitrary element in $\mathbf{m}_\alpha$ which satisfies Ψ, otherwise, if there is no element in $\mathbf{m}_\alpha$ satisfying Ψ, choose any element.*

Thus we now have that each $\mathbf{m}_\alpha$ can be considered a model of $\mathbb{L}(\mathbb{P}, \varepsilon)$. We now need to extend our definition to model expressions like $Q(\varphi)$. We regard Q as a modality

$$\mathbf{m}_\alpha \vDash Q(\varphi) \text{ iff } \alpha * (Q) \vDash \varphi$$

where $\alpha * (Q)$ is the sequence obtained from α by adding Q at the end.

For example, we have

$$\alpha \vDash P_1(P_2(P_3(\varphi))) \text{ iff } (\alpha, P_1, P_2, P_3) \vDash \varphi.$$

Remark 13.42. Note that now we have syntax and semantics. So let us see what theorems are valid.

1. $\alpha \vDash \neg\varphi(\varepsilon^\alpha x \varphi(x) \rightarrow \neg\varphi(y)$
   $\alpha \vDash \varphi(\varepsilon^\beta x \varphi(x)) \rightarrow \varphi(y)$

2. $\alpha \vDash Q(\Downarrow \{A, B\})$ iff $(\alpha, Q) \vDash\Downarrow (A, B)$
   iff $(\alpha, Q) \nvDash A$ and $(\alpha, Q) \nvDash B$
   iff $\alpha \nvDash Q(A)$ and $\alpha \nvDash Q(B)$
   iff $\alpha \vDash\Downarrow (Q(A), Q(B))$.
   We can see that the following holds

$$Q(\Downarrow \{A_i\}) \equiv\Downarrow \{Q(A_i)\}.$$

3. Note that although $\mathbf{m}_\alpha$ have the same domain V, the element which $\varepsilon^\alpha x P(x)$ picks up at each α may be different. The reason is that the extension of P at each $\mathbf{m}_\alpha$ may be different and so we will not be able to organise the same choice for $\varepsilon^\alpha x P(x)$.

4. Consider $\varepsilon^\alpha x(Q_1, Q_2 P(x))$. We are choosing an x at α, such that $P(x)$ holds at $\beta = (\alpha, Q_1, Q_2)$. This is possible because all models have the same domain V and for each Q the "modality" suggested by Q is linear discrete. Generally in modal logic when we write $t \vDash \Box \exists x P(x)$ we may have a different c_s such that $s \vDash P(c_s)$ for each s such that tRs (s accessible to t). We have

$$t \vDash \Box P(\varepsilon x P(x))$$

to deduce

$$t \vDash \exists z \Box P(x)$$

but this is not true!
So the rule in general

$$\vdash \varphi(y) \to \exists z \varphi(z)$$

becomes problematic.
We avoid this problem because all of our ε-symbols have the form $\varepsilon^\beta x$ for some β. So $\varepsilon^\beta x P(x)$ goes to $\mathbf{m}_\beta$ only.
Let us illustrate: Assume

$$\mathbf{m}_\alpha \vDash Q_1(P(a)) \wedge Q_2(P(b))$$

Thus

$$\mathbf{m}_\alpha \vDash Q_2(P(\varepsilon^{(\alpha, Q_1)} x P(x)) \wedge Q_2(P(\varepsilon^{(\alpha, Q_2)} x P(x))).$$

We have no problems with that.
However, had we not indexed the ε operator we would have got

$$\mathbf{m}_\alpha \vDash Q_1(P(\varepsilon x P(x))) \wedge Q_2(P(\varepsilon x P(x)))$$

which would have allowed us to deduce

$$\mathbf{m}_\alpha \vDash \exists y[Q_1(P(y)) \wedge Q_2(P(y))]$$

which may not be true.
Another approach is to make elements of the domain be vectors of the form $y = (a, b)$ and the semantics would be different. Which solution we choose depends on the application. Our aim is not to solve difficulties with the ε-symbol but to apply it to argumentation. For this reason we also simplify by adding the axiom

$$Q \Downarrow \{A_i\} \equiv\Downarrow \{QA_i\}.$$

Our purpose is not to develop a comprehensive theory of **HFP** $+\varepsilon$ symbol but to develop a predicate argumentation theory and so we need less of the **HFP** $+\varepsilon$-symbol.

Example 13.43. Let us revisit Figure 13.24 and see how it can be dealt with in our semantics. This discussion is only intuitive at this stage. Consider Figure 13.25. Remember that $\neg\varphi$ is $\Downarrow\varphi$ in the Dagger language.

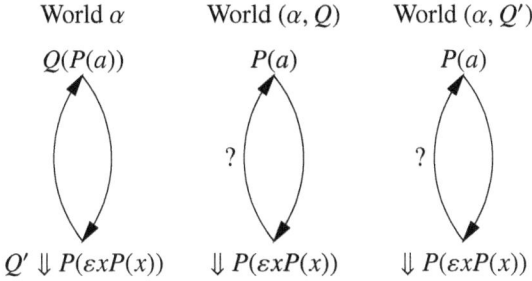

World α	World (α, Q)	World (α, Q')
$Q(P(a))$	$P(a)$	$P(a)$
	?	?
$Q' \Downarrow P(\varepsilon x P(x))$	$\Downarrow P(\varepsilon x P(x))$	$\Downarrow P(\varepsilon x P(x))$

Fig. 13.25.

We have $Q' < Q$ and α is related to (α, Q) and to (α, Q').

We would need to say that there cannot be any attack from world Q' to Q is $Q' < Q$.

We shall present a precise system later. We already see that we may need to allow attacks from one world α to another world β.

There is a way to simplify. We notice that argument $P(a)$ and argument $Q(P(a))$ are both atomic units and are independent of one another. So we can put them all in one network/model. We can recognise the world Q by taking all arguments of the form $Q(x)$. So we can have one network and the different worlds are subnetworks of it. This allows us to represent $Q' < Q$ by an attack from Q to Q' done in a certain way. The next Example 13.44 will illustrate.

Example 13.44. We consider the basic situation in Figure 13.26.

We have that argument a attacks arguments b and b'. Argument a has metalevel value Q_1, argument b has value Q and argument b' has value Q_2.

The terms $Q_1(a), Q(b)$, and $Q_2(b')$ are not part of the attack network but are metalevel to it. The information $Q_1 < Q$ and $Q_2 < Q_1$ say which value is higher than which in the Bench-Capon sense [43].

Thus the attack of a on b cannot take place because b has higher value but the attack on b' is OK because a has higher value.

Now, how are we going to represent this metalevel information in the object level? We need:

1. To include $Q_1(a), Q(b)$ and $Q_2(b')$ as arguments in the object level. We know how to do that from the previous discussion in this section.
2. We need to represent the metalevel information $Q_1 < Q$ and $Q_2 < Q_1$ in the object level. To do this we need two higher level concepts introduced in Chapter 7. These are

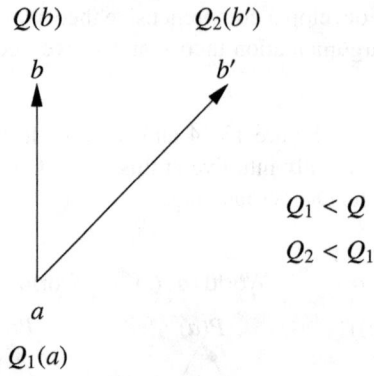

$$Q_1 < Q$$
$$Q_2 < Q_1$$

Fig. 13.26.

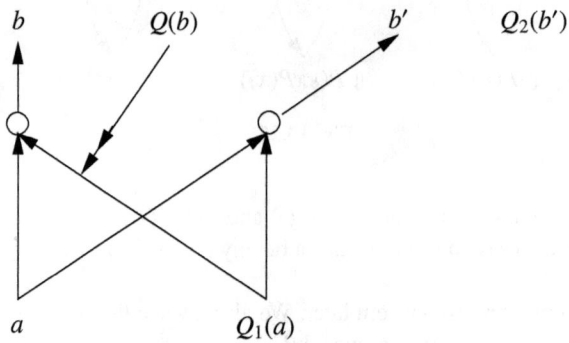

Fig. 13.27.

a) joint attacks
b) higher level attacks.
The concept of joint attacks is illustrated in Figure 13.28 and is discussed in [164].

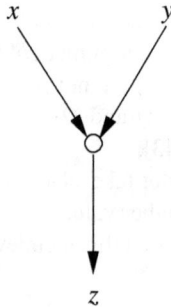

Fig. 13.28.

x and y join forces to attack z. For the attack to succeed, we need both x and y to be "in" and to be joined.

The concept of higher level attack [163], which originates in [37][7] allows arguments to attack and disconnect attacks arrows. This is illustrated in Figure 13.29

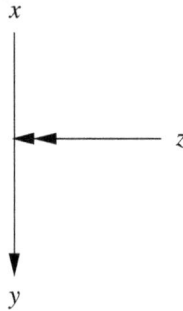

z attacks the arrow from x to y

Fig. 13.29.

If z is "in" the attack from x to y is "out" (x, y are still untouched).

Figure 13.27 represents our implementation of Figure 13.26, where we use joint attacks and higher level attacks. We have that the double arrow emanating from $Q(b)$ onto the arrow which joins $Q_1(a)$ to the attack on b, is representing the fact that $Q_1 < Q$ and therefore making sure that a cannot attack b.

Note that in Figure 13.30 the following holds:

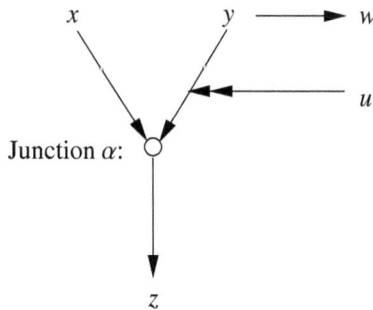

Junction α:

Fig. 13.30.

u attacks the contribution of y to the joint attack (on z), which is forming, (or getting organised or "gathering") at junction α. This means that the joint attack cannot go on, if u is in. It does not mean that x can attack alone! This is why we have the junction notation.

---

[7] This concept was later independently discovered by S. Modgil [260] and studied by Baroni and others [32]. See discussion and references in Chapter 8.

y itself is not attacked by u, so y's attack on w is valid. In Figure 13.30 we have the extension $u = $ in, $y = $ in, $w = $ out, $x = $ in, $z = $ in.

Remark 13.45 (Representation of joint attacks, and higher level attacks). In Example 13.44 and in Figure 13.27 we used joint attacks and higher level attacks to code the metalevel information that value Q_1 is lower than value Q_2, written as $Q_1 < Q_2$. We therefore need to show how to represent joint attacks and higher level attacks in our canonical model. We need to translate Figures 13.28 and 13.29 into the canonical model.

We start with joint attacks. Following [164], consider Figure 13.31

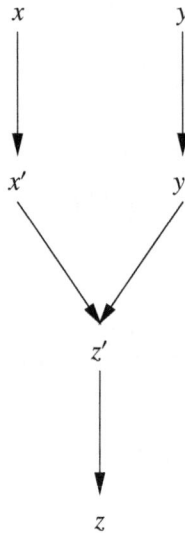

Fig. 13.31.

Note that only when x and y are "in", do we get that z is out.

Consider now Figure 13.32

Clearly this is the same figure as Figure 13.31, and we can see that the joint attack of x and y is executed by $x \wedge y$ in the canonical model.

Let us now address higher level attack. Consider Figure 13.28. This can be implemented by Figure 13.33 (which is the same as Figure 9 of [163, p. 365].

We note that this figure represents the higher level attack of Figure 13.29, as shown in [163] and as can be readily verified directly. We further observe that the part of Figure 13.33 enclosed in a circle, is the same as the representation of joint attacks in Figure 13.32. This is a joint attack of x and z' on y. We note that z' is really $\neg z$. So we get that Figure 13.29 can be represented as Figure 13.34.

Now since the case of joint attacks we already know how to represent (Figure 13.32), we can also represent higher level attacks.

Remark 13.46. Following our discussion in Remark 13.45, Figure 13.27 becomes Figure 13.31. In this figure the fact that $Q_1 < Q$ is represented by the fact that $\Downarrow Q(b)$,

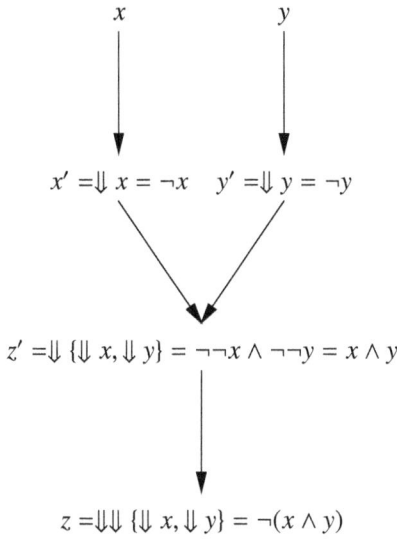

Fig. 13.32.

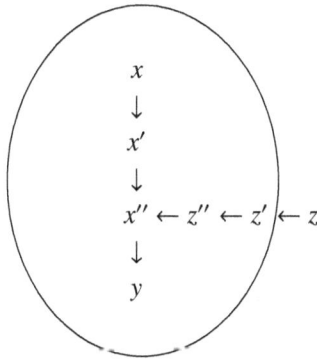

Fig. 13.33.

(i.e. $\neg Q(b)$) joins the contribution of $Q_1(a)$ to the joint attack of $\{a, Q_1(a)\}$ on b. We thus have the joint attack of $[a$ and the joint attack of $\{Q_1(a), \neg Q(b)\}]$ jointly attacking b. In symbols:

$$\text{Joint}\{a, \text{Joint}\{Q_1(a), \neg Q(b)\}\} \text{ attack on } b.$$

Note that what we have not done yet is to have a uniform representation of how $Q'(x)$ joins all the attacks of x on any y and how any $Q''(y)$ also participates but Q'' participation is effective only if $Q' < Q''$.

So far we have participation of Q'' with Q' only when $Q' < Q''$.

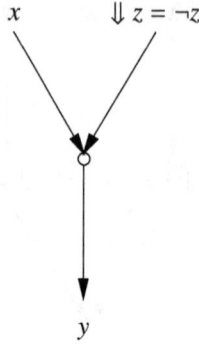

Fig. 13.34.

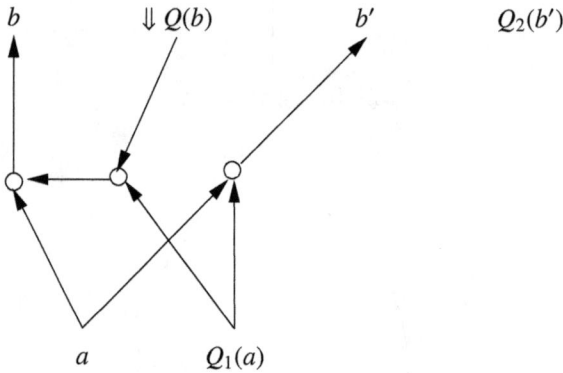

Fig. 13.35.

13.5 Resource considerations

Let us do again Definition 13.6, this time paying attention to resources.

Definition 13.47. *Let (S, R, α) be a decorated ordering. Let h be a model. Let us define the notion of resource annotated valuation of nodes in S. We write Val(h, x) = $(\mathbf{f}(x), \mathbf{F}(x))$.*

1. *Val(h, x) = $(h(\alpha(x)), \{x\})$ for x an endpoint of the ordering.*
2. *Val(h, x) = $(1 - \max_{xRy}(\mathbf{f}(y), \bigcup_{xRy} \mathbf{F}(y))$ for x which is not an endpoint.*
3. *For any $x, \mathbf{f}(x)$ is the truth value in $\{0, 1\}$ of the formula $\alpha(x)$ and $\mathbf{F}(x)$ indicates which end nodes x relies on. $\mathbf{F}(x)$ is needed for resource logic, not for classical logic.*

Example 13.48 (Resource considerations). We now explain the components of Definition 13.47.

Consider the following deduction

1. *A*, assumption
2. *A* → (*A* → *B*), assumption
3. *A* → *B*, from (1) and (2) by modus ponens.

4. B from (1) and (3) by modus ponens.

The above deduction is not valid in linear logic because (1) A is used twice (in (3) and (4).

A correct deduction for linear logic would be the following

1. A, assumption
2. A, assumption
3. $A \rightarrow (A \rightarrow B)$, assumption
4. $A \rightarrow B$, from (2) and (3)
5. B, from (1) and (4).

The above means that the formula

$$C = (A \wedge [A \rightarrow (A \rightarrow B)] \wedge \neg B)$$

is not a contradiction in resource logic as described above. While

$$D = A \wedge A \wedge (A \rightarrow (A \rightarrow B)) \wedge \neg B$$

is a contradiction.

Let us now reflect the resource idea in our $\Downarrow$ system.

Example 13.49 (Resources considerations for $\Downarrow$).

1. First recall a convention for drawing figures. When we have a binary relation $R \subseteq S \times S$ and we have xRy, we draw it as $x \leftarrow y$ (arrow going into the x). We can now build two possible decorated acyclic graphs for the formula $\neg A \wedge (\neg A \rightarrow (\neg A \rightarrow B)) \wedge \neg B$. We use the equivalent $\neg A \wedge (A \vee A \vee B) \wedge \neg B$. See Figures 13.36 and 13.37.

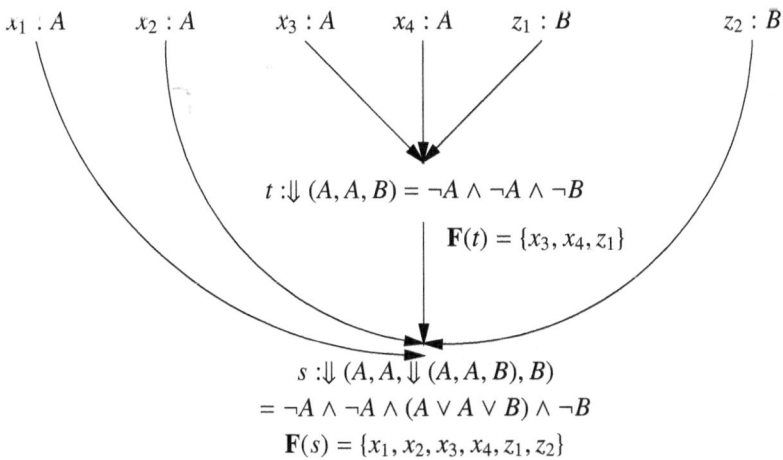

$$t :\Downarrow (A, A, B) = \neg A \wedge \neg A \wedge \neg B$$
$$\mathbf{F}(t) = \{x_3, x_4, z_1\}$$

$$s :\Downarrow (A, A, \Downarrow (A, A, B), B)$$
$$= \neg A \wedge \neg A \wedge (A \vee A \vee B) \wedge \neg B$$
$$\mathbf{F}(s) = \{x_1, x_2, x_3, x_4, z_1, z_2\}$$

Fig. 13.36.

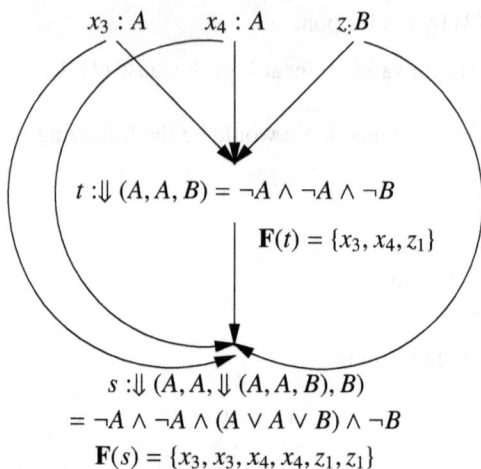

$$x_3 : A \qquad x_4 : A \qquad z:B$$

$$t :\Downarrow (A, A, B) = \neg A \wedge \neg A \wedge \neg B$$

$$\mathbf{F}(t) = \{x_3, x_4, z_1\}$$

$$s :\Downarrow (A, A, \Downarrow (A, A, B), B)$$
$$= \neg A \wedge \neg A \wedge (A \vee A \vee B) \wedge \neg B$$
$$\mathbf{F}(s) = \{x_3, x_3, x_4, x_4, z_1, z_1\}$$

Fig. 13.37.

In Figure 13.36 all the elements of $\mathbf{F}(s)$ are different. No node is used twice. In Figure 13.37 all nodes are used twice. In classical logic it does not matter. In resource logic it does matter.

Remark 13.50 (Argumentation frames based on linear logic). The connection between logic and networks, together with the discussion and figures of this section shows that for the case of acyclic networks, if we allow for each node in the network to attack only a single other node at the most, then the corresponding underlying logic is linear propositional logic based on Dagger, and the network is actually a tree.

If we want a node to attack several other nodes then we must make multiple copies of it. Thus for example Figure 13.38 becomes Figure 13.39.

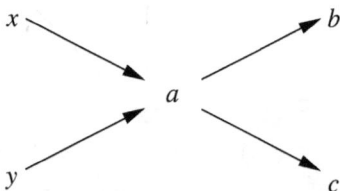

Fig. 13.38.

If we allow cycles the definitions are more complicated.

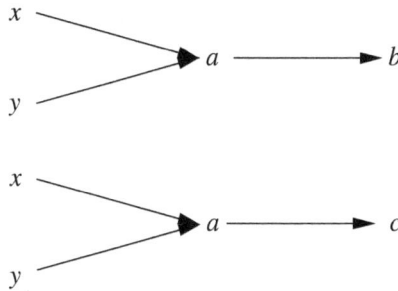

Fig. 13.39.

13.6 Conclusion: Discussing the correspondence between Dung networks and Dagger logic and comparing it with seemingly relevant literature

In this concluding Section, we would like to evaluate the correspondence between Argumentation networks and classical logic formulated with the Dagger connective. We try and answer possible criticisms from both the pure logic community and the down to earth practical argumentation community.

The title of this chapter may be seen by some as a sweeping statement. We are saying that Dung argumentation frames together with its machinery of finding extensions is essentially equivalent to classical propositional logic with the Pierce-Quine Dagger connective and its machinery for finding models for wffs.

We have discussed and demonstrated the connections under (S1) and (S2) of Section 2.2. However, since there is a general confusion in the community about what it means to be an "equivalent" system, we thought we had better clarify these concepts in this section. Furthermore, this clarification will allow us to compare this work with some seemingly related works, namely, references [47, 48, 97, 319, 208, 81].

Let us begin with a very simple example of two systems which are *not* the same. Take two strong enough computer languages. Say modern Basic and Pascal. These languages are strong enough for each to simulate the other. Give me a program in one and I can write an equivalent program in the other, where "equivalent" means doing exactly the same job!

Yet, despite the above, we will not say the two languages are the same. Yes, they have the same expressive power, Yes, they can do the same jobs. However, their basic internal constructs are different. They are not based on the same "internal movements".

We ask the reader to accept this intuitively. Don't ask us what we mean by "internal movements".

We now show you two systems which are generally considered the same.

Start with intuitionistic implication $\Rightarrow$, formulated and axiomatised as a Hilbert System, call it **I**. We have the following two axioms and the rules of Modus ponens and substitution.

1. $a \Rightarrow (b \Rightarrow a)$
2. $(a \Rightarrow (b \Rightarrow c)) \Rightarrow ((a \Rightarrow b) \Rightarrow ((a \Rightarrow c)))$

We now want to add negation to the system. We do it in two ways, creating two separate systems $\mathbf{I}_1$ and $\mathbf{I}_2$. $\mathbf{I}_1$ adds a negation unary symbol $\neg_1$ with the following additional axioms:

3. $(a \Rightarrow \neg_1 b) \Rightarrow (b \Rightarrow \neg_1 a)$
4. $a \Rightarrow (\neg_1 a \Rightarrow b)$

The second system $\mathbf{I}_2$ adds a constant to the logic, call it $\mathbf{f}$, with the additional axiom

5. $\mathbf{f} \Rightarrow a$

We now show that the two systems are really the same.

$\mathbf{I}_2$ finding itself inside $\mathbf{I}_1$

We are now in the raelm of $\mathbf{I}_1$. We want to look at $\mathbf{I}_1$ through the eyes of $\mathbf{I}_2$. $\mathbf{I}_2$ has a constant $\mathbf{f}$. Can we find it in $\mathbf{I}_1$? Yes. Let $\mathbf{f}_1$ be any $q \wedge \neg_1 q$. We must prove in $\mathbf{I}_1$ that any two $p \wedge \neg_1 p$ and $q \wedge \neg_1 q$ are equivalent (i.e. from axioms (1)–(4) we prove $\mathbf{I}_1 \vdash (p \wedge \neg_1 p) \Rightarrow (q \wedge \neg_1 q)$).

This makes $\mathbf{f}_1$ unique. We now prove in $\mathbf{I}_1$ that $\mathbf{f}_1 \Rightarrow a$ is a theorem of $\mathbf{I}_1$. Having done all that we managed to look at $\mathbf{I}_1$ through the eyes of $\mathbf{I}_2$. We found an $\mathbf{f}_1$ which is what $\mathbf{I}_2$ has.

$\mathbf{I}_1$ finding itself in $\mathbf{I}_2$

Here we want to look at $\mathbf{I}_2$ through the eyes of $\mathbf{I}_1$. $\mathbf{I}_2$ has $\neg_1$ as negation. Can we find it in $\mathbf{I}_2$? The anser is yes. Let

$$\neg_2 a = a \Rightarrow \mathbf{f}.$$

We now need to prove in $\mathbf{I}_2$ (using axioms (1), (2) and (5)) that axioms (3) and (4) hold for $\neg_2$.

This means prove in $\mathbf{I}_2$ that:

3*. $(a \Rightarrow (b \Rightarrow \mathbf{f})) \Rightarrow (b \Rightarrow (a \Rightarrow \mathbf{f}))$
4*. $a \Rightarrow ((a \Rightarrow \mathbf{f}) \Rightarrow b)$

This must be done, and can be done, using axioms (1), (2), (5) of $\mathbf{I}_2$.

Why is it that all logicians agree that $\mathbf{I}_1$ and $\mathbf{I}_2$ are the same really? It is because both formulations of negation are based on the same idea. As it says in the Bible, "Keep away from falsity".

Now let us compare the above with what we did in (S1) and (S2) of Section 2.2. The Peirce–Quine connective dagger of logic and the Dung attack relation of argumentation are based on the same idea. This we have shown. Now the question is are we doing for logic and argumentation network something similar to what we did for $\mathbf{I}_1$ and $\mathbf{I}_2$?

Logic finding itself in the argumentation world

The typical figure for logic is Figure 13.1. As discussed in (S2), of Section 2.2, logic can "find" this figure in the argumentation world as Figure 13.2.

Argumentation finding itself in the logic world

A typical argumentation network is the one of Figure 2.1. As discussed under (S1) of Section 2.2, this network can find itself in the logic world as Exercise 4 mentioned in (S1) of Section 2.2. This direction is mathematically complex, as seen from Remark 13.22 (in fact we prove a stronger mathematical theorem for this direction).

We now want to clarify the concept of one system **S1** acting as a metalevel language to describe another system **S2**. We do this by example, and once the reader understands what we mean, we can compare our chapter with the papers [47, 48, 97, 319, 208, 81].

Our starting point is classical predicate logic with term symbols (constants). We want to describe the Hilbert system of intuitionistic implication with axioms (1) and (2), modus ponens and substitution. To achieve this we must say what is an atom, a formula, an axiom and a theorem.

Let $\{q_1, q_2, \ldots\}$ be the atomic wffs of **I**. We use in predicate logic the constants $\{\mathbf{q}_1, \mathbf{q}_2, \ldots\}$ to represent them. Let the function symbol **Imp** represent $\Rightarrow$. Let the predicates **Atom** and **Formula** represent the notion of atomic formula and a general wff of **I**. Let the predicate **axiom** represent the notion of an axiom of **I** and the predicate **Theorem** represent the notion of a theorem of **I**. All the above are predicates and functions in predicate logic.

We now describe **I**. This will be a predicate logic theory $\Delta(\mathbf{I})$.

(P1)   **Atom** $(\mathbf{q}_i), i = 1, 2, \ldots$
(P2) $\forall x(\mathbf{Atom}(x) \rightarrow \mathbf{Formula}(x))$
         $\forall xy(\mathbf{Formula}(x) \wedge \mathbf{Formula}(y) \rightarrow \mathbf{Formula}(\mathbf{Imp}(x, y)))$
(P3) $\forall xy(\mathbf{Axiom}(\mathbf{Imp}(x, \mathbf{Imp}(y, x))))$
         $\forall xy(\mathbf{Axiom}(\mathbf{Imp}(\mathbf{Imp}(x, \mathbf{Imp}(y, z)), \mathbf{Imp}(\mathbf{Imp}(x, y), \mathbf{Imp}x, z))))$
(P4) $\forall xy(\mathbf{Axiom}(x) \rightarrow \mathbf{Theorem}(x))$
(P5) $\forall xy(\mathbf{Theorem}(x) \wedge \mathbf{Theorem}(\mathbf{Imp}(x, y)) \rightarrow \mathbf{Theorem}(y))$

We now got a theory $\Delta(\mathbf{I})$ of first order logic describing **I**. We can use a resolution theorem prover to ask for example does $\Delta(\mathbf{I})$ prove in classical logic **Theorem**$(\mathbf{Imp}(x, x))$? This means does **I** prove $a \Rightarrow a$?

We use this example to clarify the following concepts.

1. One logic/system **S1** talks about another system **S2** in the metalevel.
2. One system, say **S2** is used by another system **S1**, as a case study application.
   In our detailed example, classical logic was describing **I** as a case study application acting as a metalevel system when describing **I**.

Now that we have these concepts, let us compare with other papers.

Grossi's paper [208] uses modal language as a metalevel language describing argumentation.

Paper [97] is a survey paper of argumentation theory in general.

Paper [81] discusses with examples, various ways of looking at argumentation theory. It is well worth reading. It is written in the spirit of the current chapter but it was written before we discovered the "equivalence" described in the current chapter.

Paper [48] characterises extensions in set-theoretic terms. It does not deal with equivalence or connection with classical logic.

Paper [319] uses adaptive logic as a metalevel language for reasoning about argumentation and extensions.

The book [47] uses classical logic as a case study application of argumentation. It uses wffs of classical logic as arguments and defines various attack relations in terms of classical logic consistency. It is strongly related to logic but has no bearing on the question of the equivalence of logic with $\Downarrow$ and argumentation netwroks. To make the point crystal clear, note that we can take another case study where the arguments are restaurant menus for first course and main course for dinner. We have a chef who tells us what goes with which dish and this is the attack relation.

In summary, we see that none of the above papers do what we do here and they seem related to us because they have "logic" in the title.

We now proceed to address some possible criticism of our work in this chapter.

The pure logician may say that the Dagger connective is not central to logic. Much more important are the traditional intuitive connectives **and, or, not,** and **implies**. The Sheffer stroke and the dagger were studied in logic for technical reasons. The logic community was interested in connectives which are functionally complete (the Sheffer stroke and the Peirce-Quine dagger) and can define all other connectives and can be axiomatised by a minimal number of axioms containing a minimal number of letters. However, such connectives are not so important. So if the argumentation attack relation is essentially the Dagger connective, then good for argumentation. We the logicians are not necessarily interested.

My answer to this is to say that we are generalising the concept of what is a logical system, a foundational issue for logic, and advise the logician to take a look at section 3. This might help.

To the argumentation people I would say that first they should realise that they have the power of classical logic in a natural way. The correspondence shown in Table 13.2 and discussed in (S1) and (S2) in Section 2.2 and further methodologically described in Section 6 above is very natural. So they should not be surprised at what one can do with argumentation. See, for example, Section 2.4.

The argumentation researcher may come up with three objections:

1. The correspondence in Table 2 is for a very restricted argumentation frame, namely The canonical argumentation frame based on Dagger logic of Definition 13.14 which is basically a tree with small cycles at the top as in Figure 13.2. This is a very restricted argumentation frame and all the richness of argumentation is lost.

   To this we answer that this canonical frame is only our starting point, showing how logic can find itself in argumentation. We get the richness of all argumentation frames with cycles , by applying equivalence relations to this starting frame and obtaining new factor frames. This is done in Section 2.4 and also see Remark 13.45 where higher level attacks is simplified and is reduced to logic. Section 4 gives as predicate argumentation suggested by the correspondence with logic.

2. The second objection is more tricky to answer. It claims that ordinary argumentation is simple and that the connection with logic makes things more complicated. So it can't be useful. My answer is that it is not more complicated, just a new natural way of looking at argumentation. The correspondence is natural and anyone used to logic can see it. Real complications arise when the new representations

are completely different paradigm, say something like category theory. Logic is already related to and is firmly embedded in argumentation!

3. The third objection is linguistic. We talk about classical logic and yet we end up needing three valued logic, {in ,out, undecided}. My answer to this is two fold.

   First, having functions undefined on some elements of its domain, was never considered a departure from classical logic. Look at the field of partial recursive functions, it is one of the pillars of classical logic!

   We talk about three values for convenience and also recall Chapter 10, where we use continuous values in [0,1]. But the equational model of Chapter 10 is a different interpretation altogether and equally applies to Logic or to Argumentation.

   Second, from the semantic point of view it can be argued, that any non-classical logic which can be characterised by a finite matrix is essentially classical logic. This is because the matrix can be expressed in classical logic. Of course this is a semantical point of view, which includes the Semantic Tableaux formulation as well. However, when it comes to writing a Hilbert system or a Gentzen system for such logics it can be extremely difficult and much different from classical logic. See our book [257].

Further, I stress that Argumentation needs a wider theoretical basis, not just for methodological reasons but for the social cohesiveness of the community. Without a good logical connection and meta-logical foundations, there is the danger that the community will fragment and be absorbed into the variety of its application areas. Those who apply argumentation to Law will get absorbed in the law community, those who apply argumentation to Agent Theory will be absorbed in the agents community. History shows us that this is how the Logic Programming community fragmented! In fact the Logic Programming community was lucky; because of its close connection with logic, a vibrant core group did survive. I am not sure this would happen to the argumentation community. The strength of argumentation is in its applicability across many diverse disciplines (much more than Logic Programming) and that can be a problem!

Now I have something to say to both communities.

It is clear from what we are doing that the most general formal setup is that of a network (S, R) with various annotation to the elements of the network and some algorithms, relying on the geometry of the network and the annotation, for going around the network updating the annotations, possibly in loops, possibly never terminating seeking steady state solutions to the annotation function.

This is the most general case and it integrates logic, argumentation networks , neural networks, ecology networks and more.

We have done this for numerical values in Chapters 9 and 10.

The important point here is that the general integrating framework is natural in itself. It has a meaning in itself and logic and networks each can be identified (I have not written this yet) internally as specialised case of the general framework. This is an important test showing the integration is natural!

It is not the case that in this integration each component loses its identity. Not at all, they give each other ideas, being residents of a more general framework.

We saw in Sections 2 and 3 that logic can be viewed as a network. However, Logic is also strong enough to describe and talk about networks. There is a feature logic can do which is not available in networks — and this is the ease of the interaction of object level and meta-level features. Can we expand networks to be able to talk about itself?

This problem we still have to solve, namely how to embed meta-level features of networks in the object level of networks themselves. In other words how can networks talk about networks much in the same way that logic can talk about logic.

We did start to address this in the predicate logic section, Section 4. Roughly we showed a correspondence between meta-predicates in logic and attacks on one network from another network. So for a network to talk about itself it must have certain attacks from one part of itself to another part, but we leave this for future papers.

We close this section by listing what future papers we can write on topics hinted in this chapter.

1. The ideas in Section 4 can give rise to several papers on predicate argumentation and a systematic study of various papers in the literature which can be simplified and integrated using predicate argumentation.
2. The ideas of Section 5, where we replace classical propositional logic by linear logic, naturally give us a handle on resource considerations in argumentation. We do not need to define resource features ad hoc but we can be systematic. Use arguments in attack only once, or if you keep using the same argument again and again , it weakens because people get weary of it, and more.
3. The results of Section 3 can be further developed in a systematic way into a paper: "What is a logical system, version 2013".

14

Numerical Network Modalities: An Exploration

14.1 Introduction

The aim of this chapter is to bring together several kinds of formalisms widely used in artificial intelligence and computer science, which up to now have been evolved in virtual independence from one another. We have in mind, on the one hand, the communities that use a variety of network models and, on the other hand, the modal logic community. Networks have different uses. We have Bayesian networks, neural networks, inheritance networks, abstract argumentation networks, flow and transport networks, transition networks, Petri-nets and so on. We also have a variety of modal and temporal logics, whose natural semantics is a possible world semantics with accessibility relations. These are akin to networks: the worlds are the network nodes and the accessibility relations provide the network arrows. It is therefore a natural unifying step to introduce and discuss the concept of *network modalities*. This is what we shall do in this chapter.

By way of comparison, we shall show that not everything with nodes and arrows falls under the remit of our formalism. The kind of models of interest to category theory are not compatible with our point of view.

We begin by explaining our basic idea using a simple example. In Figure 14.1 we

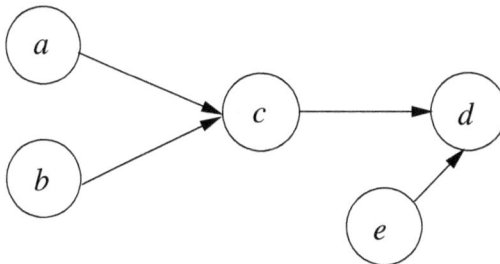

Fig. 14.1. A Simple Network

see a simple network where *a*, *b*, *c*, *d*, *e* are the nodes together with some directed arrows between some of these nodes. The nodes can be viewed as events with the arrows indicating a causal relation or as possible worlds with the arrows indicating that a and b are accessible to c and c and e are accessible to d. They can also be viewed as neurons, where a and b fire into c, and c and e fire into d.

Depending on what the network is supposed to be, we will have different annotations (labels) on the nodes and arrows. Usually for a node x we denote the value by $V(x)$. We also use the function $\mathbf{f}$ to propagate the values along the arrows. So in Figure 14.1 we propagate from a and b onto c, and onwards to d. The propagation can be done locally along each arrow or globally in parallel. In general we need a multi-variable propagation functional $\mathbf{f}$, taking as input values $x_1, \ldots, x_n$ (n arbitrary) and giving out a value $\mathbf{f}(x_1, \ldots, x_n)$. It is convenient to define $\mathbf{f}(\varnothing) = 1$ (i.e. $\mathbf{f} = 1$ for $n = 0$). In Figure 14.1 we need the use of $\mathbf{f}(V(a), V(b))$ to go to node c and $\mathbf{f}(V(c), V(e))$ to go to node d.

In looking more closely at Figure 14.1, we are able to give several interpretations for the network it schematizes.

Example 14.1 (Bernoulli Transmission). The first interpretation is that a, b, c, d, e are nodes x bearing initial values $V^0(x)$ in $[0, 1]$ and the arrows indicate how the values are transmitted in the network. Let us adopt the Bernoulli way of transmission whereby the new value $V^1(y)$ of a node y is determined as $V^1(y) = V^0(y) \prod_x (1 - V^0(x))$, where x ranges over all nodes x with arrows leading into the node y. Note that if there are no nodes leading into y we take $\prod = 1$.

Thus we have in this case $\mathbf{f}(x_1, \ldots, x_n) = \prod_i (1 - x_i)$ and the new y' is $y' = y \, \mathbf{f}(x_1, \ldots, x_n)$, where x_i range over all values of nodes with arrows leading into a target node with value y.

Thus, if we denote in Figure 14.1 the value of node x by x (i.e. $V^0(x) = x$) then we have

- $V^1(a) = a$
- $V^1(b) = b$
- $V^1(e) = e$
- $V^1(c) = c(1 - a)(1 - b)$
- $V^1(d) = d(1 - e)(1 - c)$.

We can generally define V^{n+1} in terms of V^n as follows:

- $V^{n+1}(a) = V^n(a)$
- $V^{n+1}(b) = V^n(b)$
- $V^{n+1}(e) = V^n(e)$
- $V^{n+1}(c) = V^n(c)(1 - V^n(a))(1 - V^n(b))$
- $V^{n+1}(d) = V^n(d)(1 - V^n(c))(1 - V^n(e))$

In general we have

- $V^{n+1}(y) = V^n(y)\mathbf{f}(V^n(x_1), \ldots, V^n(x_n))$, where x_i are all the nodes with an arrow leading into y.

Example 14.2 (Temporal Network). The second interpretation of the network is temporal. We read the arrow from node x to y as saying that y is a possible *immediate* future node of x, i.e. accessibility is an earlier–later relation indicated by the arrows.

The propagation function $\mathbf{g}(x_1, \ldots, x_n)$ is different in this case. If $x_1, \ldots, x_n$ are the values of all nodes leading into a target node with value y then we let $\mathbf{g}(x_1, \ldots, x_n) = \prod_i x_i$ and then the new value y' at the target node is $y' = y\, \mathbf{g}(x_1, \ldots, x_n)$.

Let $U^0(x) = x$, for the nodes x of Figure 14.1, then we have in the temporal case

- $U^1(a) = a$
- $U^1(b) = b$
- $U^1(e) = e$
- $U^1(c) = cab$
- $U^1(d) = dec$

We can generally get

- $U^{n+1}(a) = U^n(a)$
- $U^{n+1}(b) = U^n(b)$
- $U^{n+1}(e) = U^n(e)$
- $U^{n+1}(c) = U^n(c)U^n(a)U^n(b)$
- $U^{n+1}(d) = U^n(d)U^n(c)U^n(e)$.

To compare the two approaches, let us assume the values $V^0(x)$ are in $\{0, 1\}$, i.e. $a, b, c, d, e, \in \{0, 1\}$. Let q be an atomic formula and let q get the value x at node x, for $x = a, b, c, d, e$, i.e.

(1) $\qquad\qquad\qquad\qquad x \vDash q$ iff $x = 1$.

Let HA mean A holds in all immediate past points and let GA mean A holds in all immediate future points. Let us evaluate $c \vDash Hq$ and $c \vDash H\neg q$. The first holds if $a = b = 1$ and the second holds iff $a = b = 0$. It is easy to see that

$$c \vDash q \wedge Hq \text{ iff } U^1(c) = 1$$
$$c \vDash q \wedge H\neg q \text{ iff } V^1(c) = 1.$$

It is therefore sensible to define a network modality $\square q$ for the V case by

- $\mathrm{Val}(x, q) = x = V^0(x)$
- $\mathrm{Val}(x, \square^n q) = V^n(x)$.

Note the meaning of $x \vDash \square^n q$. It means "propagate the values at the network from 'distance' n in the past of node x and find the present value at the node x". In the U case the modality is a reflexive modality $q \wedge Hq$ and in the V case it is $q \wedge H\neg q$. Since our view of the modality in this example is temporal, let us look at the dual of $\square$. Call this connective $\leftarrow\!\square$. What properties do we expect of it? The connective $\square$ is viewed as 'propagating' values of $V(x)$ from nodes x in the direction of the arrows. $\leftarrow\!\square$ should therefore be read as propagating values backwards, against the direction of the arrow. Since our arrows reflect the flow of time from past to future, then $\leftarrow\!\square$ is propagating values from the future into the past. Since the future is open and unknown, $\leftarrow\!\square$ has an abductive aspect. $\leftarrow\!\square$ is going to be different from the traditional future operator G.

Let us now look at the meaning of the dual $x \vDash\, \leftarrow\!\square\, q$. This should mean propagate the value q backwards from all y such that $x \to y$ is in the network onto the node x. If we take $x = c$ in Figure 14.1 then what we need is to propagate the value from d onto c. Not all networks allow for backward propagation. It is an applicable concept for Bayesian nets for example, but not for neural nets. In general this is a question of *abduction* in the network. Thus evaluating $\mathrm{Val}(x, \leftarrow\Box\, q)$ is a form of abduction.

Let us see what this means for the temporal interpretation. Ordinary temporal logic allows for the immediate future operator G and we evaluate $c \vDash Gq$ iff for all immediate future nodes $y, y \vDash q$ and this holds iff $d = 1$ (in Figure 14.1). This is *not* the meaning of $\mathrm{Val}(c, \leftarrow\Box\, q)$. This value wants to know "what value we abduce at c, such as when propagated forward, it will give us value d at d". So let us check.

The value $V^1(d)$ is $d' = d(1 - c)(1 - e)$. We want $d' = d$.

So if $d = 0$ then any value c will do.

If $d = 1$ and $e = 1$ then no value of c will make any difference, we will have $d' \neq d$.

If $d = 1$ and $e = 0$ then the value of c is required to be 0.

Thus $\leftarrow\Box\, q$ holds at c iff

$$(d = 0 \vee (d = 1 \wedge e = 0 \wedge c = 0)) \wedge \neg(d = 1 \wedge e = 1).$$

We can try and write this condition using the traditional temporal modalities G, H for the case of Figure 14.1.

$$c \vDash \leftarrow\Box\, q \text{ iff } c \vDash [G\neg q \vee G(q \wedge H\neg q)] \wedge [q \to \neg G(q \wedge Hq)] \wedge (\neg q \to G(q \to H\neg q))$$

This comparison suggests a new way of looking at future temporal logic. Consider $\leftarrow\Box$ instead of G. $\leftarrow\Box\, q$ is an abductive request. It abduces what value we need now to ensure that q holds in all possible futures.

We note in passing that $\leftarrow\Box$ may not be used to solve Aristotle's famous puzzle of tomorrow's sea-battle. The reason for this is that Aristotle's problem bears no intrinsic tie to temporal modalities or to considerations of tense. It is easy to see that Aristotle's problematic reasoning about tomorrow's sea fight is reproducible in tenseless, temporally unspecified contexts. Consider the tenseless sentence "It is true that a sea fight is going on". If this is so, then it must be the case that a sea fight is going on. But if this is true it is a matter of necessity that the fight occurs. Therefore, it cannot have occurred by the free decision of the admiralty. The problem is not one of temporality or tense. The problem is a scope error, the confusion of $\alpha \to \Box\beta$ with $\Box(\alpha \to \beta)$.

Perhaps we could be a bit clearer about our approach to the future. Consider Figure 14.1 again and assume it describes a temporal flow of time with node c as the present. d is in the future of c and e is in the past of d but is not related to c. Thus the past is branching and future worlds may have new ideas about the past. This kind of flow corresponds to the basic temporal logic known as $\mathbf{K}_t$. When we evaluate Gq at c, we check whether $d \vDash q$ or not. There is symmetry here between the future operator G and the past operator H. We have

- $x \vDash Gq$ iff $\forall y(x < y \to y \vDash q)$
- $x \vDash Hq$ iff $\forall y(y < x \to y \vDash q)$.

where $<$ is the irreflexive earlier later than relation (usually transitive).

The view of time this model assumes is that we are outside time and history has already happened. It is like God talking all-knowingly about time. However, from the point of view of a person living at a world, say $x = c$, the future has not yet happened, so what meaning do we give to a statement saying $d \vDash q$? Our reading of this is assuming q must hold at d when can we abduce about what needs to hold now (at c)? This is exactly backward propagation.

Example 14.3 (Bayesian Nets). Bayesian nets, which are considered more closely in the next section, have a different flavour altogether. Since Figure 14.1 is an acyclic graph, it can serve as a base for a Bayesian net. If we look at $\{0, 1\}$ probabilities and read the arrows as causal connections, then we need initial probability distributions $P(a), P(b)$ and $P(e)$ and conditional distributions $P(c \mid a, b)$ and $P(d \mid c, e)$. From this a general joint probability distribution $P(a, b, c, d, e)$ can be calculated.

In a Bayesian network there is no strong notion of 'transmission'. The probabilities are fixed and are evaluated (with due caution and understanding) from the application areas. Target nodes which are not end nodes (c and d in Figure 14.1) do not get their own probabilities but only what can be calculated from the nodes leading into them. Even the simple dynamic point of view of making the probabilities of the end nodes dependent on a parameter e.g. time (for example, we write $P_t(a), P_t(b)$ and $P_t(e)$ in Figure 14.1) is not central to their considerations.

Example 14.4 (Categories). The perceptive reader might ask whether, according to us, anything with nodes and arrows can be the subject of investigation of network modalities. This question implicitly implies that maybe our concept is too general to be of any use. The answer is that this is not the case. For example, the usual concepts of categories does not fall within our range of interest. Category theory relies on the observation that many properties of mathematical systems can be unified and simplified by a presentation with diagrams and arrows. The nodes represent mathematical objects (topological spaces, groups, sets, etc.) and the arrows represent mappings. Transitivity and commutativity of diagrams is therefore a basic feature in category theory and the fundamental point of view is to look at the totality of the class of objects and the arrows between them and study its properties. This approach is diametrically opposed to the basic meaning of the modality □ which seeks to bring information from neighbouring nodes only. In fact, the very idea of transmission of items from one node to another is foreign to category theory. To quote S. Maclane in [246]:

> Since a category consists of arrows, our subject could also be described as learning how to live without elements, using arrows instead.

It is necessary at this point to take note of a possible confusion. In the literatures of computer science and philosophy of science, there is an ingrained habit of regarding any kind of backwards propagation or regressive reasoning as abductive. So conceived of, transcendental reasoning would be abductive. An example of transcendental reasoning is the inference of α from the input that β exists and that β would be impossible without α. This contrasts with a notion of abduction advanced by Gabbay and Woods [150], the so-called G-W model. On the GW-model, abduction is not only backwards propagating but is also ignorance-preserving. Abduction is seen as a response to an ignorance-problem. Problems of this sort prompt one or other of two standard responses. One is to overcome one's ignorance by getting some new knowledge, and using that knowledge as the basis for future action. The other is to acquiesce in one's ignorance, and to allow this state of affairs to inhibit new action. G-W abduction is a third response. It splits the difference between the prior two. Like the second, it does not overcome the ignorance that triggered the problem in the first place; but like the first, it serves as the basis for new action, albeit defeasibly. In our employment of it here, we use the more traditional notion of abduction, rather than GW-abduction.

14.2 Bayesian Network Modalities

In this section we will give some examples of Bayesian network modalities.

If we assume that Figure 14.1 displays a Bayesian net, then a, b, c, d and e are variables which can take values in $\{0, 1\}$ (we assume two state variables). We associate with a, b and e probability distributions $P(a)$ and $P(b)$, and $P(e)$ with values $P(a = 1)$, $P(a = 0)$, $P(b = 1)$, $P(b = 0)$, $P(e = 1)$ and $P(e = 0)$ and conditional probability for c, depending on a and b, $P(c \mid a, b)$, and similarly $P(d \mid c, e)$. We therefore have the set of numbers, for $i = 0$ and $i = 1$

$$P(c = i \mid a = 1, b = 1)$$
$$P(c = i \mid a = 1, b = 0)$$
$$P(c = i \mid a = 0, b = 1)$$
$$P(c = i \mid a = 0, b = 0)$$

Therefore we have

(∗1) $$P(c = i) = \sum_{x,y=0,1} P(c = i \mid a = x, b = y) P(a = x) P(b = y)$$

Equation (∗1) gives us $P(c)$, and of course we have $P(a, b, c) = P(c \mid a, b) P(a) P(b)$. We similarly have numbers for $P(d = i \mid c = 1, e = 1)$, $P(d = i \mid c = 1, e = 0), \ldots$ and we have

(∗2) $$P(d = i) = \sum_{x,y=0,1} P(d = i \mid c = x, e = y)$$

Equation (∗2) gives $P(d)$, and of course we have $P(a, b, c, d, e) = P(d \mid c, e) P(c) P(e)$.

The above gives us a traditional two state Bayesian net based on Figure 14.1. Our notion of a network, however, requires values $V^0(x)$ for $x = a, b, c, d, e$ and so we can let $V^0(x) = P(x)$, for $x = a, b, e$ but we must also add probabilities $V^0(c)$ and $V^0(d)$. These may be different from the calculated Bayesian probabilities $P(c)$ and $P(d)$. So our Bayesian modal network based on Figure 14.1 is the system with V^0 just defined.

Conversely, every Bayesian modal network based on an acyclic graph with assignment V^0 and with given conditional probabilities can be viewed as a traditional Bayesian network provided we consider only V^0 applied to endpoints (points with no arrows leading into them) and compute the probabilities for the rest of the network.

Let us now compute $V^1(c)$ in Figure 14.1. We need to transmit the values $P(a)$ and $P(b)$ to node c and let it interact with $V^0(c)$. We know the result of the transmission is $P(c)$. Let $V^1(c)$ be $P(c) \oplus V^0(c)$, being some combination of probabilities yet to be defined. We need to define a new pair of numbers $(z, 1 - z)$ out of $(P(c = 1), P(c = 0))$

and $(V^0(c = 1), V^0(c = 0))$. Let $\oplus$ be, for example, the Dempster–Shafer combination (*3) below.[1]

(*3) $$(u, 1 - u) \oplus (v, 1 - v) = (\frac{uv}{uv + (1 - u)(1 - v)}, \frac{(1 - u)(1 - v)}{uv + (1 - u)(1 - v)})$$

We can therefore let

(*4) $$P(c) \oplus V^0(c) = (P(c = 1), P(c = 0)) \oplus (V^0(c = 1), V^0(c = 0))$$

The above is a proper two state Bayesian modal network.

Let us now introduce a modality $\square$ into the system, as well as a stock of atomic propositional variables $\{q_1, q_2, \ldots\}$. We regard the nodes of the network as worlds, and we say that propositional variables get values in these worlds. We have to say that for each atomic q and for each node $x(x = a, b, c)$ we have an assignment $h(x, q) = q_x$ which is a probability distribution (i.e. q_x^1 for $x = 1$ and q_x^0 for $x = 0$ and $q_x^1 + q_x^0 = 1$). This assignment can now be extended to more complex formulas (built up from the atoms q_i and the connectives $\square$ and perhaps $\neg$ and $\wedge$) in the traditional way. For any Boolean combination A of $q_1, \ldots, q_n$, we can calculate $h(x, A)$ from $h(x, q_i)$.

Let us see how to calculate $h(x, \square q)$? For this, let us look at the network and calculate $h(c, \square q)$. We can compute the probability of c by using $P(c \mid a, b)$ from $h(a, q)$ and $h(b, q)$.

The value we get is $P_{\square q}(c) = \sum_{a,b=0,1} P(c \mid a, b) \cdot h(a, q) \cdot h(b, q)$.

This value is not necessarily equal to $h(c, q)$. The way we view the situation is that $h(c, q)$ is the value assigned for q at c and $P_{\square q}(c)$ is the calculated value at c. We let $P_q(c) \oplus h(c, q)$ be the value of $\square q$ at c, i.e. $h(c, \square q)$. The idea here is that $h(c, \square q)$ is calculated through the net from the immediate descendants of c.

The remaining question is now what do we take as value of $h(a, \square q)$ and $h(b, \square q)$ and $h(e, \square q)$?

---

[1] The formula for the Dempster–Shafer combination is

($\sharp 1$) $$(a, b) \oplus (c, d) = \left(1 - \frac{(1 - a)(1 - c)}{1 - (ad + bc)}, 1 - \frac{(1 - b)(1 - d)}{1 - (ad + bc)}\right)$$

We require $a + b \le 1$ and $c + d \le 1$. When $a + b = 1$ and $c + d = 1$, formula ($\sharp 1$) reduces to (*3).

There is another way of looking at (*3):

Let $x_1, \ldots, x_n$ be n Boolean variables, assuming values in $\{0, 1\}$. Let P_1, P_2 be two probability distributions over the space of elements $(x_1 = 0, 1, \ldots, x_n = 0, 1)$. Then we can have a new composite probability $P = P_1 \otimes P_2$ by letting for $e = (e_1, \ldots, e_n) \in \{0, 1\}^n$.

($\sharp 2$) $$P(x_1 = e_1, \ldots, x_n = e_n) = \frac{P_1(x_i = e_i) \cdot P_2(x_i = e_i)}{\sum_{e \in \{0,1\}^n} P_1(x_i = e_i) P_2(x_i = e_i)}$$

In the case of a single variable x which can receive values $P_1(x = 0) = u, P_1(x = 1) = 1 - u, P_2(x = 0) = v, P_2(x = 1) = 1 - v$ we get

($\sharp 3$)
$$P(x = 0) = \frac{uv}{uv+(1-u)(1-v)}$$
$$P(x = 1) = \frac{(1-u)(1-v)}{uv+(1-u)(1-v)}$$

which is the same as (*3).

We let it be the network $P(a)$ and $P(b)$ and $P(e)$ respectively.

It is clear that if we want to turn a Bayesian net into a logic, we need not assume that the net is an acyclic graph. For example we can have the situation in Figure 14.2.

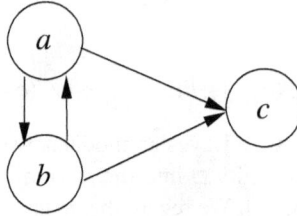

Fig. 14.2. A Cyclic Network

We need $P(c \mid a, b), P(a \mid b)$ and $P(b \mid a)$ to be given as part of the network. We start with assignments $h(a, q), h(b, q)$ and $h(c, q)$ and compute the following values:

- $h(a, \Box q) = (h(b, q)P(a \mid b) \upharpoonright a) \oplus h(a, q)$
- $h(b, \Box q) = (h(a, q)P(b \mid a) \upharpoonright b) \oplus h(b, q)$
- $h(c, \Box q) = (P(c \mid a, b)h(a, q)h(b, q) \upharpoonright c) \oplus h(c, q)$

where $P(x_1, \ldots, x_n, y) \upharpoonright y$ denotes the probability distribution P restricted to y, i.e. it is $\sum\limits_{x_1, \ldots, x_n = 0, 1} P(x_1, \ldots, x_n, y)$.

We can now compute by induction the values $\lambda x h(x, \Box^n q)$:

- $h(a, \Box^{n+1} q) = (P(a \mid b)h(b, \Box^n q) \upharpoonright a) \oplus h(a, \Box^n q)$
- $h(b, \Box^{n+1} q) = (P(b \mid a)h(b, \Box^n q) \upharpoonright b) \oplus h(b, \Box^n q)$
- $h(c, \Box^{n+1} q) = (P(c \mid a, b)h(a, \Box^n q)h(b, \Box^n q) \upharpoonright c) \oplus h(c, \Box^n q)$.

Remark 14.5. We feel we must stop here to reassure our Bayesian network researcher who probably has had his tolerance stretched to its limits. He probably feels by now that our Bayesian network modality is probably a useless generalization and might even distort the good work being done by traditional Bayesian networks in [224, 279]. The last straw is probably the network of Figure 14.2 where the acyclic condition is discarded!

Can we say anything to reassure our reader? Consider the cyclic figure 14.3 below:

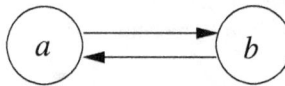

Fig. 14.3.

Imagine we are dealing with two school teachers, a and b, teaching at different schools. Such schools often have headlice epidemics. If a catches headlice she will pass it on to b. Similarly b can pass on the headlice to a. Of course, a and b get cleaned up at their respective schools, but the timings may be such that they keep infecting one another. Thus we do need in this model the probabilities $P(a), P(b), P(a \mid b)$ and

$P(b \mid a)$. The network modality $\square$ reflects one cycle of this repeated infection. The reader will notice that Figure 14.3 does not really represent a cyclic Bayesian net but rather the two traditional nets, to be used alternatively as events unfold. Perhaps this is enough to reassure Bayesians.

14.3 Fuzzy Network Modalities

Ordinary modal logic, say modal logic **K**, is complete for Kripke semantics of the form (S, R, h), where S is the set of possible worlds and $R \subseteq S^2$ is the accessibility relation. h is the assignment, giving value $0, 1$ to atoms in each world. Write h as a function: $h(t, q) \in \{0, 1\}, t \in S, q$ atomic. We can regard S as the nodes of a network where we have an arrow from t to s just in case sRt holds. Consider now a fixed atomic q and the situation in Figure 14.4. In this network, s is a possible world and $t_1, \ldots, t_n, \ldots$ are

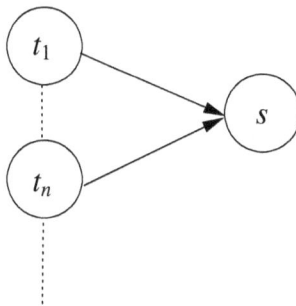

Fig. 14.4. Accessible Worlds

all the worlds accessible to it, i.e.

$$\{t_1, t_2, \ldots\} = S_s = \{t \mid sRt\}.$$

We know that

$$s \vDash \square q \text{ iff for all } t \in S_s, t \vDash q \text{ iff } \min_i h(t_i, q) = 1.$$

Let $x_i = h(t_i, q)$, $h = h(s, q)$ and let $\mathbf{f}(y, \bar{x}_i) = \min_i(x_i)$. Then $(S, \mathbf{f})$ forms a network as discussed.

We retrieve the modal logic modality as a network modality as follows:

- For every t, q, let
  - $V(t, q) = h(t, q)$.
- Proceed by induction to define $V(t, A)$ for any wff A and any t.
  - $V(t, A \wedge B) = \min\{V(t, A), V(t, B)\}$
  - $V(t, \neg A) = 1 - V(t, A)$
  - $V(t, A \vee B) = \max\{V(t, A), V(t, B)\}$
  - $V(t, A \Rightarrow B) = \min(1, 1 - V(t, A) + V(t, B))$
  - $V(t, \square A) = \mathbf{f}(V(t, A), \bar{V}_n(t_i, A))$, where $\bar{V}_n(t_i, A)$ abbreviates $(V_n(t_1, A), \ldots)$

where $t_1, t_2, \ldots$ are all the nodes in S_t, i.e. all nodes with arrows leading into t.

The beauty of the above definition is that we can change the function h into an assignment into $[0, 1]$ and give a fuzzy value $\rho(a, b) \in [0, 1]$ for every a, b such that $(a, b) \in R$, turning the logic into fuzzy modal logic. We can now change the definition of $\mathbf{f}, h$ in the network accordingly and get fuzzy modal logic. Let from the network:

- $\mathbf{f}(y, \overline{(x_i, \varepsilon_i)}) = \inf_i \{\varepsilon_i \Rightarrow x_i\} = \inf_i \{\min(1, 1 - \varepsilon_i + x_i)\}.$

We define

- $V(t, \Box A) = \mathbf{f}(V(t, A), \overline{(\rho(t, t_i), V(t_i, A))})$, for t_i such that $(t, t_i) \in R$.

Example 14.6 (General abstract argumentation network). In this example the network has the form $\mathbf{m} = (\mathcal{A}, R, V)$, where $\mathcal{A}$ is a set of arguments, $R \subseteq \mathcal{A}^2$, is the relation of what argument attacks what (we indicate the fact that $(a, b) \in R$ by $a \to b$) and V is an evaluation function into $[0, 1]$ with domain $\mathcal{A} \cup R$. V gives the strength of an argument and the transmission rate of the attack.

A typical situation is depicted in Figure 14.5.

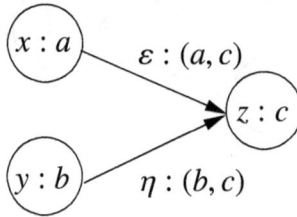

Fig. 14.5. An Argumentation Network

We have $\mathcal{A} = \{a, b, c\}$, $R = \{(a, c), (b, c)\}$ and $V(a) = x, V(b) = y, V(c) = z, V((a, c)) = \varepsilon$ and $V((b, c)) = \eta$.

In Chapter 9 we proposed several options for calculating the value of attacks, among them is one where the new value is $V'(c)$, taking into account the attack from (a) and (b), see also Chapter 10 and the later Chapter 20 below.

$$V'(c) = V(c)(1 - V(a)V((a, c)))(1 - V(b)V((b, c)))$$
$$= z(1 - \varepsilon x)(1 - \eta y).$$

The idea behind these values is that when $(u : d)$ attacks $(v : e)$ with transmission δ, as in Figure 14.6, then the value transmitted is δu and it reduces the original value v of e in proportion, so the new value is $v(1 - \delta u)$. If there are several attacks on e then they are all done in parallel. Thus if $(u_i : d_i)$ attack $(v : e)$ with strengths δ_i, respectively, the new value of e is $v \prod_i (1 - \delta_i u_i)$.

Let $\mathbf{f}(v, \overline{(u_i, \delta_i)}) = v \prod_i (1 - \delta_i u_i)$.

Clearly now if we are given a general network $(\mathcal{A}, R, V)$, we can define a modality for a language with $\Box$ using $\mathbf{f}$. See Example 14.12 below.

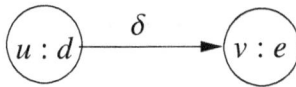

Fig. 14.6. A Transmission Factor

Here, too, it is advisable to forestall a possible confusion. According to the argumentation network approach every argument is open to attack. This is not to say that every argument that attacks an argument is successful. Openness to attack is a dialectical fact, not a normative one. The attack modelled in Figure 14.6 does have normative force, however. This is measured by the degree to which an attacking argument reduces the value of the attacked argument. However, the diminished normative force of a successfully attacked argument is only relative to the presumed normative force of the attacking argument.

14.4 Abstract Network Modality

We now introduce network modalities in an abstract setting (compare with sections 10.8.5–10.8.6).

Definition 14.7.   *1. Let S be a set of states. Let S^* be defined as follows:*
  - $x \in S^*$ *if* $x \in S$
  - *If* $x, y \in S^*$ *so is* (x, y).
 2. A network N (based on S) is a subset $T \subseteq S^*$ such that the following holds:
  - $(x, y) \in T$ *implies* $x \in T$ *and* $y \in T$
 3. A valuation V on a network T is a function from T into a space Ω of values. (Ω can be the set $\mathbb{R}$ of real numbers.)
 4. A functional $\mathbf{f}$ is a a function from pairs of the form $(x, \{(u, v)\})$, $x, u, v \in \Omega$ into Ω, i.e. $\mathbf{f} : \Omega \times 2^{\Omega \times \Omega} \mapsto \Omega$.
 5. An abstract valuation network is a system $(T, \Omega, V, \mathbf{f})$.

Example 14.8. Figure 14.7 is a network with values in some Ω.

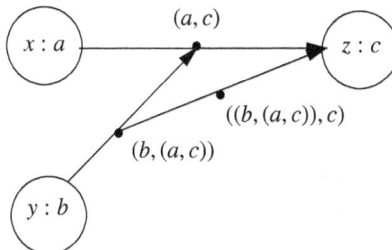

Fig. 14.7. Abstract Network

The network has the nodes a, b, c. The node (a, c) is the connection between a and c (think of it as $a \to c$) and $(b, (a, c))$ (or $b \to (a \to c)$) is a connection from b to c. We displayed the values from Ω for a, b, c but not for the other nodes.

Definition 14.9. *1. Let $(T, \Omega, V, \mathbf{f})$ be an abstract valuation network. Define $V^n(t), t \in T, n \geq 0$ as follows:*

    a) $V^0(t) = V(t)$

    b) $V^{n+1}(t) = \mathbf{f}(V^n(t), \{(V^n((s,t)), V^n(s)) \mid (s,t) \in T\})$

Definition 14.10. *Let $\mathbb{L}$ be a propositional logic with connectives $\sharp_1, \ldots, \sharp_k$ of $m_1, \ldots, m_k$ places respectively. Let $\mathbf{e}_1, \ldots, \mathbf{e}_n$ be function on Ω with $m_1, \ldots, m_k$ places, respectively. For each wff A of $\mathbb{L}$ define a valuation V_A as follows:*

    1. $V_q(t) = $ arbitrary function $h(t, q)$ for q atomic.

    2. $V_{\sharp_i(A_1, \ldots, A_{m_i})}(t) = \mathbf{e}_i(V_{A_1}(t), \ldots, V_{A_{m_i}}(t))$

Definition 14.11. *A network modal model for a logical algebra $\mathcal{A}$ with values in Ω has the form $(T, \Omega, h, \mathcal{A}, \mathbf{e})$. We define $V_a^n(t), a \in \mathcal{A}, t \in T, n \geq 0$ as follows:*

    1. $V_q^n(t)$ is defined as in Definition 14.9 for $V = V_q = \lambda th(t, q)$.

    2. $V_{\sharp(A_1, \ldots, A_k)}^0(t) = \mathbf{e}_{\sharp}(V_{A_1(t)}^0, \ldots, V_{A_k}^0(t)))$

    3. $V_{\Box A}^n(t) = V_A^{n+1}(t)$.

Example 14.12 (Argumentation networks (continued)). We continue Example 14.6 here. Recall that these networks have the form $(\mathcal{A}, R)$ with $\mathcal{A}$ a set of atomic arguments and $R \subseteq \mathcal{A}^2$, a binary relation telling us which arguments attacks which argument.

Let $\Omega = \{0, 1\}$ and let the initial value V of all arguments and all attacks be 1. Define $\mathbf{f}$ as follows

$$f(y, (x_i, \varepsilon_i)) = y \cdot \prod_i (1 - x_i \varepsilon_i)$$

$$= \begin{cases} 1 \text{ if } y = 1 \text{ and } x_i \varepsilon_i = 0 \text{ for all } i \\ 0 \text{ otherwise.} \end{cases}$$

The above definition does not allow attacks on arrows as we have in Figure 14.7 of Example 14.8, e.g. we cannot have connections of the form $a \to (b \to c)$.

If we adopt $\mathbf{f}$ and Ω for a general network as in Definition 14.11, then we can have argumentation networks like Figure 14.7. What would be the meaning of the modality? The modality tells us how far in the network we go to establish whether an argument is defeated or not. Consider Figure 14.7 and the node c.

Step 0 — no modality
Value at c is 1.

Step 1 — $\Box^1$
c is attacked by $(b, (a, c))$ and by a. The connection (a, c) is attacked by b and is therefore made invalid. However, the value of c is 0, since the value of $(b, (a, c))$ is still 1.

Step 2 — $\Box^2$
From now on, the values don't change. See [37, 41]. Recall Chapter 9.

14.5 Argumentation Fuzzy Logic Ł𝒜

We observe from previous sections the following features for network modalities.

1. We are dealing with networks that have nodes and relations between nodes that are used to propagate some values. These values require some numerical algebraic calculations.
2. We need a logic to talk about the properties of such networks.
3. The propagation aspect can be viewed as a modality.
4. Fuzzy modal logic is a logic with both numerical capability and modality.

In view of (1)–(4) above, it is reasonable for us to seek a good fuzzy modal logic with a clear proof theory/axiom system, capable of expressing the network examples of Section 1.

Let us approach this task in stages.

We begin by looking at traditional Łukasiewicz modal logic. The language has $\Rightarrow, \neg$ and $\Box$. Values are in $[0, 1]$ and the tables for $\neg, \wedge$ and $\Rightarrow$ (at a world t).

(1*) $V_t(A \Rightarrow B) = \min(1, 1 - V_t(A) + V_t(B))$
(2*) $V_t(\neg A) = 1 - V_t(A)$
(3*) $V_t(A \wedge B) = \min(V_t(A), V_t(B))$

The modality is given values through the translation into classical logic. We have generally

(4*) $t \vDash \Box A$ iff $\forall s(tRs \rightarrow s \vDash A)$.

Reading "$t \vDash B$" as "$V_t(b)$" and reading "$\rightarrow$" as "$\Rightarrow$" and "$\forall$" as "$\wedge$" we let

(5*) $V_t(\Box A) = \bigwedge_{\{s|tRs\}} (V(t, s) \Rightarrow V_t(A))$

where $V(t, s)$ is the fuzzy value of tRs holding.

If we look at our argumentation example, we need the following connective * for multiplication:

(6*) $V_t(A * B) = V_t(A)V_t(B)$

To accommodate the way values are propagated in argumentation networks we need the following:

(7*) $V_t(\Box A) = V_t(A) \prod_{\{s|tRs\}} (1 - V_s(A)V(t, s))$

This corresponds to a possible world modality of the form

$$t \vDash \Box A \text{ iff } \forall s\neg(tRs \wedge s \vDash A)$$

If we read "$\wedge$" as * "$\neg$" as $\neg$ we get

(8*) $V_t(\Box A) = \prod_{\{s|tRs\}} (1 - V_s(A)V(t, s))$ and we let $\Box A$ be
(9*) $\Box A = \text{def} A \wedge \Box A$.

Note that if R is reflexive we get the additional factor $(1 - V_t(A)V(t, t))$.

So we see from the above discussion that we need a fuzzy logic with at least $\neg$, $\wedge$ and $*$.

Such a logic exists, we quote from [98].[2]

Definition 14.13. *The propositional fuzzy logic ŁA has the connectives $\perp$ for falsity, Łukasiewicz $\Rightarrow$ and product $\to_\pi$ and product $*$ with the following tables*

    1. $V(\perp) = 0$
    2. $V(A \Rightarrow B) = min(1 - V(A) + V(B))$
    3. $V(A \to_\pi B) = \begin{cases} 1 \text{ if } V(A) \leq V(B) \\ V(B)/V(A) \text{ if } V(A) \geq V(B) \end{cases}$
    4. $V(A * B) = V(A)V(B)$

Using the above the following connectives are definable, we write the function which are their tables.

    5. $\neg_Ł x = 1 - x$
    6. $\neg_\pi x = \begin{cases} 1 \text{ if } x = 0 \\ 0 \text{ if } x > 0 \end{cases}$
    7. $\delta(a) = \begin{cases} 1 \text{ if } x = 1 \\ 0 \text{ if } x < 1 \end{cases}$
    8. $x \otimes y = max(0, x + y - 1)$
    9. $x \oplus y = min(1, x + y)$
    10. $x \ominus y = max(0, x - y)$
    11. $x \wedge y = min(x, y)$
    12. $x \vee y = max(x, y)$
    13. $x \to_G y = \begin{cases} 1 \text{ if } x \leq y \\ y \text{ if } x > y \end{cases}$

Theorem 14.14. *There exists a Hilbert axiomatisation of $Ł\prod$ with the connectives $\{\perp, \Rightarrow, \to_\pi, *\}$.*

Proof. See [98]

Remark 14.15. Quantifiers can be added. The value $V(\forall x A(x))$ is taken as $\inf_x\{V(A(x))\}$.

[98] also provides axiomatization for the predicate case. $\exists x A(x)$ is definable using Supremum. $V(\exists x A(x)) = \sup_x\{V(A(x))\}$.

14.6 Discussion

We have considered a number of examples of networks arising in computer science and related areas and introduced the notion of network modality to provide connection between, and uniform approach to computation or analysis over, such networks. This

---

[2] In a system with $\neg$, $\wedge$, $*$, the $\Rightarrow$ and $\to_\pi$ of Definition 14.13 below are definable.

is the beginnings of a unified treatment addressing the traditional formula based non-classical logics as well as reasoning directly with networks. We hope that the concept of network modality will show the respective communities the symbiotic relationship between their approaches.

The aim of this section is to persuade the reader that this is indeed the case and the new concept of network modality is indeed useful.

Whenever a new generalisation is introduced, we must ask ourselves the following methodological questions:

1. Are there enough similarities between the areas to justify at least theoretically the introduction of the new concept?
2. Is the new concept intuitive and natural? Does it match natural aspects resident in applications?
3. Is there a benefit from the point of view of each area in studying the new concept? Can new ideas be imported from neighbouring areas by looking at how this new concept manifests itself across areas?
4. Can we get new and interesting technical results using the new concept?

We can indeed answer these questions by summarising the main points of network modalities. The basic situation we study is described in Figure 14.8 $a_1, \ldots, a_n$ are

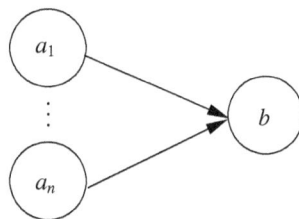

Fig. 14.8.

nodes connected to node b. There are values $x_1, \ldots, x_n, y$ associated with the nodes $a_1, \ldots, a_n, b$ respectively. We write $x_i = V^0(a_i), y = V^0(b)$. This is the initial position. A transmission occurs and a new value $V^1(b) = \mathbf{f}(y, x_1, \ldots, x_n)$ emerges at b.

We note the following:

1. The idea that there are values attached to nodes arises from modal logic or from ordinary network theory (see [13]).
2. The idea that a transmission occurs and the values change as a result of the transmission arises from transport networks or from argumentation networks or from neural networks.
3. Modal logic does not look at a possible world model as a network but mathematically it is very easy to do so. From the strictly mathematical point of view, possible worlds can be understood in any number of different ways. When we view modal logic in this way, then fuzzy values become very natural to consider. Modal logic also suggests the concept of modality as a symbol for the transmission.
4. Bayesian nets give us the idea of the function $\mathbf{f}$. All other nets transmit individually from a_i to b, for each i separately and the results aggregate. Bayesian nets

are the only ones where the transmission is done in parallel (the conditional probability $P(b \mid a_1, \ldots, a_n)$ is the transmission function). So we import from other networks the idea that P is a 'transmission' and then export back the idea that the 'transmission' should be done in parallel and not individually. Now we import back into Bayesian nets the idea that we need to give a probability value at node b and need to define how to compose an old probability value with a new one!

5. Now that we have got the basic idea of $V^1(b) = \mathbf{f}(V^0(b), V^0(a_1), \ldots, V^0(a_n))$ we can import from modal logic the idea of regarding $V^1(b)$ as $b \vDash \Box V$ and $V^0(b)$ as $b \vDash V$. We can therefore consider now $b \vDash \Box^m V$ as $V^m(b)$ which we understand as a transmission starting from distance m from b. Figure 14.9 illustrates the distance:

Fig. 14.9.

Notes (1)–(5) clearly show that the concepts are natural and fruitful and also show that the concept is just a union of features appearing naturally in the different research areas.

For those readers who are drawn to new results and new theorems, here is a list of such opportunities.

1. The Bayesian network modality presents a new way of doing probabilistic modal logic. The way in which probability and modality (and logic) have been connected to one another so far is by either making the predicates probabilistic (i.e. the probability that x is red) or making the value at a possible world probabilistic (q holds at a world t with some probability). There is nothing like what we have here. We also get various possibilities of completeness theorems and axiomatisations.

2. Abstract argumentation networks can benefit by allowing arguments to attack jointly (we learn this from Bayesian nets). This captures an important aspect of real life attack and defence. Recall Chapter 7 for a start in this direction.

3. Fuzzy logic can get new ways of introducing fuzzy modality with lots of opportunities for axiomatisations.

4. Passing information backwards is done naturally in Bayesian nets. If we try and do it systematically in neural nets it becomes abduction (the backwards propagation kind, rather than the more discriminating G-W kind). This is new to neural nets. In temporal logic it becomes planning. See Example 14.2

5. All in all we can see a lot of benefit and cross fertilisation in our point of view.

14.7 Concluding Remark

There is a joke (of sorts) about abstraction: "At the right level of abstraction anything can be interpreted as anything else." Since abstract comparisons are suppressors of difference, it is not hard to see the joke's point. This raises two questions about any proposed abstraction. One is whether it has been achieved in a technically competent way. The other is whether its gain in systematicity is worth the loss of what it suppresses. The networks sketched here operate at a fairly high level of abstraction. Of course, they are not so abstract as to fall within the ambit of the joke; we have already remarked that they won't do for category theory. Still, our fondness for networking does raise the issue of how much good it does at what cost. Perhaps it is early days for a definitive answer to this question. But, unless we are mistaken, progress with the opportunities listed just above will point us in the direction of an approving answer.

15

Logical Foundations for Bipolar Argumentation Networks

15.1 Introduction, background and orientation

15.1.1 Clarifying the context

Let us begin by highlighting the context of our work. Consider Figure 15.1. The top node **d** indicates traditional Dung networks. These are systems (S, R) where $R \subseteq S^2$ is the attack relation on a set $S \neq \emptyset$ of arguments.

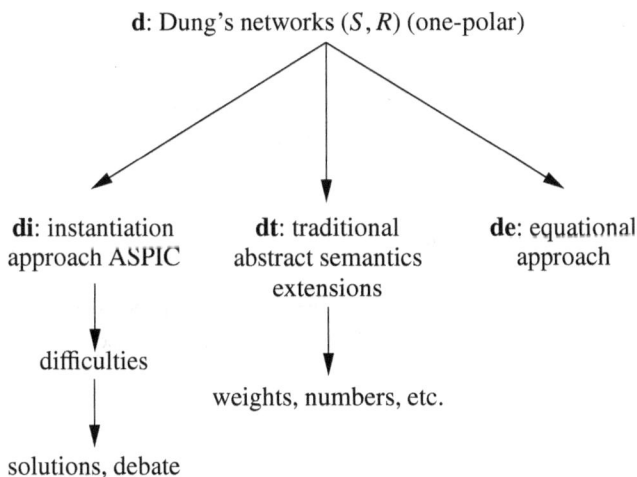

d: Dung's networks (S, R) (one-polar)

di: instantiation approach ASPIC

dt: traditional abstract semantics extensions

de: equational approach

difficulties

weights, numbers, etc.

solutions, debate

Fig. 15.1.

Given (S, R) we use traditional absract methods and define the notion of admissibility and extensions, and get various notions of semantics. For this see [35, 114], and Chapter 2.

In Figure 15.1 this is node **dt**. The instantiation approach (node **di**, ASPIC) puts forward that our starting point should be a database and a notion of deduction and arguments and attacks should be derived from the database set up. See [290, 198, 211].

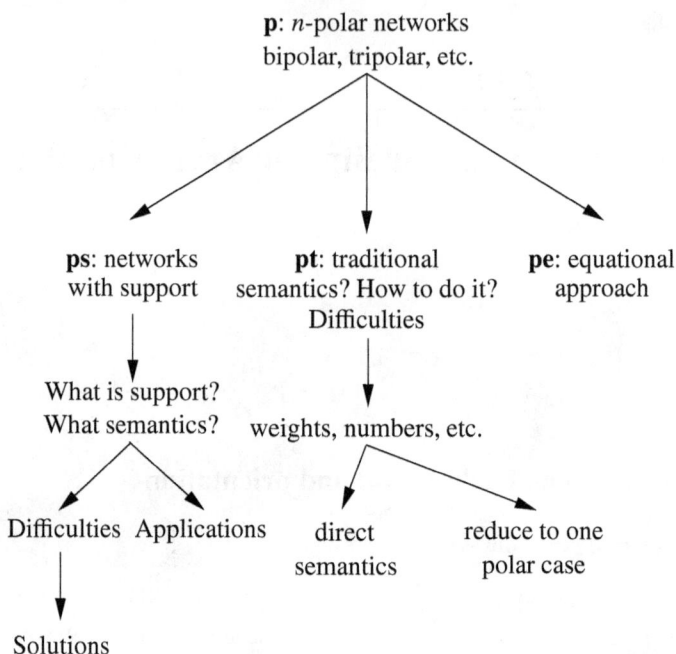

Fig. 15.2.

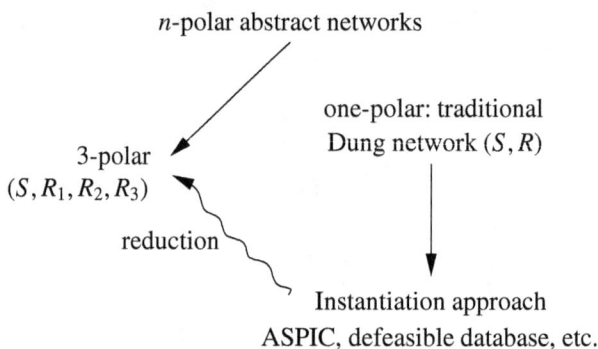

Fig. 15.3.

This approach has its difficulties and a very serious debate sometimes even emotional is currently ongoing in the community. See [79, 16, 289].

There is also the equational semantics (node **de**) for abstract argumentation networks which generalises the traditional semantics see [167, 166, 187, 37, 41].

Thus Figure 15.1 represents some existing research activity concerning traditional one-polar (S, R) networks.

We now come to the context of the results of this chapter. Consider Figure 15.2. In this figure our starting point is an n-polar abstract network of the form $(S, R_1, R_2, \ldots, R_n)$ for $n \geq 2, R_i \subseteq S^2$. The case $n = 1$ is the ordinary Dung's network for which we have Figure 15.1. We want to develop research (for Figure 15.2) similar to that for the case of $n = 1$. We thus look at Figure 15.2. What do we have known in this area? We know only some bipolar networks where we have attack and support [96, 95] which has some difficulties which require solutions, see [61, 96, 331] (node **ps**). We also have some applications [169]. We do not have anything about a general theory or about the equational approach, (nodes **pt** and **pe** in Figure 15.2).

This chapter fills this gap. We develop nodes **pt** and **pe**. We also indicate an application to the debate between [16] and [289]. We show a reduction from instantiations of a one-polar traditional networks and a 3-polar abstract networks. See Figure 15.3.

So let us summarise what we are going to do.

1. Develop general theory of extensions for bipolar, 3-polar and general n-polar networks. Show that existing published research (node **ps**) is compatible with the theory.
2. Develop the equational approach to bipolar and 3-polar network.
3. Show how certain 1-polar ASPIC instantiations can be reduced to 3-polar networks. This reduction is complex because ASPIC is a complex family. So we just give the idea in this chapter and postpone the research to [190].

15.1.2 Clarifying some concepts

We need to clarify some concepts we are going to use as we are going to have a more general point of view, and we would like our readers to see things as we see them.

Let (S, R) be a set with a binary relation $R \subseteq S \times S$. This is a mathematical structure which can be iewed in many different ways. For example, we can view S as a set of possible worlds and view R as an accessibility relation on it, or we can view S as a set of arguments and view R as the attack relation. The way we view (S, R) dictates what we do with it. So far we have no problem with these views because (S, R) as is, lends itself to these views. Now suppose we want to view (S, R) as a graph with R telling us about accessible ideal worlds, that is (S, R) is a deontic system. To do this we need the property $(\forall x \exists y)xRy$.

If this does not hold we cannot hold this view. Every world x must have at least one y which x sees as ideal for it. We can of course do something. We can modify R or understand it differently. We can say, for example, that if $(\neg \exists y xRy$ then x itself is ideal to itself. x is happy with itself, it does not need to look up to any other worlds. So if we do this we are looking at a new relation R' where:

$$xR'y = \text{def}.xRy \vee (x = y \wedge (\neg \exists z)xRz).$$

We now work with the system (S, R') derived from (S, R).

Let us now look at (S, R_1, R_2), a bi-relational structure with $R_i \subseteq S \times S, i = 1, 2$. We can view it as a (bi-modal) set of possible worlds for two modalities. Can we also view it as bipolar argumentation network? If we want to do that we have to say what our view is of R_1 and R_2. Are they two attack relations? Can we view R_1 as attack and R_2 as support? The point is, do we expect some properties or axioms on R_1 and R_2 to go with the view? Let us look at some cases:

Case 1: Let us view R_1 and R_2 as two attack relations. We have several options here. We can let $R = R_1 \cup R_2$ and proceed Dung style with (S, R).

We could also consider each attack separately and talk about R_i-admissible sets $E \subseteq S$:

- E is R_i admissible iff (1) and (2)
  1. $\neg \exists x, y(x, y \in E$ and $xR_iy)$
  2. $\forall y \forall x(x \in E \wedge yR_ix \rightarrow \exists z \in E(zR_iy))$.
  We can talk about R_1 protecting R_2:
  3. $\forall y \forall x(yR_2x \rightarrow \exists z(zR_1y))$
  Can we say here that R_1 supports R_2?

The reader can see that we can take a view and then write axioms and definitions which are forced by this view.

Case 2: We can view R_1 as attack and R_2 as support. We ask are there any special properties on R_1 and R_2 to be required by this view?

See the beginning of section 3 for further discussion of how Cayrol and Lagasquie-Schiex [96], attempted to solve this problem.

Let us take ideas from the world of real arguments. Suppose we have a database Δ in some language and two consequence relations

- A defeasible consequence $\vdash\!\!\!\sim$
- A strict consequece $\vdash$

We need $\vdash$ to be reflexive, transitive and monotonic and we need $\vdash\!\!\!\sim$ to be a nonmonotonic consequence satisfying the Gabbay 1985 axioms (reflexivity, cut and restricted monotonicity). See [129]. We now use $\vdash\!\!\!\sim$ and $\vdash$ to form argument chains. For example, as in Figure 15.4

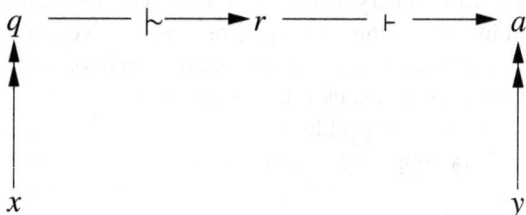

Fig. 15.4.

We have

- $q\vdash\!\!\!\sim r$ (represented by the arrow $\vdash\!\!\!\sim \rightarrow$)
- $r \vdash a$ (represented by the arrow $\vdash\rightarrow$)
- x attacks q (represented by $\twoheadrightarrow$)
- y attacks a (represented by $\twoheadrightarrow$)

Note the following

1. Since y attacks a and $r \vdash a$ (r strictly proves a) then we can say that y attacks also r. y does not attack q because the move from q to r is defeasible.
2. Since x attacks q it also attacks r because we do not have the basis to deduce r.

In view of this example we can say that R_1 is attack and R_2 support if one or both of he following conditions (i) and (ii) hold.[1]

(i) $uR_2v \wedge yR_1v \rightarrow yR_1u$

---

[1] In the sequel we refer to these conditions as "conditions (i), (ii) of Section 1.2".

(ii) $uR_2v \land xR_1u \rightarrow xR_1v$

Now just as we did in the case of deontic accessibility, where we replaced R by R' to suit our purpose, we can also, in case we have R_1, R_2 which do not satisfy conditions (i) or (ii) or both, replace R_1 by R'_1 which does satisfy (i) or (ii) or both. We get R'_1 by closing R_1 under (i) or (ii) or both.

Let us call this process *the reduction approach*. This means that we start with some (S, R_1, R_2) which is not an argumentation network and we define (from R_1 and R_2) a new R'_1 (and maybe even a new R'_2) and create an argumentation network (S, R'_1, R'_2).

Case 3: We may be given a system (S, R_1, R_2) presented semantically, with R_1 and R_2 already defined (instantiated) by some semantics. This system may not be so easily interpreted or changed into an argumentation system. This can also happen. So we would get a bipolar network with a different character. It is a network which is not exactly an argumentation network. We shall give such an example in Section 2.1.

We would like to give one more figure to show the complexities involved with the instantiation approach and explain why we are postponing the examination of [16, 289] to a later paper [190]. Consider Figure 15.5, and compare it with Figure 15.4.

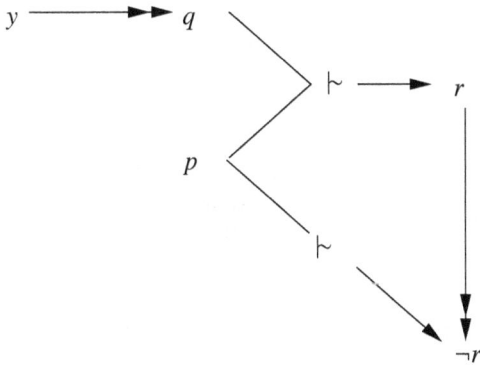

Fig. 15.5.

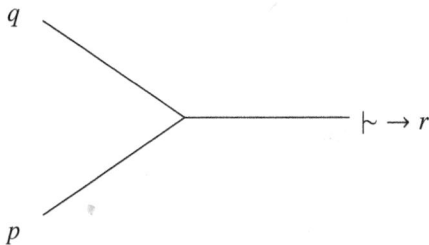

Fig. 15.6.

We have:

- $p \wedge q \vdash r$, (represented in Figure 15.5 by the subfigure 15.6)
- $p \vdash \neg r$
- $r \twoheadrightarrow \neg r$
- $y \twoheadrightarrow q$

In a Nute style defeasible system, without the attack of y on q we have that r attacks $\neg r$ because it is based on more specific information. When we add that y attacks q, we propagate the attack on the double arrow $r \twoheadrightarrow \neg r$ which should now be cancelled. This means that the reduction process in general is more complex.

Note also that the support relation is not binary but ternary. We need to write $((p, q), r)$ to represent $p \wedge q \vdash r$.

In general we need an argumentation network with joint supports and joint attacks. See a dedicated section later in Chapter 7 and a later section in this chapter.

15.2 Abstract *n*-polar argumentation networks

15.2.1 General considerations

Our starting point for addressing abstract semantics for bipolar and tripolar networks (node **pt** of Figure 15.2) are ideas presented in Chapter 10 entitled analysis of support. The approach there was to look at bipolar argumentation networks, (S, R_A, R_S) with attack and support relations and reduce the original bipolar network to a new network (T, ρ) with attack only, where T is obtained from S and where the new attack relation ρ is defined with the help of the support relation R_A, see below. We obtain extensions for the new network using ρ, (call them ρ extensions) and then somehow convert these new ρ extensions into extensions of the original bipolar network. This methodology is adopted in papers [61, 96, 95, 275]. Chapter 10 is mainly concerned with the equational approach and accepts the methodology above and only discusses possibilities for defining ρ and applies the equational approach to such possibilities. This chapter respects the above methodology but also proposes to work out the details of how to define extensions in the traditional manner directly on the original bipolar network (S, R_A, R_S), and in fact also on general multi-polar networks, as well as using the equational approach on the original bipolar network. We will compare our results with other papers (see Section 3) dealing with bipolar argumentation networks.

We start by giving a general point of view on multi-polar argumentation.

Definition 15.1 (Multi-polar pre-argumentation system). *This has the form* $(S, R_1, \ldots, R_m), i = 1, \ldots, n$ *where* $R_i \subseteq S \times S, S$ *is a non-empty set of arguments and* R_i *for* $i = 1, \ldots, n$ *are various contra arguments relations.*

Remark 15.2. We cannot define at the moment what is a multipolar argumentation system because we need to either give some axioms on the relations R_i of the multipolar pre-argumentation system or give some semantic view. There are so many diverse possibilities that there is nothing we can say in general. We can only give some examples.

Furthermore, we shall see in a later section that we need ternary relations on the set of arguments to model joint attacks and joint supports and disjunctive attacks and disjunctive support, introduced in Chapter 7 and can be used in connection with the ASPIC debate of [16, 289].

Example 15.3. Figure 15.7 gives a typical configuration around a point a, of a tripolar pre-argumentation network. We can turn this into a tripolar argumentation system by presenting a view, as discussed in Section 1.2.

$x \twoheadrightarrow y$ denotes attack
$x \rightarrow y$ denotes support
$x \Rightarrow y$ denotes a challenge resulting in confusion.

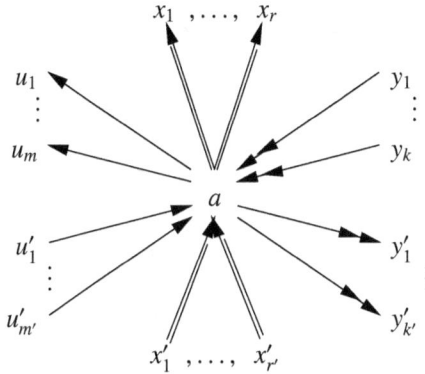

Fig. 15.7. Basic configuration tripolar network

We can use the following heuristic to explain these relations:

$x \twoheadrightarrow y$ means attack. So if x is in then y is out.
$x \rightarrow y$ means support. So depending on one's concept of support, if y is out, then x is also out. (This is condition (i) of Section 1.2.)
$x \Rightarrow y$ means x challenges and confuses the issue about y. So if x is in, then y becomes undecided.

We give an important word of caution to the reader: From the formal point of view the semantic value of the node a (in terms of {in,out,undecided}) depends in some functional way on the values given to surrounding points.

We can however give separate dependencies for each type of arrow involved, $\twoheadrightarrow$ (attack), $\rightarrow$ (support) and $\Rightarrow$ (challenge).

By giving these arrows the above names we are creating a view, as discussed in Section 1.2, and an expectation of comparison to the way these arrows are used in the literature under these names. For example the notion of support is taken in a technical sense as in (i) of Section 1.2 and as in [166, 61]. This notion of support would not be an adequate for the ASPIC interpretation, condition (ii) of section 1.2. For example, if support is the subargument relation as in ASPIC, then condition (i) does not seem to make sense. All subarguments of y can be in while yet y is out, since it is rebutted by another argument.

Example 15.4.   1. Let us offer a semantic view via a labelling discipline for a bipolar network of the form $(S, \Rightarrow, \twoheadrightarrow)$. Let a be a point in S and let $y_1, \ldots, y_k$ be all its attackers and let $x_1, \ldots, x_r$ be all of its (confusing) challengers. We have thus $y_j \twoheadrightarrow a$ and $x_i \Rightarrow a$. We write the following rules for a labelling function λ from S into {in, out, undecided}.

    a) a is labelled in if y_j, x_i are all labelled out, for $1 \le i \le r$ and $1 \le j \le k$.

    b) a is labelled out if for some j, y_j is labelled in.

    c) a is labelled undecided if each y_j is either labelled out or undecided and if all x_i are labelled out then at least one y_j is labelled undecided.

2. Let us define extensions in the traditional way for $(S, \Rightarrow, \twoheadrightarrow)$.

    a) A subset $\mathbb{E} \subseteq S$ is conflict free if for no $x, y, \in \mathbb{E}$ do we have $(x \Rightarrow y \vee x \twoheadrightarrow y)$ holds.

    b) A conflict free subset can defend itself if it defends all its members $x \in \mathbb{E}$. A node x is defended by $\mathbb{E}$ if whenever x is challenged or attacked by a y, i.e. we have that $(y \twoheadrightarrow x)$ or $(y \Rightarrow x)$ then there exists a $z \in \mathbb{E}$ such that $z \twoheadrightarrow y$.

    c) $\mathbb{E}$ is called an extension if $\mathbb{E}$ contains all the nodes it defends and satisfies (a) and (b) above.

3. We show that $\mathbb{E}$ is an extension iff there exists a labelling λ as in (1) such that $\mathbb{E} = \mathbb{E}_\lambda$, where $\mathbb{E}_\lambda = \{x | \lambda(x) = \text{in}\}$.

    a) Assume we are given a labelling λ and let $\mathbb{E} = \mathbb{E}_\lambda$. We show that $\mathbb{E}$ is conflict free, can defend itself and contains all nodes it defends.

        i. $\mathbb{E}$ is conflict free because if $x, y \in \mathbb{E}$ with $x \Rightarrow y$ or $x \twoheadrightarrow y$ holding, then by definition $\lambda(y)$ should be either out or undecided.

        ii. $\mathbb{E}$ can defend itself. Suppose y is such that $(y \twoheadrightarrow x \vee y \Rightarrow x)$ holds and $x \in \mathbb{E}$. We are looking for a $z \in \mathbb{E}$ such that $z \twoheadrightarrow y$.
Since $\lambda(x) = \text{in}$ we must have that $\lambda(y) = \text{out}$ therefore, for some $z, z \twoheadrightarrow y$ and $\lambda(z) = \text{in}$, so $z \in \mathbb{E}$ is our defender.

        iii. Let z be a node which $\mathbb{E}$ defends. We show that $\lambda(z) = \text{in}$.
Since z is defended by $\mathbb{E}$, it follows that for any y such that $y \twoheadrightarrow z$ or $y \Rightarrow z$ holds, there exists an x such that $\lambda(x) = \text{in}$ and $x \twoheadrightarrow y$ holds.
Therefore by condition (1a) on λ, we have that $\lambda(z) = \text{in}$.

    b) Given an extension $\mathbb{E}$ we need to define a labelling λ such that $\mathbb{E} = \mathbb{E}_\lambda$.

        i. Define a partial function λ_0 as follows:

          • $\lambda_0(x) = \text{in}$, for $x \in \mathbb{E}$

          • $\lambda_0(y) = \text{out}$, if for some $x_1 \in \mathbb{E}$ we have $x_1 \twoheadrightarrow y$ holds.

          • $\lambda_0(z) = \text{undecided}$, if for some $x_2 \in \mathbb{E}$ we have $x_2 \Rightarrow z$ holds, and for no $x \in \mathbb{E}$ do we have $x \twoheadrightarrow z$ holding.

        We quickly note that λ_0 is properly defined on its domain of definition and does not give to any element u more than a single value. This holds because $\mathbb{E}$ is conflict free.

        ii. We now expand λ_0 to a wider labelling λ. Assume that for $n \ge 0$, λ_n has been defined and that $\lambda_0 \subseteq \lambda_1 \subseteq \ldots \subseteq \lambda_n$. We define $\lambda_{n+1} \supset \lambda_n$ as follows. We let λ_{n+1} be the closure of λ_n under (a), (b), (c) of item (1) of this example, considered as closure conditions.
Let $\lambda_\infty = \bigcup_n \lambda_n$.
Let $\lambda(y) = \lambda_\infty(y)$ if defined, and let $\lambda(y) = \text{undecided}$ otherwise.

        iii. We now show that λ is a proper labelling satisfying conditions (a), (b), (c) and that $\mathbb{E} = \mathbb{E}_\lambda$.

We know that λ_∞ satisfies these conditions. The question is, have we affected the result of these closure conditions by moving from λ_∞ to λ and thus assigning lots of additional undecided labels.

The closure condition which may be affected is (c). But this closure condition makes new undecided labellings. So even if it is affected and it gives a new undecided label to some nodes z such that $\lambda_\infty(z)$ was undefined, we have no problem, because $\lambda(z)$ is defined as undecided anyway.

iv. Why is $\mathbb{E}_\lambda = \mathbb{E}$? We know that $\mathbb{E} \subseteq \mathbb{E}_\lambda$. Has λ_∞ labelled new in labels? Assume $m+1$ is the first time that a new element z got $\lambda_{m+1}(z) = $ in. Thus $z \in \mathbb{E}_\lambda$ and $z \notin \mathbb{E}$. This means that all attackers $y_i \twoheadrightarrow z$ and $x_j \Rightarrow z$ have $\lambda_m(x_j) = \lambda_m(y_i) = $ out. But this means that there exist x'_j and y'_i such that $\lambda_m(x'_j) = \lambda_m(y'_i) = $ in and $x'_j \twoheadrightarrow x_j$ and $y'_i \twoheadrightarrow y_i$ hold. But this implies that x'_j, y'_i are in $\mathbb{E}$. So $\mathbb{E}$ defends z and by condition (c), $z \in \mathbb{E}$. A contradiction.

Remark 15.5. Note that a bipolar network need not be just attack and support. It can be any two relations, e.g. support and challenge ($\rightarrow$ and $\Rightarrow$). The next Definition 15.6 explains this option and many more general options. Example 15.8 below gives a modal logic option and Chapter 17 below gives a contrary to duty deontic option.

Definition 15.6 (A general multi-polar multi-label network).

1. Let $\mathcal{A} = (S, R_1, \ldots, R_n)$ be a *pre-multi-polar network* with $S \neq \emptyset$ and $R_i \subseteq S \times S$. Let $\mathbb{L} = \{e_1, \ldots, e_m\}$ be a finite set of labels. Consider $\mathcal{A}$ as a classical model for n relations R_i on domain S. Let $Q_1, \ldots, Q_m$ be unary predicates on S with $Q_i(x)$ intended to say that x has label e_i.

   In the case of the bipolar network $(S, \rightarrow, \twoheadrightarrow)$ we have $R_1 = \rightarrow, R_2 = \twoheadrightarrow, \mathbb{L} = \{$in, out, undecided$\}$ and for any labelling λ on S, we have $Q_{in} = \{x|\lambda(x) = $ in$\}$ $Q_{undecided} = \{x|\lambda(x) = $ undecided$\}$ and $Q_{out} = \{x|\lambda(x) = $ out$\}$.

   Let $a \in S$ and let $x^i_1, \ldots, x^i_{k(i)}$ be all points in S such that $aR_i x^i_j$ holds and let $y^i_1, \ldots, y^i_{r(i)}$, be all points in S such that $y^i_j R_i a$ hold.

   Let $\mathbb{B}_i(a)$ be a Boolean combination of the predicates $\{Q_s(x^i_\alpha), Q_s(y^i_\beta)\}$ where $1 \leq s \leq m, 1 \leq i \leq n, 1 \leq \alpha \leq k(i), 1 \leq \beta \leq r(i)$.

   The predicate $\mathbb{B}_i$ says under what conditions a has label e_i. We have the axioms:

   (A1) $\vdash \bigvee_i \mathbb{B}_i(a)$, for $i = 1, \ldots, m$.
   (A2) $\vdash \neg[\mathbb{B}_i(a) \wedge \mathbb{B}_j(a)]$ for $i \neq j$
   (A3) $\vdash Q_i(a) \leftrightarrow \mathbb{B}_i(a)$, for $i = 1, \ldots, m$.

2. Let $\mathbb{E} \subseteq S$ be a subset.

   a) We say $\mathbb{E}$ is R_i *protected* if whenever $y \in \mathbb{E}$ and x is such that $xR_i y$ holds, then for some $z \in \mathbb{E}$ we have $zR_i x$ holds (think of $xR_i y$ as x attacks y, then this is the familiar condition of admissibility: whenever y in $\mathbb{E}$ is attacked by any x, then there is a z in $\mathbb{E}$ which attacks x).

   b) We say $\mathbb{E}$ is an R_i *extension* if $\mathbb{E}$ is R_i conflict free, R_i protected set which contains all the elements it R_i protects.

Example 15.7. Consider the network $(S, \rightarrow, \Rightarrow)$ with support and challenge. Define labelling as follows:

1. $\lambda(a) = $ in, if for all y such that $x \rightarrow y$ we have $\lambda(y) = $ in and for all z such that $z \Rightarrow x$ we have $\lambda(z) \neq $ in.

2. $\lambda(x)$ = out, if for some y such that $x \rightarrow y$. we have that $\lambda(y)$ = out.
3. $\lambda(x)$ = undecided, if we have that either

(c1): for some z such that $z \Rightarrow x$ we have that $\lambda(z)$ = in, and for all y such that $x \rightarrow y$ we have $\lambda(y)$ is not out

or

(c2): for all y such that $x \rightarrow y$ we have $\lambda(y) \neq$ out

and

for all y such that $y \Rightarrow x$ we have $\lambda(y) \neq$ in

and

for at least one such y we have that $\lambda(y)$ = undecided.

Example 15.8 (Counterpart modal semantics as a bipolar network[2]). Consider classical predicate models with domain D and binary relation R. We define a temporal model based on this language by looking at copies of models of the form (D_t, R_t) for different times $t \in T$. We assume $(T, <)$ is an acyclic temporal ordering. Figure 15.8 is an example of such a temporal model.

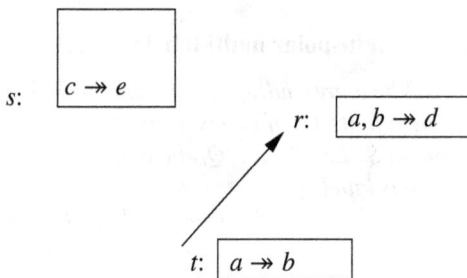

Fig. 15.8.

In the figure, t, s, r are temporal possible worlds and the boxes describe the domain and the relation R. We use "$\rightarrow$" for the accessibility "$<$" and "$\twoheadrightarrow$" for the relation R.

Thus our model has three possible worlds, with domains $D_t = \{a, b\}, D_s = \{c, e\}$ and $D_r = \{a, b, d\}$.

We want to describe this model using counterpart theory. We write (w, x) to mean the element x in world w. We can write $(w, x) \rightarrow (w', x)$ whenever $w \rightarrow w'$ holds, to say that (w', x) is the counterpart of (w, x) (i.e. the same element but considered in a different accessible world). We can also write $(w, x) \twoheadrightarrow (w, y)$ just in case that $x \twoheadrightarrow y$ holds in world w.

---

[2] In philosophy, specifically in the area of modal metaphysics, counterpart theory is an alternative to standard (Kripkean) possible-worlds semantics for interpreting quantified modal logic. Counterpart theory still presupposes possible worlds, but differs in certain important respects from the Kripkean view. So in ordinary temporal logic Dov Gabbay of 1970 is the same individual as Dov Gabbay of today. In counterpart theory Dov Gabbay of today is the counterpart of Dov Gabbay of 1970, but is not the same individual. This can happen if Dov Gabbay of next year has a Revelation and becomes a born again Christian missionary in Africa. See Wikipedia [102].

Under the above agreements, Figure 15.8 becomes Figure 15.9. Note that this figure looks like a bipolar network.

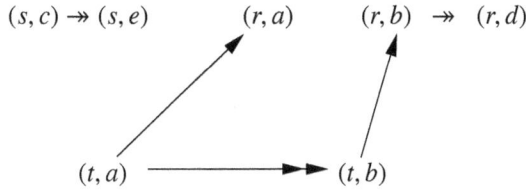

$$(s,c) \twoheadrightarrow (s,e) \qquad (r,a) \qquad (r,b) \twoheadrightarrow (r,d)$$

$$(t,a) \longrightarrow (t,b)$$

Fig. 15.9.

To make this more plausible, let us read xRy as "y owes a favour to x". So the meaning of $(t,x) \rightarrow (s,x)$ is the element x itself at different times. So $(t,x) \twoheadrightarrow (t,y)$ means at time t, y owes a favour to x. Therefore under this interpretation of R as owing a favour to and of $\rightarrow$ as being counterpart we get that we have y owes a favour to x at time s if for some earlier time t, y owes a favour to x at time t (the model has no means of describing returning the favour). This is condition (ii) from Section 1.2.

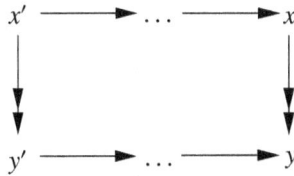

$$x' \longrightarrow \cdots \longrightarrow x$$
$$\downarrow \qquad\qquad\qquad \downarrow$$
$$y' \longrightarrow \cdots \longrightarrow y$$

Fig. 15.10.

Note also that being conflict-free for a set $\mathbb{E}$ means that nobody in $\mathbb{E}$ owes to anybody else in $\mathbb{E}$, and being able to defend oneself for a set $\mathbb{E}$ means that if $x \in \mathbb{E}$ owes a favour to a y, then y in turn owes a favour to some $z \in \mathbb{E}$. So an extension is a maximal set $\mathbb{E}$ of people who owe nothing to each other and who cannot be compromised as a group by anyone they owe anything to. This all makes sense.

Remark 15.9. For more examples of bipolar networks see Section 16.2 entitled Examples motivating the reactive idea. The examples there can easily be viewed as bipolar networks.

15.2.2 The reduction approach to bipolar and tripolar pre-argumentation networks

We briefly describe the reduction approach, following ideas of [166]. The basic concept is that of x being a threat to y.

Definition 15.10 (Threats in bipolar networks). *Let* $(S, R_A, R_S, R_1, R_1, \ldots)$ *be a multipolar pre-argumentation network with attack and support* R_A *and* R_S *respectively* *and possibly other relations.*[3] *Define two auxiliary relations for the purpose of handling support (that is we are focussing on R_A and R_S only in this definition). The definition will be schematic, dependent on a parameter relation R, which will allow us to perform iterations. Let $R \subseteq S^2$ be any relation defined already using R_A and R_S. We define $\rho_1(R)$ and $\rho_2(R)$.*

Option 1

$$u\rho_1(R)x \text{ iff def. } uRx \vee [\exists z\exists n(xR_S^n z \wedge uRz)]$$

Option 2

$$u\rho_2(R)x \text{ iff def. } uRx \vee [(\exists z\exists n(zR_S^n x \wedge uRz)]$$

The meaning of $u\rho_i(R_A)x$ is that x is under threat from u. The notation $xR^n y$ means that there exists $z_1, \ldots, z_n$ such that we have xRz_1, z_1Rz_2, and z_nRy.

Option 1 corresponds to condition (i) of Section 1.2 and Option 2 corresponds to condition (ii).

Figure 15.11 explains the situation. This is a bipolar argumentation framework. The nodes are arguments. Arguments can either attack or support other arguments. If argument x supports argument y, we write $x \rightarrow y$. If argument x attacks argument y we write $x \twoheadrightarrow y$.

In Figure 15.11, node u_0 attacks x directly, it is therefore both $\rho_1(R_A)$ and $\rho_2(R_A)$ threat to x.

u_1 attacks z which is at the end of a chain of support from x. According to $\rho_1(R_A)$, u_1 is a threat to x. It undermines x's credibility.

u_2 attacks a great grandparent supporter of x. Again it can be a threat to x.

The process can be iterated. We can look at $\rho_1(\rho_1)$ and $\rho_2(\rho_1), \rho_1(\rho_2), \rho_2(\rho_2)$. See Example 15.12.

Let ρ_e be an arbitrary $\rho_{e_1}(\rho_{e_2}(\ldots \rho_{e_n}(R_A)\ldots)$ where $\mathbf{e} = (e_1, \ldots, e_n)$ and each $e_i \in \{1, 2\}$. So ρ_e is an arbitrary iteration of ρ_1, ρ_2.

We can use ρ_e to define extensions and equations for the network. This means that we use n-level threats to define our extensions and equations. So the role of support in the network is to propagate different levels of threats on a node x through the networks by means of direct R_A attacks on various supporters or supported nodes related to x.

A policy of handling support will choose ρ_e or a combination and define the extensions and equations accordingly.

Remark 15.11. We remark about Definition 15.10 and Figure 15.11.

The question of whether u_1 really is a threat to x seems to depend on the nature of the support relations in the diagram. We use an abstract notion of support as discussed in [166, 61]. For other notions of support, say the ASPIC one, where the support relations are Pollock-style defeasible inferences and u_1's attack on z is a Pollock-style rebutting or undercutting attack, then u_1 does not seem to threaten x. Note that defeasible inference arguably does not satisfy contraposition.

Example 15.12. Consider Figure 15.12.

---

[3] We are calling the relation R_S a support relation but is is just another binary relation which can be interpreted as support. It can also be interpreted as a modal accessibility relation. It is still too general to be instantiated only as support.

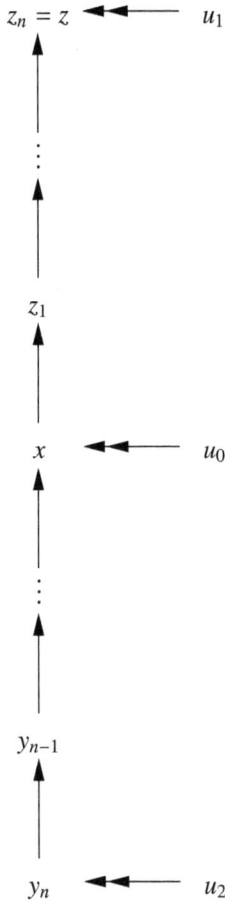

Fig. 15.11. General configuration for support and attack

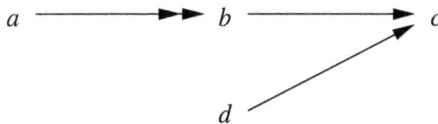

Fig. 15.12. A simple configuration

We have that $a\rho_2(R_A)c$ and $a\rho_1(\rho_2(R_A))d$, but a is not a $\rho_1(R_A)$ threat to c. So depending on how far we want to look for threats, our extensions can be either $\{a, d, c\}$ if we look at $\rho_1(R_A)$ only, or $\{a\}$ if we look at all of $\rho_1(\rho_2(R_A))$, or we can look at both $\rho_1(R_A)$ and $\rho_2(R_A)$.

Definition 15.13. *We are defining here a special case of Definition 15.10, we are taking Option 1 for the attack relation $\twoheadrightarrow$ and creating $\twoheadrightarrow_*$. Let $(S, \rightarrow, \twoheadrightarrow)$ be a bipolar argumentation network. Let $\rightarrow_*$ be the reflexive and transitive closure of $\rightarrow$. Let us seek subsets $\mathbb{E}$ of arguments defined as follows.*

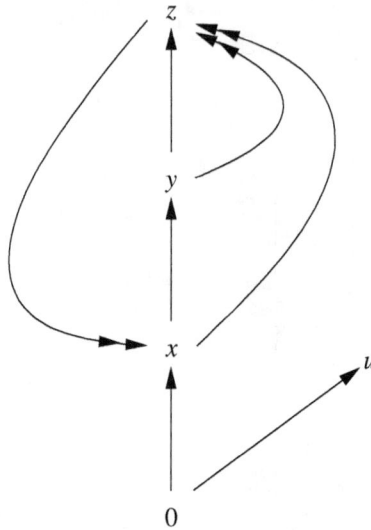

Fig. 15.13. A more complicated example

1. Let $x \twoheadrightarrow_* y$ be defined as meaning that for some x', y' we have that $y \rightarrow_* y'$ and $x \rightarrow_* x'$ and $x' \twoheadrightarrow y'$ all hold.
2. We seek subsets $\mathbb{E}$ of S, called extensions, satisfying the following:
    a) $\mathbb{E}$ is *-conflict free, namely for no $x, y \in \mathbb{E}$ do we have $x \twoheadrightarrow_* y$.
    b) $\mathbb{E}$ can *-defend itself, namely if for some x and some $y \in \mathbb{E}$, $x \twoheadrightarrow_* y$ then there exists $z \in \mathbb{E}$ such that $z \twoheadrightarrow_* x$.
    c) If $x \rightarrow_* y$ and $x \in \mathbb{E}$ hold then $y \in \mathbb{E}$.

Remark 15.14. The perceptive reader might ask why are we including item (2c) in the definition of admissible extensions (Definition 15.13), surely there would be subsets $\mathbb{E}$ which satisfy all the other conditions and since we are ignoring in the reduction the support $\rightarrow$ then why should we care?

The answer is two fold

1. We want to facilitate the equivalence of Definition 15.13 with Definition 15.23

and

2. When dealing with support one tends to view all support groups as one element. This is the notion of coalition in Cayrol and Lagasquie-Schiex's bipolar argumentation framework, discussed in Section 3.

Lemma 15.15.   *1. In the notation of Definition 15.13 we have that $y \rightarrow_* y'$ and $x \rightarrow_* x', x' \twoheadrightarrow_* y'$ imply $x \twoheadrightarrow_* y$.*
2. *Let $\mathbb{E}$ be a maximal set satisfying (a) and (b) of item (2) of Definition 15.13, then (c) also holds, namely if $x \rightarrow_* y$ and $x \in \mathbb{E}$ hold then $y \in \mathbb{E}$.*

Proof. 1. Is obvious.

2. We note that $\mathbb{E} \cup \{y\}$ is ∗-conflict free. This holds because if some $e \in \mathbb{E}$ either $e \twoheadrightarrow_* y$ or $y \twoheadrightarrow_* e$ holds then by (1) above we have that either $e \twoheadrightarrow_* x$ or $x \twoheadrightarrow_* e$. Then we note that if $z \twoheadrightarrow_* y$, then $z \twoheadrightarrow_* x$ (by (1)), and so for some $d \in \mathbb{E}$ we have $d \twoheadrightarrow_* z$. Thus $\mathbb{E} \cup \{y\}$ is ∗-conflict free and can ∗-defend itself. Since $\mathbb{E}$ is a maximal such set we have $y \in \mathbb{E}$.

Let us now check what extensions we get for Figure 15.13. We get $\mathbb{E} = \{u\}$. Certainly $\{u\}$ is ∗-conflict free and can defend itself (it is not attacked). The nodes $\{0, x, y\}$ ∗-attack themselves and so they cannot be added to $\{u\}$. The node z does not ∗-attack itself but it cannot defend itself. It is attacked by y and neither z nor u attack y. So $\mathbb{E} = \{u\}$ is maximal.

Example 15.16. Consider Figure 15.14.

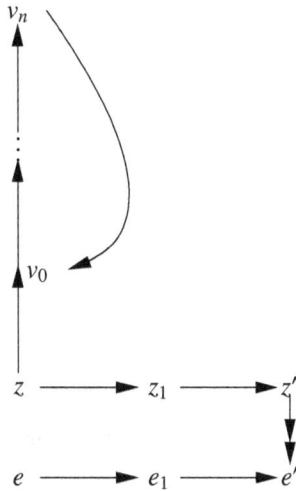

Fig. 15.14. Support loop

In this figure the set $G_1 = \{z, z_1, z', v_0, \ldots, v_n\}$ is a a set of nodes which basically support the element z' (we shall later define the concept of Support Group) which is not being attacked. So we can take them all as our extension. The group $G_2 = \{e, e_1, e'\}$ is attacked by G_1 and so it is not in the extension.

Example 15.17. We continue Example 15.8 to show how attacks can propagate along the arrows of support. Note that in this example the propagation is forward, as in condition (ii) of Section 1.2. Compare with Definition 15.13, where the propagation is backwards ,as in condition (i) of Section 1.2.

Figure 15.10 shows how owing a favour (this means being attacked) can propagate along the counterpart relation. We define

$x \twoheadrightarrow_* y$ iff for some x' and y', we have $x' \rightarrow_* x, y' \rightarrow_* y$ and $x' \twoheadrightarrow y'$ all hold. Where $\rightarrow_*$ is the transitive and reflexive closure of $\rightarrow$ and $\twoheadrightarrow_*$ is the new propagated attack relation.

Example 15.18. The following is is a completely different way of propagating attacks along the $\rightarrow$ of support. The idea comes from [37, 41]. Figure 15.15 illustrates this. The idea is that the following propagation principles are adopted:

1. If x supports y and y attacks z then this is the same as x and y attacking z.
2. If x attacks y and y supports z then this implies that x attacks z.
3. If x attacks z and y supports z and y and x are equally attacked and supported[4] then x and y cancel each other. (Note that in Figure 15.15 item (a), the points x and w do not have isomorphic configurations above them, so this rule does not apply!)

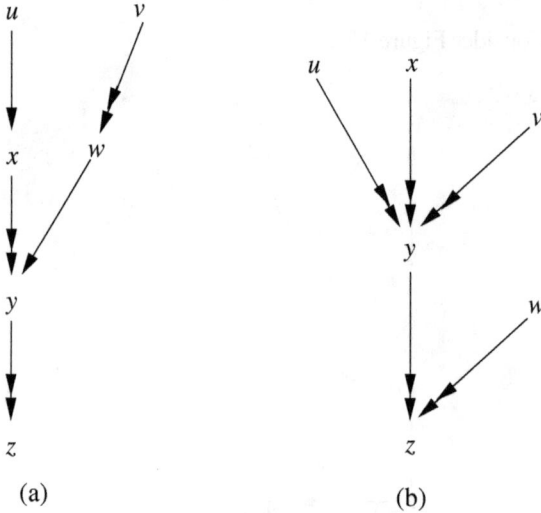

(a)   (b)

Fig. 15.15. Attack and support (a) reduces to (b)

The exact procedure is defined in [37, 41] in a numerical context.

Lemma 15.19. *The conditions of Example 15.18 are coherent.*

Proof. We provide a numerical interpretation for bipolar networks and show that this interpretation satisfies the conditions. We assume the given network has no cycles. Let x be any node. We associate a number $f(x)$ with x as follows:

1. If x has no attackers nor supporters, let $f(x) = 1$.
2. If a_i are all the attackers of x and s_j are all the supporters of x then let $f(x) = (\prod f(s_j))/(\prod f(a_i))$.

---

[4] Let w be a point in a bipolar network. Define the set S_w as the smallest set satisfying the conditions below:
a) $w \in S_w$
b) if $u \in S_w$ and v attacks or v supports u then also $v \in S_w$.
   We say that the nodes y and x are equally attacked and supported exactly when S_x and S_y are isomorphic.

It is easy to see that the rules of Example 15.18 are compatible with this numerical assignment. In the general case where we do have loops we can regard 1. and 2. above as a system of equations for f. There is always a solution.

Example 15.20 (Married John). This is example 4 from Caminada and Amgoud's paper [79], reformulated as a tripolar pre-argumentation network using our terminology. The vocabulary is

wr: John wears a ring
m: John is married
hw: John has a wife
b: John is a bachelor
go: John often goes out until late with friends.

The argumentation network is tripolar, with $\rightarrow$ indicating strict implication $\Rightarrow$ indicating defeasible implication and $\twoheadrightarrow$ indicating the attack relation. We have here two support relations, $\Rightarrow$ and $\rightarrow$, and one attack relation $\twoheadrightarrow$. We are going to apply Option 1 of Definition 15.10 to the pair $\twoheadrightarrow$ and $\rightarrow$, thus modifying the attack relative to the strict support. Note that the attack relation is defined between atoms of the form x and $\neg$x, and does not take into account how the atoms are derived. The reader should be aware that we have a very special case here, bearing in mind the discussion we had in Section 1.2.

In the general case we may have chains of several defeasible rules, where contradiction should not be propagated e.g. to the first argument. Here support is identified with the strict $\rightarrow$ inference relation. This is sufficient to solve the difficulties mentioned by Caminada and Amgoud.

A proper analysis of the general problem is postponed to [190].

The arguments are obtained from a database with strict rules S and defeasible rules $\mathcal{D}$.

$$S = \{\top \rightarrow wr, \top \rightarrow go, b \rightarrow \neg hw, m \rightarrow hw\}$$
$$\mathcal{D} = \{wr \Rightarrow m, go \Rightarrow b\}.$$

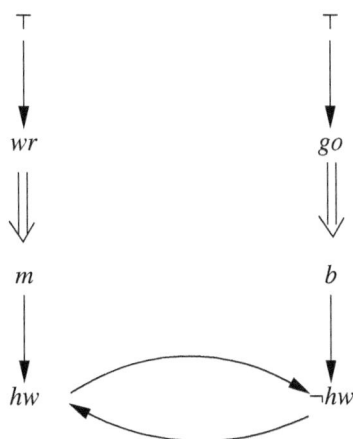

Fig. 15.16.

Caminada and Amgoud form the following arguments from the data $(S, \mathcal{D})$.

$$A_1 : [\top \rightarrow wr]$$
$$A_2 : [\top \rightarrow go]$$
$$A_3 : [A_1 \Rightarrow m]$$
$$A_4 : [A_2 \Rightarrow b]$$
$$A_5 : [A_3 \rightarrow hw]$$
$$A_6 : [A_4 \rightarrow \neg hw]$$

We form (in our own tripolar notation) the following network of Figure 15.16.

The correspondence in this case between the arguments of Caminada and Amgoud and the arguments in Figure 15.16 is to paths in this Figure as follows:

$$A_1 : (\top \rightarrow wr)$$
$$A_2 : (\top \rightarrow go)$$
$$A_3 : (\top \rightarrow wr \Rightarrow m)$$
$$A_4 : (\top \rightarrow go \Rightarrow b)$$
$$A_5 : (\top \rightarrow wr \Rightarrow m \rightarrow hw)$$
$$A_6 : (\top \rightarrow go \Rightarrow b \rightarrow \neg hw)$$

In the above set of arguments, only A_5 and A_6 attack each other and so we get the extension $\{A_1, A_2, A_3, A_4\}$. The *output* of this extension are the argument heads, namely $\{wr, go, m, b\}$.

Caminada and Amgoud's approach corresponds to our considering the network of Figure 15.16, with attack relation $\twoheadrightarrow$, and we thus get the extension $\{wr, go, m, b\}$. Caminada and Amgoud proceed to close this extension under the strict rules and they get inconsistency. They consider this a problem and offer to remedy the problem by systematically adding, with every strict rule $x \rightarrow y$, its contrapositive rule $\neg y \rightarrow \neg x$. This allows them essentially to also have that m attacks b and b attacks m, and so the extension becomes only $\{wr, go\}$, which is consistently closed under the strict rules.

Our approach already contains this remedy. We use $\twoheadrightarrow_*$ as defined in item 1 of Definition 15.13, to generate our extensions, and thus get only $\{wr, go\}$ as the extension.

The remedy that Caminada and Amgoud propose achieves the effect of taking $\twoheadrightarrow_*$ as Figures 15.17 and 15.18 show.

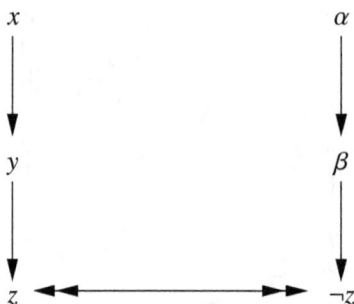

Fig. 15.17.

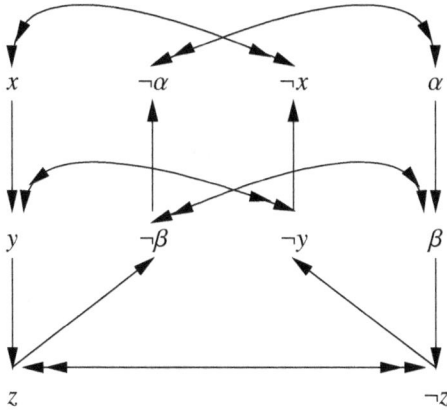

Fig. 15.18.

In Figure 15.17, our $\twoheadrightarrow_*$ would show that x is attacked by α and that α attacks x.

However, if we do not have $\twoheadrightarrow_*$ but use $\twoheadrightarrow$ only, we get that neither x nor α are attacked. In this case we would get an extension which will be inconsistent.

If we always add the contrapositives, we would add $\neg y$, $\neg x$, $\neg \beta$, $\neg \alpha$ and get Figure 15.18, then we get that x and α are attacked and will not get the inconsistent extension.

The Caminada–Amgoud solution relies on classical logic. Our approach is general and applies to any logic.

The reader is cautioned that we have only shown this remedy for the present example. It remains to be seen whether our construction always leads to extensions that are consistent and strictly closed. In fact, a careful analysis is needed because the construction is similar to the ones of Prakken and Sartor [290] and Garcia and Simari [198], which were shown by Caminada and Amgoud as not to guarantee consistency and strict closure. The present examples shows the intuition of how we want to remedy the problem, using tripolar networks. We shall address this in a separate paper on ASPIC. See [190] for a detailed analysis of the Caminada Amgoud paper [79].

15.2.3 Direct semantics for bipolar networks

This subsection gives direct ways of computing extensions for bipolar networks, without first reducing the network into an attack only network as proposed in Definition 15.13. We define exensions directly using what we call the bipolar Gabbay-Caminada labelling. In certain cases the two methods are equivalent, as they are in the traditional case (see Appendix), for example see Lemmas 15.25 and 15.27. In general, however, these are two different methodologies and may have different reach.

Definition 15.21 (Support groups).

1. *Let* $(S, \rightarrow, \twoheadrightarrow)$ *be a bipolar network. Let* $G \subseteq S$. *we say* G *is a support group if* G *is* $\twoheadrightarrow$ *conflict free and whenever* $x \in G$ *and* $x \rightarrow y$ *or* $y \rightarrow x$ *holds then also* $y \in G$.
2. *A suport group* G *is said to be unattacked if there does not exist* z *and* $a \in G$ *such that* $z \twoheadrightarrow a$ *holds.*

Remark 15.22.

1. Note that every element $x \in S$ belongs to a support group G_x. We let

$$G_x^0 = \{x\}$$
$$G_x^{n+1} = \{y | x \to y \text{ or } y \to x \text{ holds}\}$$
$$G_x = \bigcup_n G_x^n$$

2. Note that if a support group G is unattacked then it is also $\twoheadrightarrow_*$ unattacked. For if $z \twoheadrightarrow_* a, a \in G$, then for some z' and a', we have $z \twoheadrightarrow_* z', a \twoheadrightarrow_* a'$ and $z' \twoheadrightarrow a'$. But then we also have $a' \in G$ and z' attacks a', and so G is not unattacked.

Definition 15.23 (Bipolar Gabbay-Caminada labelling).

1. *Let $(S, \to, \twoheadrightarrow)$ be a bipolar network. Let $a \in S$ be a node. Figure 15.19 shows the relative situation of a in the network.*
   $y_1, \ldots, y_k$ are all the attackers of a and $u_1, \ldots, u_m$ are all the arguments which a supports.

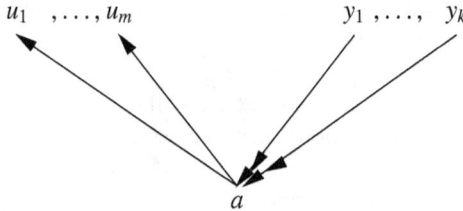

Fig. 15.19. General local configuration for a node

2. *A Gabbay-Caminada labelling of the network is a function $\lambda' : S \mapsto \{in, out, undecided\}$ such that the following conditions (a1), (a2), (b) and (c) hold for each $a \in S$ in a configuration of Figure 15.19:*
   (a1) $\lambda(a) = in$, if the support group G_a of a is *-unattacked.[5] This means that there is no y in G_a and a z and a u such that $z \twoheadrightarrow u$ and $y \to_* u$.

---

[5] Here we differ from the approach taken in our COMMA 2010 paper [61]. Suppose a supports b and b supports a, and that is all we have. Then $\{a, b\}$ form an unattacked mutual support group $\{a \to b \to a\}$ and therefore according to item (a1) of our definition here, both will be in.

According to our paper [61], when a supports b, we read it as a hidden attack of b in the direction of a, namely when b is out, a must also be out. We implement this in [61] using an auxiliary point $x_{(b,a)}$, such that b attacks $x_{(b,a)}$ and $x_{(b,a)}$ attacks a. Similarly, since b supports a, we get an auxiliary point $x_{(a,b)}$, such that a attacks $x_{(a,b)}$ and $x_{(a,b)}$ attacks b.

Thus the mutual support group $\{a, b\}$ is understood according to [61] as the attack loop $\{a \twoheadrightarrow x_{(a,b)} \twoheadrightarrow b \twoheadrightarrow x_{(b,a)} \twoheadrightarrow a\}$. This attack loop has two extensions, one in which both a and b are in and one in which both a and b are out.

We shall see later on that the equation for the mutual support group $\{a, b\}$ is $a = b$ which possible solutions agree with the interpretation of [61]. If we want to achieve the effect of the understanding of [61], we need to modify item (a1) to the new item [a1], reading as follows

[a1] $\lambda(a)$ is in, if a neither supports anything nor is attacked by anything.

[a1] may be a better option to take in view of its agreement with the equational approach.

(a2) $\lambda(a) = $ *in, if all* $\lambda(u_i)$ $i = 1, \ldots, m$ *are in and all* $\lambda(y_j)$, $j = 1, \ldots, k$ *are out.*

(b) $\lambda(a) = $ *out, if either for some* $1 \le i \le m$ *we have that* $\lambda(u_i) = $ *out or for some* $j, 1 \le j \le k$, *we have that* $\lambda(y_j) = $ *in.*[6]

(c) $\lambda(a) = $ *undecided, if the following* (c1), (c2), (c3) *all hold:*

(c1) *For all* $1 \le i \le m$, $\lambda(u_i)$ *is either in or undecided.*

(c2) *For all* $1 \le j \le k$, $\lambda(y_j)$ *is either out or undecided.*

(c3) *For some node* $x \in \{u_1, \ldots, u_m, y_1, \ldots, y_k\}$ *we have that* $\lambda(x)$ *is undecided.*

Lemma 15.24. *Let* λ *be a labelling as in Definition 15.23. Then* $\lambda(a) = $ *in and* $a \rightarrow_* b$ *imply* $\lambda(b) = $ *in.*

Proof. It is sufficient to show that $\lambda(a) = $ in and $a \rightarrow b$ imply $\lambda(b) = $ in.

Assume otherwise that $\lambda(b) \ne $ in. First note that $\lambda(b)$ cannot be out because this will force $\lambda(a)$ to be out.

Consider now the situation of Figure 15.19. We have that $b = u_i$ for some i and $\lambda(b) = $ undecided. None of $\lambda(u_i)$ can be out, $1 \le i \le m$. How about $\lambda(y_i), 1 \le j \le k$? None of $\lambda(y_j)$ can be in because this would force $\lambda(a) = $ out.

We are thus in situation (c) of Definition 15.23 and so $\lambda(a)$ should be undecided, not in. A contradiction.

Lemma 15.25. *Let* λ *be a labelling as in Definition 15.23. (Note we have not proved yet that such* λ *exist.) Let* $\mathbb{E}_\lambda = \{x \in S | \lambda(x) = $ in$\}$. *Then* $\mathbb{E}_\lambda$ *is an extension in the sense of Definition 15.13.*

Proof. We show that $\mathbb{E}_\lambda$ satisfied condition 2 of Definition 15.13.

1. We show that $\mathbb{E}_\lambda$ is *-conflict free. Assume that $e_1, e_2 \in \mathbb{E}_\lambda$ and that $e_1 \twoheadrightarrow_* e_2$. This means that for some $a_1, a_2 \in S$ we have

$$e_1 \rightarrow_* a_1$$
$$e_2 \rightarrow_* a_2$$
$$a_1 \twoheadrightarrow a_2$$

all hold.

This means that for some $a_1^1, \ldots, a_{n(1)}^1$ and $a_1^2, \ldots, a_{n(2)}^2$, we have

$$e_1 \rightarrow a_1^1 \rightarrow \ldots \rightarrow a_{n(1)}^1 = a_1$$
$$e_2 \rightarrow a_1^2 \rightarrow \ldots \rightarrow a_{n(2)}^2 = a_2.$$

Since $\lambda(e_1) = \lambda(e_2) = $ in, we must have also, in view of the above, that $\lambda(a_1) = \lambda(a_2) = $ in. But this is a contradiction since $a_1 \twoheadrightarrow a_2$. Condition (c2) on λ requires $\lambda(a_2) = $ out!

2. We show $\mathbb{E}_\lambda$ can defend itself.

Assume $e \in \mathbb{E}_\lambda$ and $z \twoheadrightarrow_* e$. We seek an $a \in \mathbb{E}_\lambda$ such that $a \twoheadrightarrow_* z$.

Note that since we have shown that $\mathbb{E}_\lambda$ is conflict free, we have that $\lambda(z) \ne $ in.

Since $z \twoheadrightarrow_* e$, we have for some z' and e' that the following holds:

---

[6] This definition is not compatible with Pollock-style defeasible inference, as it implicitly assumes contraposition. Same applies to Lemma 15.24 below. The definition is compatible with Chapter 10 and [61, 95, 61] and uses condition (i) of Section 1.2.

$$z \to_* z'$$
$$e \to_* e'$$
$$z' \twoheadrightarrow e'.$$

We know that $\lambda(e) = $ in. Hence $\lambda(e') = $ in. Hence $\lambda(z') = $ out. Hence $\lambda(z) = $ out. For $\lambda(z)$ to be out, we must have that G_z, the support group of z, is attacked by some node which is in, for otherwise $\lambda(z)$ will be either in or undecided, and also we must have one of the cases below, case $\alpha 0$ or else we have case $\beta 0$ to hold.

Case $\alpha 0$

For some w_0 such that $w_0 \twoheadrightarrow z$ we have $\lambda(w_0) = $ in. In this case let w_0 be the node a we seek.

Case $\beta 0$

For some v_0 such that $z \to v_0$ we have $\lambda(v_0) = $ out. We distinguish two cases.

Case $\alpha 1$ For some w_1 such that $w_1 \twoheadrightarrow v_0$ we have $\lambda(w_1) = $ in. In this case let w_1 be the a we seek. We have $w_1 \to v_0, z \to v_0$ which imply $w_1 \twoheadrightarrow_* z$ and $w_1 \in \mathbb{E}_\lambda$.

Case $\beta 1$

For some v_1 such that $v \to v_1$ we have $\lambda(v_1) = $ out.

We continue by induction and get case βn, for $n \geq 1$ as in Figure 15.20. We also know that none of cases $\alpha 0, \alpha 1, \ldots, \alpha(n-1)$ hold.

If we end up that case αn for $n = 0, 1, 2, \ldots$ never hold, for any n then we get that the support group G_z of z is unattacked by any point which is in and so $\lambda(z) = $ either in or undecided, which is impossible.

So for some n, case αn does hold. This means that we have the situation in Figure 15.21, where w_{n+1} attacks v_n and $\lambda(w_{n+1}) = $ in and $z \to_* v_n$. These conditions show that $w_{n+1} \in \mathbb{E}_\lambda$ and $w_{n+1} \twoheadrightarrow_* z$.

We have thus shown that $\mathbb{E}_\lambda$ defends itself.

3. Condition (2c) of Definition 15.13 follows from Lemma 15.24.
4. We show that if the group G_a is unattacked then $a \in \mathbb{E}_\lambda$. This holds because of condition (a1) of Definition 15.23.

Lemma 15.26. *Let $(S, \to, \twoheadrightarrow)$ be a bipolar network and let a be in S with Figure 15.19 describing its immediate situation. Then (1) is equivalent to (2):*

1. *$z \twoheadrightarrow_* a$ (as defined in Definition 15.13).*
2. *Either (a) or (b) holds:*
   a) $z \twoheadrightarrow_* u_i$ for some $1 \leq i \leq m$
   b) $z \to_* y_j$ for some $1 \leq j \leq k$.

Proof. $z \twoheadrightarrow_* a$ means that for some z' and a' we have

$$z \to_* z'$$
$$a \to_* a'$$
$$z' \twoheadrightarrow a'.$$

Case 1

If $a' = a$, then we have that $z' \twoheadrightarrow a$ and hence $z' = y_j$ for some j and so $z \to_* y_j$.

Case 2

If $a' \neq a$, then for some $i, y_i \to_* a'$ and so by definition of $\twoheadrightarrow_*, z \twoheadrightarrow_* u_j$.

The other direction is clear.

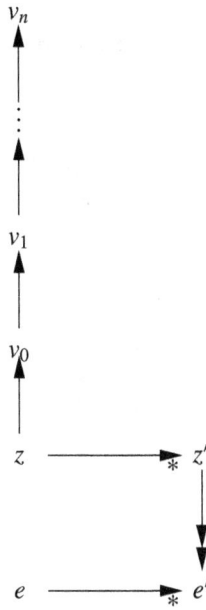

$$v_n$$

$$\vdots$$

$$v_1$$

$$v_0$$

$$z \xrightarrow{\quad\quad}_* z'$$

$$e \xrightarrow{\quad\quad}_* e'$$

$\lambda(e) = $ in. $\lambda(e') = $ in. $\lambda(z') = $ out.

$\lambda(z) = $ out. $\lambda(v_r) = $ out, $0 \leq r \leq n$.

Fig. 15.20. Case βn for Lemma 15.25

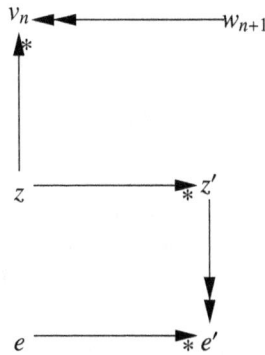

$$v_n \xleftarrow{\quad\quad} w_{n+1}$$

$$z \xrightarrow{\quad\quad}_* z'$$

$$e \xrightarrow{\quad\quad}_* e'$$

Fig. 15.21. Case αn for Lemma 15.25

Lemma 15.27. *Let $(S, \rightarrow, \twoheadrightarrow)$ be a bipolar network and let $\mathbb{E}$ be an extension defined as in Definition 15.13. Then there exists a Gabbay-Caminada labelling λ such that $\mathbb{E} = \mathbb{E}_\lambda$.*

Proof. $\mathbb{E}$ is an extension of $(S, \twoheadrightarrow_*)$ and so there exists a Caminada labelling λ for $\mathbb{E}$ as such. We show that this very λ also satisfies the conditions of a Gabbay-Caminada labelling for $(S, \rightarrow, \twoheadrightarrow)$ as given in Definition 15.23.

(a1) This condition holds by virtue of item (2) of Remark 15.22.

(a2) Assume $a \in S$ is in a situation as described in Figure 15.19 and assume that $\lambda(y_j)$ out for all $1 \le j \le k$ and $\lambda(u_i) = $ in for all $1 \le i \le m$. We want to show that $\lambda(a) = $ in.

Let us show that all $\twoheadrightarrow_*$ attackers of a are out. Let z be a candidate for being attacker. We have for some z' and a' that

$$z \rightarrow_* z'$$
$$a \rightarrow_* a'$$
$$z' \twoheadrightarrow a'$$

From the fact that $a \rightarrow_* a'$ we deduce that either $a = a'$ or for some $u_i, u_i \rightarrow_* a'$. If $a \ne a'$, then since $\lambda(u_i) = $ in, we get by condition (c) of Definition 15.13 that $\lambda(a') = $ in and hence $\lambda(z') = $ out and again by condition (c) of Definition 15.13, we get that $\lambda(z) = $ out. The other possibility is that $a' = a$. But then we have $z' \twoheadrightarrow a$ and so $z' = y_j$ for some j and since we are given that $\lambda(y_j) = $ out, we deduce again from condition (c) of Definition 15.13 that $\lambda(z) = $ out. Therefore we showed that all $\twoheadrightarrow_*$ attackers of a are out and so $\lambda(a) = $ in.

(b) Assume that either for some i, $\lambda(u_i)$ is out or for some j, $\lambda(y_j) = $ in. We show that we must have $\lambda(a) = $ out.

If $\lambda(u_i) = $ out, then there must be a z such that $\lambda(z) = $ in and $z \rightarrow_* u_i$. Therefore for some z', u_i' we have $z \rightarrow_* z', u_i \rightarrow_* u_i'$ and $z' \twoheadrightarrow u_i'$. But since $a \rightarrow u_i$ holds, we have $a \rightarrow_* u_i'$ and the above situation means $z \twoheadrightarrow_* a$ and so $\lambda(a)$ must be out. If for some $y_j, \lambda(y_j) = $ in then since $y_j \twoheadrightarrow a$, we immediately get that $\lambda(a)$ must be out.

(c) Assume that (c1) and (c2) and (c3) of Definition 15.23 all hold. We show that $\lambda(a) = $ undecided. To achieve this we show that in $(S, \twoheadrightarrow_*)$, we have that all $*$-attackers of a are either out or undecided, with at least one of them being undecided.

By Lemma 15.26 we have that $z \twoheadrightarrow_* a$ iff either $z \twoheadrightarrow_* y_i$ for some j or $z \twoheadrightarrow_* u_i$ for some i.

However we do know the value of λ on y_j and u_i. In all of the cases (c1), (c2) and (c3), such $*$-attackers z of u_i or y_j are either out or undecided with at least one of them undecided. Therefore all $*$-attackers of a in $(S, \twoheadrightarrow_*)$ are either out or undecided, with at least one of them being undecided. Therefore $\lambda(a) = $ undecided.

Example 15.28. Let us calculate the extensions for some examples from the literature. See the networks in Figure 15.22.

In (a) the extension is $\{a, c, d\}$. a is not attacked so a is in. So b is out and now $\{c, d\}$ is an unattacked support group.

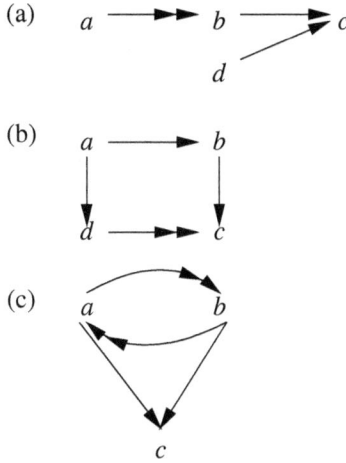

Fig. 15.22. Sample bipolar networks

In (b) the extension is just $\{d\}$ as it $*$-attacks all the other nodes.

In (c), the extensions are $\{a, c\}$ and $\{b, c\}$.

Example 15.29 (Provability networks[7]). We want to give a credible example of attack and support bipolar network. For this we use the Lambek calculus with right hand arrow $\Rightarrow$. This logic is widely used in formal linguistics. It contains atomic propositions including a falsity constant $\perp$ and the implication symbols $\Rightarrow$.

Formulas are built in the traditional manner. The data structures are sequences of formulas and we allow the rule of right hand modus ponens:

$$A \Rightarrow B, A \vdash B$$

(the minor premise A has to be immediately to the right of the ticket $A \to B$) and the right hand deduction theorem.

We have

$$(A_1, \ldots, A_n) \vdash A \Rightarrow B$$

iff

$$(A_1, \ldots, A_n, A) \vdash B.$$

A has to be the right most element of the sequence and B the only element to the right of the turnstile $\vdash$.

Consequence is defined between data structures as follows: Let

$$\mathbf{A} = (A_1, \ldots, A_n)$$
$$\mathbf{B} = (B_1, \ldots, B_n)$$

Then

$$(\mathbf{A}, X \Rightarrow Y, X.\mathbf{B}) \vdash (\mathbf{A}, Y, \mathbf{B})$$

---

[7] See Section 21.2.1 of Chapter 21 for related details.

we form modus ponens between the correctly adjacent components $X \Rightarrow Y, X$ in the sequence and replace the pair in the sequence by the result of the modus ponens Y.

Thus for example we have

$$A, A \Rightarrow \bot \nvdash \bot$$

becausewe cannot do modus ponens when minor assumption A is to the left of the ticket but

$$A \Rightarrow \bot, A \vdash \bot.$$

Define inconsistency of **A** to mean:

$$\mathbf{A} \vdash \bot.$$

We now turn the set W of all data structures of the logic into a bipolar network $(W \rightarrow, \twoheadrightarrow)$, with support $\rightarrow$ and attack $\twoheadrightarrow$ to be defined using the provability of the logic as follows:

Let

1. $\mathbf{A} \rightarrow \mathbf{B}$ means $\mathbf{A} \vdash \mathbf{B}$
2. $\mathbf{A} \twoheadrightarrow \mathbf{B}$ means $\mathbf{B}, \mathbf{A} \vdash \bot$.

Note that $\twoheadrightarrow$ is not symmetrical. We have $A \twoheadrightarrow (A \Rightarrow \bot)$ but not $(A \Rightarrow \bot) \twoheadrightarrow A$.

Note also that $\rightarrow_*$ (the reflexive and transitive closure of $\rightarrow$) is the same as $\vdash$, i.e. as $\rightarrow$).

We have $\mathbf{A} \rightarrow_* \mathbf{B}$ and $\mathbf{C} \rightarrow \mathbf{D}$ and $\mathbf{B} \twoheadrightarrow \mathbf{D}$ imply $\mathbf{A} \twoheadrightarrow \mathbf{C}$.

15.3 Cayrol and Lagasquie-Schiex's bipolar argumentation framework

In this section we summarize the definitions of bipolar argumentation frameworks with the terminology used by Cayrol and Lagasquie-Schiex [96].

We begin with a discussion of what is support. Given a multipolar argumentation network and, focussing on one of the relations, say R, in this network, we ask when can we say that this R is an attack relation and when can we say that this R is a support relation?

Of course an essential part of the presentation of the network is a view or semantics for it. Our answer to the question depends on the view. Suppose the view we have is equational. This means that nodes (see Figure 15.7 for illustration) get values in [0, 1] (where 1 is considered crisp "in" and 0 is considered crisp "out") and with each node a we associate an equation of the form

$$a = \mathbf{f}_a \text{ (surrounding nodes)}.$$

So if for all R related surrounding nodes x (i.e. nodes for which we have xRa holding) we have that the value of a increases with the increase in the value of x, then R is support. If it decreases then R is attack.[8] See Example 15.18 and note that there are lots of numerical examples in Chapter 9.

---

[8] Note that since the equations are local to the nodes a, a relation R can be support for one node a and attack for another node b.

If our view is instantiation ASPIC type semantics then we have the base proof system to tell us what is support. Since the relation R is instantiated, it has a meaning in some defeasible system, and we can tell from this instantiation what R is.

What if we have a system which is abstract, with one relation R_1 which we call attack (this we can always do) and another relation R_2 which we would like to call support. We have no additional view on R_1 and R_2 beyond their property of being general binary relations on the set of arguments (i.e. we have no semantics, no equations, no axioms — we have only our desire to treat R_2 as some sort of support). What can we do, can we make R_2 into some sort of support?

Cayrol and Lagasquie-Schiex are trying to do that in their paper [96]. They correctly realise that if R_2 is supposed to be support then we need to take maximal sets of mutually supporting elements and handle them as our "arguments". They get into difficulties, as discussed in [61], and [61] also offers a solution. This section reproduces and examines what Cayrol and Lagasquie-Schiex do and shows that indeed condition (i) of Section 1.2 is what is needed to turn R_2 into support. This is indeed the remedy offered [61]].

We take up the Cayrol and Lagasquie-Schiex approach also in the next section where we show that the equational approach can also solve the problem, if the right conditions are available. Furthermore there is more to the Cayrol and Lagasquie-Schiex approach than originally intended. They connect with argumentation modelling of Contrary to Duty obligations, see [169], which is another reason for discussing their approach in this chapter.

Definition 15.30 (Bipolar argumentation framework BAF). *A bipolar argumentation framework* $(A, \rightarrow, \twoheadrightarrow)$ *consists of a finite set A called arguments and two binary relations on A called support and attack respectively.*

Definition 15.31 (Conflict free). *Given an argumentation framework* $AF = (A, \twoheadrightarrow)$ *a set* $C \subseteq A$ *is conflict free, denoted as* $cf(C)$, *iff there do not exists* $\alpha, \beta \in C$ *such that* $\alpha \twoheadrightarrow \beta$.

The union of elementary coalitions in [96] is defined as follows:

Definition 15.32 (Elementary coalitions). *An elementary coalition of BAF is a subset* $EC = \{a_1, \ldots, a_n\}$ *of A such that*

1. *there exists a permutation* $\{i_1, \ldots, i_n\}$ *of* $\{1, \ldots, n\}$ *such that the sequence of support* $a_i \rightarrow a_{i_1} \rightarrow a_{i_n}$ *holds;*
2. *cf (EC);*
3. *EC is maximal (with respect to $\subseteq$) among the subsets of A satisfying (1) and (2).*

EC denotes the set of elementary coalitions of BAF and $ECAF = (EC(A), c\text{-attacks})$ is the elementary coalition framework associated with BAF. Cayrol and Lagasqie-Schiex define a conflict relation on $EC(A)$ as follows:

Definition 15.33 (c-attacks relation). *Let* EC_1 *and* EC_2 *be two elementary coalitions of BAF.* EC_1 *c-attacks* EC_2 *if and only if there exists an argument* a_1 *in* EC_1 *and an argument* a_2 *in* EC_2 *such that* $a_1 \twoheadrightarrow a_2$.

Remark 15.34. Extensions can now be defined in the Dung traditional manner on ECAF. We get sets of coalitions on A. Our interest stops at this point. However [96] continued to the next definition 15.35.

Definition 15.35 (Acceptability semantics).

- S is a ecp-extension of BAF if and only if there exists $\{EC_1, \ldots, EC_p\}$ a preferred extension of ECAF such that $S = EC_1 \cup \ldots \cup EC_p$.
- S is an ecs-extension of BAF if and only if there exists $\{EC_1, \ldots, EC_p\}$ a stable extension of ECAF such that $S = EC_1 \cup \ldots \cup EC_p$.
- S is a ecg-extension of BAF if and only if there exists $\{EC_1, \ldots, EC_p\}$ a grounded extension of ECAF such that $S = EC_1 \cup \ldots \cup EC_p$.

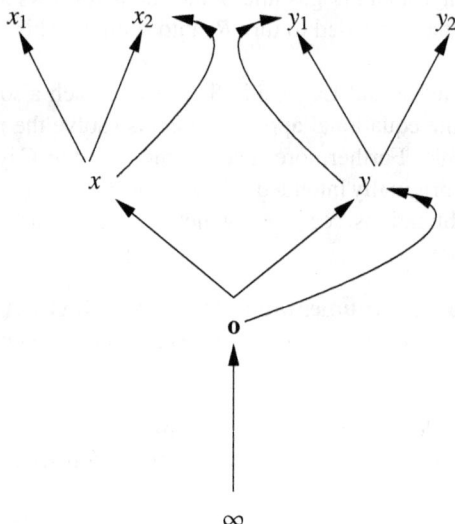

Fig. 15.23. Bipolar configuration coming from the Chisholm paradox, see [3]

Example 15.36. Consider the bipolar network of Figure 15.23. Figure 15.23 arises from the modelling of the Chisholm paradox, see [169].

Fig. 15.24. Simultaneous support and attack

Note that we allow nodes to both attack and support other nodes at the same time. This is allowed in [96].[9]

---

[9] If the reader is uncomfortable with this we can add auxiliary points and the problem disappears. We can replace any occurrence of Figure 15.24 by Figure 15.25.

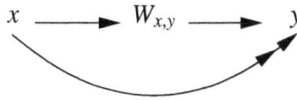

Fig. 15.25. Normalising Figure 15.24

The maximal support paths of Figure 15.23 are

$$\beta = (\infty, \mathbf{0}, y, y_2)$$
$$\alpha = (\infty, \mathbf{0}, y, y_1)$$
$$\delta = (\infty, \mathbf{0}, x, x_2)$$
$$\gamma = (\infty, \mathbf{0}, x, x_1)$$

The maximal coalitions according to Definition 15.32 are

$$\{\infty, \mathbf{0}, x, x_1\}, \{y, y_2\}.$$

The attack relation among coalitions is Figure 15.26.

$$\gamma = \{\infty, \mathbf{0}, x, x_1\} \quad \beta' = \{y, y_2\}$$

Fig. 15.26. Maximal coalitions for Figure 15.23

Thus the extension is empty since they do attack each other (**0** attacks y).

If we want to proceed according to Definition 15.35, we get the extension for Figure 15.23 to be $\gamma \cup \beta' = \{\infty, \mathbf{0}, x, x_1, y, y_2\}$.

Note that if we compute the extensions for Figure 15.23 directly, we get the same set, $\{\infty, \mathbf{0}, x, x_1, y, y_2\}$.

Note that the reason we cannot take the maximal paths $\alpha, \beta, \gamma, \delta$ as our "coalition" network elements is that α, β and δ are not conflict free, and Definition 15.32 requires conflict freeness. If we ignore the conflict free restriction we can consider the network with $\alpha, \beta, \gamma, \delta$.

The attack relation among the paths is as in Figure 15.27.

Example 15.37. Another example from [61] is Figure 15.28.

The coalitions are as in Figure 15.29. The extension of Figure 15.29 is $\{(d), (e)\}$, and according to Definition 15.35, this yields the extension $\{d, e\}$ for Figure 15.28. The $\{d, e\}$ extension is not a correct extension when viewed according to Dung's theory. Strictly speaking Dung's theory deals with networks with attack relation only, while here we also have support. However, if we follow the intuition behind Dung's theory, we have here an extension containing the node d which is attacked by node b which is not attacked by anything. This is counter intuitive. The only way out is to say that node

---

Since the bipolar semantics deals with coalitions, and $W_{x,y}$ is between x and y, it will appear in any coalition containing x and y and there will be no technical consequences to using Figure 15.24 instead of Figure 15.25.

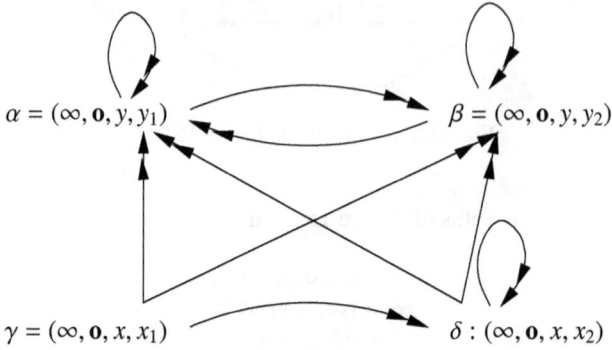

Fig. 15.27. Attack relation among maximal paths of Figure 15.23

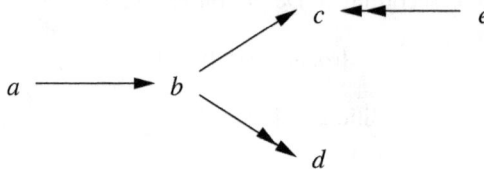

Fig. 15.28. Problematic network for [2]

e, which is definitely in, also attacks b by proxy, since it attacks c which is supported by b. Paper [61] offers a remedy.[10] We are going to offer an equational approach remedy.

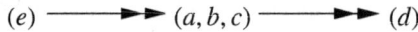

Fig. 15.29. Coalitions for Figure 15.27

15.4 The equational approach

Let $\mathcal{A} = (S, R)$ be an argumentation framework, $S \neq \varnothing$ is the set of arguments and $R \subseteq S \times S$ is the attack relation. The equational approach views (S, R) as a bearer of equations with the elements of S as the variables ranging over $[0, 1]$ and with R as the generator of equations. We use the letter $\mathbf{f}$ to denote a solution to the equations. Let $x \in S$ and let $y_1, \ldots, y_k$ be all of its attackers. Viewing the letters x and $y_1, \ldots, y_k$ as equational variables, we write two types of equations $Eq_{\max}(\mathcal{A})$ and $Eq_{\text{inverse}}(\mathcal{A})$.[11] For $Eq_{\max}$ we write:

- $x = 1 - \max(y_1, \ldots, y_k)$;

---

[10] In [61] we extend the notion of attack. If a supports b and c attacks b we consider a as being under attack as well. See Chapter 10, Section 3, Analysis of Support, for an extensive discussion.

[11] In Chapter 10 there are more Eq options.

- $x = 1$ if the letter x when viewed as an argument is (S, R) has no attackers.

For $Eq_{inverse}$ we write:

- $x = \prod_{i=1}^{k}(1 - y_i)$;
- $x = 1$, if the letter x when viewed as an argument is (S, R) has no attackers.

We seek solutions $\mathbf{f}$ for the above equations. In Chapter 10 we proved the following:

Theorem 15.38. *(1) There is always at least one solution in $[0, 1]$ to any system of continuous equations $Eq(\mathcal{A})$.*
(2) If we use $Eq_{max}(\mathcal{A})$ then the solutions $\mathbf{f}$ correspond exactly to the Dung extensions of A. This correspondence is as follows:
(a) Let E be an extension given to us through a Caminada labelling of each node in S, using the labels {in, out, undecided}, (for Caminada labelling and its equivalence to the traditional way of defining extensions see Chapter 2 and [35]). Let $\mathbf{f}$ be defined on the letters $x \in S$ regarded as variables ranging over $[0, 1]$, by letting $\mathbf{f}(x) = 1$ (resp. 0, resp. $\frac{1}{2}$) if the letter x regarded as an argument in (S, R) is given the Caminada label in (resp. out, resp. undecided). Then the function $\mathbf{f}$ is a solution to the system of Equations Eq_{max}.
(b) Let $\mathbf{f}$ be a solution to the system Eq_{max} generated by (S, R) as explained above. We can generate a Caminada labelling on S by the following correspondence between the values of $\mathbf{f}$ and the labels {in, out, undecided}:

- $\mathbf{f}(x) = 1$ *corresponds to x = in*
- $\mathbf{f}(x) = 0$ *corresponds to x = out*
- $0 < \mathbf{f}(x) < 1$ *corresponds to x = undecided.*
  The actual value in $[0, 1]$ reflects the degree of odd looping involving x.

(3) If we use $Eq_{inverse}$, we give more sensitivity to loops. For example the more undecided elements y attack x, the closer to 0 (out) its value gets.

Proof. See [166].

Definition 15.39. *We now define equations for bipolar networks. Let $(S, \rightarrow, \twoheadrightarrow)$ be a bipolar network and let a be an element whose configuration is as in Figure 15.19. We define the equations for a as follows: This definition is compatible with condition (i) of Section 1.2: if x supports y and z attacks y then z attacks x.*

- $x = 1 - \max(y_1, \ldots, y_k, 1 - u_1, \ldots, 1 - u_n)$
- $x = 1$ *if the support group G_a of a is unattacked.*

Theorem 15.40.

1. *There is always at least one solution to any system of equations of Definition 15.39.*
2. *Let $\mathbf{f}$ be a solution of the equations. Define a function $\lambda_{\mathbf{f}}$ as follows:*
   - $\lambda_{\mathbf{f}}(x) = in,$ *if $\mathbf{f}(x) = 1$*
   - $\lambda_{\mathbf{f}}(x) = out,$ *if $\mathbf{f}(x) = 0$*
   - $\lambda_{\mathbf{f}}(x) = undecided,$ *if $0 < \mathbf{f}(x) < 1$.*

Then λ_f is a Gabbay-Caminada labelling.

3. Let λ be a Gabbay-Caminada labelling and define $\mathbf{f}_\lambda$ as follows

- *$\mathbf{f}_\lambda(x) = 1$, if $\lambda(x) = in$*
- *$\mathbf{f}_\lambda(x) = 0$, if $\lambda(x) = out$*
- *$\mathbf{f}_\lambda = \frac{1}{2}$, if $\lambda(x)$ is undecided.*

Then $\mathbf{f}_\lambda$ is a solution to the equations.

Proof.   1. holds from Brouwers fixed point theorem.[12]

2. Let $\mathbf{f}$ be a solution to the equations and consider λ_f. We show it satisfies the conditions (a1), (a2), (b), (c) of Definition 15.23

   (a1) holds by the definition of the equations

   (a2) holds because we get $\mathbf{f}(a) = 1 - \max(0, \dots, 0, 1 - 1, \dots, 1 - 1)$

   (b)   If any $\lambda_f(y_j) = in$ or $\lambda_f(u_i) = out$, this means that $\max(\mathbf{f}(y_j), 1 - \mathbf{f}(y_i)) = 1$ and hence $\mathbf{f}(a) = 0$, i.e. $\lambda_f(a) = out$.

   (c)   If $0 < \mathbf{f}(u_i), i = 1, \dots, m$, $\mathbf{f}(y_j) < 1$ for $j = 1, \dots, n$ and for at least one x of $\{u_i, y_j\}$ we have $0 < \mathbf{f}(x) < 1$ then $0 < \max(\mathbf{f}(y_j), 1 - \mathbf{f}(u_i)) < 1$ and so $0 < \mathbf{f}(a) < 1$, i.e. $\lambda_f(a) = $ undecided.

3. Assume λ is a Gabbay-Caminada labelling and consider $\mathbf{f}_\lambda$. We show it satisfies the equations. We check the equation $\mathbf{f}_\lambda(a) = 1 - \max(\mathbf{f}_\lambda(y_j), 1 - \mathbf{f}_\lambda(u_i))$. We distinguish three cases, and verify the equation holds for each case. We have that $\mathbf{f}_\lambda(a)$ is either 0, $\frac{1}{2}$ or 1.

Case 1

For the right hand max side to be 1, we must have that all $\mathbf{f}_\lambda(y_j) = 0$ and all $\mathbf{f}_\lambda(u_i) = 1$. This means that all $\lambda(y_j) = out$ and all $\lambda(u_i)$ are in. But then $\lambda(a) = in$ and so the equation holds

Case 0

For the right hand max side to be 0, we must have that either some $\mathbf{f}_\lambda(y_j) = 1$ or some $\mathbf{f}_\lambda(u_i) = 0$. But then we have either some $\lambda(y_j) = in$ or some $\lambda(u_i) = out$. But this implies $\lambda(a) = out$, i.e. $\mathbf{f}_\lambda(a) = 0$ and the equation holds.

Case $\frac{1}{2}$

For the right hand max side to be $\frac{1}{2}$ we must have all $\mathbf{f}_\lambda(y_j) \leq \frac{1}{2}$, all $\mathbf{f}_\lambda(u_i) \geq \frac{1}{2}$ and for at least one $x \in \{y_j, u_i\}$, $\mathbf{f}_\lambda(x) = \frac{1}{2}$. But this implies $\lambda(a) = $ undecided and so $\mathbf{f}_\lambda(x) = \frac{1}{2}$ and the equation holds.

Corollary 15.41. *There exist Gabbay-Caminada labellings on bipolar networks.*

Proof. Follows from Theorem 15.40.

Example 15.42. Let us check the equations for some of the networks we have so far.

The equations are written from the point of view that if x supports y and y is attacked by z then z also attacks x.

If we look, for example, at Figure 15.12, we see that b supports c, and so the equation for b should reflect the possibility of a hidden attack on c which will become an attack on b, given our view. In labelling numerical terms this view manifests itself in the fact that if c is labelled 0 then b must also be labelled 0. This explains equations

---

[12] See http://en.wikipedia.org/wiki/Brouwer_fixed-point_theorem.

(1b) and (1c) below for Figure 15.12, and equations (2a) and (2b) for item c of Figure 15.22.

The equations for Figure 15.23 are put forward for two reasons. One is the connection of this figure with Contrary to Duty obligations and two that we can have a node x both supporting and attacking the same node y. The equations do not care, they just need to be solved. Note that we are not using the reduction method on the figure itself . Node $\mathbf{o}$, for example, because it both supports and attacks y, also attacks itself. The reduction method would have added a double arrow $\mathbf{o} \twoheadrightarrow \mathbf{o}$ to the figure. The equational approach does not do that; the view is built into the equations. Indeed equation (3b) incorporates an attack from y on $\mathbf{o}$.

It is also important to note that we are definitely not first reducing the graph in our mind and then writing equations for it. Had we done so the equation for $\mathbf{o}$ would have been $\mathbf{o} = 1 - \max(x, \mathbf{o}, y)$. This point is analysed explicitly in Example 15.30 below.

1. **Network of Figure 15.12**

   The equations are:
   a) $a = 1$
   b) $b = 1 - \max(a, 1 - c)$
   c) $d = 1 - \max(1 - c)$
   We get
   d) $b = 0$, from (1)
   e) $d = c$
   Since $\{d, c\}$ is now an unattacked cycle, we get
   f) $d = c = 1$
   The extension is $\{a, d, c\}$.

2. **Network of Figure 15.22, item (c):**

   The equations are
   a) $a = 1 - \max(b, 1 - c)$
   b) $b = 1 - \max(a, 1 - c)$
   c) $c = 1$
   The solutions are $a = r, b = 1 - r, c = 1$ where r is any number $0 \le r \le 1$.

3. **Network of Figure 15.23**

   The equations are:
   a) $\infty = \mathbf{o}$
   b) $\mathbf{o} = 1 - \max(1 - x, 1 - y)$
   c) $x = 1 - \max(1 - x, 1 - x_2)$
   d) $y = 1 - \max(\mathbf{o}, 1 - y_2, 1 - y_2)$
   e) $x_2 = 1 - x$
   f) $x_1 = 1$
   g) $y_2 = 1$
   h) $y_1 = 1 - y$
   Simplifying, we get
   i) $\mathbf{o} = 1 - \max(x_2, y_1)$ from (b), (e), (h)
   j) $x_2 = 1 - x_2$, from (c), (e), (f)
   k) $y_1 = 1 - y_1$, from (d), (g), (h)
   Therefore we have

   $$x_1 = 1, x_2 = \tfrac{1}{2}, y_1 = \tfrac{1}{2}, y_2 = 1, \mathbf{o} = \tfrac{1}{2}, \infty = \tfrac{1}{2}, y = \tfrac{1}{2}, x = \tfrac{1}{2}.$$

4. **Equations for Figure 15.28**
   This is the figure causing difficulty to Cayrol and Lagasquie-Schiex [96].
   a) $a = b$
   b) $b = c$
   c) $c = 1 - e$
   d) $e = 1$
   e) $d = 1 - b$.
   The solution is
   $$a = b = c = 0, d = e = 1.$$

Example 15.43 (Reduction approach to Figure 15.23). Continuing Example 15.42, let us apply the reduction method to Figure 15.23, get the new Figure 15.30 and then apply the equational method to it an see what we get.

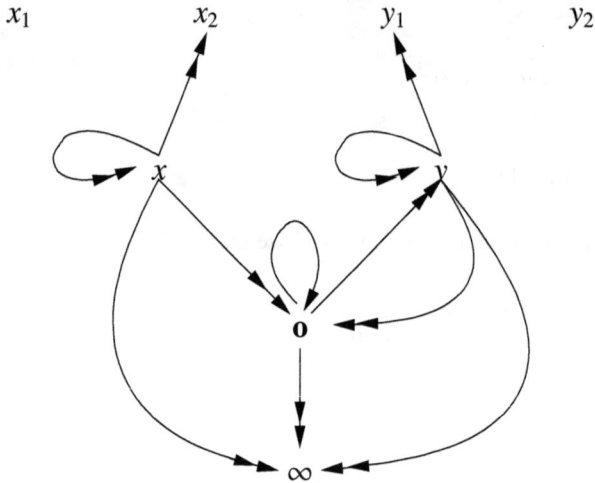

Fig. 15.30.

We remind the reader that we used the principle of reduction (ii) of Section 1.2.

(ii) $x \rightarrow y$ and $z \twoheadrightarrow y$ imply $z \twoheadrightarrow x$.

We applied this to Figure 15.23 and got Figure 15.30. The support arrows $\rightarrow$ are now ignored.

We now write equations for Figure 15.30. These are:

$\infty = 1 - \max(\mathbf{o}, x, y)$
$\mathbf{o} = 1 - \max(x, \mathbf{o}, y)$
$y = 1 - \max(y, \mathbf{o})$
$x = 1 - x$
$x_2 = 1 - x$
$y_1 = 1 - y$
$x_1 = 1$
$y_2 = 1$

The solution to these equations is the same as in Example 15.42 for Figure 15.23 namely

$$\infty = \mathbf{0} = y = x = x_2 = y_1 = \tfrac{1}{2}$$
$$x_1 = y_2 = 1$$

We clearly see that the equations for the bipolar case as done in Example 15.42 for Figure 15.23 contain implicitly the information of the reduction Figure 15.30.

Example 15.44 (Equations for challenge). We apply the equational approach to networks of the fom $(S, \rightarrow, \Rightarrow, \twoheadrightarrow)$. The novelty here is that we have two kinds of support and so two kinds of equations are involved. Let a be a node and let $u_1, \ldots, u_m$ be all nodes such that $a \rightarrow u_i$, $1 \le i \le m$ holds. Let $y_1, \ldots, y_k$ be all nodes such that $y_j \twoheadrightarrow a$ holds, $1 \le j \le k$. Let $x_1, \ldots, x_n$ be all nodes such that $x_r \Rightarrow a$ holds, $1 \le r \le n$.

We regard the network, as we did before, as a carrier of equations, and we regard the nodes as variables in these equations. The nodes take values in $[0, 1]$. We regard the value 0 as corresponding to out, 1 as corresponding to in and any other value $0 < v < 1$ as corresponding to undecided.

The Eq_{max} for this case are:

- $a = [1 - \max(y_j, u_i)] \dfrac{(1 - \max(x_r))}{2}$, where $1 \le i \le m, 1 \le j \le k, 1 \le r \le n$.

15.5 Joint support, joint attack and equations

In Chapter 7 we introduced the notions of joint attacks and disjunctive attacks. In this section we recall these notions and introduce the notions of joint support and disjunctive support and develop the equational approach to these concepts. These concepts are needed for the analysis of various defeasible bases involved in the ASPIC debate [16, 289] because we have in such systems proofs from several assumptions and this translates into joint support. We need semantics and the most general immediately usable semantics is the equational one. We then illustrate our system by looking at more examples from [79].

The idea of joint attack is not new. It arose naturally in the context of fibring argumentation networks of Chapter 7, but was used before (see references and comparisons in Chapter 7). Figure 15.31 introduces the notation we use.

Definition 15.45. *Consider Figure 15.31. In this figure the element a is attacked by n groups of joint attackers, where group i has n(i) attacking members. If the figure shows all attackers of a, then we have*

1. *a is out if for at least one i, all of $x^i_1, \ldots, x^i_{n(i)}$ are in.*
2. *a is in if for each i there exists a j such that x^i_j is out.*
3. *a is undecided if neither (1) nor (2) hold and for at least one i we have that all x^i_j are either in or undecided.*

The Eq_{max} equation for Figure 15.31 is

$$a = 1 - \max_i \min_j (x^i_j)$$

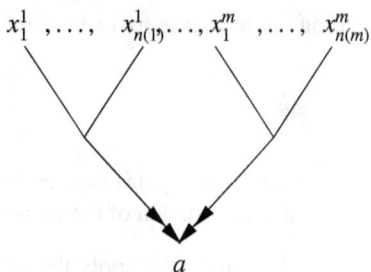

Fig. 15.31.

Remark 15.46. It was shown in Chapter 7 that joint attacks can be implemented using ordinary single attacks if we allow for auxiliary points. Thus Figure 15.32 can be implemented as Figure 15.33. $\{x, y, a\}$ are the original points and $\{\alpha_x, \alpha_y, \beta_a\}$ are the auxiliary points. Note the Eq_{max} equations for a in terms of x, y are the same as before, with the auxiliary nodes not appearing in the equations. This shows they play no role as far as the extensions are concerned. The relevant concepts and theorems are developed in Chapter 7.

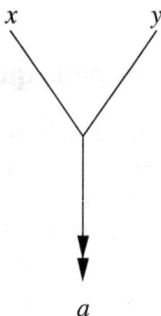

Fig. 15.32.

We have

$$\alpha_x = 1 - x$$
$$\alpha_y = 1 - y$$
$$\beta_a = 1 - \max(1 - x, 1 - y)$$
$$a = \max(1 - x, 1 - y)$$
$$= 1 - \min(x, y)$$

We now define the notion of joint support. The motivation are examples where we have

$$\bigwedge A_i \vdash B$$

We can think of it as $\{A_i\}$ jointly support B (compare with Example 15.29).

The notation is as in Figure 15.31, except that we use single arrow $\rightarrow$ instead of double arrow $\twoheadrightarrow$.

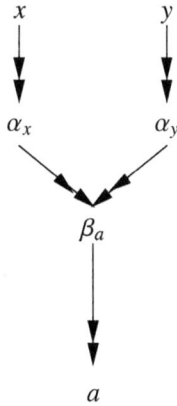

Fig. 15.33.

The understanding of support means that if all of A_i are in then we expect B to be in. Alternatively, if B is out and $A_1, \ldots, A_{n-1}$ are in then A_n must be out. So we can reduce joint support to joint attacks. This corresponds to contraposition.

$$A_1, \ldots, A_n \vdash B$$

iff

$$A_1, \ldots, A_{i-1}, \neg B, A_{i+1}, \ldots, A_n \vdash \neg A_i.$$

Definition 15.47. *Let* $(S, \rightarrowtail, \twoheadrightarrow)$ *be a bipolar network with joint attacks and joint support. So* $\rightarrowtail$ *(joint support and* $\twoheadrightarrow$ *(joint attacks) are subsets of* $2^S \times S$.

Consider the general configuration of a node $a \in S$, *described in Figure 15.34. In this figure* $\{x_1^i, \ldots, x_{n(i)}^i\}$ *are all the joint groups attacking* $a, i = 1, 2, \ldots$ *and* $\{y_1^k, \ldots, y_{m(k)}^k\}$ *and* z_k *are all the group for which* a *is part in supporting, as shown in the figure.*

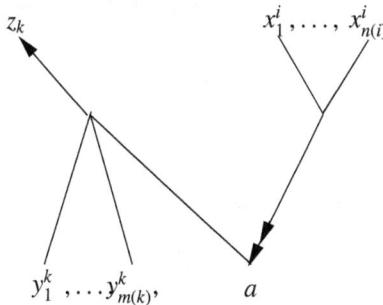

Fig. 15.34.

The equation for a is then the following:

$$a = 1 - \max_{i,k} \min_{j,r}(x^i_j, y^k_r, 1 - z_k)$$

Compare this equation with the equation in Definition 15.39.

Example 15.48. Consider the situation in Figure 15.35

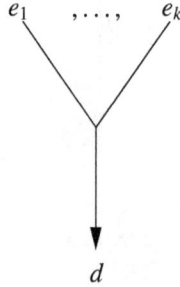

$e_1 \quad , \ldots, \quad e_k$

d

Fig. 15.35.

We have, according to Definition 15.47, the following equations:

$$e_i = 1 - \min(1 - d, e_j, j \neq i).$$

We claim that if we want for all of these equations to hold in the case that $d = 0$, and we further insist that $e_i \in \{0, 1\}$ then we have that exactly one of e_i is 0. Let us check. If all $e_i = 1$, we get a contradiction and if $e_i = e_j = 0$ for some $i \neq j$ we get $e_i = 1 - 0 = 1$, again a contradiction.

Note that if we regard the situation described in Figure 15.35 as a logic Programming Horn clause of the form $\bigwedge e_i \to d$ then by the equational approach of Chapter 10, the equation for this clause should be $d = \prod e_i$. We are not adopting in this chapter the Logic programming equation of Chapter 12, because it allows for more than one e_i to be 0, when $d = 0$. The reason being that in the context defeasible logic, we want minimal corrections to the database. So when a strict rule is contradicted or when a defeasible rule is defeated, we want to implement a remedy with minimal change only![13]

Example 15.49. We consider Example 5 from [79]. We have a database with strict $\to$ and defeasible $\Rightarrow$ rules, where[14]

$$S = \{\top \to a, \top \to d, \top \to c, b \wedge e \to \neg c\}$$
$$D = \{a \Rightarrow b, d \Rightarrow e\}.$$

We construct the arguments as in Example 15.20. In [79] they consider the following arguments.

---

[13] In the ASPIC approach (and similar approaches, such as Defeasible Logic, Defeasible Logic Programming and Pollock's work) strict rules are by definition beyond defeat. They can however be revised or updated, in which case we want minimal change.

[14] When x is a strict assumption we write $\top \to x$, (in [79] they just write $\to x$), and when y is a defeasible assumption we write $\top \Rightarrow y$ (in [79] they just write y).

$$A = [[\top \rightarrow a] \Rightarrow b]$$
$$B = [[\top \rightarrow d] \Rightarrow e]$$
$$C = [\top \rightarrow c]$$

These arguments have no defeaters according to [79]. Their *output* is the set $\{b, e, c\}$ which is not consistent in view of the strict rule $b \wedge e \rightarrow \neg c$.

Let us see what happens if we use our joint attack and support system. Figure 15.36 describes the situation. We show no attack from $\neg c$ to c because c is strict consequence and $\neg c$ is defeasible consequence and so the convention is that a defeasible consequence cannot attack a strict consequence.

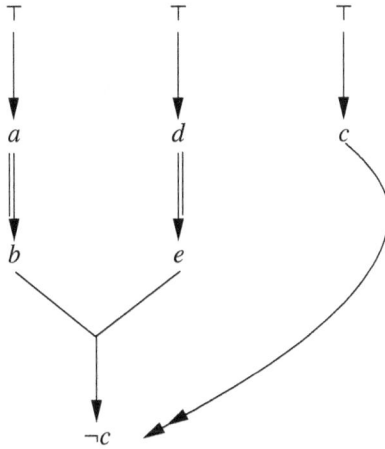

Fig. 15.36.

We have not said formally how the equational approach treats $\Rightarrow$. So let us leave aside for the moment the equational connection between a and b and the connection between d and e. We do know however, how to compute the connection between b, e and $\neg c$. The equations are

$$e = 1 - \min(b, 1 - \neg c)$$
$$= 1 - \min(b, c)$$

and

$$b = 1 - \min(e, 1 - \neg c)$$
$$= 1 - \min(e, c)$$

From $\top \rightarrow c$, we get $c = 1$ and thus get that $b = 1 - e$. If we insist on $\{0 = \text{out}, \frac{1}{2} = \text{undecided}, 1 = \text{in}\}$ values, we have two extensions

1. $a = d = c = 1$, $b = 1$ and $e = 0$.
2. $a = d = c = 1$, $b = 0$ and $e = 1$.

If we allow also $\neg c$ to attack c, we get one more solution

3. $a = d = 1, b = e = c = \frac{1}{2}$

This is a consistent situation, since $a \Rightarrow b$ and $d \Rightarrow e$ are defeasible rules, and we do not insist on $a = b$ and on $d = e$.

We remarked before that joint support can be rewritten as joint attack and that joint attack can be reduced to ordinary attack with the help of auxiliary points. Let us do this to Figure 15.36 and see what we get. We first turn Figure 15.36 into Figure 15.37 and then transform the latter into Figure 15.38.

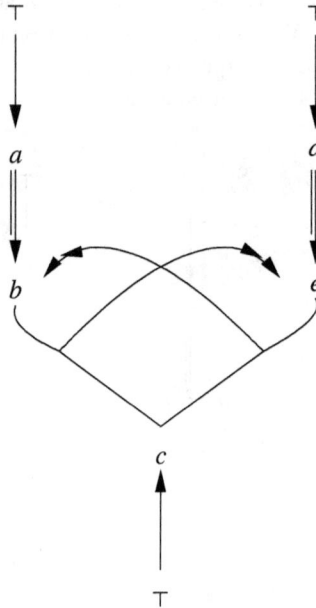

Fig. 15.37.

We clearly see that Figure 15.38 allows for the same extensions as before.

Remark 15.50. We shall treat the equational approach to argumentation networks with defeasible rules in another paper [190]. Here we mention the idea only. Consider a system of equations which has no solutions. Since each equation is a demand on the variables, the lack of solutions means the demands are not consistent. So to get a solution we drop some equations. So in a network with defeasible rules, we get some equations with low priority (these are the equations arising from the defeasible rule), which we drop if there is no solution.

Let us consider the defeasible rules $a \Rightarrow b$ and $d \Rightarrow e$ in Figure 15.36 and let us offer the equations $a = b$ and $d = e$ for them (which means treating $\Rightarrow$ as $\rightarrow$ and applying the equation for $\rightarrow$), but give them low priority. The resulting system of equations will have no solution and so we get solutions by dropping out one of the equations.

Example 15.51. We consider example 6 of [79]. This example was also considered in Chapter 4. We have a network obtained from strict and defeasible rules as follows

$$S = \{\top \rightarrow a, \top \rightarrow d, \top \rightarrow g, b \wedge c \wedge e \wedge f \rightarrow \neg g\}$$
$$\mathcal{D} = \{a \Rightarrow b, b \Rightarrow c, d \Rightarrow e, e \Rightarrow f\}.$$

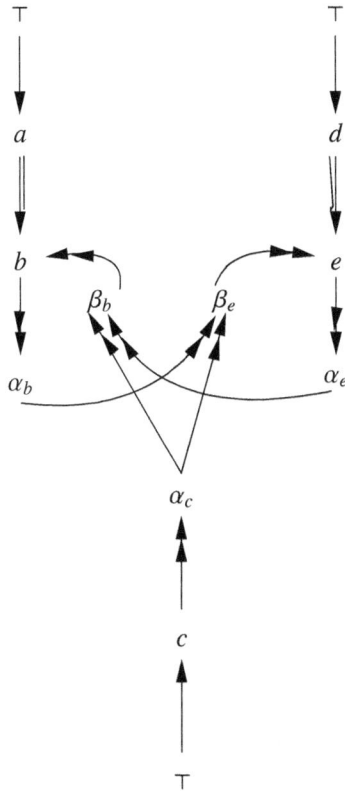

Fig. 15.38.

In [79] they consider the following arguments

$$A = [[\top \rightarrow a] \Rightarrow b]$$
$$B = [[\top \rightarrow d] \Rightarrow e]$$
$$C = A \Rightarrow c$$
$$D = [B \Rightarrow f]$$

The arguments A, B, C, D have no defeaters, according to [79]. This yields an *output* with $\{b, c, e, f, a, g\}$ which is, however, inconsistent because of the strict rule $b \wedge c \wedge e \wedge f \rightarrow \neg g$.

Using our methods we get Figure 15.39 for this example. We cannot write equations for $a \Rightarrow b, b \Rightarrow c, d \Rightarrow e, e \Rightarrow f$ but we can write equations for $b \wedge c \wedge e \wedge f \rightarrow \neg g$. These are

$$b = 1 - \min(c, g, f, e)$$
$$c = 1 = \min(b, g, f, e)$$
$$f = 1 - \min(b, g, c, e)$$
$$e = 1 - \min(b, g, c, f)$$

Since $g = 1$, the solutions are that exactly one of $b = 0 \bigvee c = 0 \bigvee f = 0 \bigvee e = 0$ is true.

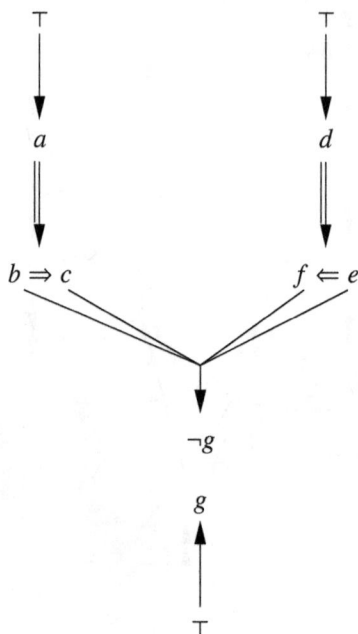

Fig. 15.39.

Again, because we have some defeasible rules, there is no inconsistency.

This example gives us an opportunity to illustrate how the priorities for the defeasible equations can be obtained from the geometry of the network. We start by treating $\Rightarrow$ like $\rightarrow$ and get the equations

$$a = 1 \quad d = 1$$
$$b = a \quad e = d$$
$$c = b \quad f = e$$

We can drop out only either $c = b$ or $f = e$.

So, for example, we can get $c = 0, a = d = g = 1, b = e = 1, f = 1$.

Note that the equations *do not* allow us to drop the middle equation, say $b = a$. This will make two nodes equal 0.

Definition 15.52. *We now define disjunctive support and disjunctive attack. Figure 15.40 shows the basic configuration and notation. The disjunctive attack was introduced in [164], see Chapter 7. If b is in, then at least one of $y_1, \ldots, y_m$ must be out. The disjunctive support is a new concept. a supports the group $\{x_1, \ldots, x_n\}$. It is like giving money to a charity agent, who will pass on the money in due course to one of the charities $x_1, \ldots, x_n$.*

So for disjunctive attack, if all of $y_1, \ldots, y_m$ are in then b must be out. For disjunctive support, if all $x_1, \ldots, x_n$ are out, then a must be out.

Note that logically disjunctive support is not

$$A \vdash B_1 \vee \ldots \vee B_n$$

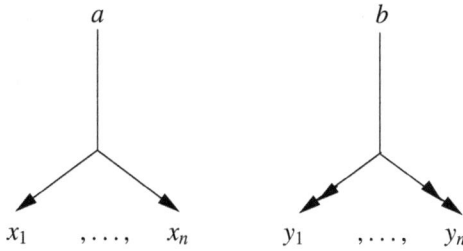

Fig. 15.40.

because we requrie it to mean $\bigvee_i[A \vdash B_i]$. *This is the disjunction property for* A, B_i. *we do not have it in classical logic. We have* $\top \vdash p \vee \neg p$ *but* $\top \nvdash p$ *and* $\top \nvdash \neg p$.

15.6 Conclusion and future research

The most important conclusion to be drawn from our studies and examples in this paper is that the ASPIC approach can most likely be modelled as a tripolar network with three relations, $\rightarrow$ for strict implication, $\Rightarrow$ for defeasible implication and $\twoheadrightarrow$ for the attack relation. These relations are joint, i.e. they are subsets of $2^S \times S$, where S is the set of arguments. The examples in Section 6 above show the way, but the details and technical machinery have to be worked out in a future paper [190]. We give here one example, just a taste.

Example 15.53 (Equations and specificity). Consider the following two defeasible rules

- $\bigwedge a_i \wedge \bigwedge b_i \Rightarrow c$
- $\bigwedge a_i \Rightarrow \neg c$

Defeasible logic would give priority to the first, more specific rule.

Let us see what our equational approach has to say. How does the specificity condition manifest itself in the equations?

Consider the rule

$$\bigwedge x_i \Rightarrow y.$$

The equations for it are

$$x_i = 1 - \min(1 - y, x_j, j \neq i).$$

Suppose we want the principle of forcing as many arguments z to be in (i.e. $z = 1$) as we can. If we choose $y = 1$, then $1 - y = 0$, so each x_i will be 1.

If we choose $y = 0$, then $x_i = 1 - \min(x_j, j \neq i)$ and exactly one of x_i will be 0. Obviously we want $y = 1$.

Now out of the two contradictory rules above, one for c and one for $\neg c$, which one of $\{c, \neg c\}$ do we choose to make in (i.e. do we choose $c = 1$ or do we choose $c = 0$?).

Obviously the best choice that maximises the number of in arguments is to give 1 to the head of the rule which is more specific and has more arguments in the body. So the specificity rule can be derived from the equational principle of finding a solution which maximises the number of variables with value 1.

To complete the picture we remark that as the equation corresponding to a strict rule

$$\bigwedge x_i \to y$$

we take as

$$y = \prod x_i.$$

Let us now look at a clash between a strict rule and a defeasible rule.

If we make the head of a strict rule 0, we can potentially make all the elements in the body 0. The equations do not forbid that. In comparison, if we make the head of a defeasible rule 0, only one more element from the body has to be 0. So if we give preference to strict rules over defeasible rules, we still, in principle, maximise the number of elements which get value 1 (in as much as we are limiting the potential of getting values 0).

Our future research plan in this area is as follows:

1. Develop equational semantics for defeasible systems for strict rules and defeasible rules.
2. Implement such systems in tripolar joint attacks argumentation networks with equational semantics.
3. Since the ASPIC approach instantiates arguments in defeasible systems of (1), we can implement ASPIC argumentation system in (2).

The above has direct bearing on the problem of instantiation of abstract argumentation, as it shows that the ASPIC instantiation is actually an abstract tripolar joint attack argumentation network.

To be absolutely clear, let (S, R) be a traditional argumentation network, with a set S of atomic letters standing for arguments and an attack relation $R \subseteq S^2$. We can take a defeasible system $\mathbb{D}$ with database A of atoms in the form q or the form $\neg q$ and strict rules (written using $\to$) and defeasible rules (written using $\Rightarrow$) and construct a set of arguments ARG and an attack relation among these arguments (denoted by $\twoheadrightarrow$ and defined appropriately using $\neg$) as done in [79] and as quoted in our Examples 15.49 and 15.51. We thus get a specific argumentation pair (ARG, $\twoheadrightarrow$). If we have that the pair (S, R) can be isomorphically mapped onto the pair (ARG, $\twoheadrightarrow$), then we say that (S, R) is instantiated by $\mathbb{D}$. What we can show is that $\mathbb{D}$ can be isomorphically represented as a tripolar abstract argumentation system with attacks and joint defeasible supports and joint strict supports of the form $(A, \to, \Rightarrow, \twoheadrightarrow)$ where $A, \to, \Rightarrow, \twoheadrightarrow$ are the same as in the rules of $\mathbb{D}$ now viewed as relations on $2^A \times A$.

We do not know yet what conditions to impose on a tripolar joint abstract (A, R_1, R_2, R_3) for it to be of the above instantiated form obtained from some $\mathbb{D}$.

This is a representation theorem.

16

Reactive Kripke Models and Argumentation Networks

16.1 Reactive arrows and modalities

Modal logic **K** is complete for tree Kripke frames. These have the form $(S, R, 0)$, where S is a non-empty set of worlds, and $R \subseteq S^2$ is a tree relation on S with root point 0. Being a tree means that for every point $x \in S$, there exists a unique path $(0, x_1, \ldots, x_n)$ leading from 0 to $x_n = x$, such that $0Rx_1, x_1Rx_2, \ldots, x_{n-1}Rx_n$ hold. The option $n = 1$ means $0Rx$.

We add reactivity to such a model by adding a further set of relations $R' \subseteq S^2 \times S^2$.

We write $x \to y$ when $(x, y) \in R$ and we write $(x \to y) \twoheadrightarrow (u \to v)$ when $((x, y), (u, v)) \in R'$. The first are called arrows and the second are called double arrows.

Figure 16.1 is an example of a reactive frame $(S, R, R', 0)$. Since $(S, R, 0)$ is a tree, Figure 16.1 can be represented as Figure 16.2.

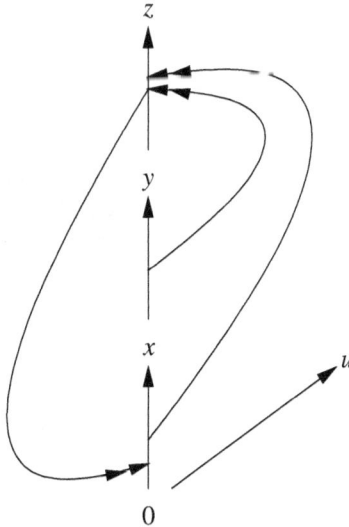

Fig. 16.1.

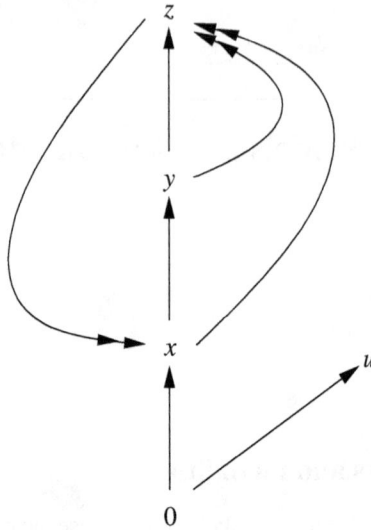

Fig. 16.2.

How do we interpret the double arrows? We imagine an agent traversing the tree from point 0 along the arrows. This agent may be moving along the arrows in the course of evaluation of a modal formula. Let us take, for example, the formula $\lozenge\lozenge\lozenge\Box\bot$, and evaluate it at node 0.

We need to find a path $(0, x_1, x_2, x_3)$ such that $x_3 \vDash \Box\bot$ holds and $0Rx_1, x_1Rx_2, x_2Rx_3$ hold. Such a path exists; it is $(0, x, y, z)$. Let the operators $\lozenge$ and $\Box$ look only at the relation R. $\lozenge$ and $\Box$ move along the arrows. They disregard the double arrows.

We now have to say what the double arrows do. We consider two options.

Option 1. Ordinary reactivity

When we evaluate at a node x_m such that $(0, x_1, \ldots, x_m)$ is the path leading to x_m, then we disconnect any arrow of the form $u \to v$ such that $(x_i \to x_{i+1}) \twoheadrightarrow (u \to v)$ holds for some $0 \leq i < m$, (i.e. we take $u \to v$ out of R when the above double arrow is in R'). So, for example, when we move along the path $(0, x, y)$ the arrow $y \to z$ gets disconnected because it is being hit twice; once from $(0 \to x) \twoheadrightarrow (y \to z)$ and once from $(x \to y) \twoheadrightarrow (y \to z)$. Let $\lozenge'$ be a modality which respects the effects of the double arrows. Then we have

$$0 \vDash \lozenge\lozenge\Box'\bot$$
$$0 \nvDash \lozenge\lozenge\Box\bot$$

Note that the double arrow $(y \to z) \twoheadrightarrow (0 \to x)$ has no effect on the evaluation because it is triggered after we pass through the arrow $0 \to x$. It will have an effect if we have some operators which go backwards.

Option 2. Switch reactivity

Switch reactivity counts the number of hits by double arrows which impact on an arc. If it is even, the arc remains connected. If it is odd then the arc is disconnected. Thus a switch double arrow switches a connection "on" if it is "off" and "off" if it is "on". So it acts like a switch. The default condition for arcs is that they are "on".

Thus in Figure 16.1 the arc $y \to z$ shall remain on because it is hit twice by double arrows. So for the case of switch double arrows, $0 \nVdash \Diamond\Diamond\Box' \bot$.

Note that the reactivity idea does not require $(S, R, 0)$ to be a tree. We can add reactivity and move to $(S, R, R', 0)$ for any Kripke frame. The tree property is used only in reducing $R' \subseteq S^2 \times S^2$ to an $R_1 \subseteq S \times S$, since in a tree, the arc $x \to y$ is uniquely determined by the node y. In fact, we can allow for reflexivity at the root 0 as well if we understand $0 \twoheadrightarrow 0$ as $(0 \to 0) \twoheadrightarrow (0 \to 0)$. We can have this understanding only for $x = 0$. If $x \neq 0$ then $y \to x$ exists for $y = $ predecessor (x) and $x \twoheadrightarrow x$ must mean $(y \to x) \twoheadrightarrow (y \to x)$.

Of course if there is no double arrow $(y \to x) \twoheadrightarrow (y \to x)$ then $x \twoheadrightarrow x$ can mean $(x \to x) \twoheadrightarrow (x \to x)$. So in principle we can reduce arc reactivity R' to point to point relation R_1 in many more cases if we have the right conventions.

Consider Figure 16.3.

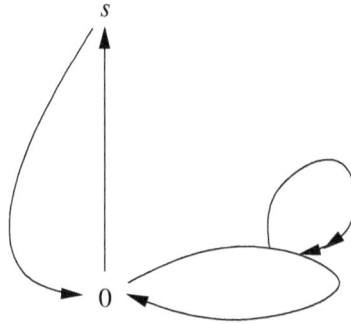

Fig. 16.3.

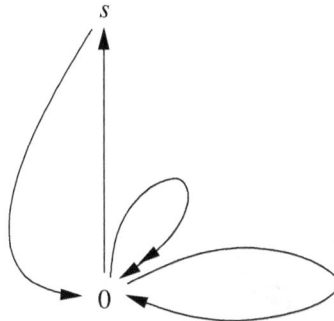

Fig. 16.4.

The above figure gives rise to a modal logic with a single modality $\Box'$ which is not complete for any ordinary class Kripke frames but is complete for the above reactive frame.

The above discussion places reactive Kripke structures somewhere in the area of bimodal logics. It yields some sort of a modality $\Diamond'$ which is more than a single modality and less than an independent additional modality.

This gives us scope for further research. We can consider the following questions:

1. Expressive power of $\Diamond, \Diamond'$
2. Applications of the idea of reactivity
3. Axiomatisation, proof theory and completeness proofs for these modalities.

For references see [147, 157, 158]. The option which connects with argumentation is mainly Option 1.

16.2 Examples motivating the reactive idea

Before we continue with reactivity and argumentation, let us motivate our idea of reactive semantics and consider some examples. The reader should note that all of these examples can also be viewed as bipolar argumentation networks!

16.2.1 Airline example

We begin with a very simple and familiar example. Consider Figure 16.5

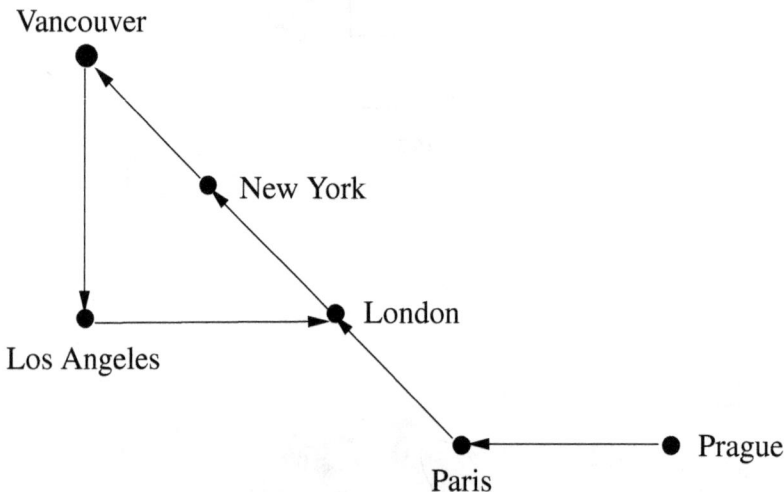

Fig. 16.5.

Figure 16.5 gives the possible flight routes for the aeroplanes of TUA (Trans Universal Airlines). It is well known that many features of a flight depend on the route. These include the cost of tickets, as well as the right to take passengers at an airport. The right to take passengers at an airport depends on the flight route to that airport and on bilateral agreements between the airlines and governments. Thus, for example, flights to New York originating in London, may take on passengers in London to disembark in New York. However, a flight starting at Paris going to New York through London may not be allowed to pick up passengers in London to go onto New York. It is all a matter of agreements and landing rights. It is quite possible, however, that on the route Prague–Paris–London–New York, the airline is allowed to take passengers in London to disembark in New York. We can describe the above situation in Figure 16.6.

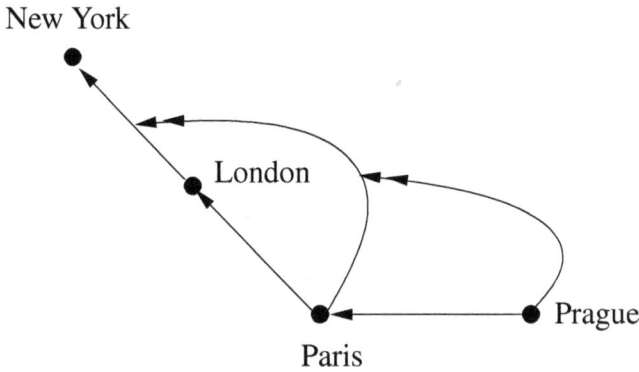

Fig. 16.6.

The double-headed arrow from Paris to the arc London→New York indicates a cancellation of the 'passenger' connection from London to New York. The double-headed arrow from Prague to the double arrow arc emanating from Paris indicates a cancellation of the cancellation.

Figure 16.6 looks like a typical reactive Kripke model, where we have arcs leading into arcs.

Let us see more examples of this.

16.2.2 Inheritance networks example

This example offers a different point of view of arc semantics, coming from the non-monotonic theory of inheritance networks. Consider Figure 16.7.

In Figure 16.7, the circular nodes are predicates, such as Fly, Birds, etc. The arrows indicate inheritance, so for example, we have $\forall x(\text{Bird}(x) \rightarrow \text{Fly}(x))$. The arrows with a bar indicate blockage, for example $\forall x(\text{Penguin}(x) \rightarrow \neg\text{Fly}(x))$. The square nodes indicate instantiation, so son of Tweetie is a special penguin.

Figure 16.7 is the kind of figure one finds in papers on inheritance networks. The figure indicates that Penguins are Birds, that Birds Fly but that Penguins do not Fly. However, special Penguins do Fly and the son of Tweetie, a rare bird, is a special Penguin, and therefore does Fly. The arrow with the bar on it blocks the information from flowing from the Penguin node to the Fly node. The theory of inheritance networks spends a lot of effort on algorithms that allow us to choose between paths in the network so that we can come up with the desired intuitively correct answers. In the case of Figure 16.7 we want to get that the son of Tweetie does fly, since we have the most

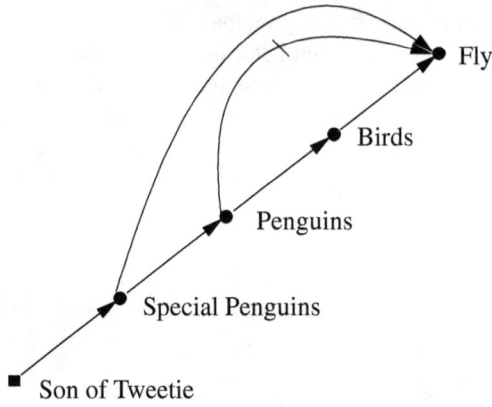

Fig. 16.7.

specific information about him. It is not important to us in this paper to take account of how inheritance theory deals with this example. We want to look at the example from our point of view, using our notation, as in Figure 16.8.

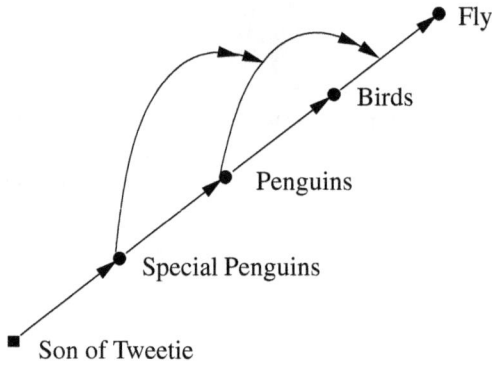

Fig. 16.8.

In Figure 16.8 the double headed arrow ↠ emanating from Penguins attacks the arrow from Birds to Fly, and the double arrow emanating from Special penguins attacks the double arrow emanating from Penguins and attacking the arrow from Birds to Fly. This is not how inheritance theory would deal with this situation but we are not doing inheritance theory here. Our aim is to motivate our approach and what we need from the inheritance example is just the idea of the algorithmic flow of information during the dynamic evaluation process.[1]

We have already put forward the reactive and dynamic idea of evaluation in earlier papers and lectures (see [141]). A typical example we give is to consider $t \vDash \Diamond A$. In

---

[1] It is our intention to explore whether our idea of double headed arrows cancelling other arrows can simplify inheritance theory algorithms.

modal logic this means that there is a possible world s such that we have $s \vDash A$ we take a more dynamic view of it.

We ask: where is s? How long does it take to get to it? and how much does it cost to get there?

The reader should recall the way circumscription theory deals with the Tweetie example, see [241, section 4.1, especially page 324]. We write

- Birds $(x) \wedge \neg Ab_1(x) \rightarrow$ Fly(x)
- Penguins$(x) \rightarrow$ Birds(x)
- Penguins$(x) \wedge \neg Ab_2(x) \rightarrow Ab_1(x)$
- Special Penguins$(x) \rightarrow$ Penguins(x)
- Special Penguins (son of Tweetie)
- Special Penguins $(x) \rightarrow Ab_2(x)$.

"$Ab(x)$" stands for "x is abnormal". If the clause $C(x) \rightarrow B(x)$ represents the arc $C \rightarrow B$ then $C(x) \wedge \neg Ab(x) \rightarrow B(x)$ represents the situation in Figure 16.9

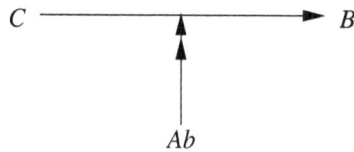

Fig. 16.9.

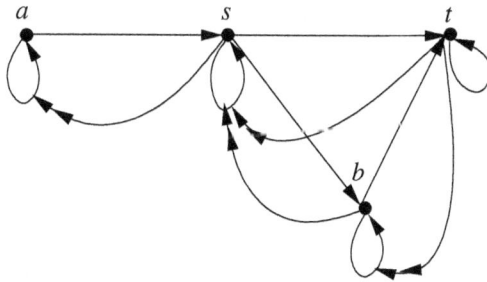

Fig. 16.10.

16.2.3 A technical example

It is now time to give a technical example. Consider Figure 16.10. This figure displays a past flow of time. The node t is the present moment and a single headed arrow from one node to another, say from s to t, means that t is in the immediate future of s. We use the modality $\square$ to mean 'always in the immediate past'. Thus the accessibility relation R of figure 16.10 is as follows:

- $tRs, tRb, bRs, sRa, tRt, bRb, sRs$ and aRa.

The double-headed arrows cancel the accessibility relation.

Let us calculate $t \vDash \square^3 q$ in Figure 16.10.

Initial Position: Starting point is t and all arrows are active.

Step 1: Send double arrow signal from t to all destinations inverting the active/inactive status of all destination arrows. Then go to all accessible worlds (in this case s and b) and evaluate $\square^2 q$ there. If the result is positive 1 and at all nodes, then send 'success' back to node t.

Step 2: Evaluate $\square^2 q$ at nodes b and s.

Subcase 2.b. Evaluation at b: First we send a double arrow signal from b to all destinations reversing the activation status of these destinations. Thus the single arrow from s to s will be re-activated and we will evaluate $\square q$ at s at the next step 3 (with s accessible to itself). b is not accessible to itself because its arrow has been deactivated by t at Step 1.

Subcase 2.s. Evaluation at s: First we send a double arrow signal to reverse the status arrow from a to a. Then we evaluate $\square^2 q$ at s with s not accessible to itself, since the arrow from s to s was deactivated by t at step 1.

This can go on, but we shall not continue as we trust that the reader has got the idea by now.

Note that if we start at t and evaluate $B = \square^3 q \wedge \square^2 q$, we will get that $\square q$ must be evaluated at s in two ways. One with s accessible to itself (coming from b via $t \vDash \square^3 q$) and once with s not accessible to itself (coming from t via $t \vDash \square^2 q$).

Let us now calculate $t \vDash \lozenge^2 q$ in Figure 16.10.

Initial Position: Starting point is t and all arrows are active.

Step 1: Send double arrow signal from t to all destinations inverting the active/inactive status of all destination arrows. Then go to one of the accessible worlds (in this case s or b) and evaluate $\lozenge^2 q$ there. If the result is positive 1 at this node, then send 'success' back to node t.

Step 2: Evaluate $\lozenge^2 q$ at nodes b or s.

Subcase 2.b. Evaluation at b: First we send a double arrow signal from b to all destinations reversing the activation status of these destinations. Thus the single arrow from s to s will be re-activated and we will evaluate $\lozenge q$ at s at the next step 3 (with s accessible to itself). b is not accessible to itself because its arrow has been deactivated by t at Step 1.

Subcase 2.s. Evaluation at s: First we send a double arrow signal to reverse the status arrow from a to a. Then we evaluate $\lozenge^2 q$ at s with s not accessible to itself, since the arrow from s to s was deactivated by t at step 1.

This can go on, but we shall not continue as we trust that the reader has got the idea by now. For the case of $\lozenge$ we make a non-deterministic choice. The model is not sensitive to whether we come to a point because we are evaluating $\square$ or a $\lozenge$. If we want this kind of sensitivity we can have arrows of the form $\twoheadrightarrow_\square$ and $\twoheadrightarrow_\lozenge$.

Such distinctions may be desirable in dealing with quantifier games, where changes may be different for the cases of $\forall$ and $\exists$.

16.2.4 Tax example

Having explained the technical side of our reactive (changing) semantics, let us give some real examples.

House prices in London have gone up a great deal. An average upper middle class family is liable to pay inheritance tax on part of the value of their house (if the house is valued over £500,000, for example, then there is tax liability on £250,000). Some parents solved the problem by giving the house as a gift to their children. If at least one of the parents remains alive for seven years after the transaction, then current rules say that there is no tax. Consider therefore the following scenario:

1. current date is April 2004
2. parents gave house as a gift to children in 1996
3. parents continued to live in house as guests of the children

(1)–(3) above imply that (4):

4. if parents both die in March 2004, then no tax is liable.

To continue the story, there were rumours that the tax people were going to change the rules in April 2004, declaring that if parents remain living in the house after it was given as a gift, then the gift does not count as such an there is tax liability. The rumours also said that this law is going to apply *retrospectively*.[2]

Thus we have that (5) holds:

5. If parents both die on March 2005, then tax is liable.

We assume that (4) still holds even after the new law as we cannot imagine that the UK tax inspector would be opening closed old files and demanding more tax.

The way to represent (4) and (5) is to use two dimensional logic. We write $t \vDash_s A$ to mean at time t A is true given the point of view proposed or held at time s.

Thus $2005 \vDash_{2003} \neg(5)$ holds, because from the 2003 laws point of view (before legislation) no tax is liable ((5) says tax is liable).But $2005 \vDash_{2004} (5)$ also holds, because according to 2004 legislation tax is liable.

So far we have no formal problem and no need for our new semantics, because we can write

- $t \vDash_t \Box A$ iff for all future s, $s \vDash_s A$.

In other words we evaluate sentences at time t according to the point of view held at the very same time t.

The problem arises when we want to formalise the following scenario. The parents die in 2003. The lawyer is dealing with the estate. We do not know when he is going to finish. When he submits the paperwork then the tax liability at 2003 is judged according to the time of submission. Now the second index s in $t \vDash_s A$ behaves like a reactive model as we are evaluating

$$2003 \vDash_{\text{time lawyer submits}} (4).$$

16.2.5 Salesman example

Consider the simple graph of Figure 16.11

---

[2] Some countries, like Austria, for example, would *never* legislate retrospectively. They regard this as a cultural taboo.

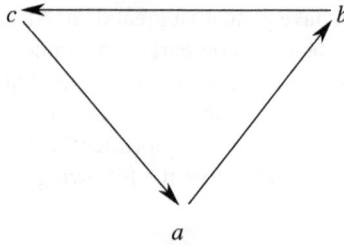

Fig. 16.11.

A salesman wants to traverse this graph in such a way that he doesn't pass through the same edge twice. Such problems are very common in graph theory. The simplest way of implementing this restriction is to cancel an edge once it has been used. Figure 16.12 will do the job.

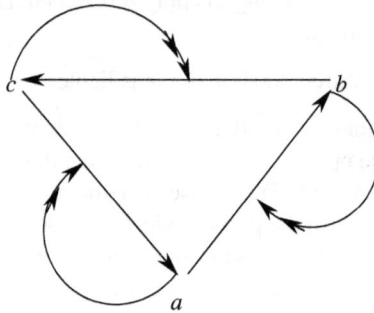

Fig. 16.12.

One cannot always implement the salesman problem in this way. It depends on the graph we deal with. However, it is one more reason for considering arc accessibility.[3]

16.2.6 Resource example

Consider a road system as in Figure 16.13.

Assume that whenever we drive through a road we need to pay a fee to pass. Payment must be in cash and it is very expensive. We can withdraw money from cash machines along the way to pay for our passage as follows:

Point a: Cash withdrawal $100 is available.
              Road a to b: costs $ 70.
Point b: Cash withdrawal $50 available.
              Road b to c: costs $50

---

[3] For the salesman example we need accessibility from arcs to arcs. So in Figure 16.12, rather than have the double arrow $(b, (a, b))$, we need the double arrow $((a, b), (a, b))$. The arc cancels itself as we go through it.

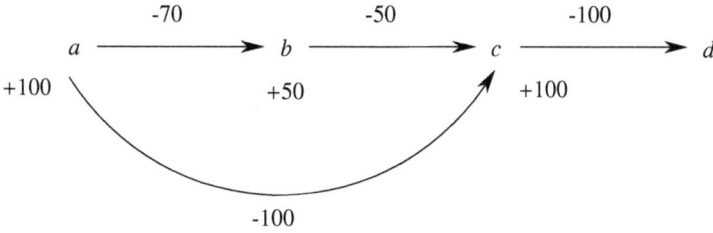

Fig. 16.13.

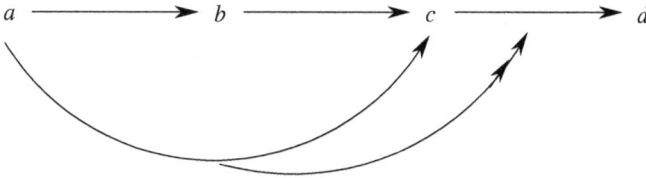

Fig. 16.14.

Point c: Cash withdrawal $100 is available.
             Road c to d: costs $120.

We start with no money at all at point a. If we drive from a to c directly, we can get $100 from the cash machine at a, and this will get us to c with no money left and so we cannot get at c enough cash to pay passage from c to d. However, if we pass through b, we can withdraw money at b and later at c and we will have enough to pay for the passage from c to d.

Figure 16.13 is essentially an ordinary annotated graph describing the resource situation but the qualitative situation (where money considerations are hidden) can be described in Figure 16.14. The upshot of Figure 16.14 is that the arc from a to c sends a signal to cancel the arc from c to d. This is also the first case where we have double arrows going from arc to arc.

16.2.7 Flow products example

Ordinary products of Kripke frames are defined in a straightforward manner. Given two frames (S_1, R_1), and (S_2, R_2), we form the product space $S = S_1 \times S_2$, and define two modalities $\square_1, \square_2$ on pairs as follows:

- $(a, b) \vDash \square_1 A$ iff $\forall x(aR_1x \rightarrow (x, b) \vDash A)$
- $(a, b) \vDash \square_2 A$ iff $\forall y(bR_2y \rightarrow (a, y) \vDash A)$.

Figure 16.15 shows the configuration for the case of $S_1 = S_2 = N$, the set of natural numbers
    $\square_1$ shifts the x coordinate, leaving b fixed and $\square_2$ shifts the y coordinate, leaving a fixed.

We have, for example, among other things that $\square_1\square_2 = \square_2\square_1$. Products spaces are used whenever we deal with two independent modal or temporal aspects. This is a

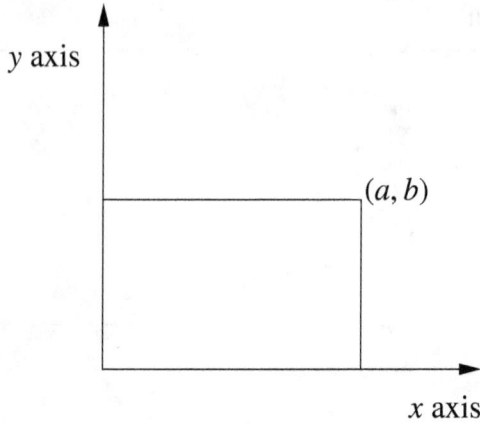

Fig. 16.15.

very active area of many dimensional modal logics, also related to classical predicate logic with a fixed number of variables. See our book [144].

We introduce in [141] the concept of flow products. Imagine the x axis is space (measured in kilometres) and the y axis is time measured in hours. Any shift in space will necessarily cause a shift in time. Assuming speed of 1 mile per hour, we get

- $(a, b) \vDash \Box_1 A$ iff $\forall u((a + u, b + u) \vDash A)$
- $(a, b) \vDash \Box_2 A$ iff $\forall u((a, b + u) \vDash A)$.

When we move in space time shifts.

We can view this as a reactive model. Imagine for any point (a, b) the following double arrows exist, see Figure 16.16

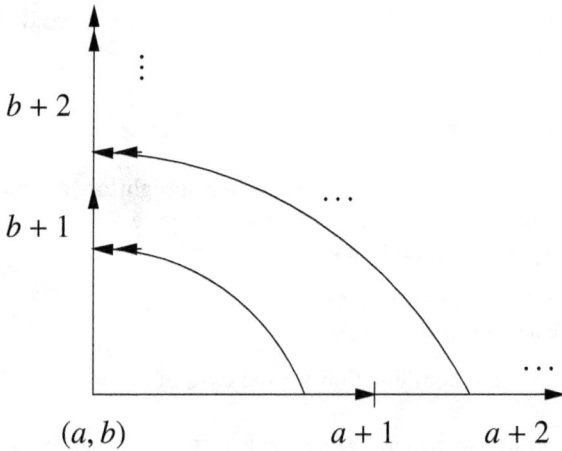

Fig. 16.16.

as we move from a to $a+1$, the connection from b to $b+1$ is switched off. Therefore we get

$$(a, b) \vDash \Box_1 A$$

in the reactive model iff $\forall u((a + u, b) \vDash A$ in the new model after firing the reaction of moving through $a+1, a+2, \ldots, a+u$, which disconnects b from $b+1, b+2, \ldots, b+u$). Hence $(a + u, b) \vDash \Box_2 B$ is now connected to $(a + u + 1, b + u + 1) \vDash B$, i.e. it is as if we are at the point $(a + u, b + u)$.

Of course a general flow product model will require a shift function $f(a, b, x)$ on the x-coordinate, satisfying

$$(a, b) \vDash \Box_1 A \text{ iff } \forall x(a R_1 x \rightarrow (x, f(a, b, x)) \vDash A.$$

$f(a, b, x)$ tells us how much the y coordinate shifts. f must satisfy some additivity properties, such as

- $a R_1 x \rightarrow b R_2 f(a, b, x)$
- $a R_1 x_1 \wedge x_1 R_1 x_2 \wedge a R_1 x_2 \rightarrow f(a, b, x_2) = f(x_1, f(a, b, x_1), x_2)$.

16.3 Reactivity and argumentation

We now motivate the need to use reactive double arrows in argumentation. We have already seen in Chapter 8 the use of double arrows attacking arrows, and seen some examples using them. The reader might have got the impression that reactive arrows are mainly a technical device, very useful, but not conceptually essential to argumentation. This is not the case, as the following story shows.

Example 16.1 (The politician). John is running for Mayor in a small town. The population is divided over some serious moral issues. John is an opportunist and would like to present a coherent view which would appeal to a maximal number of voters. To achieve this he wants to lay out the arguments involved in these issues in a nice argumentation network, look up the extensions and compare with the statistical distribution of voters' opinions and choose the extension which gains the support of a maximal number of voters. This would give John a coherent view. The problem of calculating the best extension from the point of view of numbers of voters need not concern us here. What we want to examine is how to build an argumentation network which represents the issues and contains in it as an identifiable subnetworks the various views of the different groups of voters. We shall show that to do this well we need reactive double arrows.

Let us assume that the issues are

1. Abortion A
2. Adultery D
3. Punishment P

We have the following groups of voters classified by their beliefs

4. Religious fundamentalists F
   They believe ($\rightarrow$ means attack!).

a) No abortion $F \rightarrow A$
b) Committing abortion should be punished $A \rightarrow \neg P$
c) No adultery $F \rightarrow D$
d) Committing adultery should be punished $D \rightarrow \neg P$

5. Conservatives C

These believe only in
a) = 4a: $C \rightarrow A$
b) = 4b: $A \rightarrow \neg P$
c) = 4c: $C \rightarrow D$

They do not believe that committing adultery should be punished. So they *do not* have $D \rightarrow \neg P$.

6. Liberals L

Liberals accept 4a and 4c, i.e. no abortion $L \rightarrow A$ and no adultery $L \rightarrow D$ but do not believe in punishment.

The question is how do we represent these views in a network. First let us try to represent the fundamentalists. We can do this in Figure 16.17

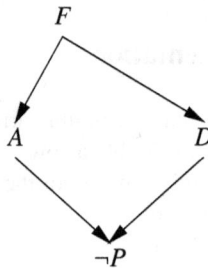

Fig. 16.17.

There is a problem with that because the extension is $\{F, \neg P\}$. It is coherent but somehow loses the information in the network. We really want two networks for the fundamentalists, see Figure 16.18

Perhaps Figure 16.19 will do the job, where we split P into two:

P_1:   Punishment after having committed abortion
P_2:   Punishment after having committed adultery.

The advantage of this approach is that we can represent the information. The disadvantage is that it is not intuitive. We do not think in terms of P_i

P_i: Punishment after having committed crime number i.

Let us reserve judgement on this and try not to combine in one network the views of both F and C.

Let us start with the representation for C as Figure 16.20:
How do we combine the two views, of F and C in one network?
Try Figure 16.21. The problem is the dotted attack arrow

$$D - - - - \rightarrow \neg P$$

a)

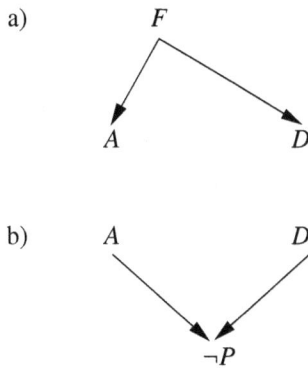

b)

Fig. 16.18.

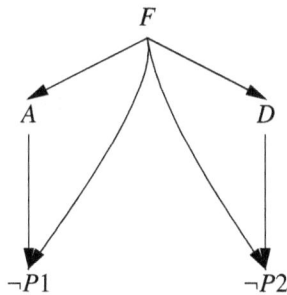

Fig. 16.19.

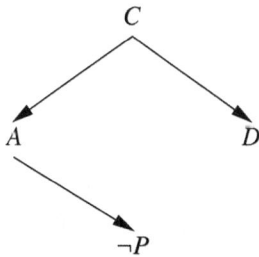

Fig. 16.20.

F wants this arrow and C does not want it. If we split P into $P1$ and $P2$, it could help. See Figure 16.22.

We do not have the attack $C \rightarrow \neg P2$. However, the split $P1$ and $P2$ and the solution in Figure 16.22, is not a natural one. A more natural way is to have only one P and look at Figure 16.21 and say: If we come to D from F (i.e. point of view of F) then we have the arrow $D \rightarrow \neg P$, and if we come to D from C (point of view of C) then we do not have the arrow $D \rightarrow \neg P$.

But this is reactivity!

To represent this we need double arrows.

Consider Figure 16.23 and look at extensions:

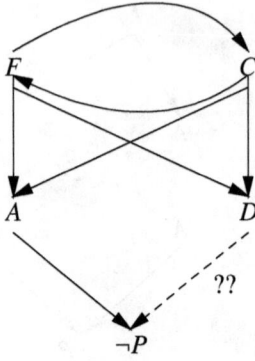

Fig. 16.21.

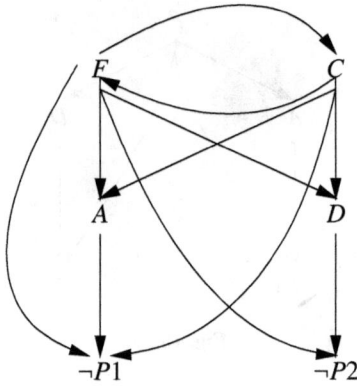

Fig. 16.22.

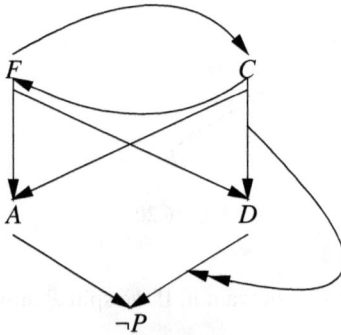

Fig. 16.23.

Case *F* (extension representing *F*)
This is obtained by having in Figure 16.23 *F* = in, so *C* = out. We delete *C* from the figure and all arrows emanting from it and get in fact Figure 16.17.

Case *C* (extension representing *C*)
This is obtained by having in Figure 16.23 *F* = out, *C* = in. We delete *F* from the figure and all arrows emanating from it. We get Figure 16.24 which is equivalent to Figure 16.23, because the double arrow cancels the arrow *D* → ¬*P*.

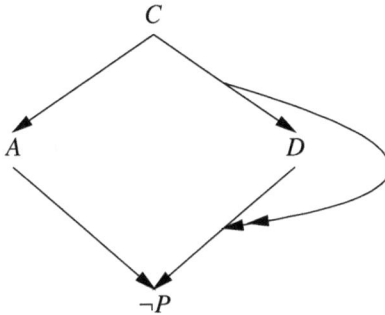

Fig. 16.24.

There is no way of getting from *C* to ¬*P* by passing through *D*.
Thus the network representing all views, *FCL* would be in Figure 16.25.

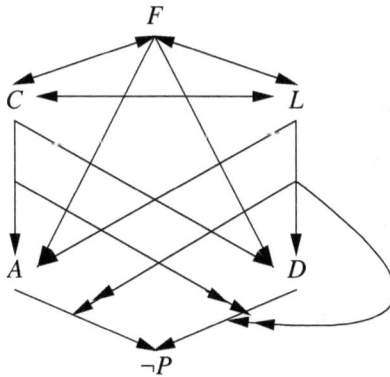

Fig. 16.25.

Note that we have used three principles here:

*Π*1:   The reactive double arrows
*Π*2:   The idea that arguments are paths and not just nodes. So for example the node *D* is not just a node we distinguish as two nodes:
*D*1:   *D* after coming from *C* (i.e. (*C*, *D*))

and

D2:   *D* after coming from *F* (i.e. (F, D)).
      This is similar to splitting *P* into *P1* and *P2*.
¬*P1*: ¬*P* after coming from *A*.
¬*P2*: ¬*P* after coming from *D*.
Π3:   The principle of how to additively combine the information contained in several
       networks based on the same *S*. We use reactivity as follows:
       Let $N_i = (S, R_i)$ be several networks. Add the information by letting

$$S^+ = S \cup \{N_i\}$$
$$R^+ = \{(N_i, N_j)|i \neq j\} \cup \bigcup_i R_i \cup \{(N_i, (x, y))|(x, y) \in (R_j - R_i) \text{ for } j \text{ different from } i\}$$

To conclude our example, if number of voters of *F*, *C* and *L* are more or less the same,
I think John should say he does not believe in punishment of any sort.

Remark 16.2. Note that additive combination of networks is not merging. Compare
with Chapter 20.

   The above discussion in Example 16.1 showed the use of two types of attack →
and ↠. this is in line with Chapter 8 and in line with Figures 16.1 and 16.2.
   We read → and ↠ as two types of attack. One can read Figure 16.1 as containing
support → and attack ↠ and maybe read it in this case as Figure 16.2. How do we
deal with attack ↠ and support →? Let us briefly look at the next Definition. Chapters
15 and 17 have lots of examples and discussions.
   Let us take another look at Figure 16.2. This can be viewed as a bipolar argumentation frame. The nodes are arguments. Arguments can either attack or support other
arguments. If argument *x* supports argument *y*, we write $x \to y$. If argument *x* attacks
argument *y* we write $x \twoheadrightarrow y$. Let $\to_*$ be the reflexive and transitive closure of →. The
way we play some of the game in argumentation theory is to seek maximal subsets $\mathbb{E}$
of arguments defined as follows.

1. Let $x \twoheadrightarrow_* y$ be defined as meaning that for some x', y' we have that $y \to_* y'$ and
   $x \to_* x'$ and $x' \twoheadrightarrow y'$ all hold.
2. We have $y \to_* y'$ and $x \to_* x', x' \twoheadrightarrow_* y'$ imply $x \twoheadrightarrow_* y$.
3. We seek maximal subsets $\mathbb{E}$ of *S*, called preferred extensions, satisfying the following:
   a) $\mathbb{E}$ is *-conflict free, namely for no $x, y \in \mathbb{E}$ do we have $x \twoheadrightarrow_* y$.
   b) $\mathbb{E}$ can *-defend itself, namely if for some *x* and some $y \in \mathbb{E}$, $x \twoheadrightarrow_* y$ then there
      exists $z \in \mathbb{E}$ such that $z \twoheadrightarrow_* x$.

Lemma 16.3.
Let $\mathbb{E}$ be an extension and let $x \to_* y$ and $x \in \mathbb{E}$ hold then $y \in \mathbb{E}$.

Proof. First we note that $\mathbb{E} \cup \{y\}$ is *-conflict free. For if some $e \in \mathbb{E}$ either $e \to_* y$ or
$y \twoheadrightarrow_* e$ holds then by (2) above we have that either $\twoheadrightarrow_* x$ or $x \twoheadrightarrow_* e$. Then we note that
if $z \twoheadrightarrow_* y$, then $z \twoheadrightarrow_* x$ (by (2)), and so for some $e \in \mathbb{E}$ we have $e \twoheadrightarrow_* z$. Thus $\mathbb{E} \cup \{y\}$ is
*-conflict free and can *-defend itself. Since $\mathbb{E}$ is a maximal such set we have $y \in \mathbb{E}$.

Let us now check what preferred extensions we get for Figure 16.2. We get $\mathbb{E} = \{u\}$. Certainly $\{u\}$ is *-conflict free and can defend itself (it is not attacked). The nodes $\{0, x, y\}$ *-attack themselves and so they cannot be added to $\{u\}$. The node z does not *-attack itself but it cannot defend itself. It is attacked by y and neither z nor u attack y. So $\mathbb{E} = \{u\}$ is maximal.

When the argumentation network is not a tree, we need to use double arrows attacking arrows or nodes attacking arrows. These are known as higher level attacks. We studied them in Chapter 8.

17

Bipolar Argumentation Frames and Contrary to Duty Obligations

This chapter shows a connection between deontic contrary to duty obligations [188, 189] and bipolar argumentation networks [61, 96], recall Chapter 15. We need to give a short introduction to each area.

17.1 Bipolar argumentation networks with a starting node

Our starting point is the Cayrol and Lagasquie-Schiex's bipolar argumentation framework described in Chapter 15, Section 15.3 above. We need to modify the definition of bipolar networks to enable us to compare with contrary to duty obligation networks. The modification is nothing essential, we just ask for a starting point.

For example, we prefer to take a modified form of Figure 15.28. Namely, Figure 17.1. We add a new node ∞ and add $\infty \to x$, to any x such that there is no y with $y \to x$, i.e. to all $\to$ minimal xs.

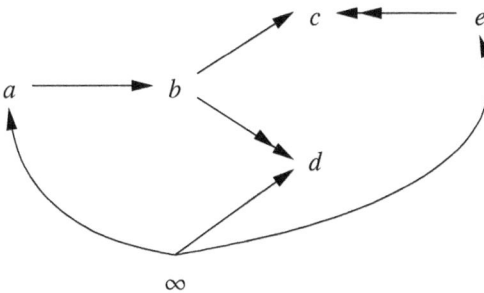

Fig. 17.1.

This helps the bipolar network to be similar to models of contrary to duty obligations.

Definition 17.1. *Let $(S, \to, \twoheadrightarrow)$ be a bipolar network. Let ∞ be a new point. Define $S' = S \cup \{\infty\}$ and extend $\to$ on S' as follows.*

For any $x \in S$ such that for no $y \in S$ do we have that $y \to x$ holds, let $\infty \to x$ be added.

We call $(S', \to, \twoheadrightarrow)$ the deontic friendly extension of $(S, \to, \twoheadrightarrow)$.

Proposition 17.2. *The mapping*

$$(\infty, x_1, \ldots, x_n) \leftrightarrow (x_1, \ldots, x_n)$$

between the coalitions in S' to coalitions in S is an isomorphism of the coalition graph of S (as in Definition 15.33) and the coalition graph of S'.

Proof. Obvious.

17.2 Reactive semantics for contrary to duty obligations

We now very quickly present the problems of contrary to duty paradoxes and outline our reactive models for their solution.

Consider the following set of obligations, known as the Chisholm set. Notation: $p = $ go, $q = $ tell.

1. It is obligatory to go and help your friend.
2. If you go, you ought to tell him you are coming.
3. If you do not go, you ought to tell him you are not going.
4. Fact: you do not go.

A proper modelling of these clauses requires that these clauses be independent and consistent. Standard deontic logic **SDL** cannot do the job and in our papers [188, 189] we offer a reactive variant of **SDL**. **SDL** is the modal logic **K** with the operator O and the additional axiom $\neg O\bot$. The English statements (1)–(4) are formalised (at best) as (1a), (2b), (3a), (4a) below. If read as wffs of **SDL** they are still problematic, but if modelled by reactive frames like Figure 17.2, we avoid difficulties and have a solution, see [188, 189].

Let us take the translation into **SDL** (1a), (2b), (3a), (4a).

(1a) Op
(2b) $p \to Oq$
(3a) $\neg p \to O\neg q$
(4a) $\neg p$

The problem with this translation when taken as wffs of **SDL** is that (4a) implies logically (2b). We lose independence.

Let us look at Figure 17.2, which gives a graphical representation of the linguistic clauses (1a), (2b), (3a), and (4a).

Figure 17.2 expresses exactly the same as the linguisitic clauses. However, it is clear that this is a different representation and we can see from the figure at which node each of these clauses is associated with.

(1a) is associated with node **o**
(2b) is associated with node x
(3a) is associated with node y
(4a) is associated with node y

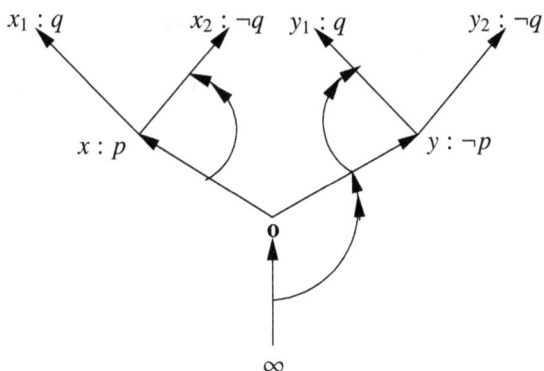

Fig. 17.2. Representation of the Chisholm set

Before we explain how we understand Figure 17.2, let us compare it, purely as a graph, with Figure 15.23. First note that since the $\rightarrow$ part of this figure is a tree, we can represent the double arrows

$$(\infty \rightarrow \mathbf{0}) \twoheadrightarrow (\mathbf{0} \rightarrow y)$$
$$(\mathbf{0} \rightarrow y) \twoheadrightarrow (y \rightarrow y_1)$$
$$(\mathbf{0} \rightarrow x) \twoheadrightarrow (x \rightarrow x_2)$$

respectively by

$$\mathbf{0} \twoheadrightarrow y$$
$$y \twoheadrightarrow y_1$$
$$x \twoheadrightarrow x_2$$

This can be done because each point in the tree (except ∞) has a unique $\rightarrow$ predecessor, and so we can identify any $u \rightarrow v$ by v alone.

If we do this identification we see that Figure 17.2 becomes Figure 15.23.

This is good, however, in bipolar argumentation we read and do extensions with the figure while in deontic logic we read contrary to duty obligations from the figure. Are these two "readings" related?[1] Let me tell you first the deontic reading and then we compare the way we read the figure. Our deontic reading is as follows. ∞ is our starting point. We go to node $\mathbf{0}$ (say office), where our obligations begin. Obligation 1 says we ought to go and help.

This means we need to travel to node x. This is why we have the double arrow from $(\infty \rightarrow \mathbf{0})$ to $(\mathbf{0} \rightarrow y)$. Double arrows in the reactive semantics post a warning sign

"Do not pass through this arc"

If we indeed go through $\mathbf{0} \rightarrow x$, we need to continue to point x_1 and so another double arrow $(\mathbf{0} \rightarrow x) \twoheadrightarrow (x \rightarrow x_2)$ tells our traveller not to go through $x \rightarrow x_2$. Similarly if we ignore the warning sign on the arc $\mathbf{0} \rightarrow y$ and go across it to y, then the contrary to

---

[1] Note that this is a key question. If what we do with the basic figure is different then there is no relation between the approaches. This point will arise also in Remark 17.10 when we compare our approach with other papers, for example with paper [322].

duty says do not go to y_1, do go to y_2. It put a warning sign on $y \rightarrow y_1$. This is done by the double arrow $(\mathbf{o} \rightarrow y) \twoheadrightarrow (y \rightarrow y_1)$.

Our deontic perception of Figure 17.2 and its equivalent (as a graph only) Figure 15.23 is that we walk along maximal paths following the arrows.

So let us ask ourselves, what are the possible paths? These are

$$\beta = (\infty, \mathbf{o}, y, y_2)$$
$$\alpha = (\infty, \mathbf{o}, y, y_1)$$
$$\delta = (\infty, \mathbf{o}, x, x_2)$$
$$\gamma = (\infty, \mathbf{o}, x, x_1)$$

The facts in the contrary to duty set, is information about the path we actually took. The full information is a maximal path, and partial information is a family of (possible) paths. In our example the fact is that he went to node y but we do not know whether he continued to y_1 or to y_2. So the possible paths are α and β.

The $\rightarrow$ are the attacks. We can perceive them as attacks in the contrary to duty deontic case because they represent instructions to block some paths. This is their function.[2]

Now we ask, if we walk according to all obligations and obey all double arrows (obey all signs which say **do not pass**), which paths are Kosher and OK?

By looking at the figure, we see it is

$$(\infty, \mathbf{o}, x, x_1)$$

Here we obey all our obligations. We are really good.

It can also be (y, y_2) if we obey the contrary to duty. Having violated the Ox and gone to y, we no longer commit violations and obey $y \rightarrow Oy_2$.

Now we ask, doing the bipolar extensions according to [96], as in Definition 15.35, what do we get? The answer is $\{\infty, \mathbf{o}, x, x_1\}$, and $\{y, y_2\}$.

So we are getting all the paths in which we are somewhat obedient.

We now get a suggested correspondence as follows:

Support = possible path connections (taking the view that allowing you to go whenever you want is being supportive)
Attack = Obligation (taking the view that restricting where you want to go is a form of attack)
Extensions = sets of maximal paths without violation (that is moving around without being attacked)

We shall see later in Example 17.5, that we recommend to change the notion of attack in bipolar argumentation and take paths instead of sets as our elements. This would be our message to [96].[3]

---

[2] The idea of nodes or arrows attacking other arrows was introduced in [147] and gave rise to the notion of Reactive Kripke Semantics, which was widely applied. In our papers [37] and its expanded version [41], these ideas were used in argumentation. See also [170] for details and discussions. [32, 263] independently considered this idea. See also [61]. In our paper [188] we used reactive arrows to model contrary to duty obligations, and this use allows for the connection between argumentation and normative systems. This material is included of course in this book.

[3] The reader familiar with Abstract Dialectical frameworks of Brewka and Woltran [69] will recall that they claim in their paper, that they do better analysis of support than [96]. Our aim

Example 17.3. Let us look at Figure 17.1 from the deontic contrary to duty point of view. First we comment that the family of examples (paradoxes) of contrary to duty sets is limited and this graph (given in [61] as an example to discuss the adequacy of [96]) does not fit any known deontic example. However, we do have an interpretation for it. We have in Figure 17.1 several options for paths beginning at ∞ and we would like to offer a maximal package of paths for the righteous people to follow, without any risk of violations. Looked at the figure in this way we want extensions in the sense of Definition 15.35 for paths. The answer is

$$\{(\infty, d), (\infty, e)\}$$

Note that is is meaningless for the deontic case to look at $\{d, e\}$ as suggested in Definition 15.35, as an extension of the points graph. We want only coalition extensions.

17.3 New ideas arising from the connection between argumentation and deontic logic

The previous two sections presented the connections between bipolar networks and contrary to duty obligations. This section will describe new ideas arising from these connections. We discuss

1. Bipolar argumentation semantics for contrary to duty (looping) obligations
2. The equational approach to contrary to duty
3. New path semantics for bipolar argumentation networks

We now know how to turn every reactive tree frame into a bipolar system, and hence get argumentation semantics for contrary to duty sets. Let us get some mileage out of this correspondence. In [189] there is a loop example whose semantics was left as an open problem. Can we make use of our new insight?

Example 17.4 (Loop). Consider the following 3-loop.

1. $\neg x \rightarrow$ Obligatory y
2. $\neg y \rightarrow$ Obligatory z
3. $\neg z \rightarrow$ Obligatory x.

We cannot build a tree graph for it. We have a loop.

The coalitions in this case are

$$\{\infty, a, x\}$$
$$\{\infty, a, \neg x, y\}$$

---

here is not to do better than [96] but to show the interaction of BAF with Deontic reasoning. By the way, we use in Remark 17.13 the notion of disjunctive attacks, introduced in [164], see Chapter 7, to model multiple support. The logical machinery of Abstract Dialectical Framework is implicit in [164], as recognised by Brewka and Woltran in their paper. This means that they are also capable of doing what we are doing in Remark 17.13.

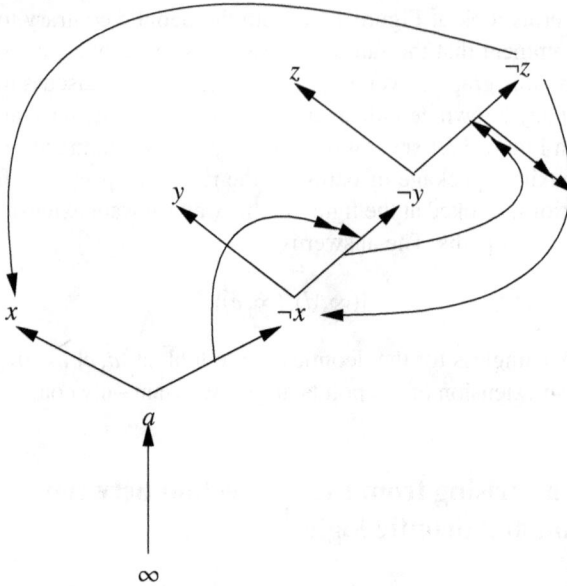

Fig. 17.3. 3-level loop

These are the only conflict free coalitions in Figure 17.3 and are therefore the bipolar extensions according to Definition 15.33 and Remark 15.34.[4]

We now check the maximal path point of view for Figure 17.3.

The finite paths are as follows:

$$\alpha = (\infty, a, x)$$
$$\beta = (\infty, a, \neg x, y)$$
$$\gamma = (\infty, a, \neg x, \neg y)$$
$$\delta = (\infty, a, \neg x, \neg y, \neg z, x)$$
$$w = (\neg y, \neg z, \neg x)$$

Then we loop into the infinite number of sequences

---

[4] The perceptive reader will note that Figure 17.3 is not a kosher figure of bipolar network as defined by Cayrol and Lagasquie-Schiex's in Definition 15.30. The attack arrows do not emanate from nodes but from arcs. Such attacks were never considered in the argumentation networks, except in [37] and [41]. The attacks arrows also terminate in arcs. The idea of attacking arcs was also first introduced in [37] and was followed up, sometimes independently, in the literature. Note that Figure 17.3 is not a tree and so we cannot convert it to an equivalent figure with double arrows emanating from nodes and terminating at nodes. We can and should, however, generalise the notion of bipolar argumentation networks by saying that attacks should emanate from support arrows, and terminate at other support arrows, using the rationale that when we support an argument , part of our support strategy is to initiate attacks on other arguments and their supports. We shall later develop these concepts. Meanwhile let us execute the obvious steps for Figure 17.3.

$$\zeta_n = \alpha(w)^n(y)$$
$$\eta_n = \alpha(w)^n(\neg y, z)$$
$$\rho_n = \alpha(w)^n(\neg y, \neg z, x).$$
$$n = 1, 2, 3, \ldots$$

The path attack relation is as follows:

1. α attacks nobody
2. $\beta \twoheadrightarrow$ all except α, β
3. All nodes except α, β attack each other.

Clearly the extension there for the path network is: $\{\alpha, \beta\}$.

These are also the righteous paths with no violations, and they correspond to the bipolar extensions.

Example 17.5 (Gabbay's first proposal for the concept of bipolar network). This example imports concepts from contrary to duty obligations into bipolar argumentation. Consider the network of Figure 17.4. This represents the contrary to duty

1. You ought to either work or rest
2. If you work you ought to rest
3. If you rest you ought to work.

What these rules say is that you ought to start working or get yourself ready by resting, and then alternate between work and rest.

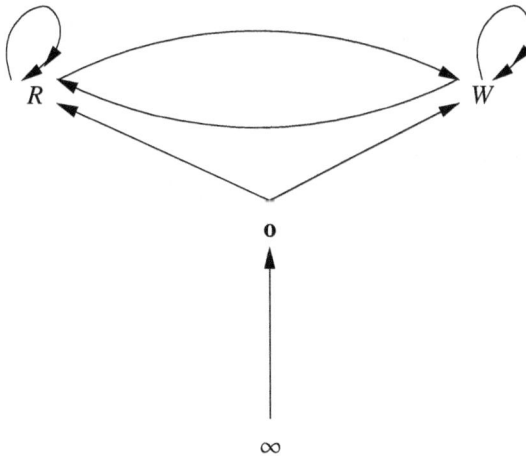

Fig. 17.4.

The paths here are
$$\Pi_1 = (\infty, \mathbf{o}, W, R, W, R, \ldots)$$
$$\Pi_2 = (\infty, \mathbf{o}, R, W, R, W, \ldots)$$

According to the bipolar definitions, Definitions 15.32 and 15.33 we take subsets as maximal conflict free coalitions and the attack relation is done element-wise. So there is only one coalition $\{\infty, \mathbf{o}\}$. This does not mean much in our context.

This example shows a sharp difference between the path approach and the coalition bipolar approach. We did have good correspondence between the two approaches when the graphs were trees (Examples 18.103 and 15.37 and the analysis of Figure 17.2).

The correspondence still worked even in the looping example 17.4, but not for Figure 17.4.

We need now to recommend changes in the abstract machinery of Definitions 15.30–15.33.

This is our message of change to C. Cayrol and M. C. Lagasquie-Schiex's bipolar concept.

The following table, Table 17.1 outlines the differences, which is then followed by a discussion.

We assume in this table that we have as given a bipolar graph $(S, \rightarrow, \twoheadrightarrow)$.

Table 17.1.

Concept	Cayrol and Lagasquie-Schiex	Gabbay
Path	A sequence $(x_1, \ldots, x_n)$ such that $x_1 \rightarrow x_2 \rightarrow \ldots \rightarrow x_n$ holds	same
conflict free path	For no $x, y, \in \{x_1, \ldots, x_n\}$ do we have $x \twoheadrightarrow y$	For no $1 \leq i \leq n-1$ do we have $x_i \twoheadrightarrow x_{i+1}$
Attack between paths	$(x_1, \ldots, x_n)$ attacks $(y_1, \ldots, y_k)$ if for some x_i and y_j we have $x_i \twoheadrightarrow y_j$	$(x_1, \ldots, x_n)$ attacks $(y_1, \ldots, y_k)$ if for some $r < \min(k, n)$ we have $x_r \twoheadrightarrow y_{r+1}$.
Coalition	Maximal conflict free set which is a path (Definition 15.32)	Maximal Gabbay conflict free path.

According to Gabbay's path semantics, the paths Π_1 and Π_2 are conflict free each and do not attack each other. They are the two violation free courses of action for the contrary to duty Figure 17.4. So we get the right result here.[5]

Example 17.6 (Makinson Moebius example). Consider the loop

1. $a \rightarrow Ox$
2. $x \rightarrow Oy$
3. $y \rightarrow O\neg a$

Figure 17.5 draws this set

The only coalition or path coalition which is not self attacking in Figure 17.5 is $(\infty, a, x, , \neg a)$. Neither Cayrol and Lagasquie-Schiex (Definition 15.33) nor Gabbay (Example 17.7) is bothered by the fact that we have both a and $\neg a$ in this coalition because there is no formal attack $a \twoheadrightarrow \neg a$ or $\neg a \twoheadrightarrow a$ n the Figure. One must not

---

[5] Note that sequences can be proofs and so we may be able to deliver an abstract bipolar network in between the abstract Dung networks and the fully instantiated ASPIC networks. In fact in Chapter 15 we use tripolar networks, where we have attacks and two types of support one strict and one defeasible.

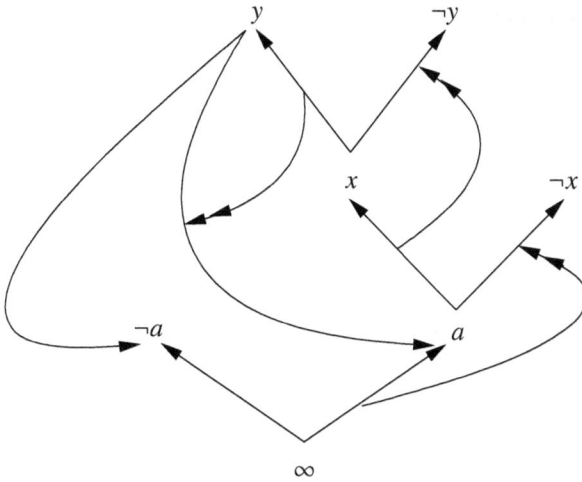

Fig. 17.5.

criticise Cayrol and Lagasquie-Schiex for not requiring that $a \twoheadrightarrow \neg a$ an $\neg a \twoheadrightarrow a$ because "¬" is not assumed in the language.

In the case of Gabbay the definition is in terms of paths and although we pass through node a we do follow instructions and end up at $\neg a$. So Gabbay would not need to stipulate $a \twoheadrightarrow \neg a$ and $\neg a \twoheadrightarrow a$.

Example 17.7 (The equational approach to contrary to duty obligations). We can import more from argumentation to deontic logic. We can import the equational approach. See [166].

Consider Figure 17.6.

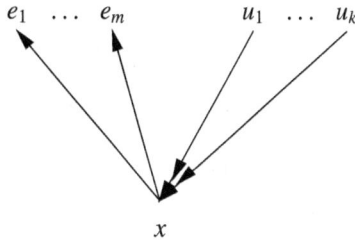

Fig. 17.6.

x is attacked by $u_1, \ldots, u_k$ and is supporting $e_1, \ldots, e_m$. The equations are:

$Eq_{\max} : x = 1 - \max\{u_1, \ldots, u_k, 1 - e_1, \ldots, 1 - e_m\}$
$Eq_{\text{inverse}} : x = e_1 \times \ldots \times m \times (1 - u_1) \times \ldots \times (1 - u_k).$

Let us check the equations for the Chisholm paradox in Figure 15.23.

Note that the figure has the node **o** to distinguish it from the numeral 0. We use Eq_{max}:

1. $\infty = \mathbf{o}$
2. $\mathbf{o} = 1 - \max\{1 - y, 1 - x\}$
3. $y = 1 - \max\{1 - y_2, \mathbf{o}, 1 - y_1\}$
4. $x = 1 - \max\{1 - x_1, 1 - x_2\}$
5. $x_1 = y_2 = 1$
6. $y_1 = 1 - y$
7. $x_2 = 1 - x$

Solving the equations we get:
From (6) and (3) we get

8. $y = 1 - \max\{y, \mathbf{o}, 1 - y_2\}$

From (7), (5) and (4) we get

9. $x = 1 - x = \frac{1}{2}$

From (7) and (9) we get

10. $x_2 = x = \frac{1}{2}$

From (8), (5), (9) and (2) we get

11. $y = 1 - \max\{y, 1 - y\}$ and hence $y = \frac{1}{2}$

We compare with equation 2 and get that the solution is

$$y = \mathbf{o} = \frac{1}{2}$$

This will solve the equations.
We get

$$\infty = \mathbf{o} = x = y = x_2 = y_1 = \frac{1}{2}$$

and

$$x_1 = y_2 = 1.$$

The result we got is not yet satisfactory. We hope for a solution

$$\infty = \mathbf{o} = x = x_1 = 1$$

which gives the correct

$$(\infty, \mathbf{o}, x, x_1)$$

path (no violations on this path) and all the other paths have 0 in them. Instead we got a lot of $\frac{1}{2}$ values. However, we can still continue, because $\frac{1}{2}$ means undecided.

Indeed, in Figure 15.23 we have three loops with $\frac{1}{2}$ = undecided values. These are (the loop is that a node both supports and attacks the other node).

$$L_1 = \{x, x_2\}, L_2 = \{\mathbf{o}, y\} \text{ and } L_3 = \{y, y_1\}$$

We can use loop busting methods to break these loops, see [170] and get

$x = 1, x_2 = 0$, for L_1
$\mathbf{o} = 1, y = 0$, for L_2
$y_1 = 1, y = 0$, for L_3

We now have the solution we want!

Let us use the equational approach for this example in a different way and compare. Let us look at Figure 15.27. This is the coalitions attack figure used to compute the bipolar extension. The winning extension was $\{(\infty, \mathbf{o}, x, x_1)\}$.

Let us give values 1 to all nodes in this winning extension and 0 to all other nodes. Did we get the same result as before? The answer is yes.

We can follow this method in general. Start with a bipolar argumentation network. Obtain the extensions as in Remark 15.34 and then form the set S as in Definition 15.35.

Now give numerical value 1 to members of S and 0 to other nodes.

We now want to define a bipolar argumentation network associated with a general arbitrary set of contrary to duty obligations. Such obligations have the form $x \rightarrow$ Obligatory y, which we can also write as (x, y) or the form Obligatory z, which we can also write as $(\top, z)$.

We look at all the atomic letters $x, y, z, \ldots$ and their negations $\neg x, \neg y, \neg z, \ldots$ and take these as our nodes in the bipolar argumentation network, together with the additional node ∞.

We let $x \rightarrow y$ and $x \rightarrow \neg y$ be support arrows in the network whenever (x, y) is in the contrary to duty set. We also let the attack $x \twoheadrightarrow \neg y$ be in the network if (x, y) is in the set and $x \twoheadrightarrow y$ be in the network if $(x, \neg y)$ is in the contrary to duty set. If x is $\top$ we use ∞ instead of x. This is basically the idea. We now give the definition.

Definition 17.8 (Bipolar argumentation semantics for a general contrary to duty set).

1. Let Q be a set of atomic letters and let

$$Q^{\pm} = Q \cup \{-x \mid x \in Q\} \cup \{\top\}.$$

A set $\mathbb{C}$ of pairs of the form (x, y) where $x, y \in Q^{\pm}$ and $y \neq \top$ is called a contrary to duty obligations set. We can write $(x, y) \in \mathbb{C}$ in a linguistic semi-formal language as $x \rightarrow$ Obligatory y. If $x = \top$, we can write Obligatory y.

2. Let $\mathbb{C}$ be a contrary to duty obligations set. We define a bipolar argumentation network $(S, \rightarrow, \twoheadrightarrow)$ as follows. Let Q_0 be the set of all $y \in Q$ such that y or $\neg y$ appear in $\mathbb{C}$.

Let

$$S = Q_0 \cup \{\neg y \mid y \in Q_0\} \cup \{\infty\}.$$

We define $\rightarrow$ and $\twoheadrightarrow$ on S as follows

a) Let $\infty \rightarrow x$ and $\infty \rightarrow -x$ hold whenever $(\top, x)$ or $(\top, \neg x)$ are in $\mathbb{C}$ for $x \in Q_0$.

b) Let $\infty \rightarrow x$ hold whenever (x, y) is in $\mathbb{C}$ and there is no z such that (z, x) is in $\mathbb{C}$.

c) Let $x \rightarrow y$ and $x \rightarrow -y$ hold, whenever either (x, y) or $(x, -y)$ are in $\mathbb{C}$ for $y \in Q_0$.

d) *Let $x \twoheadrightarrow y$ hold whenever $(x, -y)$ is in $\mathbb{C}$ for $y \in Q_0$, and let $x \twoheadrightarrow -y$ hold whenever (x, y) is in $\mathbb{C}$ for $y \in Q_0$.[6]*

3. *A complete fact is a maximal path of the form $(\infty, x_1, \ldots, x_n, \ldots)$, such that $\infty \to x_1$ holds and for all i, $x_i \to x_{i+1}$ holds. A fact is a set of complete facts.*

4. *The path semantics outlined in Definition 17.5 would give us bipolar path argumentation semantics for contrary to duty obligations.[7]*

Example 17.9 (Chisolm set). Let us use Definition 17.8 on the Chisholm set

1. $\top \to$ Obligatory g
2. $g \to$ Obligatory t
3. $\neg g \to$ Obligatory $\neg t$
4. $\neg g$

Figure 17.7 describes this set, according to Definition 17.8.

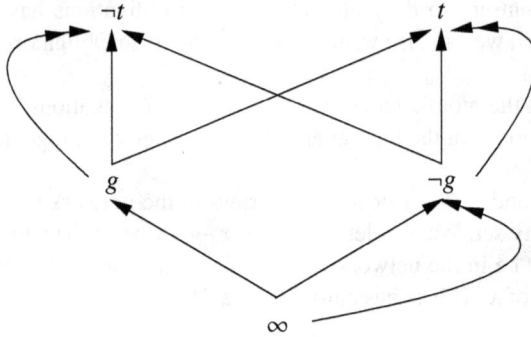

Fig. 17.7.

Figure 17.7 should be compared with Figure 17.2. They are essentially the same. The latter is the unfolding of the former, with an additional intermediary point **o**. The complete facts are paths in Figure 17.7. These are:

$$(\infty, g, t), (\infty, g, \neg t), (\infty, \neg g, t), (\infty, \neg g, \neg t).$$

The fact g corresponds to the set of two paths

$$\{(\infty, g, t), (\infty, g, \neg t)\}.$$

Figure 17.8 shows the path coalitions and the attacks among them.

Obviously there is exactly one winning coalition, namely (∞, g, t).

---

[6] Note that we defined attacks from node to node and not from arc to arc, as remarked in footnote 4. We adopt the view that the contrary to duty obligation $x \to Oy$ actually means that once you are at x you should continue to y and this obligation is independent of how (by what route) you got to x.

[7] This definition should be compared with Definition 2 of [322].

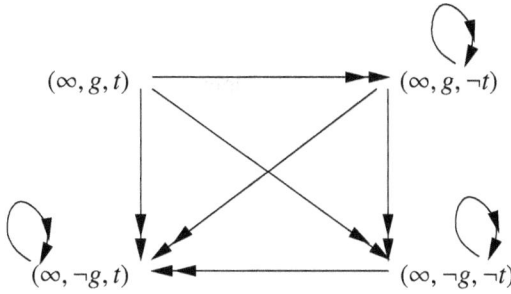

Fig. 17.8.

Remark 17.10 (Comparison with [322]). We take this opportunity to compare our analysis of the Chisholm set in Example 17.9, with example 6 of [322]. In [322], Tosatto, Boella, Torre and Villata use input-output logic to model abstract normative systems. Example 6 and Figure 6 in their paper addresses the Chisholm set. Our Figure 17.9 reproduces their Figure 6.

The arrows are input-output connections.

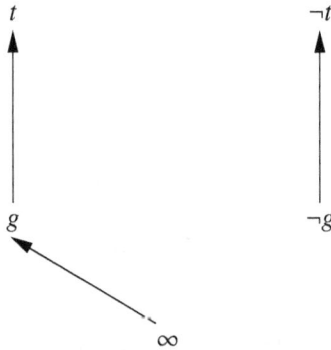

Fig. 17.9.

In Figure 6 of [322] they use a for our g and $\top$ for our ∞. Their representation of facts (i.e. $\neg g$) is a set A of nodes. Their choice is $A = \{\infty, \neg g\}$. They apply a variety of input-output operations to A (called deontic operations) to obtain families of sets $\odot_i A$, where $\odot_i$ is one of twelve possible operations.

Let us consider their operation $\odot_3$. We have

$$\odot_3(\{\infty, \neg g\}) = \{\neg t\}.$$

My understanding of this in terms of our representation is that $\odot_3$ says if you go down the path $(\infty, \neg g)$ it is recommended by the norms that you continue to $\{\neg t\}$.

So in general for a set of nodes A, the operations $\odot_i(A)$ are various recommendations of where else to "continue". The approach in [322] is essentially proof theoretic whereas our approach is essentially semantic. The two approaches of course can be combined and benefit one another to obtain better tools for modelling. We shall pursue this in a future paper. We note in passing that the Chisholm set has a temporal aspect to it, the tell comes before go. Our approach can represent the temporal order by using Figure 17.10 instead of Figure 17.7.[8] The approach of [322] cannot be modified so easily.

We need to systematically compare the two approaches, and this we shall do in the aforementioned future paper.

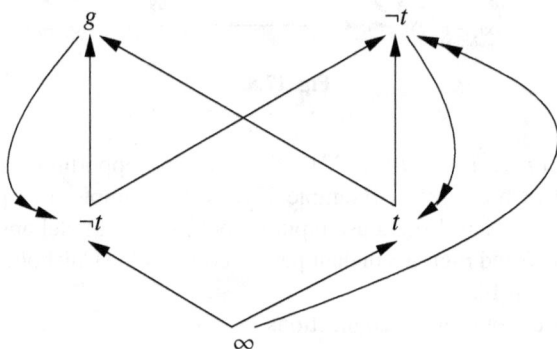

Fig. 17.10.

Definition 17.11 (Gabbay's second proposal for the concept of bipolar network).

1. *Let* $\mathcal{A} = (A, \rightarrow, \twoheadrightarrow)$ *be a bipolar network as in Definition 15.30. We define the notion of tree coalitions as follows:*
   A subset TC of A is a tree coalition iff the following holds
   a) *TC is conflict free*
   b) *If* $x \in TC$ *and* $y \rightarrow x$ *holds then* $y \in TC$
   c) *Let* $\rightarrow_*$ *be the transitive closure of* $\rightarrow$. *Then if* $x, y \in TC$ *then for some* $z, y \rightarrow_* z$ *and* $x \rightarrow_* z$ *hold.*
2. *Let S be the set of all tree coalitions. Define a notion of attack on S by letting* $TC_1 \twoheadrightarrow TC_2$ *iff for some* $x \in TC_1$ *and* $y \in TC_2$ *we have* $x \twoheadrightarrow y$ *holds. Then* $\mathcal{TA} = (S, \twoheadrightarrow)$ *is a traditional Dung argumentation network. It is the derived network of* $\mathcal{A}$.

Example 17.12 (Proof theory example). We can give a proof theory interpretation to the concepts defined in Definition 17.11.

Let Q be a set of atoms q and their negations $\neg q$. Construct formulas of the form φ, where

---

[8] See paper [188], where temporal sequences are addressed. Note that we would need to change the notion of attack as defined in Table 17.1. We would need to adopt Cayrol and Lagasquie-Schiex's definition and let:

$(x_1, \ldots, x_n)$ attacks $(y_1, \ldots, y_k)$ if for some r, j we have that $x_r \twoheadrightarrow y_j$.

$$\varphi = (\alpha_1 \Rightarrow (\alpha_2 \Rightarrow \ldots \Rightarrow (\alpha_n \Rightarrow \beta)\ldots))$$

where $\Rightarrow$ is logical implication, and $\alpha_i, \beta \in Q$.

Let A be the set of all such formulas. Define $\to$ (support) on A by letting

$$\varphi_1 \to \varphi, \psi \to \varphi$$

where ψ is

$$\psi = (\alpha_2 \Rightarrow (\alpha_3 \Rightarrow \ldots (\alpha_n \Rightarrow \beta)\ldots))$$

and φ is as above.

Define $x \twoheadrightarrow y$, for $x, y \in A$ iff x and y are inconsistent together, i.e. can prove in the logic $\neg q$ and q.

Consider $\mathcal{A} = (A, \to, \twoheadrightarrow)$ and consider $\mathcal{T}\mathcal{A}$. The latter is an abstract argumentation network which can be instantiated in the ASPIC [288] sense. In fact it was built that way.

Thus in fact we can offer a notion of abstract instantiation which is still abstract and is an intermediate concept between the Dung and the ASPIC concepts.

A tree bipolar argumentation network $\mathcal{A}$ is an instantiation of an argumentation network $\mathcal{B}$ if we have $\mathcal{B} = \mathcal{T}\mathcal{A}$.

Remark 17.13. Note that Figures 15.28 and 15.29 fall under Definition 17.11 and the problem that $\{d, e\}$ is not a valid extension of the network of Figure 15.28 still exists. We need to offer a solution.

Our solution is similar to the solution of [61]. We need to consider the situation in Figure 17.11.

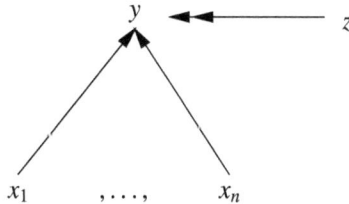

Fig. 17.11.

If we understand $x_1, \ldots, x_n$ as the essential assumptions for proving y then a successful attack of z on y must reflect an attack on at least one of the supportive x_i, because we understand the support illustrated in Figure 17.11 as $\bigwedge_i x_i \vdash y$.

This is a disjunctive attack of z on $\{x_1, \ldots, x_n\}$.

Disjunctive attacks were considered in [164], where we used the notation of Figure 17.12.

Disjunctive attacks can be realised in ordinary argumentation networks by adding new variables and replacing the structure of Figure 17.12 by the structure of Figure 17.13.

Thus the original $\mathcal{A} = (A, \to, \twoheadrightarrow)$ can be reduced to a $(B, \twoheadrightarrow)$ with additional new nodes. See [164].

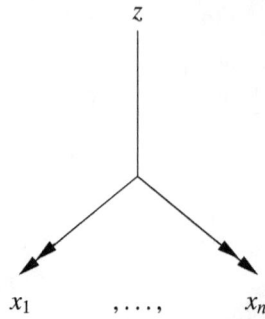

Fig. 17.12.

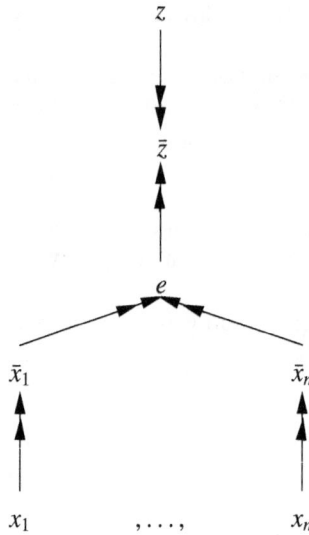

Fig. 17.13.

Example 17.14 (Support only network). We illustrate our methods by applying them to an example given to us by Martin Caminada. Let x = rent an apartment and thus have an address, z = open a bank account.

It is the case that you need x to achieve z and you need z to achieve x. We thus have the mutual support situation of Figure 17.14.

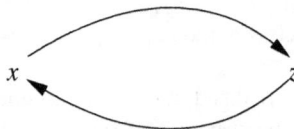

Fig. 17.14.

According to our theory in Remark 17.13, Figure 17.14 is equivalent to Figure 17.15, which contains auxiliary points

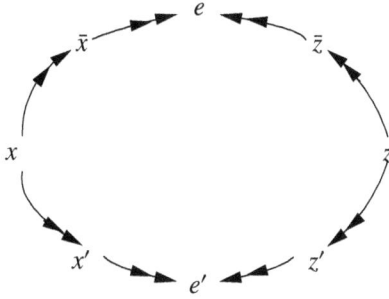

Fig. 17.15.

We can also write directly equations for Figure 17.14, as proposed in Example 17.7:

$$x = 1 - \max(1 - z) = z$$
$$z = 1 - \max(1 - x) = x.$$

Note the following:

1. Common sense requires a solution of two extensions, either both x and z are in or both x and z are out.
2. the equational approach, as well as the situation in Figure 17.15, do give such two solutions: ϕ and $\{x, z\}$. The coalition approach of Cayrol and Lagasquie-Schiex (Definition 15.35) give only one extension, namely $\{x, z\}$.

Example 17.15. Consider Figure 17.16

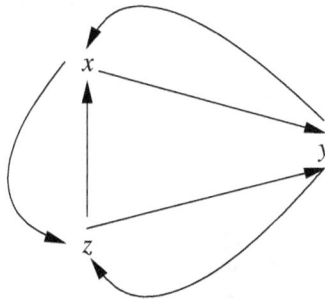

Fig. 17.16.

Here, any two nodes support the third. The equations are as follows:

$$x = 1 - \max(1 - y, y - z)$$
$$y = 1 - \max(1 - x, 1 - z)$$
$$z = 1 - \max(1 - x, 1 - y)$$

There are two solutions: $x = y = z = 1$ and $x = y = z = 0$. So there are two extensions $\{x, y, z\}$ and ϕ. As it should be!

17.4 Concluding Discussion

In this chapter, we took the concept of bipolar abstract argumentation network and modified it a bit. This change enabled us to connect to contrary to duty obligations and give them argumentation semantics (using Gabbay's reactive modelling of contrary to duty). We were also able to extend our equational approach to bipolar argumentation as well as hint at instantiations for abstract argumentation in the ASPIC spirit, by using a proof theoretic interpretation of support (this is further developed in [171]).

The connection with input-output logic allows us to give argumentation semantics for input-output logic, and more interestingly, equational semantics. This connection will be addressed in a future paper.

To wrap up this section we would like to give a modal logic which generalises both contrary to duty obligations and bipolar argumentation networks.

So far we have translated one into the other. Showing a modal logic in which both can reside is a nice conclusion to our present chapter. More details on such modal logics will be given in our general paper [172].

Our starting point is a bimodal logic with two disjoint **KD** modalities, $\Box_1$ and $\Box_2$. Its semantics has frame models of the form (S, R_1, R_2, ∞) where S is the set of possible worlds, $\infty \in S$ is the actual world and $R_i \subseteq S \times S, i = 1, 2$. Let $R = R_1 \cup R_2$ and $\Box = \Box_1 \wedge \Box_2$. We can assume that (S, R, ∞) is a tree with the following properties:

1. Each point $x \neq \infty$ has a unique R predecessor and furthermore for $x \neq \infty$ there exists a unique sequence $(\infty, x_1, \ldots, x_n = x)$ such that $\infty R x_1 \wedge x_1 R x_2 \wedge \ldots \wedge x_{n-1} R x_n$.
2. R_1 and R_2 are disjoint, i.e. $x R_1 y$ iff $\neg x R_2 y$ holds.
3. We also require the conditions $\forall x \exists y x R_i y, i = 1, 2$.

If we look at Figure 15.23, then we see an example of the beginning of such a frame, where $x \rightarrow y$ means $x R y$ and $x \twoheadrightarrow y$ means $x R_2 y$. R_1 can be obtained as

$$x R_1 y = \text{Def}.x R y \wedge \neg x R_2 y.$$

The full figure is given in Figure 17.17

The paths in the tree indicate obligation choices taken by an agent starting at ∞ and proceeding along a path π, say $\pi = (\infty, o, z_1, z_2, z_3, \ldots)$ with ∞R_0 and $o R_{z_1}$ and $z_1 R_{z_2}$ and $\ldots$. We have a violation at point z_{i+1} along the path if $z_i R_2 z_{i+1}$.

So if our modal logic with $\Box_1$ and $\Box_2$ is augmented with the possibility of talking about paths then we can express whatever is relevant to say on contrary to duty obligations.

In our paper [172] we used the following modality for obligation

$$\bigcirc A = \text{Def}.\Box_1 A \wedge \Box_2 \neg A.$$

We can also regard Figure 17.17 as an argumentation network. We consider the two modalities $\Box = \Box_1 \wedge \Box_2$ and $\Box_2$ as representing support ($\Box = \Box_1 \wedge \Box_2$ with relation R) and attack ($\Box_2$ with relation R_2) respectively.

The figure is a very special bipolar network with it being a tree with properties (1), (2), (3) above. Furthermore, in the modal logic which corresponds to the figure we again need to talk about paths, with one path π_1 attacking another π_2 if for some z in π_1 and z' in π_2 we have zR_2z'.

So the common logic which can host both notions is modal logic with two modalities and some ability to talk about paths.

Note that since we are dealing with trees a backward temporal additional modality may suffice.

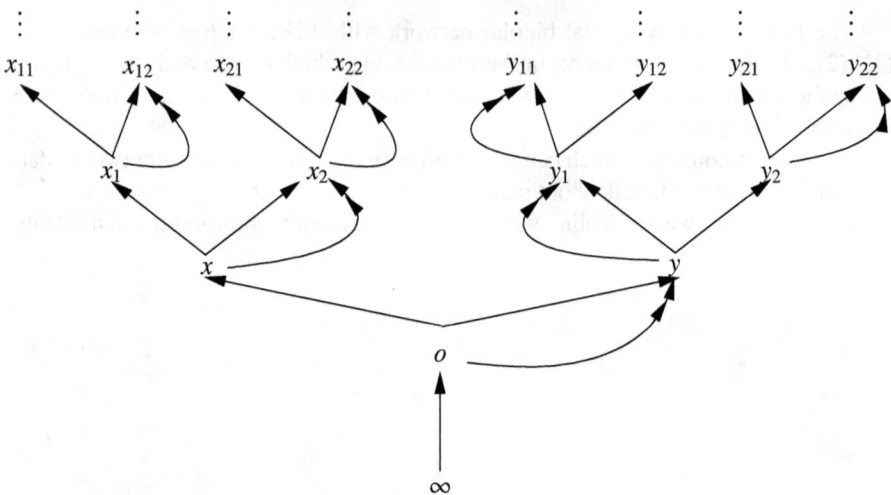

Fig. 17.17.

18

The Handing of Loops in Argumentation Networks

18.1 Background and orientation

There are two aspects to handling loops in argumentation networks. These aspects apply to any formal system based on any application area. These are:

1. The conceptual side of explaining and handling loops from the point of view of the application area. In our case it is dealing with the loop in the context of the instantiation of the argumentation system. The relevant questions are: why does the loop arise, what to do with it, and how to break/prevent it, etc.

2. The algorithmic aspect. We have various algorithms for calculating various properties/sets in the formal system and such algorithms can loop. Some of the looping comes from conceptual loops from the intended application areas and some loops may come from the nature of the algorithm used. We need technical means for handling such loop situations. In the case of argumentation, the loops in the network can trivialise our quest for extensions and we need ways and means to handle such loops.

The two aspects are connected, of course.

We shall discuss options of how to handle loops in the abstract (for several sample systems) in the concluding section. In this background and orientation section, let us focus on loops in argumentation networks.

Consider Figure 18.64. Here we have a floating attack from the cycle $\{a, b, c\}$ on the node d. Because of the cycle, the only traditional extension we get is all "undecided", but conceptually we do want node d to be definitely "out". We need a loop handling method to address this.

Then two aspects of loops mentioned above manifest themselves in this network as follows: The conceptual aspect will ask how does this loop arise in practice. What are the instantiations of arguments a, b, and c and why do they loop. Knowing this and knowing the nature of the attacks on node d (and what d actually means) will allow us to resolve the situation. The reader will see in Chapter 19 a legal Talmudic example. It is a real life example which could arise today. Let me just hint that formally what happens in this example is that the arguments are time-stamped (they get into the network at different times) and the Talmud rejects the temporally last argument which creates (closes) the loop. This is called the Rabbi Shkop method. So on the conceptual side this method handles loops in

(Talmudic) argumentation. It simply does not allow for the creation of loops. This would equally apply to 2 cycles (even though 2 cycles do have stable extensions). The Talmud does not think in terms of Dung extensions. It does not like cycles. The algorithmic aspect wants to break the 3-cycle loop in order to obtain meaningful extensions. So for example the Baroni et al. [30] CF2 semantics will take maximal conflict-free subsets of $\{a, b, c\}$ and proceed from there.

Baroni et al. [30] devote a long discussion about the inadequacy of the traditional semantics in handling odd and even loops. They say, and I quote:

> the length of the leftmost cycle should not affect the justification states [of an argument]. More generally, it is counter-intuitive that different results inconceptually similar situations depend on the length of the cycle. Symmetry reasons suggest that all cycles should be treated equally and should yield the same results.

We agree with [30] on the need for a new approach but we feel that the CF2 semantics offered as a solution requires further independent methodological justification.The notion of conflict freeness is a neutral notion and does not use the central notion of "attack" of the Dung semantics. When we get a loop like a three cycle $\{\alpha, \beta, \gamma\}$, in a real life application, there are good reasons for the loop in the context of the application area where it arises, and we want a decisive solution to the looping terms of {"in", "out"}, which makes sense in the application area. We do not want just a technical, non-decisive choice of maximal conflict free sets, a sort of compromise which involves no real decision making. Imagine we have a loop with $\{\alpha, \beta, \gamma\}$, and we go to a judge and we expect some effective decision making. We hope for something like "I think γ is not a serious argument".

These considerations are still part of the conceptual side. Conflict free subsets of $\{a, b, c\}$ are nevertheless still conceptually meaningful and can be considered.

We now explain the technical side of loops. Assume we have an algorithm $\mathcal{A}$ for calculating extensions. This is technical. The loop in Figure 18.64, for example, (which has a conceptual meaning in the appropriate application area) would technically cause a loop in the algorithm $\mathcal{A}$. This algorithmic loop (call it $\mathcal{A}$-loop) has to be resolved in the context of $\mathcal{A}$. This $\mathcal{A}$-loop is no different from any loop in any software program such as a logic programming program, database query program, graph traversing program, etc., etc. The most naural action of a programmer when detecting a loop is to instruct the program to skip and backtrack if possible. Our question is, what loop policy do we install in $\mathcal{A}$ so that the result will have meaning in argumentation? Just skipping may not be the right (sound) policy.

Let us summarise our overview of what is involved in handling loops in argumentation:

a) There is a conceptual framework for argumentation. Loops can be handled at this level. This conceptual framework takes a view on what to do with an abstract argumentation graph of the form (S, R), $S \neq \varnothing, R \subseteq S^2$, where S is the abstract set of arguments and R is the attack relation. Part of this framework is the traditional quest for extensions.

There are other approaches to (S, R), e.g. the equational approach, viewing (S, R) as a bearer of equations to be solved. Whatever the view is, it gives rise to (b).

b) Algorithms $\mathcal{A}$ for computing various aspects in (a). Loops in (S, R) cause loops in $\mathcal{A}$. We need loop checkers which are meaningful and are sound for (a).

We conclude this section by describing an algorithm for Figure 18.64. The states of the algorithm have the form $?x = r$, where x is a node in Figure 18.64 and $r \in \{0, 1\}$. The starting state can be any state, for example $?d = 0$ and the transitions of the algorithm from state to state is governed by the graph connections of Figure 18.64. The loop $\{a, b, c\}$ in this figure generates a loop in the algorithm. The computation tree of this algorithm is described in Figure 18.65.

Whatever happens to this algorithm is discussed in Section 5. At this point the reader is just invited to look at the figure, and not analyse it.

The material is structured as follows:

Section 2 deals with annotated computation trees for argumentation networks. If we want to analyse loops and handle them we need the machinery of computation trees with annotations, which record and lay before us all the relevant information about the loops. Given the trees we can introduce various approaches for handling loops.

Section 3 introduces the annihilator approach to handling loops. This approach is very simple: if an argument is involved in a loop then take it out of the network and deal with the rest. Of course we need the correct sound policies to do this, and this is what section 3 does. We also show how to get the CF2 semantics in this way (using annihilators).

Section 4 introduces the impulse method. It is equivalent to the annihilator method of Section 2, but is more general as a methodology, and does not rely on argumentation concepts. A loop can be a cycle of arguments attacking in a loop or it can be a loop in the algorithm computing something for example computing extensions for a networks which itself does not have any attacking loops (but the algorithm loops) or it can be in an algorithm simulating argumentation (e.g. in a logic program) looping back into an earlier state. A general algorithm, when in a loop, can receive an impulse to do something else. The most common impulse is to abandon that branch which is looping and move on. This has nothing to do with argumentation. We can ask, however, what impulse corresponds to the argumentation concept of annihilators? This correspondence is the topic of section 4.

Section 5 introduces the linear logic like approach which can be really generalised to any intelligent algorithm, such as Logic Programming, database queries, Tableaux, Intuitionistic Logic, theorem proving, and more. The basic idea is quite intuitive: first identify and define the notion of an item of resource (argument, logic formula, clause of a logic program) as "being used in a computation", and then restrict its use to a fixed number of times. This will stop all loops. The question is: What happens when the algorithm stops after having used all available resources (i.e. what is the declarative meaning of what is computed) and how to continue the computation after that with a view of trying to get sound results. This is done in this section for argumentation.

Section 6 handles loops through the equational approach. An argumentation networks gives rise to a system of equations whose variables are the arguments themselves. The extensions are obtained from solutions to the equations. Loops in the argumentation network become dependencies of variables on each other in the equations. Handling loops in argumentation become problems of dealing with implicit systems of equations.

Section 7 studies the adjustment method. It specifically gives a representation theorem for CF2 semantics. Its basic idea is that given an argumentation network which has loops, we adjust the attack relation a bit and look for extensions from the new network. It does not get into computation at all , not conceptually. We show that the CF2 extensions of the original network become ordinary stable extensions of the adjusted network.

Section 8 generalises the linear logic like approach which we introduced in section 6 and applies it to loops in logic programming.

Section 9 concludes with a discussion and comparison with the relevant literature.

18.2 Annotated computation trees

This section develops the notion of a computation tree for a given argumentation network (S, R). Take a point x in S and ask is it "in" in some extension $(?x = 1)$ or is it "out" in some extension $(?x = 0)$. The computation tree for this query traces what can happen. We can read cases of looping from it and we can also manipulate it to resolve loops. These basic notions are introduced in this section.

Definition 18.1 (Complete extensions).

1. *An argumentation network has the form (S, R) where S is a finite non-empty set of nodes and $R \subseteq S \times S$*
2. *A subset $E \subseteq S$ is conflict free if for no $x, y \in E$ do we have xRy*
3. *A subset $E \subseteq S$ protects an element x if whenever zRx then there exists a $y \in E$ such that yRz*
4. *A subset $E \subseteq S$ is a complete extension if the following three conditions hold:*
    a) E is conflict free
    b) E contains all the elements it protects
    c) E protects its own elements

Definition 18.2. *A tree $(T, \rho, 0)$ is a triple with T finite and $0 \in T$ is the top element and $\rho \subseteq T \times T$ is the tree binary relation satisfying the following:*

1. *Every $t \in T$ has a unique set of predecessors $s_1, \ldots, s_k$ (we allow also that t has no predecessors) such that $s_i \rho t, i = 1, \ldots, k$.*
2. *If $t \neq s$ then the sets of predecessors of t and s are disjoint.*
3. *Every point s in T $s \neq 0$ has a unique path*

$$S = s_1, \ldots, s_n = 0$$

Such that s_i is a predecessor of $s_{i+1}, i = 1, \ldots, n - 1$.

Definition 18.3. *Let (S, R) be an argumentation network.*

1. *A goal has the form $?x = 0$ or $?x = 1$, where $x \in S$. Think of $?x = r$ as Òwe want a complete extension with x in (resp. x out)Ó.*
2. *A sequence of goals of the form*

$$H = (?x_1 = r_1, \ldots, ?x_n = r_n)$$

where $x_i \in S, r_i \in \{0, 1\}$ is said to be a legitimate history of goals if the following holds:

a) For no $i \neq j$ do we have $?(x_i = r_i) = (?x_j = r_j)$ (i.e. $x_i = x_j$ and $r_i = r_j$)

b) For all i, $x_{i+1}Rx_i$ and $r_{i+1} = 1 - r_i$ hold.

3. It will be convenient to represent a history sequence with unique names for each position. The simplest choice of names is numerals. So the sequence history $(?x_1 = r_1, \ldots, ?x_n = r_n)$ will be presented as the sequence $(1 : ?x_1 = r_1, \ldots, n : ?x_n = r_n)$. This allows us to say statements like "$?x = r$ appears in position 5 of the history". Sometimes when we use expressions of the form $?x = r$ to annotate elements of a path $(t_1, \ldots, t_n)$ in some tree, we can write the history sequence as $(t_1 : ?x_1 = r_1, \ldots, t_n : ?x_n = r_n)$, provided all items t_i are pairwise different. What we do will be clear from the context.

Definition 18.4 (Computation trees). Let (S, R) be an argumentation network and let H be a legitimate history sequence of goals, presented as $(e_1 : ?x_1 = r_1, \ldots, e_n : ?x_n = r_n)$ where $e_i \neq e_j$ for $i \neq j$ and e_i not in S.

Let $?x_0 = r_0$ be a goal. Let $(T, \rho, 0)$ be a tree with T not containing any of the $\{e_i\}$.

We say that a function $\mathbf{f}$ makes $(T, \rho, 0, \mathbf{f})$ into a computation tree over (S, R) with initial history sequence $(H_0, ?x_0 = r_0)$ iff $\mathbf{f}$ is a function giving each $t \in T$ a pair $(H(t), g(t))$ such that the following holds:

1. For each t, $H(t)$ is a legitimate history and $g(t)$ is a goal and $\mathbf{f}(0) = (H_0, (?x_0 = r_0))$

2. For any $t \in T$, such that $g(t) = (?x(t) = r(t))$, let $y_1, \ldots, y_k$ be all the elements in S such that $y_i Rx(t)$ hold. Then the predecessors of t in T can be written as $s_1, \ldots, s_k$ (same k as above) and can be correlated with $y_1, \ldots, y_k$ in such a way that the following holds:

   a) $H(s_i) = H(t) * (t : g(t))$

   b) $g(s_i) = (?y_i = (1 - r(t))), i = 1, \ldots, k$

   (note that $*$ is concatenation of sequences).

3. If t is a point in T with no predecessors in T then either (a) or (b) hold:

   a) $x(t)$ has no attackers in (S, R), i.e. $\neg \exists zzRx(t)$.

   b) There is a loop, that is (formally) $g(t)$ appears in the sequence $h(t)$. Note that we can identify exactly the position of $g(t)$ in the sequence. Let this position be s (i.e. $g(t) = g(s)$). Then we say $t : g(t)$ is in a loop with $s : g(s)$.

Definition 18.5 (Annotation function).

1. Let $(T, \rho, 0, \mathbf{f})$ be an annotated computation tree over an argumentation network (S, R). We add to it an annotation function V giving to each node t an annotation value $V(t)$ of the form "success" or "failure" or "loop".

   The definition yields an algorithm for computing V which is bottom up in the tree, and is executed in several passes, giving a sequence of partial functions $V_1, V_2, \ldots$, with V being the last in the sequence.

   The first pass will annotate the endpoints (bottom of the tree) with values "success", "failure" or "loop" and later passes will propagate the values "success" and "failure" up the tree and also replace some "loop" values by "success" or by "failure".

   The first pass is done using group (1) of rules. It will annotate the bottom nodes t by the values $V(t)$ being "failure", "success" or "loop". The value "loop" will *be given because the query t :?x = r, x ∈ S and r ∈ {0, 1} has been asked before in the history at a node say s :?x = r and so it is no use to continue. So we stop and the node t becomes an endpoint at the bottom of the tree. We say that the source of the loop at t is the position s in the history. However, the loop value may be temporary only, because the loop may clear later on at position s in a later pass, and the information will be passed on to t.*

The next pass is using the rules of groups 2 and 3 which propagate the "success", "failure" values up the tree. We execute group 2 rules and group 3 rules again and again until no change occurs (no new values are propagated) and then we execute the group 4 rules. The result of the final pass is the final value of V.

2. *Let t ∈ T and let the annotation* $\mathbf{f}(t)$ *be* $(H(t), g(t))$.
   We give the following groups of rules for defining V(t):

(Group 1) Case of $t \in T$ has no predecessors

a) *If x(t) has no attackers in (S, R) and r(t) = 0, then let V(t) = "failure".*
b) *If x(t) has no attackers in (S, R) and r(t) = 1, then let V(t) = "success".*
c) *If x(t) does have attackers in (S, R) then let V(t) = "loop".*[1]

(Group 2) Case of $t \in T$ has predecessors in (T, ρ)

Let the predecessors be $s_1, \ldots, s_k$. *Then by construction x(t) ∈ S has the attackers* $y_1, \ldots, y_k$ *(i.e.* $y_i R x(t)$ *holds i = 1, ..., k) and we can assume that they are correlated with* $s_1, \ldots, s_k$ *(same k!) and that* $g(s_i) = ?y_i = 1 - r(t)$ *and* $H(s_i) = H(t) * g(t)$. *We also assume that some of* $V(s_i), i = 1, \ldots, k$ *are defined, maybe not all but enough are defined to allow us to define V(t).*

a) *If r(t) = 1 and for some i, V(s_i) = "failure" then V(t) = "failure".*
b) *If r(t) = 1 and for all i, V(s_i) = "success" then let V(t) = "success".*
c) *If r(t) = 0 and for some i, V(s_i) = "success" then let V(t) = "success".*
d) *If r(t) = 0 and for all i, V(s_i) = "failure" then let V(t) = "failure".*
e) *If none of the above then V(t) gives no value. These are cases (e1) and (e2) below:*

(e1) *r(t) = 1 and for some i, V(s_i) is not yet defined and for no i do we have that V(s_i) = "failure", then we leave V(t) undefined.*
(e2) *r(t) = 0 and for some i, V(s_i) is undefined and for no i do we have that V(s_i) = "success" then we leave V(t) undefined.*

(Group 3) Replacing "loop" by "success" or by "failure"

We consider any endpoint labelled "loop". We look up the history H(t) and by definition there is a node s above with the same query ?x = r which is the source of the loop. If this node is not labelled by V we do nothing. If this node is labelled "success" or "failure" we change the label "loop" to the same label as the one of the node s above (i.e. to "success" or "failure" according to the label of the node s above)

(Group 4) Propagating the "loop" value

a) *If r(t) = 1 and for some i, V(s_i) = "loop" and for no i do we have that V(s_i) = "failure" then let V(t) = "loop".*
b) *If r(t) = 0 and for some i, V(s_i) = "loop" and for no i do we have that V(s_i) = "success" then let V(t) = "loop".*

---

[1] If there were no loop, i.e. $g(t) \notin H(t)$ then the tree would have continued.

3. **The algorithm for defining** V

Step 1. First pass

Compute V_1, a partial annotation function defined using group 1 of rules. This will annotate only the end points of the tree.

Step 2. Several additional passes

a) *Use group 2 rules to extend V_1 to V_2.*
b) *Use group 3 rules to extend V_2 to V_3.*
c) *Go to (a) to extend V_3 to V_4 and then go to (b) to extend V_4 to V_5 and keep this execution loop until $V_{2n+1} = V_{2n+3}$.*

Step 3. Final step

Use group rules 4 to propagate the "loop" value in V_{2n+3} up the tree to obtain the final V. This final V is always defined on all points of T.

Example 18.6. Consider Figure 18.1

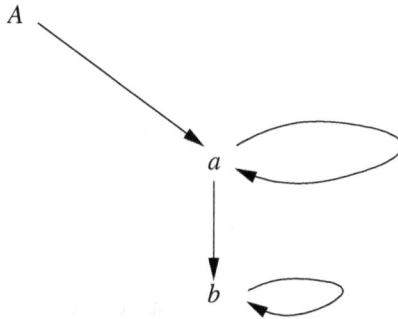

Fig. 18.1.

1. Let us ask $?a = 1$ and construct the tree for this query: we get Figure 18.2. We use the notation $x \leftarrow y$ for $x\rho y$ in the tree.

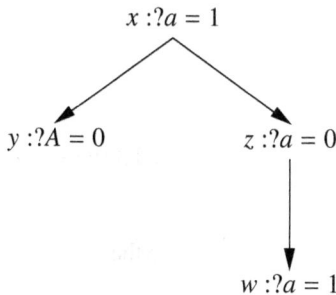

$x : ?a = 1$

$y : ?A = 0$ $z : ?a = 0$

$w : ?a = 1$

Fig. 18.2.

Let us now apply our algorithm to define the annotation V to the nodes. First we look at Group 1 of rules as in Definition 18.4. This gives us the annotation "loop" at node w, and "failure" at node y. We then apply Group 2 rules and get the annotation "failure" at node x. We then realise that further applying Group 2 rules does not give us anything new and so we now apply group 3 rules and get "failure" to annotate node w instead of "loop", because the source of the loop, node x is annotated "failure" by now.

Note that argument b of Figure 18.1 did not participate in the computation tree.

2. Let us now ask the query $?b = 0$ and see what happens. See Figure 18.3.

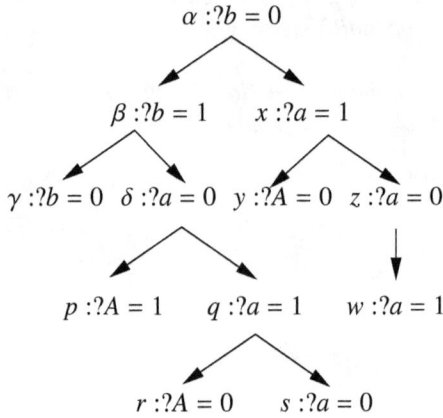

$$\alpha : ?b = 0$$

$$\beta : ?b = 1 \qquad x : ?a = 1$$

$$\gamma : ?b = 0 \quad \delta : ?a = 0 \quad y : ?A = 0 \quad z : ?a = 0$$

$$p : ?A = 1 \qquad q : ?a = 1 \qquad w : ?a = 1$$

$$r : ?A = 0 \qquad s : ?a = 0$$

Fig. 18.3.

The initial annotation of endpoints by V is as follows:

$V(2) = $ "loop", from node x
$V(y) = $ "failure"
$V(s) = $ "loop", from node δ
$V(r) = $ "failure"
$V(p) = $ "success"
$V(\gamma) = $ "loop", from node α.

We now propagate the "failure" and "success" values up the tree using Groups 2 and 3 rules and get

$V(q) = $ "failure"
$V(x) = $ "failure"
$V(\delta) = $ "success"

We make another pass using Groups 2 and 3 rules and get

$V(w) = $ "failure".

This is as far as we can go.
We now use Group 4 rules and propagate the "loop" value and get

$V(\beta) = $ "loop"
$V(\alpha) = $ "loop"

Therefore the computation tree for $\alpha : ?b = 0$ gave us "loop".
We are now ready for applying loop busting procedures. This, however, is the job for Section 4.

$$t : ?x(t) = r(t)$$

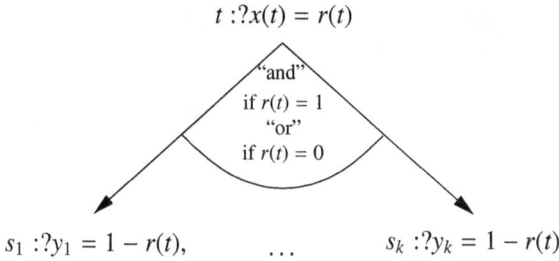

"and"
if $r(t) = 1$
"or"
if $r(t) = 0$

$$s_1 : ?y_1 = 1 - r(t), \qquad \ldots \qquad s_k : ?y_k = 1 - r(t)$$

Fig. 18.4.

Remark 18.7. To explain the meaning of Definition 18.5, consider Figure 18.4

Remember the query $?x(t) = r(t)$ must always succeed: if $r(t) = 1$, we want $x(t)$ to succeed (be in) and so all the attackers s_i must be 0 (be out) and so $?y_i = 0$ must all succeed. So if $r(t) = 1$ the branching is an "and" branching. If $r(t) = 0$, we want $x(t)$ to be out and so at least one of $?y_i = 1$ must succeed. So if $r(t) = 0$ the branching is an "or" branching.

To summarise, the branching at the tree is an "and" branching if $r(t) = 1$ and an "or" branching if $r(t) = 0$.

Proposition 18.8. *Let (S, R) be an argumentation network and let $x_0 \in S$ and $r_0 \in \{0, 1\}$. Let H_0 be a legitimate history. Then there exists a unique (up to renaming nodes) annotated computation tree $(T, \rho, 0, \mathbf{f})$ such that $g(0) = (H_0, ?x = r)$.*

Proof. By induction on the depth of the tree.
Case 1: The tree has only one point $T = \{0\}$. This case can arise only if (a) or (b) hold:

(a) x_0 has no attackers in (S, R).
(b) x_0 does have attackers but $?x_0 = r_0$ is in H_0.

In either case the tree is unique.

Case n: Let $(T, \rho, 0, \mathbf{f})$ be a tree for $(H_0, ?x_0 = r_0)$ of depth > 1. Let $s_1, \ldots, s_k$ be the predecessors of x_0. Let $y_1, \ldots, y_k$ be the attackers of x_0 in (S, R). Then by the induction hypothesis, the trees $(T_i, \rho_i, s_i, g \upharpoonright T_i)$ are unique up to renaming, for $i = 1, \ldots, k$. Now since by construction the correlation between s_i and y_i is fixed, we get that $(T, \rho, 0, g)$ is also unique.

Proposition 18.9. *Let $(T, \rho, 0, \mathbf{f}, V)$ be an annotated tree, for $(H_0, ?x = r)$. Change every value appearing in the tree from "0" to "1", from "1" to "0", from "success" to "failure" and from "failure" to "success". Then we get a correctly defined dual tree.*

Proof. We really need to check the correctness of the propagation of V, after we swap values. Consider Definition 18.5. We note that condition (1a) becomes (1b) and (1b) becomes (1a) and (1c) remains the same.

Further we get that (2) swaps with (2c) and (2b) swaps with (2d). Also (2e) swaps with (2f).

Remark 18.10. Proposition 18.9 is important for our loop checking. It means that for the purpose of loops and computation it does not matter if we ask $?a = 1$ or $?a = 0$, we get looping just the same. Also note that

- $?x = r$ is annotated by "failure" iff $?x = 1 - r$ is annotated by "success"

holds.

Example 18.11. Consider Figure 18.5^2 and the goal $?a = 0$, with the empty starting history.

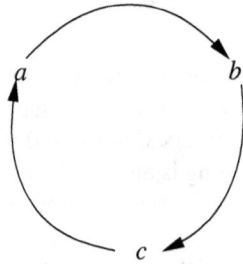

Fig. 18.5.

The following Figure 18.6 is a tree for it:

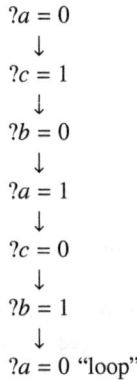

$$?a = 0$$
$$\downarrow$$
$$?c = 1$$
$$\downarrow$$
$$?b = 0$$
$$\downarrow$$
$$?a = 1$$
$$\downarrow$$
$$?c = 0$$
$$\downarrow$$
$$?b = 1$$
$$\downarrow$$
$$?a = 0 \text{ "loop"}$$

Fig. 18.6.

In Figure 18.6 we did not explicitly name the nodes, but their position in the Figure identifies them. The history of a node can be read from the figure, it is all the queries above the node. We stopped the tree at the bottom node "$?a = 0$" because of a loop.

The evaluation function V will propagate the value "loop" to all the nodes above.

Example 18.12. Consider Figure 18.7.

A tree for this figure is Figure 18.8, where the query is $(\emptyset, ?a = 1)$.

---

^{2} Our notation is $x \to y$ for xRy. Recall that $x \leftarrow y$ means $x\rho y$. Whether we mean R or ρ will be clear from context and arrow direction.

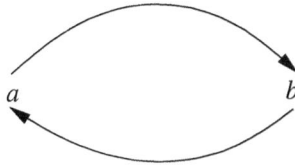

Fig. 18.7.

$?a = 1$
$\downarrow$
$?b = 0$
$\downarrow$
$?a = 1$ "loop"

Fig. 18.8.

Example 18.13. Consider Figure 18.9 Let us offer a tree for it, in Figure 18.10, with

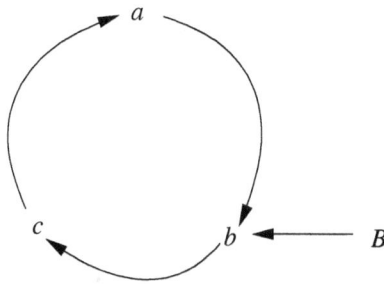

Fig. 18.9.

query $(\varnothing, ?a = 0)$.

Example 18.14. Consider Figure 18.11. Let us do 4 trees for it. We do only the first pass for V. We get Figures 18.12 to 18.15. It will be important to have these trees when we deal with loop checking.

Note that according to the definition of the propagation of the value function V, all nodes in Figures 18.12 to 18.15 will eventually be annotated by "loop".

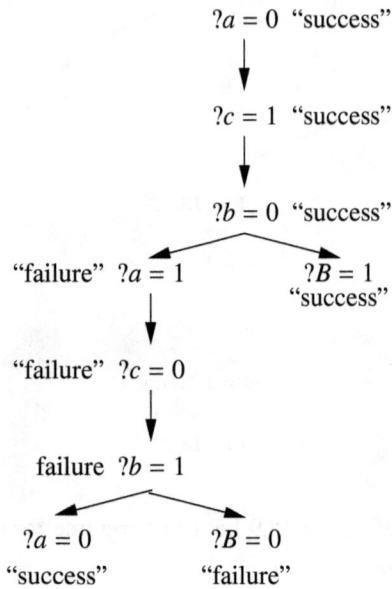

Fig. 18.10.

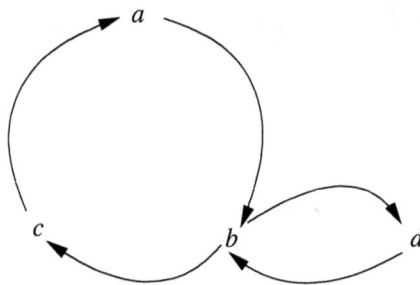

Fig. 18.11.

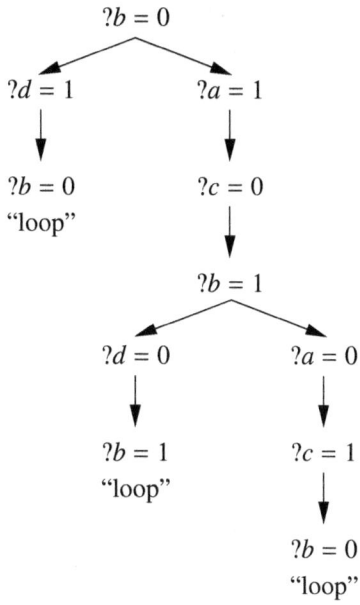

Fig. 18.12.

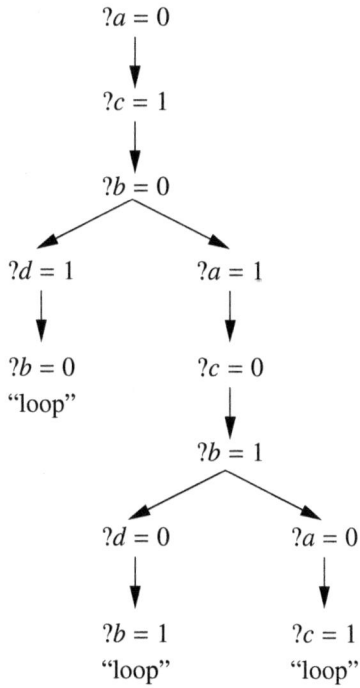

Fig. 18.13.

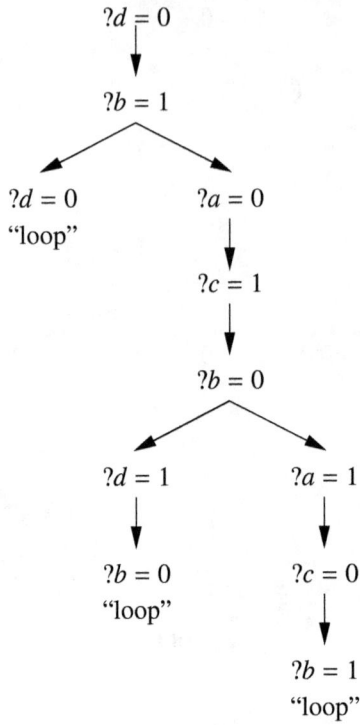

Fig. 18.14.

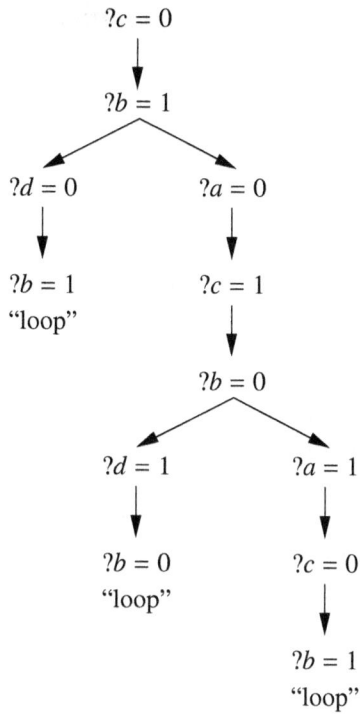

Fig. 18.15.

18.3 The annihilators approach to handling loops

This section introduces annihilators and defines the directional LB1 and LB2 semantics. We then show that the LB2 semantics is equivalent to the CF2 Semantics.

18.3.1 LB extensions and annihilators

This subsection identifies looping points in computation trees and introduces annihilators to deal with them. The various definitions and theorems study the interplay between loop points and annihilators.

Definition 18.15 (Annihilators).

1. *Let (S, R) be an argumentation network. Let $x \in S$. Let $\alpha(x)$ be a new atom depending on x, which we call the annihilator of x. We assume that all $\alpha(x)$, $x \in S$ are pairwise different and are all different from all elements of S.*
   It is convenient to denote the elements of S by lower case letters x and the annihilators $\alpha(x)$ by the respective capital letters. Thus, for example

   $$\alpha(a_8) = A_8$$
   $$\alpha(b) = B$$

   and so on.
2. *Let (S, R) be an argumentation network. Let $b \in S$. Then the b-annihilator expansion network is the network (S_B, R_B) where*

   $$S_B = (S \cup \{B\})$$

   and

   $$R_B = R \cup \{(B, b)\}$$

3. *Similarly we define the network $(S_{\{B_i\}}, R_{\{B_i\}})$, for b_i in S, as*

   $$S_{\{B_i\}} = S \cup \{B_i\}$$
   $$R_{\{B_i\}} = R \cup \{(B_i, b_i)\}.$$

Example 18.16. Consider the network of Figure 18.5. The *b*-annihilator expansion network associated with it is Figure 18.9.

Theorem 18.17.

1. *Let (S, R) be a network and let E be a complete extension and let $b_i \in S$ be elements such that $\exists x \in E(xRb_i)$ holds.*
   Then $E' = E \cup \{B_i\}$ is a complete extension of the annihilator $(S_{\{B_i\}}, R_{\{B_i\}})$.

Proof. 1. We show E' is conflict free. Let $x, y \in E'$ such that x attacks y, then if $x = B_i$ for some i, then y must b_i, and $b_i \in E$. But $\exists y \in E(yRb_i)$ which contradicts the fact that E is conflict fee. If $x \in E$ then y also is in E since B_i are not attacked by anything. Again a contradiction.
2. We show that E' contains what it protects. Assume x is protected by E'. We need to show that $x \in E'$. If $x = B_i$ for some i then $x \in E'$.
   Assume x is not attacked then x is not attacked in (S, R) and so $x \in E \subseteq E'$.
   Assume x is attacked by y. There are two cases

a) $y = B_i$ for some i. Then x is not protected by E' because B_i is not attacked by anything.

b) So since we assume that E' protects x we must assume that $y \neq B_i$. Therefore for some $z \in E'$ we have that z attacks y. If $z = B_i$, then $y = b_i$ and b_i is attacked by E by assumptions on b_i. If z is not B_i, then $z \in E$. Thus either way E protects x from any attacking $y \in S$. Hence $x \in E \subseteq E'$.

3. E' protects its elements. This holds because E protects its elements and nobody attacks any B_i.

Example 18.18. There is no converse to Theorem 18.17. Consider Figure 18.16

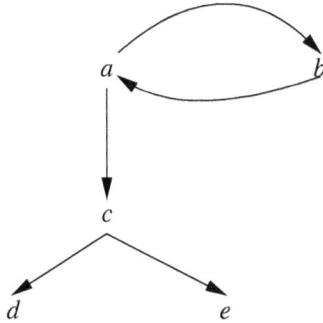

Fig. 18.16.

There is a complete extension $\{b, c\}$. In this extension both d and e are out. Consider now the annihilator Figure 18.17. It does have the extension $\{b, c, D, E\}$ as predicted by Theorem 18.17. But it also has the extension $\{a, D, E\}$ which has no counterpart in Figure 18.16.

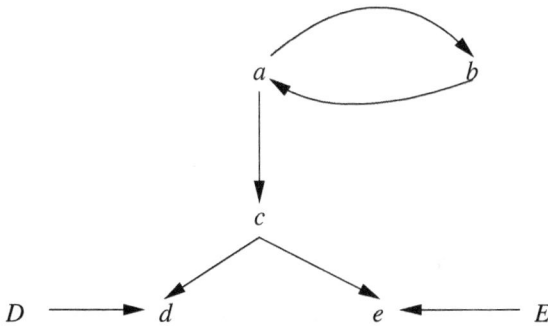

Fig. 18.17.

Let us go back to the extension $\{b, c\}$ of Figure 18.16. We have that c is out in this extension. Let us consider Figure 18.18.

In this case there is correlation between the extensions containing $\{b, c\}$ and $\{b, c, A\}$.

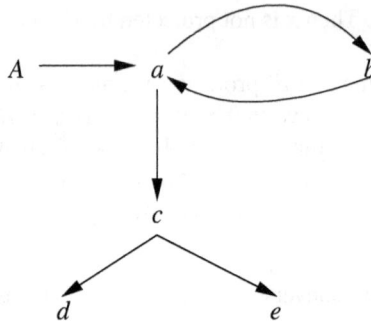

Fig. 18.18.

Our intention is to use annihilators to resolve loops. Perhaps we can find a way of choosing the annihilators in such a way as to give us the correct correlations when the loop can be resolved by the traditional semantics, for example the case of even cycles.

Definition 18.19 (Top loop-nodes).

1. *Let (S, R) be an argumentation network. Let $x \in S$. x is said to be a top loop-node iff the following holds:*
   a) $\exists y (y R x)$
   b) *If for some $y, u_1, \ldots, u_k$ we have $R u_1 \wedge u_1 R u_2 \wedge \ldots \wedge u_k R x$ then for some $v_1, \ldots, v_n$ we have $x R v_1 \wedge v_1 R v_2 \wedge \ldots v_n R y$.*
2. *Let $R^{-1}(x)$ be defined as $\{y | y R x\}$. Let $R^{-m}(x) = \{y | \exists z \in R^{-m+1}(x)$ such that $y R z\}$.*

Lemma 18.20. *If x is a top loop-node then for some m, $x \in R^{-m}(x)$.*

Proof. Since part of the definition of top loop-node is $\exists y (y R x)$ we get that $R^{-1}(x) \neq \emptyset$. Since also we have that for some $v_1, \ldots, v_k$ we have $x R v_1 \wedge \ldots \wedge v_k R y$. We get that $R^{-k-1}(x)$ is nonempty and contains x.

Definition 18.21 (Strongly connected subsets). *Let (S, R) be an argumentation network.*

1. *A sequence of nodes $(x_1, \ldots, x_n)$ is a cycle iff the following holds:*
   a) $x_i R x_{i+1}, 1 \leq i < n - 1$
   b) $x_n R x_1$
2. *Let $x \equiv y$ hold by definition iff both x and y belong together to some same cycle.*
3. *$\equiv$ is an equivalence relation and it divides S into equivalence classes. These are called maximal strongly connected sets by Baroni et al, [30]. Let $S/\equiv$ be the set of such equivalence classes and define $R/\equiv$ on $S/\equiv$ by $x/\equiv R/\equiv y/\equiv$ iff for some $u \equiv x$ and $v \equiv y$ we have $u R v$.*
4. *We say that $x/\equiv$ is a top loop-class, iff whenever $u R v$ and $v \equiv x$, we also have $u \equiv x$.*

Lemma 18.22. *Let (S, R) be an argumentation network. Then an element x is a top loop-node iff $x/\equiv$ is a top loop-class.*

Proof. Condition (3) of Definition 18.21 holds for top loop-nodes.

Theorem 18.23. *Let (S, R) be an argumentation network. Let $a_i \in S$ for $i = 1, \ldots, m$ be top loop-nodes. Assume that for each i, there exists a complete extension E_i such that a_i is attacked by E_i (i.e. $\exists y \in E_i y R a_i$). Consider the annihilator extension $(S_{\{A_i\}}, R_{\{A_i\}})$ for $a_i \in S$, as defined in Definition 18.15, namely $S_{\{A_i\}} = S \cup \{A_i\}$ and $R_{\{A_i\}} = R \cup \{(A_i, a_i)\}$. Let E'_* be a complete extension in $(S_{\{A_i\}}, R_{\{A_i\}})$ and assume that $E_* = E'_* \cap S$ attacks each of the a_i for $i = 1, \ldots, m$, then $E_* = E'_* \cap S$ is a complete extension in (S, R).*

Proof. We show that Definition 18.1 holds for E_*.

1. E_* is conflict free. This holds because E'_* is conflict free.
2. We prove that E_* contains all elements $x \in S$ which it protects.

   Assume x is protected by E_*. We want to show that $x \in E_*$. We claim that x is protected by E'_* in $(S_{\{A_i\}}, R_{\{A_i\}})$ and therefore $x \in E'_*$ and so $x \in E_*$. We shall prove that the assumption that x is not protected by E'_* leads to a contradiction.

   Let y attack x. If $y \in E_*$ then since E_* protects x, there will be a $z \in E_*$ such that zRy. So the only way that E'_* does not protect x in $(S_{\{A_i\}}, R_{\{A_i\}})$ is that $y = A_i$, for some i, and of course there is no attacker to A_i. In this case we get $x = a_i$. We can assume that $i = 1$.

   We now show that this situation is impossible. The situation we have is as follows:
   a) a_1 is neither in E_1 (a_1 not in E_1 because E_1 attacks a_1) nor in E_* (a_1 not in E_* because E'_* attacks a_1).
   b) $\exists y \in E_1$ such that $y R a_1$
   c) Both E_1 and E_* protect a_1.

   From (b) we get for some $y_1, y_1 R a_1$ holds. Thus $R^{-1}(a_1)$ is not empty. Consider any $y_1 \in R^{-1}(a_1)$. Now since E_* protects a_1 it follows that for some $y_2 \in E_*$ we have $y_2 R y_1$.

   Thus we have $R^{-2}(a_1) \neq \emptyset$. Note that y_2 is not in E_1 because E_1 is conflict free and $y_1 \in E_1$.

   We thus got the following hierarchy so far:
   Level 0: $a_1 \notin E_1, a_1 \notin E_*$
   Level 1: $R^{-1}(a_1) \neq \emptyset$ elements
   $y_1 \in R^{-1}(a_1)$ are in E_1 but not in E_*.

   We continue with our recursive considerations with an arbitrary $y_2 \in R^{-2}(a_1)$. Since E_1 is a complete extension and it protects its members, and $y_2 R y_1$, we have a $y_3 \in E_1$ such that $y_3 R y_2$ holds. $y_3 \in E_1$ and so y_2 is not in E_* because E_* is conflict free.

   We thus get level 3:
   Level 3: $R^{-3}(a_1) \neq \emptyset$
   Elements $y_3 \in R^{-3}(a_1)$ are in E_1 and not in E_2.
   We continue by induction and get
   Level 2r: $R^{-2r}(a_1) \neq \emptyset$
   Elements $y_r \in R^{-2r}(a_1)$ are in E_* but not in E_1.
   Level 2r + 1: $R^{-2r-1}(a_1) \neq \emptyset$
   Elements $y_{2r+1} \in R^{-2r-1}(a_1)$ are in E_1 but not in E_*.
   Figure 18.19 illustrates the situation we get.

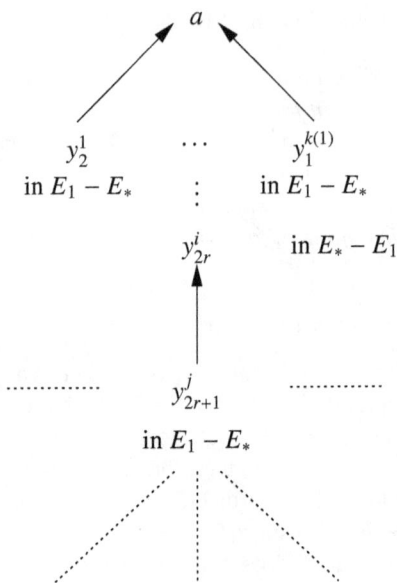

Fig. 18.19.

We now have a contradiction, because a_i is a top loop-node and by Lemma 18.20 for some m, $a_1 \in R^{-m}(a_1)$, but a_1 is neither in E_1 nor in E_*.

This situation arose because we assumed that E_* protects x but E'_* does not protect x and so $A_1 R_{\{A_i\}} x$ and so $x = a_1$. Thus E'_* does protect x and hence $x \in E_*$. This proves the condition that E_* contains the elements it protects.

3. We now prove that E_* protects its own members. Let $x \in E_*$ and assume y attacks x. We seek a $z \in E_*$ such that z attacks y. We know that E'_* is a complete extension and protects y and so there is such a $z \in E'_*$. If for all i we have that $z \neq A_i$ we are finished because $z \in E_*$.

What if $z = A_i$ for some i?

In this case $x = a_i$ and this is impossible because we assumed that E_* attacks a_i. This completes the proof of the theorem.

Example 18.24. Consider the network (S, R) of Figure 18.20. This network has the extension $E_1 = \{y, x\}$.

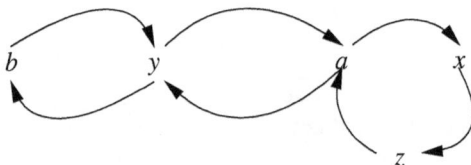

Fig. 18.20.

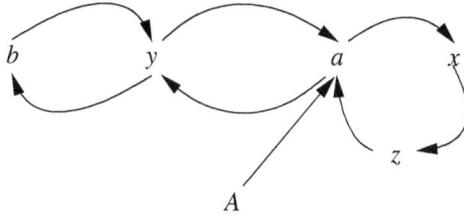

Fig. 18.21.

a is attacked by this extension. When we move to the annihilators (S_A, R_A) of Figure 18.21 we also get the extension $E'_2 = \{A, x, b\}$. E_2, however is not an extension of (S, R). The condition of Theorem 18.23 is not satisfied.

The proof that E_2 protects its own members does not go through. E_2, however, does contain whatever it protects. $E_2 = \{x, b\}$ does not protect anything.

Consider the extension $E'_3 = \{A, y, x\}$. This does satisfy the condition of the theorem. The extension $E'_4 = \{A, x\}$ also does not satisfy the conditions of the theorem.

Remark 18.25. Consider Figure 18.20. This is a loop (a strongly connected set), and so let us see how we can break this loop. If we use the CF2 semantics, we will be looking for maximal conflict-fee sets.[3] We get the following options in CF2:

$\{b, x\}$
$\{b, z\}$
$\{b, a\}$
$\{y, x\}$.

the last one, $\{y, x\}$ is available as a stable extension anyway, without any loop checking mechanism. The problem with any loop checking and loop busting method is the tension between the following two needs:

1. Apply the method to any loop (e.g. odd length cycle) and get all the extensions. For example apply CF2 to Figure 18.20 and get all the extensions.
2. It may be the case that the loop does have traditional extensions already (e.g. even length cycles) and in this case we want *local means* to ensure that we identify the extensions that already exist.

In the case of the annihilator approach we do have the heuristic means required by (2) courtesy of Theorem 18.23. We proceed as follows:

1. Identify a top loop-node in the network (S, R) (in our example we identified a in Figure 18.20).
2. Move to the annihilator network (S_A, R_A) (in our case Figure 18.21).
3. Compute all extensions E' of (S_A, R_A) and take all $E = E' \cap S$ as the extension of (S, R).
4. Identify those existing extensions by checking whether E attacks a. (This is a geometrical test, no need to compute extensions in (S, R).)

---

[3] The CF2 semantics is defined later on in Definition 18.32. for strongly connected sets the CF2 extensions are all maximal conflict free subsets.

Thus in our case after identifying the top loop-node a and using the annihilator method, we get the extensions $\{b, x\}$ and $\{y, x\}$ and only $\{y, x\}$ was identified as existing.

So the CF2 extensions $\{b, z\}$ and $\{b, a\}$ are not obtained.

We can ask whether choosing another annihilator will give us different extension? After all, all nodes in Figure 18.20 are top loop-nodes!

Let us check what happens for the different choices of annihilators:

1. **Case a is annihilated**
   We get extenssions $\{b, x\}$ and $\{y, x\}$. The latter is identified as existing.
2. **Case y is annihilated**
   We get the extension $\{b\}$ and the loop $\{a, z, x\}$ remains. Further annihilators are required. For example we can next take out a and get the extension $\{b, x\}$, or next take out z and get the extension $\{b, a\}$ or next take out x and get the extension $\{b, z\}$.[4]
3. **Case b is annihilated**
   Still loops remain, more annihilators are required.
4. **Case x is annihilated**
   We get $\{z\}$ as an extension. The loop $\{b, y\}$ remains. More annihilators are required. This case does not satisfy theorem 18.23.
5. **Case z is annihilated**
   Loop remains.

Remark 18.26. One may ask of the loop checker to give only the extensions which exist when there are such traditional extensions, (for example the loop checker should break odd loops but give us all and only existing extensions for even loops) but we are not so strict. Take, for example (see Example 18.43), an even 6-cycle with $S = \{a_1, a_2, a_3, a_4, a_5, a_6\}$ and the attack relation with n attacking $n + 1$, for $1 \leq n < 6$ and a_6 attacking a_1. CF2 (which takes as extensions all maximal conflict free sets) allows for the extension $E_2 = \{a_3, a_6\}$. This is not an existing extension, the existing traditional extensions are $E_1 = \{a_1, a_3, a_5\}$ and $E_3 = \{a_2, a_4, a_6\}$. The point is, do we want to modify CF2 to exclude $E_2 = \{a_3, a_6\}$ or do we want to modify CF2 with a local quick test enabling it to recognise $E_2 = \{a_3, a_6\}$ as a non-existent new extension, without fully computing all existing extensions and comparing?

We shall see later (Theorem 18.33) that we can get CF2 extensions by taking annihilators for the complements of a maximal conflict-free sets. So the extension $E_1 = \{a_1, a_3, a_5\}$ is obtained by annihilating $\{a_2, a_4, a_6\}$ and the extension $E_2 = \{a_3, a_6\}$ is obtained by annihilating $\{a_1, a_2, a_4, a_5\}$. Similarly the extension $E_3 = \{a_2, a_4, a_6\}$ is obtained by annihilating $\{a_1, a_3, a_5\}$.

We ask can we recognise by some local condition that $E_2 = \{a_3, a_6\}$ is not an existing complete extension. Theorem 18.23 can help us . It is true that each of the annihilated $\{a_1, a_2, a_4, a_5\}$ is attacked by some extension, but it is not true that our candidate extension $\{a_3, a_6\}$ attacks all of them. So an extension obtained via annihilators has to attack by itself all of its annihilated nodes.

Remark 18.36 continues this discussion.

---

[4] We shall prove a theorem (Theorem 18.33) that if we simultaneously annihilate the complements of the maximal conflict free sets we get all the CF2 extensions.

Example 18.27. Let us connect the considerations of this section with those of the previous section. What is the connection between annihilators, procedures, and goal directed computation? Let us apply our goal directed procedure of Section 2, to the network of Figure 18.20.

Suppose we are interested in an extension containing *a*. So we ask $?a = 0$ and follow up the goal directed computation. This is done in Figure 18.22.

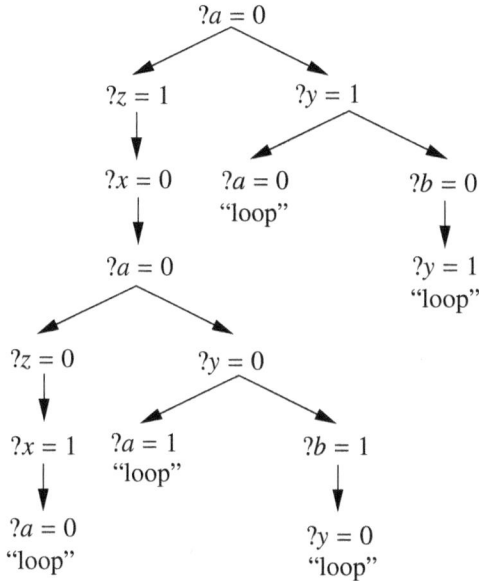

Fig. 18.22.

We propagate the loop values and see that the query $?a = 0$ loops. We now check whether the element *a* is a top loop-node. The answer is yes and so we move to the annihilator network of Figure 18.21 and ask the same question. Figure 18.23 shows what we get.

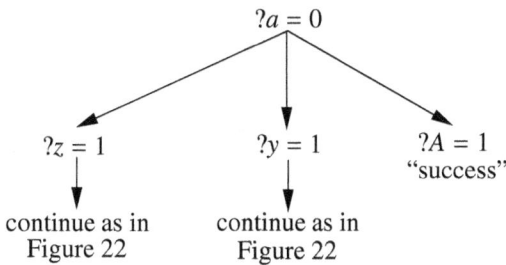

Fig. 18.23.

We should be able to read the extension from Figure 18.23. Let us do it intuitively. The formal definition was given in the previous section, Definition 18.5. Obviously

?$a = 0$ is labelled "success" as seen from the top of the tree. Therefore any node where ?$a = 0$ loops, it loops with "success" and therefore ?$a = 1$ is a "failure" and so ?$y = 0$ is a "failure" and therefore ?$b = 1$ is a "failure" and also ?$y = 1$ is a "success". Also ?$z = 0$ and ?$x = 1$ are "success".

So we read the extension from the tree $a = 0, y = 1, b = 0, x = 1$, and $z = 0$.

We now want to define procedures which will give us our first loop busting approach (which we shall call LB1, or the top loop-nodes approach) and we will also define a more refined method (which we call LB2, or the top loop-sets approach) and compare with the CF2 extensions of Baroni *et al.*, [30].

Definition 18.28 (Residuals, left-overs). *Let $\mathcal{A} = (S, R)$ be a network and let E be its grounded extension. Let $\mathbf{f}_1$ be a partial $\{0, 1\}$ valued function defined possibly only on some of the points of S as follows.*
$\mathbf{f}_1(x)$ *is defined to be 1 if x in E.*
$\mathbf{f}_1(x)$ *is defined to be 0 if x is attacked by some y in E.*
$\mathbf{f}_1(x)$ *is undefined otherwise.*
Let S_1 be the set of all x in S on which $\mathbf{f}_1$ is not defined. Let R_1 be R restricted to S_1. Let $\mathcal{A}_1 = (S_1, R_1)$.

We say that $\mathcal{A}_1$ is the residual, or the left-over from $\mathcal{A}$ after the grounded extension and its effects (i.e. the domain of $\mathbf{f}_1$) is taken out.

Example 18.29. Consider the network of Figure 18.24. Begin the procedure with ground extension. We get $\mathbf{f}_1(a) = 1, \mathbf{f}_1(b) = 0$ and the remaining network $\mathcal{A}_1$ of Figure 18.25.

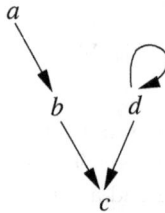

Fig. 18.24.

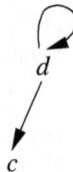

Fig. 18.25.

We can now continue and observe that in Figure 18.24, the node d is a top loop-node and we want to continue and annihilate d. So we follow the procedure we just about to define in Definition 18.30 below for $\mathcal{A}_1$ and get $d = 0, c = 1$. Thus the solution yields the function $\mathbf{f}_2(a) = 1, \mathbf{f}_2(b) = 0, \mathbf{f}_2(d) = 0, \mathbf{f}_2(c) = 1$.

The corresponding extension is $\{a, c\}$.

Definition 18.30 (General loop busting).

1. *We are going to define the following procedure to find* $\{0, 1\}$ *functions* $\mathbf{f}$ *on an argumentation network* $\mathcal{A}_0 = (S_0, R_0)$.

   Step 1.
   Choose a top loop-node a_1 *in* S_0, *introduce the annihilator* A_1 *for it and consider the annihilator network* (S_{0,A_1}, R_{0,A_1}).

   Step 2.
   Apply the procedure of Definition 18.28, to the annihilator network (S_{0,A_1}, R_{0,A_1}), *and get* $\mathbf{f}_1$ *and* $\mathcal{A}_1$.
   We say that all elements in $S_0 - S_1$ *have rank 1. These are the elements instantiated to numerical values at Step 1. Furthermore* $\mathbf{f}_1$ *is a* $\{0, 1\}$ *function. Also note that* $\mathcal{A}_1 = (S_1, R_1 = R_0 \cap S_1 \times S_1)$. *Note that the annihilator* A_1 *is thrown out anyway because it is in the grounded extension of the annihilator network* (S_{0,A_1}, R_{0,A_1}).

   Step 3.
   Choose a new top loop-node a_2 *of* $\mathcal{A}_1$.

   Step 4.
   Go to apply step 2 to $\mathcal{A}_1$, a_2 *and obtain* $\mathbf{f}_2$ *and* $\mathcal{A}_2$.
   Also identify the elements of $S_2 - S_1$ *as the elements of rank 2.*

   Step $n + 2$.
   Continue until you get $\mathcal{A}_{n+3} = \varnothing$.
   The function $\mathbf{f}_{n+2}$ *will be total on* S_0 *and will give you the extension*

   $$E(a_1, a_2, \ldots, a_{n+2}) = \{x | \mathbf{f}_{n+2}(x) = 1\}.$$

   All elements in the network have a clearly defined rank, it being the step in which they were instantiated to numerical value in $\{0, 1\}$.

2. *We define the semantics LB1, or the top loop-nodes semantics as the family of all the extensions of the form* $E(a_1, \ldots, a_{n+2})$, *for all possible choices of* a_i *allowable by the procedure.*

18.3.2 Comparison with CF2 semantics

We now show that our LB1 top loop-nodes semantics, if refined to a special version which we call the LB2 semantics, defined using certain parallel annihilators, (as in Definition 18.31 below), can be shown to be the same as CF2.

Definition 18.31 (Top loop-sets, LB semantics).

1. *Let* (S, R) *be an argumentation network. Move to the equivalence classes as defined in Definition 18.21. Choose a top equivalence class* C, *(as defined in item 4 of Definition 18.21). Let* M *be a maximal conflict free subset of* C. *Then the set* $B = (C - M)$ *is referred to as a top loop-set in* (S, R).

2. *To define the LB2 semantics (the top loop-sets semantics), we modify the process of Definition 18.30 as follows:*
   In step 1 we choose a top loop-set $B_1 = \{a_i\}$ in S_0, introduce the annihilators A_i for each a_i. We let $\mathbf{B}_1 = \{A_i\}$ and we consider the annihilator network $(S_{0,\mathbf{B}_1}, R_{0,\mathbf{B}_1})$. We proceed in step 2 as described.
   In step 3 we proceed as in step 1 with the network (S_1, R_1) and choose another top loop-set B_2 in S_1, introduce the respective annihilators set $\mathbf{B}_2$ and we consider the annihilator network $(S_{1,\mathbf{B}_2}, R_{1,\mathbf{B}_2})$.
   Step 4 is the same as step 2, for this network $(S_{1,\mathbf{B}_2}, R_{1,\mathbf{B}_2})$, and we carry on until we get to the empty network as in Definition 18.30. We end up with an extension denoted by $E(B_1, B_2, \ldots, B_{n+2}) = \{x | f_{n+2}(x) = 1\}$.

The following is a definition of CF2 extensions, as given in [30, Definition 4]. The notion of strongly connected set used by Baroni is the same as our notion of equivlance class as was defined in Definition 18.21.

Definition 18.32 (CF2 semantics). *Let $\mathcal{A} = (S, R)$ be an argumentation network and let $E \subseteq S$ be a set of arguments, then E is a CF2 extension of $\mathcal{A}$ iff*

1. *If (S, R) itself is a strongly connected set then E is a maximal conflict free subset of S.*
2. *Otherwise, for every C where C is a maximal strongly connected subset of S, we have that the set $C \cap E$ is a CF2 extension of the network (T_1^C, R_1), where*

$$T_1^C = C - \{x | \exists y \in E((y, x) \in R \wedge y \notin C\}$$
$$R_1 = R \cap (T_1^C \times T_1^C).$$

Theorem 18.33. *The semantics CF2 is the same as the LB2 semantics (obtained by using the process of Definition 18.30 modified by the requirement to choose top loop-sets at each stage, as in Definition 18.31).*

Proof.   1. We start by showing that every top loop-sets extension E of a network $\mathcal{A}_0 = (S_0, R_0)$ is also a CF2 extension as defined in Definition 18.32. To achieve this goal we need to follow closely how the extension E was defined in top loop-sets semantics for $\mathcal{A}_0$.
Let us list the way the top loop-sets extension E of $\mathcal{A}_0$ is defined.

   a) E is defined according to item 2 of Definition 18.30. The extension is obtained in the form $E = E(B_1, \ldots, B_{n+2})$, where each B_{i+1} is a top loop-set of the network $\mathcal{A}_i = (S_i, R_i)$.
   b) The top loop-sets was chosen according to the protocol of Definition 18.31.
   c) The elements of $S_i - S_{i+1}$ are of rank i, where i is the step in which they got a numerical value in $\{0, 1\}$ by the function f_i.
   The function f_{n+1} gives numerical values in $\{0, 1\}$ to all the elements of S_0 and we have
$$E = \{x \in S_0 | f_{n+1}(x) = 1\}.$$

We are now going to use the rank i to show that E is a CF2 extension according to Definition 18.32. We need a bit more preparation.

   d) Let $\mathcal{A}_0^* = (S_0^*, R_0^*)$ be the equivalence classes ordering without loop derived from (S_0, R_0) as in Definition 18.21. The element classes of S_0^* are all the maximal strongly connected subsets of S_0

We continue the proof by induction on $n + 2$, being the number of steps required to define E.

e) Case $n + 2 = 2(n = 0)$.

In this case we have that (S_0, R_0) itself is an equivalence class. Then E is obtained from B_1 which satisfies the CF2 condition as in Step 1 of Remark 18.30. Using the notation of Step 1, we have $E = \{x | \mathbf{f}_1(x) = 1\}$. In this case E is also a CF2 extension.

f) Case $n > 0$

Take any equivalence class subset C of (S_0, R_0). Let $k + 1$ be the step at which all elements of C get a numerical value. There are two possibilities:

i. At step k some maximal subset $C' \subseteq C$ does not yet have numerical values. C' is a top loop in $\mathcal{A}_k$. In this case B_k is an annihilator set C' and makes it get a numerical value in $\mathcal{A}_{k+1}$.

ii. C is not a top loop in $\mathcal{A}_k$, in which case B_k busts some other loops and in the process of obtaining $\mathcal{A}_{k+1}$, C' disappears as all these elements get a numerical value. In fact, in this case it is Step k which gives all elements of C a numerical value.

Let us now look at the set T_1^C, as defined in item 2 of Definition 18.32.

$$T_1^C = C - \{\exists y \in (E - C)((y, x) \in R)\}$$

The set T_1^C is comprised from two parts, $T_{1,k}^C$ and $T_{1,k+1}^C$. Part $T_{1,k}^C$ are all points $z \in T_1^C$ that get numerical value at step k by $\mathbf{f}_k$ and the set $T_{1,k+1}^C$ is the set of all points that get numerical values at step $k + 1$ by the function $\mathbf{f}_{k+1}$.

In case (i) above, $T_{1,k+1}^C$ is still the loop C', but still $T_{1,k}^C$ may be $\neq \varnothing$.

In case (ii), $T_{1,k+1}^C = \varnothing$. We ask is $E \cap T_1^C$ a CF2 extension of T_1^C according to Definition 18.32? The answer is yes. The part $T_{1,k}^C$ is calculated traditionally and if there is a loop $C' = T_{1,k+1}^C$, it will be busted by B_k which was chosen in top loop-sets semantics to yield a maximal conflict free set.

We thus see through considerations (a)–(e) that LD2 $\subseteq$ CF2.

The reader should see Remark 18.34, to appreciate the difference between the way top loop-sets semantics and CF2 calculate their extensions.

2. We now prove the other direction, namely that every CF2 extension is also a top loop-sets semantics extension.

Let E_0 be a CF2 extension of the network $\mathcal{A}_0 = (S_0, R_0)$. We would like to define a sequence of top loop-sets semantics and sets $B_1, B_2, \ldots, B_{n+2}$ such that $E_0 = E(B_1, \ldots, B_{n+2})$. We choose B_i by looking at E_0.

Step 1.

Look at the top loops of (S_0, R_0). Use E_0 to choose the set B_1.

Let us look at top strongly connected sets of (S_0, R_0). These are either single unattacked points x (in agreement with CF2) or a loop C, for which CF2 gives a choice of maximal conflict-free subset $E_0 \cap C$. We can now choose our set B_1 to be

$$B_1 = \bigcup_{\text{top loops } C \text{ of } \mathcal{A}_0} (C - E_0).$$

We now apply step 1 of the top loop-sets procedure and get $\mathcal{A}_1 = (S_1, R_1)$. Again consider the top loops of $\mathcal{A}_1$. Let

$$B_1 = \bigcup_{\text{top loops } C \text{ of } \mathcal{A}_1} (C - E_0)$$

We carry on in this manner and get

$$E(B_1, B_2, \ldots).$$

To show that $E_0 = E$ is not difficult. This is done along the lines of the proof of (1) above.

Remark 18.34.

1. The way we compute the extensions of top loop-sets semantics is not synchronised with the way CF2 itself works. This can be seen from the network of Figure 18.26. In this figure the top loop is $\{a, b, c\}$. By choosing the maximal conflict free set $\{c\}$, we are led to the set $B = \{a\}$.

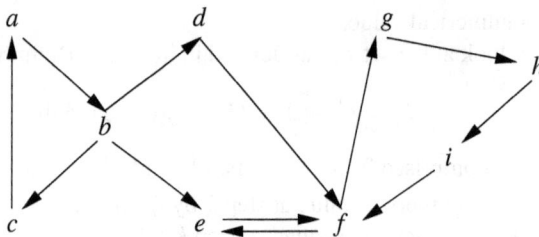

Fig. 18.26.

In step 1 we propagate the attacks to get "in" and "out", (1 and 0 respectively) numerical values for as many variables as we can.
We get in top loop-sets semantics

$$a = 0, c = 1, b = 0, d = 1, e = 1, f = 0, g = 1, h = 0 \text{ and } i = 1.$$

We get the extension in one step

$$E = \{c, d, e, g, i\} = E(\{b\}).$$

In comparison, when we follow the CF2 procedures, we look at two loops, the equivalence classes $\{a, b, c\}$ and $\{e, f, g, h, i\}$.
Step 1 of the LB2 procedure corresponds to what the CF2 definition does, namely treat the loop $\{a, b, c\}$. This we do by choosing $c = 1$ and calculating $a = 0, b = 0$ and $d = 1$.
We now look at the loop $\{e, f, g, h, i\}$ and take from it the elements attacked from outside it. In this case we take the element f attacked by d. Thus we are left with $S_1 = \{e, g, h, i\}$ and R_1 being $\{(g, h), (h, i)\}$. The CF2 extension for (S_1, R_1) is $\{e, g, i\}$. So the extension we get finally is $E = \{c, d, e, g, i\}$.

2. It is important for our discussion of CF2 in a later section, to highlight here the directionality in which the LB2 semantics works (LB2 gives the same extensions as CF2, and so imposes the directionality on CF2).

a) . Given a network (S, R), identify, in view of Definition 18.21, the top [strongly connected sets].

b) . Break these top loops by choosing a respective maximal conflict free subset of each.

c) Propagate the attacks from the chosen sets of item 2 on the rest of (S, R), and identify the residual left over network (in the sense of Definition 18.28. Call this network (S_1, R_1).

d) If the left over network of item 3 is non empty, then do a recursion, go to item 1. and repeat the process for this leftover network. If it is empty then stop (and we have got our LB2/CF2 extension).

e) The reader should compare with Gaggl and Woltran's impressive paper [192]. As far as I can see, there is no directionality there either.

Remark 18.35 (Comparison of LB1 top loop-node semantics with the CF2 semantics). Let us make a quick comparison.

1. The top loop-node give the correct exact extension for even loops. CF2 gives more extensions.

2. The top loop-node semantics does bust odd loops but gives less extensions than CF2.

3. Most importantly, and this has been pointed out to me by Martin Caminada, CF2 is not robust against conceptual extensions, such as joint attacks. This is shown in the next remark, 18.39.

Remark 18.36. We continue our discussion of Remark 18.26 and compare the LB1 semantics with CF2. Consider Figure 18.27. In this figure the CF2 extensions are the same as the traditional extensions, namely $\{b\}$ and $\{a, c\}$.

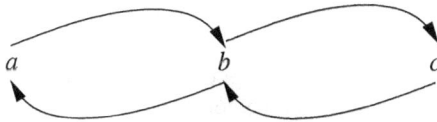

Fig. 18.27.

However, the annihilator method can get more extensions as follows:

1. Annihilate node b and get the extension $\{a, c\}$.

2. Annihilate node a first and get the secondary loop $\{b, c\}$. So further annihilation is required. If we now annihilate node b in the secondary loop we get the extension $\{c\}$ and if we annihilate node c we get the extension $\{b\}$.

3. Annihilate node c first and get the secondary loop $\{a, b\}$. Further annihilation of node a will yield the extension $\{b\}$ and the annihilation of node b will yield the extension $\{a\}$.

To summarise, the possible LB1 extensions in this case are $\{a\}, \{b\}, \{c\}, \{a, c\}$.

So, depending on the network CF2 might give the correct results while LB1 does not, and vice versa, LB1 might give the correct result (see Remark 18.26), while CF2 does not.

We might be able to rescue LB1 as well as CF2 by putting more restrictions on them. For example W. Dvorak and S. Gaggl [119] suggested to modify CF2 by requiring not just maximal conflict free sets E but such sets for which $E' = E \cup \{y|$ for some $x \in E, xRy\}$ is maximal. Such a restriction (called CF2-stage) will make CF2-stage give the correct results on even cycles.

LB1 and LB2 can restrict the choice of annihilators as well. LB2 can mirror the CF2-stage semantics just as easily as it mirrors CF2, and become LB2 stage. Similarly LB1 can also restrict the choice of single annihilators by putting geometrical restrictions on the choice of annihilated nodes. For example we can choose nodes which break as many (a maximal number of) cycles as possible. Adopting this restriction for LB1, for example, will force us to annihilate node b in Figure 18.27, yielding the extension $\{a, c\}$ but we will not be able to get the extension $\{b\}$ unless we annihilate in parallel both node a and node c. So we need to formulate a policy for choosing sequences of nodes to annihilate.

The next remark 18.37 is a sample proposal and a discussion for such a policy.

Going back to the network of Figure 18.27, note also that the extensions $\{a\}$ and $\{c\}$ do not comply with the heuristic suggested by Theorem 18.23, because the extension does not attack the annihilators (a does not attack c and c does not attack a).

Remark 18.37. The following is a proposal for a policy for generating sequences of nodes to annihilate. It needs to be theoretically investigated.

Given a complete loop (S, R), i.e. all the nodes in S are top loop-nodes, the problem is to identify which nodes are best to annihilate, for example, identify the odd cycles in (S, R) and annihilate nodes in the odd cycles, or identify nodes which can break a maximal number of cycles and annihilate them.

The brute force approach is to calculate all cycles in (S, R) and then make a choice of which nodes to annihilate. There are known algorithms from graph theory which output all such cycles. See, for example, [253]. Once we have identified these cycles we can also see which nodes appear in a maximal number of cycles.

The perceptive reader may not like this approach. He might say that we might as well calculate all extensions and then see how to break odd loops by annihilating which points. This in fact might even be computationally cheaper than outputting all cycles in (S, R). This my be true, but as a methodology it is faulty. We do not want to compute extensions in order to find extensions. The cycle generating algorithms have nothing to do with argumentation. They are part of graph theory. So methodologically, we are using tools from graph theory to solve argumentation problems.

So, for example, consider Figure 18.26. The graph cycle algorithm will output the cycles $\{a, b, c\}, \{e, f\}$ and $\{f, g, h, i\}$ and will also tell us that f is involved in a maximal number of cycles. We can see immediately that we need to annihilate a node in $\{a, b, c\}$ in order to break the odd cycle $\{a, b, c\}$.

If we decide to annihilate the node which can break the most loops, it would be node f. With $f = $ "out", we get $g = $ "in", $h = $ "out", $i = $ "in" and we are left with the residual left over Figure 18.28

Now we can deal further with this left-over figure and break the loop $\{a, b, c\}$.

Note that had we dealt with the loop $\{a, b, c\}$ first, we could have got an extension by using only one annihilator.

The different policies for sequencing annihilators will give us different semantics for argumentation networks. Let us illustrate this in the next example 18.38.

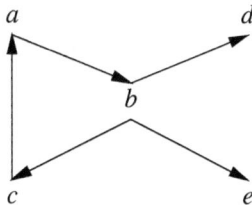

Fig. 18.28.

Example 18.38. This example illustrates and compares the various approaches for obtaining extensions by breaking loops. Consider Figure 18.29.

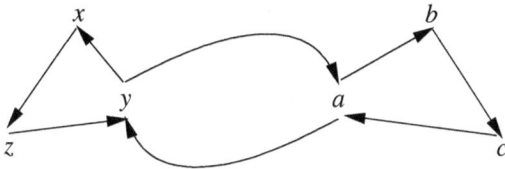

Fig. 18.29.

Let us compute the various extensions obtained by various policies for the network of this figure.

1. **CF2 (same as LB2) extensions**
   These are maximal conflict free sets. We get all pairs, one from each odd loop, except the pair $\{y, a\}$, namely:

   $$\{x, a\}, \{x, b\}, \{x, c\}, \{y, b\}, \{y, c\}, \{z, a\}, \{z, b\}, \{z, c\}.$$

2. **CF2-stage (same as LB2 stage) extensions**
   The extensions are the same as in (1).
3. **LB1 extensions**

   a) General loop busting policy, no restriction on the choice of annihilators. This is the policy of Defiition 18.30.
      We get the following extensions:
      i. $\{z, a\}$, obtained by annihilating x then annihilating c
      ii. $\{z, b\}$, annihilate first x and then a
      iii. $\{z, c\}$, annihilate x then b
      iv. $\{x, a\}$, annihilate y then c
      v. $\{x, b\}$, annihilate y then a
      vi. $\{x, c\}$ annihilate y then b
      vii. $\{y, b\}$, annihilate z. So y is "in", x is "out", a is "out", b is "in" and c is "out".

      To make it clear what happens here, when we annihilate z we get the network of Figure 18.30, which has the only stable extension $\{y, b\}$.

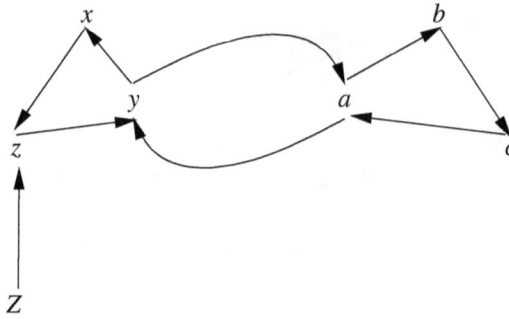

Fig. 18.30.

By symmetry, we can compute the further LB1 extensions:

viii. $\{c, y\}$, annihilate b then z

ix. $\{c, x\}$, annihilate b then y. This is the same as (vi)

x. $\{c, z\}$, annihilate b then x. This is the same as (iii)

   Similarly

xi. $\{b, y\}$, annihilate a then z. This is another way of getting (vii).

xii. $\{b, x\}$, annihilate a then y. Same as (v).

xiii. $\{b, z\}$, annihilate a then x. Same as (ii),

and the mirror of (vii) is (xiv):

xiv. $\{a, x\}$, annihilate c. This is another way of getting (iv).

Thus we get all the CF2 extensions in different ways.

The method of choice of annihilating is purely geometrical no argumentation considerations like "conflict freeness".

4. **Modified LB1 extensions**

Let us modify our choice policy as follows:

Step 1 Compute all geometric cycles using graph theory algorithms. We get $\{x, y, z\}$, $\{a, b, c\}$ and $\{y, a\}$.

Step 2 Annihilate only nodes that participate in maximal number of cycles.

   If the maximal number of cycles is 1 then choose nodes from odd cycles. If all cycles are even then proceed as in traditional argumentation.

According to this policy we can annihilate only either a or y initially and then go on to annihilate the remaining odd loop. So we get the extensions $\{b, x\}$, $\{b, y\}$, $\{b, z\}$ and $\{x, a\}$, $\{x, c\}$.

Remark 18.39 (CF2 and joint attacks).

1. In my paper [164] fibring argumentation frames, see Chapter 7, I introduced the notion of joint attack of say two arguments a and b on a third argument c. This means that c is out only when both a and b are in. I used the notation of Figure 18.31.

   We have that c is "out" (i.e. $c = 0$) exactly when both a and b are "in" (i.e. $a = b = 1$) namely

$$c = 1 - ab.$$

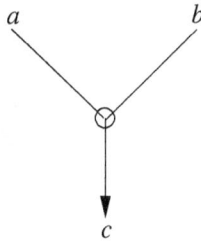

Fig. 18.31.

I also showed in the paper how to interpret joint attacks within ordinary argumentation networks, using for each node e the new auxiliary points $x(e)$ and $y(e)$. The joint attack of Figure 18.31 can be represented faithfully by Figure 18.32.

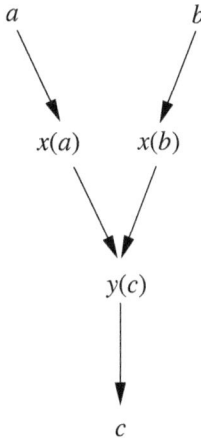

Fig. 18.32.

It is important to note that if we calculate 0 and 1 values (in and out) of Figure 18.32 we get

$$x(a) = 1 - a$$
$$x(b) = 1 - b$$
$$y(c) = ab$$
$$c = 1 - ab$$

It is clear that the value for c does not depend on the auxiliary points. Therefore any extensions of the enlarged network when restricted to the original network will not be affected by the auxiliary points.

2. Let us now take a loop involving joint attacks. Suppose we have 3 items, a, b and c and enough money to buy only two. Thus buying any two items attacks jointly the buying the third item. We get the loop of Figure 18.33.

The representation of this figure using the auxiliary point into an ordinary argumentation network is presented in Figure 18.34.

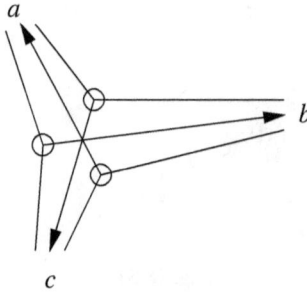

Fig. 18.33.

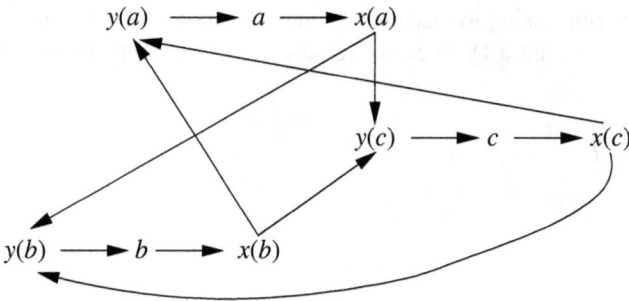

Fig. 18.34.

In this figure, the set $\{a, b, c\}$ is conflict free in the expanded network with the auxiliary points but it is not so in the original network, contrary to intuition. So CF2 messes up the concept of joint attack.

In comparison, the LB1 semantics is not affected by the auxiliary points as we have already seen from the equations. Not being affected means that if we start with a network (S, R) and move for whatever reason to an expanded network (S', R'), containing additional auxiliary points , then any extension E' obtained traditionally for (S', R') will endow a correct and acceptable extension $E = E' \cap S$ of (S, R), and furthermore all such correct extensions E are so obtained.

Let us find a set of top loop-nodes to break the loop for Figure 18.34. Note that this system is a 9-point loop.

A computational such set would be, for example, $\{y(a), y(b)\}$.

This gives $a = b = 1$ and $c = 0$. $\{y(a), y(b), y(c)\}$ is not minimal so we cannot get the extension $\{a, b, c\}$.

It is clear that our LB1 machinery works correctly here.

18.4 Impulse inputs approach to handling loops

In previous sections we identified top loop-node elements causing loops and introduced annihilators to deal with them; and we defined the semantics LB1. In this section we use the new impulse approach to deal with looping elements. The two methods are different but give the same results.

Remark 18.40. Consider the networks of Figures 18.20 and 18.21. The only difference between the two networks is the annihilator A attacking node a.

Similarly, if we look at the difference between their respective computation trees as represented in Figures 18.22 and 18.23, we see that the effect of the addition of the node

$?A = 1$, "success"

is to turn the root node

$?a = 0$

into a "success" node. So in Figure 18.22 the root node $?a = 0$ will be annotated "loop" and in Figure 18.23 the root node $?a = 0$ will be annotated "success".

If we ignore the process of going through the use of annihilators, what we have done in effect, is to take Figure 18.22 and introduce an *impulse input* operation which changed the annotation "loop" of the root node $?a = 0$ into the new annotation "success".

Of course, this kind of impulse input has to do with the definition of the annotation function V on a computation tree. So the proper formal way to define impulses and how to use them is to modify Definition 18.5. We can add to the groups of rules of Definition 18.5 a new impulse group of the following form:

(Group I): Introduce the following impulse input value to extend the domain of definition of the annotation function V (so far defined), namely $[\ldots$ DO THIS$\ldots]$.

Definition 18.41 (Ready for impulse). *Let* $(T, \rho, 0, \mathbf{f})$ *be an annotated computation tree over an argumentation network* (S, R). *A pair* $(t, \mathbf{f}(t)), t \in T$ *is said to be* ready for impulse *if the following holds:*

1. *$g(t) = (?x(t) = r)$.*
2. *There is no s in T such that s is higher in the tree than t (i.e. s is nearer the root 0) with $g(s) = g(t)$.*
3. *The argument $x(t)$ is a top loop-node in (S, R) as defined in Definition 18.19*

Definition 18.42 (Impulse procedure). *Let* $(T, \rho, 0, \mathbf{f})$ *be an annotated computation tree over an argumentation network* (S, R). *We modify the procedure for defining the annotation function V given in Definition 18.5 into a new procedure called impulse procedure as follows:*

1. *Add a new group of rules, (Group I) below*
   *(**Group I**)*
   Let $(t, \mathbf{f}(t))$ be a ready for impulse node as defined in Definition 18.41. Let $g(t) = (?x(t) = r)$. then make the new $V(t)$ to be "success" if $r = 0$ and "failure" if $r = 1$. This rule will be applied only when the existing value $V(t)$ is "loop". We do not change "failure" into "success" or "success" into "failure".[5]

---

[5] If we do that we will be doing abduction. We shall discuss this later. We have to be careful with the case where there is no value , because it might really be "success" or "failure" and we do not want to change it by executing an impulse action. So an impulse action really just annihilates a top loop-node from the network.

2. *Use (Group I) rules for one ready for impulse node after group rules (1)–(3) have been exhausted. Once you use Group I rule, start using again group rules (2) and (3) until there is no change in the annotations. You can iterate the process several times with several $(t_i, \mathbf{f}(t_i))$ until you wish to stop. You can then move on to use group 4 rules.*

Example 18.43. Consider the 6-loop of Figure 18.35

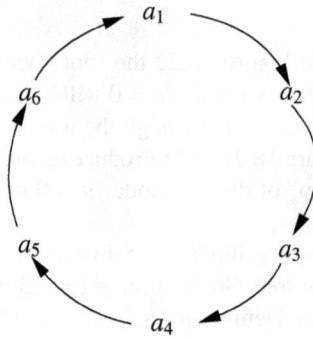

Fig. 18.35.

Consider the tree of Figure 18.36

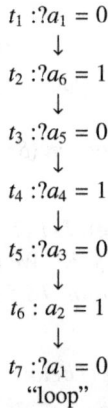

$$t_1 : ?a_1 = 0$$
$$\downarrow$$
$$t_2 : ?a_6 = 1$$
$$\downarrow$$
$$t_3 : ?a_5 = 0$$
$$\downarrow$$
$$t_4 : ?a_4 = 1$$
$$\downarrow$$
$$t_5 : ?a_3 = 0$$
$$\downarrow$$
$$t_6 : a_2 = 1$$
$$\downarrow$$
$$t_7 : ?a_1 = 0$$
$$\text{"loop"}$$

Fig. 18.36.

1. Apply impulse to $t_1 : ?a_1 = 0$, the root of the tree. We now execute group (2) and group (3) rules and get the annotations $V(t_i) = $ "success" for all $t_i, i = 1, \ldots, 7$. This gives us the extension $\{a_2, a_4, a_6\}$.

2. Alternatively, apply (Group I) rule to node t_4 and annotate $V(t_4)$ = " failure". Apply group rules (2) and (3) to get this new V and get $V(t_i)$ = "failure" for $i = 1,\ldots,7$. This gives us the extension $\{a_1, a_3, a_5\}$.

3. Note that we cannot get the CF2 extension $\{a_1, a_4\}$. To get this we need to apply annihilators to $\{a_2, a_3, a_5, a_6\}$.

   In (Group I) terminology, we want to send impulses *simultaneously* to $\{a_2, a_3, a_5, a_6\}$. Namely

   $V(t_2)$ = "fail"
   $V(t_3)$ = "success"
   $V(t_5)$ = "success"
   $V(t_6)$ = "fail".

   So we need to allow simultaneous impulses!

Example 18.44. Consider Figure 18.37.

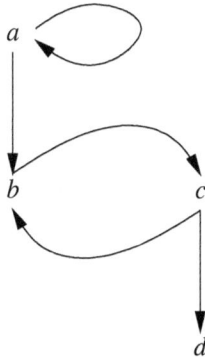

Fig. 18.37.

Let us construct the computation tree for the query $?d = 0$. we get Figure 18.38. In this figure node t_2 is not ready for impulse, because c is not a top loop-node. The node t_4, however, is ready for impulse. We can (execute an impulse and) annotate it with "failure". If we do that we can propagate the annotation and also annotate node t_7 with "failure" and node t_6 with "success".

After that, we are not able to continue and further propagate our annotations. So nodes t_2 and t_3 cannot be further annotated.

We get Figure 18.39.

We have already remarked that inputting the impulse to node t_4 amounts to (has the same effect as) moving to the annihilator network of Figure 18.40 and constructing the computation tree for that figure.

The tree for this network is Figure 18.41.

In Figure 18.41 the value "failure" propagates from node s to node t_4 and because of the loop (t_7 loops with t_4), the "failure" propagates to t_7 and from there to value "success" at t_6.

We get the same annotations in Figure 18.41 as in Figure 18.39, for the t-nodes. However after the first impulse, the loop at t_5 :$?c = 1$ is now ready for a further

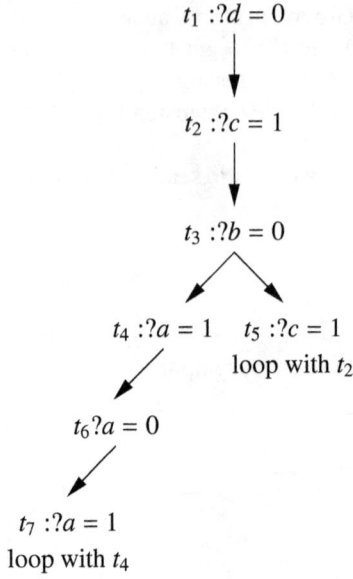

$t_1 :?d = 0$

$t_2 :?c = 1$

$t_3 :?b = 0$

$t_4 :?a = 1$ $t_5 :?c = 1$
                 loop with t_2

$t_6?a = 0$

$t_7 :?a = 1$
loop with t_4

Fig. 18.38.

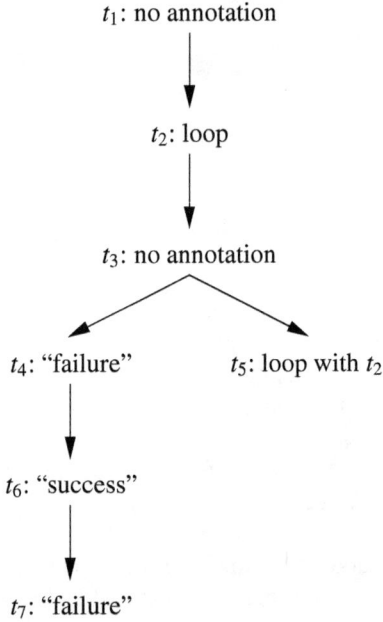

t_1: no annotation

t_2: loop

t_3: no annotation

t_4: "failure" t_5: loop with t_2

t_6: "success"

t_7: "failure"

Fig. 18.39.

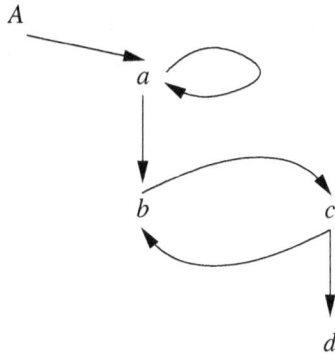

Fig. 18.40.

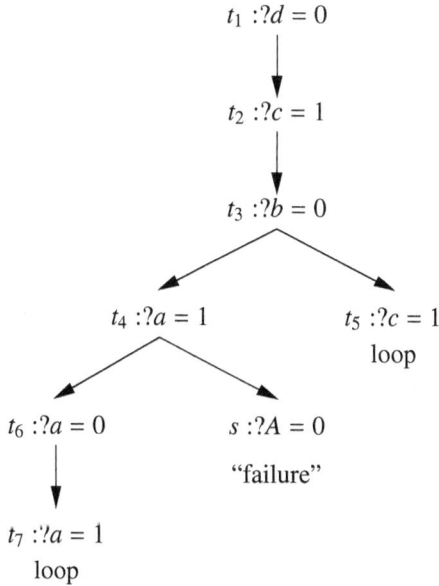

Fig. 18.41.

impulse because c is now a top loop-node in the network of Figure 18.40. We can see it clearly in this figure, but not so clearly in Figure 18.39 because the node s is not there.

So realy we need rules telling us how to apply a sequence of impulses correctly (without having to construct the corresponding annihilator figures). This is a methodological requirement, if not a complexity one.

My guess is that it may be quicker, to construct the annihilators figures and thus identify top loop-nodes, than to follow some algorithm on the construction tree (because all we need to do is to add single annihilator arrows to the original figure each time we add an impulse).

The reader might ask why bother with the requirement of top loop-node for being ready for impulse? Why not just impulse any loop? The problem is that this policy can give wrong results.

Consider Figure 18.42.

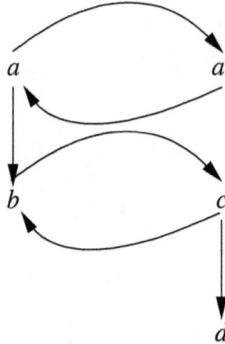

Fig. 18.42.

The construction tree we get for $?d = 0$ is Figure 18.43.

If we impulse node t_5 :$?c = 1$, we get the equivalent annihilator network of Figure 18.44. This allows for the extension $a' =$ "out", $a =$ "in", $b =$ "out", $c =$ "out", $d =$ "in" which is not correct as this network does have traditional extensions.

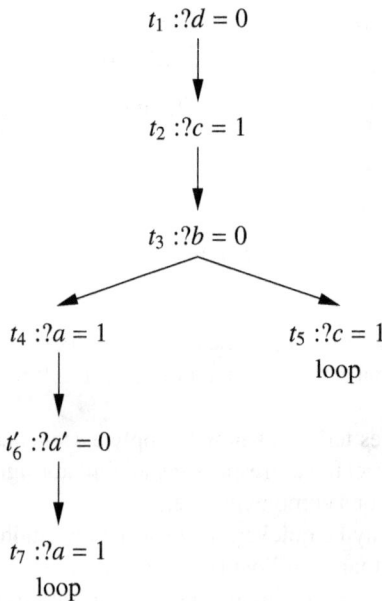

$$t_1 :?d = 0$$

$$t_2 :?c = 1$$

$$t_3 :?b = 0$$

$t_4 :?a = 1$ $t_5 :?c = 1$
                                     loop

$t_6' :?a' = 0$

$t_7 :?a = 1$
loop

Fig. 18.43.

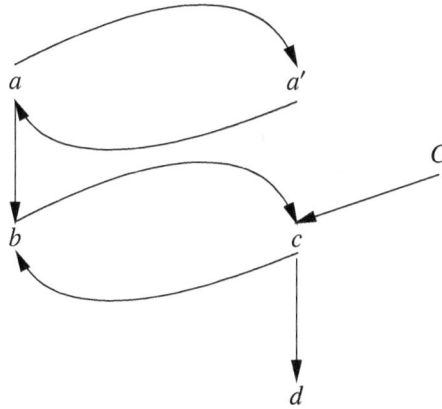

Fig. 18.44.

We might think that perhaps we can solve the problem of how to handle a sequence of impulses by using Definition 18.21 and ordering the looping elements according to the ordering in the equivalence classes network. We can then impulse the looping elements respecting this order.

In our case, the loop of element a will come before the loop of element c because of the ordering. This does not work, however, because after the impulse the ordering would change. Consider Figure 18.44 and let us add to this figure an attack arrow $d \to a'$. Now we have a network which is one big loop and we would have no ordering on the nodes. If we take out by impulse the node d, we are still left with a network with two looping elements, say $?c = 0$ and $?a = 0$ with no inherited ordering among them.

We shall address this problem in Section 5. Note that in Figure 18.39, if we delete the nodes which have a clear "success" or "failure" annotation (nodes t_4, t_6, t_7) we are left with a tree for the $?c = 1$ loop. This will be our strategy in Section 5.

Lemma 18.45. *Let (S, R) be an argumentation network and let $a \in S$ be a top loop-node. Let $x \in S$ and $r \in \{0, 1\}$ and let $\mathbb{T}_1$ be the construction tree for the query $?x = r$. Let A be the annihilator for a and let $\mathbb{T}_2$ be the construction tree for the same query $?x = r$, but this time for the network (S_A, R_A). Assume $\mathbb{T}_1$ and $\mathbb{T}_2$ have no node names in common. Further, assume that the nodes in $\mathbb{T}_2$, in which $?A = 0$ or $?A = 1$ occur, are $E = \{s_1, \ldots, s_m\}$. These are all endnodes of the tree $\mathbb{T}_2$, and are annotated by "success", if the query at the node is $?A = 1$ and "failure" if the query at the node is $?A = 0$. Let $\mathbb{T}_3$ be the tree obtained by deleting all the nodes of E from $\mathbb{T}_2$.*
Then $\mathbb{T}_1$ and $\mathbb{T}_3$ are isomorphic.

Proof. Clear from our discussions.

The impulse methodology is more powerful and general than the annihilator approach. We now illustrate this point in an abduction example.

Example 18.46. Consider the network of Figure 18.45. We want the extension $\{a, b\}$. I.e. we want a and b to be "in" and x and y to be "out".

The only way to achieve that is a higher order annihilator, as in Figure 18.46.

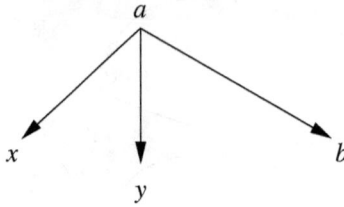

Fig. 18.45.

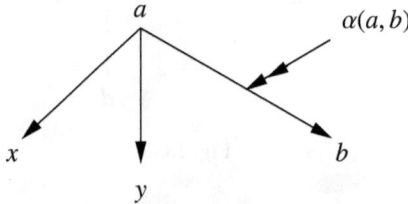

Fig. 18.46.

The annihilator $\alpha(a, b)$ attacks the connection attack arrow $a \to b$. This higher order attack is described in the figure by a double arrow

$$\alpha(a, b) \twoheadrightarrow (a \to b)$$

Let us see what kind of impulse input we need to model this type of attack. See Figure 18.47, it being a computation tree for Figure 18.45. The node t is already annotated

$$t : ?b = 1$$
"failure"

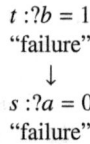

$$s : ?a = 0$$
"failure"

Fig. 18.47.

with "failure". Our impulse rules do not allow us to change the annotation to success. If we introduce a new group (GA group) which allow us to overturn "success" to "failure" and "failure" to "success", then we can do it! This would be abduction.

Example 18.47 (Abduction example). Consider the knowledge base:

1. If you have a cold you cough a bit.
   We observe fact 2:
2. You cough a bit.
   We want to explain our observation. We ask the database

? Cough

↓

? Have a cold

"failure" (because have a cold is not in the database)

Abduction allows us to *abduce* the fact:
3. You have a cold.

We can form the following network, see Figure 18.48, which represents data item 1 (which we read as the contrapositive ¬cough → ¬cold).

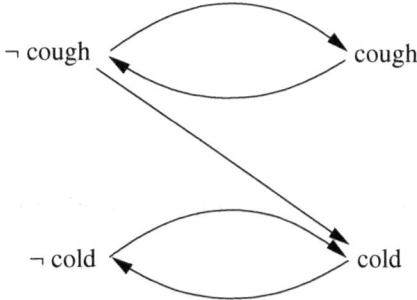

Fig. 18.48.

The fact that we observe cough is represented by an annihilator. This is an annihilator ¬COUGH to ¬cough.

We get Figure 18.49

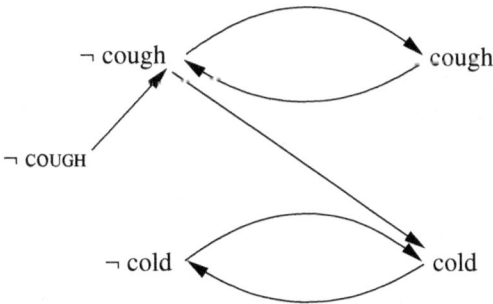

Fig. 18.49.

Let us now ask ?cold=1. See Figure 18.50.

If we propagate the annotations we get that cough is successful but we still have the loop with {cold, ¬cold}. If we want to do the abduction we annihilate ¬cold. If we do not want to abduce we do not annihilate.

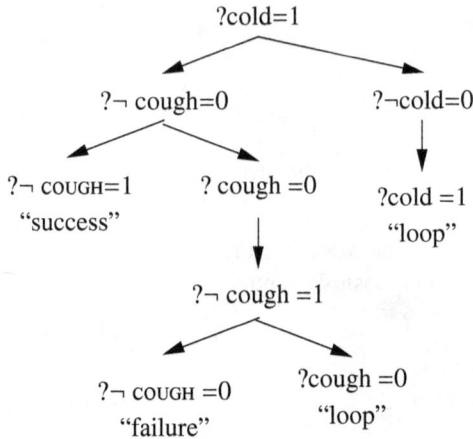

$$?cold=1$$

$$?\neg\, cough=0 \qquad\qquad ?\neg cold=0$$

$$?\neg\, \text{COUGH}=1 \qquad ?\,cough =0 \qquad ?cold =1$$
$$\text{``success''} \qquad\qquad\qquad\qquad \text{``loop''}$$

$$?\neg\, cough =1$$

$$?\neg\, \text{COUGH} =0 \qquad ?cough =0$$
$$\text{``failure''} \qquad\qquad \text{``loop''}$$

Fig. 18.50.

18.5 Linear logic like (LLL) approach to handling loops

We now introduce a diminishing resource linear logic like approach for dealing with looping elements. This approach is more general than the annihilator or the impulse approaches. It is more efficient and can be generalised to other looping algorithms in other areas, for example Logic Programming.

Consider again Figure 18.22. This is a computation tree for the query $?a = 0$, following the network of Figure 18.20. In this tree we stopped developing the tree at points where we got a loop. These nodes are the end nodes of the tree. We ask ourselves what happens if we do not stop? Well, the obvious will happen, namely we will keep on looping and looping. Now suppose we say that we can use each node of the argumentation network (in this case Figure 18.20) at most twice. What will happen?

First note that if the procedure for constructing the computation tree requires us to use a node three times then we have a loop, because we have only two possible questions involving a node x, $?x = 1$ and $?x = 0$. Thus we can check for loops with x by restricting the use of the node x.

We might get a slightly bigger computation tree, but in principle the properties of the bigger tree are the same as the shorter tree. The following may happen: we start with $?x = r$ and loop along the path in the tree and ask again $?x = r$. The procedure for constructing trees in section 2 will stop at this point. The linear logic procedure will continue and repeat another cycle. What are the advantages of generating the construction tree and identifying the looping point using the linear logic method?

We list several:

1. We need not record history in the tree in order to identify loop points, all we need to do is to just let a diminishing resource program run its course.
2. Once we stop and choose a looping endpoint in the tree, say $t : ?x = r$, which we want to deal with, we have more options of how to proceed. We can either annihilate the point or we can decide the point must be "in". Note that according to Sections 2–4, our only option is to annihilate the point x. If we do that, we will

be operating in the annihilator network of Sections 2–4 and may need to construct trees again. We certainly need to spend effort on propagating annotations V. In the diminishing resource approach for the case of annihilating x, we can carry on with the tree with all other nodes $s :?y = r'$, for all y different from x, by giving them more resources to continue while bearing in mind that if the query $?x = r$ occurs under the computation of the query $?y = r'$, we do know the value. So we do not need the annotation function V, only a simple list of what to bear in mind! Furthermore, under the diminishing resource approach we may decide to annotate the looping $?x = r$ as success, and carry on with the rest of the nodes $?y = r'$. Of course in this case we need a soundness theorem to characterise what this choice means (are we annihilating all the attackers of x or are we annihilating the attack arrows only?).

3. The approach itself is new. It allows us to deal with arguments which can be used only a certain number of times. This is a new area which we may call the area of resource argumentation theory. This approach is not just a tool for handling loops. There are variations on this approach, we can say let us use some of the nodes, say only nodes $y_1, \ldots, y_n$, only $k_1, \ldots, k_n$ times, respectively.[6]

4. The approach can be applied in other areas, for example to Logic Programming.

18.5.1 Motivating examples

The next example illustrates the use of diminishing resource for a 2-loop. We could use the network of Figure 18.20, but the tree will be too big for an illustration.

Example 18.48. Consider the two loop of Figure 18.7. It has the computation tree of Figure 18.8. The tree stops because of a loop. According to what we have learnt so far we have two approaches to such loops: we can either introduce an annihilator $A \to a$ and get the extension $\{b\}$ or introduce an impulse action annotating the first node $?a = 1$ as "failure" and again get the extension $\{b\}$.

Similarly we can do this to node b and get the extension $\{a\}$. Figures 18.51–18.52 illustrate these options for node a. We now use the approach where we ignore the loop and just carry on with the computation one more round. We get Figure 18.53. Our starting query is $?a = r, r \in \{0, 1\}$. This figure has information annotated on the right hand side of nodes taking the view of us "walking" along the attacking arcs in Figure 18.7.

So we start at node a and ask $t_1 :?a = r$. We "walk" to node b along the attack arc $a \leftarrow b$ and ask $t_2 :?b = (1 - r)$. We "walk" back along the attack arc $b \leftarrow a$ and ask $t_3 :?a = r$.

The node a says to us, "Hello, you have been here before". We "walk" back to b along the attack arc $b \leftarrow a$ and ask $t_4 :?b = (1 - r)$. The node b says to us "Hello, you have been here before". We "walk" back along the attack arc $a \leftarrow b$ and ask $t_5 :?a = r$. The node a says to us "Hello again. This is the third time you came here. Two times is is all you get, get lost!". So the query $t_5 :?a = r$ has no access. We ask is "no access" to be considered a "failure" or is it to be considered a "success"?

Let us get a clue for what to do from the basic network of Figure 18.54.

---

[6] There are intuitive reasons for using an argument a limited number of time. People get fed up of hearing it and it loses its potency. This has nothing to do with the technical use of dealing with loops in abstract argumentation.

(i)

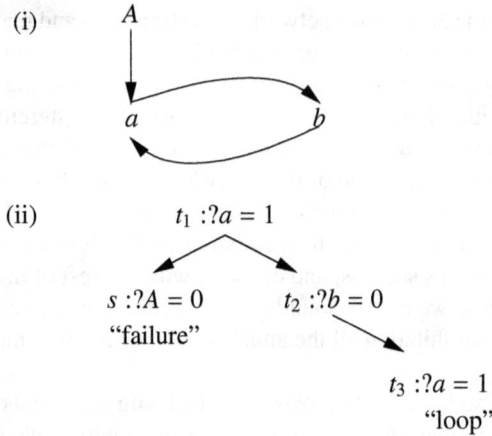

(ii)

$t_1 : ?a = 1$

$s : ?A = 0$
"failure"

$t_2 : ?b = 0$

$t_3 : ?a = 1$
"loop"

(i) is the annihilator figure
(ii) is the computation tree for (i) with initital annotation.

The final annotation for the tree is "failure" at all points.

Fig. 18.51.

$t_1 : ?a = 1$ ⟸   Impulse action
annotate "failure"

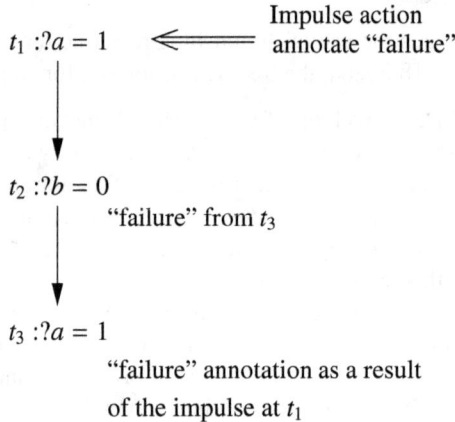

$t_2 : ?b = 0$
"failure" from t_3

$t_3 : ?a = 1$
"failure" annotation as a result
of the impulse at t_1

Fig. 18.52.

$a_1, \ldots, a_k$ are all the attackers of b in a network (S, R). If we delete all of a_i from the network then b will be "in", because it will be unattacked. So if we ask $?b = 1$ and go to each a_i, we will need to ask $a_i = 0$. If a_i give no access, then we have two options, the skeptical and the credulous options. The credulous option will say no access to a_i is as if a_i are not there. So $?a_i = 0$ must be a success, because b has to end up "in". The skeptical option will say let us assume that at least one of the a_i is there and so $?a_i = 0$ must be a failure because b has to end up "out".

We therefore adopt the following two policies for handling "no access", the skeptical and the credulous:

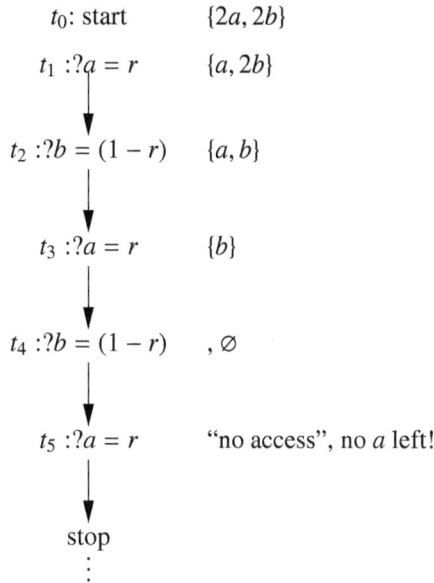

Fig. 18.53.

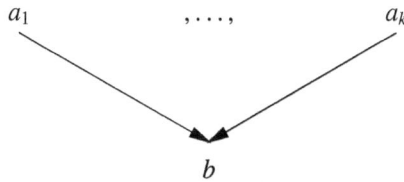

Fig. 18.54.

For the credulous case, if $r = 1$, i.e. we want to ask $?a = 1$ and we get no access, we assign (report) "failure" (of the query $?a = 1$) and so a is "out". If $r = 0$, i.e. we want to ask $?a = 0$, and we have no access, we report "success" (of the query $?a = 0$) and so again a is "out". This approach corresponds to annihilating top loop-nodes in the argumentation network.

For the skeptical option we do the opposite, if $r = 1$ we replace "no access" by "success" and if $r = 0$ we replace "no access" by "failure". This corresponds to deciding that the top loop-node should be "in".

Note that for the credulous approach we have relative soundness, we know what it does to the original network; it moves us to the corresponding annihilator network. For the skeptical approach we do not know yet what it does to the original network.

Let us assume we started with $r = 1$, i.e. we asked $t_1 :?a = 1$. What computation tree do we get?

We get Figure 18.55

Note that this is a similar annotation as what we get from the annotation we give for loops in Section 3, except that now we have a straightforward program to do it. If we propagate the annotation, say in the credulous case, we get that $?b = 0$ is annotated

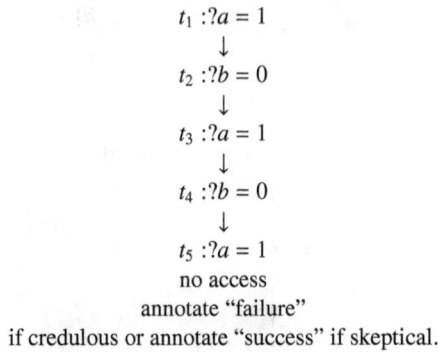

$$t_1 : ?a = 1$$
$$\downarrow$$
$$t_2 : ?b = 0$$
$$\downarrow$$
$$t_3 : ?a = 1$$
$$\downarrow$$
$$t_4 : ?b = 0$$
$$\downarrow$$
$$t_5 : ?a = 1$$
no access
annotate "failure"
if credulous or annotate "success" if skeptical.

Fig. 18.55.

"failure" and we get the extension a = "out", b = "in". By symmetry if we start by asking $?b = 1$ we get the extension a = "in", b = "out".

Definition 18.49 (LLL argumentation networks).

1. *Let S be a set of arguments. We define a multiset $\mathbb{S}$ based on S, as a multiset containing elements of the form !x or nx where n is a natural number. !x in $\mathbb{S}$ means that $\mathbb{S}$ contains an infinite number of copies of x and nx in $\mathbb{S}$ means that $\mathbb{S}$ contains n copies of x.*
2. *Let $\mathbb{S}$ be a multiset. Let $nx \in \mathbb{S}, n > 0$. Then $\mathbb{S} - \{x\}$ is obtained by replacing nx in $\mathbb{S}$ by $(n - 1)x$. Note that if $!x \in \mathbb{S}$ then $\mathbb{S} - \{x\} = \mathbb{S}$.*
3. *A linear logic argumentation network (LLL) has the form $(S, R, \mathbb{S})$ where (S, R) is an argumentation network and $\mathbb{S}$ is a multiset based on S. Let $(S, R, \mathbb{S})$ be a network and let $x \in S$. We define the computation rules for the query:*

$$?x = r, r \in \{0, 1\}, \text{ from } \mathbb{S}.$$

   a) *$?x = r$ is "success" (respectively "failure") from $\mathbb{S}$ if $r = 1$ (resp. $r = 0$) and there are no attackers of x in $\mathbb{S}$. (Note the meaning of "no attackers in $\mathbb{S}$". There may be an attacker y of x in (S, R) but in $\mathbb{S}$ we have 0y. So we have no access to y. This definition is what we called the credulous approach. For the skeptical approach we swap "success" and "failure").*
   So to be crystal clear about how this works, suppose we ask $?x = 1$ and y is the only attacker of x. Then we should ask $?y = 0$, and this should succeed. But if we have no access to y, i.e. 0y is in $\mathbb{S}$, then the query $?y = 0$ is a success, i.e. y is "out", and so $?x = 1$ is also a success. This is the credulous view. But 0y in $\mathbb{S}$ means $?y$ is in a loop, and so the credulous approach annihilates looping elements. By mirror symmetry the skeptical approach insists on looping elements being "in".
   b) *$?x = r$ is "success" from $\mathbb{S}$ iff for all (resp. some) attackers y of x we have $?y = (1 - r)$ is "success" (resp. "failure") from $\mathbb{S} - \{x\}$.*
   c) *$?x = r$ is "success" from $\mathbb{S}$ iff $?x = 1 - r$ is "failure" from $\mathbb{S}$.*

Definition 18.50 (LLL computation tree).

1. *Let $(S, R, \mathbb{S})$ be a linear logic argumentation network as defined in item 3 of Definition 18.49. Assume that $\mathbb{S} = \{2x | x \in S\} = 2S$. This assumption is all we need to define the notion of a computation tree for linear logic argumentation networks. It does not restrict the generality of the notion of a computation tree!*

2. *Let $(S, R, 2S)$ be a linear argumentation network. Let $(T, \rho, 0)$ be a tree and let $?x(0) = r(0), x(0) \in S$ and $r(0) \in \{0, 1\}$ be a goal. A function $\mathbf{f}$ makes $(T, \rho, 0, \mathbf{f})$ into a computation tree for the goal $?x(0) = r(0)$ from $(S, R, 2S)$ if the following holds:*

   a) *$\mathbf{f}(0) = 2S\,?x(0) = r(0)$*

   b) *For any $t \in T$ such that $\mathbf{f}(t) = \mathbb{S}_t ?x(t)?r(t)$, and such that $x(t) \in \mathbb{S}_t$ let $y_1, \ldots, y_k$ be all the elements in S such that $y_i R x(t)$ hold. Then the predecessors of t in S can be written as $s_1, \ldots, s_k$ (same k as above) and can be correlated with $y_1, \ldots, y_k$ in such a way that*

   $$\mathbf{f}(s_i) = (\mathbb{S}_t - \{x(t)\}?y_i = 1 - r(t).$$

   c) *If t is an endpoint in the tree T (i.e. no predecessors) then either (i) or (ii) hold:*

      i. *$x(t)$ is not a member of $\mathbb{S}_t$*

      ii. *$x(t)$ has no attackers in (S, R), i.e. $(\neg \exists z) z R x(t)$.*

3. *Let $(T, \rho, 0, \mathbf{f})$ be a computation tree. A function V is a partial annotation for the computation tree if the following holds:*

   a) *If t is an endpoint where $x(t) \notin \mathbb{S}_t$ then $V(t) = $ "no access". Furthermore, for the credulous (resp. skeptical) approach we add to the annotation $V(t)$ "failure" (resp. "success") if $r(t) = 1$ and we add to $V(t)$ "success" (resp. "failure") if $r(t) = 0$.*

   b) *If t is an endpoint and $x(t) \in \mathbb{S}_t$, then if $r(t) = 0$, we have $V(t) = $ "success" and if $r(t) = 1$, we have $V(t) = $ "failure".*

   c) *Let s be any point in the tree and let t be any endpoint. Then if $x(s) = x(t)$ then $V(s)$ is defined and its value is as follows:*
   Let $V(s) = $ "success", if $V(t) = $ "success" and $r(t) = r(s)$ or if $V(t) = $ "failure" and $r(t) = 1 - r(s)$.
   Let $V(s) = $ "failure", if $V(t) = $ "failure" and $r(t) = r(s)$ or if $V(t) = $ "success" and $r(t) = 1 - r(s)$.

Remark 18.51 (LLL annotated computation tree). Let $(S, R, 2S)$ be a linear argumentation network and let $x(0) \in S$ and let $?x(0) = r(0)$ be a goal. We can inductively construct an annotated computation tree $(T, \rho, 0, \mathbf{f}, V)$ for this goal which is compatible with Definition 18.50 as follows:

We construct the tree in steps. In step n we have the tree $(T_n, \rho_n, 0, \mathbf{f}_n, V_n)$.

Step 0.   *Let $T_0 = \{0\}$. $\rho_0 = \emptyset$, $\mathbf{f}_0(0) = 2S\,?x(0) = r(0)$, $V_0 = \emptyset$*

Step $n + 1$.   *Let $t_1, \ldots, t_m$ be all endpoints of T_n*

Let $t \in \{t_i | i = 1, \ldots, m\}$ be any such an endpoint. Assume $\mathbf{f}(t) = \mathbb{S}_t ?x(t) = r(t)$. We distinguish seveal cases for t and in each case we might add new point to the tree T_n and thus form the new tree T_{n+1}.

Case 1.   *$x(t) \in \mathbb{S}_t$ and $x(t)$ has no attackers in (S, R). In this case we do nothing.*

Case 2. $x(t)$ is not in $\mathbb{S}_t$, then do nothing.

Case 3. $x(t) \in \mathbb{S}_t$ and t has the attackers $y_1^{(t)}, \ldots, y_{k(t)}^{(t)}$ in S. In this case we add to T_n the new points $s_1(t), \ldots, s(t)_{k(t)}$ (i.e. the points $s_j(t)$ are all different for different ts and different from each other and completely new to T_n).

Let $\rho_{n+1} = \rho_n \cup \{(s_j(t), t)|x(t)$ is in $\mathbb{S}_t$ and $x(t)$ has attackers in S and $j = 1, \ldots, k(t)\}$.

$T_{n+1} = T_n \cup \{s_j(t)|x(t)$ is in $\mathbb{S}_t$ and $x(t)$ has attackers in S and $j = 1, \ldots, k(t)\}$.

Let $\mathbb{S}_{s_{j(t)}} = \mathbb{S}_t - \{x(t)\}$.

Let $x(s_j(t)) = y_j^t, j = 1, \ldots, k(t)$.

Extend $\mathbf{f}_n$ to be $\mathbf{f}_{n+1}$ by defining $\mathbf{f}_{n+1}$ on the new points $s_j(t)$ by letting

$$\mathbf{f}_{n+1}(s_j(t)) = \mathbb{S}_{s_{j(t)}}?x(s_j(t)) = 1 - r(t).$$

Let $V_{n+1} = V_n$.

Since at each step the multisets $\mathbb{S}_t$, for endpoints t, become smaller, then at some final $n = N$, only cases 1 and 2 would apply and the process would stop.

Step final N. At this stage we define V_N using the clauses of item 3 of Definition 18.50. We first define V_N on the endpoints of T_N as in item 3(a) and 3(b) and then propagate it to some other internal points as suggested by item 3(c).

Example 18.52. Let us consider again the network of Figure 18.11, and use the diminishing resource linear logic approach. We adopt the policy where all elements of the network can be used at most twice. The computation tree for this network is in Figure 18.56. Compare it with Figure 18.12. We also adopt the credulous approach. We use the diminishing resource for all the nodes of S in parallel.

If we propagate the annotations up the nodes of Figure 18.56 using group rules (2)–(3) of Definition 18.5, we get the annotations in Figure 18.57.

We can read the extension we get from Figure 18.57: $b = $ "out", $c = $ "in", $a = $ "out", $d = $ "in".

Note that the annotation in Figure 18.57 is consistent. If $x = r$ is annotated "success" (resp. "failure") then $x = 1 - r$ is annotated "failure", (resp. "success').

The reader might think now that perhaps the right loop checking policy for a network (S, R) is to do a diminishing resource computation for $\mathbb{S} = \{2x \mid x \in S\}$. However, the next few examples show that this policy is not sound. The reason being the parallel nature of the policy. We shall use the examples to get ideas on how to modify the policy.

Example 18.53. Consider Figure 18.58

The construction tree is for the query $?b = 1$ is Figure 18.59, where we adopt the credulous approach to no-access. Figure 18.60 gives the annotations for the nodes for both the credulous and skeptical approaches.

The credulous annotation is the top annotation and below it is the skeptical annotation. As you can case both give the same extension $b = $ "in", $a = $ "out".

Example 18.54. Let us check the 3-cycle of Figure 18.5. We use Figure 18.61 and compare with Figure 18.6.

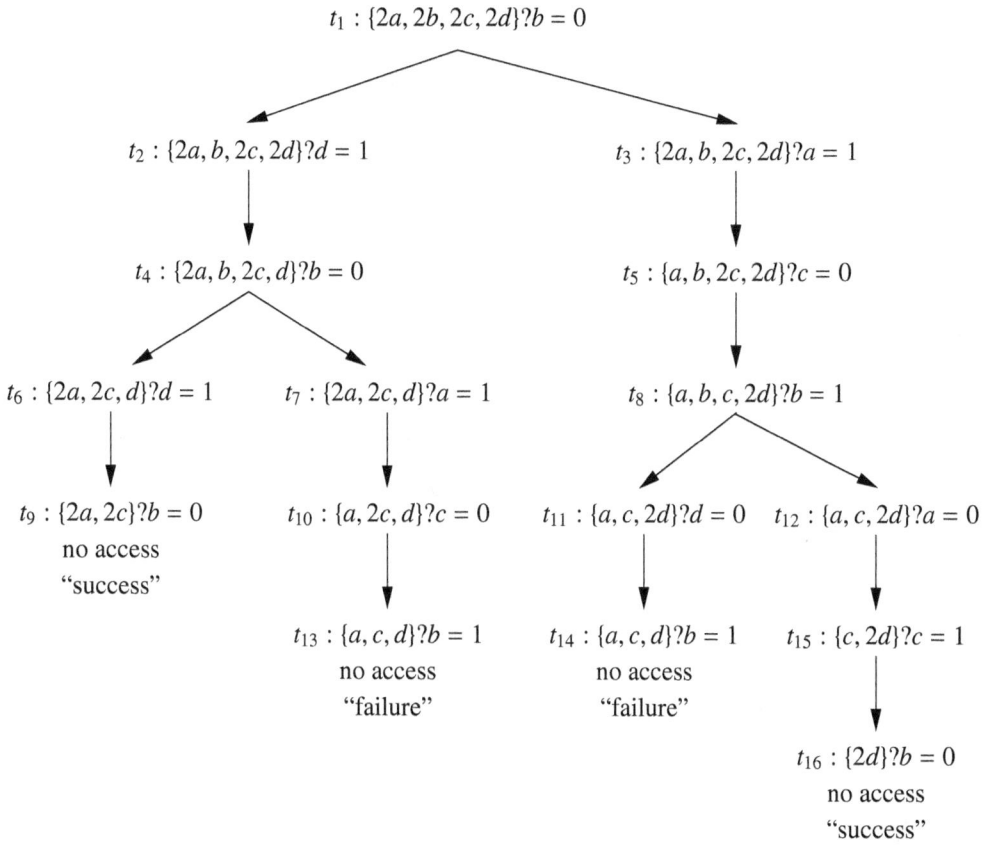

$$t_1 : \{2a, 2b, 2c, 2d\}?b = 0$$

$$t_2 : \{2a, b, 2c, 2d\}?d = 1 \qquad\qquad t_3 : \{2a, b, 2c, 2d\}?a = 1$$

$$t_4 : \{2a, b, 2c, d\}?b = 0 \qquad\qquad t_5 : \{a, b, 2c, 2d\}?c = 0$$

$$t_6 : \{2a, 2c, d\}?d = 1 \qquad t_7 : \{2a, 2c, d\}?a = 1 \qquad t_8 : \{a, b, c, 2d\}?b = 1$$

$$t_9 : \{2a, 2c\}?b = 0 \qquad t_{10} : \{a, 2c, d\}?c = 0 \qquad t_{11} : \{a, c, 2d\}?d = 0 \quad t_{12} : \{a, c, 2d\}?a = 0$$

t_9: no access "success"

$$t_{13} : \{a, c, d\}?b = 1 \qquad t_{14} : \{a, c, d\}?b = 1 \qquad t_{15} : \{c, 2d\}?c = 1$$

t_{13}: no access "failure"

t_{14}: no access "failure"

$$t_{16} : \{2d\}?b = 0$$

t_{16}: no access "success"

Fig. 18.56.

Figure 18.62 shows the propagation of annotations in Figure 18.61, using the credulous approach. t_7 gets "success", (with the skeptical approach t_7 gets "failure").

The annotation is not consistent. t_4 gives $a = 1$ "success" while t_7 gives $a = 0$ "success". We can ignore this and simply take the value at t_1 as the value of a. Thus a = "out" and by symmetry we get also $b = c$ = "out". The skeptical approach will have all of $a = b = c$ = "in".

This means the entire odd loop is thrown out by the credulous approach or kept in by the skeptical approach. Another possible interpretation is to say that the diminishing resource linear logic approach does not bust odd loops. This is fine. We can still ask does the approach give us all existing extensions? Let us look at more examples before we decide on what view to take.

We need, however, to clarify how to interpret the result of the skeptical approach which says $a = b = c$ = "in". How can this be? We can understand this as a joint "in", meaning they can only attack jointly. See Figure 18.63

Compare with Example 18.55 below.

Example 18.55. Consider the network in Figure 18.64.

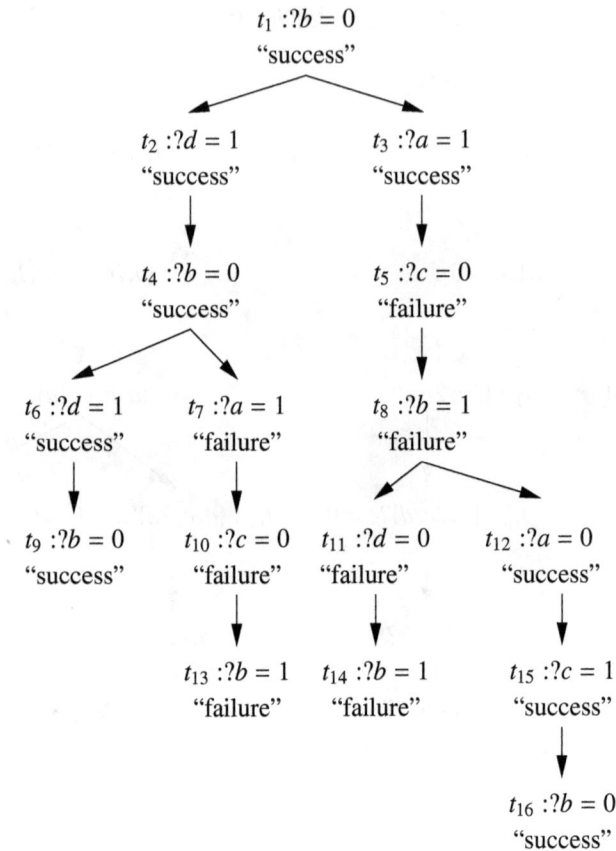

$t_1 : ?b = 0$
"success"

$t_2 : ?d = 1$
"success"

$t_3 : ?a = 1$
"success"

$t_4 : ?b = 0$
"success"

$t_5 : ?c = 0$
"failure"

$t_6 : ?d = 1$
"success"

$t_7 : ?a = 1$
"failure"

$t_8 : ?b = 1$
"failure"

$t_9 : ?b = 0$
"success"

$t_{10} : ?c = 0$
"failure"

$t_{11} : ?d = 0$
"failure"

$t_{12} : ?a = 0$
"success"

$t_{13} : ?b = 1$
"failure"

$t_{14} : ?b = 1$
"failure"

$t_{15} : ?c = 1$
"success"

$t_{16} : ?b = 0$
"success"

Fig. 18.57.

Fig. 18.58.

In this figure, d is a floating point "out". Figure 18.65 shows the computation for $?d = 0$.

The credulous approach will give the extension $a = b = c =$ "out", $d =$ "in".

The skeptical approach will give the extension $a = b = c =$ "in", $d =$ "out". This result interprets Figure 18.64 as being Figure 18.63.

Before we congratulate ourselves on the successful treatment of the floating node, let us look at the next example where a two-cycle attacks a third node. We get the wrong answers!

Example 18.56. Consider the network of Figure 18.66.

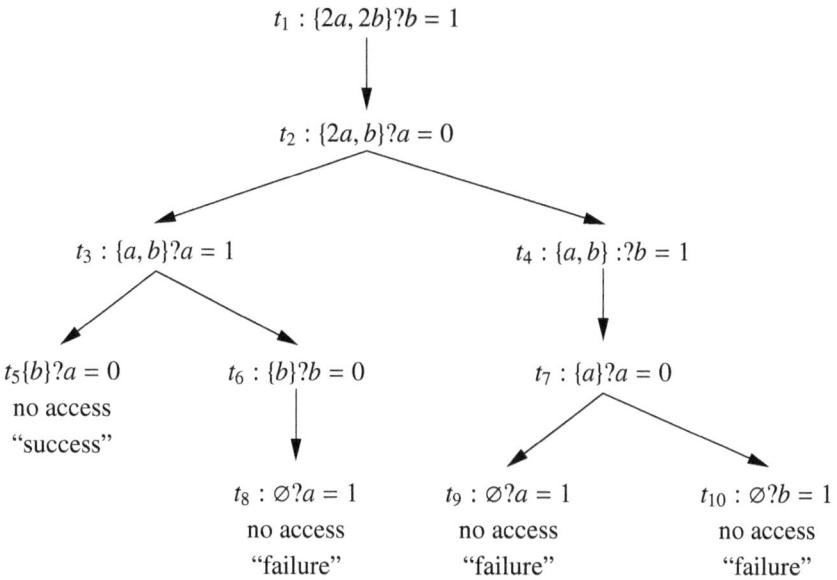

$$t_1 : \{2a, 2b\}?b = 1$$

$$t_2 : \{2a, b\}?a = 0$$

$$t_3 : \{a, b\}?a = 1 \qquad\qquad t_4 : \{a, b\} :?b = 1$$

$$t_5\{b\}?a = 0 \qquad t_6 : \{b\}?b = 0 \qquad\qquad t_7 : \{a\}?a = 0$$
no access
"success"

$$t_8 : \varnothing?a = 1 \qquad t_9 : \varnothing?a = 1 \qquad t_{10} : \varnothing?b = 1$$
no access   no access   no access
"failure"   "failure"   "failure"

Fig. 18.59.

We ask the goal $?z = 0$. We get the tree of Figure 18.67.

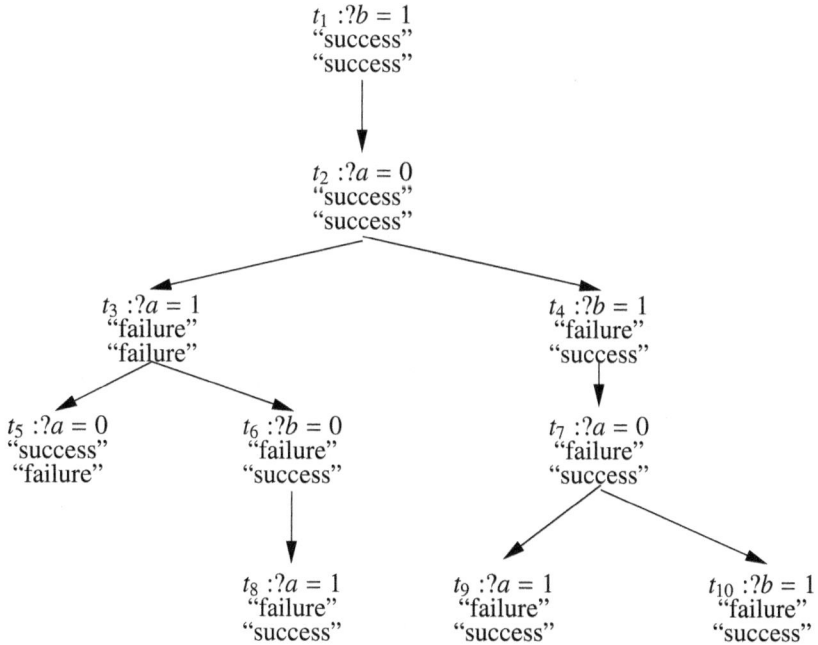

$$t_1 :?b = 1$$
"success"
"success"

$$t_2 :?a = 0$$
"success"
"success"

$$t_3 :?a = 1 \qquad\qquad t_4 :?b = 1$$
"failure"   "failure"
"failure"   "success"

$$t_5 :?a = 0 \qquad t_6 :?b = 0 \qquad t_7 :?a = 0$$
"success"   "failure"   "failure"
"failure"   "success"   "success"

$$t_8 :?a = 1 \qquad t_9 :?a = 1 \qquad t_{10} :?b = 1$$
"failure"   "failure"   "failure"
"success"   "success"   "success"

Fig. 18.60.

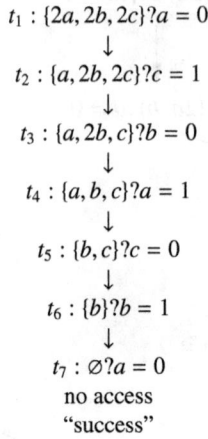

$t_1 : \{2a, 2b, 2c\}?a = 0$

$\downarrow$

$t_2 : \{a, 2b, 2c\}?c = 1$

$\downarrow$

$t_3 : \{a, 2b, c\}?b = 0$

$\downarrow$

$t_4 : \{a, b, c\}?a = 1$

$\downarrow$

$t_5 : \{b, c\}?c = 0$

$\downarrow$

$t_6 : \{b\}?b = 1$

$\downarrow$

$t_7 : \varnothing?a = 0$

no access

"success"

Fig. 18.61.

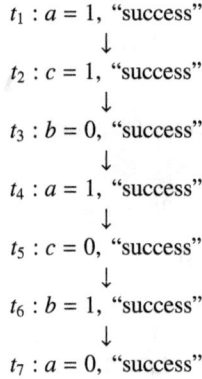

$t_1 : a = 1,$ "success"

$\downarrow$

$t_2 : c = 1,$ "success"

$\downarrow$

$t_3 : b = 0,$ "success"

$\downarrow$

$t_4 : a = 1,$ "success"

$\downarrow$

$t_5 : c = 0,$ "success"

$\downarrow$

$t_6 : b = 1,$ "success"

$\downarrow$

$t_7 : a = 0,$ "success"

Fig. 18.62.

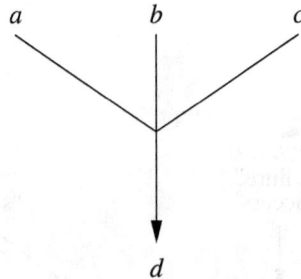

d is out only if $a = b = c =$ "in".

Fig. 18.63.

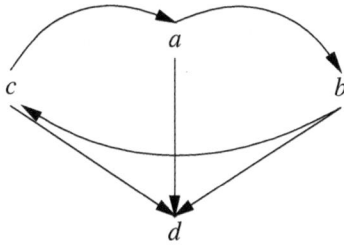

Fig. 18.64.

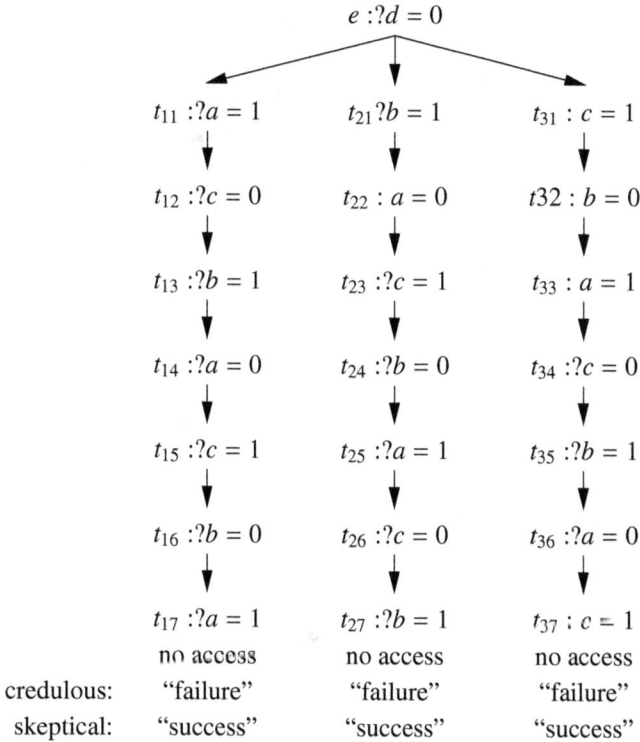

Fig. 18.65.

The next step is to replace the "no access" in nodes t_5 and s_5 by the annotations "success" or "failure", depending on the policy we want to adopt. If we adopt the credulous policy, we replace "no access" by "success" and we get the extension $x = y = $ "out" and $z = $ "in". If we adopt the skeptical approach, we replace "no access" by "failure" and get the extension $x = y = $ "in" and $z = $ "out".

So what have we done wrong?

The answer is that we replaced "no-access" for all nodes in parallel at once.

Let us do the replacement in sequence. There are two possibilities.

1. First x then y,

and

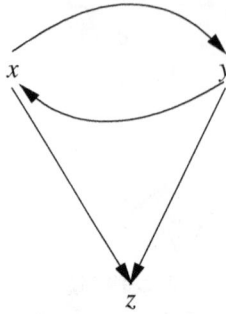

Fig. 18.66.

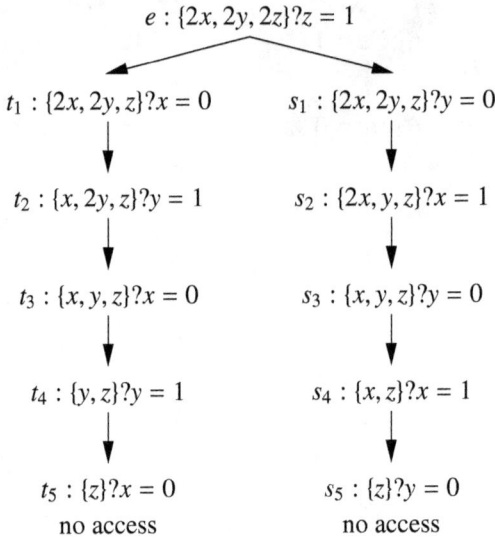

Fig. 18.67.

2. First y then x.

Let us do the first possibility and the credulous policy. This means that node t_5 is annotated "success". Thus $?x = 0$ is considered "success". We now annotate everywhere in the construction tree any node with $?x = 0$ by "success' and any node with $?x = 1$ by "failure". We get the tree of Figure 18.68.

Clearly by our propagation rules we need to annotate $?y = 0$ by "failure" and $?y = 1$ by "success". Given that, we must continue and annotate $?z = 1$ by failure.

So the extension we get is $y = $ "in" and $x = z = $ "out".

By symmetry, if we replace first $?y = 0$ by "success" we get the other extension, namely $x = $ "in" and $y = z = $ "out".

Example 18.57 (3-cycle revisited). Let us check again the diminishing resource computation tree for Example 18.55. This is Figure 18.65. Let us replace only one of the "no access" annotations by "failure" or "success" annotations. Let us choose

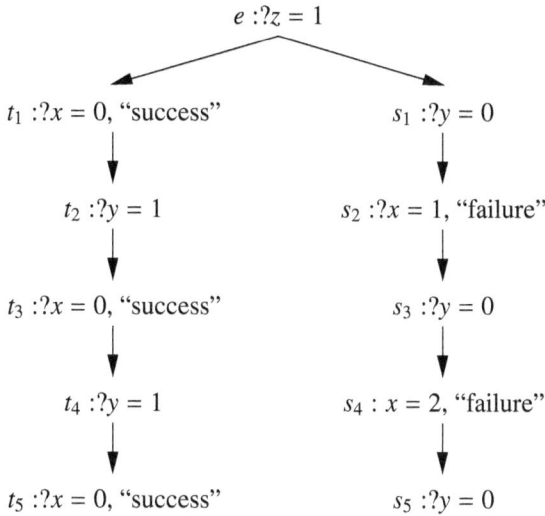

$e : ?z = 1$

$t_1 : ?x = 0,$ "success" $s_1 : ?y = 0$

$t_2 : ?y = 1$ $s_2 : ?x = 1,$ "failure"

$t_3 : ?x = 0,$ "success" $s_3 : ?y = 0$

$t_4 : ?y = 1$ $s_4 : x = 2,$ "failure"

$t_5 : ?x = 0,$ "success" $s_5 : ?y = 0$

Fig. 18.68.

$t_{17} : ?a_1 = 1$ and use the credulous approach and give it "failure". We then annotate any node with $?a = 1$ by "failure" and any node with $?a = 0$ by "success'.

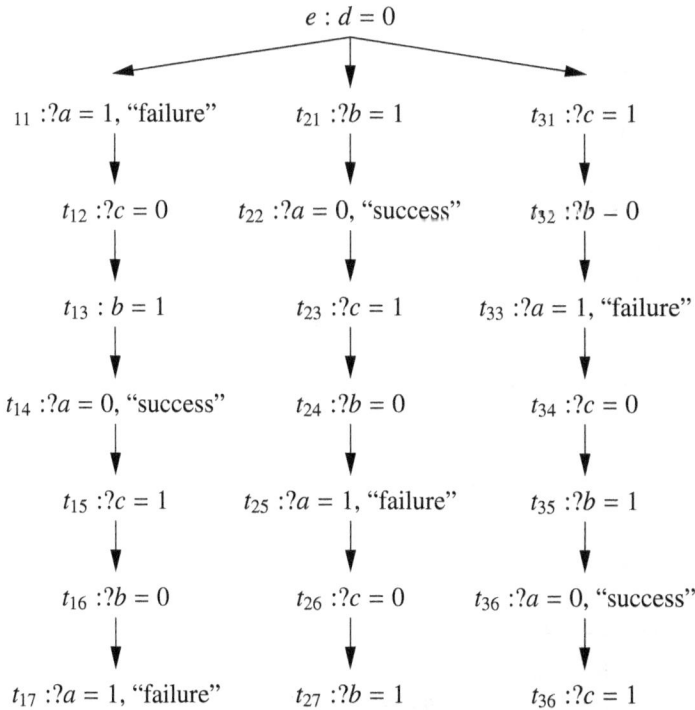

$e : d = 0$

$_{11} : ?a = 1,$ "failure" $t_{21} : ?b = 1$ $t_{31} : ?c = 1$

$t_{12} : ?c = 0$ $t_{22} : ?a = 0,$ "success" $t_{32} : ?b - 0$

$t_{13} : b = 1$ $t_{23} : ?c = 1$ $t_{33} : ?a = 1,$ "failure"

$t_{14} : ?a = 0,$ "success" $t_{24} : ?b = 0$ $t_{34} : ?c = 0$

$t_{15} : ?c = 1$ $t_{25} : ?a = 1,$ "failure" $t_{35} : ?b = 1$

$t_{16} : ?b = 0$ $t_{26} : ?c = 0$ $t_{36} : ?a = 0,$ "success"

$t_{17} : ?a = 1,$ "failure" $t_{27} : ?b = 1$ $t_{36} : ?c = 1$

Fig. 18.69.

We now propagate the given values up the tree. We use the group rules (1)–(3) of Definition 18.5 to propagate value sup the tree with one restriction. If a node with $?a = 0$ or $?a = 1$ already has a value, we stop. We do not change the value. (So we treat the annotations of Figure 18.69 as impulses or equivalently as the result of annihilators. So we are operating as if in Figure 18.70.)

So the value "failure" of node t_{17} is propagated upwards to t_{16} and t_{15} and stops, because the value for t_{14} is fixed already. Similarly the "success" value from t_{14} is propagated upwards to t_{13} and t_{12} but stops at t_{11} and does not change the value of t_{11}.

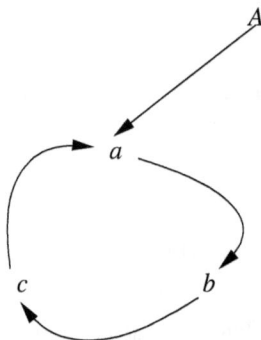

Fig. 18.70.

We do this everywhere in the tree and get the following Figure 18.71.

The figure is consistent. It gives the values b = "in", $a = c$ = "out".

By symmetry, starting with t_{27} :$?b = 1$ "failure"or t_{37} :$?c = 1$, "failure" will yield the other two extensions, namely a = "in', $b = c$ = "out" and c = "in", $a = b$ = "out".

Note that these 3 extensions are exactly the maximal conflict free sets adopted by the CF2 semantics for this case.

18.5.2 The LLL computation

We are now ready to describe the linear logic like diminishing resource compuration.

Definition 18.58 (LLL computation procedure). *Let* (S, R) *be an argumentation network. Let* $\mathbb{E}$ *be a partial function on* S, *giving for each* x *in its domain a value of either 1 (meaning "in") or 0 (meaning "out"). We assume that* $\mathbb{E}$ *is coherent and respects the* (S, R) *attack relation. This means that*

1. *If* x *attacks* y *and* $\mathbb{E}(x) = 1$ *then* $\mathbb{E}(y) = 0$, *if it is defined on* y.
2. *If* $\mathbb{E}$ *gives value 0 to all attackers of* y *in* (S, R), *then* $\mathbb{E}$ *gives value 1 to* y, *if defined on* y.
3. *If* y *has no attackers in* (S, R), *then* $\mathbb{E}$ *gives value 1, if defined on* y.

Let $?x(0) = r(0)$ *be a goal, with* $x(0) \in S$ *being a top loop-node and* $r(0) \in \{0, 1\}$. *Further assume that* $x(0) \notin \mathbb{E}$. *Further assume that we have a loop handling policy* $\mathbb{P}$ *which decides what to do if we find a loop in the middle of a computation. The exact way the loop policy operates will be clear when we describe our computation.*

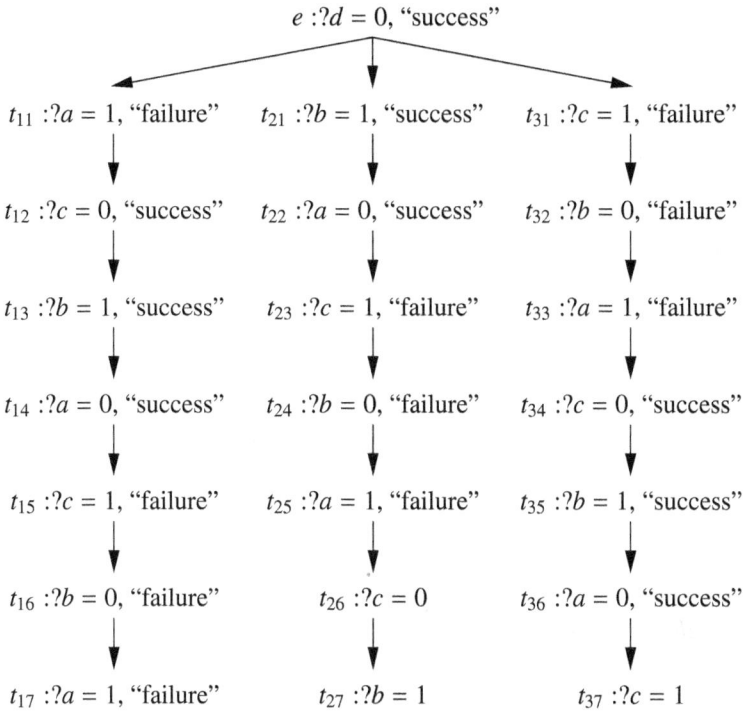

$$e :?d = 0, \text{“success”}$$

$t_{11} :?a = 1, \text{“failure”}$	$t_{21} :?b = 1, \text{“success”}$	$t_{31} :?c = 1, \text{“failure”}$
$t_{12} :?c = 0, \text{“success”}$	$t_{22} :?a = 0, \text{“success”}$	$t_{32} :?b = 0, \text{“failure”}$
$t_{13} :?b = 1, \text{“success”}$	$t_{23} :?c = 1, \text{“failure”}$	$t_{33} :?a = 1, \text{“failure”}$
$t_{14} :?a = 0, \text{“success”}$	$t_{24} :?b = 0, \text{“failure”}$	$t_{34} :?c = 0, \text{“success”}$
$t_{15} :?c = 1, \text{“failure”}$	$t_{25} :?a = 1, \text{“failure”}$	$t_{35} :?b = 1, \text{“success”}$
$t_{16} :?b = 0, \text{“failure”}$	$t_{26} :?c = 0$	$t_{36} :?a = 0, \text{“success”}$
$t_{17} :?a = 1, \text{“failure”}$	$t_{27} :?b = 1$	$t_{37} :?c = 1$

Fig. 18.71.

We now give a computation procedure for the goal from (S, R) using $\mathbb{E}$ as a basis and giving a new strictly larger function $\mathbb{E}_1 \supseteq \mathbb{E}$. When this procedure is repeated again and again we end up with a total function $\mathbb{E}_M$ on S, namely a $\{0, 1\}$ extension. Thus our overall loop handling policy will depend on our procedure and on our policy $\mathbb{P}$. This will be done in three stages.

First Stage *Construct a tree and get some additional new values for nodes of S.*
Second Stage *Propagate values up the tree of the first stage to get an annotation function V*
Third stage *Use V to define a new $\mathbb{E}_1$.*

So let us assume we have $(S, R), \mathbb{P}$ and $\mathbb{E}$ and let us start the first stage.
We define the computation tree of the form $(T, \rho, 0, \mathbf{f}, V, \mathbb{E})$ of the goal $?x(0) = r(0)$, based on $\mathbb{E}$ as follows (compare with remark 18.51). We construct the tree in steps.
Recall that ρ is a binary relation on $T, 0 \in T, \mathbf{f}$ is a function on T giving values of the form $\mathbb{S}'?x(t)?r(t)$, where $\mathbb{S}' \subseteq 2S, x(t) \in S$ and $r(t) \in \{0, 1\}$ and V is a function on T giving values in $\{\text{“success”}, \text{“failure”}\}$.
In step n we have the tree $(T_n, \rho_n, 0, \mathbf{f}_n, V_n, \mathbb{E})$.

Step 0 *Let $T_0 = \{0\}, \rho_0 = \varnothing, \mathbf{f}_0(0) = 2S ?x(0) = r(0), V_0 = \varnothing$.*
Step n+1 *Let $t_1, \ldots, t_m$ be all endpoints of T_n. Let $t \in \{t_1, \ldots, t_n\}$ be any such endpoint. Let $\mathbf{f}(t) = \mathbb{S}_t ?x(t) = r(t)$.*
We distinguish several cases for t and in each case we might add new points to the tree T_n and thus form the new tree T_{n+1}.

Assume we have defined $(T_n, \rho_n, 0, \mathbf{f}_n, V_n, \mathbb{E})$.

Case 1a $x(t) \in \mathbb{S}_t$ *and t has no attackers in* (S, R). *In this case do nothing.*

Case 1b $x(t) \in \mathbb{S}_t$ *and* $\mathbb{E}$ *is defined on* $x(t)$. *In this case do nothing.*

Case 2 $x(t)$ *is not in* $\mathbb{S}_t$, *then do nothing. (Note that this is a case of a loop. Had* $\mathbb{E}$ *been defined on* $x(t)$, *we would have stopped earlier by case 1b).*

Case 3 $x(t) \in \mathbb{S}_t$ *and* $\mathbb{E}$ *is not defined on* $x(t)$ *and* $x(t)$ *has the attackers* $y_1^t, \ldots, y_{k(t)}^t$. *In this case we add to* T_n *the new points* $s_1(t), \ldots, s_{k(t)}(t)$. *We make sure these points are pairwise different from each other, from all other points* $s_j(t')$ *for other* t' *and from all other points in* T_n.

Let $\rho_{n+1} = \rho_n \cup \{(s_j(t), t) | t$ *an endpoint and* $j = 1, \ldots, k(t)\}$.

$T_{n+1} = T_n \cup \{s_j(t) | t$ *an endpoint* $j = 1, \ldots, k(t)\}$.

Let $\mathbb{S}_{s_j(t)} = \mathbb{S}_t - \{x(t)\}$.

Let $x(s_j(t)) = y_j^t, j = 1, \ldots, k(t)$.

Extend $\mathbf{f}_n$ *to be* $\mathbf{f}_{n+1}$ *where we define* $\mathbf{f}_{n+1}$ *on the new points by* $\mathbf{f}_{n+1}(s_j(t)) = \mathbb{S}_{s_j(t)}?x(s_j(t)) = 1 - r(t)$.

Let $V_{n+1} = V_n$.

Since for each n the multisets $\mathbb{S}_t$ *for endpoints t become smaller, there will be an N such that at Step N we do nothing on all endpoints of* T_N.

Step final N+1 *At this step define* V_{N+1} *as folows. First define* V_{N1+1} *on the endpoints using the following cases:*

Case 1a *t is an endpoint and* (t) *has no attackers in* (S, R), *then let* $V_{n+1}(t) = $ *"success" if* $t(t) = 1$ *and "failure" if* $t(t) = 0$.

Case 1b *t is an endpoint and* $x(t) \in \mathbb{S}_t$ *and* $\mathbb{E}$ *is defined on* $x(t)$, *then let* $V_{N+1}(t) = $ *"success" if* $r(t) = \mathbb{E}(x(t))$ *and "failure" otherwise. Note that because* $\mathbb{E}$ *is coherent, this case agrees with case 1a.*

Case 2 $x(t)$ *is not in* $\mathbb{S}_t$. *This is the case of a loop. We note that* $x(t)$ *is also a top loop-node.[7] Our loop checking policy will tell us what value to give to* $V_{N+1}(t)$. *Our loop checking policy is assumed to be committed to doing the following steps:*

Let E_N *be the set of all endpoints of* T_N *falling under Case 2. The loop checking policy has three options:*

1. *Abstain option: given no value*
2. *Credulous option. give value* $V_{N+1}(t) = $ *"success" if* $r(t) = 0$ *and "failure" other-wise. This is the annihilation option, saying let* $x(t)$ *be "out".*
3. *Skeptical option. Give value* $V_{N+1}(t) = $ *"failure" if* $r(t) = 0$ *and "success" other-wise. This says let* $x(t)$ *be "in".*

We know what the credulous option does. It is as if we are operating in the annihilator network $(S_{x(t)}, R_{x(t)})$. *We have not investigated yet what the skeptical option means. So*

---

[7] $x(t)$ is a top loop-node in this case. It attacks $x(0)$ and loops with itself, so we have a situation like this:

$$x(t) \to \ldots \to x(t) \to \ldots \to x(0).$$

If any y transitively attacks $x(t)$ it also transitively attacks $x(0)$ and so $x(0)$ being a top loop-node will attack back. So we have

$$y \to \ldots \to x(t) \to \ldots \to x(t) \to \ldots \to x(0) \to \ldots \to y.$$

Clearly then $x(t)$ also transitively attacks y.

let us not use this option. Let us assume that our policy chooses the credulous option for at least one point in E_N and abstains from giving values to the other points. Let $E'_N \subseteq E_n$ be the set of points chosen by the policy $\mathbb{P}$.

Policy considerations of what to choose may involve some meta-level algorithms. From the point of view of the tree construction process we want to know for which points $x(t)$ in E_N we let V_{N+1} give values to t.

We have now completed the first stage of constructing the tree and we can start the second stage of propagating values to continue to define V_{N+1}.

Now that we have defined V_{N+1} on some of the endpoints we can propagate these values to higher points of the tree, using the protocol given in item 3 of Definition 18.50.

We further extend the definition of V_{N+1} using the protocol of Group 2 of Definition 18.5.

Since we started with a coherent $\mathbb{E}$ and defined V_{N+1} on the endpoints using it and then propagated th evalues of V_{N+1} from the endpoints again in a coherent way, we are not going to have the incoherent situation where $V_{N+1} = $ "success" (resp. "failure") and $V_{N+1}(s) = $ "success" (resp. "failure") and $x(t) = x(s)$ and $r(t) = 1 - r(s)$ or the incoherent situation where $x(t) = x(s), r(t) = r(s)$ and V_{N+1} is defined on both t and s and $V_{N+1}(t) \neq V_{N+1}(s)$.

We now proceed to the third stage.

We can therefore coherently define the set $\mathbb{E}_1 \supseteq \mathbb{E}$ by letting for $x \in S$

$$
\begin{aligned}
\mathbb{E}_1(s) \quad &= 1, \quad \textit{if for some } t \in T_{N+1}, r(t) = 1 \textit{ and } V_{N+1}(t) = \textit{ "success" or} \\
&\qquad r(t) = 0 \textit{ and } V_{N+1}(t) = \textit{ "failure"} \\
&= 0, \quad \textit{if for some } t \in T_{N+1}, r(t) = 1 \textit{ and } V_{N+1}(t) = \textit{ "failure" or} \\
&\qquad r(t) = 0 \textit{ and } V_{N+1}(t) = \textit{ "success"} \\
&\textit{undefined, otherwise}
\end{aligned}
$$

Remark 18.59. We make some comments about the algorithm in Definition 18.58 before we go on to the next definition.

1. Since we use the credulous policy, when we move from $\mathbb{E}$ to $\mathbb{E}_1$, through the policy $\mathbb{P}$ of credulously handling elements in $E'_N \subseteq E_N$, we note that this move is equivalent to moving from (S, R) to the annihilator network $(S_{E'_N}, R_{E'_N})$. This also means that $\mathbb{E}_1$ is coherent in $(S_{E'_N}, R_{E'_N})$.
2. The elements in $E_N - E'_N$ are top loop-nodes in $(S_{E'_N}, R_{E'_N})$ if $E_n - E'_N$ is not empty. We can start a new computation for a new point $y(0) \in E_N - E'_N$ based on $(S_{E'_N}, R_{E'_N})$.
3. Note that we never refer to the network beyond looking at the attackers of a node. So let us check what exactly happens to points in E'_N. Let x be a node and consider Figure 18.72

$b_1, \ldots, b_n, y_1, \ldots, y_k$ are all the attackers of x in $(S_{E'_N}, R_{E'_N})$. b_j are in E'_N and B_i are the annihilators. When we are at a node t in the tree with $x(t) = b_j$, we stop, either because $x(t) \notin S_t$ or because $\mathbb{E}_1(b_j)$ is defined.

This is case 1b or case 2. In either case the annihilator B_j and the attackers z_j of b_j are not consulted and not involved. So from the point of view of the computation, it is as if we are still working with (S, R) and $\mathbb{E}_1$, and everything is sound.

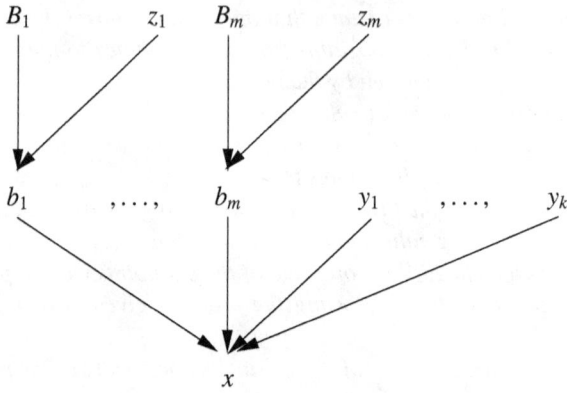

Fig. 18.72.

The view involving the annihilators is just to assure us of the soundness of the function $\mathbb{E}_1$.

So we can continue to use the tree T_N by considering the nodes in $E_N - E'_N$ and giving them the new resource $2S$ again and using $\mathbb{E}_1$ instead of $\mathbb{E}$.

Example 18.60. To illustrate this key remark of item 3 of Remark 18.59, let us look at the network of Figure 18.11. The computation tree for this network is described in Figure 18.56. In this tree we have $\mathbb{E} = \emptyset$, $x(0) = b$, $r(0) = 0$. In the tree the maximal length of a path is 6.

So the tree describes a correct computation according to Definition 18.58 which stops at $N = 6$. There is no more change in the tree T_N. The final looping nodes E_N are $\{t_9, t_{13}, t_{14}, t_{16}\}$. The policy $\mathbb{P}$ would annihilate node b and this would give us the extension for this choice of initial goal $?b = 0$.

Now imagine that we would start the initial goal $?a = 1$ (same network of Figure 18.11). We would get the tree of Figure 18.73 ($\mathbb{E} = \emptyset$, as before. The endpoints are $\{t_{11}, t_{15}, t_{16}, t_{14}\}$. The computation stops at $N = 7$. The set E_N is $\{b, d, a\}$. Our credulous loop policy can either annihilate b or d or a or several of them in parallel.

Depending on what we choose we get different extensions. Let us annihilate d. we get $d = $ "out". We take $\mathbb{E}_1$ as the partial function giving d value 0. We give nodes t_{14} and t_{16} the original resource of $2S$ back and continue the computation with $\mathbb{E}_1$. So actually what we do in the algorithm is to impulse to all endnotes the command "change resource to 2S again" and carry on with $\mathbb{E}_1$.

What would happen now is that the algorithm will continue. Nodes $t_{11}, t_{15}, t_{14}, t_{16}$ will continue. Really there is no need to deal with nodes t_{11} and t_{15} because their goal is $?b = 1$ or $?b = 0$ and the job is already done at node t_{16} and we need not duplicate. However it is simpler to say continue with all endnodes.

Figures 18.74 and 18.75 show how the computation continues for nodes t_{14} and t_{16}. The tree for $?a = 1$ is easy to reproduce by looking at Figure 18.56 with the knowledge that $\mathbb{E}_1(d) = 0$.

The tree for t_{16} and $?b = 0$, we can get by looking at the tree for $?b = 0$ of Figure 18.56. We denote the nodes by t' with the appropriate subscripts and use the knowledge that $\mathbb{E}_1(d) = 0$.

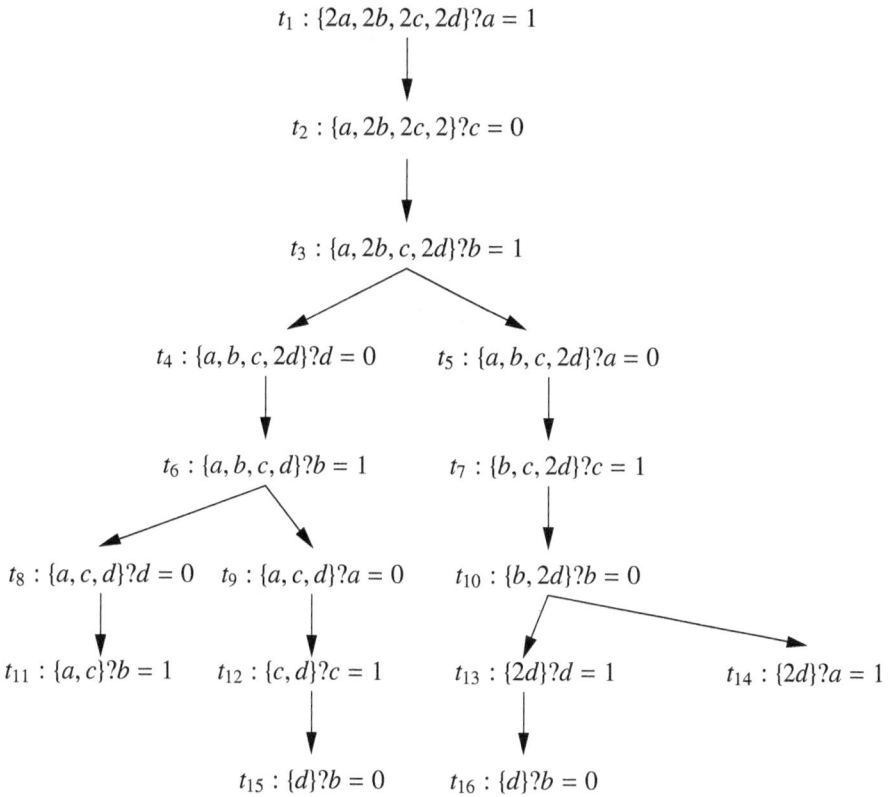

$t_1 : \{2a, 2b, 2c, 2d\}?a = 1$

$t_2 : \{a, 2b, 2c, 2\}?c = 0$

$t_3 : \{a, 2b, c, 2d\}?b = 1$

$t_4 : \{a, b, c, 2d\}?d = 0$ $t_5 : \{a, b, c, 2d\}?a = 0$

$t_6 : \{a, b, c, d\}?b = 1$ $t_7 : \{b, c, 2d\}?c = 1$

$t_8 : \{a, c, d\}?d = 0$ $t_9 : \{a, c, d\}?a = 0$ $t_{10} : \{b, 2d\}?b = 0$

$t_{11} : \{a, c\}?b = 1$ $t_{12} : \{c, d\}?c = 1$ $t_{13} : \{2d\}?d = 1$ $t_{14} : \{2d\}?a = 1$

$t_{15} : \{d\}?b = 0$ $t_{16} : \{d\}?b = 0$

Fig. 18.73.

We get Figure 18.75.

This ends the second stage of the comptuation.

The tree we get is shown in Figure 18.76. We indicate only the schematic view

We now have the looping endpoints of Figure 18.76 to deal with. These are s_9 : $?a = 1$ and $t'_{16} : ?b = 0$. Our policy has a choice of either credulously annihilating b or annihilating a or both. This will give us $\mathbb{E}_2$. Let us choose to have $b = 0$. This completes the first stage and we can proceed to the second stage and propagate this value in order to see what other variable arguments get values (in this case does a get a value). This is according to the protocol of Definition 18.58.

We look at the tree of Figure 18.74 and get that $c = 1$ and $a = 0$. We can finish here. However, rather than propagate up the tree and (use the second stage) and get the values, it is much simpler to just continue the computation tree and not use the second stage. This as a policy (no second stage) will give us the values for a and c anyway.

So we can now continue the computation from all endpoints of the tree of Figure 18.76 (including s_8 and t'_{16}) with resource $2S$ and $\mathbb{E}_2$.

We carry on and on in this manner until an iteration M where $\mathbb{E}_M$ is defined on all of S.

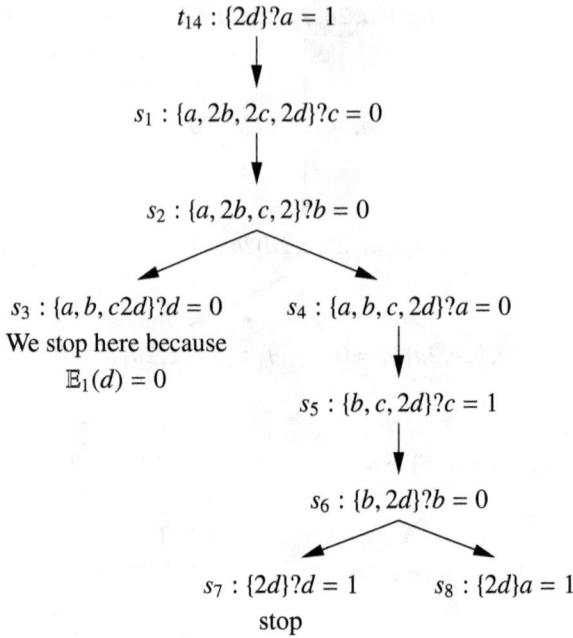

$t_{14} : \{2d\}?a = 1$

$s_1 : \{a, 2b, 2c, 2d\}?c = 0$

$s_2 : \{a, 2b, c, 2\}?b = 0$

$s_3 : \{a, b, c2d\}?d = 0$
We stop here because
$\mathbb{E}_1(d) = 0$

$s_4 : \{a, b, c, 2d\}?a = 0$

$s_5 : \{b, c, 2d\}?c = 1$

$s_6 : \{b, 2d\}?b = 0$

$s_7 : \{2d\}?d = 1$
stop

$s_8 : \{2d\}a = 1$

Fig. 18.74.

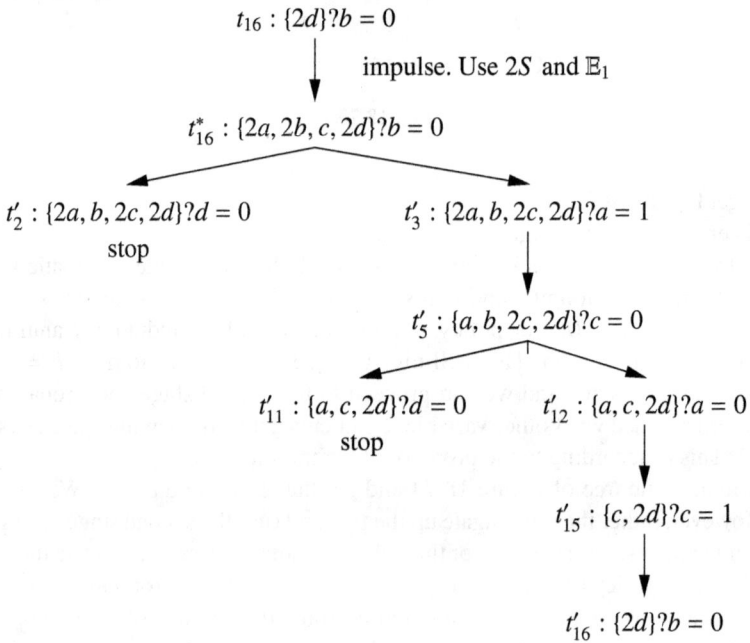

$t_{16} : \{2d\}?b = 0$

impulse. Use $2S$ and $\mathbb{E}_1$

$t_{16}^* : \{2a, 2b, c, 2d\}?b = 0$

$t_2' : \{2a, b, 2c, 2d\}?d = 0$
stop

$t_3' : \{2a, b, 2c, 2d\}?a = 1$

$t_5' : \{a, b, 2c, 2d\}?c = 0$

$t_{11}' : \{a, c, 2d\}?d = 0$
stop

$t_{12}' : \{a, c, 2d\}?a = 0$

$t_{15}' : \{c, 2d\}?c = 1$

$t_{16}' : \{2d\}?b = 0$

Fig. 18.75.

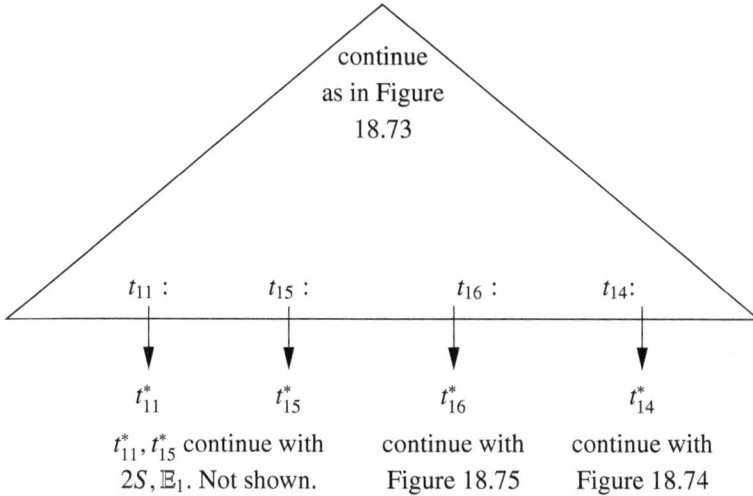

Fig. 18.76.

18.5.3 Formal LLL algorithm

We are now ready to put forward the formal LLL algorithm for finding all possible LLL extensions for an argumentation network (S, R).

Definition 18.61 (Formal LLL algorithm). *Let (S, R) be an argumentation network. Let $\mathbb{E}_0$ be the empty function from S to $\{0, 1\}$. Let there be a complete ordering of S (arbitrary, just for us to keep track of what we are doing). Let $x(0) \in S$ be the first element in the ordering which we have not dealt with yet. Let $r(0) \in \{0, 1\}$. We are now going to deal with $?x(0) = r(0)$. We are going to operate in waves.*

Wave 0. *Construct the tree $(T_{N_0}, \rho_{N_0}, 0, \mathbf{f}_0, \mathbb{E}_0)$ as described in Definition 18.58. We use only the first and third stages and skip the second stage. We use the first stage to define $\mathbb{E}_1$.*

Wave m+1. *Assume that at wave m, we extended the tree to $(T_{N_m}, \rho_{N_m}, 0, \mathbf{f}_{N_m}, \mathbb{E}_m)$ and that we identified the endpoints of T_{N_m} and we defined $\mathbb{E}_{N_{m+1}}$. We continue the computation from these endpoints with new resource $2S$ and $\mathbb{E}_{N_{m+1}}$. If t is an endpoint we add a t^* with $t^* \rho_{N_{m+1}}(t)$ and carry on with the same query but with $2S$ and $\mathbb{E}_{N_{m+1}}$. Note that some of these endpoints have been annihilated by our policy and so the new computation will automatically not start there. The other points will continue with the computation.*

We carry on until such N_M such that the domain of $\mathbb{E}_{N_M}$ equals all of S.

We thus get an extension. If we move along the ordering of S and do this for all its elements, we get all the extensions of the LLL semantics for (S, R).

18.6 The equational approach to handling loops

In our paper [166], see Chapter 10, we introduced the equational approach to argumentation. Loops in argumentation becomes looping dependencies of variables in equations and the equational environment suggests policies for handling loops. This section studies this option. We first have to introduce the Equational approach.

18.6.1 The equational approach

Let $\mathcal{A} = (S, R)$ be an argumentation frame $S \neq \varnothing$ is the set of arguments and $R \subseteq S \times S$ is the attack relation. The equational approach views (S, R) as a bearer of equations with the elements of S as the variables ranging over $[0, 1]$ and with R as the generator of equations. Let $x \in S$ and let $y_1, \ldots, y_k$ be all of its attackers. We write two types of equations $Eq_{max}(\mathcal{A})$ and $Eq_{inverse}(\mathcal{A})$.

For Eq_{max} we write

- $x = 1 - \max(y_1, \ldots, y_k)$
- $x = 1$ if it has no attackers.

For $Eq_{inverse}$ we write

- $x = \prod_{i=1}^{k}(1 - y_i)$
- $x = 1$, if it has no attackers.

We seek solutions $\mathbf{f}$ for the above equations. In [166], see Chapter 10, we prove the following:

Theorem 18.62. *1. There is always at least one solution in $[0, 1]$ to any system of continuous equations $Eq(\mathcal{A})$.*
   2. If we use $Eq_{max}(\mathcal{A})$ then the solutions $\mathbf{f}$ correspond exactly to the Dung extensions of A. Namely
   - $\mathbf{f}(x) = 1$ *corresponds to* $x = in$
   - $\mathbf{f}(x) = 0$ *corresponds to* $x = out$
   - $0 < \mathbf{f}(x) < 1$ *corresponds to* $x = undecided$.
      The actual value in $[0, 1]$ reflects the degree of odd looping involving x.
   3. If we use $Eq_{inverse}$, we give more sensitivity to loops. For example the more undecided elements y attack x, the closer to 0 (out) its value gets.

In the context of equations, a very natural step to take is to look at *Perturbations*. If the equations describe a physical or economic system in equilibrium, we want to change the solution a bit (perturb the variables) and see how it affects the system. For example, when we go to the bank to negotiate a mortgage, we start with the amount we want to borrow and indicate for how many years we want the loan and then solve equations that tell us what the monthly payment is going to be. We then might change the amount or the number of years or even negotiate the interest rate if we find the monthly payments too high.

In the equational system arising from an argumentation network we can try and fix the value of some arguments and see what happens. In the equational context, this move is quite natural. We shall see later, that fixing some values to 0 in the equations of $Eq(\mathcal{A})$, amounts to adopting the CF2 semantics, when done in a certain way. When done in other ways it gives the new loop-busting semantics LB.

Example 18.63. Consider Figure 18.86. The equations for this figure are (we use $Eq_{inverse}$)

1. $\alpha = 1 - \phi$
2. $\beta = 1 - \alpha$
3. $\phi = 1 - \beta$
4. $\gamma = (1 - \alpha)(1 - \beta)(1 - \phi)$
5. $\delta = 1 - \gamma$

The solution here is

$$\alpha = \beta = \phi = \tfrac{1}{2}$$
$$\gamma = \tfrac{1}{8}$$
$$\delta = \tfrac{7}{8}$$

Let us perturb the equation by adding an external force which makes a node equal zero. The best analogy I can think of is in electrical networks where you make the voltage of a node 0 by connecting it to earth.

Let $Z(x)$ be the "earth" connection for node x. We now do several perturbations as examples

(a). Let's choose to make $\phi = 0$.
    We replace equation 3 by
$3^*a.\ \phi = (1 - \beta)Z(\phi)$
$3^*b.\ Z(\phi) = 0.^8$
    The equations now solve to

$$\phi = 0, \alpha = 1, \beta = 0$$
$$\gamma = 0$$
$$\delta = 1.$$

This gives us the extension $\{\alpha, \delta\}$
(b). If we try to make $\alpha = 0$, we replace equation (1) by
$1^*a.\ \alpha = (1 - \phi)Z(\alpha)$
$1^*b.\ Z(\alpha) = 0$
    We solve the equations and get

$$\alpha = 0, \beta = 1, \phi = 0$$
$$\gamma = 0$$
$$\delta = 1$$

This corresponds to the extension $\{\beta, \delta\}$.
(c). Now let us make $\beta = 0$. We replace equation (1) by
$2^*a.\ \beta = (1 - \alpha)Z(\beta)$
$2^*b.\ Z(\beta) = 0$
    Solving the new equations gives us

$$\beta = 0, \phi = 1, \alpha = 0$$
$$\gamma = 0$$
$$\delta = 1$$

This gives us the extension $\{\phi, \delta\}$.

---

[8] We use Z and write 3^*a and 3^*b, rather than just writing $3^*\ a = 0$ because of algebraic considerations. The current equations can be manipulated algebraically to to prove $a = b = c$. By adding a fourth variable $Z(\phi)$ we prevent that.

If we compare these extensions with the CF2 extensions, we see that they are the same.

Example 18.64. Let us see what happens with Figure 18.85(b). Here we have a well behaved even loop. Let us write the equations

1. $\alpha = 1 - \gamma$
2. $\delta = 1 - \alpha$
3. $\beta = 1 - \delta$
4. $\gamma = (1 - \beta)(1 - \phi)$
5. $\phi = 1 - \gamma$

Let us do some perturbations:

(a) Let us make $\gamma = 0$. We change equation 4 to

$4^*a.$ $\gamma = (1 - \beta)(1 - \phi)Z(\gamma)$
$4^*b.$ $Z(\gamma) = 0$

We solve the new equations and get

$$\gamma = 0, \alpha = 1, \delta = 0, \text{ and the two options, } \gamma = 0 \text{ and } \phi = 1 \text{ or } \beta = 1, \phi = 1.$$

The extension is $\{\alpha, \beta, \phi\}$.

(b) Let us try $\alpha = 0$. we replace equation 1 by

$1^*a.$ $\alpha = (1 - \gamma)Z(\alpha)$
$1^*b.$ $Z(\alpha) = 0$

We solve the new equations and get

$$\alpha = 0, \delta = 1, \beta = 0, \gamma = 1, \phi = 0$$

The extensions we get are $\{\delta, \phi\}$ or $\{\delta, \gamma\}$.

(c) Let us make $\delta = 0$. We replace equation 2 by

$2^*a.$ $\delta = (1 - \alpha)Z(\delta)$
$2^*b.$ $Z(\delta) = 0$

The solution is

$$\delta = 0, \alpha = 1, \beta = 1, \gamma = 0 \text{ and } \phi = 1$$

This gives the extension

$$\{\alpha, \beta, \phi\}$$

(d) Let us make $\beta = 0$. the new equations for β are

$3^*a.$ $\beta = (1 - \delta)Z(\beta)$
$3^*b.$ $Z(\beta) = 0$

We solve the new set of equations and get the two options, either

$$\beta = 0, \gamma = 1, \phi = 0, \alpha = 0, \delta = 1 \text{ or } \beta = 0, \gamma = 0, \phi = 1, \alpha = 1, \delta = 0.$$

The extensions are $\{\phi, \alpha\}$ and $\{\gamma, \delta\}$.

(e) Let us make $\phi = 0$. We change equation 5 to

$5^*a.$ $\phi = (1 - \gamma)Z(\phi)$
$5^*b.$ $Z(\phi) = 0$

We solve the new equations.
From (3) and (4) we get

5. $\delta = \gamma$

From (1) and (2) we get

7. $\alpha = \beta$.

Let $\alpha = \beta = x$. Then $\gamma = \delta = 1 - x$.

If we want $\{0, 1\}$ extensions, i.e. $x \in \{0, 1\}$, then we get the extensions

$\{\alpha, \beta\}$, case $\{x = 1, \phi = 0\}$

$\{\gamma, \delta\}$, case $\{x = 0, \phi = 0\}$.

(f) Let us make $\alpha = \gamma = 0$. The new equations are

1*a. $\alpha = (1 - \gamma)Z(\alpha)$

1*b. $Z(\alpha) = 0$

  2. $\delta = 1 - \alpha$

  3. $\beta = 1 - \delta$

4*a. $\gamma = (1 - \beta)Z(\gamma)$

4*b. $Z(\gamma) = 0$

  5. $\phi = 1 - \gamma$.

The solution is

$$\alpha = \gamma = 0$$
$$\delta = 1$$
$$\beta = 0$$
$$\phi = 1$$

The extension we get is $\{\delta, \phi\}$.

(g) Let us summarise in Table 18.1.

Case	Set B of points made 0	Corresponding extensions
(a)	$B_a = \{\gamma\}$	$\{\alpha, \beta, \phi\}$
(b)	$B_b = \{\alpha\}$	$\{\delta, \gamma\}$
(c)	$B_c = \{\delta\}$	$\{\alpha, \beta, \phi\}$
(d)	$B_d = \{\beta\}$	$\{\gamma, \delta\}$
(e)	$B_e = \{\phi\}$	$\{\alpha, \beta\}, \{\gamma, \delta\}$
(f)	$B_f = \{\alpha, \gamma\}$	$\{\delta, \phi\}$

Table 18.1.

18.6.2 The equational loop-busting semantics LB for complete loops

We now introduce our loop busting semantics, the LB semantics for complete loops. We need a series of concepts leading up to it.

Definition 18.65 (Loops). *Let $\mathcal{A} = (S, R)$ be an argumentation network.*

1. A subset $E = \{x_1, \ldots, x_n\} \subseteq S$ is a loop cycle, (or a loop set, or a loop) if we have

$$x_1 R x_2, x_2 R x_3, \ldots, x_{n-1} R x_n, x_n R x_1$$

(S, R) *is said to be a* complete loop *if every element of S is an element of some loop cycle.*[9]

2. *A set $B \subseteq S$ is a* loop-buster *if for every loop set E we have $E \cap S \neq \emptyset$*

3. *Let $B \subseteq S$ be a loop-buster and let M be a meta-predicate describing properties of B. We can talk about the semantics LBM, where, (when we define it later), we use only loop-busters B such that M(B) holds. Criteria for adequacy for LBM are*

    a) It busts all odd numbered loops

    b) It busts all even numbered loops and yields all allowable Dung extensions for such loops.

4. *Our first two proposals for conditions M on loop-busters is minimality. The idea is the smaller B is, the more options we have.*

    Therefore, we define: A loop-buster set B is minimal absolute *if there is no loop-buster set B' with a smaller number of elements (we do not require $B' \subseteq B$!).*

5. *A loop-buster set B is* minimal relative *if there does not exist a $B' \subsetneq B$ which is a loop-buster set.*

Example 18.66 (Loop-buster 1). Consider Figures 18.77 and 18.78.

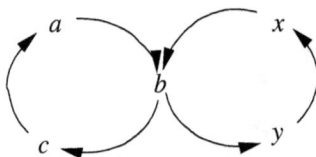

Fig. 18.77.

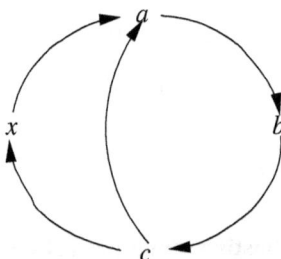

Fig. 18.78.

1. In Figure 18.77 there are two loop sets, $\{a, b, c\}$ and $\{b, x, y\}$. The loop-buster $\{b\}$ is minimal absolute and $\{y, c\}$ is minimal relative. The loop set $\{y, b\}$ is not minimal absolute.

---

[9] Comparing with the terminology of [30], a complete loop is a union of disjoint strongly connected sets. Compare with Definition 18.21.

2. Consider Figure 18.78. There are two loops $\{a, b, c\}$ and $\{a, b, c, x\}$. The minimal absolute loop-buster sets are $\{c\}, \{a\}$. $\{x, b\}$ is not minimal relative.

Example 18.67 (Loop-buster 2).
   Consider Figure 18.79.

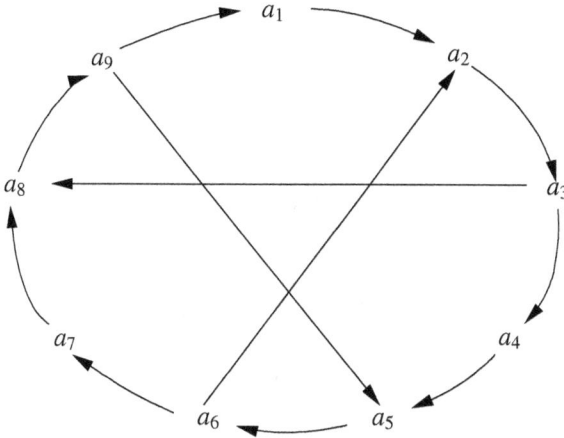

Fig. 18.79.

The loops in this figure are many. For example, we list some

1. $\{a_1, a_2, a_3, a_4, a_5, a_6, a_7, a_8, a_9\}$
2. $\{a_6, a_2, a_3, a_4, a_5\}$
3. $\{a_3, a_8, a_9, a_1, a_2\}$
4. $\{a_9, a_5, a_6, a_7, a_8\}$

Consider the loop-buster

$$\{a_2, a_5, a_8\}$$

This is not a minimal absolute set but if we delete one of its elements we get a minimal absolute set. No one element is a loop-buster.

Definition 18.68 (The loop-busting semantics LB$\mathcal{M}$ for complete loops). *Let $\mathcal{A} = (S, R)$ be an argumentation network. Assume that (S, R) is a complete loop, namely that each of its elements belongs to some loop cycle, as defined in item 1 of Definition 18.65. We define the LB$\mathcal{M}$ extensions for $\mathcal{A}$ as follows.*

1. Let B be a loop-buster for $\mathcal{A}$ satisfying $\mathcal{M}$.
2. Let $Eq_{max}(\mathcal{A})$ be the system of equations generated by $\mathcal{A}$. These have the form

$$(\mathbf{eq}(x)) : x = \mathbf{h}_x(y_1, \ldots, y_{k(x)})$$

where $x \in S$, and $y_1, \ldots, y_{k(x)}$ are all the attackers of x. If x has no attackers then $h_x \equiv 1$.
3. For each $x \in B$ replace the equation $\mathbf{eq}(x)$ by the two new equations

- $(\mathbf{eq}_a^*(x)) : x = \mathbf{h}_x(y_1, \ldots, y_{k(x)})Z(x)$
- $(\mathbf{eq}_b^*(x)) : Z(x) = 0$

where $Z(x)$ is a new variable syntactically depending on x alone.

4. *Solve the equations in (3) and let $\mathbf{f}_B$ be any solution.*
   Then the set

$$E_{\mathbf{f},B} = \{x \in S \,|\, \mathbf{f}_B(x) = 1\}$$

   is an LBM extension.

5. *Thus the set of all LBM extensions for $\mathcal{A} = (S, R)$ is the set*

$$\{E_{\mathbf{f},B} \,|\, B \text{ is as in (1), } \mathbf{f}_B \text{ is as in (4) and } E_{\mathbf{f},B} \text{ is as in (4)}\}$$

Note that our definition of extension for a general network will be given in the next section.

Before we prove soundness of LBM relative to the traditional Dung semantics and compare LBM with CF2 semantics, let us do some examples. We use Figures 18.86 and 18.85(b).

Example 18.69. Consider Figure 18.86. The only loop here is $\{\alpha, \beta, \phi\}$. There are three minimal absolute loop-busting sets, $B_\alpha = \{\alpha\}$, $B_\beta = \{\beta\}$ and $B_\phi = \{\phi\}$.

For each one of these sets we need to modify the equations of Figure 18.86 and solve them and see what extensions we get. This has already been done in Example 18.63, parts (a), (b) and (c).

In (a) we made $\phi = 0$, i.e. we used the loop-busting set B_ϕ. We solved the modified equations and got the extension $\{\alpha, \delta\} = E_\phi$. In (b) we made $\alpha = 0$, i.e. we used the set B_α, solved the modified equations and got the extension $E_\alpha = \{\beta, \delta\}$.

In (c) we made $\beta = 0$, i.e. we used the set B_α, solved the modified equations and got the extension $E_\beta = \{\phi, \delta\}$.

Let us now compare with the CF2 extensions for the figure (Figure 18.86). The maximal conflict free sets of the first loop $\{\alpha, \beta, \phi\}$ are $C_\alpha = \{\alpha\}$, $C_\beta = \{\beta\}$ and $C_\phi = \{\phi\}$. They are the same as our loop-busting sets, but they are used differently. They are supposed to be in (i.e. value 1) not out (value 0). We use $C_\alpha, C_\beta, C_\phi$ to calculate the CF2 extensions and get $\{\alpha, \delta\}, \{\beta, \delta\}$ and $\{\phi, \delta\}$, indeed the same as the LB extensions.

Example 18.70. We now consider Figure 18.85(b). The only minimal absolute loop-buster set here is $B_\gamma = \{\gamma\}$. We have three more minimal relative sets, $B_1 = \{\beta, \phi\}$, $B_2 = \{\delta, \phi\}$ and $B_3 = \{\alpha, \phi\}$.

We refer the reader to Example 18.64, where some equational calculations for this figure are carried out.

1. In (a) of Example 18.64, we make $\gamma = 0$, we solve the modified equation and get the extension $E_\gamma = \{\alpha, \beta, \phi\}$.
   This takes care of the case $B_\gamma = \{\gamma\}$.
2. Let us address the case of $B_3 = \{\alpha, \phi\}$. We use (b) of Example 18.64, where we make $\alpha = 0$. We modify the equation for α and get a solution $\alpha = 0, \delta = 1, \beta = 0, \gamma = 1$ and $\phi = 0$.
   We thus get the extension $E_{\alpha,\phi} = \{\delta, \gamma\}$.
3. Let us address the case of $B_1 = \{\beta, \phi\}$. This corresponds to case (d) $\beta = 0$ of Example 18.64. We modify the equations and solve them and get $\beta = 0, \gamma = 1, \phi = 0, \alpha = 0$ and $\delta = 1$.
   The extension is $\{\gamma, \delta\}$.

4. We now check the case of $B_2 = \{\delta, \phi\}$.

The modified equation system for $B_2 = \{\delta, \phi\}$ is the following:

1. $\alpha = 1 - \gamma$
2*a. $\delta = (1 - \alpha)Z(\delta)$
2*b. $Z(\delta) = 0.$
3. $\beta = 1 - \delta$
4. $\gamma = (1 - \phi)(1 - \beta)Z(\gamma)$
5*a. $\gamma = \phi = (1 - \gamma)Z(\phi)$
5*b. $Z(\phi) = 0.$

We solve the equations and get $\phi = 0, \delta = 0, \beta = 1, \gamma = 0, \alpha = 1$.
The extension is $\{\alpha, \beta\}$.

Example 18.71 (CF2 and the LB minimal absolute semantics). The LB minimal absolute semantics does not give all the CF2 extensions in the case of even loops. Consider Figure 18.87. The set $B = \{a_{10}\}$ yields the extension $E_B = \{b, a_1, a_3, a_5, a_7, a_0\}$. B is minimal absolute. Consider now B' being $B' = \{a_{10}, a_3, a_6\}$.

This yields

$$E_{B'} = \{a_1, a_4, a_7, a_9\}$$

However, B' is not minimal absolute. $E_{B'}$ is a CF2 extension. B' is a minimal relative set.

What happens here is that the minimal absolute semantics gives the same extensions for even loops as the traditional Dung extensions, but the CF2 semantics gives more. This is a weakness of CF2.

Remark 18.72 (CF2 and the minimal relative extensions). Let us discuss the results of Example 18.70 calculated for Figure 18.85(b) and compare them with the CF2 extensions of Figure 18.85(b). This will give us an idea about the relation of CF2 to the minimal relative semantics. We use Example 18.64, where all the extensions were calculated and especially refer to Table 18.1, given in item (g) of Example 18.64, which summarises these calculations.

The CF2 extensions are all the conflict free subsets. These are $\{\delta, \phi\}, \{\delta, \gamma\}, \{\alpha, \beta, \phi\}$.

Comparing with the semantics of Table 18.1, we get the following: the LB minimal absolute extensions are one only, namely $\{\delta, \gamma\}$. The LB minimal relative extensions are $\{\delta, \gamma\}$ and $\{\alpha, \beta\}$.

We see that LB minimal absolute gives less extensions (but breaks loops) while LB minimal relative gives more extensions. Obviously we need to identify a policy $\mathcal{M}$ which will yield exactly the CF2 extensions.

We now need to demonstrate the soundness of the LB semantics. The perceptive reader will ask himself, how do the LB extensions relate to the extensions of traditional Dung semantics? After all, we start with the standard equational semantics, which for the case of Eq_{max} is identical with the Dung semantics, but then using a loop-busting set B of one kind or another, we get a new set of equations and call the solutions LB extensions. What are these solutions and what meaning can we give them?

Obviously, we need some sort of soundness result. This is the job of the next theorem.

Theorem 18.73 (Representation theorem for LB semantics using annihilators).
Let $\mathcal{A} = (S, R)$ be an argumentation net being a complete loop as in Definition 18.68 and let B be a loop-busting subset of S (of some sort M). Let $\mathbf{E}(B, \mathcal{A})$ be the family of LB extensions obtained from $\mathcal{A}$ and B by following the procedures of Definition 18.68. Then $\mathbf{E}(B, \mathcal{A})$ can be obtained also following the procedure below

1. *For each $x \in B$, let $\mathbf{z}(x)$ be a new point not in S. Let $\mathbf{z}(x)$ be all different for different xs.*
2. *Define (S_B, R_B) as follows:*

$$S_B = S \cup \{\mathbf{z}(x) | x \in B\}$$
$$R_B = R \cup \{(\mathbf{z}(x), x) | x \in B\}.$$

3. *The network (S_B, R_B) is an ordinary Dung network and has traditional Dung extensions. We have (for $Eq_{\max}$):*

$$\mathbf{E}(B, \mathcal{A}) = \{E \cap S | E \text{ is an extension of } (S_B, R_B)\}$$

Proof. The new equations for each $x \in B$ in (S_B, R_B) are

$$(\mathbf{eq}_a^*(x)) : x = 1 - \max(y_1, \ldots, y_{k(x)}, \mathbf{z}(x))$$
$$(\mathbf{eq}_a^*(\mathbf{z}(x))) : \mathbf{z}(x) = 1$$

where $y_1, \ldots, y_{k(x)}$ are all the attackers of x in (S, R).
   Since $\mathbf{z}(x) = 1$, we get that

$$1 - \max(y_2, \ldots, y_{k(x)}, \mathbf{z}(x)) = (1 - \max(y_1, \ldots, y_{k(x)}))(1 - \mathbf{z}(x))$$
$$= (1 - \max(y_1, \ldots, y_{k(x)}))Z(x)$$

provided $Z(x) = 1 - \mathbf{z}(x)$.
   Of course $\mathbf{z}(x) = 1$ means $Z(x) = 0$.
   So we get the same modified equations as required by the LB semantics in Definition 18.68.

Example 18.74. Let us represent the cases of Example 18.70, which dealt with Figure 18.85(b). See Figures 18.80, 18.81, 18.82, 18.83 corresponding to cases (1)–(4) of Example 18.70.
   We are also adding Figure 18.84, describing the situation for $B_2' = \{\delta\}$ as discussed in Remark 18.72 in item (b).

18.6.3 The equational semantics LB and its connection with CF2

We now define the family of LB semantics and identify the loop-busting counterpart of CF2. We need to develop some concepts first. We begin with a high school example.

Example 18.75 (High school example).

1. Solve the following equations in the unknowns x, y, z.
    a) $x - y = 1$
    b) $x + y = 5$

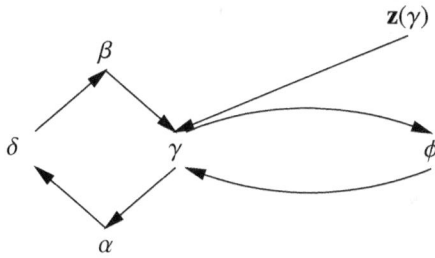

Fig. 18.80. Case (1): $\gamma = 0$ for $B_\gamma = \{0\}$

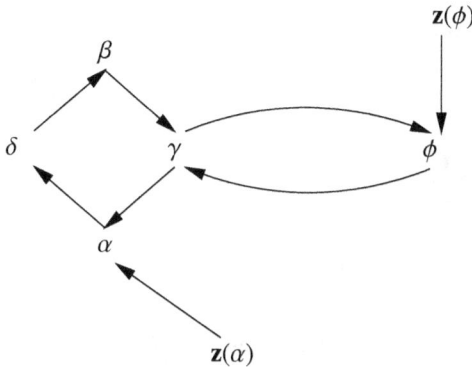

Fig. 18.81. Case (2): $B_3 = \{\alpha, \phi\}$

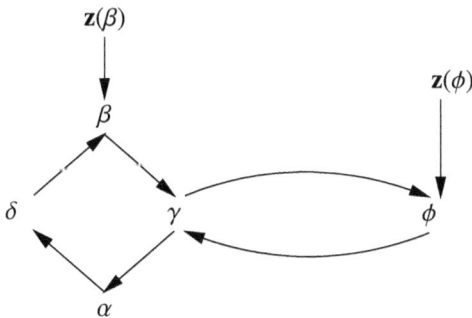

Fig. 18.82. Case (3): $B_1 = \{\beta, \phi\}$

c) $z^2 - 4yz + x + 1 = 0$

The point I want to make is that we solve the equations directionally. We first find the values of x and y from equations (a) and (b) to be $x = 3$ and $y = 1$ and then substitute in equation (c) and solve it. We get

c) $z^2 - 4z + 4 = 0$

$z = 2$

2. Let us change the problem a bit. We have the equations

a) $x - \sin y = 2.99$

b) $x + \sin y = 3.01$

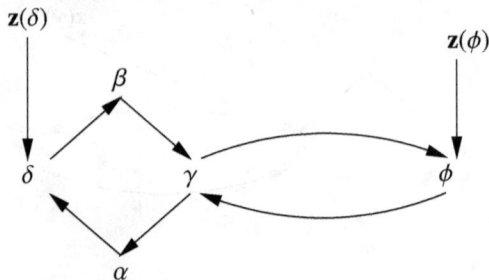

Fig. 18.83. Case (4): $B_2 = \{\delta, \phi\}$

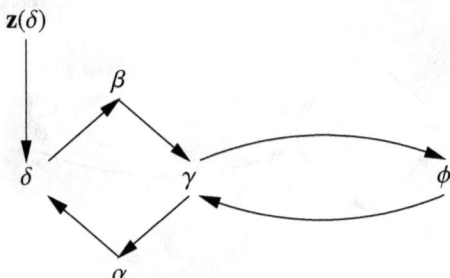

Fig. 18.84. Case (b): $B_2' = \{\delta\}$

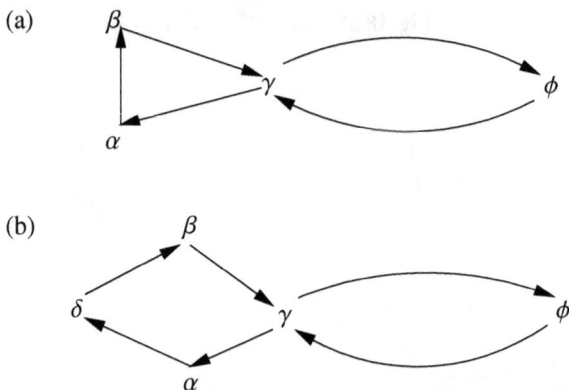

Fig. 18.85. Figure 8 of [30]. Problematic argumentation of frameworks

c) $z^2 - 400yz + x + 1 = 0$

Here we may again consider equations (a) and (b) first but also use the approximation $y \approx \sin y$. We find $x = 3, y \approx 0.01$ and solve the third to get $z = 2$.

3. A third possibility is to look at equations (a) and (b) and decide to ignore them altogether,[10] and substitute $x = y = 0$. We get

(c*). $z^2 + 1 = 0$

4. Another example is the equation

---

[10] Of course, ignoring (a) and (b) needs to be justified.

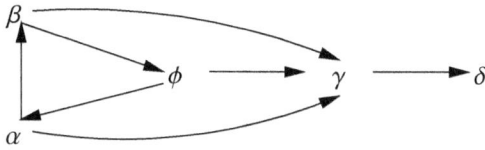

Fig. 18.86. Figure 9 of [30]: Floating defeat and floating acceptance

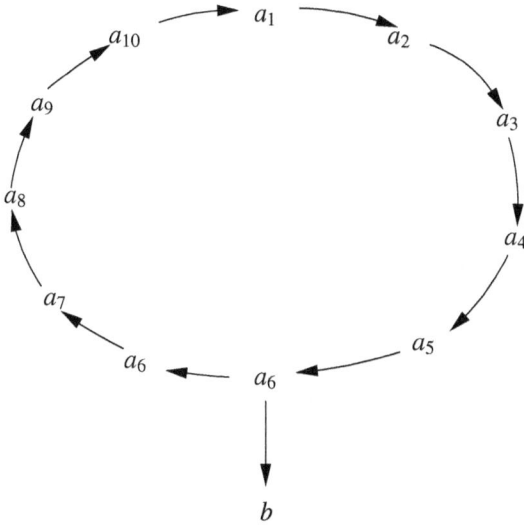

Fig. 18.87.

$$x^4 - 2x^2 + \frac{x}{1000} + 1 = 0.$$

To solve this equation we decide on the perturbation which ignores $\frac{x}{1000}$ on account of it being relatively small. We solve

$$x^4 - 2x^2 + 1 = 0$$

we get $x = \pm 1$.

Definition 18.76. *We present a perturbation protocol for solving equations of the form*

$$x = \mathbf{h}_x(v_1, \ldots).$$

1. Let V be a set of variables and $\mathbb{E}$ be a set of equations of the form $x = \mathbf{h}_x(V_x)$, where $V_x \subseteq V$ are the variables appearing in $\mathbf{h}_x$, and x ranges over V. We seek solutions to the system $\mathbb{E}$ with values hopefully in $\{0, 1\}$. If $\mathbf{h}_x$ are all continuous functions in $[0, 1]$, then we know that there are solutions with values in $[0, 1]$, but are there solutions with values in $\{0, 1\}$?

Even if we are looking for and happy with any kind of solution, we may wish to shorten the computation by starting with some good guesses, or some approximation or follow any kind of protocol $\mathbb{P}$ which will enable us to perturb the equations *and get some results which we would find satisfactory from the point of view of our application area.*

In the case of equations arising from argumentation networks, we would like perturbations which help us overcome odd-numbered loops.

Note that in numerical analysis such equations are well known. If $x_1, \ldots, x_m$ are variables in $[0, 1]$ and $\mathbf{h}_1, \ldots, \mathbf{h}_m$ are continuous functions in $[0, 1]$, we want to solve the equations

$$x_i = \mathbf{h}_i(x_1, \ldots, x_m), i = 1, \ldots, m.$$

One well known method is that of successive approximations. We guess a starting value

$$x_1 = a_1^0, \ldots, x_m = a_m^0$$

and continue by substituting

$$a_i^{j+1} = \mathbf{h}_i(a_1^j, \ldots, a_m^j).$$

Under certain conditions on the functions $\mathbf{h}_i$ (Lipschitz condition), the values $a_i^j, = 1, 2, \ldots$ converge to a limit $a_i^\infty, i = 1, 2, \ldots, m$ and that would be a solution. What we are going to do in this chapter is in the same spirit.

2. *Let us proceed formally adopting a purely equational point of view and take a subset $B_1 \subseteq V$ of the variables and decide for our own reasons to substitute the value 0 for all the variables in B_1 in the equations $\mathbb{E}$.*

How we choose B_1 is not said here, we assume that we have some protocols for finding such a B_1. In the application area of argumentation, these protocols will be different loop-busting protocols $LB(\mathcal{M})$.

For the moment, formally from the equational point of view, we have a set of equations $\mathbb{E}$ with variables V and a $B_1 \subseteq V$, which we want to make 0. How do we proceed?

This has to be done carefully and so we replace for each $u \in B_1$, the equation

$$\mathbf{eq}(u) : u = \mathbf{h}_u(V_u)$$

by the pair of equations

$$\mathbf{eq}^*(u) : \begin{cases} u = \mathbf{h}_u(V_u)Z(u) \\ Z(u) = 0 \end{cases}$$

We now propagate these values through the new set of equations, solve what we can solve and end up with new equations of the form

$$\mathbf{eq}^1(x) : x = \mathbf{h}_x^1(V_x^1)$$

for $x \in V$, where $\mathbf{h}_x^1$ is the new equation for x and V_x^1 are its variables. We have

$$V_x^1 \subseteq V - B_1.$$

The variables of B_1 get all value 0 and maybe more variables solve to some numerical values. Note that we can allow also for the case of $B_1 = \varnothing$.

We always have a solution because the functions involved are all continuous. Let $U_1 = \{x | V_x^1 = \varnothing\}$. U_1 is the set of x which get a definite numerical value, for which V_x, the set of variables they depend on, is empty. We have $B_1 \subseteq U_1 \subseteq V$. Let $\mathbf{f}_1$ be a function collecting these values on U_1, i.e. $\mathbf{f}(x) = \mathbf{h}_x^1$, for $x \in U_1$.

3. We refer to U_1 as the set of all elements instantiated to numerical values at step 1. We declare all variables of U_1 as having rank 1.
4. Let $\mathbb{E}^1$ be the system of equations for the variables in $V - U_1$.

   We now have a new system of variables and we can repeat the procedure by using a new set B_2 chosen to make 0.

   We can carry this procedure repeatedly until we get numerical values for all variables. Say that at step n we have that the union of all sets $U_1, U_2, \ldots, U_n$ equals V. Then also each element of V has a clear rank k, the step at which x was instantiated. Call this procedure Protocol $\mathbb{P} = (B_1, B_2, \ldots)$. Note that we did not say why and how we choose the sets $B_1, B_2, \ldots$. In the case of equations arising from argumentation networks, these sets B_i will be loop-busting sets.
5. Note that the equations initially give variables either 0 or 1 and our loop busters also give variables o, and Eq_{max} and $Eq_{inverse}$ are such that they keep the variables in $\{0, 1\}$, then all the functions $\mathbf{f}$ involved are $\{0, 1\}$ functions

Example 18.77. We now explain why we use Z in our perturbation. Consider the equations

1. $a = 1 - c$
2. $b = 1 - a$
3. $c = 1 - b$.

These equations correspond to a 3-element argumentation loop.

We take the $B_1 = \{a\}$ and want to execute a perturbation. If we do just substitute $a = 0$, we get a contradiction because the equations prove algebraically through manipulation that

$$a = b = c = 1 - a = 1 - b = 1 - c$$

So we need to change the equation governing a. We write

$1^*a.$ $a = (1 - c)Z(a)$
$1^*b.$ $Z(a) = 0$

Algebraically we now have 4 equations in 4 variables

$$a, b, c, Z(a)$$

The solution is

$$Z(a) = 0, a = 0, b = 1, c = 0.$$

We cannot any more execute an algebraic manipulation to get $a = b = c$!

Example 18.78. Let us recall Example 18.63, manipulating the equations arising from Figure 18.86. This is an illustration of our procedure. We used the loop-busting sets $B_\alpha = \{\alpha\}, B_\beta = \{\beta\}$ and $B_\phi = \{\phi\}$, and followed the procedure as described in Definition 18.76.

Let us now proceed with more concepts leading the way to the full definition of our loop-busting LB semantics.

We saw how to get a set of equations $Eq(\mathcal{A})$ from any argumentation network $\mathcal{A}$. Now we want to show how to get an argumentation network $\mathcal{B}_{\mathbb{E}}$ from any set of equations $\mathbb{E}$.

Furthermore, once we have a set of equations $\mathbb{B}_{\mathbb{E}}$, we can perturb it to get a new set of equations $\mathbb{E}_B$ using some perturbation set B and then from the equations $\mathbb{E}_B$ get a new argumentation network $\mathcal{A}_{\mathbb{E}_B}$. The net result of all these steps is that we start with a network $\mathcal{B} = (S, R)$ and a perturbation set of nodes $B \subseteq S$ and we end up with a new network which we can denote by $\mathcal{A} = \mathcal{B}_B$. If B is a loop-busting set, then $\mathcal{A}$ is the loop-busted result of applying B to $\mathcal{B}$.

Definition 18.79. *1. Let V be a set of variables and let $x = \mathbf{h}_x(V), x \in V, V_x \subseteq V$ be a system of equations $\mathbb{E}$, where V_x is the set of variables actually appearing in $\mathbf{h}_x$. We now define the associated argumentation network $\mathcal{A}_{\mathbb{E}} = (S_{\mathbb{E}}, R_{\mathbb{E}})$ as follows:*
   a) Let $S_{\mathbb{E}} = V$
   b) Let $yR_{\mathbb{E}}x$ hold iff $y \in V_x$.[11]
2. Let $\mathcal{B} = (S, R)$ be a network and let $\mathbb{E} = Eq(\mathcal{B})$ be its system of equations. $\mathbb{E}$ is a system of equations as in (a) above. Let $B \subseteq V$ be some of the variables in V. Let $\mathbf{f}$ be a function giving numerical values 0 to the variables in B. Let $\mathbb{E}_{\mathbf{f}}$ be the system of equations obtained from $\mathbb{E}$ by substituting the values $\mathbf{f}(u)$ in the equations for the variables of B. The variables of $\mathbb{E}_{\mathbf{f}}$ are $V - B$. Consider now the argumentation network

$$\mathcal{A}_{\mathbb{E}_{\mathbf{f}}} = (S_{\mathbb{E}_{\mathbf{f}}}, R_{\mathbb{E}_{\mathbf{f}}}).$$

We say that $\mathcal{A}_{\mathbb{E}_{\mathbf{f}}}$ was derived from $\mathcal{B}$ using $\mathbf{f}$. We can also use the notation $\mathcal{B}_{\mathbf{f}}$ or $\mathcal{B}_B$.

---

[11] The definition of yRx as $y \in V_x$ is a very special definition, making essential use of the fact that the equation

$$x = \mathbf{h}_x(V_x)$$

is of a very special form of either

$$x = 1 - \max V_x$$

or

$$x = \prod_{y \in V_x}(1 - y)$$

The real definition, which is more general, should be

$$yRx \text{ iff when we substitute } y = 1 \text{ in } \mathbf{h}_x, \text{ we get that } \mathbf{h}_x = 0.$$

This definition is good in a more general context.
   Suppose y_1, y_2 attack x *jointly*. This means that $x = 0$ only if both $y_1 = y_2 = 1$. See Chapter 7 for a discussion of joint attacks.
   The equation for that is

$$x = 1 - y_1 y_2$$

Given a general equation

$$x = \mathbf{h}_x(V_x)$$

for example

$$x = \mathbf{h}_x(y_1, y_2, z) = (1 - z)(1 - y_1 y_2)$$

We define the notion for $V_x^0 \subseteq V_x$ of joint attack as follows.
   V_x^0 attack x jointly if the substitution of $u = 1$ for all variables in V_x^0 makes $\mathbf{h}_x = 0$ and for no proper subset of V_x^0 do we have this property.
   So in the above example, y_1, y_2 attack x jointly and z attacks x singly.

Example 18.80. Let us use the network of Figure 18.26 to illustrate the process outlined in Definition 18.76.

The variables of this figure are

$$V = \{a, b, c, d, e, f, g, h, i\}$$

The equations are, using Eq_{inverse} as follows:

1. $a = 1 - c$
2. $b = 1 - a$
3. $c = 1 - b$
4. $d = 1 - b$
5. $e = (1 - b)(1 - f)$
6. $f = (1 - e)(1 - d)(1 - i)$
7. $g = (1 - f)$
8. $h = (1 - g)$
9. $i = (1 - h)$

Let us take $B = \{a\}$ and let $\mathbf{f}$ be the function making $a = 0$ (i.e. $\mathbf{f}(a) = 0$). (This is a loop-busting move, breaking the loop $\{a, b, c\}$).

The new equations for a are

$1^*a.$ $a = (1 - c)Z(a)$
$1^*b.$ $Z(a) = 0$

or we can simply write

$1^*.$ $a = 0$

Substituting this value in the equations and solving we get the new system of equations for the unknown variable as follows

1. $b = 1$, known value
2. $c = 0$, known value
3. $d = 0$, known value
6. $f = 1 - i$
7. $g = 1 - f$
8. $h = 1 - g$
9. $i = 1 - h$

We get the solution function $\mathbf{f}_1$ giving the known values to the variables $a = 0, b = 1, c = 0, d = 0, e = 0$ (these are the variables of rank 2) and the new system of equations (6), (7), (8), (9). Eliminating the known variables from the original network, we get the derived network in Figure 18.88.

We can continue now with this loop and choose a loop-busting variable say $B' = \{i\}$. We substitute $i = 0$ in the equations and get $f = 1, g = 0, h = 1, i = 0$ (these are the variables of rank 1). We extend the function $\mathbf{f}_1$ to be $\mathbf{f}_2$ giving these values.

We thus get the extension $\{b, f, h\}$ (these are the variables which get value 1 from $\mathbf{f}_2$). We also get clear ranks for the variables for the particular protocol $\mathbb{P} = (B, B') = (\{a\}, \{i\})$.

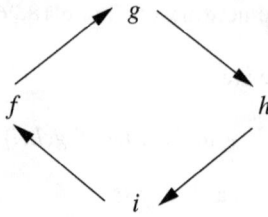

Fig. 18.88.

Example 18.81. Let us do another example. We use Figure 18.85 item b. We note that in Example 18.64, item (e), we make $\phi = 0$. This means we start with $B_1 = \{\phi\}$.

We manipulated the equations in item (e) of Example 18.64 and got the remaining equation (1)–(4), namely

1. $\alpha = 1 - \gamma$
2. $\delta = 1 - \alpha$
3. $\beta = 1 - \delta$
4. $\gamma = 1 - \beta$

We proceeded in item (e) of Example 18.64 to find two solutions to these equations. However, if we follow the procedures of Definition 18.76, we need now to extract a new network out of equations (1)–(4), and keep in mind the partial function $\mathbf{f}_1(\phi) = 0$. The new network is presented in Figure 18.89.

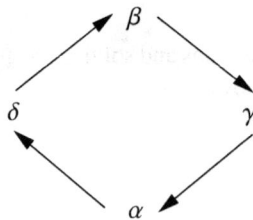

Fig. 18.89.

We can now proceed by choosing a new loop-busting set B_2. We have four options here. Choosing B_2 to be α or β will make the extension $\alpha = \beta = 0, \delta = \gamma = 1$ and choosing B_2 to the γ or δ will give the extension $\alpha = \beta = 1, \gamma = \delta = 0$.

These are also the solutions we got in item (e) of Definition 18.64.

18.7 The adjustment approach to handling loops

Previous sections offered a method of handling loops in networks (S, R), which were external. They did not change the attack relation R. The methods basically tried to identify extensions in various ways, using various algorithms and to the extent that we moved to another network, it was to show soundness; to explain what the algorithm was doing in terms of another network. The adjustment method actually says that since (S, R) has problematic loops, let us consider another network $(S, \check{R})$ instead. This approach is risky because it may seem that we are evading the problem, rather than solving it.

The perceptive reader might say that we are not dealing with (S, R) but with a different network $(S, \check{R})$! Whereas previous sections tried to compute the (S, R) extensions and when presented with a loop tried to fix the algorithm and get results, the adjustment method abandons (S, R) and goes to $(S, \check{R})$ instead. The perceptive reader might think that this method is methodologically faulty.

The answer is that it all depends on how near $(S, \check{R})$ is to (S, R). To take a family example, if I offer you soup and you don't like it and I give you a sandwich instead, then I have abandoned the soup in favour of a sandwich. However, if I put pepper in the soup and thus make it nicer and more acceptable to you, then I have fixed it. Let us illustrate this idea by an example which leads to a nice theorem.

Example 18.82. Consider a 6-cycle (S, R) as in Figure 18.35. This cycle has the traditional non-empty extensions $\{a_1, a_3, a_5\}$ and $\{a_2, a_4, a_6\}$. (It also has the all undecided empty extension.) It also has the CF2 extensions $\{a_1, a_4\}, \{a_2, a_5\}, \{a_3, a_6\}$.

The LB1 annihilator approach, as well as the impulse and linear logic like (LLL) approaches can get these extensions, depending on their loop handling policy. Let us focus our attention on the CF2 approach. Let us change R into $\check{R}$, being its symmetric closure. We have

$$x\check{R}y \text{ iff } xRy \vee yRx$$

We get Figure 18.90

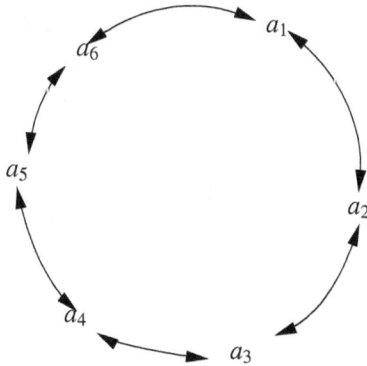

Fig. 18.90.

It is easy to see that for any strongly connected network (S, R) the maximal conflict free subsets of (S, R) and the maximal conflict free subsets of its symmetric closure $(S, \check{R})$ are the same. So they both have the same CF2 extensions. There is, however a fundamental difference between the way we compute the extensions for (S, R) and $(S, \check{R})$. In $(S, \check{R})$ we do the CF2 approach, not the traditional Dung style approach.

For $(S, \check{R})$, on the other hand, we can take all the traditional Dung style non-empty extensions!

Thus the CF2 semantics becomes (at least for the case of strongly connected sets) an adjustment approach semantics.

Start with (S, R). Move to $(S, \check{R})$, take all non-empty extensions for $(S, \check{R})$ and these will be your CF2 extensions for (S, R).

It seems this may lead us to some sort of a general theorem. To find out we need more examples.

We have to be careful here at the boundaries, i.e. when E is empty. Take (S, R) to be $(\{a\}, \{((a, a)\})$, i.e. a single point attacking itself. The symmetric closure is (S, R) itself and the CF2 and traditional extensions are empty. So we had better talk only about (S, R) not containing self attacking elements!

We can prove that if we delete from a network (S, R) all the self attacking elements, then the CF2 extensions (which do not contain self attacking elements anyway) remain the same.

We are aiming at something like Conjecture 18.83.

Conjecture 18.83. There exists an algorithm which gives us for any (S, R) without self attacking elements, another network (S, R_1) such that for any non-empty subset $E \subseteq S$, (1) and (2) are equivalent:

1. E is a CF2 extension in (S, R)
2. E is an ordinary traditional complete extension in (S, R_1).

This conjecture is certainly true if (S, R) is a strongly connected set. So maybe we can take R_1 to be the result of adding to R all the symmetric closures of all maximal strongly connected subsets in (S, R).

Example 18.84. Consider Figure 18.91. The top maximal strongly connected set is $\{\alpha, \beta, \gamma\}$ and the bottom one is $\{x, y, z\}$. Our conjecture says that we should take the symmetric closure of each cycle and get Figure 18.92 and that on this new figure the traditional extensions would give all CF2 extensions on Figure 18.91.

This is not the case!

Let us check Figure 18.91. We can start by letting (CF2 style) $\beta = $ "in", and $\gamma = \alpha = $ "out". Therefore $x = $ "out" and $y = $ "in" and $z = $ "out".

If we do the same with Figure 18.92, i.e. start with $\beta = $ "in", we get Figure 18.93.

The CF2 extensions of Figure 18.91 do not allow for $\beta = $ "in" and $z = $ "in", while Figure 18.93 does allow for that.

By symmetry we would have the same problem if we had started with the extension $\gamma = $ "in", $\alpha = \beta = $ "out".

We ask ourselves, is there another formulation of the conjecture which will be a general theorem? The idea is, after all, very attractive.

We want something like the following:

Let (S, R) be a network. Then there exists an adjustment of R into R_{CF2} (same S) where R_{CF2} may be possibly a higher level attack relation, such that for any set $E \subseteq S, E \neq \varnothing$ we have (1) iff (2)

1. E is a CF2 extension of (S, R).

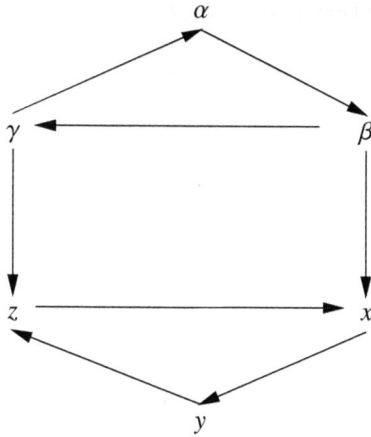

Fig. 18.91.

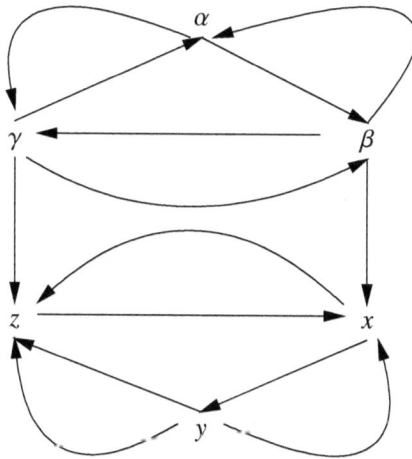

Fig. 18.92.

Fig. 18.93.

2. E is a Dung traditional extension of (S, R_{CF2}).[12]

The next example analyses the situation, with a view to formulating a general theorem.

Example 18.85. Consider Figure 18.94. This is a schematic figure of what can happen in our directional computation of C2 extensions, if we follow the procedure of item 2 of Remark 18.34.

C_1 and C_2 are maximal strongly connected sets attacking the maximal strongly connected set C. We are in the middle of the construction of the CF1 extension and we have several possibilities to continue the construction by choosing extensions in C_1 and in C_2. In Figure 18.94 we show two for each. E_1^1 and E_2^1 and E_1^2 and E_2^2.

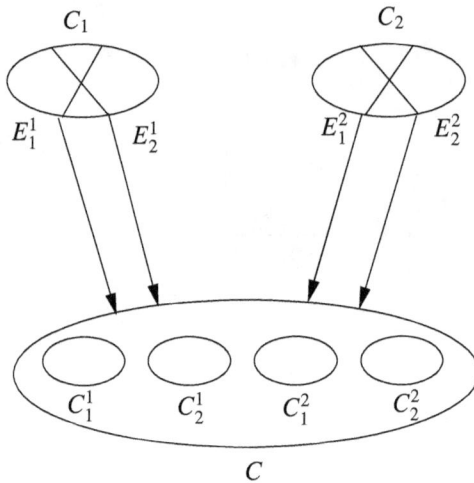

Fig. 18.94.

Suppose we choose E_1^1 and E_1^2 as the extensions. These attack C and propagate and we are left with the strongly connected subset C_1^1 of C, which we can use to continue our inductive constructions.

Had we chosen a different pair, say E_2^1 and E_2^2, we would have been left with a possibly different strongly connected subset C_2^2. We know that to continue the construction we can take the symmetric closure on C_1^1 and this will work for the current inductive construction which chose the extensions E_1^1 and E_1^2 but it will not work had we made a different choice and taken the extensions E_2^1 and E_2^2. So we do not know beforehand what symmetric closure to take of what parts of C. The choice depends on the inductive process itself. More explicitly, let xy be in C and assume x attacks y. For some choice of extensions, say E_1^1 and E_1^2, after they attack C and propagate, it may be the case that x and y will remain in a loop and so we can add to the loop that y attacks x. However, had our process made a different choice, say E_2^1 and E_2^2, x and y

---

[12] We need to show how traditional extensions can be defined for networks with higher level attacks. See [163] and Chapter 8, as well as [32, 33].

could be out of the loop and so we must not add that y attacks x. So what do we do? We must take a higher level approach. We use higher level joint attacks. Recall Figure 18.31. This figure shows a joint attack on a node c. We have $c =$ "out" if both $a = b =$ "in".

We use this device. Let E_1 be CF2 extension of C_1 and E_2 of C_2. Take the symmetric closure of C and let x attacks y be at C. then we have added that y attacks x. It may be that the result of the combined attack of E_1 and E_2 on C requires that y does not attack x. In this case we form a joint attack from E_1 and E_2 on the arrow $y \to x$, to cancel it. Figure 18.95 shows this situation.

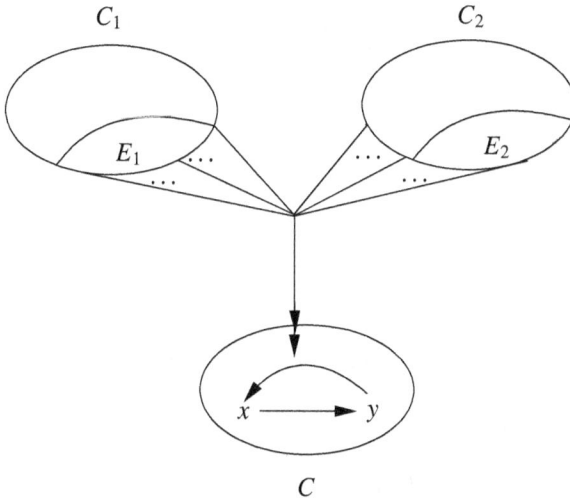

Fig. 18.95.

Thus for any CF2 extension E_1 and E_2 we mount joint attacks from E_1 and E_2 onto any arrow $y \to x$ that should not be there (i.e. was added when we took the symmetric closure of C). If the CF2 construction process chooses a different set of extensions, say E_2', E_2', then these joint attacks will not fire and the joint attacks from E_1' and E_2' will fire.

So really the embedded joint attacks in the higher level network ensures that the process of construction proceeds as if we have never added any arrows to the original network (S, R).

Figure 18.96 shows how to modify Figure 18.92 in the spirit of what we have just discussed. We add the double arrow attacks from β and from γ.

Our disucssion shows that the following theorem is true.

Theorem 18.86. *Let (S, R) be an argumentation network without self attacking elements. Then there exists a higher level argumentation network (S, R_{CF2}) (same S, different attacks) such that (1) and (2) are equivalent for any subset $E \subseteq S$.*

1. E is a CF2 extension in (S, R)
2. E is a traditional Dung extension in (S, R_{CF2}).

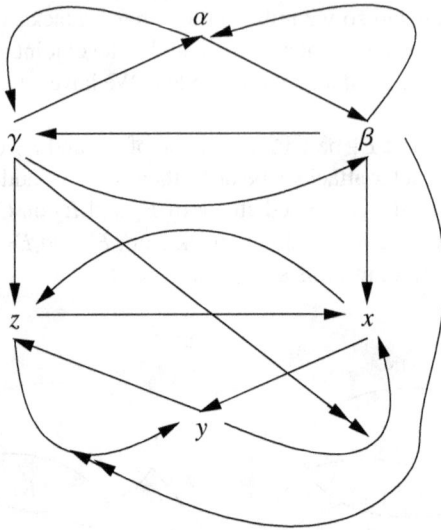

Fig. 18.96.

Proof. As discussed and illustrated in Example 18.85.

Remark 18.87. We need to figure out what to do with (S, R) which contains self attacking elements. My guess is that we need to adjust it to a network (S', R') without any self attacking elements and then apply Theorem 18.86.

Theorem 18.88. *Let $S, R)$ be a network and let D be the subset of all self attacking elements. Then the network with $S - D$ and R restricted to $S - D$ has the same CF2 extensions as (S, R).*

Proof. By induction of the directional construction of the CF2 semantics as given in item 2 of remark 18.34. It is sufficient to prove this for strongly connected sets, since the CF2 extensions of such sets drive the induction process. Clearly, if any maximal conflict free subset of S would be maximal in $S - D$ but not in S, then it can be extended in S with at least one more element x. x is not in D because the elements of D are self attacking. So x is in $S - D$ and so E is not maximal in $S - D$. The other direction is even more obvious.

Remark 18.89. Note for example that the theorem holds for the network pair of Figure 18.91 and Figure 18.96.

The perceptive reader might raise some objections to Theorem 18.86 as follows:

Question 1: You are using higher level networks. I don't like it.
Answer 1: Higher level networks are well accepted and used by now. They make sense and arise in many contexts. See Chapter 8 and [163, 32].

Question 2: To find R_{CF2} you need to follow the LB2/CF2 inductive construction process and compute all the CF2 extensions of (S, R) in order to define the double arrows of R_{CF2}. So you give us R_{CF2} which yields all extensions only after you have already computed all of them.

Answer 2: We are giving an existence representation theorem of conceptual value. It is not a computational complexity theorem. When you give a representation theorem you assume you know what you have and you want to represent it in a different way. In many cases it is true, that the computation on the representation is faster than on the original (see Remark 18.89, for example) but this would be only a bonus.

Furthermore, the basic idea is to take the symmetric closure of the attack relation. We can use only this idea as follows:

Begin with a network (S, R)

1. Identify the top loop maximal strongly connected sets. Delete all self attacking elements.
2. Make the attack relation symmetric on these sets.
3. Find all traditional extensions on the resulting sets.
4. Propagate the attack in the network for each extension.
5. Take out all points with clear $\{0, 1\}$ value. Let the residual network be (S_1, R_1).
6. If (S_1, R_1) is empty stop, otherwise go to 1. and proceed with (S_1, R_1).

Compare with item 2 of Remark 18.34.
See also our answer to Question 3.

Question 3: Are there any advantages to your theorem?

Answer 3: Yes, there are. Higher level networks have a simple and natural equational semantics. This is an advantage, as the equations involve only the nodes in S as variables. So, as a result, we have the following theorem.

Theorem 18.90. *There exists an algorithm yielding for any network (S, R) a system of equations $\mathbb{E}$ involving the elements of S as variables, such that all the $\{0, 1\}$ solutions of $\mathbb{E}$ give exactly the CF2 extensions of (S, R).*

Proof. The CF2 extensions of (S, R) can be obtained as ordinary Dung extensions of (S, R_{CF2}). The latter has equational semantics involving elements of S only as variables. So this is the system of equations on S all of whose solutions give the CF2 extensions for (S, R).

Also using a general theorem which says that we can eliminate higher level attacks, we can get another theorem:

Theorem 18.91. *Let (S, R) be a network. Then we can effectively construct a network (S_1, R_1), such that $S \subseteq S_1$, $R \subseteq R_1$ and the following holds:*

1. *Any CF2 extension E of (S, R) can be uniquely extended to a traditional Dung extension E_1 of (S_1, R_1).*
2. *For any traditional extesnion E_1 of (S_1, R_1) the set $E = S \cap E_1$ if non-empty is a CF2 extension of (S, R).*

Proof. Follows from reductions in [164] and Chapter 7.

Example 18.92. To see how the equational approach works, let us compare the equations for 6 cycle (Figure 18.35 with its symmetric closure (Figure 18.90).

The equations for 6-cycle are:

1. $a_1 = 1 - a_6$
2. $a_2 = 1 - a_1$
3. $a_3 = 1 - a_2$
4. $a_4 = 1 - a_3$
5. $a_5 = 1 - a_4$
6. $a_6 = 1 - a_5$.

There are three solutions:

1. $a_1 = a_2 = a_3 = a_4 = a_5 = a_6 = \frac{1}{2}$ (this is all undecided solution.)
2. $a_1 = a_3 = a_5 = 1$
   $a_2 = a_4 = a_6 = 0$
3. $a_1 = a_3 = a_5 = 0$
   $a_2 = a_4 = a_6 = 1$

The symmetric closure of Figure 18.90 has the following equations:

1. $a_1 = 1 - \max(a_6, a_2)$
2. $a_2 = 1 - \max(a_1, a_3)$
3. $a_3 = 1 - \max(a_2, a_4)$
4. $a_4 = 1 - \max(a_3, a_5)$
5. $a_5 = 1 - \max(a_4, a_6)$
6. $a_6 = 1 - \max(a_5, a_1)$

The solutions (2) and (3) are also available here.

18.8 Handling loops in Logic Programming

This section shows how the linear logic LLL approach applies to Logic Programming loop checking. There is a close connection between argumentation and logic programming (see [164] and Chapter 7) and so it is worthwhile for us to do this.

For the purpose of orientation we assume some familiarity with logic programming. Formal definitions will be given in later sections.

We begin with several simple examples.

Example 18.93. Consider the looping logic program P_0:

(i) $\neg a \rightarrow b$
(ii) $\neg b \rightarrow a$.

Let us use the traditional goal directed computation, see [140]. In this computation we loop. We can use a traditional loop detector (a loop occurs when the same query is asked again, with the same annotations, from the same database, see [140]) and get

1. $?a$ original query (to succeed, we also write $?a = 1$)
2. $?\neg b$ (to succeed), from (i)
3. $?b$ (to fail, we also write $?b = 0$)
4. $?\neg a$ (to fail) from (ii)
5. $?a$ (to succeed), loop because it is the same as 1.
6. abandon the computation, in view of the loop.

An ordinary loop policy would give up and probably say that a does not succeed. Similarly we will have to say that b does not succeed. So we get $\varnothing$.

The above problem is solved by answer set programming. We choose a subset of $X \subseteq \{a, b\}$ and see if it is a fixed point under P_0 acting as a closure operator.[13]

We get two answer sets in this case: $\{a\}$ and $\{b\}$. $\varnothing$ is not an answer set.

Note that the above logic program corresponds to the argumentation network of item (b) of Figure 18.97. The extensions are $\varnothing, \{a\}$ and $\{b\}$.

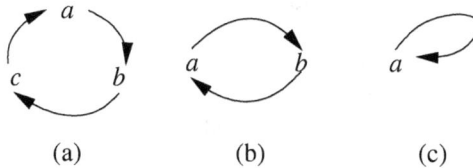

(a)   (b)   (c)

Fig. 18.97.

Remark 18.94. We need to clarify the methodology involved here; lay down the conceptual framework in which our loop checking is going to operate. This is as follows

1. We are given a logic program P, built up using a set of literals and clauses (this will be defined in Definition 18.98 below)
2. We are interested in identifying subsets of literals, generated using P and some additional algorithmic mechanisms. We call such subsets extensions. We have seen in Example 18.93 several such mechanisms
   a) Ordinary logic programming computation
   b) ordinary logic programming computation augmented by a traditional loop checker
   c) Answer set programming
   d) For certain logic programs arising from argumentation we have the various semantics generating extensions as done in argumentation.

In this section we want a more sophisticated loop checker, and we want to get the extensions not by guessing answer sets, but by following ordinary goal directed computation with a clever loop checker.

---

[13] The way it works is as follows: Given a set of literals X, we create a new program P^X out of P_0 in the following way: We look at each clause in the program (clauses are defined in Definition 18.98). The body of the clause has positive literals of the form a_i and negative literals of the form $\neg b_j$. If all of the b_j of this clause are not in X, then we delete them from the clause and put the new clause we thus obtained into P^X. If this is not the case we skip and look at another clause. By going through the clauses of the program one by one we thus define P^X. We now say that X is an answer set for P_0 if X is a fixed point of P^X regarded as a closure operator. See [199]. Note that our view here of answer set programming is very restricted, we look only at the narrow aspect of this area as a mechanism to finding extensions.

Example 18.95 (Example 18.93, continued). Let us do the computation of Example 18.93 again. We follow lines (1) to (5) of Example 18.93, but instead of abandoning the computation in line (6), we delete the clause involved in the loop, in this case the clause $\neg b \to a$, and continue with the original goal involved in the loop.

So let us continue

(6*) Delete clause $\neg b \to a$ and start again.
(7*) ?a, fail because there is no clause with head a.

We now come to a key point: We asked ?a from the original program P_0 and using our new loop checking algorithm we failed, and in the process we got a reduced smaller program. Now we want to ask ?b but we need to decide from which program do we ask ?b from? Do we ask ?b from the original one or from the reduced one? If we ask ?b from the original program, it will also fail and the extension we get is the empty set of literals. However, if we ask ?b from the reduced program it will succeed, because we are computing

$$\{\neg a \to b\}?b.$$

Thus if we ask the sequence (?a, ?b) we get the extension $\{b\}$ and by symmetry if we ask the sequence (?b, ?a) we get the extension $\{a\}$. So the basic idea is to devise a clever loop checker that detects loops and tells you what to do next in view of what it detected, and hope to get the right results.

Remark 18.96. The considerations in Example 18.95 motivate the following sequence of assumptions for our quest for extensions:

A1.   Assume we are given a finite logic program P based on a set of literals Q.
A2.   Assume we devised a new dynamic loop checker algorithm that whenever we ask a literal query ?$x = 1$ or ?$x = 0$ gives a definite success or failure answer. Of course ?$x = 1$ succeeds (resp. fails) iff ?$x = 0$ fails (resp. succeeds)
A3.   Assume that during the application of the loop checker in A2, certain clauses are thrown out from the program P. Let the resulting reduced program be denoted by P_x.
A4.   Assume that the process of A2. and A3. not only gives a definite answer to ?$x = 1$ but also indicates (exports) that some other literals (e.g. $Y_x = \{y_j\}$) should be failing (that is, the answer to ?$y_j = 1$ should be fail). This will happen, for example, if there is only one clause with head y_j and this clause is not in P_x.

Given the above assumptions on the loop checking process we can describe how to find extensions for our program P based on our loop checking process

E1.   Arrange all literals of Q in a sequence. We find an extension relative to this sequence. If we arrange the literals in a different sequence the extension may be different. The family of all extensions is obtained by looking at all possible sequences.
E2.   Given a sequence **s**, let x be its first element. Ask the query P?$x = 1$. Use the loop checker. You will get a definite answer success or failure. In the process we get a new reduced program P_x and a set Y_x as in A3. Let $\mathbf{s}_x$ be the sequence obtained from **s** by deleting x and the elements of Y_x. Then continue by applying E2 to P_x and $\mathbf{s}_x$

E3.   Continue until you get success or failure answers to all elements of **s**. This is your extension denoted by Extension(**s**)

E4.   Restart the process with another sequence to obtain another extension. Carry on for all possible sequencing of Q to obtain all possible extensions for P based on our loop checking algorithm.

E5.   The big question is of course what is the semantical meaning of these extensions which we can get algorithmically using our loop checker.

We hope to formulate the process in A3 in a careful way so that we get a good semantical meaning.

Example 18.97. Consider now the program $\neg a \to a$.

1. ?a (to succeed)
2. ?$\neg a$ to succeed
3. ?a to fail
4. ?$\neg a$ to fail
5. ?a to succeed. Loop
6. Delete $\neg a \to a$ and continue by asking the original goal from the smaller program
7. $\varnothing$?a, fail.

This result is different from what we get in argumentation. $\neg a \to a$ corresponds to the argumentation network of item (c) of Figure 18.97. Argumentation wants a as undecided. The CF2 semantics will throw a out and the extension will be the empty set. The moral of the story is that we need to define our loop checking procedure carefully.

We have shown in [164] and Chapter 7 that argumentation networks can be represented faithfully in the object level by logic programs.[14] Therefore problems with loops in argumentation networks such as the problems arising from the treatment of odd and even loops (see [30]) may be treated using loop checkers as used in logic programming. This section uses linear logic loop checking techniques of Section 6 to solve the problems of loops in argumentation networks through the translation into logic programs.

We begin with some definitions and examples.

Definition 18.98. *A logic program P in the atoms Q is a set of clauses of the form*

$$\bigwedge a_i \wedge \bigwedge \neg b_j \to x, \text{ where } a_i \text{ and } b_j \text{ are atoms.}$$

x is the head of the clause, the rest is the body of the clause. We say a_i appear positively in the body of the clause and b_j appear negatively. Note that we may not have any a_i or b_j or both appear in the clause.

---

[14] The logic programs we get (corresponding to argumentation networks) have clauses of special form (compare with Definition 18.98 below):

1. Each literal appearing in the program is the head of exactly one clause
2. The clauses in the program do not have any positive literals in their body.

Let us write the basic metalevel computation predicate as $P?x = 1$ or $P?x = 0$, for x an atom. $P?x = 1$ corresponds to the traditional $P?x$ and $P?x = 0$ corresponds to the traditional $P?\neg x$.

The computation rules are as follows:

1. $P?x = 1$ *(resp. $P?x = 0$) succeeds if for some clause C in P, of the form $\bigwedge a_i \wedge \bigwedge \neg b_j \to x$, we have that all of (resp. some of) $P?a_i = 1, P?b_j = 0$ succeed (resp. fail).*
2. $P?x = 0$ *(resp. $P?x = 1$) fails if $P?x = 1$ (resp $P?x = 0$) succeeds.*
3. $P?x = 1$ *succeeds if $x \in P$*
4. $P?x = 0$ *succeeds if x is not the head of any clause in P.*

Example 18.99 (Loop checking). Consider the program $P = \{1.\neg a \to b, 2.a \to a\}$. In loop checking we may get different results, depending on our loop checker. We shall see that the results depend on how we order the queries.

Loop checking with $?b = 1$

$P?b = 1$, from first clause, if $P?\neg a = 1$, if $P?a = 0$ from second clause. We are in a loop. We stop the computation and say that the original query $?b = 1$ loops or fails (if we consider looping as failure).

Loop checking with $?a = 1$.
$P?a = 1$, if $P?a = 1$ from second clause, we are in a loop.
We consider the query $?a = 1$ as looping or as failed.

Loops checking with both $?a = 1$ and $?b = 1$ in parallel single steps

1.	$?b = 1$	$?a = 1$
2.	$?\neg a = 1$	$?a = 1$
3.	$?a = 0$	$?a = 1$

We get opposing requirements on a from line 3. We need a policy of what to do. In the above, we followed the traditional policy 1.

Policy 1: (traditional)
If we start a computation with original goal G and all possible computation paths either finitely fail or loop then we can consider the original goal either as an unsuccessful looping goal or as a failed goal.

This is the traditional loop checking policy, where we abandon a computation path option if we detect a loop on that option.

Policy 2: (Gabbay[15])
If we detect a loop in a computation path, we stop and examine the loop. We identify the looping elements (current goal and previous goal which create the loop) and depending on the nature of the loop we decide what to do, by looking at an already prepared loop resolution table. Every computation algorithm must contain a loop resolution table.

Example 18.100 (Linear loop checking). Let $P = \{\neg a \to a\}$. 1. $P?a = 1$, if 2. $P?\neg a = 1$, using the clause $\neg a \to a$, if 3. $P?a = 0$. 4. We detect a loop with $a = 1$ and $a = 0$.

---

[15] Introduced in 1985, mentioned orally in various lectures at the time.

We look up our loop checking table and let us assume it says for this case we have to delete the clause used at the origin of the loop, i.e. the clause $\neg a \to a$. We continue with $P' = \varnothing$ by asking the query at the origin of the loop again, but this time from P'.

5. $P' = \varnothing ?a = 1$, repeating (1) above. fail.

Thus $P?a = 1$ fails. Therefore $a = 0$ (if we seek $\{0, 1\}$ solutions).

Note that our loop checking table might have said that we ask the current query, at the end of the loop, in which case we would ask

5. $P'?a = 0$, and succeed, since P' is empty.

We need, of course, to make the concepts involved precise. We will do this later in the section.

The perceptive reader might ask why are we calling this type of loop checking "linear loop checking"? The answer is clear, we are allowing for any clause to be used no more than once in a loop and we delete it, if used. The meaning of "linear" here is as in "Linear Logic".

Example 18.101 (Odd loop of 3 elements). Consider the program $P = \{1.\neg c \to a, 2.\neg a \to b, 3.\neg b \to c\}$. We want to ask for a sequence of three goals $?(a, b, c)$ in that order. Let us ask $P?a = 1$

Step 1. $P?a = 1$
Step 2. $P?c = 0$, from clause 1
Step 3. $P?b = 1$, from clause 3
Step 4. $P?a = 0$, from clause 2
Step 5. $P?c = 1$
Step 6. $P?b = 0$
Step 7. $P?a = 1$.

We get a loop with $P?a = 1$. Its origin is step 1, using clause 1.

Our loop checker says delete the clause used in the origin of the loop. Our new database is $P' = \{\neg a \to b, \neg b \to c\}$.

We continue the query at origin of the loop from the new database.

Step 8. $P'?a = 1$ fail (no clause with head a).

We now continue and ask the next goal b:

Step 9. $P'?b = 1$
Step 10. $P'?a = 0$, from clause 2. Success as a fails in line 8

We now continue and ask the third goal c:

Step 11. $P'?c = 1$
Step 12. $P'?b = 0$, from clause 3. We fail, since $P'?b = 1$ succeeded, (respectively failed) in line 7.

Example 18.102. Let $P = \{1.\neg a \wedge \neg b \to c, 2.\neg a \to a, 3.\neg a \to b\}$. We ask $P?c = 1$.

1. $P?c = 1$
2. (a) $P?b = 0$ and (b) $P?a = 0$, from clause 1
3. (a) $P?a = 1$ and (b) $P?a = 1$, from clauses 3 and 2 respectively.

4. (a) $P?a = 0$ and (b) $P?a = 0$, both from the second clause.

We get a loop at 4, with $a = 0$ originating at 2b from clause 2.
  We therefore delete clause 2 and continue with $P' = \{\neg a \wedge \neg b \rightarrow c, \neg a \rightarrow b\}$.
  We continue and ask the original goals 2b and 3a again from P'

5. (a) $P'?a = 1$ (this is a repetition of 3a) and (b) $P'?a = 0$ (this is a repetition of 2b).
   4a fails and 4b succeeds for the same reason, namely that a is not the head of any
   clause in P'.
6. $P?c = 1$ fails because 5a fails.

Example 18.103 (Non-contradictory loop). Let $P = \{1.\neg a \rightarrow b, 2.\neg b \rightarrow a\}$. Let us ask
for $P?b = 1$

1. $P?b = 1$
2. $P?a = 0$, from clause 1
3. $P?b = 1$, from clause 2.

This is a loop with same value, same goal at origin step 1 with $b = 1$ and end step 3
with $b = 1$. We follow the same procedure and delete the clause used in the origin of
the loop.
   Let $P' = \{\neg b \rightarrow a\}$. We ask the query origin of the loop (which is the same as the
current query) again from P'.

4. $P'?b = 1$ fails. b is not the head of any clause in P'!
5. We ask now $P'?a = 1$
6. $P'?b = 0$ succeeds.

We thus get $a = 1$ (from steps 6 and 7) and $b = 0$ from step 5.
   To get the other "extension", $a = 0, b = 1$, we start by asking $?a = 1$ first, then
$?b = 1$.

Definition 18.104. *Let P be a logic program and let Q_P be the set of literals involved
in P. Define a binary relation R on Q_P as follows:*

  *xRy iff there is a clause C in the program P with head y in whose body the
  literal x appears.*

*The above means that to compute the success or failure of y we need to know the
success or failure of x.*

Remark 18.105. Now that we have defined (Q_P, R) in Definition 18.104, we can proceed as in Definition 18.19 and define the notion of top loop-literal (node) for P (i.e. a
top loop-node in (Q_P, R)). We can also define the notion of a cycle and the notion of a
maximal strongly connected class as in Definition 18.21.
   These concepts we are going to use in our loop checking for logic programs.

We are ready now to define the loop-checking mechanism in a precise way.

**Definition 18.106 (Prolog computation with loop checking, with the policy of ask-
ing the original query from a reduced program in the case of a loop).**

1. *Let P be a program with clauses of the form*

$$\text{clause } C_i : \bigwedge_j a_{i,j} \wedge \bigwedge_k \neg b_{i,k} \rightarrow c_i, \text{ for } i = 1, \ldots, n.$$

2. *Let $(T, <, 0)$ be a finite tree with root $0 \in T$ and for each $x \in T$, let $S_x \subseteq T$ be the set of its immediate successors. If x has no immediate successors, then we say x is an endpoint.*

   Let $y \in T$, then there exists a unique path $(y_0 = 0, y_1, \ldots, y_k = y)$, leading from 0 to y such that for each j, y_{j+1} is an immediate successor of y_j. We call any sequence $(y_j, y_{j+1}, \ldots, y_k = y)$ an end path leading from y_j to y.

   Let $x \in T$. Let T_x be the set defined by $x \in T_x$ and if $y \in T_x$ then $S_y \subseteq T_x$. Then $(T_x, <, x)$ is a subtree, called the truncation at x.

3. *A function $\mathbf{F}$ is an annotation function of nodes $x \in T$, if $\mathbf{F}(x)$ has the form $\mathbf{F}(x) = (P', a, r, X)$ where $P' \subseteq P$, a is a literal of P, $r \in \{0, 1\}$ and X is either a clause in P with head a, or $X = $ loop, or $X = $ success or $X = $ failure.*

4. *Let $(T, <, 0, \mathbf{F})$ be an annotated tree. We say that it is a computation tree with loop checking for $P?a_0 = r_0$ iff the following holds:*

   a) *$\mathbf{F}(0) = (P_0, a_0, r_0, C_0)$, where C_0 is our beginning clause, used in the computation and $?a_0 = r_0$ is the beginning goal.*

   b) *Let $(x_1, \ldots, x_k)$ be a path in the tree leading from x_1 to x_k. We say $\mathbf{F}(x_1)$ and $\mathbf{F}(x_k)$ are in a loop if $\mathbf{F}(x_1) = (P_1, a_1, r_1, C_1)$, $\mathbf{F}(x_k) = (P_1, a_1, r_1, loop)$, and for no $x_j, 1 < j < k$ do we have the word "loop" in $\mathbf{F}(x_j)$.*

   We say that node x_1 is the origin of the loop and x_k is the end of the loop and clause C_1 is the cause of the loop.

   c) *If $\mathbf{F}(x) = (P', c, r, C)$ and C is the clause $\bigwedge_j a_j \wedge \bigwedge_k \neg b_k \rightarrow c$. Then $S_x = \{t_j, s_k\}$ with $\mathbf{F}(t_j) = (P', a_j, r, X_j)$ and $\mathbf{F}(s_k) = (P', b_k, 1 - r, X'_k)$.*

   d) *If $\mathbf{F}(x) = (P', c', r', loop)$ then for some y above x in the tree we have that $\mathbf{F}(y)$ and $\mathbf{F}(x)$ are in a loop and we have two possibilities:[16]*

      1. *The literal c' is not a top loop-literal in P' and x is an endpoint, or*
      2. *c' is a top loop literal in P' and $S_x = \{t\}$ and $\mathbf{F}(t) - (P' - \{C'\}, a_0, r_0, X)$ where $\mathbf{F}(y) - (P', c', r', C')$, and a_0 and r_0 are the original beginning query of the computation and X is the original clause C_0 of the computation. If it happens the case that $C_0 = C'$, i.e. C_0 is deleted because it is involved in the loop, then X is another clause with head a_0. If C_0 happens to be the only clause with head a_0, then $X = $ "failure" and this branch of the computation stops.*

   e) *If x is an endpoint then we have that $F(x) = (P', c, r, X)$.*

      i. *if $r = 1$ then either $X = $ success and $c \in P'$ or $r = 0$ and c is not the head of any clause in P'.*

      ii. *if $r = 0$ then either $X = $ success and c is not the head of any clause in P' or $X = $ failure and $c \in P'$.*

      iii. *the words "failure" and "success' appear only in endpoints.*

5. *We define the concept of $P?a = r$ succeeds (resp. fails) by the computation tree $(T, <, 0, \mathbf{F})$:*

   a) *If T is a one point tree, i.e. $T = \{0\}$, then $\mathbf{F}(0) = (P, a, r, X)$ and $X = $ success (resp. failure).*

---

[16] See Remark 18.107.

b) Assume $\mathbf{F}(0) = (P, a, r, C)$ *and C is the clause*

$$\bigwedge_{j=1}^{m} a_j \wedge \bigwedge_{k=m+1}^{n} \neg b_k \to a$$

Let $t_1, \ldots, t_n$ *be all immediate successors of* 0.

Then by the definition of the computation tree we can assume that $\mathbf{F}(t_j) = (P, a_j, r, X_j)$, *for* $1 \le j \le m$ *and that* $\mathbf{F}(t_k) = (P, b_k, 1 - r, X_k)$ *for* $m < k \le n$.

Then we say that $P?a = r$ *succeeds (resp. fails) if all (resp. some)* $P?a_j = r$ *and* $P?b_k = 1 - r$ *succeed (resp. fail) by their respective subtrees, namely* $(T, <, t_j, \mathbf{F})$.

c) Assume $\mathbf{F}(x) = (P, a, r, \text{loop})$ *and let y above x be the origin of the loop, with* $\mathbf{F}(y) = (P, a, r', C)$ *and C is a clause. Then the node x has exactly one successor t with* $\mathbf{F}(t) = (P - \{C\}, a, r', Z)$, *and the success or failure of* $\mathbf{F}(x) = (P, a, r, \text{loop})$ *is the same as the success or failure of* $\mathbf{F}(t) = (P - \{C\}, a, r', Z)$.

6. *A query* $?x = r$ *succeeds (resp. fails) from P if there is a (resp. every) computation tree for it succeeds (resp. fails).*

Remark 18.107. This remark refers to item 4(d) of Definition 18.106. At this case (item 4(d)) we know we are in a loop.

We need to decide whether to continue the computation on this branch or to abandon it and move on. The decision is based on the nature of the literal and clause involved in the loop.

If the literal is not a top loop-literal, then we stop.

If the literal involved is a top loop-literal, we continue but delete the clause involved in the loop.

This is condition 2 in item 4(d).

There is room for variation of policy here. If we follow condition 2. for any top loop-literal, then we would be deleting (in the overall computation) all top looping clauses in parallel.

We could however, as another option for a policy, linearly order all top loop-literals before the start of the computation and delete during the computation only the clause belonging to the first top loop-literal in the list (which has not been addressed yet). We wait for the computation to stop and then restart it as a new computation without the looping clause. This way we would be breaking the loops (by deleting top loop-nodes) not in parallel but one at a time. As an example consider the argumentation network of Figure 18.66.

The logic program corresponding to it is the following:

1. $\neg x \wedge \neg y \to z$
2. $\neg x \to y$
3. $\neg y \to x$.

In this program (and in the argumentation network) the top loop-literals (nodes) are $\{x, y\}$.

If we start the computation with $?z = 1$, we will continue to the queries $?x = 0$ and $?y = 0$. These will loop.

If we follow the parallel policy, both clauses 2. and 3. will be deleted and $?z = 1$ would succeed. However if we order the top loop-literals and delete them one at a time, we get the extensions $\{x\}$ and $\{y\}$, with z always "out".

Example 18.108 (Even loop of 4 points). Let $P = \{1.\neg a \rightarrow b, 2.\neg b \rightarrow c, 3.\neg c \rightarrow d, 4.\neg d \rightarrow a\}$. We ask $P?a = r$, for $r \in \{0, 1\}$. See Figure 18.98.

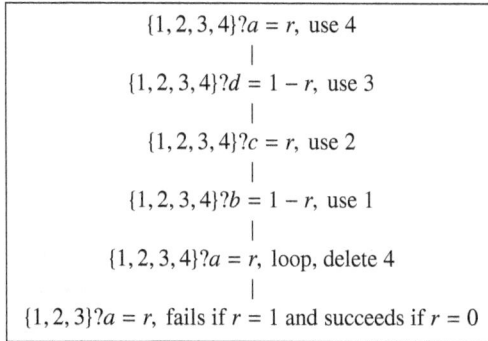

$\{1, 2, 3, 4\}?a = r$, use 4
|
$\{1, 2, 3, 4\}?d = 1 - r$, use 3
|
$\{1, 2, 3, 4\}?c = r$, use 2
|
$\{1, 2, 3, 4\}?b = 1 - r$, use 1
|
$\{1, 2, 3, 4\}?a = r$, loop, delete 4
|
$\{1, 2, 3\}?a = r$, fails if $r = 1$ and succeeds if $r = 0$

Fig. 18.98.

For $r = 1$, we get that the query fails. From the subtree for $d = 0$ we get that $d = 1$ succeeds. From the subtree for $c = 1$ we get that $c = 1$ fails. Similarly from the subtree for $b = 0$ we get that $b = 0$ fails, i.e. $b = 1$ succeeds.

So for $r = 1$, the "extension" is $a = 0, b = 1, c = 0$ and $d = 1$.

For the case $r = 0$, we get mirror image extension, namely $a = 1, b = 0, c = 1$ and $d = 0$.

Remark 18.109 (Soundness).

1. We note that Definition 18.106 presented a Prolog computation with loop checking. We ask ourselves, what is the logical meaning of this loop checking? The answer is that we are doing a version of goal directed linear logic with negation as failure. If we examine clause (4d) of Definition 18.106, we see that in case of a loop the computation continues with the original goal but with the new database which is the old database without the clause causing the loop. This is equivalent to saying "use each clause at most once, on each computation path if it is involved in a top maximal strongly connected loop". This answer however, is not satisfactory. Linear logic is a-priori resource bounded. It says right at the start, use any clause exactly once. We can be happy with a resource logic which fixes an integer k and says that we are allowed to use any clause at most k times. But to say use a clause at most once if it is involved in a loop is not a-priori declaration, it is run-time trick. Can we do better? The answer is yes. We can say that we are adopting a resource policy of using each clause at most twice. Now we have a legitimate a-priori resource logic programming computation. Does it do the same job as our loop checker? The answer is yes. Why is that true?

   What happens is that we use the original clause C at say point x in the path to ask for the goal say $a = r$?, then we get into a loop and reach the end of the loop at point say z, where we ask $a = r$ again? Then we would be using the goal and clause twice and so we cannot use it any more. This is exactly our loop checker policy.

So this is equivalent to deleting the clause and asking for the original goal at point z, the end of the original loop, as prescribed in item (4d) of Definition 18.106.

On the other hand if we use a clause and do not loop, this means we do not need to use it again with the same goal and therefore we can throw it out anyway. The reader might say that our notion of resource logic does not count the uses of the clause but the uses of the clause with the goal. So a clause with head c cannot be used twice with the goal $?c = r$, same r. So we are still describing how you use it and not simply counting the number of uses of the clause. This is no problem to us. If we say do not use the clause more than three times then certainly two out of three times will be with the same r and so we still have our correct loop checker, (but waste some computation time).

So let us summarise: Our loop checking method is to turn the traditional prolog computation into a resource logic where we say do not use a clause more than twice, on each computation path.

2. Thus going back to Definition 18.98, we can modify the definition of the computation there to reflect this idea of "use each clause at most twice, on each computation path" and suit our purpose as follows: Let P be a program. Turn P into a multi-set $\mathbb{P}$ containing each clause twice. Now compute like in linear logic, where whenever you use a clause of $\mathbb{P}$, immediately delete it.

The clauses of this new linear logic computation become the following:

1*. $\mathbb{P}?x = 1$ (resp. $\mathbb{P}?x = 0$) succeeds if for some clause C in $\mathbb{P}$, of the form $\bigwedge a_i \wedge \bigwedge \neg b_j \to x$, we have that all of (resp. some of) $\mathbb{P} - \{C\}?a_i = 1, \mathbb{P} - \{C\}?b_j = 0$ succeed (resp. fail).

2*. $\mathbb{P}?x = 0$ (resp. $\mathbb{P}?x = 1$) fails if $\mathbb{P}?x = 1$ (resp $\mathbb{P}?x = 0$) succeeds.

3*. $\mathbb{P}?x = 1$ succeeds if x is in $\mathbb{P}$

4*. $\mathbb{P}?x = 0$ succeeds if x is not the head of any clause in $\mathbb{P}$.

3. Of course the linear logic deletion of used clauses acts as a loop checker. It is actually more efficient to use the linear logic computation, (even though we are duplicating the database), than to use the computation of Definition 18.106, because we do not need to record the history of the computation to detect loops. Such logics were investigated in my papers from 20-30 years ago, see my book with Olivetti, [140].

18.9 Concluding discussion

Given an argumentation network (S, R), we would like to have algorithms which will give us only total $\{0, 1\}$ extensions, i.e. stable extensions in which every $x \in S$ is either "in" or "out". This means we must be able to break odd loops in some sound reasonable manner.

Our methods were in principle of two types:

1. Loop checking methods which eliminate the loops which gave rise to the value "undecided".

2. Other methods which reduce the problem to a neighbouring area (1: equations or 2: higher level networks) and solve the problem in the neighbouring area and import the solution back.

All approaches we studied were general and could be applied not only to argumentation but also to looping problems in other intelligent systems areas. So we developed general methodologies and not necessarily relied on specific argumentation concepts.

The relationship with CF2 semantics is of special interest to us.

We showed in Section 7 that any argumentation network (S, R) can be translated/reduced into a similar network (S_1, R_1) such that the ordinary Dung extension of (S_1, R_1) can generate all the CF2 extensions of (S, R). If E_1 is a non-empty ordinary extension of (S_1, R_1) then $E_1 \cap S$ is a CF2 extension of (S, R).

Our algorithms used the directionality of the attack relation in the family of equivalence classes of the maximal strongly connected subsets of (S, R). Our algorithms proceeded, each in its own characteristic way, from top loop-nodes to other lower nodes in algorithmic waves.

The idea of directionality came from our use of the equational approach, see Gabbay's [170], where all the equations had the form $x = \mathbf{f}(y_1, \ldots, y_k)$, giving direction from y_i to x. At the time (2012) the only algorithm for CF2 extensions was that of Gaggl and Woltran [192]. Their algorithm is not directional but is direct and is not implicit as was the original CF2 definition of [30].

Our approach can be viewed as a prelude to a general method of loop checking for general intelligent algorithms. In Section 8 we showed how our ideas can be used for logic programs. This was only a taste. We need to address the implications fully in a subsequent paper. We also aim to look at loops in planning (a big problem in the planning area) and loops in logic theorem proving.

19

The Handling of Loops in Talmudic Logic

19.1 Background

The Talmud is a body of arguments and discussions about all aspects of the human agent's social, legal and religious life. It was completed over 1500 years ago and its argumentation and debates contain many logical principles and examples very much relevant to today's research in logic, artificial intelligence, law and argumentation.

In a series of books on Talmudic Logic, the authors have studied the logical principles involved in the Talmud, one by one, devoting a volume to each major principle

We have just finished writing Volume 5, entitled *Resolution of Conflicts and Normative Loops in the Talmud*, and the present chapter describes how the Talmud deals with even and odd loops and compares the results with open issues in argumentation.

For other English papers corresponding to previous books, see [2, 5, 3, 9, 4, 10], and see our new book [8].

We start by looking at two typical loops, as in Figures 19.1 and 19.2.

Fig. 19.1.

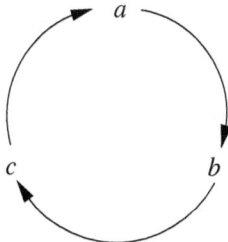

Fig. 19.2.

We need to give some definitions.

An abstract network has the form (S, R), where S is a set of abstract nodes (arguments) and $R \subseteq S^2$ is the attack relation. Traditional research looks at extensions, these are subsets of S satisfying certain conditions (formulated in terms of R). Given (S, R) there may be several possible extensions of several types. In our case, for example, Figure 19.1 has three complete extensions $\{a\}, \{b\}$ and $\varnothing$, and Figure 19.2 has only one extension $\varnothing$.

Current research in argumentation, which relates to such loops and which connects with Talmudic logic, has two aspects:

1. Giving new definitions of extensions which can apply to abstract argumentation networks containing loops and allow us to get some new extensions other than "all undecided".
2. Adding extra information to the argumentation network which helps resolve the loops or help choose an extension.

The extra information one can add to the nodes of the network can be valuations or preferences among nodes. Mathematically one can look at valuations only, as preferences can be derived from them.

When we add valuations, we add a function $V : S \mapsto U$ where U is a value domain, giving some value to each $x \in S$.

V can be used in two extreme ways:

(a) Use V in the definition of extensions, by modifying the network or by disregarding and removing attacks, etc.
(b) Calculate the extensions without using V (i.e. ignoring V) and then using V to choose one's favourite extension or modify existing extensions and create new modified extensions.
(c) There is a third way, highly recommended by some members of the community, which is to use V in combination with the internal structure of the argument. (Note that V is not definded on arguments here but on components of arguments).

(a) is supported by Leila Amgoud and Trevor Bench-Capon.

(b) is supported by the 1500 years old Talmudic logic and recently by a 2010 paper by Toshiko Wakaki, [335].

(c) is supported by Henry Prakken in a 2010 paper [288].

The (b) and (c) approaches maintain consistency while (a) is problematic. See a critique by Martin Caminada [78]. We are grateful to Martin Caminada for providing us with the above information, as well as sending us his critique of approach (a).

Our plan for this chapter is very simple. In Section 2 we present the notion of Shkop extension to an abstract network (S, R) and compare it with Baroni's and his colleagues [31, 30] CF2 extensions.[1]

In Section 3 we discuss some counter examples by Martin Caminada. In Section 4 we conclude the chapter. In a follow-up paper, yet to be written, we give examples of how the Talmud offers valuations to resolve loops of odd and even types and how the Talmud chooses extensions.

---

[1] Rabbi Shimon Shkop, 1860–1930. A Talmudic scholar analysing many logical principles in the Talmud.

19.2 Shkop extensions

We begin with a motivating Talmudic example, the dates are all in the same year, say 2010.

Example 19.1 (The divorce). Jane is married to John. She develops some feelings for Frank and wants a divorce from John. Frank is a rich man and promises to compensate John generously if he cooperates. We now have the following temporal sequence:

Jan 01:   John gives divorce papers to Jane. The divorce is conditional on Jane marrying Frank by the 31st of March. Such conditional divorces are allowed in the Talmud. If Jane marries Frank before 31st March then all is well. If Jane does not marry Frank by 31st March then the divorce papers, from the beginning, from January 01, are nullified and the divorce is not valid from Jan 01. This is Talmudic legal backwards causality.

Feb 01:   Jane takes her divorce papers and marries Terry. This marriage is valid because Jane's divorce papers are valid. Jane can still potentially fulfill the condition mentioned in the divorce papers; she can still divorce Terry and marry Frank.

31st March: Jane, without getting a divorce from Terry, goes and marries Frank.

There is no doubt that Jane is a naughty girl! Frank is a bit paranoid, asking John to give Jane a conditional divorce.

Now we seem to have landed in a logical loop.

Let us build up an argumentation network based on this story.

The base logic is classical temporal logic. The base theory in the logic is the following:

1. If x is married to someone then x cannot marry someone else.
2. If x is married to y at time t then x continues to be married to y until there is a divorce or death.
3. a) A divorce can be given at time t, conditional on an action taken at time $s > t$.
   b) If the action is not taken at time s then there is backward causality and the divorce is not valid from time t.
   c) If the action is taken at time s then the divorce is valid at time t.
   d) At any time $t', t \le t' < s$, the divorce given at time t on a condition to be fulfilled at time s, is considered valid at time t' as long as there is the reasonable possibility, as seen from time t', that the condition will be fulfilled at time s.
4. **Fact**: John gave a divorce to Jane on January 01, conditional on Jane marrying Frank by March 31.
5. **Fact**: Jane married Terry on Feb 01.
6. **Fact**: Jane married Frank on March 31, without ever getting a divorce from Terry.
7. Note: It is possible for x to give a divorce to y at time t on the condition that y marries z ($z \ne x$) at time $s > t$.

   One might argue that at time s, we have a problem:

   y is still married to x therefore y cannot marry z. It is only when y marries z at time s that y is divorced from x at time t and is therefore able to marry z at time s. Since we allow for such conditions, we regard marrying z at s and enabling the divorce at t as simultaneous.

The answer is that the condition of marrying z is not an enabling condition for the divorce papers but a nullifying condition. If it is not fulfilled the divorce papers are nullified.

We now consider the following arguments seen from the temporal point of view of March 31.

DJJ-   John's divorce from Jane on January 01 is not valid.

The reasoning in this argument from base data goes as follows:

On February 01, the divorce was valid because there was the possibility of fulfilling the condition of the divorce, from Rule (3d). Therefore the marriage to Terry (Fact (5)) is valid and does not nullify the divorce, since Jane can still divorce Terry and marry Frank (Rule (3d)).

Therefore at the time March 31, when Jane married Frank without divorcing Terry, her marriage to Frank was not valid (Rules (1) and (2)). Hence, since the condition of the divorce was not fulfilled, the divorce is not valid.

MJT+   Jane's marriage to Terry on Feb 01 is valid.

The argument goes as follows:

Since Jane got a conditional divorce from John and the condition can still be fulfilled her divorce stands and she can marry Terry.

MJF+   Jane's marriage to Frank is valid.

the argument for that is as follows:

Assume the marriage to Frank is not valid. Then Jane's divorce from John is not valid. Hence her marriage to Terry is not valid. But then Jane has a conditional divorce from John and she is not married to Terry, therefore she is free to marry Frank and the marriage is valid. Therefore since $\neg$**MJF+** $\rightarrow$ **MJF+**, we conclude **MJF+**.

We now get the argumentation loop presented in Figure 19.3.

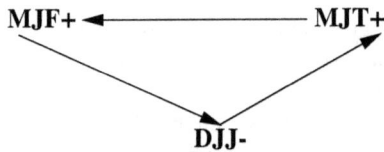

Fig. 19.3.

It is clear that we have an odd loop here and the only Dung extension is ∅, being all undecided. However, life must go on and we need a resolution as to whom Jane is married to! Is she married to John, to Terry or to Frank?!

Here we introduce the intuitive Rabbi Shkop principle:

Shkop principle

If by assuming $x =$ **in, we deduce that** $x =$ **out, then surely** x **must be out.**

Let us apply this to our example. We have three possibilities for the choice of x, see Figure 19.3.

1. $x = $ **DJJ–**
2. $x = $ **MJT+**
3. $x = $ **MJF+**

We reason against the direction of the attack arrows. This reasoning is done later on, see Example 19.6 below for the calculation.

We get three extensions for each one of the choices of x:

1. Marriage to Frank is valid.
2. Divorce not valid — Jane is married to John.
3. Marriage to Terry is valid.

Common sense dictates that we should not test the validity of the divorce because at the time (and here we make use of the temporal sequence) we did not know what Jane was going to do. Similarly we should not test the validity of the marriage to Terry because Jane could still have divorced him. So the only test is that of the validity of marriage to Frank. This test gives by the Shkop principle that **MJF+** = out and therefore the network looks like Figure 19.4, (see also Example 19.6 below for a detailed analysis).

a is an annilhilator node

making sure **MJF+** is out

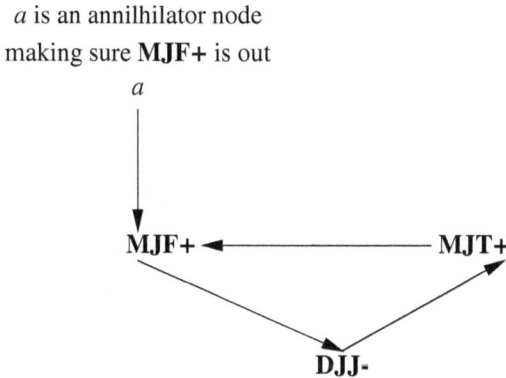

Fig. 19.4.

Figure 19.4 has the extension:

$$
\begin{aligned}
a &= \text{in} \\
\textbf{DJJ–} &= \text{in} \\
\textbf{MJT+} &= \text{out} \\
\textbf{MJF+} &= \text{out}
\end{aligned}
$$

In the above considerations we kept the temporal aspects in the metalevel. We can include these aspects in the object level. We time stamp each argument and each attack arrow, according to the way the story unfolds. If we do this we get Figure 19.5.

Obviously the loop occurs on March 31. So we have to do the Shkop test on the March 31 argument, which is **MJF+**.

In general we can talk about Shkop temporal argumentation frames of the form $\mathbf{N} = (S, R, \mathbf{T})$, where (S, R) is an ordinary network and $\mathbf{T}$ is a time stamping function:

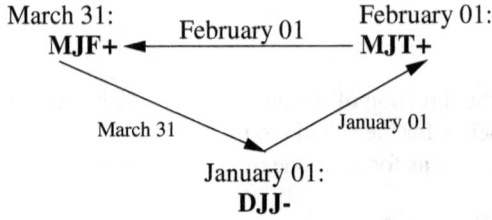

Fig. 19.5.

$$\mathbf{T} : S \cup R \mapsto \text{ Time axis.}$$

For any choice of time t we look at the network

$$\mathbf{N}_t = (S_t, R_t),$$

where

$$S_t = \{a \in S \,|\, \mathbf{T}(a) \leq t\}$$
$$R_t = \{(x, y) \in R \,|\, \mathbf{T}(x, y) \leq t\}$$

Given $a \in S$ with $\mathbf{T}(a) = s$, we check according to Shkop the test $a = 1$? in the network $\mathbf{N}_s = (S_s, R_s)$.

Let us now be a bit more formal about Shkop extensions. Our aim is to offer the argumentation community the notion of Shkop semantics, and compare it with CF2 or Stage semantics. To do that, we need to generalise the intuitive Shkop principle in a sensible way.

For reasons of clear exposition, we find it advantageous to actually start from a recent paper of Martin Caminada, entitled Preferred semantics as Socratic discussion [77].

Caminada sets himself to give a game theoretic answer to the question:

Q: Given (S, R) and $a \in S$, can a be an element of some admissible extension?

His method is to assume that $a = $ in and see by Socratic discussion whether such a position can be maintained. The method is best explained by two examples.[2]

Example 19.2. Consider Figure 19.6
    We ask can we have $c = $ in in some extension? We proceed as follows:

1. $c = $ in, assumption
2. $b = $ out, from (1)
3. $a = $ in, from (2)
4. $a = $ out, from (1)

We get a contradiction. The assumption $c = $ in, lead us, using the attack rules and the geometry of the figure, that both $a = $ out, and $a = $ in.

Thus the answer the question about c is that it cannot be in, it must be out.

Example 19.3. Consider Figure 19.2. Ask the question can $c = $ in? Let us check:

---

[2] Appendix A offers a Tableaux algorithm for this test.

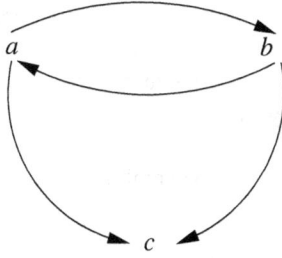

Fig. 19.6.

1. c = in, assumption
2. b = out, from (1)
3. a = in, from (2)
4. c = out, from (3)

So again the answer is negative, there is no extension in which c = in.

Remark 19.4. Note that the proofs in the Caminada Socratic discussion obtain a contradiction by using the direction in the graph against the arrow. Thus if we have

$$x \to y \to z$$

and we assume y = in, Caminada is allowed to deduce x = out, going against the arrow, but is not allowed to deduce z = out going with the arrow.

It seems that even with this restriction, the Socratic discussion is strong enough to identify all nodes a in the network for which a = in is impossible.

Caminada's paper stops when we get our answers to the question of whether a = in is possible or not.

Now let us use these two examples to explain what Shkop does. Shkop introduced a principle for resolving loops:

Shkop's original principle
If the test assumption a = in leads to the conclusion that a = out, then a must be annihilated and be out.

To implement such a principle we need some notation. Let (S, R) be a network and let the elements of S be denoted by lower case letters. Let us add for any $a \in S$ a new annihilator letter, capital A.

With the above notation, let us redo Examples 19.2 and 19.3 according to Shkop.

Example 19.5 (Doing Example 19.2 according to Shkop). We start by testing c = in in Figure 19.6.

1. c = in, test assumption
2. b = out, from (1)
3. a = in, from (2)
    The Caminada Socratic discussion goes against the arrow and would continue

4*. a = out, from (1), a contradiction, because we get both a = in and a = out.

The Shkop original principle requires us to get c = out for a contradiction, because our original test was for c = in?. Therefore we need to go forward with the arrow using (3), as this is the only way to get back to c, and get (4) below. Going forward:

4. c = out, from (3)

5. From (1)–(4) we get that c must be annihilated by the Shkop principle.

This means that we replace Figure 19.6 by Figure 19.7.

We may now feel comfortable, allowing ourselves to go both backwards and forwards with the arrow, and thus maintaining the intuitive spirit of the Shkop principle. This, however, is problematic. Caminada has shown a counter example which is problematic. We discuss this later in Section 3. So we cannot allow ourselves to prove forward with the arrow. So we need to modify the Shkop principle.

Our choice of modifying the Shkop principle is to state:

- If a = in leads to a contradiction then a must be out. In deriving the contradiction, we use reasoning going backwards with the arrow only, see Appendix A. Once the contradiction is derived we introduce an annihilator for a.

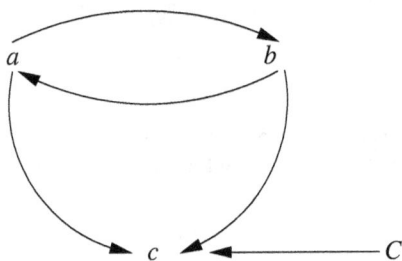

Fig. 19.7.

Coming back to the argumentation network of Figure 19.6, having tested c = in?, we can continue to test a = in and test b = in but this will not require any more annihilators.

The Shkop extensions for Figure 19.6 are obtained by taking ordinary extensions for Figure 19.7 and ignoring the annihilators. In the case of Figure 19.6 the Shkop procedure made no difference but for Figure 19.2 it does as it resolves loops.

Example 19.6 (Doing Example 19.3 according to Shkop). We have three tests to conduct:

Test 1: c = in
Test 2: b = in
Test 3: a = in

Test 1

1. c = in, test assumption
2. b = out, from (1)
3. a = in, from (2)

4. c = out, from (3)
5. Using the Shkop principle c must be annihilated and Figure 19.2 replaced by Figure 19.8.

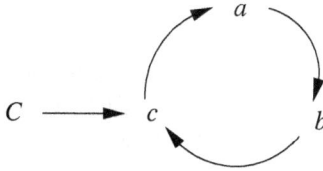

Fig. 19.8.

Figure 19.8 is a new network and we can apply the Shkop test to it. We will get no more contradictions. The Dung extension for it is $\{C, a\}$.

The other tests will give us Figures 19.9 and 19.10.

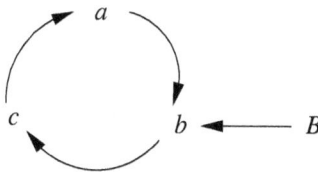

Fig. 19.9.

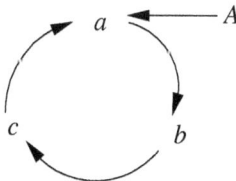

Fig. 19.10.

The normal Dung extensions for Figures 19.9 and 19.10 are respectively $\{B, c\}$ and $\{A, b\}$.

According to Shkop, the Shkop extensions for Figure 19.2 are obtained from the normal extensions of Figures 19.8, 19.9 and 19.10 by ignoring the annihilators letters.

Thus we get the extensions $\{a\}, \{b\}, \{c\}$. Notice that these are the conflict free sets of Figure 19.2.

We ask the reader to remember this because we shall compare the Shkop extensions with Baroni's CF2 extensions.

Remark 19.7. The reader should note that the Shkop procedure was originally intended for elements x of a network which are part of an odd loop, see Example 19.1.

Once the element is found to be out by the Shkop principle, we move to a new network containing the annihilator X of x and we deal with the new network only. Shkop would never test $c = $ in ? immediately (at that moment, if we take into account the temporal aspect, see Section 4) in Figure 19.1 because c is not part of a loop. He would test $a = $ in? and $b = $ in? and find no contradiction. For the sake of mathematical completeness and generalising Shkop, we can allow the use of the Shkop principle to any x in the network. The test is similar to the Caminada Socratic discussion (see Appendix A for a Tableaux algorithm doing the same as Caminada's Socratic discussion), and if $x = $ in is found contradictory, this means that x must be out. Thus adding the annihilator X with $X \rightarrow x$ to the network will make no difference and we get an equivalent network.

We therefore put forward the Generalised Shkop Principle:

Generalised Shkop Principle

Let (S, R) be a network and let $a \in S$. If the assumption $a = $ in leads to a contradiction (i.e. for some $x \in S$, we get both $x = $ in, and $x = $ out) by reasoning only backward against the direction of the arrow (as Caminada does in his Socratic discussion, or as we do in Appendix A using Tableaux) then a must be out. To ensure that a is out, we move to a new network $(S \cup \{A\}, R \cup \{A, a)\})$, where A is a new letter, being the annihilator of a.

Note that Caminada proved in his Socratic paper that for a network (S, R) and $a \in S$ the condition:

- The assumption $a = $ in leads to a contradiction by correctly reasoning backwards against the direction of the arrow.

is equivalent to the declarative condition

- a is not a member of any admissible set.

We can therefore formulate the Generalised Shkop principle in an equivalent declarative way as follows:

Generalised Shkop principle (declarative)

Let (S, R) be a network and let $a \in S$. If a is not a member of any admissible set, then a must be out. To ensure that a is visibly out we move to a new network $(S \cup \{A\}, R \cup \{(A, a)\})$, where A is a new letter, being the annihilator of a.[3]
We are now ready to define the notion of Shkop extensions.

Definition 19.8. *1. Let* $\mathbf{N} = (S, R)$ *be a finite argumentation network. Assume elements* $y \in S$ *are denoted by lower case letters. For each such* y *let* Y *be the annihilator of* y.
We define by induction the notions of
   a) $(y_1, \ldots, y_k), y_i \in S$ *is a legitimate Shkop sequence.*
   b) $\mathbf{N}_{(y_1, \ldots, y_k)}$ *is a Shkop model dependent on* $(y_1, \ldots, y_k)$.

---

[3] So for example the Liar paradox network $(\{a\}, \{a \rightarrow a\})$ becomes the network $(\{A, a\}, \{a \rightarrow a, A \rightarrow a\})$.

Case $k = 1$

y_1 *is a legitimate Shkop sequence if* y *is not a member of any admissible set of* (S, R) *(or equivalently by Caminada [77], if the assumption* $y = in$, *in* (S, R) *leads to a contradiction using Caminada Socratic discussion). In this case let*

$$\mathbf{N}_{y_1} = (S \cup \{Y_1\}, R \cup \{(Y_1, y_1)\})$$
$$= (S_{y_1}, R_{y_1}).$$

Case $k + 1$

Assume $(y_1, \ldots, y_k)$ *is a legitimate sequence and assume* $\mathbf{N}_{(y_1,\ldots,y_k)}$ *is well defined. Let* $y_{k+1} \in S$ *be a point such that* y_{k+1} *is different from all* $y_1, \ldots, y_k$. *Assume that* y_{k+1} *is not a member of any admissible set in* $\mathbf{N}_{(y_1,\ldots,y_k)}$, *(or equivalently the assumption* $y_{k+1} = in$, *in the network* $\mathbf{N}_{(y_1,\ldots,y_k)}$ *leads to a contradiction using Caminada Socratic discussion). Then* $(y_1, \ldots, y_{k+1})$ *is a legitimate sequence and let* $\mathbf{N}_{(y_1,\ldots,y_{k+1})}$ *be* $(S_{(y_1,\ldots,y_{k+1})}, R_{(y_1,\ldots,y_{k+1})})$, *where*

$$S_{(y_1,\ldots,y_{k+1})} = S_{(y_1,\ldots,y_k)} \cup \{Y_{k+1}\}$$
$$R_{(y_1,\ldots,y_{k+1})} = R_{(y_1,\ldots,y_k)} \cup \{(Y_{k+1}, y_{k+1})\}.$$

2. *Let* $(y_1, \ldots, y_k)$ *be a legitimate sequence. Let* n *be the number of elements of* S. *Then we say the rank of* $\mathbf{N}_{(y_1,\ldots,y_k)}$ *is* $n - k$.
3. *Let* $(y_1, \ldots, y_k)$ *be a legitimate sequence. Let* $\mathbf{N}_{(y_1,\ldots,y_k)}$ *e its associated Shkop network. We say* $\mathbf{N}_{(y_1,\ldots,y_k)}$ *or equally* $(y_1, \ldots, y_k)$ *is* clean *iff there are no legitimate sequences extending* $(y_1, \ldots, y_k)$. *Alternatively, iff for any* $y \in S, y \neq y_i, i = 1, \ldots, k$, *we have that the test* $y = in$ *does* not *lead to a contradiction.*
4. *Let* $\mathbf{N}_{(y_1,\ldots,y_k)}$ *be clean. Then we define the set of Shkop extensions of* $\mathbf{N} = (S, R)$ *as derived from* $(y_1, \ldots, y_k)$.
   Notation
   $$\mathbb{B}^{\text{Shkop}}_{(y_1,\ldots,y_n)}$$
   to be defined as follows.
   Let E *be any ordinary Dung extension of* $\mathbf{N}_{(y_1,\ldots,y_k)}$ *or equivalently let* λ *be any Caminada labelling for* $\mathbf{N}_{(y_1,\ldots,y_k)}$, *then* $E \cap S$ *(or equivalently)* $\lambda \upharpoonright S$ *be an element of* $\mathbb{B}^{\text{Shkop}}_{(y_1,\ldots,y_k)}$.
5. *We now define the notion of all Shkop extensions of a finite network* $\mathbf{N} = (S, R)$. *We define the set of all Shkop extensions of* $\mathbf{N}$ *to be*

$$\mathbb{B}^{\text{Shkop}}_{\mathbf{N}} = \bigcup_{\substack{(y_1, \ldots, y_k) \\ \text{clean}}} \mathbb{B}^{\text{Shkop}}_{y_1,\ldots,y_k)}$$

Remark 19.9. Note that this is our definition based on the generalised Shkop principle. We can give restricted variations of it. For example, following Baroni *et al.* in their paper [30] of SCC recursiveness, we can first rewrite (S, R) as an acyclic ordering of maximal loops and then apply the Shkop procedure to loop elements starting from the top loops. This is like the way the CF2 extensions are calculated. We shall give a substantial example below to show you what happens.

It is now time to give some more Shkop examples.

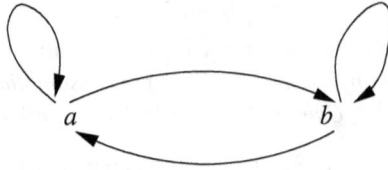

Fig. 19.11.

Example 19.10. Consider Figure 19.11

Testing *b* and then testing *a* or testing *a* and then testing *b* will lead to the same Figure 19.12.

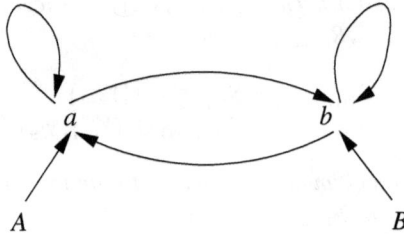

Fig. 19.12.

Therefore the Shkop extension of Figure 19.11 is {*a* = out, *b* = out}.
This does not contradict the usual Dung extension of all undecided!

Example 19.11 (Shkop compared with CF2). Consider the network in Figure 19.13. This figure appears in [191] as an example of how Baroni's CF2 semantics works. Gaggl and Woltran have a program which can compute the CF2 extensions. See also [192, 191].

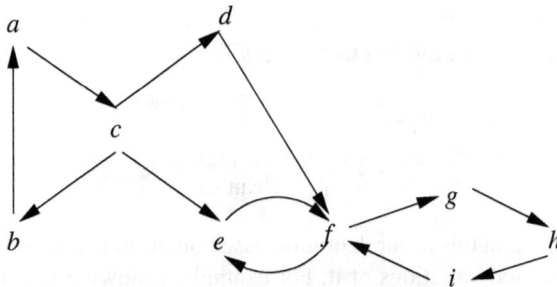

Fig. 19.13.

The CF2 extensions for Figure 19.13 are the following:

$$E_1 = \{c, f, h\}$$
$$E_2 = \{c, g, i\}$$
$$E_3 = \{b, d, e, g, i\}$$
$$E_4 = \{a, d, e, g, i\}.$$

The CF2 semantics would start with the top cycle $\{a, b, c\}$. They would take maximal conflict free subsets which are in this case $\{a\}, \{b\}, \{c\}$ and then arbitrarily decide on the three assignment:

1. $c = $ in, $b = $ out, $a = $ out
2. $b = $ in, $c = $ out, $a = $ out
3. $a = $ in, $c = $ out, $a = $ out

Having now given values to a, b and c, one can propagate the values to the rest of the network and get extensions.

For example:

If $c = $ in, then $d = e = $ out.

Therefore $f = $ in and hence $g = $ out, $h = $ in and $i = $ out.

We got ourselves an extension by breaking the loop $\{a, b, c\}$. The alternative, if we follow traditional Dung style approach is to have one extension only = all undecided.

The method makes sense, it is not arbitrary, it is not just a technical device to generate extensions. Consider Figure 19.14 for example.

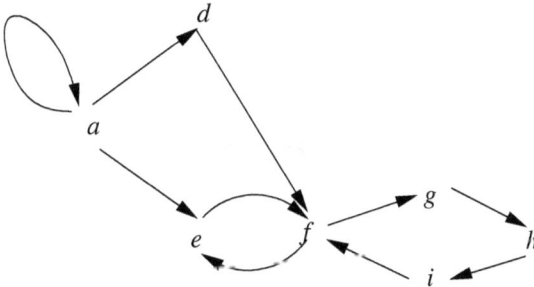

Fig. 19.14.

CF2 would take maximal conflict free subsets of the loop $\{a\}$, which is the empty set, therefore d and e are in and so f is out, g is in, h is out and i is in.

Now let us look at Shkop extensions of Figure 19.13.

Option 1

Accept the procedure where we start from the top loops. Call this top-down Shkop procedure. In this case we start from $\{a, b, c\}$ and ask, as in Example 19.6,

Test 1: $a = $ in
Test 2: $b = $ in
Test 3: $c = $ in

This will yield Shkop figures 19.15, 19.16 and 19.17.

From Figure 19.15 we get the extensions E_1 and E_2. From Figure 19.16 we get Extension E_4 and from Figure 19.17 we get extension E_3.

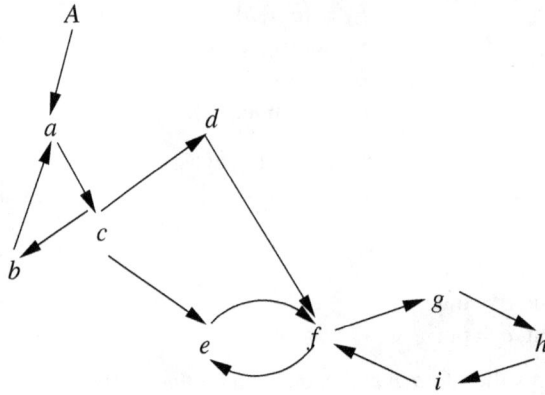

Fig. 19.15.

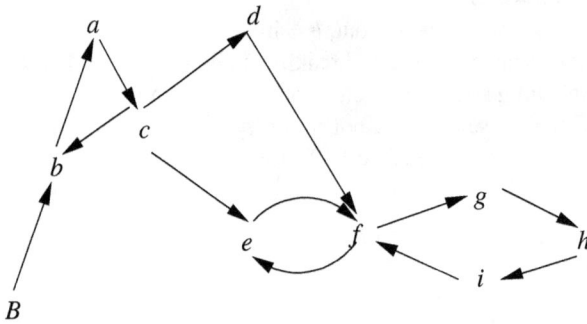

Fig. 19.16.

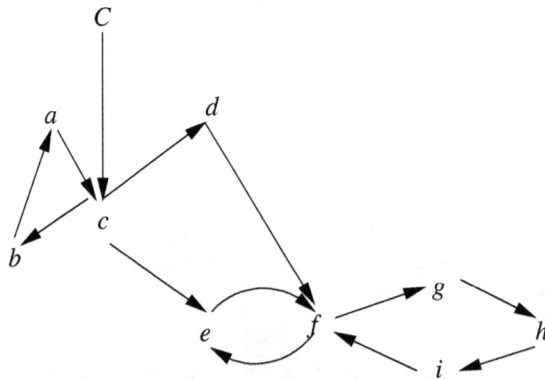

Fig. 19.17.

In the case of Figure 19.14, using the Shkop procedure on $a = $ in will give Figure 19.18, and we get the extension $\{d, e, g, i\}$..

Let us now check what happens if we allow the Shkop process to start from any point. Let us start with $d = $ in? and then $e = $ in?. We will get that both need to be

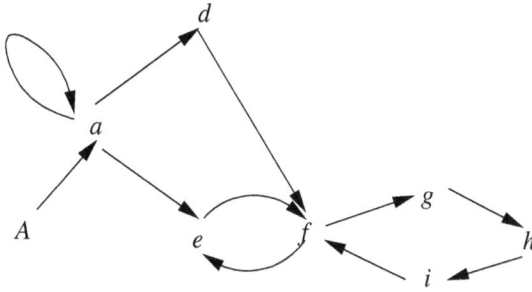

Fig. 19.18.

annihilated. If we carry on asking $a = $ in? or $b = $ in? or $c = $ in? we get the extensions

$$\{c, f, h\}$$
$$\{c, g, i\}$$
$$\{a, f, h\}$$
$$\{a, g, i\}$$
$$\{b, f, h\}$$
$$\{b, g, i\}$$

This chapter is mainly qualitative. A more mathematical exposition will need to address some open problems.

Problem 1
Under what circumstances is the top down Shkop process the same as CF2?

Problem 2
Is there a set of equations in the equational approach of Chapter 10 characterising say the top down Shkop extensions?

Note that all extensions obtained by the Shkop procedure are stable. There is no undecided. Shkop kills all undecided!

Remark 19.12. The reader may seek some meaning to the Shkop algorithm. For this, see the conclusion Section 4. The reader must remember that Talmudic logical argumentation and debate was conducted from the first to the end of the fifth centuries and was used in Jewish communities in the world during the following 1500 years.

Rabbi Shkop just explained the principles involved and we in this chapter are formally modelling them in terms of known abstract argumentation methods.

The principle works!

Remark 19.13 (Comparison with stage semantics). Stage semantics is discussed in detail in [82]. It has similarities with the Shkop extensions but it is not the same. Both ignore self loops but stage semantics may ignore arguments which are not attacked by any other argument. Shkop extensions never do that.

We take our examples from [82]. Consider Figure 19.19.

Stage semantics will ignore the self looping a and will have the extension $\{b\}$. The same is the case with the Shkop semantics. They both agree on $b = $ in. Stage will say $a = $ undecided while Shkop will say $a = $ out.

Fig. 19.19.

As a second example from [82], take Figure 19.20.[4] Shkop will not agree here with the traditional extension. According to Shkop we have

a = in, b = out, c = out.

The traditional extension will have

a = in, b = out, c = undecided.

Stage semantics allows for two extensions: the first one is the same as the traditional one

a = in, b = out, c = undecided.

The second one is

a = undecided, b = in, c = out.

Note that in the second stage a is not in, even though it has no attackers. This is rather strange. Caminada has proved, however, that every argumentation network has at least one stage extension which contains its ground extension. So it can be well behaved. Compare the stage semantics result for the network of Figure 19.20 with Example 19.14 and the considerations leading to Figure 19.21. We get the stage semantics if we go forward. Is this a coincidence? We think it is.

19.3 Caminada counter examples: A discussion

Martin Caminada read an earlier version of Section 2 and gave us penetrating comments and devastating counter examples. The aim of this Section is to put forward an alternative formulation of the Shkop principle which maintains the spirit of Shkop while avoiding the counter examples of Caminada.

We need to summarise the intellectual chain of reasoning events.

(1)   The original Shkop principle, as formulated by Shkop, says as follows:
(*1)   Let $\mathbf{N}$ = (S, R) be a network. Let $x \in S$. Assume (test) x = in. If one can prove that this entails x = out, then surely x must be out.
          Our modelling of this principle was to move to the network $\mathbf{N}_x$, as defined in Definition 19.8.
          Shkop does not specify what it means "to be able to prove that x = in entails x = out". We adopted the Caminada Socratic method to give meaning to this notion.

---

[4] In fact Pietro Baroni and Massimilano Giacomin invented this figure in order to show that CF2 semantics has some advantages over stage semantics.

(2)   Here we had a problem. Caminada's method uses reasoning against the direction of the arrow. So if we have, for example

$$y \rightarrow x \rightarrow z$$

and we test the assumption $x = 1$, then Caminada allows us to deduce $y = 0$, but we are not allowed to deduce $z = 0$.
The difficulty with this is that Shkop formulated his principle by saying "$x =$ in can prove $x =$ out".
It is the same x.
The "same x" restriction is OK for cases of pure loops of the form

$$x \rightarrow a_1 \rightarrow a_2 \rightarrow \ldots \rightarrow a_k \rightarrow x$$

We can prove $x =$ in implies $x =$ out by going backwards, but for cases like Figure 19.6 (the test case assuming $c =$ in) we cannot get $c =$ out by going against the arrow only, as discussed in Example 19.5.
Our original modification of Shkop principle was to allow forward reasoning with the arrow. However, Caminada landed a devastating counter example on this attempt (see Example 19.14 below).
We therefore reformulated the generalised Shkop principle in a safe way, as follows.

(*2)   Let $N = (S, R)$ be a network. Let $x \in S$. Assume (test) $x =$ in. If one can prove a contradiction from this assumption, say that for some $y \in S$, both $y =$ in and $y =$ out are derivable, then surely x must be out, and move to N_x
The above is equivalent to the following (in view of Caminada's Socratic paper).

(*3)   Let $N = (S, R)$ and let $x \in S$. If x is not part of any extension (equivalently if there is no Caminada labelling λ with $\lambda(x) =$ in), then surely x must be out and we move to N_x

So the Shkop extensions and Shkop semantics are obtained by systematically annihilating all points which cannot be part of an extension, as defined in Definition 19.8. This is a Draconian instrument. Note that it needs to be done in sequence, one node at a time.

Example 19.14 (Caminada's counter example). Consider Figure 19.20

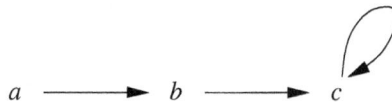

Fig. 19.20.

Let us test $a = 1$? allowing reasoning both with and against the arrow. We reproduce Caminada's reasoning

1. $a =$ in, assumption

2. b = out, from (1)
3. We now do case analysis for c.

Case 3a c = in
In this case we continue
(4a) c = out, since c attacks itself.
(5a) b = in, from (4a)
(6a) a = out, from (5a)

Case 3b c = out

(4b) b = in, since c = out
(5b) a = out, from (4b)
4. Since in both cases we get a = out, then by the Shkop principle surely a = out and we move to Figure 19.21.

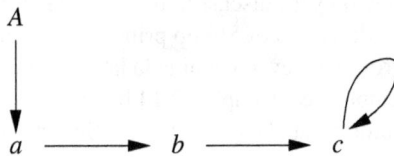

Fig. 19.21.

Clearly this is not acceptable.

Later on we shall modify the forward proof procedures by means of Labelled Deductive Systems and hopefully avoid the Caminada counter example.

We shall now show the idea behind this modification. Let us do the proof again, using our idea:

1. a = in, assumption
2. b = out, from (1)
3. we now have a node c which is not attacked by any node which is in, and instead of doing a case analysis, let us ask, by way of a subcomputation, can c = in?

Subcomputation

•   Given assumptions: a = in, b = out
•   we test: c = in.
(3.1) c = in, assumption
(3.2) c = out, from (3.1)
       Therefore using the Shklop principle c surely must be out, and we move to Figure 19.22.
(4)   We now continue the original computation with Figure 19.22.

To make our idea crystal clear, let us present the reasoning structure as follows: (Note the network changes as we reason, so each line has to indicate which network we are dealing with).

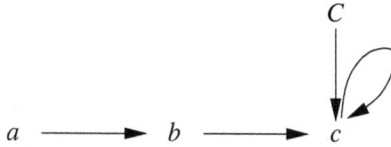

Fig. 19.22.

1. Figure 19.20, a = in, assumption
2. Figure 19.20, b = out, from (1)
3. subcomputation in Box1

Box 1:
3.1 Figure 19.20, c = in, assumption
3.2 Figure 19.22, c = out, from (3.1)
3.3 Use (3.1) and (3.2) and the Shkop principle: c must be out
3.4 Exit subcomputation with Figure 19.22

4. We are now in Figure 19.22 from Box 1: we continue reasoning.
   Thus the network changes as we reason along the arrows!

At present we do not know if this new computation is sound. It may be that counter examples can be found. Even if it is sound, we do not know exactly what it does. Our conjecture is that it just forces us to consider the loops first and eliminate them. At any rate, this is not crucial to our chapter, since we are happy with the General Shkop Principle and the algorithm we have in the Appendix.

19.4 Conclusion

The original Shkop principle is given in temporal context: Imagine a group of agents operating in time and taking actions. To execute an action **a** the pre-condition of the action α_a needs to be fulfilled and then after the action is taken, the post-condition β_a of the action holds.

So if we start at time $t = 0$ in a certain state s and let our agents proceed with their actions then we move from state to state without any trouble and no argumentation networks arise and no loops arise.

The difference between ordinary actions and Talmudic actions is that the Talmud allows for the pre-conditions to contain future conditions and actions. Thus the enabling conditions of the actions can depend on the future and this can create loops. The Talmud also says that if the future condition is not fulfilled, then the action is nullified backwards in time (backward causality). See our papers [9] and [10].

To give a simple example, suppose that on Monday John orders a new laptop to be delivered on Friday. John gives his old laptop on Monday to a student named Tracy, free of charge, on the condition that on Friday, Tracy will configure his new laptop. Call this action **a**.

On Tuesday Tracy is ready to sell the laptop she got from John to a new buyer, Mary, for a good price, but Mary insists that on Friday Tracy transfers the contents of her old computer to the laptop she is buying. Call this action **b**.

The pre-condition for action **b** is that Tracy owns the laptop she is selling. For this to hold she must configure John's new laptop on Friday. However, if we allow action **b**, then Tracy will not be able to configure John's new laptop on Friday, because she will be busy transferring Mary's old data. If Tracy does that for Mary, then action **a** is nullified and so Tracy will not be the owner of the laptop she wants to sell and therefore action **b** is nullified.

What we get here is that if action **b** is allowed then it is nullified. The Shkop principle says that in this case do not allow action **b**.

We see here the context in which the Shkop principle operates. It is a time action model with future pre-conditions and backward causality, which progresses in time. Shkop says that any action which is about to be taken at time t which causes a chain reaction which cancels its own pre-condition at the same time t, should not be taken at time t.

We used the idea of Shkop to suggest and create the Shkop extensions for argumentation networks. These networks are not temporal but are static. We get them from the temporal action model by looking at what is happening at any certain fixed time.

This initial chapter is mainly qualitative and a more detailed modelling of the temporal aspects is forthcoming.

19.5 Appendices

19.6 Tableaux for Caminada Socratic discussion

We offer here a tableaux method designed to test, for an element x in a finite argumentation network, whether x is an element of any admissible extension. Compare also with the Verheij paper [330].

Definition 19.15. *Let* $\mathbf{N} = (S, R)$ *be a finite argumentation frame.*

1. A tableaux for $\mathbf{N}$ *has the form*

$$\tau = (\mathbb{A}_\tau, \mathbb{B}_\tau, \mathbb{D}_\tau)$$

where $\mathbb{A}_\tau \subseteq S$ *is the left inside of* τ *and* $\mathbb{B}_\tau \subseteq S$ *is the right outside of* τ, *and* $\mathbb{D}_\tau$ *is the set of elements marked to be treated in* τ. $\mathbb{D}_\tau$ *will be treated in the next tableau derived from* τ. *We have either* $\mathbb{D}_\tau \subseteq \mathbb{A}_\tau$ *(left treatment) or* $\mathbb{D}_\tau \subseteq \mathbb{B}_\tau$ *(right treatment).*

2. A tableau τ *is said to be closed if one or more of the following holds:*
- $\mathbb{A}_\tau \cap \mathbb{B}_\tau \neq \emptyset$
- *For some* $y \in \mathbb{B}_\tau$, *we have* $\{x \in S \mid xRy\} = \emptyset$.

Definition 19.16. *Let* $\mathbf{N} = (S, R)$ *be finite argumentation frame and let* $x \in S$. *We define a tree* $\mathbb{T}$ *of tableaux for testing whether* $x = in$ *is possible at all, i.e. whether* x *can be a member of any admissible extension. The tree of tableaux will have tree relation* ρ.

Step 1
Form the tableau $\tau_1 \in \mathbb{T}$, *where*

$$\tau_1 = (\{x\}, \varnothing, \{x\})$$

say $\{x\}$ is marked to be dealt with at this stage.

Step 2
Form the tableau $\tau_2 \in \mathbb{T}$, where

$$\tau_1 = (\{x\}, \{y|yRx\}, \{y|yRx\})$$

Say $\{y|yRx\}$ are marked to be dealt with at this stage and that $\{x\}$ has been dealt with. Let $\tau_1\rho\tau_2$ hold.

If for some y such that yRx we have $\{z|zRy\} = \varnothing$ or if xRx then this tableau is closed. Otherwise we move to Step 3.

Step 3
Let $\mathbf{f}$ be any choice function such that for each y to be dealt with in the tableaux τ_2 of the previous step, (i.e. $y \in \mathbb{D}_{\tau_2}$), it chooses an element $\mathbf{f}(y) \in S$ such that $\mathbf{f}(y)Ry$. Form the tableaux, $\tau_3^{\mathbf{f}} \in \mathbb{T}$:

$$\tau_3^{\mathbf{f}} = (\mathbb{A}_3^{\mathbf{f}}, \mathbb{B}_3^{\mathbf{f}}, \mathbb{D}_3^{\mathbf{f}})$$

for each such an $\mathbf{f}$, where

$$\mathbb{A}_3^{\mathbf{f}} = \mathbb{A}_2 \cup \{\mathbf{f}(y)|y \in \mathbb{B}_2\}$$
$$\mathbb{B}_3^{\mathbf{f}} = \mathbb{B}_2$$
$$\mathbb{D}_3^{\mathbf{f}} = \{\mathbf{f}(y)|y \in \mathbb{B}_2 \text{ and } \mathbf{f}(y) \notin \mathbb{A}_2\}.$$

Say that all elements of $\mathbb{B}_2$ (all the ys) have been dealt with and all elements of $\mathbb{D}_3^{\mathbf{f}}$ are marked to be dealt with.

Let $\tau_2\rho\tau_3^{\mathbf{f}}$, for all $\mathbf{f}$.

Note that $\mathbb{D}_3^{\mathbf{f}}$ may be empty.

Step 4
Let $\tau_3^{\mathbf{f}}$ be any tableau of Step 3. Construct the tableau $\tau_4^{\mathbf{f}} \in \mathbb{T}$ as follows:

$$\mathbb{A}_4^{\mathbf{f}} = \mathbb{A}_3^{\mathbf{f}}$$
$$\mathbb{B}_4^{\mathbf{f}} = \mathbb{B}_3^{\mathbf{f}} \cup \{z| \text{ for some } u \in \mathbb{D}_3^{\mathbf{f}} \text{ we have } zRu\}$$
$$\mathbb{D}_4^{\mathbf{f}} = \{z| \text{ for some } u \in \mathbb{D}_3^{\mathbf{f}} \text{ we have } zRu \text{ and } z \notin \mathbb{B}_3^{\mathbf{f}}\}.$$

We say the elements of $\mathbb{A}_3^{\mathbf{f}}$ have been dealt with and the elements of $\mathbb{B}_4^{\mathbf{f}}$ are marked to be dealt with.

Let $\tau_3^{\mathbf{f}}R\tau_4^{\mathbf{f}}$.

Inductive step type odd
We assume by induction that we have $\tau = (\mathbb{A}, \mathbb{B}, \mathbb{D})$ and the elements marked to be dealt with are all in $\mathbb{A}$, i.e. $\mathbb{D} \subseteq \mathbb{A}$ and $\mathbb{D} \neq \varnothing$. In this case proceed as in Step 3 and create τ' and let $\tau' \in \mathbb{T}$ and let $\tau R\tau'$.

Inductive step type even
We assume by induction that we have $\tau = (\mathbb{A}, \mathbb{B}, \mathbb{D})$ and all the elements to be dealt with are from $\mathbb{B}$ (i.e. $\mathbb{D} \subseteq \mathbb{B}$), and that $\mathbb{D} \neq \varnothing$.

Then proceed as in Step 4.

Lemma 19.17. *If* $\mathbf{N} = (S, R)$ *is finite then after a finite number of steps the Tableaux process terminates. We reach tableaux at the bottom of the* ρ*-tree such that they are either closed or their* $\mathbb{D}$ *is empty.*

Proof. Since $\mathbb{D}$ always adds new elments either to $\mathbb{A}$ or to $\mathbb{B}$ and $\mathbb{A}$ and $\mathbb{B}$ do not decrease, and S is finite, sooner or later $\mathbb{D} = \varnothing$.

Lemma 19.18. *Let* (S, R) *be a finite argumentation network and let* $(\mathbb{T}, \rho)$ *be the tableaux for it.*
Then there exists a maximal path $\tau_1 \rho \tau_2 \rho \ldots \rho \tau_n$ *of non-closed tableaux in* $\mathbb{T}$*, if and only if x is a member of some admissible extension E.*

Proof.

1. Assume $x \in E$ and E is an admissible extension. We will define a maximal path $\tau_1 \rho \tau_2 \rho \ldots \rho \tau_n$ of non-closed tableaux in $(\mathbb{T}, \rho)$.
   Let $\tau_1 = (\{x\}, \varnothing, \{x\})$ as in Step 1 of the inductive definition of $(\mathbb{T}, \rho)$.
   Let τ_2 be as in Step 2. τ_2 is not closed, because if xRx holds, then x cannot be in any admissible extension, and if for some y, yRx and $\neg \exists z(zRy)$ hold, then x is out.
   Assume by induction that we have defined a chain $\tau_1 \rho \tau_2 \rho \ldots \rho \tau_k$ of non-closed tableaux such that for each $1 \leq i \leq k$ we have
   - If $y \in \mathbb{A}_{\tau_i}$ then $y \in E$
   - If $y \in \mathbb{B}_{\tau_i}$ then for some $z \in E, zRy$ holds.
   We now define τ_{k+1}.

   Case k is odd
   In this case we have
   $$\mathbb{D}_{\tau_k} \subseteq \mathbb{A}_{\tau_k}$$
   Let τ_{k+1} be defined in Inductive Step type odd (same as Step 3). Clearly $\tau_k \rho \tau_{k+1}$ holds. We want to show that τ_{k+1} is not closed. Since $\mathbb{A}_{\tau_k} \subseteq E$ and $\mathbb{D}_{\tau_k} \subseteq \mathbb{A}_{\tau_k}$ we have that any yRu for $u \in \mathbb{D}_{\tau_k}$ is atatcked by E and hence is out. Thus
   $$\mathbb{A}_{\tau_{k+1}} \cap \mathbb{B}_{\tau_{k+1}} = \varnothing.$$
   Also every such y is attacked by something and so τ_{k+1} is not closed.

   Case k is even
   In this case we have $\mathbb{D}_{\tau_k} \subseteq \mathbb{B}_{\tau_k}$. This means that all points of $\mathbb{D}_{\tau_k}$ are out. Moreover by construction, $\mathbb{D}_{\tau_k}$ are points attacking points in $\mathbb{A}_{\tau_{k-1}}$, and so by the admissibility of E each such point y has an attacker $\mathbf{f}(y) \in E$. Then let τ_{k+1} be $\tau_{k+1}^{\mathbf{f}}$ for this function $\mathbf{f}$. we have that $\tau_k \rho \tau_{k+1}^{\mathbf{f}}$ and $\tau_{k+1}^{\mathbf{f}}$ is non-closed.
   We carry on until such an n that $\mathbb{D}_{\tau_n} = \varnothing$.

2. Assume there exists a maximal path of non-closed tableaux $\tau_1 \rho \tau_2 \rho \ldots \rho \tau_n$ in $(\mathbb{T}, \rho)$. Then clearly
   $$\mathbb{D}_{\tau_n} = \varnothing.$$
   Let $E = \mathbb{A}_{\tau_n}$. We show that E is conflict free and self-defending. If xRy holds for $x, y \in E$, then at some $\tau_i, y \in \mathbb{A}_{\tau_i}$ and so $x \in \mathbb{B}_{\tau_{i+1}}$ and so τ_j will be closed, for some $j \geq i$ (the j in which x gets into $\mathbb{A}_{\tau_j}$).
   Assume for some z that $zRx, x \in E$. We need to show a $u \in E$ such that uRz. Since $x \in E$ then $x \in \mathbb{A}_{\tau_i}$ for some i. Then $z \in \mathbb{B}_{\tau_{i+1}}$ and so in $\mathbb{B}_{\tau_{i+1}} = \mathbb{B}_{\tau_i} \mathbf{f}$ we have $\mathbf{f}(z) \in \mathbb{A}_{\tau_i} = \mathbb{A}_{\tau_{i+1}}$ and $\mathbf{f}(z)Rz$.

This completes the proof.

Example 19.19. Let us check again whether $c = $ in is possible in Figure 19.6, this time using tableaux.

$$\tau_1 : (\{c\}, \varnothing, \{c\})$$
$$\tau_2 : (\{c\}, \{a, b\}, \{a, b\})$$
$$\tau_3^f : (\{c, a, b\}, \{a, b\}, \{a, b\}).$$

Here $\mathbf{f}(a) = b$ and $\mathbf{f}(b) = a$. τ_3^f is closed.

Remark 19.20. Note that the tableaux method works for the query for several points, namely

- Can $c_1, \ldots, c_n$ all be together in some admissible set?

We simply start our tableaux with

Step 1:

$$(\{c_1, \ldots, c_n\}, \varnothing, \{c_1, \ldots, c_n\})$$

19.7 Shkop principle in temporal context

It would be helpful to the reader if we present the Shkop principle in its natural temporal context. Imagine a linear flow of time of the form $(N, <)$ where N is the set of natural numbers $\{1, 2, 3, \ldots\}$ and $<$ is smaller than relation. We associate with each $n \in N$, a state of the world, which we denote by Δ_n, being a classical propositional logical theory in the language with the atoms $Q = \{q_1, \ldots, q_k\}$. We imagine history as evolving. At step 1 we have only state Δ_1 as given and state Δ_2 has not been created yet. The future states are created by actions. An action has the form $\mathbf{a} = (\alpha_{\mathbf{a}}, \beta_{\mathbf{a}})$, where $\alpha_{\mathbf{a}}$ is the precondition of the action and $\beta_{\mathbf{a}}$ is the post condition, all in the same classical langauge of the states Δ.

So at state Δ_1 we might wish to take action $\mathbf{a}$. We can do that if the precondition holds, i.e. $\Delta_1 \vdash \alpha_{\mathbf{a}}$. If this is the case, then we take the action and we move to state Δ_2, which is the state at time 2. Δ_2 is connected with Δ_1 via a revision process, denoted by "$\circ$". Thus $\Delta_1 = \Delta_1 \circ \beta_{\mathbf{a}}$.

The exact nature of the revision process is not relevant to our purpose. It is sufficient to know that for any Δ and any action $\mathbf{a} = (\alpha_{\mathbf{a}}, \beta_{\mathbf{a}})$, such that $\Delta \vdash \alpha_{\mathbf{a}}$ we get a new state $\Delta' = \Delta \circ \beta_{\mathbf{a}}$.

This is a simple model which can easily be made richer and more complicated. The Talmudic twist to this model is that the Talmud allows for future *conditional actions*. Part of the precondition for allowing action $\mathbf{a}$ to take place at time n is that a related action $\mathbf{a}'$ be taken at future time $n + n'$.

For example: I give you this computer to be yours now on the condition that you clean my garden in a week's time. We have

$\mathbf{a} = $ (I own computer, you own computer)
$\mathbf{a}' = $ (Truth, you have cleaned my garden)

So if you do not clean my garden in a week, then the original action is cancelled. This is *backward causality*. We denote these conditional actions by **a** if a' within n' days.

Consider now Figure 19.23.

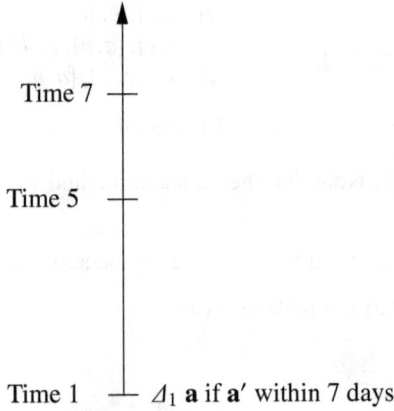

Time 7 ┼

Time 5 ┼

Time 1 ── Δ_1 **a** if **a**$'$ within 7 days

Fig. 19.23.

So action **a** can be taken at Δ_1 on the condition that action **a**$'$ is taken at time 7. Note that time 7 has not yet happened.

Now suppose we continue to take actions and at time 5 we want to take action **b**. The precondition of action **b** holds and so we want to proceed. It could be the case that if we take action **b** at time 5, then a situation is created where action **a**$'$ cannot be taken at time 7. If action **a**$'$ is not taken at time 7, then action **a** at time 1 is not valid and past history is affected to the extent that at time 5 in the new history, the precondition for action **b** does not hold. The Shkop prinicple says that any action **b**, which when taken, changes history backwards in such a way that it cannot be taken (cancelling its own precondition) then **b** should not be taken!

A simple example will illustrate the idea:

Example 19.21. On Monday, John buys a new computer to replace his old one. He gives the old computer (which is still good and fast) to his student Terry, on the condition that on Saturday, Terry comes to John's home and installs the new computer.

Terry decides to sell the computer he was given to a housewife neighbour called Mary. The precondition for the sale is that Terry owns the computer. This is OK because there is still the possibility for Terry to fulfil the condition to John and go on Saturday and install John's new computer.

Mary is prepared to buy the computer from Terry but she has her own condition. She wants Terry to come on Saturday and teach her how to use it.

We ask: can Terry sell the computer to Mary? We reason, following Shkop, that if Terry does sell the computer to Mary, he will have to spend Saturday with her and would not be able to go to John and install John's new computer. Failing to go to John on Saturday would nullify the gift of John giving Terry the old computer, which would nullify the precondition of the sale of this computer by Terry to Mary, namely Terry is not the owner of this computer.

So by selling the computer to Mary, Terry is nullifying the legitimacy of the sale!
So the Shkop prinicple applies and the sale cannot be permitted.

Let us now give another temporal argumentation model in which the Shkop principle can apply for resolving loops.

Suppose we have a sequence of argumentation networks of the form $(S_n, R_n), n = 1, 2, 3, \ldots$ such that $S_n \subseteq S_{n+1}$ and $R_n \subseteq R_{n+1}$. Thus as time passes on, (i.e. $n = 1, 2, \ldots$) we get more and more arguments and more and more attacks.

Consider time $n + 1$ and let $x \in S_{n+1} - S_n$. So x is a new argument added at time $n + 1$. So if x causes an odd loop and cannot be part of any extension, then we apply the Shkop principle, as detailed in Section 2, and annihilate it. To understand the usefulness of this principle and the temporal setup, consider Figure 19.13. We have many options for resolving the loops there. Our task is made easier if we have a temporal sequence of when each argument was put into the figure. We can follow the temporal sequence and use the Shkop principle to incrementally in time resolve the loops.

A Numerical Approach to the Merging of Argumentation Networks

20.1 Introduction

An argumentation system is a tuple $\langle S, R \rangle$, where S is a non-empty set of *arguments* and R is a binary relation on S representing *attacks* between the arguments [114]. One may argue that the main objective of an argumentation system is to identify sets of *winning* arguments in S, based on the interactions represented by R and an appropriate semantics determining which subsets of S can be taken as a coherent view. Such subsets are called extensions.

This chapter concerns the merging of argumentation systems. From the methodological point of view of this chapter follows from Chapters 9 and 10. We imagine a family of k agents and a large set of possible arguments. Each agent a_i can see a subset S_i of these arguments and in her opinion, the attack relation should be $R_i \subseteq S_i^2$. Each agent a_i further adopts a set of winning arguments $E_i \subseteq S_i$. The agents form a community and a consensus is required. Thus our problem is to merge these k systems $\langle S_i, R_i, E_i \rangle$ into a single system representing the views of the community and draw appropriate conclusions from it.

At first, one may think that the merging process can be done at the *meta level*, i.e., by considering only the winning arguments E_i in each local system. However, as pointed out in [101], this not only will sometimes produce unintuitive results, but will also fail to simultaneously satisfy well-known social choice properties [296]. The reasons have to do with loss of information during the merging process. In particular, the extensions of winning arguments do not carry the information about the local attacks, and as it turns out, these may well be relevant to the collective decision making process. If we want to take both the local preferences for winning arguments and the local topologies of the various systems into account, we need a framework that can deal with all this information.

Our starting point is an augmented argumentation system containing the arguments and attacks of all individual networks. We approach the merging problem from a voting perspective: agents put forward a vote on the components of the augmented system depending on how they perceive these components locally. However, the votes are not used as in an usual voting procedure such as majority voting, etc. For us, votes are used to support the idea of *reinforcement*: the more a component appears in individual networks, the more it is represented collectively. We aggregate the votes of the components resulting in an augmented argumentation system in which both arguments and attacks have weights with values in the interval $U = [0, 1]$. Thus, we get a network of the form $\langle S, R, V \rangle$, where $\langle S, R \rangle$ is a traditional network and V is a function from $S \cup R$ into U. Such augmented systems can be seen a special case of *support and attack networks* Chapter 9. We believe that the merging of argumentation systems is a scenario that naturally justifies the employment of weights in attacks and arguments.

We now have a situation whereby each agent has a traditional argumentation system, they all vote, and as a result we get a merged argumentation system with weights combined numerically. There is a mismatch between the "type" of the original networks and that of the network resulting from their merging, and thus we need to explain how we understand the numerical weights and how we can extract/project a set of winning arguments from the merged system. Had we started working from the outset with numerical weighted systems, we would have more choice on how to perform the merging because we could use the original weights in the computation of the overall result, e.g., by constructing a new weighted argumentation system representing the group as a whole.

Given an augmented argumentation system with weights constructed as described above, we see the weights of the nodes as the overall initial level of support for the arguments in the community and the weights of the edges as the intensity with which the attacks between the arguments are carried out.

It is natural to expect that the overall support for an argument will decrease in proportion to the strength of its attacking arguments and the intensity with which these attacks are carried out. However, since the attacking arguments may themselves be attacked, we need to find a way to systematically propagate the values in the network and determine *equilibrium* values for the nodes based on their interactions, much in the spirit of an *interaction-based valuation* [86]. This is akin to finding the extensions in a traditional network. However, our work has two important differences: *1)* we allow both arguments and attacks to have weights; and *2)* we calculate the equilibrium values using the equational approach of Chapter 10: the augmented system works as a generator of numerical equations whose solutions correspond to the equilibrium values.

Argumentation systems using weights in one form or another have been studied before. One of the first approaches using them was proposed by Besnard and Hunter who suggested a *categoriser* function assigning values to trees of arguments [46]. Subsequently, arguments were presented with weights used to express their relative strength within a particular audience [43]. Cayrol and Lagasquie-Schiex introduced the concept of *graduality* in the valuation of arguments in [86]. Since then, other frameworks using weights include the ones proposed in [37, 18, 50, 118, 336, 237].

The novelty of our approach is in the use of the weights to represent the support of the community for both arguments and attacks and in the way that equilibrium values for these components are calculated using a system of equations.

The rest of the chapter is structured as follows. In Section 20.2, we introduce some basic concepts and the equational approach. In Section 20.3, we show how the merging process is done. We then show how to calculate equilibrium values in Section 20.4 and illustrate the idea with many examples in Section 20.6. Some comparisons with related work are done in Section 20.7 and we finish with a discussion and some conclusions in Section 20.8.

20.2 Background

As mentioned in the previous section, given an argumentation system $\langle S, R \rangle$, one is generally interested in finding the *winning* arguments in S according to a particular semantics.

One way of doing this is to look at subsets $E \subseteq S$ that are as large as possible and yet whose arguments are *compatible* with each other. Two common notions of compatibility require E to be *conflict-free*, i.e., for all arguments X and Y in E, it is not the case that $(X, Y) \in R$; and that all arguments $X \in E$ are *acceptable*, i.e., for every argument Y in S, if $(Y, X) \in R$, there exists an argument Z in E such that $(Z, Y) \in R$. If E is conflict-free and only contains acceptable arguments, then we say that E is *admissible*. An admissible set $E \subseteq S$ that is also maximal with respect to set inclusion amongst all admissible sets is called a *preferred extension* of $\langle S, R \rangle$.

A preferred extension can be defined in terms of a complete labelling of the set of arguments that assigns *in* to arguments that are accepted; *out* to those that are rejected; and *undec* to those that are neither [81, Theorem 2]. Such labelling is called a *Caminada labelling* [81, Definition 5] and has advantages over the extension approach, because the latter only identifies the set of arguments that are accepted. We will return to this type of labelling later in the section.

In traditional argumentation systems, there is no notion of weight associated to an argument or attack. However, there are scenarios in which this association seems natural. In the case of arguments, the weights may come, for instance, from an underlying many-valued logic; as the normalised result of a vote put to a community of agents; or as the result of interactions between the arguments in a network (as in [86]). In the first case, the values are intrinsic to the arguments whereas in the last two, the values are conceptually *external* to the argumentation framework. Mixed approaches are also possible. We may start with each agent assigning numerical values via considerations which are conceptually connected to the arguments and their meaning and end up with merged values obtained during a voting procedure. The application area can dictate the most appropriate approach.

For similar reasons, an attack between arguments X and Y may also be given varying degrees of strength rather than just 0 or 1. Again, the strength may have conceptually related, internal, argumentation meaning or may be conceptually external to the arguments themselves. For example, it may be obtained from the statistics about the correlation between X and Y; or calculated from the proportion of members of a community supporting the attack of X on Y (as in [88]). It may even come from considerations about the geometry of the network itself.

An even more compelling scenario for the use of extended values is because they arise naturally in formalisms that are concerned with the problem of *merging* of argumentation systems, which we consider here. The concept was introduced by Coste-Marquis et. al. in [101].

It may be wise when presenting a numerical argumentation network to provide not only the numerical values themselves but also to give their origin, internal or external, etc, because the origin provides the context that supports a proper interpretation of the weights.

Now, given the numerical network $\langle S, R, V \rangle$ we need to somehow figure out what the various values mean. We can regard the values given by V as *start-up values* that we may want to adjust depending on how the components interact in the network. The adjustment corresponds to the *valuation step* in Cayrol and Lagasquie-Schiex's terminology [86]. However, in our case we want arguments to be *weakened* in proportion to the strength of the attacks on them as well as the intensity with which these attacks are carried out. Ideally, we want to find *equilibrium* values for all arguments, i.e., stable values for the arguments resulting from their start-up values and the interactions with the equilibrium values of the arguments that attack them.

An interesting methodology for calculating these values is the so-called *equational approach* of Chapter 10, which sees a numerical network as a generator of equations. Each argument is associated with a variable and an equation is generated for each argument taking into account the attack relation. Provided the equations respect the meanings of the weights of the arguments and attacks an "evaluation" of the network can be done by solving the system of equations. For an argument X, the equilibrium value 1 means that X definitely "in"; 0 means that it is definitely "out"; and any other value in between means how close to "in" (or "out") X is. If we want, we may be more flexible and settle for a threshold value for the acceptance of arguments that is lower than 1.

An example of how such equations can be generated is given by the schema Eq_{max} below. The symbol $V_e(X)$ will be used to denote the *equilibrium* value of a node X. Now let $Att(Y)$ denote the set of all arguments attacking Y, i.e., $Att(Y) = \{X_i \in S \mid (X_i, Y) \in R\}$. We can define the equilibrium value of Y through the Eq_{max} schema using the equation below.

(Eq_{max}) $V_e(Y) = 1 - \max_{X_i \in Att(Y)}\{V_e(X_i)\}$

Notice that for a node Y, $V_e(Y) = 1$ if and only if $V_e(X) = 0$ for all $X \in Att(Y)$ and $V_e(Y) = 0$ if and only if $V_e(X) = 1$ for some $X \in Att(Y)$.

Thus, the network of Fig. 20.1. generates the following system of equations:

$$V_e(X) = 1$$
$$V_e(Z) = 1$$
$$V_e(W) = 1 - \max\{V_e(Z)\} (= 0)$$
$$V_e(Y) = 1 - \max\{V_e(X), V_e(W)\} (= 0)$$

We now interpret these values as "accept X and Z" and "reject W and Y". This gives the extension $\{X, Z\}$, as expected from a traditional argumentation system.

Generally speaking, Gabbay has shown that the totality of the solutions of the equations generated from a network using Eq_{max} corresponds to the totality of Caminada labellings of that network [166].

Unfortunately, Eq_{max} does not take into account the start-up value of a node or the intensity with which the attacks on it are carried out. In order to take these into account, we will consider a more sophisticated equation schema in Section 20.4.

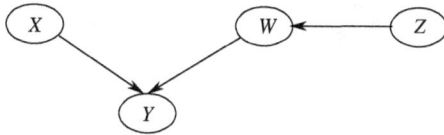

Fig. 20.1. A simple argumentation system.

20.3 Merging argumentation networks

In this section, we discuss some intuitions underpinning our proposed method of merging argumentation networks. Our first goal is to show how to combine a collection of networks into a single weighted argumentation network.

As discussed in Section 20.1, we start by associating each network with an agent who "votes" for components of another possibly larger network. Obviously, the more interesting scenarios involve distinct networks being merged. Consider the networks in Fig. 20.2 and each agent's chosen extension of winning arguments in these networks.

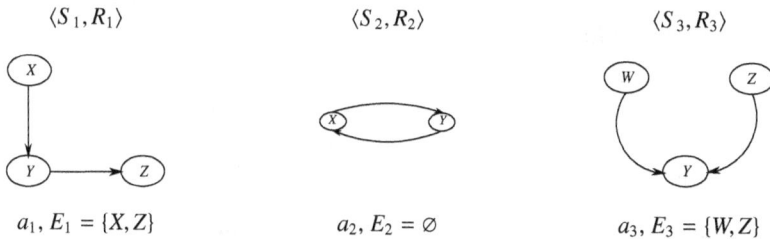

$$\langle S_1, R_1 \rangle \qquad\qquad \langle S_2, R_2 \rangle \qquad\qquad \langle S_3, R_3 \rangle$$

$$a_1, E_1 = \{X, Z\} \qquad\qquad a_2, E_2 = \varnothing \qquad\qquad a_3, E_3 = \{W, Z\}$$

Fig. 20.2. Argumentation networks of three different agents.

One can easily see that the three agents have different sets of arguments, and even in the case where some arguments coincide, the agents may disagree with respect to the attack relationship between them. For instance, argument W is only known to agent a_3, and in her network, Z attacks Y, whereas in the network of agent a_1, Y attacks Z.

There are many reasons why agents may have different points of view regarding arguments and their attack relationship. They may use different knowledge bases; they may have different resources for inference at their disposal; they may use different inference systems; they may have different preferences for arguments; etc. All of these may result in the non-availability of some arguments to some agents as well as the generation of disagreements with respect to the direction of the attacks between them (this is not very dissimilar to the existence of cycles in a single network).

A simple way of harmonising the differences is to consider expansions to the networks. However, unlike in [101], we do not expand each network individually and then combine the expansions. Instead, we consider the single augmented network that includes the components of all other networks and then analyse the representation of the components of this network with respect to the community of agents.

Since some components appear in more networks than others, the augmented network alone is not sufficient to represent the community. In order to do that we use

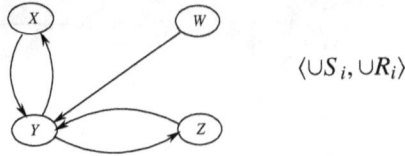

$$\langle \cup S_i, \cup R_i \rangle$$

Fig. 20.3. Augmented network containing all components of $\langle S_1, R_1 \rangle$, $\langle S_2, R_2 \rangle$ and $\langle S_3, R_3 \rangle$

weights, but let us first introduce the notion of a profile of (traditional) argumentation systems.

Definition 20.1 (Profile of argumentation systems). *A profile of argumentation systems is a tuple $P = \langle AN_1, \ldots, AN_k \rangle$ where each $AN_i = \langle S_i, R_i, E_i \rangle$ is an argumentation system $\langle S_i, R_i \rangle$ for agent a_i provided with a set $E_i \subseteq S_i$ of the winning arguments in S_i selected according to some local semantics.*

Definition 20.2 (Augmented weighted network for a profile of argumentation systems). *Let $P = \langle AN_1, \ldots, AN_k \rangle$ be a profile of argumentation systems. The* weighted augmented network *for P is a tuple $AWN_P = \langle S, R, V_0, \xi \rangle$ where*

- $S = \cup_i S_i$ *and* $R = \cup_i R_i$
- $V_0 : S \rightarrow [0,1]$ *represents the start-up values of the arguments in S and is to be interpreted as the initial level of support for an argument $X \in S$ within P*
- $\xi : R \rightarrow [0,1]$ *represents the intensity of an attack $(X, Y) \in R$ within P*

The values for V_0 and ξ need to be computed from the argumentation systems in the profile P. The basis for this is a policy interpreting each agent's perception of the arguments and attacks in AWN_P depending on the agent's own original network. For simplicity, we will refer generally to the arguments and attacks of a network as its "components".

In agreement with [101] we believe that there is an intrinsic difference between supporting a component; rejecting it and being ignorant about its existence (in which case a decision for or against it is impossible). In order to distinguish these attitudes, we let agents vote for components by assigning to them one of the three values below.

0: the agent does not know about the component
1: the agent knows about the component and supports it
−1: the agent knows about the component but does not support it

Definition 20.3 (Attitude with respect to a component). *Let P be a profile of argumentation systems and AWN_P the weighted augmented network for P. The attitude of an agent a_i towards the component c of AWN_P, in symbols $v_i(c)$, is represented in the following way.*[1]

$v_i(X)$ *(arguments)*	$v_i((X, Y))$ *(attacks)*
0: *if* $X \notin S_i$	0: *if either* $X \notin S_i$ *or* $Y \notin S_i$ *(or both)*
1: *if* $X \in E_i$	1: *if* $(X, Y) \in R_i$
−1: *if* $X \in S_i - E_i$	−1: *if* $X, Y \in S_i$, *but* $(X, Y) \notin R_i$

---

[1] To simplify notation we use the same function symbol v_i for nodes and edges.

That is, the agent a_i votes with 0 for *argument* X, if a_i has no knowledge about X. Otherwise, a_i will vote with 1 or −1 depending on whether X is amongst the winning arguments of S_i. Of course, X may be one of the winning arguments of S_i only because a_i is unaware about some other argument Y that defeats it. However, if that is the case, both Y and its supposed attack on X will be represented in the augmented network and this will have an effect on X's equilibrium value, as we shall see later.

The case of an attack from an argument X to an argument Y is similar, except that an attack may not exist because the agent is unaware of one or both of the arguments. Hence, the agent a_i will vote with 0 if at least one of X and Y is not known to her (in which case a judicious decision of a_i about the attack from X on Y is not possible). Otherwise, if both X and Y are known to a_i, she will vote with −1 if $(X, Y) \notin R_i$ and with 1 if $(X, Y) \in R_i$. Notice that the vote 1 for an attack (X, Y) by an agent a_i depends only on the existence of the attack in a_i's local network. Even if $Y \in E_i$ and $X \notin E_i$, a_i must still vote with 1 if $(X, Y) \in R_i$, since she knows about it. The fact that a_i chooses Y over X in spite of the attack of X on Y in this case is already taken into account in the agent's votes for the arguments X and Y.

The above voting strategy requires that there is a local semantics for deciding the winning arguments in each network but does not make any assumptions on what the semantics should be. In fact, the group as a whole may have several different local semantics. If local preference orderings are used to decide about the winning arguments of each agent, then it would make sense to aggregate all of these preference orderings and take the aggregated result into account when deciding on the extensions of the aggregated network.

In either case, we find it reasonable to assume that an agent will be able to make some decisions about her winning arguments and these preferences should somehow be used during the merging. This is especially important when it concerns the representation of unanimity of local decisions. In order to illustrate this, suppose that agents a_1 and a_2 share the same argumentation network given in Figure 20.4.

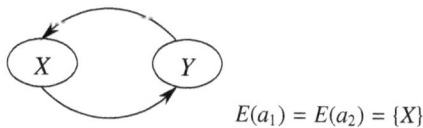

$$E(a_1) = E(a_2) = \{X\}$$

Fig. 20.4. Argumentation networks for agents a_1 and a_2.

Under the preferred semantics, $E(a_1) = \{X\}$ or $E(a_1) = \{Y\}$ (a_2 has the same choices). For the sake of discussion, let us assume that $E(a_1) = E(a_2) = \{X\}$, i.e., both agents have a preference for X over Y, and, as a result, they both take the extension $\{X\}$ to be their set of winning arguments. We would expect that the result of the merging of these networks would contain X only.[2] However, this is only possible if these prefer-

---

[2] This expectation does not hold if we consider simply the merging of *graphs*, but the objective of the merging of argumentation systems is to provide the collective view of a community of agents with respect to a set of arguments and therefore everything that involves the way the agents see the arguments and how they interact with each other must play a part in the process.

ences are somehow taken into account in the merging (as is the case in our approach). Thus, a formalism that does not consider such local preferences (such as [101]) and which would still agree that merging the networks for a_1 and a_2 would result in a_1 ($= a_2$) has no recourse when it comes to deciding between $\{X\}$ and $\{Y\}$, even though all agents involved in the merging prefer the former to the latter (we will come back to this example in Section 20.6).

If the local networks are themselves numerical, then a number of alternatives arise. One could compute each network individually, decide on the winning arguments and apply the same technique given above; or one could feed the equilibrium values of each network into the augmented one, normalise the values as appropriate, generate the equations and then compute the overall equilibrium values as before; or one could choose a combination of these ideas. However, in this work we want to keep our assumptions to a minimum so we only require that the agents can provide a set of winning arguments.

We now need to generate the initial weights for the augmented network based on each agent's attitude to its components. Again, because some components are only known to some agents, the community as a whole may take two different approaches when considering the overall level of support of a component:

- in the *credulous* approach, the weights are calculated based on the total number of agents *that know about a component*
- in the *sceptical* approach, the weights are calculated taking into account the total number of agents in the profile P

We will associate the credulous approach with the superscript $^+$ and the sceptical one with the superscript $^-$ in the definitions of the initial values V_0 and ξ below. Whenever the distinction is not important we will simply omit the superscripts.

Definition 20.4 (Initial values for nodes and attacks). *Let* $P = \langle AN_1, \ldots, AN_k \rangle$ *be a profile of argumentation systems and* AWN_P *the weighted augmented network for P. Let* $v^+(c) = |\{i \mid v_i(c) = 1\}|$ *and* $v^-(c) = |\{i \mid v_i(c) = -1\}|$. *We define*

$$V_0^+(X) = \frac{v^+(X)}{v^+(X)+v^-(X)} \qquad\qquad V_0^-(X) = \frac{v^+(X)}{k}$$

$$\xi^+((X,Y)) = \frac{v^+((X,Y))}{v^+((X,Y))+v^-((X,Y))} \qquad \xi^-((X,Y)) = \frac{v^+((X,Y))}{k}$$

where $V_0^+(X)$ *(resp.,* $V_0^-(X)$) *is the start-up value for the argument X under the credulous (resp., sceptical) approach and* $\xi^+((X,Y))$ *(resp.,* $\xi^-((X,Y))$) *is the intensity of the attack from X to Y within* AWN_P *under the credulous (resp., sceptical) approach.*

Notice that we have purposefully excluded the agents who do not know about a component c in the definitions of $V_0^+(c)$ and $\xi^+(c)$ above. These agents vote with 0 for c according to Definition 20.3 and hence are not counted in either $v^+(c)$ or $v^-(c)$. $V_0^-(c)$ and $\xi^-(c)$ on the other hand look at the components more sceptically and consider their representation across all voters.

For the example in Fig. 20.2 we get the weights shown in Fig. 20.5 for the components of the augmented networks under both approaches. Given these weights, we then need to calculate equilibrium values for the nodes, which will be done in Section 20.4.

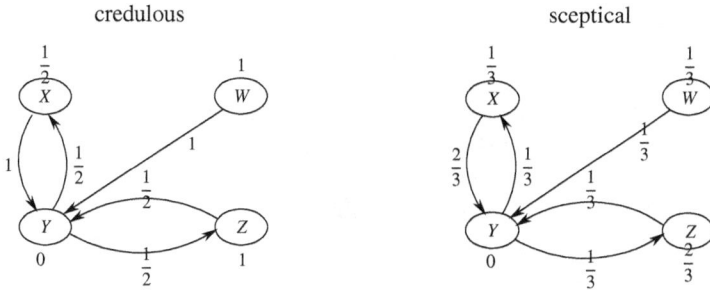

Fig. 20.5. Merged networks of Fig. 20.2 under the credulous and sceptical approaches.

By referring to Fig. 20.2, where the winning arguments in each network are given, one can easily see why the initial weight of the argument Y is 0 under both approaches (see Fig. 20.5). This is because Y is not a winning argument in any of the initial networks (see Proposition 20.5).

Analogously, the initial value of Z in Fig. 20.5 is 1 only under the credulous approach. This is because even though Z is a winning argument in every network in which it is known, it is not known in every network in the profile. Similarly, W's initial weight is 1 under the credulous approach, but only $\frac{1}{3}$ under the sceptical one. This is reasonable, since it is only known by one out of the three agents, but for that agent (a_3), it is one of the winning arguments. The weights for the attacks are assigned following the same pattern.

Generally speaking, we have the following.

Proposition 20.5. *Let* $P = \langle AN_1, \ldots, AN_k \rangle$ *be a profile of argumentation systems where each* $AN_i = \langle S_i, R_i, E_i \rangle$ *and let* $AWN_P = \langle S, R, V, \xi \rangle$ *be the weighted augmented network for P. The following hold for all arguments* $X \in S$.

1. *if* $X \in \cap_i E_i$, *then* $V_0^+(X) = V_0^-(X) = 1$
2. *if* $V_0^-(X) = 1$, *then* $X \in \cap_i E_i$
3. *if* $X \in E_i$ *for all i such that* $X \in S_i$, *then* $V_0^+(X) = 1$
4. *if* $X \notin \cup_i E_i$, *then* $V_0^+(X) = V_0^-(X) = 0$

Proof. 1. and 4. follow directly from Definitions 20.3 and 20.4. For 2., note that if $V_0^-(X) = 1$, *then* $v^+(X) = k$, *and hence* $X \in E_i$ *for every agent* a_i. *For 3., note that if* $X \in E_i$ *for all i such that* $X \in S_i$, *then* $v^-(X) = 0$, *and hence* $V_0^+(X) = 1$.

The situation with attacks is similar, but simpler.

Proposition 20.6. *For all attacks* $(X, Y) \in R$.

1. *if* $(X, Y) \in \cap_i R_i$, *then* $\xi^+((X, Y)) = 1$ *and* $\xi^-((X, Y)) = 1$.
2. *if* $\xi^-((X, Y)) = 1$, *then* $(X, Y) \in \cap_i R_i$.
3. $(X, Y) \in \cup_i R_i$ *if and only if* $\xi^+((X, Y)) > 0$ *(resp.* $\xi^-((X, Y)) > 0)$

Proof. These follow directly from Definitions 20.3 and 20.4.

Proposition 20.6 states the following. Attacks appearing in all networks are transmitted with full intensity (i.e., 1) in both approaches. If the intensity of an attack is 1 under the sceptical approach, then the attack appears in all networks. Notice that under the credulous approach, an attack may be transmitted with full intensity even if it does not appear in all networks, as long as the networks in which it does not appear do not have at least one of the nodes involved in the attack. The corresponding agents will vote with 0, which under the credulous approach indirectly means that they vote with the networks that support the attack. Under the credulous approach, the value of the weight of a component can only be reduced by the agents that vote against it (i.e., vote with −1). Finally, any attack appearing in at least one network is carried out with some non-null intensity (and vice-versa).

We now turn to the problem of calculating equilibrium values for the arguments of a weighted augmented network.

20.4 Equilibrium values in a weighted augmented network

One important aspect in the calculation of the equilibrium values of the arguments in a weighted augmented network is the decision of how the attacks to an argument should affect its initial support value.

As in an usual argumentation system, arguments may be attacked by any number of arguments. Since we work with numerical values, we want to reduce the strength of the attacked node by a measure that is the result of the aggregation of the strength of the attacking nodes. The strength of a single attack itself depends on the strength of the attacking node and the intensity with which the attack is carried out. However, the attacking nodes may be themselves attacked, so we need to perform the aggregation systematically.[3] Let us start by analysing the effect of attacks in general.

Consider the network in Fig. 20.6, in which x, y and z are the initial values of the arguments X, Y and Z, respectively. Let us for a moment ignore these values.

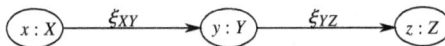

$$\boxed{x:X} \xrightarrow{\xi_{XY}} \boxed{y:Y} \xrightarrow{\xi_{YZ}} \boxed{z:Z}$$

Fig. 20.6. A typical weighted argument network.

If we want to mimic the behaviour of attacks in a standard argumentation system [114], we need to accept arguments X and Z and reject argument Y. The reasoning is as follows. Since no arguments attack X, it *persists*. X then attacks Y, which is *defeated*, and hence no persisting arguments attack Z, which then consequently also persists. In our numerical semantics, persistence is associated with the values $[t, 1]$ (for some $t > 0$) and (strong) defeat with the value 0. Of course there is a grey area for the values $0 < v < t$, which show *some* support for arguments whose values did not meet the threshold t for acceptance. In other words, acceptance in Dung's sense for us means to have an equilibrium value e equal or higher than a minimum acceptance level $t > 0$. If

---

[3] If there are no loops, the problem is simpler because we can always start with the nodes that are not attacked by any nodes and propagate the values up the graph.

we want to be strict, we can set $t = 1$. Otherwise, we may settle for any value greater than 0 (up to 1).

Ideally, we would like to remain close to the basic semantics, taking care of the arguments' start-up values (which are all in the unit interval U) and the intensity with which the attacks between them are carried out. Hence, our objective in Fig. 20.6 is to calculate the values $V_e(X)$, $V_e(Y)$ and $V_e(Z)$, based on x, y, z, ξ_{XY} and ξ_{YZ}. Arguably, since X is not attacked by any node, its equilibrium value $V_e(X)$ can be calculated directly by some manipulation on the value x alone. The simplest procedure is to make $V_e(X) = x$, its initial value. On the other hand, the value of $V_e(Y)$ depends both on $V_e(X)$ and the *intensity* ξ_{XY} with which the attack from X to Y is carried out. Once $V_e(Y)$ is calculated, the equilibrium value for $V_e(Z)$ can be calculated using ξ_{YZ} in the same way. If there are cycles, the equations get more complex, but they are solvable, as long as the functions involved are all continuous.[4]

Now suppose we give initial value 1 to all arguments and consider all attacks being transmitted with full intensity. Since X has initial value 1 and it is not attacked by any arguments, its equilibrium value is the same as its initial value, i.e., 1. It then attacks Y with full intensity (i.e., $\xi_{XY} = 1$), which means that Y's initial value, $y = 1$, is weakened by 1 and its equilibrium value is set to 0. Effectively, this annihilates Y's attack on Z, which then gets as its equilibrium value the same value as its initial one, i.e., 1. As a result, we end up with the acceptance of X (because of its equilibrium value 1); the rejection of Y (because of its equilibrium value 0); and the acceptance of Z (also because of its equilibrium value 1), as expected.

We stress that, in general, we are free to decide on the minimum value we require for considering an argument as being accepted. As we mentioned, we may decide this to be the value 1 itself, leaving all values $0 < x < 1$ to represent *undecided* arguments; or we may even do away with the notion of undecidedness altogether and divide the interval in two halves only: acceptance or rejection of an argument depends on which interval its equilibrium value falls into.

In terms of the effect of the attacks on some argument X, our problem is to determine a factor $0 \leq \pi(X) \leq 1$ representing the combined strength of the attacks on it. The equilibrium value for X can then be calculated by multiplying X's initial value by this factor, i.e., $V_e(X) = V_0(X) \cdot \pi(X)$. The function π must *aggregate* the value of the attacks on an argument. In order to remain close to the standard argumentation semantics, we want π to satisfy at least the three conditions below.

(SSC1) $\pi(X) = 1$, if $max_{Y \in Att(X)}\{\xi((Y, X))V_e(Y)\} = 0$
(SSC2) $\pi(X) = 0$, if $max_{Y \in Att(X)}\{\xi((Y, X))V_e(Y)\} = 1$
(SSC3) π is continuous

(SSC1) says that if all arguments attacking X are fully defeated or transmitted with null intensity, then X should retain its initial value fully. (SSC2) says that if any argument that attacks X has full strength *and* the attack is carried out with full intensity, then X should be fully defeated. (SSC3) ensures that the considerations about the interactions between the nodes are robust, i.e., that small changes in the initial values do not cause sudden variations in the equilibrium ones.

---

[4] This and some other related issues will be explored in more detail in a forthcoming paper dealing with the more mathematical issues of this approach.

Thus, the basic idea is that the stronger an attack is, the closer its value gets to 1 and hence the closer we want π to get to 0 so that the equilibrium value of the attacked argument can decrease proportionally (since its initial value is multiplied by π). In the case of a single attack of strength u to node X with transmission factor κ, one possibility is to make $\pi(X) = 1 - \kappa u$. In the network of Fig. 20.6 above, this would make $\pi(Y) = 1 - \xi((X, Y))V_e(X)$ and hence Y's equilibrium value would be $V_e(Y) = V(Y) \cdot (1 - V_e(X)) = 1 \cdot 0 = 0$, as expected.

Besnard and Hunter's *categoriser* function [46] is an example of a function satisfying (SSC1)–(SSC3) (more on this in Section 20.7).

We still have to tackle the problem of a node being attacked by multiple arguments, i.e., what can we say about $\pi(X)$ when X is attacked by multiple arguments?

As usual, attacking arguments combine via *multiplication*, which is compatible with the behaviour of conjunction in Boolean logic and in probability. The equations for the equilibrium values of the nodes of a weighted augmented network are defined below.

Definition 20.7 (Equilibrium value of an argument). *Let* $P = \langle AN_1, \ldots, AN_k \rangle$ *be a profile of argumentation systems and* $AWN_P = \langle S, R, V, \xi \rangle$ *the weighted augmented network for P. The equilibrium value of an argument* $X \in S$ *is defined be the equation:*

$$(Eq_{inv}) \qquad V_e(X) = V_0(X) \cdot \prod_{Y_i \in Att(X)}(1 - \xi((Y_i, X))V_e(Y_i))$$

One can choose V_0 and ξ to be V_0^+ and ξ^+ or V_0^- and ξ^- depending on whether a credulous or sceptical approach is desired (this will be explored further in Section 20.6). Notice that the highest possible intensity of the attack by an argument Y is $V_0(Y)$ itself. This happens when the attack is carried out with full intensity and Y is not itself attacked by any node — in this case it retains its initial value fully, i.e., $V_e(Y) = V_0(Y)$. Because we take the complement of this attack to 1, in such circumstances this is sufficient to make the equilibrium value of the attacked argument 0.

Eq_{max} decreases the initial support value of an argument according to the value of the strongest attack on it. Eq_{inv} on the other hand is *cumulative*: it aggregates the strength of the attacking nodes. The intuition is that each challenge to an argument contributes to decreasing the argument's overall credibility.

Henceforth, we formally set the value $\pi(X)$ to $\prod_{Y_i \in Att(X)}(1 - \xi((Y_i, X))V_e(Y_i))$.

Proposition 20.8. π *satisfies (SSC1)–(SSC3).*

Proof. If $\max_{Y \in Att(X)}\{\xi((Y, X))V_e(Y)\} = 0$, *then by Definition 20.7,* $\prod_{Y_i \in Att(X)}(1 - \xi((Y_i, X))V_e(Y_i)) = 1$. *Therefore, (SSC1) is satisfied. If* $\max_{Y \in Att(X)}\{\xi((Y, X))V_e(Y)\} = 1$, *then by Definition 20.7, for some* $Y' \in Att(X)$, $1 - \xi((Y', X))V_e(Y') = 0$, *and then* $\prod_{Y_i \in Att(X)}(1 - \xi((Y_i, X))V_e(Y_i)) = 0$. *Hence (SSC2) is also satisfied. (SSC3) is trivially satisfied.*

Combining attacks in this way was initially proposed in [37].

It is easy to see that when all attacks are carried out with full intensity, $\pi(X)$ can be defined in a simpler way as

$$\pi(X) = \prod_{Y \in Att(X)} (1 - V_e(Y))$$

which is equivalent to

$$\pi(X) = 1 - \vee_{Y \in Att(X)} V_e(Y) \qquad (20.1)$$

where $a \vee b = a + b - a.b$ and for $\Delta = \{a_1, \ldots, a_k\}$, $\vee \Delta = ((a_1 \vee a_2) \vee \ldots \vee a_k)$. The expression in (20.1) corresponds to the complement of the probabilistic sum t-conorm used by Leite and Martins in [237]. In probability theory, the probabilistic sum expresses the probability of the occurrence of independent events. Since we want to weaken the value of the attacked node, we take the complement of this sum to 1.

It is worth emphasising that the equilibrium value of a node can never be higher than its initial value.

Proposition 20.9. *For arguments X, $V_e(X) \le V_0(X)$.*

Proof. Straightforward. Note that $V_e(X) = V_0(X) \cdot \pi(X)$. By Definition 20.4, for all arguments Y, $0 \le V_0(Y) \le 1$. By Definition 20.7, $0 \le \pi(X) \le 1$ and hence $V_e(X) \le V_0(X)$.

Proposition 20.10 (Unanimity of acceptance). *Let $P = \langle AN_1, \ldots, AN_k \rangle$ be a profile of argumentation systems where each $AN_i = \langle S_i, R_i, E_i \rangle$ and let $AWN_P = \langle S, R, V, \xi \rangle$ be the weighted augmented network for P. If each E_i is conflict-free and $X \in \cap_i E_i$, then $V_e(X) = 1$.*

Proof. By Proposition 20.5, if $X \in \cap_i E_i$, then $V_0^+(X) = V_0^-(X) = 1$. Suppose $(Y, X) \in R_i$, for some argumentation framework AN_i. Since each E_i is conflict-free, then $Y \notin E_i$ and hence $Y \notin \cup_i E_i$. By Proposition 20.4, $V_0^+(Y) = V_0^-(Y) = 0$ and by Proposition 20.9, $V_e(Y) = 0$. It follows that $\pi(X) = 1$ and hence $V_e(X) = 1$.

If each E_i is conflict-free and $V_e(X) = 1$ under the sceptical approach, then it follows that $X \in \cap_i E_i$. This is not necessarily the case under the credulous approach, because under the latter the initial support value of an argument is set to 1 as long as it wins in every argumentation system *in which it is known*. Consequently, as long as there are no attacks on X its equilibrium value will also be 1 (see Example 2. in Section 20.6). It is worth emphasising that the flipside of this credulity is that attacks (which also influence the equilibrium value of an argument) are treated in the same way. This is illustrated in Example 1. of Section 20.6, where the equilibrium value of the argument Y is lower in the credulous approach than in the sceptical one as the result of credulously accepting an argument that attacks it.

Proposition 20.11 (Unanimity of rejection). *Let $P = \langle AN_1, \ldots, AN_k \rangle$ be a profile of argumentation systems where each $AN_i = \langle S_i, R_i, E_i \rangle$ and let $AWN_P = \langle S, R, V, \xi \rangle$ be the weighted augmented network for P. If $X \notin \cup_i E_i$, then $V_e(X) = 0$.*

Proof. By Proposition 20.5, if $X \notin \cup_i E_i$, then $V_0^+(X) = V_0^-(X) = 0$. By Definition 20.7, $V_e(X) = 0$.

The equilibrium values are largely (but not solely) dependent on the initial values, which in turn depend on the level of acceptance of each argument in the profile. If an argument is not accepted in any of the networks in the profile, then its initial support value will be null and, consequently, so will its equilibrium value. The converse is not necessarily true though, since $V_e(X)$ is also null when $\pi(X) = 0$.

20.5 Thresholds for acceptance

The equilibrium values simply represent how the initial overall level of support for a component is affected by the interactions with the other components in the network. Propositions 20.10 and 20.11 go some way into explaining what to expect from the equilibrium values in special cases. If one wants to make a decision on what arguments to accept overall, an appropriate threshold for acceptance for the network at hand must be decided.

The value 1 represents the strongest possible level of acceptance. We have seen that under the sceptical approach an equilibrium value of 1 means that the argument belongs to the sets of winning arguments of all agents, but setting 1 as the minimum acceptance level could prove too strict. The concept of *majority* is sometimes used in some voting systems. Acceptance of an argument by a clear majority of the networks in a profile produces a start-up value for the argument strictly greater than $\frac{1}{2}$, as can be seen below.

Proposition 20.12 (Majority of acceptance). *Let* $P = \langle AN_1, \ldots, AN_k \rangle$ *be a profile of argumentation systems where each* $AN_i = \langle S_i, R_i, E_i \rangle$ *and let* $AWN_P = \langle S, R, V, \xi \rangle$ *be the weighted augmented network for P. If* $|\{i \mid X \in E_i\}| > \frac{k}{2}$, *then* $V_0^+(X) > \frac{1}{2}$ *and* $V_0^-(X) > \frac{1}{2}$.

Proof. This comes straight from Definition 20.4. $v^+(X) = |\{i \mid v_i(X) = 1\}| = |\{i \mid X \in E_i\}|$. Notice that $v^+(X) + v^-(X) \le k$. If $v^+(X) > \frac{k}{2}$, then $V_0^+(X) > \frac{1}{2}$. The same applies to $V_0^-(X)$, since $v^+(X) = \frac{k}{2} + \epsilon$, for some ϵ and $\frac{\frac{k}{2}+\epsilon}{k} > \frac{1}{2}$.

However, having a start-up value greater than $\frac{1}{2}$ is not sufficient to guarantee an *equilibrium* value greater than $\frac{1}{2}$, because there could be arguments that attack X in those networks in which it was not accepted and if the equilibrium values of those arguments is greater than 0, they will contribute to bring X's equilibrium value down. We argue that this is a good feature and addresses the problem of "voting relying only on the selected extensions" (this is presented as "Problem 2." in [101]). Majority of acceptance alone does not guarantee overall group acceptance, precisely because of the attack relations and this is reflected in the calculation of the equilibrium values.

Alternatively, one could define the acceptance value as the maximum of the equilibrium values (if that is greater than 0). A simpler approach (adopted here) is to set the threshold for acceptance as the average value of the equilibrium values and accept the arguments whose equilibrium values are greater or equal to it. This is akin to deciding the grades in an exam by looking at the distribution of the marks, dividing it into clusters according to the average, and associating the top cluster with the best grade.

In specific scenarios a different threshold value can be defined through a more sophisticated analysis of the networks in the profile in a similar way to how it is done in [18] (which is itself based on the notion of the "inconsistency degree" of a knowledge base). This investigation itself is quite complex and left for future work.

20.6 Worked examples

We now illustrate our technique with a few examples. In each example, we show the input networks and the resulting augmented (merged) network with its components annotated with the initial weights for each approach in the form (credulous,sceptical). The equations for all nodes are given as well as the corresponding equilibrium values calculated under both the credulous and sceptical approaches. The accepted arguments are indicated within a shadowed box (they have equilibrium values greater or equal than the average). Each example is followed by a discussion of the results.

1. **Input networks**

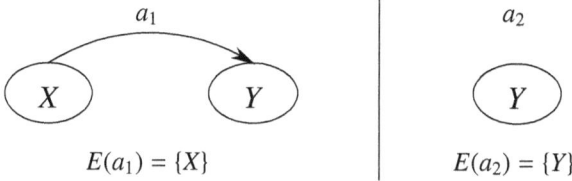

$$E(a_1) = \{X\}$$ $$E(a_2) = \{Y\}$$

Merged network

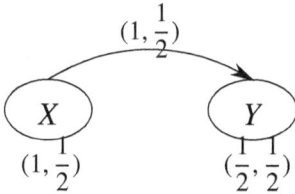

Credulous

$$V_e(X) = V_0(X)$$
$$V_e(Y) = V_0(Y)(1 - V_0(X))$$

$$V_e(X) = 1$$
$$V_e(Y) = 0$$
$$avg = \tfrac{1}{2}$$

Sceptical

$$V_e(X) = V_0(X)$$
$$V_e(Y) = V_0(Y)(1 - \tfrac{1}{2} \cdot V_0(X))$$

$$V_e(X) = \tfrac{1}{2}$$
$$V_e(Y) = \tfrac{3}{8} = 0.375$$
$$avg = 0.437$$

In this example, under the credulous approach $V_0(X) = 1$ and since X has no attacks, $V_e(X) = 1$. Its attack on Y is transmitted with full intensity. $V_0(Y) = \tfrac{1}{2}$. Therefore, $V_e(Y) = \tfrac{1}{2} \cdot (1 - 1) = 0$. Under the sceptical approach $V_0(X) = \tfrac{1}{2}$ and hence $V_e(X) = \tfrac{1}{2}$. Its attack on Y is transmitted with intensity $\tfrac{1}{2}$. Therefore, $V_e(Y) = \tfrac{1}{2} \cdot (1 - \tfrac{1}{2} \cdot \tfrac{1}{2}) = \tfrac{3}{8}$. Note that the sceptical approach produces a higher equilibrium value for Y because under the credulous approach X is fully accepted and its attack on Y fully defeats it. The only argument with equilibrium value above the average of the values is X in both approaches and therefore it is the only one accepted. As for the credulous approach, it is easy to see why X should be accepted and Y rejected. There is no dispute about the acceptance of X, but under the credulous approach so is the case about its attack on Y, which is then defeated. Under the sceptical approach, the results can be explained as follows. Both agents know about Y, so its equilibrium value should not be null. However, even though X is not unanimously known (and hence accepted), there is no knowledge of any attacks on it. X's equilibrium value of $\tfrac{1}{2}$ reflects its support across the community.

Y's initial support value has to be adjusted down by the fact that one of the agents supports an attack on it. Taking the average of the two values means that Y ends up being rejected.

2. **Input networks**

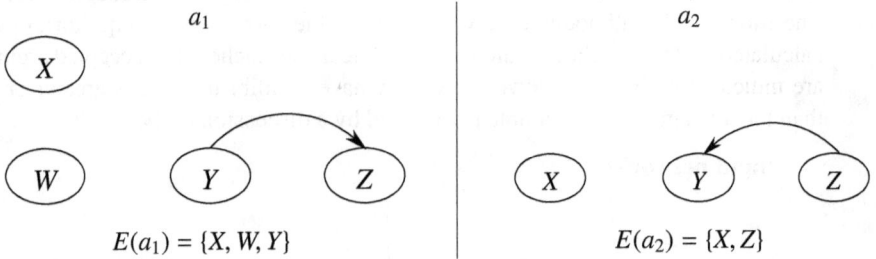

$$E(a_1) = \{X, W, Y\} \qquad\qquad E(a_2) = \{X, Z\}$$

Merged network

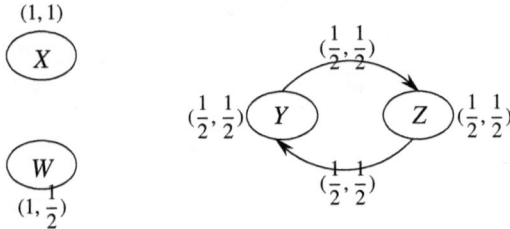

Credulous	**Sceptical**
$V_e(X) = V_0(X)$	$V_e(X) = V_0(X)$
$V_e(W) = V_0(W)$	$V_e(W) = V_0(W)$
$V_e(Y) = V_0(Y)(1 - \frac{1}{2} \cdot V_e(Z))$	$V_e(Y) = V_0(Y)(1 - \frac{1}{2} \cdot V_e(Z))$
$V_e(Z) = V_0(Z)(1 - \frac{1}{2} \cdot V_e(Y))$	$V_e(Z) = V_0(Z)(1 - \frac{1}{2} \cdot V_e(Y))$
$V_e(X) = 1$	$V_e(X) = 1$
$V_e(W) = 1$	$V_e(W) = \frac{1}{2}$
$V_e(Y) = \frac{2}{3}$	$V_e(Y) = \frac{2}{3}$
$V_e(Z) = \frac{2}{3}$	$V_e(Z) = \frac{2}{3}$
$avg = 0.7$	$avg = 0.575$

In this example, both agents accept argument X and there are no attacks on it in any network. Thus, regardless of the approach, the equilibrium value of X is 1. This value under the sceptical approach indicates unanimity of acceptance of X. Compare this with the previous example, where X's equilibrium value smaller than 1 under the sceptical approach signals some disagreement about its universal acceptance. In spite of there not being any attacks on W, this argument is only known by agent a_2. Under the credulous approach $V_e(W) = 1$, but under the sceptical approach $V_e(W) = \frac{1}{2}$, since it is accepted by only half of the community. Arguments Y and Z are also accepted by half of the community, but in each case, the other half supports a complementary attack of one on the other. As a result, their equilibrium values are both reduced from $\frac{1}{2}$ to $\frac{2}{3}$. Arguments X and W have equilibrium values above the average under the credulous approach and hence are accepted there, but under the sceptical approach only X is accepted, since W's equilibrium value takes into account the fact that it is not known by a_2. Notice that our approach provides some indication about the relative acceptability levels of all arguments, including those that are not accepted. In this case, Y and Z are at the same level but below W. This reflects the fact that no attacks on W are known.

3. **Input networks**

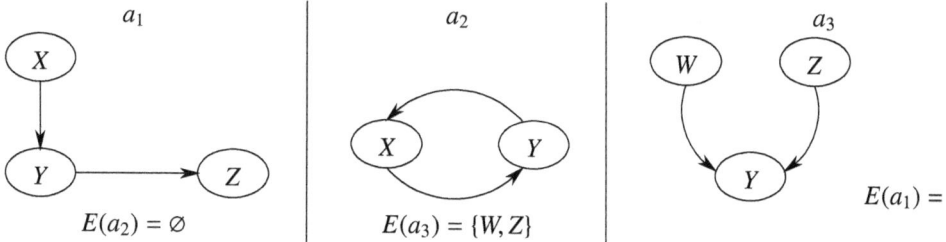

$$E(a_2) = \varnothing \qquad E(a_3) = \{W, Z\} \qquad E(a_1) =$$

Merged network

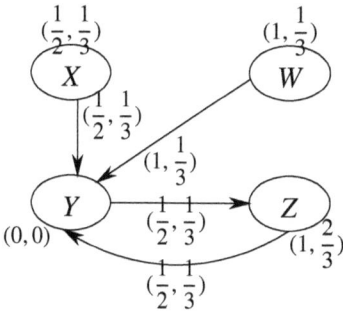

Credulous

$V_e(X) = V_0(X)$

$V_e(W) = V_0(W)$

$V_e(Y) = V_0(Y)(1 - \frac{1}{2} \cdot V_e(X))$
$\qquad (1 - \frac{1}{2} \cdot V_e(W))$
$\qquad (1 - \frac{1}{2} \cdot V_e(Z))$

$V_e(Z) = V_0(Z)(1 - \frac{1}{2} \cdot V_e(Y))$

$V_e(X) = \frac{1}{2}$

$V_e(W) = 1$

$V_e(Y) = 0$

$V_e(Z) = 1$

$avg = \frac{5}{8}$

Sceptical

$V_e(X) = V_0(X)$

$V_e(W) = V_0(W)$

$V_e(Y) = V_0(Y)(1 - \frac{1}{3} \cdot V_e(X))$
$\qquad (1 - \frac{1}{3} \cdot V_e(W))$
$\qquad (1 - \frac{1}{3} \cdot V_e(Z))$

$V_e(Z) = V_0(Z)(1 - \frac{1}{3} \cdot V_e(Y))$

$V_e(X) = \frac{1}{3}$

$V_e(W) = \frac{1}{3}$

$V_e(Y) = 0$

$V_e(Z) = \frac{2}{3}$

$avg = \frac{1}{3}$

This is the example appearing in Fig. 20.2. We start with argument Y, which does not feature in any of the sets of the agents' winning arguments. Its initial support value is null and hence its equilibrium value is also null. This leaves X's initial support values unchanged in both approaches. Under the credulous approach both W and Z get equilibrium value 1 (there are no attacks on W and Z is accepted by the majority). Under the sceptical approach Z's equilibrium value is the highest, because it is accepted by $2/3$ of the agents (as opposed to X and W which are accepted by only $1/3$ of them). Both W and Z have equilibrium values equal to

1 under the credulous approach and hence are accepted, but under the sceptical approach only Z is accepted (note that it is the only argument accepted by the majority of the agents).

4. **Input networks**

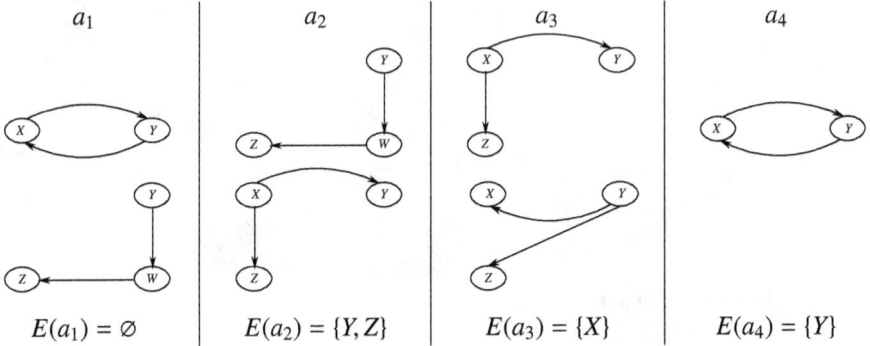

$$E(a_1) = \varnothing \qquad E(a_2) = \{Y, Z\} \qquad E(a_3) = \{X\} \qquad E(a_4) = \{Y\}$$

Merged network

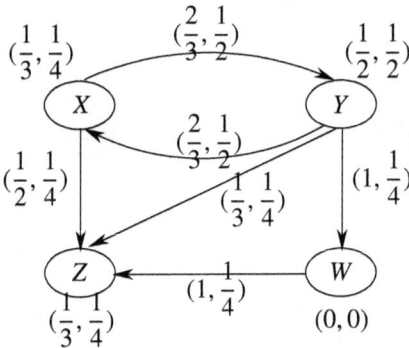

Credulous

$$V_e(X) = V_0(X)(1 - \tfrac{2}{3} \cdot V_e(Y))$$
$$V_e(W) = V_0(W)(1 - V_e(Y))$$
$$V_e(Y) = V_0(Y)(1 - \tfrac{2}{3} \cdot V_e(X))$$
$$V_e(Z) = V_0(Z)(1 - \tfrac{1}{2} \cdot V_e(X))$$
$$(1 - \tfrac{1}{3} \cdot V_e(Y))$$
$$(1 - V_e(W))$$

$$V_e(X) = \tfrac{12}{50} = 0.24$$
$$V_e(W) = 0$$
$$V_e(Y) = \tfrac{21}{50} = 0.42$$
$$V_e(Z) = \tfrac{473}{1875} \approx 0.252$$
$$avg \approx 0.228$$

Sceptical

$$V_e(X) = V_0(X)(1 - \tfrac{1}{2} \cdot V_e(Y))$$
$$V_e(W) = V_0(W)(1 - \tfrac{1}{4} \cdot V_e(Y))$$
$$V_e(Y) = V_0(Y)(1 - \tfrac{1}{2} \cdot V_e(X))$$
$$V_e(Z) = V_0(Z)(1 - \tfrac{1}{4} \cdot V_e(X))$$
$$(1 - \tfrac{1}{4} \cdot V_e(Y))$$
$$(1 - \tfrac{1}{4} \cdot V_e(W))$$

$$V_e(X) = \tfrac{6}{31} \approx 0.194$$
$$V_e(W) = 0$$
$$V_e(Y) = \tfrac{14}{31} \approx 0.452$$
$$V_e(Z) \approx 0.211$$
$$avg \approx 0.214$$

This example was introduced in [101] and is further discussed in Section 20.7. It shows a high level of disagreement between all agents. In particular, no single argument is accepted by all of these agents. The argument which is accepted by the highest number of agents is Y (accepted by half of the agents), followed by X and Z (accepted by one agent only). Y obtains the highest equilibrium values in both the credulous and sceptical approaches. This value is well above the average values of the arguments in each approach and as a result Y is accepted under both. X and Z are accepted under the credulous approach with equilibrium values just above the average of the values there. However, they fail to be accepted under the sceptical approach because their values are below the average of all values. W is not accepted by any of the agents and as a result it is not accepted in the result of the merging under either approach.

5. This example was introduced in Section 20.3. We consider the case of two agents sharing the same argumentation network in scenarios in which they sometimes agree and sometimes disagree with respect to the winning arguments in the network.

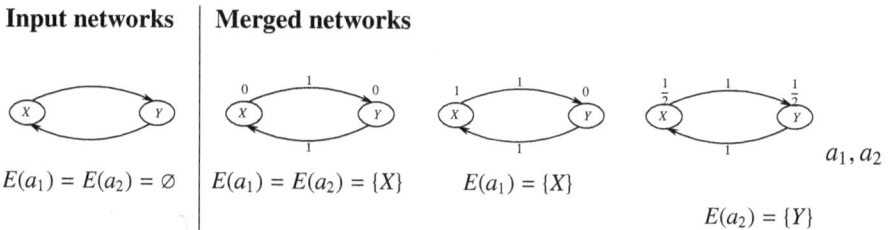

Input networks | **Merged networks**

$E(a_1) = E(a_2) = \varnothing$ | $E(a_1) = E(a_2) = \{X\}$ | $E(a_1) = \{X\}$

$E(a_2) = \{Y\}$

a_1, a_2

Credulous and Sceptical

$V_e(X) = V_0(X)(1 - V_e(Y))$
$V_e(Y) = V_0(Y)(1 - V_e(Y))$

(a)	**(b)**	**(c)**
$E(a_1) = E(a_2) = \varnothing$	$E(a_1) = E(a_2) = \{X\}$	$E(a_1) = \{X\}\ E(a_2) = \{Y\}$
$V_e(X) = 0$	$V_e(X) = 1$	$V_e(X) = \frac{1}{3}$
$V_e(Y) = 0$	$V_e(Y) = 0$	$V_e(Y) = \frac{1}{3}$
$avg = \frac{1}{2}$	$avg = \frac{1}{2}$	$avg = \frac{1}{3}$

This example includes three sub-cases. The equations are the same because the attacks are the same, only the start-up values need to change and they correspond to the sub-cases. (a) If both agents do not accept any of the arguments, then no arguments will be accepted in the result of the merging. This is because the equilibrium value of a node is always less than or equal its initial value and the initial value of the arguments in this case are all null since they are not supported by any of the agents (cf. Proposition 20.11). (b) On the other hand, if they all accept the same arguments (for instance, $\{X\}^5$) and provided their extensions are conflict-free (which is the case here), these arguments will get equilibrium value 1 and consequently will be accepted (cf. Proposition 20.10), agreeing with their own local preferences. These local preferences are not taken into account in [101] and, consequently, some other decision mechanism would have to be employed to deal with the results of the merged network there. (c) Finally, if the agents are divided with respect to the winning arguments, the resulting equilibrium values

---

[5] Choosing $\{Y\}$ would give symmetrical results.

will reflect that. In this case, the agents are equally split. Even though each argument starts with value $\frac{1}{2}$, they end up with the lower value $\frac{1}{3}$ due to the attacks these arguments make on each other. Notice that the use of *avg* causes all of the arguments to be accepted (since they have equal equilibrium values). Setting the threshold at $\frac{1}{2}$ would cause them to be rejected.

20.7 Comparisons with other work

As mentioned in Section 20.1, many frameworks consider extensions to Dung's argumentation systems that are capable of representing in one way or another the notion of strength of arguments or attacks. In this section, we compare some of these approaches with ours. There are two main aspects to consider: the numerical nature of the networks itself (independently of any merging) and the methodology for merging.

In terms of numerical merging, the formalism that most resembles ours is the one proposed in [88], which uses a weighted argumentation system. The idea is also based on the combination of all networks into a single augmented one in which attacks are assigned weights that correspond to ours under the credulous approach. However, the similarities stop there. In particular, there is no notion of sceptical support; no mechanism to consider the weights of arguments; and the concept of acceptance is based on the notion of "various-strength defence": an argument X defends an argument Y against an argument Z, if the weight of the attack of X on Z is greater than the weight of the attack of Z on Y. This notion of defence is then used in the definition of *admissibility*. We believe that once we are prepared to associate strengths to the attacks based on the opinions of the agents, we should also be prepared to take into account the opinions of these agents about the arguments themselves during the merging process.

Bistarelli and Santini also propose a numerical approach to merging in [50], but as in the formalism above, weights are given only to the attacks.

Amgoud and Kaci take a different approach to merging by considering the merging of knowledge bases whose underlying formalism is a possibilistic logic [18]. This allows for the calculation of the *inconsistency degree* of a base, which in turn can be used to determine its "plausible" consequences. The inconsistency degree of a knowledge base is an interesting concept and something we would like to investigate in the future to help us to provide a more robust definition of the threshold of acceptance of arguments.

Given an adequate meaning for the initial weights, the equational approach can be used for a single weighted network independently of the merging process. Leaving considerations about the merging aside, it is possible to compare our formalism with other weighted argumentation systems. One of the first systems of this type was proposed in [46]. In that formalism, the weight of an argument is calculated by a so-called *categoriser* function, an example of which is the **h-categoriser**. For an argument X, this is defined as follows

$$h(X) = \begin{cases} 1, & \text{if } Att(Y) = \emptyset; \text{ or} \\ 1/(1 + \sum_{Y \in Att(X)} h(Y)), & \text{otherwise} \end{cases}$$

We can think of h as the function that calculates the equilibrium values of the arguments. Let us now analyse what happens with these values in a sequence of attacks like the one below. For comparison, we assume that all nodes have the same initial value v and that the intensity with which all attacks are carried out is also 1.

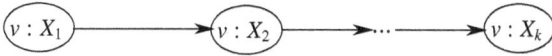

Given that X_1's initial value is 1 in the example above, we would have that $h(X_1) = 1$; $h(X_2) = 0.5$; $h(X_3) = 0.66$; and so forth. This obviously does not agree with Dung's semantics. Using the equational approach, we get that $V_e(X_1) = v$, $V_e(X_2) = v(1 - v)$, $V_e(X_3) = v(1 - (v(1 - v)))$, ... If $v = 1$, then $V_e(X_1) = 1$, $V_e(X_2) = 0$, $V_e(X_3) = 1$, and so forth, agreeing with Dung's semantics as expected. If $v = 0$, then $V_e(A_i) = 0$ for all i. This is as expected, since in this case no arguments have any initial support and we have seen that $V_e(X) \leq V_0(X)$ (cf. Proposition 20.9). If $v = 0.5$, we get $V_e(X_1) = 0.5$, $V_e(X_2) = 0.25$, $V_e(X_3) = 0.375$, ...[6]

In [237], Leite and Martins proposed the so-called *social abstract argumentation frameworks* (SAAFs), which can be seen as an extension of Dung's abstract argumentation frameworks to allow the representation of information about votes to arguments. The motivation for a SAAF is to provide a means to calculate the result of the interaction between arguments using approval and disapproval ratings from users of news forums. The idea is that when a user sees an entry in a forum, she may approve it, disapprove it, or simply abstain from expressing her opinion. The 'weights' associated with the arguments in this case can also be seen as being generated by how the users perceive the arguments. However, the initial support level for an argument is calculated differently in their formalism and there is no notion of strength of attack, even though, as in our case attacks are aggregated using the probabilistic sum t-conorm.

In terms of the merging methodology itself Coste-Marquis et. al.'s work [101] has a number of similarities with ours that are worth discussing. As in our case, their starting point is an augmented network containing all nodes in the profile to be merged. However, they consider the expansions to *each* of the networks being merged as an intermediate step. The expansions are called *partial argumentation frameworks* (PAFs). The process can be described as follows. Nodes unknown to an agent are simply added to her intermediate network. For every pair of nodes in an agent's own (non-expanded) network, the underlying assumption about the absence of an edge is that of a *non-attack* between the nodes. Therefore, an agent is never "ignorant" about attacks within her own arguments. However, as the network is expanded to include all arguments in the profile, a decision has to be taken with respect to the attacks between the arguments the agent does not know about (as is the case in our own formalism when we consider the attitude of an agent with respect to components of the augmented network) and she is left with the choice of either supporting the attack; or rejecting it; or neither (i.e., to assert her ignorance about it).

In a single agent's (smaller) network, the attack, non-attack and ignorance relations above can be reduced to two only if one assumes omniscience of the agent about attacks between her own arguments. This renders a traditional argumentation system as a special case of a PAF in which the ignorance relation is empty and the non-attack

---

[6] We can think of an infinite sequence of this kind as a node with an attack on itself. In the limit $k \to \infty$, for $V_0(X_1) = 0.5$, $V_e(X_k) = \frac{1}{3}$. Further considerations of this kind are left for a forthcoming paper on the numerical aspects of these networks.

relation is the complement of the attack relation with respect to the Cartesian product of the set of arguments.

Any PAF can be "completed" by moving all edges from the ignorance relation into one of the other two.[7]

Now given a network and a profile of networks, we want to consider the expansion containing all arguments in the profile, and hence we need to decide what to do with the attacks between arguments not present in the original network. Coste-Marquis et. al. advocate for a "consensual" expansion: if an attack that is unknown to an agent appears in the profile the agent will add the edge to her PAF in the attack relation only if all agents that know about the two arguments have that attack; otherwise the agent will add the edge to her ignorance relation.

For comparison, the idea is illustrated by the example below taken from [101]. In the example, the four networks AF_1–AF_4 are to be merged.

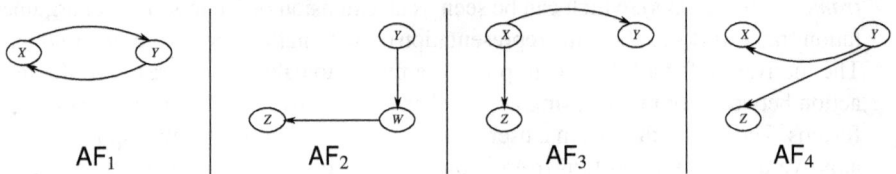

The first step is the expansion of each individual network, which results in the four partial argumentation frameworks PAF_1–PAF_4 below.

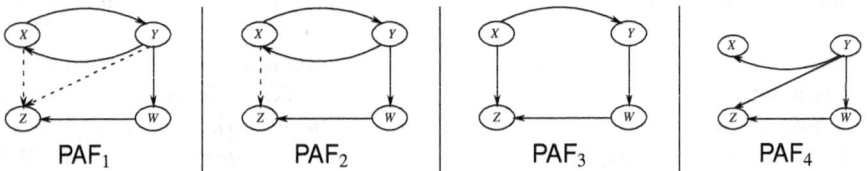

Notice that edges not appearing in any network are therefore (implicitly) added as non-attacks. Each network maintains its original edges in the expansion. Take for instance the network PAF_1. It contains the edges originally present in AF_1, but also the attacks from Y to W and W to Z, because they appear in AF_2, the only network that contains W (as a result, all expanded networks will contain all attacks to and from W — see our discussion later). An ignorance edge is added from X to Z, because even though X attacks Z in AF_3, this is not supported by AF_4, and hence there is no "consensus" on the attack (the absence of the edge between X and Z in AF_4 is treated explicitly as a non-attack).

Once the networks of all agents are expanded, the merging is computed by selecting the "super-networks" with minimal aggregated distance to all of the expanded networks. These super-networks are networks with all arguments, where the absence/presence of an edge between arguments is taken into account by a distance function. They consider the distances *sum*, max, and *leximax*.

---

[7] It is sufficient to have only the attack relation, since assuming the ignorance relation is empty, the non-attack relation can be defined as the complement of the attack relation as explained above.

Our aggregation is done implicitly through the equations, so a direct comparison is not possible. However, we can interpret each individual expansion as the attitude of the agent with respect to the augmented network containing all components, since the expanded networks are what will be used in the aggregation. With this in mind, we can compare the two approaches. In our case, each agent would "vote" for each component of the augmented network as depicted below (remember that the absence of an edge in a PAF means a non-attack – effectively a vote against the attack, which is what a "−1" vote means in our methodology).

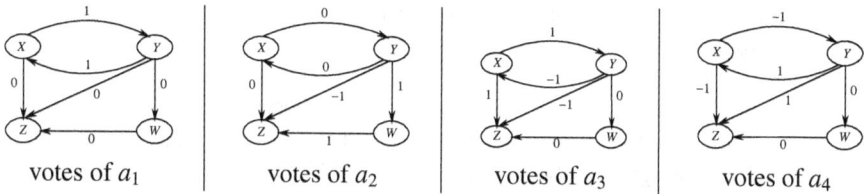

| votes of a_1 | votes of a_2 | votes of a_3 | votes of a_4 |

The attitudes are similar, except that in their formalism an agent adopts the views of the others with respect to an attack between unknown arguments as long as no agent explicitly disagrees about the attack. Our approach is more cautious about this acceptance. For instance, in our case, a_1, a_3 and a_4 remain neutral with respect to the attacks from Y to W and from W to Z suggested by AF_2, since none of a_1, a_3 and a_4 know about W. We argue that this difference is crucial. In their methodology, if there are only two agents a_1 and a_2 who disagree about everything they know in common, they would still forcibly agree with each other on everything they do not know in a consensual expansion.[8] In this sense, consensual expansion is always biased towards reaching an agreement, but this may not always be appropriate. In contrast, the votes of the agents themselves are completely independent of how the other agents vote in our framework, which seems more natural. However, our credulous approach exhibits the same bias towards agreement at the time of merging even though other considerations can also potentially be taken at that stage based, e.g., on the overall level of disagreement in the community. We feel that if an agent is to lean towards the attitude of some other agents with respect to a component, at the very least the agent should lean more towards the agents whose opinions are more in agreement with her own.

Furthermore, notice that the merging of the PAFs can still produce more than one network as the result, since there could be more than one candidate aggregation with the minimal distance. And of course there is also the matter of choosing the winning arguments from the result.

Finally, another non-numerical formalism for the merging of argumentation systems was proposed in [296].

---

[8] It is sufficient for the set of non unanimous arguments to be partitioned between disagreeing agents for them all to take the same view on those arguments.

20.8 Conclusions and future work

In this chapter, we showed how to merge a profile of argumentation systems through the use of an augmented argumentation network provided with weights for the arguments and the attacks between them.

The weights of the augmented network are calculated based on how representative each component features in the profile and are independent of the local semantics of each network. We proposed two approaches for calculating these weights. In the *credulous approach* agents vote for components and the votes generate weights that take into account only the number of agents that know about a component. The *sceptical approach* considers the votes with respect to the total number of networks in the profile.

Weighted argumentation networks have been proposed before. Sometimes weights have been assigned to the arguments (e.g., as in [18, 37, 46, 86, 237]) and sometimes they have been assigned to the attacks (e.g., as in [37, 50, 118, 336]). In our approach, both arguments and attacks have weights and the network is seen as a generator for equations. The idea is to calculate equilibrium values for the arguments based on their initial support value within the profile and on how they interact with other arguments through the attack relation. The equilibrium values can be calculated by solving a system of equations generated by the augmented network, following [167]. Once calculated, the notion of acceptance can be defined in terms of reaching a minimum threshold value for acceptance. A strict definition of this threshold is the value 1. However, our framework is flexible in the sense that a particular application is free to partition the unit interval in different ways and give an interpretation for values occurring within each of these segments. For instance, one could associate 0 with rejection; 1 with acceptance and consider anything else in between as being undecided.

We can interpret the initial values in our augmented networks as coming from an extended form of approval voting in which voters can also express ignorance and rejection for some components. There are variations of this idea that are worth investigating, including giving different degrees of preference to the components depending on the expertise level of the agents supporting them. Furthermore, there are interesting connections with several other areas of research. From the aggregation perspective, it is worth exploring similarities with other mechanisms for voting and formalisms for merging of knowledge bases as in [92, 151, 232, 233]. Some similarities also exist in the way the interactions are calculated with the approaches taken in the areas of network flows [13], belief propagation and Bayesian networks [278]. We hope to explore these issues in more detail in future work.

The merging of argumentation systems is an application that leads naturally to the employment of weights in a network. However, one need not restrict its use to such scenarios only. All that is required is a suitable interpretation for the weights; an adequate schema for generating the equations; and an interpretation for the equilibrium values. This chapter paves the way for a new type of research in argumentation networks not only because its approach is numerical, but also because it is an initial study of *vector evaluations*. We can see this work as a preliminary investigation on how to aggregate many-dimensional values of the components of a network and propagate the aggregated values through the network taking the network's interactions into account.

To realise the potential, consider the very well developed area of many-dimensional temporal logics. In these logics, a formula is evaluated at several indices. As a complex formula is evaluated in the model we move from one set of indices to another. The analogous movement in the case of argumentation is that of an attack. One can move from one node to the next evaluating and propagating the values on the way.

The equational approach can also be used in a more general context. For instance, if the underlying representation is itself based on a fuzzy or possibilistic logic, the initial weights can be obtained from the computations in the logic themselves, in the spirit of Prakken [288] or Amgoud-Kaci's "force of an argument" [18]. The weights can then be subsequently combined taking the topology of the network into account as done here.

The Law of Evidence and Labelled Argumentation Networks

21.1 Background: logic and law

The purpose of this chapter is to reveal, through examples, the potential for collaboration between the theory of legal reasoning on the one hand, and some recently developed instruments of formal logic. Three zones of contact are highlighted.

1. The law of evidence, in the light of labelled deductive systems (LDSs) and labelled argument, discussed through the example of the admissibility of hearsay evidence.
2. The give and take of legal debate in general, and regarding the acceptability of evidence in particular, represented using the abstract systems of argumentation developed in logic, notably the coloured argumentation graphs of Bench-Capon. This is considered through an imaginary example.
3. The use of Bayesian networks as tools for analysing the effects of uncertainty on the legal status of actions, illustrated via the same example

These three kinds of technique do not exclude each other. On the contrary, many cases of legal argument will need the combined resources of all three.

In the past thirty years major changes in logic have taken place. Whereas in the first half of the last century logic was mainly applied to mathematics and philosophy, the rise of computer science, artificial intelligence, computational and logical linguistics, logic in engineering, and quantum computation have given logic a big push and accelerated its evolution. All of this is well known and has been discussed in various places [148, 145, 338]. What has not been sufficiently discussed is the influence and interaction of these developments on the area of logic and law.

Consider, for example, the way logic has evolved in response to the needs of computer science, AI and theories of language. They have to do with daily human behaviour, reasoning and action. These areas deal with devices and artefacts that help and or replace the human in his daily activities.

Logic is needed partly as the underlying formal language and partly to model and analyse the human in these daily activities, with a view to producing better devices to serve, regulate or understand him.

Once logic has evolved in this direction and has developed new logical tools for this purpose, these same kind of new logics and new tools can usefully be adapted to the consideration of similar issues in the law.

Here lies the connection between logic and law. We can say without serious exaggeration that the interface of logic and law is going to be central to the further advancement of logic in the next twenty years.

We envisage the following main benefits to the law community, in addition to the benefits from existing logical tools and aids available from Artificial Intelligence.

- The proper labelled argumentation system tailored for law of evidence and other judicial arguments can help articulate and clarify (hidden) intuitive common sense principles behind existing practices.
- The labelled argumentation methodology includes a system of labelling and stylised hierarchical movements which have logical content. This kind of hierarchy can be added to legal specification formats thus giving a better specification language for law without sacrificing the use of ambiguities and variety of interpretations.

It is astonishing to realize that very few people are aware of the true potential of the interaction of the new logics and law. There are many reasons for that, most of them social. The new developments in logic are slow to spread around even among logicians, and certainly among researchers in legal reasoning and legal theory, many of whom still think of "logic" as "Aristotlian syllogism".[1]

Some bridging work between law and logic has been done by C.H. Perelman [280], who kept in touch with both logicians and judges and lawyers, arguing that logic should play a different — more restricted — role. But when Perelman wrote, the new logical tools were not as available as they are now; and such as were available, Perelman made no use of.

The rise of Horn clause logic programming in the 1980s has helped turn some logicians in the direction of the law, but early attempts to apply logic to law, such

---

[1] It is instructive to read the following passage on legal reasoning from the July 2003 edition of a basic textbook on legal philosophy, widely taught in the UK (J. W. Harris, *Legal Philosophies*, p 213):

"It is far from easy to get a comprehensive view of the subject [of legal reasoning]. Most writers who have discussed legal reasoning have either concentrated on the form as distinct from the substance of justificatory arguments, or else dealt with only part of the subject. Two forms of argument, the deductive and the inductive, have generally been considered inapposite characterizations of legal argument. Some take the view that deductive argument – from major and minor premises to a logically necessary conclusion – is inappropriate even in clear cases. This may be asserted on the general ground that deductive arguments only hold true of factual propositions not of norms; or on the more specific ground that even the clearest rule may be held not to apply to a case where that would frustrate the purpose of the law or produce absurd consequences, and the decision whether this so or not cannot be dictated by logic. On the other hand, reasoning in clear cases seems very close to deductive reasoning – here is a speed-limit rule applying to all car drivers, I am a car driver, so it applies to me. Even in unclear cases, it can be contended that the form of the argument is deductive, since what is at issue is which of competing rulings should be adopted, granted that the winner will be applied deductively in all cases of the present type – although here our major concern will be with the substantive arguments which dictate choice among the rulings.

as the formalisation of the British Nationality Act [304], has drawn a strong critical reaction from the law community on the ground that Horn clause logic is not rich enough to allow for the wealth of nuances and interpretations/explanation/ revision so common in legal reasoning. See also [14] by Judge Ruggero J. Aldisert.

This criticism may have been valid in 1980, it is no longer valid now, especially in view of many advances made in logics of practical reasoning and argumentation.

Logic programmers and deontic logicians have had a somewhat earlier interest in law, have their own conferences and journals [108]. But we doubt if they are aware as a community of all relevant developments in logic. They appear not to realize (or believe) that law is an area of potentially evolutionary significance to logic.

We recommend to the reader survey works by two key researchers in the area, Trevor Bench-Capon's [45] survey article for the *Encyclopaedia of Computer Science and Technology* and Henry Prakken's book [286], *Logical Tools for Modelling Legal Argument*. Prakken's book, especially, takes note of many of the new developments in logic, and argues very strongly in favour of the theoretical connectedness of logic and law. He especially highlights the new developments in defeasible and non-monotonic logics and reasoning from inconsistent data. However, he is unaware of the methodology of labelled deductive systems which subsumes the logic of legal reasoning, among many others, as a special case. More importantly, Prakken believes that 'logic should be regarded as a tool rather than as a model of reasoning', [286, Section 1.4]. Furthermore, the entire approach to date of the community to logic and law is further restricted by the view that [286, p. 6]:

> To understand the scope of the present investigations it is important to be aware of the fact that the information with which a knowledge-based system reasons, as well as the description of the problem, is the result of many activities which escape a formal treatment, but which are essential elements of what is called 'legal reasoning'. In sum, the only aspects of legal reasoning which can be formalised are those aspects which concern the following problem: *given* a particular interpretation of a body of information, and *given* a particular description of some legal problem, what are then the general rational patterns of reasoning with which a solution to the problem can be obtained? With respect to this question one remark should be made: I do not require that these general patterns are deductive; the only requirement is that they should be formally definable.

Thus modelling the legal theory of evidence (which decides what 'body of information' we are 'given') is beyond the horizon of current research in logic and law. In what follows, on the contrary, we shall develop a case study that will show just how important this area is.

A recent key collection of papers by Marylin MacCrimmon and Peter Tillers [245] indicates very lively activity in law and logic. However, most of the papers take a fuzzy logic, uncertainty and probabilistic approach (in the sense of [306, 209]. See also [287] and the references there.

We must here add that the Bayesian reasoning community is actively involved in (Bayesian) logic and law. This is because of several high visibility court cases and evidence where probabilities are used. Part of the problem is that the probabilistic reasoning community is not so interactive with the ordinary logic communities (and so we also need to bring logic and probability together as part of our own ongoing work). The theory of Labelled Deductive Systems is fully compatible with probabilistic reasoning and networks.

In the sections to come we examine some case studies to show how the new logics can play a role in the area of evidence and legal reasoning.

21.2 Legal theory of evidence and the new logics

Our purpose here is to show how the new labelled logics, arising from research in computer science, can be applied to the legal theory of evidence. For a sample of Labelled Deductive Systems, see [143]. For the original monograph, see [136].

21.2.1 Some labelled logic

We start with logic. One of the most well known resource logics is linear logic [202]. In this logic, the databases are multisets of formulas and each item of data must be used *exactly once*. So, for example, we have

$$A, A \rightarrow B \vdash B$$

But

$$A, A \rightarrow (A \rightarrow B) \nvdash B$$

This is because two copies of A are needed here, and we have only one. The proof would run as follows:

1. $A \rightarrow (A \rightarrow B)$, assumption
2. A, assumption
3. $A \rightarrow B$, from 1 and 2 using the rule of modus ponens.
4. B, from 1 and 3, using the rule of modus ponens.

In this proof, 2. is used twice.

To make this example more concrete, let

- A = having a drunken driving conviction
- B = driving licence suspended.

Then $A \rightarrow (A \rightarrow B)$ means that two convictions entail suspension (and of course you cannot count the same conviction twice!).

Linear logic allows for the connective $!A$, which means that A can be used as many times as needed.

Thus

$$!A, A \rightarrow (A \rightarrow B) \vdash B.$$

Let us modify the logic a bit[2] and add the connective ❤A: ❤A means that we can use A if we ask and get permission from some meta-level authority. So we can write

$$❤A, A \rightarrow (A \rightarrow B), \text{ permission given } \vdash B.$$

---

[2] See footnote 17 for an anagram example.

There is a mixing here of object level and meta-level features. Such logics are best expressed as labeled deductive systems (LDS) [136, 143]. A labelled system is comprised of formulas and labels. The labels contain additional information relating to the formulas. For example an item of data (called a *declarative unit*) may have the form

$$\Delta : \text{John has cancer.}$$

Δ can be a medical file with data confirming the fact that John has cancer. This fact can be used in certain situations of legal argument; e.g. to attempt to release John from prison. The reasoning governing Δ is medical, while the reasoning governing the release from prison is legal. Labelled logic is the methodology of how to use such mixed reasoning.

We have in LDS the following form of modus ponens:

$$\frac{t : X, s : X \to Y, \varphi(s, t)}{f(s, t) : Y}$$

Here t, s are labels (their nature and mode of handling are defined in the system), which can be themselves entire databases; φ is meta-predicate indicating that there is the permission to apply modus ponens (φ is called the compatibility predicate); and f is a function giving the new label of the result Y.

Going back to our example, we write

1. $s : (A \to (A \to B))$, where s represents here a body of legal background data on how the substantive law of
   " two drunken driving convictions $\to$ licence suspended"
   has been established.
2. $t : A$, where t is a file indicating the data establishing the facts of the drunken driving incident.
3. $\varphi(s, t)$ is a meta-level argument looking into s and t and arguing that, although we have here only *one* incident of drunken driving, the intention of law (see file s) and the severe circumstances of the incident (see file t) call for suspension (that is, permission to count as two incidents is granted).
4. $f(s, t) : A \to B$, by modus ponens from (1), (2), (3).
   $f(s, t)$ is a file containing the arguments present in granting permission, i.e.
   $f(s, t) = t + s + \varphi$
5. $f(s, t) : B$, by modus ponens from (4) and (5).
   So formally, we have $f(f(s, t), t) = f(s, t)$.

We now show a further connection with the law of evidence.

One important feature of LDS is that it regulates the admissibility of data into the database together with the label it is permitted to have. In fact, using φ we can diplomatically admit a datum D into the database with a label "don't touch", with the effect that φ will never give permission to use it.

These kinds of logics were developed to accommodate the needs of artificial intelligence and the logic ofin language. It is surprising how well these logics fit the needs of theories of evidence.

Imagine a database (Barclays Bank) containing data about a customer. One kind of data includes home telephone number, mobile telephone number, etc. Assume hat a security protocol will allow only certain individuals at the Bank to enter such data and it is up to them to decide whether to 'admit' an additional number. Suppose I call Barclays bank, identify myself and ask the representative to add my mobile number to the database. The representative will ask me some questions (usually mother's maiden name). If correct answers are given, he will add (admit) the additional telephone number. If he is still uncomfortable with my identity (for whatever reasons) he can refuse to do so. We doubt, however, that he has the authority to decide to accept the phone number even if we fail to answer the questions correctly. In other words, security protocols allow the representative to refuse admissible data but do not allow him to overrule and accept non-admissible data!

21.2.2 What Some Books on Evidence Say

Let us go to the website and to the book of Professor Steve Uglow. He teaches evidence at the University of Kent (www.kent.ac.uk/law/spu/), established the law school there, and is actively involved with the community and its problems.

In his web course notes, right at the beginning, he says:

> "Evidence is about regulating the information produced at a trial.
> - What are the general principles regarding this?
> - What are exclusionary rules?
> - What logical processes are involved?"

In our labelled logic we can phrase these points as

- With what label do we insert the new data (evidence) in our database?

The challenge of this area to the research community is made clear at the very first paragraph of Uglow's 725-page book on evidence [325] (*Textbook on Evidence*, 1997)

> "The law relating to evidence is a strange and unruly beast. It is unruly because, first, it refuses to fit into any easy structure for analysis and exposition and, second, it often adopts the characteristics of an uncharged minefield, by which is meant that any set of facts has the potential of throwing up evidential problems, not just of one but of several types, often unforeseen. It is strange because it fulfils different functions than the familiar areas of substantive law. It is in such areas that we see legal rules at their most visible, dealing with the *consequences* of facts – if a contract is broken, damages are paid; if a theft is committed, punishment is imposed. Damages, imprisonment and other civil and criminal remedies are the sanctions accompanying rules which require or prohibit certain types of conduct or which lay down conditions under which that conduct can take place. These rules are often referred to as the substantive law. Within most contested trials, such rules form the background to the case but play little part since there is no conflict over the substance of the rule. We know what the rule says and what the consequences of a breach will be: if there has been a road accident and a driver has been negligent, damages for personal injuries will be paid to any plaintiff; if a sane defendant intentionally kills another person, he or she will be prosecuted and generally receive a life sentence.

But the real conflict in a court, before any substantive rule is brought to bear, is about establishing the facts: was the driver negligent? Did the defendant cause the victim's death? What happened? The law of evidence is not about determining the consequences of facts but about establishing those facts. In a contested trial, under the common law system of justice, the opposing parties will present differing, sometimes diametrically opposed, views of the same event. Having listened to these accounts, the trier of fact must decide what the facts are. It is this problem as to how 'facts' are established with which the law of evidence is concerned: what information can be presented to the court' through what means; how does a court decide whether that information proves whether an event happened in a particular way or not? Such rules, alongside the rules of civil and criminal procedure, can be described, not as *substantive*, but as *adjectival* law.[3] This means that these rules attach themselves to and qualify the operation of a substantive rule but never, by themselves, directly decide the rights and wrongs of any issue. The law of evidence qualifies the operation of a substantive rule because it controls the flow and nature of the information which can be presented to the court. Indirectly, of course, the law of evidence can be decisive since the outcome of a case can depend on whether a particular item of evidence is allowed to be presented to the court or not. For example, a guilty verdict or an acquittal can hang on whether the prosecution can meet the preconditions for the admissibility of a confession in a criminal trial; in a civil case where the weight of the evidence is evenly balanced, the decision may hinge on the question as to where the burden of proof rests.

---

[3] This is our footnote.

Note that a substantive law in labelled logic looks like $s : A \rightarrow (A \rightarrow B)$. Facts look like $t : A$. We can also have other testimony allowing for $t' : \neg A$. The rule that decides in *LDS*, whether to deduce A or $\neg A$ given say, $t_1 : A, t_2 : A, t_3 : \neg A$ is called a *flattening rule*. More precisely, a flattening rule tells us, given $t_i : A$ and $s_j : \neg A$, what is the resultant labels $t : A$ and $s : \neg A$. So, for example, if t_i, s_j are reliability measures of various sources supporting A and $\neg A$ respectively, t and s might be some averages.

What Professor Uglow calls here *Adjectival Law*, means in LDS the logic for reasoning *inside* the label t. For example, t may contain medical evidence and a lawyer may attack that!

If we take our example

Δ : John has cancer,

Δ may be a medical file about John. Δ may contain among other things an expert opinion of a certain Dr. Smith, giving a statement $\Gamma : X$, there X is the Doctor's statement and Γ is another file showing Dr.Smith is a world expert on this kind of cancer. A lawyer wishing to attack Δ might choose to attack Γ (i.e. Dr. Smith's credentials are false), thus weakening the value of X and overall weakening Δ. So we have a structure like

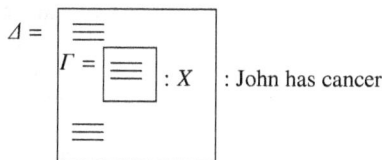

Many of these evidential issues seem very technical to a layperson and, especially in criminal trials, to exclude relevant and important information from the proceedings. Examples might be given of the rule against the admission of hearsay evidence – a witness would be usually prevented from testifying that the victim, now dead, had identified the accused as the assailant; similarly the jury would rarely be allowed to hear about any previous convictions of the defendant. But these are not technicalities for their own sake and reflect the nature and characteristics of the common law trial."

Put in the language our own LDS, what Uglow is saying right at the beginning of his textbook is that:
Given the situation

$$t : A, s : A \rightarrow B$$

he calls $A \rightarrow B$ "substantive law", (in logic it is called a "rule" or a "ticket"), and calls A the facts (called minor premises in logic), then the main part of the theory of evidence is whether to admit A into the database (i.e. establish A as a fact) and with what label t? t may be a label supporting A and what the book calls "adjectival law" is the theory (logic) of evidence.

There is no doubt that the new labelled logics have a role to play in taming this "strange and unruly beast".

Here now is another basic textbook on evidence [107], I. H. Dennis, *Law of Evidence*, 1999.[4] He says (pages 4–6)

B. Concepts and Terminology

The law of evidence uses a number of concepts which are fundamental to an understanding of the subject. This section attempts to introduce these concepts by stating a number of general propositions about them and about their relationships. The propositions are stated in summary form, with more detailed explanation given later.

1. Evidence must be *relevant* in order for a court to receive it. This means that it must relate to some fact which is a proper object of proof in the proceedings.[5] The evidence must relate to the fact to be proved in the sense that it tends to make the existence (or non-existence) of the fact more probable, or less probable, than it would be without the evidence. A simple example is a case where a fact to be proved is the identity of the accused as the person who stole certain goods. Evidence that the goods were found in the accused's house is relevant because it makes the existence of the fact that he is the thief more probable.

2. Evidence must also be *admissible*, meaning that it can properly be received by a court as a matter of law. The most important rule of admissibiltiy is that the evidence must be relevant; irrelevant evidence is always inadmissible. Generally speaking evidence that is relevant is also admissible, but certain

---

[4] Professor Dennis teaches at University College London. He also says in his introduction "Evidence is a notoriously difficult subject to organize in any logical basis".

[5] The facts which are proper objects of proof are sometimes called material facts, but materiality is a slippery term which can be used with more than one meaning. See the discussion in the text below.

rules of law prohibit the reception of certain types of evidence, even though the evidence is relevant. An example is the rule against hearsay evidence, which, broadly speaking, forbids the reception of evidence of a statement made by a person on another occasion when the purpose of adducing[6] the evidence is to ask the court to accept that the statement was true. These rules are often called the *exclusionary rules*, to indicate their function of excluding certain evidence from the court's consideration. The rules are complex because they are often accompanied by exceptions, some of which may be narrow and precisely defined, others may be in broad and flexible terms.

3. In criminal cases, in addition to exclusionary rules, there is also *exclusionary discretion*. A trial judge may exclude prosecution evidence that is relevant and admissible (in the sense that it is not excluded by an exclusionary rule) in the exercise of a discretion conferred on him by the common law or by section 78 of the Police and Criminal Evidence Act 1984 (PACE). The statutory discretion is to prevent the admission of the evidence from adversely affecting the fairness of the proceedings. The main application of the common law discretion is to exclude evidence the prejudicial effect of which outweighs its probative value. Probative value refers to the potential weight of the evidence (see next paragraph), whereas prejudicial effect refers to the tendency of evidence to prejudice the court against the accused, so as to lead the court to make findings of fact against him for reasons not related to the true probative value of the evidence.

4. At the end of a contested trial the court will have to evaluate the relevant and admissible evidence that it received. The *weight* of the evidence is the strength of the tendency of the evidence to prove the fact or facts that it was adduced to prove. This is a matter for the tribunal of fact to decide. In civil cases the judge who tries the case is generally the judge of issues of both law and fact. In criminal cases the tribunal of fact is different according to whether the case is tried on indictment or summarily. The jury is the tribunal of fact for cases tried on indictment. In summary trial the magistrates (justices) deal with issues of both law and fact; lay magistrates have the guidance of their clerk on questions of law. This book uses the term "factfinder" to refer generally to a tribunal of fact, unless the context requires a specific reference to a judge, jury or magistrate. When a factfinder has to determine the weight of evidence it will examine carefully, amongst other things, the *credibility* and *reliability* of the evidence. These terms are not always used with a consistent meaning. Credibility is most commonly used in connection with the testimony of a witness and refers to the extent to which the witness can be accepted as giving truthful evidence in the sense of honest or sincere testimony. Reliability refers most commonly to the truthfulness of testimony in the sense of its accuracy. Honest witnesses may sometimes give evidence that is inaccurate; mistaken evidence of identification by eyewitnesses is a classic example.

---

[6] "Adducing" evidence is a term often used to denote the process of presenting evidence to a court in one of the approved forms, most commonly in the form of the testimony of a witness.

Note here the central role played by the notion of *relevance*. This is also an AI and natural language concept. It is no accident that the first book of our series of books on cognitive systems is a book on relevance [146].

21.3 Case study: hearsay case, *Myers v DPP*

We begin by quoting from [15, p. 133].

A good statement of the hearsay rule was given originally in *Cross on Evidence*, [104].

> "An assertion other than one made by a person while giving oral evidence in the proceedings is inadmissible as evidence of any fact asserted".

Allen continued on page 135:

> "Hearsay law has been described as 'exceptionally complex and difficult to interpret' [297]. What we need is a method of approach to the subject which will enable us to understand why some cases were decided as they were and why others are open to criticism. Above all, we need a technique [our comment: i.e. logic] for thinking about hearsay, ...".

We now examine a key case, which seems to be quoted in every textbook on Evidence (and hearsay). This is a case of *written statements*, which may fall under hearsay law.

We quote two descriptions of this case, one from [227] and one from [325], and then we model the arguments as quoted in [325].

We begin with [227, pp. 250–252]

(b) Written statements

The leading case on written hearsay is *Myers v DPP* ([1965] AC 1001). The appellant was convicted of offences relating to the theft of motor cars. He would buy a wrecked car, steal a car resembling it, disguise the stolen car so that it corresponded with the particulars of the wrecked car as noted in its log book, and then sell the stolen car with the log book of the wrecked one. The prosecution case involved proving that the disguised cars were stolen by reference to the cylinder-block numbers indelibly stamped on their engines. In the case of some cars, therefore, they sought to adduce evidence derived from records kept by a motor manufacturer. An officer in charge of these records was called to produce microfilms which were prepared from cards filled in by workmen on the assembly line and which contained the cylinder-block numbers of the cars manufactured. The Court of Criminal Appeal held that the trial judge had properly allowed the evidence to be admitted because of the circumstances in which the record was maintained and the inherent probability that it was correct rather than incorrect. The House of Lords held that the records constituted inadmissible hearsay evidence. The entries on the cards and contained in the microfilms were out-of-court assertions by unidentifiable workmen that certain cars bore certain cylinder-block numbers. The officer called could not prove that the records were correct and that the numbers they contained were in fact the numbers on the cars in question. Their

Lordships, however, were divided as to whether the evidence should be admitted by the creation of a new exception to the hearsay rule.[7] Lords Pearce and Donovan were in favour of such a course, but the majority, comprising Lords Reid, Morris and Hodson, declined to do so, being of the opinion that it was for the legislature and not the judiciary to add to the classes of admissible hearsay.[8] It was argued before the House that the trial judge has a discretion to admit a record in a particular case if satisfied that it is trustworthy and that justice requires its admission. Lord Reid, while acknowledging that the hearsay rule was 'absurdly technical', held that 'no matter how cogent particular evidence may seem to be, unless it comes within a class which is admissible, it is excluded ... '

The actual decision in *Myers v DPP* was reversed by the Criminal Evidence Act 1965, which provided for the admissibility of certain hearsay statements contained in trade or business records. Although the 1965 Act was repealed by the Police and Criminal Evidence Act 1984, ss 23 and 24 of the Criminal Justice Act 1988 are wider in scope than the provisions of the 1965 Act and provide for the admissibility of first-hand hearsay statements in documents generally as well as hearsay statements contained in documents created or received by a person in the course of, inter alia, a trade or business. The principles enunciated in *Myers v DPP*, however, remain of importance in relation to hearsay statements falling outside the statutory exceptions. Over 25 years later, another majority of the House of Lords, in *R v Kearley*,[9] although of the opinion that there may be a case for a general relaxation of the hearsay rule, affirmed the majority view in *Myers v DPP* that the only satisfactory solution is legislation following on a wide survey of the whole field.

Patel v Comptroller of Customs[10] also illustrates the application of the hearsay rule to written statements. The appellant was convicted of making a false declaration in an import entry form concerning certain bags of seed. Evidence was admitted that the bags of seed bore the words 'Produce of Morocco'. The Privy Council held that the evidence was inadmissible hearsay and advised that the conviction be quashed. The decision may be usefully compared with that in *R v Lydon*.[11] The appellant, Sean Lydon, was convicted of robbery. His defence was one of alibi. About one mile from the scene of the robbery, on the verge of the road which the getaway car had followed, were found a gun and, nearby, two pieces of rolled paper on which someone had written 'Sean rules' and 'Sean rules 85'. Ink of similar appearance and composition to that on the

---

[7] The Lords were unanimous in dismissing the appeal on the grounds that the other evidence of guilt being overwhelming, there had been no substantial miscarriage of justice.

[8] The minority view, that it was within the provenance of the judiciary to restate the exceptions to the hearsay rule, was adopted by the Supreme Court of Canada in *Ares v Venner* [1970] SCR 608. See also per Lord Griffiths in *R v Kearley* [1992] 2 All ER 345, HL at 348.

[9] [1992] 2 All ER 345, HL, per Lords Bridge, Ackner and Oliver at 360–361, 366 and 382–383 respectively.

[10] [1966] AC 356, PC. See also *R v Sealby* [1965] 1 All ER 701 and *R v Brown* [1991] Crim LR835, CA (evidence of a name on an appliance inadmissible to establish its ownership); and cf *R v Rice* [1963] 1 QB 857, below.

[11] [1987] Crim LR 407, CA.

paper was found on the gun barrel. The Court of Appeal held that evidence relating to the pieces of paper had been properly admitted as circumstantial evidence: if the jury were satisfied that the gun was used in the robbery and that the pieces of paper were linked to the gun, the references to Sean could be a fact which would fit in with the appellant having commited the offence. The references were not hearsay because they involved no assertion as to the truth of the contents of the pieces of paper: they were not tendered to show that Sean ruled anything.[12]

If we go 480 pages into Steven Uglow's book [325], we find his account of the same case.

> "*written statements*: the classic case here is *Myers v DPP* ([1964] 2 All E.R. 877) where the defendant bought wrecked cars for their registration certificates. He would then steal a similar car and alter it to fit the details in the document. He would sell the disguised stolen car along with the genuine log book of the wrecked car. The prosecution sought to show that the cars and registration documents did not match up by reference to the engine block numbers and introduced microfilm evidence kept by the manufacturer, showing that this block number did not belong in a car of this registration date. The microfilm was prepared from cards which were themselves prepared by workers on the assembly line. Lord Reid in the House of Lords held that the microfilm was inadmissible since it contained the out-of-court assertions by unidentified workers."

The labelled structure of the above is as follows.

Let

- $t : C$ The numbers assigned to the cars by the manufacturers are $x_1, x_2, \ldots$
- $t' : C'$ The numbers in the cars' logbook are $y_1, y_2, \ldots$.

If $x_i \neq y_i$, then we get:

- $t + t' : C''$ = the numbers on the cars and numbers on the registration documents do not match

where

- t = description of how the microfilm supporting C was obtained and compiled.

---

[12] See also *R v McIntosh* [1992] Crim LR 651, CA (calculations as to the purchase and sale prices of 12 oz of an unnamed commodity, not in M's handwriting but found concealed in the chimney of a house where he had been living, admissible as circumstantial evidence tending to connect him with drug-related offences); and cf *R v Horne* [1992] Crim LR 304, CA (documents of unknown authorship, referring to H, containing calculations possibly relating to the cost of importing drugs, and found in the flat of a co-accused to which H was supposed to deliver the drugs, inadmissible against H). *R v McIntosh* was applied in *Roberts v DP* [1994] Crim LR 926, DC: documents found at R's offices and home, including repair and gas bills and other accounts relating to certain premises, were admissible as circumstantial evidence linking R with those premises, on charges of assisting in the management of a brothel and running a massage parlour without a licence.

- $t' =$ the cars' logbooks.

The candidate item of data for admissibility is

- $t : C$.

The following passage is Lord Reid's argument that $t : C$ should be inadmissible, i.e. Lord Reid wants to argue that t should also contain the phrase "do not use me".

This is done in the logic of the labels. In other words, Lord Reid's argument has to do with the data inside t.

Here is Lord Reid's argument (technically it is part of t). It also quotes the arguments given in favour of admitting $t : C$.

Myers v DPP [1964] 2 All E.R. 877 at 886b–887h, *per* Lord Reid

It is not disputed before your Lordships that to admit these records is to admit hearsay. They only tend to prove that a particular car bore a particular number when it was assembled if the jury were entitled to infer that the entries were accurate, at least in the main; and the entries on the cards were assertions by the unidentifiable men who made them that they had entered numbers which they had seen on the cars. Counsel for the respondents were unable to adduce any reported case or any textbook as direct authority for their submission. Only four reasons for their submission were put forward. It was said that evidence of this kind is in practice admitted at least at the Central Criminal Court. Then it was argued that a judge has a discretion to admit such evidence. Then the reasons given in the Court of Criminal Appeal were relied on. And lastly it was said with truth that common sense rebels against the rejection of this evidence.

At the trial counsel for the prosecution sought to support the existing practice of admitting such records, if produced by the persons in charge of them, by arguing that they were not adduced to prove the truth of the recorded particulars but only to prove that they were records kept in the normal course of business. Counsel for the accused then asked the very pertinent question — if they were not intended to prove the truth of the entries, what were they intended to prove? I ask what the jury would infer from them: obviously that they were probably true records. If they were not capable of supporting an inference that they were probably true records, then I do not see what probative value they could have, and their admission was bound to mislead the jury.

The first reason given by the Court of Criminal Appeal for sustaining the admission of the records was that, although the records might not be evidence standing by themselves, they could be used to corroborate the evidence of other witnesses.[13] I regret to say that I have great difficulty in understanding that ... Unless the jury were entitled to regard them, I can see no reason why they should only become admissible evidence after some witnesses have identified the cars for different reasons ... [14]

---

[13] This is our footnote. "corroborate evidence of other witnesses" means in our LDS language "help with the flattening process".

[14] Our footnote: i.e. $u_1 : X$ is admissible only if some other $u_2 : X$ is already admissible. See objection $s_{3,2}$ below. LDS allows formally for putting item $u_1 : X$ in the database in such

At the end of their judgement, the Court of Criminal Appeal gave a different reason. 'In our view the admission of such evidence does not infringe the hearsay rule because its probative value does not depend upon the credit of an unidentified person but rather on the circumstances in which the record is maintained and the inherent probability that it will be correct rather than incorrect.' That, if I may say so, is undeniable as a matter of common sense. But can it be reconciled with the existing law? I need not discuss the question on general lines because I think that this ground is quite inconsistent with the established rule regarding public records. Public records are prima facie evidence of the fact which they contain but it is quite clear that a record is not a public record within the scope of that rule unless it is open to inspection by at least a section of the public. Unless we are to alter that rule how can we possibly say that a private record not open to public inspection can be prima facie evidence of the truth of its contents? I would agree that it is quite unreasonable to refuse to accept as prima facie evidence a record obviously well kept by public officers and proved never to have been discovered to contain a wrong entry though frequently consulted by officials, merely because it is not open to inspection. But that is settled law. This seems to me to be a good example of the wide repercussions which would follow if we accepted the judgement of the Court of Criminal Appeal. I must therefore regretfully decline to accept this reason as correct in law.

In argument, the Solicitor-General maintained that, although the general rule may be against the admission of private records to prove the truth of entries in them, the trial judge has a discretion to admit a record in a particular case if satisfied that it is trustworthy and that justice requires its admission. That appears to me to be contrary to the whole framework of the existing law. It is true that a judge has a discretion to exclude legally admissible evidence if justice so requires, but it is a very different thing to say that he has a discretion to admit legally inadmissible evidence. The whole development of the exceptions to the hearsay rule is based on the determination of certain classes of evidence as admissible or inadmissible and not on the apparent credibility of particular evidence tendered. No matter how cogent particular evidence may seem to be, unless it comes within a class which is admissible, it is excluded. Half a dozen witnesses may offer to prove that they heard two men of high character who cannot now be found discuss in detail the fact now in issue and agree on a credible account of it, but that evidence would not be admitted although it might be by far the best evidence available.

It was admitted in argument before your Lordships that not every private record would be admissible. If challenged it would be necessary to prove in some way that it had proved to be reliable, before the judge would allow it to be put before the jury. And I think that some such limitation must be implicit in the last reason given by the Court of Criminal Appeal. I see no objection to a judge having a discretion of this kind though it might be awkward in a civil case; but it appears to me to be an innovation on the existing law which decides inadmissibility by categories and not by apparent trustworthiness ...

---

a way that it can be used only in the flattening process to support other items but not in deduction.

Structure of Lord Reid's argument

$\Delta_1 : N =$ number on car A is a, (when assembled), and Δ_1 is the support of this claim.
$\Delta_1 =$ description of procedures of entering numbers during assembly.

We also have a common sense metalevel persistence principle: numbers on cars persist (don't fade away or change).

$$N \rightarrow \textbf{Always } N.$$

Thus, according to Lord Reid, t is equal to:

$$t = \{\Delta_1 : N, N \rightarrow \textbf{Always } N\}.$$

He wants to block the use of t by attacking the admissibility of Δ_1.

Four reasons were quoted for the admissibility of Δ_1 and three reasons for non-admissibility:

r_1:   Evidence of this kind is admitted in Central Criminal Court.
r_2:   Judge has discretion to admit such evidence.
r_3:   This is a list of reasons given in Court of Criminal Appeal, namely:

  $r_{3,1}$:   The records were produced to show that the records were kept in the normal course of business (but not to prove the truth of the recorded particulars).

  $r_{3,2}$:   Although the record may not be evidence by themselves, they may be used to corroborate other evidence.

  $r_{3,3}$:   We do not have dependency on the credit of an unidentified person but rather on a probably reliable process of record maintenance, and can therefore admit them.

r_4:   Common sense rebels against rejection of such evidence.
s_0:   No reported case or any textbook as direct authority for admission.

It seems at this point that r_1–r_4 are stronger than s_0.[15] So Lord Reid is trying to weaken the force of r_3 and r_2 by attacking them logically with s_3 and s_2:

s_2:   Judges do not have the discretion to admit legally inadmissible evidence.
s_3:   Counter argument to r_3 comprising of:

  $s_{3,1}$;   If the records are not intended to prove the truth of their entries, what are they intended to prove? (I.e. they are irrelevant!)

  $s_{3,2}$:   Either the records are admissible or not. There is no sense in which they can become admissible only after some other evidence to the same conclusion becomes admissible (see Footnote 14).

  $s_{3,3}$:   Such records are not public records which are admissible for reasons that they are open to the public for inspection and correction. The current law therefore does not support their admissibility.

Figure 21.1 shows the form of t, where $E =$ admit evidence or 'use me'.

---

[15] In other words, it seems that a reasonable flattening process, weighing $\{r_1, r_2, r_3, r_4\}$ against $\{s_0\}$ will decide in favour of the former and thus admit the records. Note that no rules are given at this stage of how the decision is made. In some logics, where labels are confidence numbers, we can give a rule; e.g. admit iff $r_1 + r_2 + r_3 + r_4 > s_0$, but not here.

$t =$

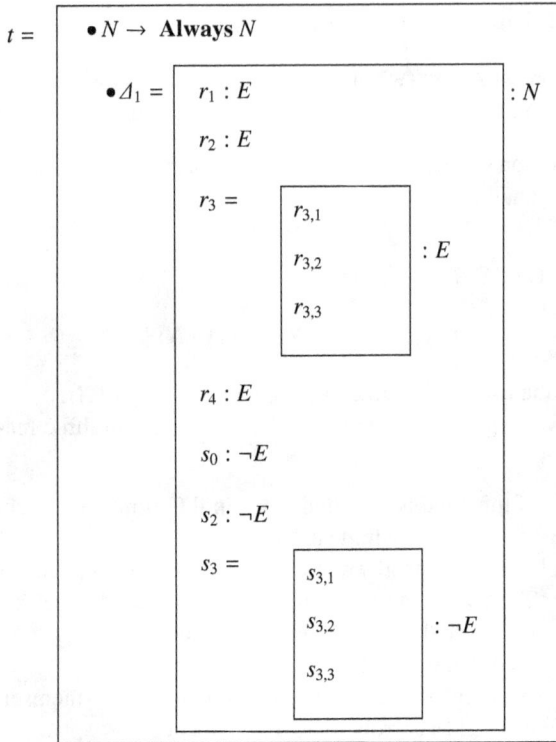

Fig. 21.1.

To strengthen his case (i.e. strengthen the overall labels for $\neg E$, Lord Reid is attacking the label r_3 by putting forward $s_{3,1}$, $s_{3,2}$ and $s_{3,3}$. Note that the reasoning in the different boxes can be of different kinds!

Note that one of the points Lord Reid is making is s_2, namely that trial judges do not have discretion to 'admit legally inadmissible evidence'.

Compare this with the Barclays Bank example. So the force of the argument is to influence the flattening process: we have r_1–r_4 : E and s_0, s_2, s_3 : $\neg E$, which one wins?

In this case the evidence was not admitted.[16]

Uglow continues:

The House of Lords recognized the absurdity of their position but felt strongly that it was for the legislature to reform the law and create new exceptions. Parliament dealt with the problem of documentary hearsay with the Criminal Evidence Act 1965 which created an exception for trade and business records This was later extended by section 68 of the Police and Criminal Evidence Act 1984 and now by sections 23 and 24 of the Criminal Justice Act 1988. Such records have all been admissible in civil proceedings since the Civil Evidence Act 1968.

---

[16] This decision was made by vote as described in the quote from [227] on our page 713.

Myers has been regularly followed in such cases as *Patel v Comptroller of Customs* ([1965] 3 All E.R. 593) where the appellant was convicted of making a false declaration to customs, having stated that the bags of seed were originally from India. The prosecution sought to prove that the seed originated in Morocco and adduced evidence that the bags were stamped with 'Produce of Morocco'. The Privy Council, following *Myers* held that these words were hearsay and inadmissible. Unlike *Myers*, there was no evidence that the writing was at all reliable, there being no testimony as to how or by whom the bags were marked."

The reader should note that the main thrust of the argument and logic of the Lord Reid example is in weakening and strengthening labels. Put schematically we have a master argument, say E which can prove a conclusion on D. E is a labelled argument containing various labels within labels. Among this maze of labels there is a label t containing another argument, say Δ. To attack E we can attack Δ. Our argument attacking Δ can itself be attacked by attacking some label s in it and so on. This is reminiscent of systems of abstract argumentation theory. Bench-Capon [43] has a paper on graphs of arguments and counterargunents, but his model is schematic. We can give actual proof rules and labelling disciplines so that questions like export from one label to another can also be considered. For example:

"If you weaken t then D will not follow from E, and that would be a bad precedent."

One cannot argue in this way unless a specific labelled model is available. We shall examine the Bench-Capon paper in the next section. For the time being, we think that we have seen enough to be convinced that labelling logics and labelled argumentation networks can play a central role here, though we would understand if the cautious reader would prefer to reserve judgement until more case studies are presented.

21.4 Value-based argument framework

The purpose of this section is to compare our approach with that outlined in Bench-Capon [43] and to show how labels can be used more effectively. We also give a Bayesian approach and a neural nets approach. In a future paper we hope to offer an LDS mix of all approaches. We believe any realistic model needs to do that!

We can indicate at this stage how the abstract argumentation model can relate to LDS. Consider the Lord Reid argument as presented in Figure 21.1. It has arguments $r_1, \ldots, r_4$ in favour of E and counter arguments $s_0, \ldots, s_3$ in favour of $\neg E$, essentially attacking $r_1, \ldots, r_4$. LDS requires in this case a flattening function (or a process) to tell us which arguments win and at what strength we can use E or $\neg E$.

This flattening process can make use of abstract argumentation theory, either in its Bench-Capon form, or modified with probability or implemented in neural nets. A taste of these options is given in this section.

21.4.1 The Framework

We begin by discussing and highlighting our method of modelling. The first principle is to work bottom up from the application area into the formal model, trying to reflect in the formal model more and more key properties of the application area. In the case of evidence this means we need to see and study many examples/case studies/debates about evidence and then try to construct a suitable logic for it. Chances are that existing logics, constructed for some other purpose, may not be the most suitable. Our starting formal system for this purpose is LDS. The theory of LDS was developed from the bottom up point of view, especially to model aspects of human behaviour, reasoning and action, and is very comprehensive, adaptable and incremental. It contains a large variety of existing logical systems as special cases. What is more important is that LDS is not a single system but a methodology for building *families* of systems, ready to be adapted to the needs of various application areas, in our case to the theory of evidence.

One very important side effect of this approach is that the logic can be worked up directly from the day-to-day activity of the practitioner of the laws of evidence, without necessarily forcing him to study logic. The 'logic' will be hidden in the stylised movements he will be asked to make, and the interplay between the labels and comments and arguments he will be using.[17]

In contrast to our approach, in a good deal of applicational work in logic, a logic is applied to various areas and tend to force the application area into a form suitable for its existing formalism. This tends to produce results intelligible mainly to the logician, ignoring that the ordinary human/lawyer/judge already knows intuitively how to handle his daily life, and that all he needs is some bottom up additional organisation of his activities which will enable him to understand it better and possibly solve some of his outstanding puzzles.[18]

---

[17] Consider the widespread use of anagrams. Take as an example the pair of words 'read on'. We can rearrange the letters (including the space between the words) into 'no dear'. Let us write this as

$$\text{read on} \vdash \text{no dear}$$

We can also write equivalently

$$\text{space, a, d, e, n, o, r} \vdash \text{read on}$$
$$\text{space, a, d, e, n, o, r} \vdash \text{no dear}$$

where on the left we just listed the basic blocks we can use, including the space.

Now suppose we allow you some 'wildcard' of the form

$$\text{space} \mapsto \text{any other already listed letter}$$

Then we get

$$\text{space, a, d, e, n, o, r, (space} \mapsto \text{any other already listed letter)} \vdash \text{adorned}$$

We chose here space $\mapsto$ d.

What we have been doing here was linear logic!

So anagrams with wildcards is linear logic.

The idea that logic can be 'translated' into stylised proof movements was put forward in the Gabbay 1984 logic lectures at Imperial College, London published as part of [152]. See the first chapter of [136] and see [180]. Peter Tillers says similar things in his paper in [245, pp. 2–11]. We assume the word 'dynamics' in the title of [245] is significant.

[18] The modelling practices of the social sciences generally are adaptations of the modelling paradigms of physics (rather than, say, biology), and are a reflection of the primacy of logical

The difference in this point of view is apparent when we look again at Prakken's book. The book does realise the potential in the interaction of logic and law. It also recognises some of the kinds of logics needed to model some aspects of the law. But having made and argued all of these points, the main part of the book gives an exposition of the relevant parts of the logic in a way that only a logician can understand. This is also true at the moment of the current chapter, but we hope in a future paper based on the ideas of this chapter we will be able to show how to do logic directly in the legal evidence application area. See Footnote 17.

Having said all that, we can now look at some specific model, namely that of abstract argumentation systems. These were put forward as a response to the realisation that no argument or proof is conclusive in real life, and that arguments have counterarguments. The argument framework has the form $AF = (AR, Attacks)$ where AR is a set of objects called arguments and *Attacks* is a binary relation (usually irreflexive), saying which arguments x attacks which argument y. The following Figure 21.2 is an example

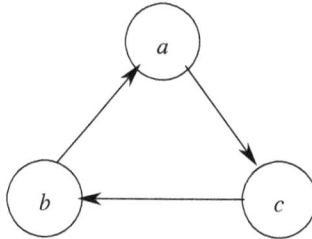

Fig. 21.2.

a attacks c, c attacks b and b attacks a.

There are no winning arguments here. This framework is too abstract to be of specific use. It equally applies to circuits and impending circuits, credits and debits, neural nets and counterweights or any system involving x and anti-x, whatever x is.

To apply such a system successfully we need to go into the structure of the arguments and analyse the mechanics of one argument attacking another.

Bench-Capon tried to improve upon such systems by introducing a clever idea; the value-based argumentation framework. In this framework we are given a set of colours (values) and a colouring of the arguments. The values are partially ordered and an argument of strictly lesser value cannot now attack an argument of stronger value.

So following Bench-Capon in the previous figure, if we make b red and a and c blue then

1. If blue is stronger than red, then b cannot attack and defeat a, a can attack c and the winning arguments are $\{a, b\}$, because c is out.

---

positivism as the social sciences were in process of articulating its philosophical presumptions. But it is almost never satisfactory to abstract from the data of human interactions in the same way that one abstracts from the interactions of physical particles.

2. If red is stronger than blue then the winning arguments are $\{b, c\}$.

Certainly this colouring with values is an intuitively welcome improvement. However, this model is still too abstract. Real life has arguments within arguments in different levels and interconnections between the levels. We can extend the Bench-Capon model by using our technique of self-fibring of networks of Chapter 7. This method allows for the recursive substitution of networks inside nodes of other networks. Thus when node (argument) x attacks argument y and y is coloured as having higher preference than x, then the attack is blocked. We offer to label x and label y and compare the labels. The labels can be entire logical systems of reasoning or entire argumentation frames for each argument or temporal stamping or a database and the comparison of labels can be done in a separate logic. We therefore think using LDS is a much better option. Compare also with Chapter 9, Section 9.6.1, comparison with other papers. The them there is to label with numerical fuzzy values. This is not the spirit here, where we label with logical information.

In LDS, this situation will arise if we have a labelled database which includes items such as $t : a$, $s : b$ and $r : c$ and some additional data, say $u_i : X_i$, such that the following can be proved, among others:[19]

- $\gamma(t) : \neg c$
- $\beta(r) : \neg b$
- $\alpha(s) : \neg a$.

α, β, γ are the labels of $\neg a, \neg b$ and $\neg c$ respectively and t, r, s are mentioned in the respective labels to indicate that e.g. $t : a$ is used in the proof of $\gamma(t) : \neg c$ (a with label t attacks c, by proving $\neg c$ with label $\gamma(t)$). The label $\gamma(t)$ shows exactly what role a plays in this attack.

The flattening process acts here as value judgement of what can win, $r : c$ or $\gamma(t) : \neg c$, by comparing r and $\gamma(t)$.

Obviously the value based argumentation machinery can be utilised as part of our flattening mechanism.

The following LDS model will reflect the Bench-Capon coloured diagram:

> red: b
> blue: a
> blue: c
> red to blue: $b \rightarrow \neg a$
> blue to blue: $a \rightarrow \neg c$
> blue to red: $c \rightarrow \neg b$

Using modus ponens in the form

$$\frac{\alpha : X, \beta : X \rightarrow Y, \varphi(\beta, \alpha)}{\alpha \cup \beta : Y}$$

---

[19] Note that we are assuming here that to defeat x we must put forward an argument for $\neg x$. This is only a simplifying assumption. In LDS, x comes with a label t and so to weaken $t : x$ we can attack t. For example suppose the label for x is an argumentation network t and a preferred extension E for t which contains the argument x. An attack on x can be a simple question why do you take the extension E why not take the extension E', which does not contain x?

We can prove:

> red:¬a if red to blue is allowed
> blue:¬c if blue to red is allowed
> blue:¬b if blue to blue is allowed.
> The flattening function has to flatten:

{red: b, blue: ¬b}
{blue: a, red: ¬a}
{blue: c, (blue: ¬c is not allowed!)}

Case 1.

red stronger than blue i.e. *blue to red* not allowed.
     We get b and ¬a and c.

Case 2.

Blue stronger than red (*red to blue* not allowed)
     We get
{blue: a, (red: ¬a not allowed)}
{blue: c, blue: ¬c}
{red: b, blue: ¬b if c is available}

We cannot decide between c and ¬c since both are blue. If we leave them both out or take ¬c then ¬b will not be obtainable and hence we will have {a, b}.

We see that in the labelled formulation we have more options

1. We can have X, ¬X or neither as choices
2. The label colour (value) can itself be a whole database and so arguments about the values and their strengths can also be part of the system.

The Bench-Capon system is only one level.
     The following Figure 21.3 shows the abstract argumentation structure of Lord Reid's arguments.

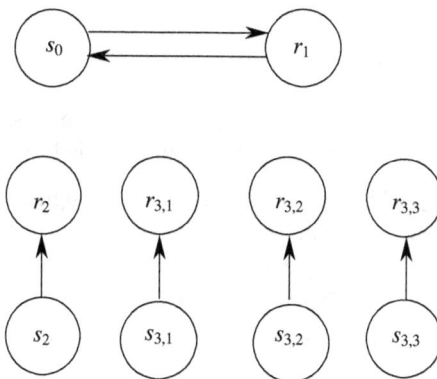

Fig. 21.3.

Accordingly, $\varDelta_1$ in Figure 21.1 can be better rewritten as Figure 21.4 below

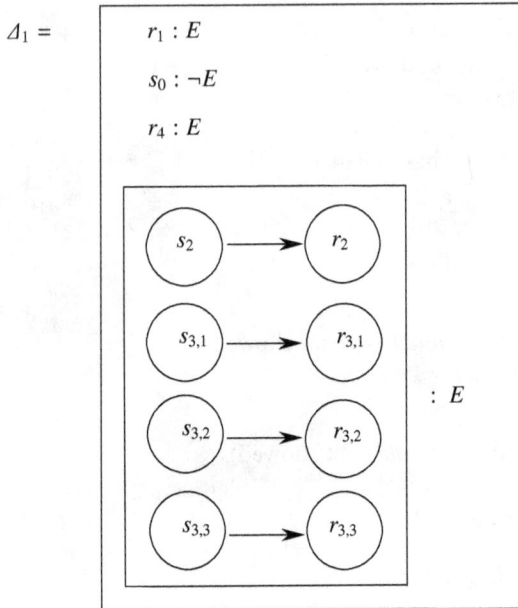

$$\Delta_1 =$$

Fig. 21.4.

Assuming that the attack of Lord Reid is successful, then Figure 21.4 reduces to $\{r_1 : E, r_4 : E$ and $s_0 : \neg E\}$. The Lords indeed decided that s_0 was stronger, but they were uncomfortable about it and decided to recommend new legislation.

Note Lord Reid's argument $s_{3,2}$. This is a metalevel value argument like "you cannot colour something red".

Also note that s_0 and s_2 can be further counter-argued if possible by other Lords. The formal labelling of these additional arguments may require self-fibring. See section 21.4.5.

21.4.2 Moral debate example

This section also follows Bench-Capon [43, p. 442]. We consider an example cited by Bench-Capon, attributed to Coleman in [100] and Christie [93].

> "Hal, a diabetic, loses his insulin in an accident through no fault of his own. Before collapsing into a coma, he rushes to the house of Carla, another diabetic. She is not at home but Hal enters her house and uses some of her insulin. Was Hal justified, and does Carla have a right to compensation?"

The following are the arguments involved as presented in the Bench-Capon paper:

A =   Hal is justified, since a person has a privilege to use the property of others to save their life - the case of necessity.

B =   It is wrong to infringe the property rights of another.

C =   Hal compensates Carla.

Bench-Capon [43] quotes that Christie [93] adds:

D₁ = If Hal is too poor to compensate Carla, he should nonetheless be allowed to take the insulin, as no one should die because they are poor.

D₂ = Moreover, since Hal would not pay compensation if too poor, neither should he be obliged to do so even if he can.[20]

Bench-Capon further suggests:

E = Poverty is no defence for theft.
F = Hal is endangering Carla's life.
G = Fact: Carla has abundant insulin.
H = Fact: Carla does not have ample insulin.

Figure 21.5 now represents the situation. Note that H = ¬G.

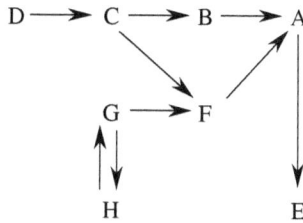

Fig. 21.5.

Bench-Capon gives the following value properties to the arguments:

Life: A, D, F
Property: B, C, E
Fact: G, H

He says one might argue whether life is stronger than property or not but facts are always the strongest.

Since H = ¬G, and since we cannot have both facts, he regards that part of Figure 21.5 as a case of uncertainty.

We cite this example because we want to analyse what is needed for a better representation of it.

We begin by listing the points:

1. The model needed for a proper analysis of this kind of problem in general (though maybe not necessarily the Hal problem) is a time/action model. There is a difference of values depending at what stage of the action sequence we are at. Has Hal entered Carla's house? Has he checked for insulin? Is it all over and Carla is dead? Each of these cases may have a different argument diagram, possibly with values depending on the previous one! We might add at this point that the need for time/action models has already been strongly emphasised in Gabbay [180] in connection with puzzles involved in the logical analysis of conditionals. This is

---

[20] Christie puts D₁ + D₂ = D together as D. The division into D₁ and D₂ is ours, for later discussion.

factors of connected to contrary-to-duty models[21] and also needed to incorporate uncertainty. We can get a quite complicated (but highly intuitive) model.[22]

2. We require a better metalevel hierarchy of values and rules, as are available in Labelled Deduction. Possibly such options can also be made adequately available to the abstract argumentation model via self-fibring.

3. The links $(X \to Y)$ should be given strength labels to help us model more realistic cases where an argument X is attacked by arguments $Y_1, \ldots, Y_k$ with strength measuring $m_1, \ldots, m_k$.

This is an essential generalisation. One of the quotes we cited from the car case study was (see footnote 7) had the Lords rejecting the written evidence because there was other ample evidence to the same effect (and they didn't want to create a precedent by admitting it).[23]

4. We can read the link $X \to Y$ as preventative action of X to stop Y and thus by giving probability of success turn any acyclic network into a Bayesian one. This will introduce uncertainty into the framework. Actually the probability of success is inversely proportional to the conditional probability of Y on X.

21.4.3 Bayesian aspects of the moral debate example

We begin this section with a closer look at Figure 21.5. We require a time/action model and contrary-to-duty considerations. We shall explain these features as we model the example.

We imagine an agent, such as Hal, who has available a stock of optional actions. These actions have the form $\mathbf{a} = (A, (B^+, B^-))$ where A is the precondition of the action and B^+, B^- are the post-conditions. A must hold in order for Hal to be allowed to perform the action, in which case the resulting state is guaranteed to satisfy B^+. However, the agent may take the action anyway, without permission (i.e. A does not hold), in which case the post-condition is B^-. Note that in most cases $B^- = B^+$.

---

[21] See the authoritative survey [84] of A. Jones and J. Carmo in the *Handbook of Philosophical Logic*, 2nd edition.

[22] We take this opportunity to reinforce our methodological remark of footnote 18. In modelling human practical reasoning, actions and general behaviour it is often a disadvantage and a deficiency to try and use a stylised model and abstract too much from the actual reality (in contrast possibly with modelling physical nature). Often the details of the reality to be modelled suggests the solution to what otherwise is a puzzle. Let us look at the story and focus on the part which assumes Hal is too poor to replace Carla's insulin. We can ask how is he getting his insulin? Is he getting it on National Health Service? If yes, can't he call the NHS and try to get a replacement? So surely the question of replacement is not 'whether' but 'when', i.e. can he get a replacement in time before Carla runs out of insulin? If life is more important than property this is a good question. If property is more important, then we know he can replace it! Another question, if Hal steals the insulin from Carla and then calls for a replacement, would it not be more difficult to get a replacement (as opposed to calling the NHS first)? We need more details. We are *not* transforming the problem to one more suited to our framework. There are many other examples in other areas which need more details.

[23] This is a mixture of metalevel/strength/proof argument that only LDS can model. We shall address this kind of argument later.

We imagine we are at a state (or time) T_0, described by a logical theory Δ. The actions available to us to perform are $\mathbf{a}_1, \mathbf{a}_2, \ldots \mathbf{a}_i = (A_i, (B_i^+, B_i^-)), \ldots$. If $\Delta \vdash A_i$, then action $\mathbf{a}_i$ is allowalbe at time (state) T_0, otherwise not. If we perform the action $\mathbf{a}$, with post-condition B (B is either B^+ or B^-) then we move to time T_1, with state $\Delta_\mathbf{a} = \Delta \circ B$ where $\Delta \circ B$ is the revision of Δ by B. We have $\Delta \circ B \vdash B$.

So to have time action model we need

1. A language for the theories Δ to describe states
2. A language for pre-condition and a language for post-conditions for actions
3. A logic or algorithm for determining when $\Delta \vdash A$ holds, where A is a pre-condition.
4. A revision algorithm giving for each Δ and post-condition B a new theory $\Delta' = \Delta \circ B$. This algorithm can satisfy some reasonable axioms.

Note that the languages for Δ, the pre-conditions and the post-conditions need not be the same!

The flow of time is future branching and is generated by the actions. So if for example our agent can perform actions $\mathbf{a}_1, \ldots, \mathbf{a}_k$ as options then after two steps in which he performs say $\mathbf{a}_1$ first and then say $\mathbf{a}_3$, we may get a situation as in figure 21.6

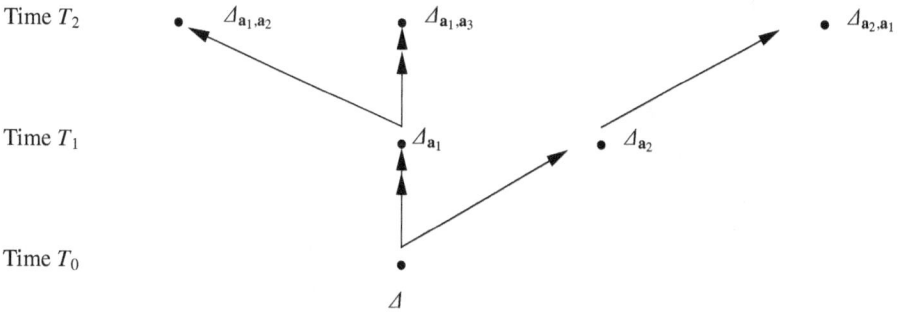

Fig. 21.6.

The real history at time T_2 is $(\Delta, \Delta_{\mathbf{a}_1}, \Delta_{\mathbf{a}_1, \mathbf{a}_3})$. The states $\Delta_{\mathbf{a}_1, \mathbf{a}_2}$ and $(\Delta_{\mathbf{a}_2}, \Delta_{\mathbf{a}_2, \mathbf{a}_1})$ are hypotheticals.

At time T_0, our agent chose to take action $\mathbf{a}_1$ moving onto state $\Delta_{\mathbf{a}_1}$, but he could have chosen to take action $\mathbf{a}_2$ and done action $\mathbf{a}_1$ afterwards, ending up at state $\Delta_{\mathbf{a}_2, \mathbf{a}_1}$ at time T_2. In reality, however, he chose to perform $\mathbf{a}_1$ and then $\mathbf{a}_3$.

The pre-conditions of actions can talk about states and hypotheticals. They need not be in the same language as Δ or the same language as the post-conditions. What is important are the algorithms for '$\vdash$' and '$\circ$'.

We are now ready to analyse the moral debate example. First we tell the story in a more realistic way (see footnote 22!). Then we propose some probabilities as an example and we conclude by translating the Bench-Capon statements A–H (page 724) into our time/action set up.

Our story goes as follows. Hal needs insulin. So does Carla. Both are poor and get their insulin from the Health Service. They get it in batches, though not at the same time. So the question whether Carla has spare insulin (*G*) depends on the time, and is a matter of probability.

Hall loses all his insulin and would need to break into Carla's property to get hers. He has the option of calling the NHS and asking for replacement, which he can use either for himself if it arrives immediately or to replace Carla's if necessary. He might get some money from friends. One thing is clear to him. If he steals Carla's insulin, it will complicate matters; it might be more difficult to find a replacement. So the question of compensation *C* is also a matter of probability. The following are the possible scenarios.

If property is valued more than life, then if Hal steals Carla's insulin, the probability of getting a replacement is lower in the case where Carla's life is not threatened.

If life is valued more than property, his chances of obtaining replacement is higher in case Carla's life is threatened.

We must clarify what 'getting a replacement' means. Hal will probably start a process for getting insulin for himself immediately at start time T_0. Since it might not arrive in time, he will break into Carla's home and use hers, and hope to use the insulin he 'ordered' to replace Carla's. If Carla has ample insulin, there is a higher chance or that the replacement will arrive in time before Carla's life is threatened. If Carla does not have ample insulin, Hal can use this as a further reason to rush the process of replacement. This further reason might be counterproductive if property is valued above life.

So the statement

C = Hal gets a replacement

should be taken as (see footnote 22):

Hal gets a replacement before Carla is in need of it.

We may then have the following scenarios (*P* stands for Probability *P*(*x*) and it should be indexed by case and time, i.e. $P_{1,a}, P_{1,b}, P_{2,a}$ and $P_{2,b}$:

Case 1. Property stronger than life

(a) Time = Before Hal breaks into Carla's house.
$P(G) = \frac{2}{3}$
$P(\neg G) = \frac{1}{3}$
$P(C/G) = 0.9$
$P(\neg C/G) = 0.1$
(Since Carla does have ample insulin, Hal has more time to replace what he might take.)

$P(C/\neg G) = 0.5$
$P(\neg C/\neg G) = 0.5$
(Admittedly, Carla's life is in danger but there may not be enough time to get a replacement. On the other hand, this very fact might help get the insulin more quickly. Note that the event *C* means 'getting replacement in time'.)
(b) Time = After Hal breaks into Carla's house.
At this stage the value of *G* is known: either *G* = 1 or *G* = 0. We get
$P(C/G = 1) = 0.7$
$P(\neg C/G = 1) = 0.3$

(less than before breaking into the house, because Hal committed a serious crime. He may not be favourable with the authority.)

$P(C/G = 0) = 0,4$

$P(\neg C/ G = 0) = 0.6$

Again, less than before.

Case 2. Property not stronger than life[24]

(a) Time = Before Hal breaks into Carla's house

$P(G) = \frac{2}{3}$

$P(\neg G) = \frac{1}{3}$

$P(C/G) = 0.9$

$P(\neg C/G) = 0.1$

$P(C/\neg G) = 0.9$

$P(\neg C/\neg G) = 0.1$

(b) Time = After Hal breaks into Carla's house

$P(G) = \frac{2}{3}$

$P(\neg G) = \frac{1}{3}$

$P(C/G = 1) = 0.9$

$P(\neg C/G = 1) = 0.1$

$P(C/G = 0) = 0.7$

$P(\neg C/G = 0) = 0.3.$

Let us now translate the arguments involved in the original moral debate example of Section 21.4.2.

When is Hal justified in breaking into Carla's home? The answer is yes only in the case that life is stronger than property and he can reasonably say he is not risking her life. That depends on finding a replacement. We therefore have to calculate the probability of C given all the data we have.

Thus our time/action axis has the form of Figure 21.7:

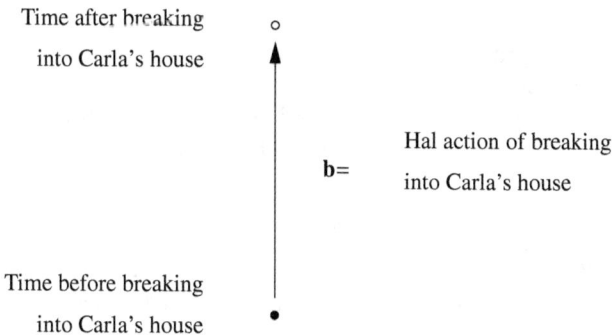

Fig. 21.7.

---

[24] Jon Williamson reminded us that it is reasonable to assume that the legal process does not make general value judgements like this, nor can a legal argument appeal to such judgements. Instead much more specific 'mitigating circumstances' can be used to reduce the length of a sentence on conviction ('I did it to save my life, guv').

The actions available to Hal are:

1. **b** = breaking into Carla's house. The post-condition is breaking in and taking the insulin. The pre-condition of **b** is high probability of replacing Carla's insulin (in time before she needs it) in case *life is stronger than property* and $\perp$ (falsity i.e. no permission to do the action) in case *life is not stronger than property*.
2. **r** = actions having to do with getting a replacement of insulin. We assume he can perform these actions at any time but the post-conditions are not clear.[25]

We need also agree the value of the threshold probability, e.g. only if there is at least 0.9 chance of replacement can Hal break into Carla's home to take the insulin. Consider now:

B = It is wrong to infringe the property of others.

B is an argument reflected in the pre-condition of the action **b**, it can be done when B satisfied otherwise not. I would write it as

$$\mathbf{b} = (\text{Justification, Break in and taking insulin}).$$

Let us now model the chain of events as a Bayesian network. The story is clear. Depending on the probability $P(G)$, Hal decides whether he wants to break into Carla's house **b** (no use breaking into her house if she does not have enough insulin). He is justified J in breaking **b** into Carla's house if there is high probability of compensation C. Thus C depends both on **b** and G, and **b** also depends on G. We have the following network, Figure 21.8.

There are two problems with this representation.

1. The dependency of **b** on G is not on $G = 1$ or $G = 0$ but on $P(G)$. Say if $P(G) < 0.1$ then maybe **b** = 0.
   This is OK because the probabilities can be made to take account of that. This is allowed in the theory of Bayesian nets.
2. The probabilities in Figure 21.8 depend on whether property is stronger than life or not. The best way to represent this is to have a Bayesian net with one variable only, *Case*.
   Case =1 means property stronger than life and *case* =0 means property is not stronger than life.
   For each case we get a different copy of Figure 21.8 with different probabilities. So we get a substitution of the network of Figure 21.8 into a one point network:

   • *Case*. This operation is in accordance with the ideas in [337].

   We can also allow for several justification variables to make it more realistic.

It is not difficult to work out the details of the rest of C–H, but the reader can already see that in the simple minded model there is lack of sensitivity to a variety of metalevels.

---

[25] We may need a temporal language for the post-conditions so that we can say something like 'insulin will be delivered in two days'.

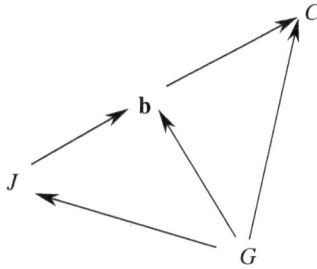

Fig. 21.8.

21.4.4 Neural representation of argumentation frameworks

This subsection, based on [185] will outline how to represent (in neural nets) any value-based argumentation framework involving x and anti-x (i.e. arguments and counter-arguments). For instance, it can be implemented in neural networks with the use of Neural-Symbolic Learning Systems [197]. A neural network consists of inter-connected neurons (or processing units) that compute a simple function according to the weights (real numbers) associated to the connections. Learning in this setting is the incremental adaptation of the weights [217]. The interesting characteristics of neural networks do not arise from the functionality of each neuron, but from their collective behaviour, thus being able to efficiently represent (and learn) multi-part, cumulative argumentation, as exemplified below.

Cumulative behaviour can be encoded in Neural-Symbolic Learning Systems with the use of a hidden layer of neurons in addition to an input and an output layer in a feedforward network. Rules of the form $A \wedge B \rightarrow C$ can be represented by connecting input neurons that represent concepts A and B to a hidden neuron, say h_1, and then connecting h_1 to an output neuron that represents C in such a way that output neuron C is activated (true) if input neurons A and B are both activated (true). If, in addition, a rule $B \rightarrow C$ is also to be represented, another hidden neuron h_2 can be added to the network to connect input neuron B to output neuron C in such a way that C is now activated also if B alone is activated.[26] This is illustrated in Figure 21.9. The network can be used to perform the computation of the rules in parallel such that C is true whenever B is true [197].

In a neural network, positive weights can represent the support for an argument, while negative weights can be seen as an attack on an argument. Hence, a negative weight from a neuron A to a neuron B can be used to implement the fact that A attacks B. Similarly, a positive weight from B to itself can be used to indicate that B supports itself. Since we concentrate on feedforward networks, neuron B will appear on both the input and the output layers of this network as shown in Figure 21.10, in which dotted lines are used to indicate negative weights.

---

[26] In the general case, hidden neurons are necessary to implement the following conditions: **(C1)** The input potential of a hidden neuron (N_l) can only exceed N_l's threshold (θ_l), activating N_l, when all the positive antecedents of r_l are assigned the truth-value *true* while all the negative antecedents of r_l are assigned *false*; and **(C2)** The input potential of an output neuron (A) can only exceed A's threshold (θ_A), activating A, when at least one hidden neuron N_l that is connected to A is activated.

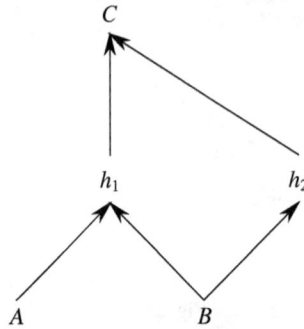

Fig. 21.9. A simple example of the use of hidden neurons

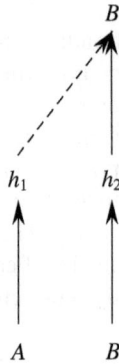

Fig. 21.10. A simple example of the use of negative weights for counter-argumentation

In Figure 21.10, A attacks B via h_1, while B supports itself via h_2. Suppose now that, in addition, B attacks C. We need to connect input neuron B to output neuron C via a new hidden neuron h_3. Since B appears on both the network's input and output, we also need to add a feedback connection from output neuron B to input neuron B such that the activation of B can be computed by the network according to the chain 'A attacks B', 'B attacks C', etc. As a result, in Figure 21.11 (in which we do not represent B's feedback connection for the sake of clarity), if the attack from A on B is stronger (according to the network's weights) than B's support to itself, then A will block the activation of (output) B, and (input) B will not be able to block the activation of C. In this case, the network's final computation will include C and not B in a stable state. If, on the other hand, A is not strong enough to block B, then B will be activated and block C.

Let us take the example in which an argument A attacks an argument B, and B attacks an argument C, which in turn attacks A in a cycle. In order to implement this in a neural network, we need positive weights to explicitly represent the fact that A supports itself, B supports itself and so does C. In addition, we need negative weights from A to B, from B to C and from C to A (see Figure 21.12) to implement attacks. If all the weights are the same in absolute terms, no argument wins, as one would expect, and the network stabilises with none of $\{A, B, C\}$ activated. If, however, the value of A (i.e. the weight from h_1 to A) is stronger than the value of C (the weight from h_3 to

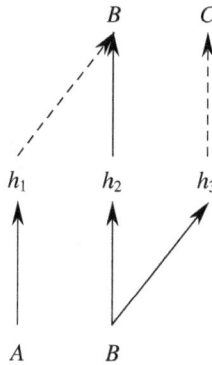

Fig. 21.11. The computation of arguments and counter-arguments

C, which is expected to be the same in absolute terms as the weight from h_3 to A), C cannot attack and defeat A. As a result, A is activated. Since A and B have the same value (as e.g. in the previous case of an unspecified priority), B is not activated, since the weights from h_1 and h_2 to B will both have the same absolute value. Finally, if B is not activated then C will be activated, and a stable state $\{A, C\}$ will be reached in the network. In Bench-Capon's model [43], this is exactly the case in which colour blue is assigned to A and B, and colour red is assigned to C with blue being stronger than red. Note that the order in which we reason does not affect the final result (the stable state reached). For example, if we started from B successfully attacking C, C would not be able to attack A, but then A would successfully attack B, which would this time round not be able to successfully attack C, which in turn would be activated in the final stable state $\{A, C\}$. This indicates that a neural (parallel) implementation of this reasoning process could be advantageous also from a purely computational point of view.

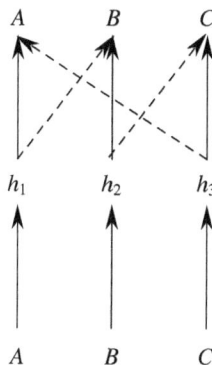

Fig. 21.12. The moral-debate example as a neural network

Note that (as in the general case of argumentation networks) in the case of neural networks, we can extend Bench-Capon's model with the use of self-fibring neural networks, which allow for the recursive substitution of neural networks inside nodes of other networks [149].

The implementation of the network's behaviour (weights and biases) must be such that, when we start form a number of positive arguments (input vector $\{1, 1, \ldots 1\}$), weights with the same absolute values cancel each other producing zero as the output neuron's input potential. A neuron with zero or less input potential is then deactivated, while a neuron with positive input potential is activated. This allows for the implementation of the argumentation framework in neural-symbolic learning systems, in the style of the translation algorithms developed at [186].

21.4.5 Self-fibring of argumentation networks

We will conclude this section by recalling from Chapter 7 how to do self-fibring of argument networks. The mechanics of it is simple. We begin with one network, say the one in Figure 21.2. We pick a node in it, say node a, and substitute another network for that node, say we substitute the network of Figure 21.5. We thus get the 'network' of figure 21.13.

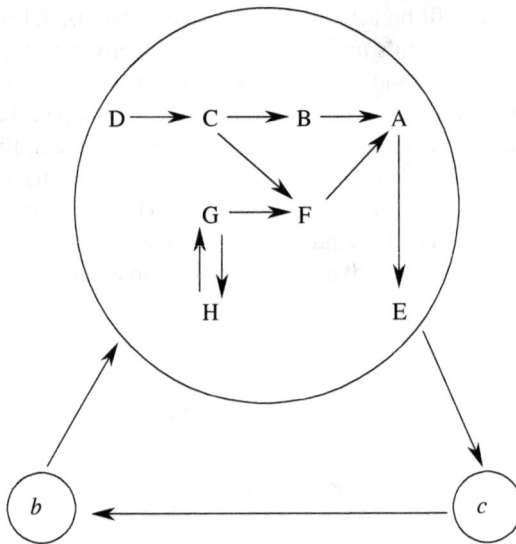

Fig. 21.13.

The need of self-fibring may arise if additional arguments are available supporting the contents of the node.

The self-fibring problem has three aspects:

Aspect 1: Intuitive Meaning
What is the intended interpretation/meaning of this substitution? This can be decided by the needs of the application area. Here are some options:

(1.1) a is supposed to be an argument, so Figure 21.5 can be viewed as delivering some winning argument (A of Figure 21.5) which can combine/support a.

(1.2) Figure 21.5 is a network so *b* of Figure 21.2 can plug into it. We can connect *b* to all (or some) members of Figure 21.2 and similarly connect all (or some) members of Figure 21.5 into *c* of Figure 21.2.

For various options see Chapter 7 as well as [337, 186, 139] and [85].

Aspect 2: Formal aspect

(2.1) *Syntactical substitution*
Formally the node *a* is supposed to be an argument. So we need a fibring function **F**(node, network) = *e* yielding a node *e* and so we end up with Figure 21.14

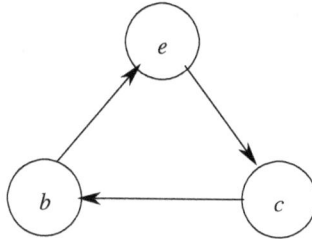

Fig. 21.14.

F might do, for example, the following: **F** can use the colour of node *a* to modify the colours of the nodes in Figure 21.5 (the substituted network), and maybe also modify some connections in Figure 21.5, and then somehow emerge with some winning argument *e* and a colour to be substituted/combined with *a* and its colour.

(2.2) *Semantic substitution*
If the original network has an interpretation, then the node *a* can get several possible semantic values. We can make the definition of the substitution context sensitive to those values. We may even go to the extent of substituting different networks for different options of values.

Aspect 3: Coherence
To enable successful repeated recursive substitution of networks within networks, we have to modify our definition of the original network. For example:

(3.1) Possibly extend the notion of network and allow arrows to either support or defeat arguments.

(3.2) Restrict the substitution of networks for nodes by compatibility/consistency conditions.

Example: Self-fibred argumentation network
We have a set of nodes and links of the form (*a*, *b*) meaning *a* attacks *b*. We also have valuation colours. A weaker colour cannot attack a stronger colour. So far this is the Bench-Capon definition.

Let *a* be a node. Define the notion of *x* is a supportive (resp. attacking) node for *a* as follows:

- *a* is supportive of *a*

- if x is supportive (resp. attacking) node of a and y attacks x then y is an attacking (resp. supportive) node of a.

Now let a be a node in a network A and suppose we have another network N which we want to substitute for a. We must assume a appears in N with the same colour value as it is in A. We substitute N for a and make new connection as follows:

- Any node x of A which attacks a in A is now connected to any node y in N which supports a in N.
- Any node y in N which supports a in N is now made connected to any node x of A which a of A is attacking.

This definition is reasonable. a is an argument in network A. N is another network which is supposed to support a (a in N). Thus anything which attacks a in A will attack of all a supporters in N and these in turn will attack whatever nodes a attacks in A. Note that he may be attacking facts in N by this wholesale connection of arrows. However, Bench-Capon has already remarked that facts should get the strongest colour and so the colours will take care of that!

Compare with Chapter 7 Fibring Argumentation Frames. See reference [164].

References

1. S. Abiteboul, R. Hull, and V. Vianu. *Foundations of Databases*, Addison-Wesley Pub. Co., 1996.
2. M. Abraham, D. M. Gabbay, and U. Schild. Analysis of the Talmudic Argumentum A Fortiori Inference Rule (Kal-Vachomer) using Matrix Abduction. *Studia Logica*, vol 92 , No 3, pp 281–364, 2009.
3. M. Abraham, D. M. Gabbay, and U. Schild. Obligations and Prohibitions in Talmudic Deontic Logic. In G. Governatori and G. Sartor, eds., *DEON 2010, LNAI 6181*, pp. 166–178, 2010.
4. M. Abraham, D. M. Gabbay, and U. Schild. Obligations and Prohibitions in Talmudic Deontic Logic. *Journal of Artificial Intelligence and Law*, 19:2, 117–148, 2011.
5. M. Abraham, G. Hazut, D. M. Gabbay, Maruvka and U. Schild. Logical Analysis of the Talmudic Rule of General and Specific (Klal-u-Prat). Special issue on Judaic Logic edited by A Schumann, *Journal of the History and Philosophy of Logic*, Vol 32 issue 1, pp 47–62, 2011.
6. M. Abraham, D.Gabbay and U. Schild. *Resolution of conflicts and normative loops in the Talmud*, College Publications, 2011.
7. M. Abraham, D. M. Gabbay, and U. Schild. *The handling of loops in talmudic logic, with application to odd and even loops in argumentation.* Expanded version of paper published in 2011 as part of the monograph [6].
8. M. Abraham, D.Gabbay and U. Schild. *Principles of Talmudic Logic*, 300pp. College Publications, 2013.
9. M. Abraham, D. M. Gabbay, and U. Schild. Contrary to Time Conditionals. *J of Artificial Intelligence and Law*, 20(2), 145–179, 2012.
10. M. Abraham, I. Belfer, D. M. Gabbay, and U. Schild. Future determination of entities in Talmudic Logic. *Journal of Applied Logic*, 11(1), 63–90, 2013.
11. W. Ackermann. Untersuchungen uber das eliminationsproblem der mathematischenlogik, *Mathematische Annalen*, 110, 390–413, 1935.
12. C. Adler. *Modern Geometry*. Second edition, McGraw Hill, 1967.
13. R. K. Ahuja, T. L. Magnani and J. B. Orlim. *Network Flow*, Prentice-Hall, 1993.
14. Ruggero J. Aldisert. *Logic for Lawyers: A Guide to Clear Legal Thinking*, Clark Boardman, 1989.
15. Christopher Allen. *Practical Guide to Evidence*, 2nd ed. Cavendish Pub, 2001.
16. Leila Amgoud. Five weaknesses of ASPIC+.+. In *Proceedings of the 14th International Conference on Information Processing and Managing of Uncertainty in Knowledge-Based Systems*. LNAI, Springer Verlag, 2012.

17. L. Amgoud, N. Maudet, and S. Parsons. Modelling dialogues using argumentation. In *Proceedings of the Fourth International Conference on MultiAgent Systems (ICMAS-00)*, pages 31–38, Boston, MA, 2000.

18. L. Amgoud and S. Kaci, 'An argumentation framework for merging conflicting knowledge bases', *Journal of Approximate Reasoning*, **45**, 321–340, (2007).

19. L. Amgoud, C. Cayrol, M. C. Lagasquie-Schiex, and P. Livet. On bipolarity in argumentation frameworks. *Int J Intell Syst* 23(10):1062–1093, 2008.

20. R. M. Anderson, B. D. Turner and L. R. Taylor, eds. *Population Dynamics*. Blackwell, 1979.

21. G. Antoniou. A tutorial on default logics. *ACM Computing Surveys*, 31(3), September 1999.

22. K. R. Apt and M H. van Emden. Contributions to the Theory of Logic Programming, *J. Assoc. Comput. Much.* 29:841-862 (1982).

23. K. R. Apt and R. N Bol. Logic programming and negation: A Survey. *J. Logic Programming*, 19/20, 9–72, 1994.

24. J. C. Augusto, G. R. Simari: A temporal argumentative system. AI Communications 12 (1999) 237Đ257 ISSN 0921-7126, IOS Press, 1999.

25. ASPIC-consortium. Deliverable D2.5: Draft formal semantics for ASPIC system, June 2005.

26. F. Baader, D. Calvanese, D. L. McGuinness, D. Nardi, and P. F. Patel-Schneider, eds. *Description Logic Handbook*, Cambridge University Press, 2002.

27. P. Balbiani. Modal logic and negation as failure. *J Logic Computation*, 11, 331-356, 1991.

28. Pietro Baroni and Massimiliano Giacomin. Comparing argumentation semantics with respect to skepticism. In *Proc. ECSQARU 2007*, pages 210–221, 2007.

29. Pietro Baroni and Massimiliano Giacomin. On principle-based evaluation of extension-based argumentation semantics. *Artificial Intelligence*, 171(10-15):675–700, 2007.

30. P. Baroni, M. Giacomin, and G. Guida. SCC-recursiveness: a general schema for argumentation semantics. *Artificial Intelligence*, 168 (1-2):162–210, 2005.

31. P. Baroni and M. Giacomin. Solving semantic problems with odd-length cycles in argumentation. In *Proceedings of the 7th European Conference on Symbolic and Quantitative Approaches to Reasoning with Uncertainty (ECSQARU 2003)*, pp. 440–451. LNAI 2711, Springer-Verlag, Aalborg, Denmark, 2003.

32. P. Baroni, F. Cerutti, M. Giacomin and G. Guida. Encompassing attacks to attacks in abstract argumentation frameworks. In Proceedings of the 10th European Conference on Symbolic and Quantitative Approaches to Reasoning with Uncertainty. pp. 83–94, 2009.

33. P. Baroni, F. Cerutti, M. Giacomin and G. Guida. Afra: Argumentation framework with recursive attacks. *International Journal of Approximate Reasoning*, 52(1): 19-37 (2011).

34. P. Baroni, F. Cerutti, P. E. Dunne, and M. Giacomin. Computing with Infinite Argumentation Frameworks: the Case of AFRAs. In *TAFA 11*, 2011.

35. Pietro Baroni, Martin Caminada and Massimiliano Giacomin. An introduction to argumentation semantics, *The Knowledge Engineering Review*, 26(4), 365-410, 2011.

36. Pietro Baroni, Martin Caminada and Massimiliano Giacomin. An introduction to argumentation semantics, *The Knowledge Engineering Review* (November 2011), 26 (4), pg. 365-410

37. H. Barringer, D. M. Gabbay and J. Woods. Temporal dynamics of argumentation networks. In D. Hutter and W. Stephan, eds., *Mechanising Mathematical Reasoning*, pp. 58–98. LNCS 2605, Springer, 2005.

38. H. Barringer and D. M. Gabbay. Modal and temporal argumentation networks. In the Amir Pnueli Memorial Volume *Time for Verification*, D. Peled and Z. Manna, eds., pp. 1–25. LNCS, Springer, 2010.

39. H. Barringer D. M. Gabbay and J .Woods. Modal and temporal argumentation networks, *Argument and Computation*, 3(2-3), 203–227, 2012.

40. H. Barringer, D. M. Gabbay, and J. Woods. Network modalities. In G. Gross and K. Schulz, eds., *Linguistics, Computer Science and Language Processing. Festschrift for Franz Guenthner on the Occasion of his 60th Birthday*, pp. 79–102. College Publications, 2008.

41. H. Barringer, D. M. Gabbay and J. Woods. Temporal, Numerical and Metalevel Dynamics in Argumentation Networks. *Argument and Computation*, 3(2-3), 143–202, 2012.

42. H. Barringer, D Gabbay, A. d'Avila Garcez, O. Rodrigues, and J. Woods. *Neuro-fuzzy Argumentation Networks*. Monograph, in preparation for Springer.

43. T. Bench-Capon. Persuasion in practical argument using value based argumentation framework. *Journal of Logic and Computation*, **13**, 429–448, 2003. See also http://www.csc.liv.ac.uk/tbc/FTP/kings2.ppt

44. T. J. M. Bench-Capon and K. Atkinson. Abstract argumentation and values. In I. Rahwan and G. Simari, editors, *Argumentation in AI*, Chapter 3, pages 45–64. Springer-Verlag, 2009.

45. T. Bench-Capon. Knowledge based systems in the legal domain. A survey article in *Encyclopaedia of Computer Science and Technology*, available on the web at http://www.csc.liv.ac.uk/~lial/lial/tut.html

46. P. Besnard and A. Hunter. A logic-based theory of deductive arguments. *Articial Intelligence*, 128(1-2):203–235, 2001.

47. P. Besnard and A. B. Hunter. *Elements of Argumentation*, 300pp. MIT Press, 2008.

48. Ph. Besnard and S. Doutre. Checking the acceptability of a set of arguments. In *Proceedings NMR-2004*, pp. 59–64, 2004.

49. Philippe Besnard and Sylvie Doutre. Characterization of semantics for argument systems. In *Proceedings of the Ninth International Conference on the principles of Knowledge Representation and reasoning*, pages 183–193. AAAI Press, 2004.

50. S. Bistarelli and F. Santini, 'A common computational framework for semiring-based argumentation systems', in *Proceedings of the 2010 conference on ECAI 2010: 19th European Conference on Artificial Intelligence*, pp. 131–136, Amsterdam, The Netherlands, The Netherlands, (2010). IOS Press.

51. P. Blackburn and J. Seligman. Hybrid languages. *Journal of Logic, Language and Information*, 4, 41–62, 1995.

52. A. Bochman. *Explanatory Nonmonotonic Reasoning*. Advances in Logic, World Scientific, 2005.

53. Gustavo A. Bodanza and Fernando A. Tohmé. Two approaches to the problems of self-attacking arguments and general odd-length cycles of attack. *Journal of Applied Logic*, 7, 403–420, 2009.

54. G. Boella, J. Hulstijn, and L. van der Torre. A logic of abstract argumentation. In *Proceedings of the Workshop on Argumentation in Multi-Agent Systems (ArgMAS)*, 2005.

55. G. Boella, D. M. Gabbay, L. van der Torre and S. Villata. Argumentation modelling of the Toulmin scheme. *Studia Logica*, 93(2-3):297-354, 2009.

56. Guido Boella, Souhila Kaci, Leendert van der Torre: Dynamics in Argumentation with Single Extensions: Attack Refinement and the Grounded Extension (Extended Version).*ArgMAS 2009*: 150-159

57. G. Boella, L. van der Torre and S. Villata. Social viewpoints for arguing about coalitions. In T. D. Bui, T. V. Ho, and Q. T. Ha, eds., *PRIMA*, Springer, Lecture Notes in Computer Science, vol 5357, pp 66–77

58. G. Boella, D. M. Gabbay, L. van der Torre and S. Villata. Meta-argumentation part 1, *Studia Logica*, 93 (2-3): 297–354, 2009,

59. G. Boella, L. van der Torre, and S. Villata. On the acceptability of meta-arguments. In *Proc. of the 2009 IEEE/WIC/ACM International Conference on Intelligent Agent Technology, IAT 2009*, IEEE, pp 259–262

60. G. Boella, L. van der Torre, and S. Villata. Analyzing cooperation in iterative social network design. *Journal of Universal Computer Science*, 15(13):2676–2700, 2009.

61. G. Boella, D. M. Gabbay, L. van der Torre and S. Villata. Support in abstract argumentation In *Computational models of Argument, COMMA 2010*, P. Baroni, F. Cerutti, M. Giacomi and G. Simari , eds., pp. 111–122. IOS press, 2010.

62. G. Boella, D. M. Gabbay, L. van der Torre, and S. Villata. Argumentative agents negotiating on potential attacks. In *5th Internatioal KES Conference on Agents and Multi-agent Systems — Technologies and Applications* (KES-AMSTA 2011). LNCS Vol 6682, pp. 280–290, 2011.

63. A. Bondarenko, P. M. Dung, R. A. Kowalski, and F. Toni. An abstract, argumentation theoretic approach to default reasoning. *Artificial Intelligence*, 93:63–101, 1997.

64. G. Boole. *The Mathematical Analysis of Logic*, Cambridge and London, 1847.

65. G. Boella, D. M. Gabbay, L. van der Torre and S. Villata. An argumentation based approach to coalition formation using voluntary attacks. In preparation.

66. A.J. Bonner and L.T. McCarty. Adding Negation-as-Failure to Intuitionistic Logic Programming. Technical report, Department of Computer Science, Rutgers University, New Brunswick, NJ 08903, 1990.

67. G. Boolos. *The Logic of Provability*. Cambridge University Press, 1993.

68. G. Brewka. Dynamic argument systems: A formal model of argumentation processes based on situation calculus. *JLC*, 11(2):257–282, 2001.

69. G. Brewka and S. Woltran. Abstract dialectical frameworks. In *Proc. of the 20th International Conference on the Principles of Knowledge Representation and Reasoning (KR 2010)*, pages 102–111, 2010.

70. G. Brewka, P. Dunne, and S. Woltran. Relating the Semantics of Abstract Dialectical Frameworks and Standard AFs. In *Proc. IJCAI-11*, pp. 780–785, 2011.

71. For Brouwer Fixed Point Theorem, see `http://en.wikipedia.org/wiki/Brouwer_fixed_point_theorem` and Sobolev, V. I., "Brouwer theorem" in Hazewinkel, Michiel, *Encyclopaedia of Mathematics*, Springer, 2001.

72. M. Cadoli. *Tractable Reasoning in Artificial Intelligence*. Vol 941 of em LNCS, Springer-Verlag, 1995.

73. M. Cadoli, and F. M. Donini. A survey on knowledge compilation. *AI Communications*, 10(3–4), 137–150, 1997.

74. M.W.A. Caminada. Semi-stable semantics. In P.E. Dunne and T.J.M. Bench-Capon, editors, *Computational Models of Argument; Proceedings of COMMA 2006*, pages 121–130. IOS Press, 2006.

75. M.W.A. Caminada. On the issue of reinstatement in argumentation. In M. Fischer, W. van der Hoek, B. Konev, and A. Lisitsa, editors, *Logics in Artificial Intelligence; 10th European Conference, JELIA 2006*, pages 111–123. Springer, 2006. LNAI 4160.

76. M.W.A. Caminada. An algorithm for computing semi-stable semantics. In *Proceedings of the 9th European Conference on Symbolic and Quantitalive Approaches to Reasoning with Uncertainty (ECSQARU 2007)*, number 4724 in Springer Lecture Notes in AI, pages 222–234, Berlin, 2007. Springer Verlag.

77. M. Caminada. Preferred Semantics as Socratic Discussion, 2011.

78. M. Caminada. On the Limitations of Abstract Argumentation. Submitted to *BNAIC*, 2011.

79. Martin Caminada and Leila Amgoud. On the evaluation of argumentation formalisms. *Artificial Intelligence*, 171(5-6):286–310, 2007.

80. M. Caminada. A gentle introduction to argumentation semantics. Lecture material, Summer 2008.

81. M. Caminada and D. M. Gabbay. A logical account of formal argumentation. *Studia Logica*, 93(2-3):109-145, 2009.

82. M. Caminada. A labelling approach for ideal and stage semantics. *Argument and Computation*, 2(1):1–21, 2011.

83. M. Caminada and Yining Wu. On the limitations of abstract argumentation. In *Proceedings of BNAIC 2011; the 23rd Benelux Conference on Artificial Intelligence* 3-4 November

2011, Ghent, Belgium. Editors, Patrick De Causmaecker, Katja Verbeeck, Joris Maervoet, Tommy Messelis, pp. 69-66, 2011.

84. J. Carmo and A. Jones. Deontic logic and contrary-to-duties. *Handbook of Philosophical Logic*, Volume 8, 2nd edition, pp. 265–345, D. M. Gabbay and F. Guenthener, eds. Kluwer, 2002.

85. W. Carnielli, M. Coniglio, D. M. Gabbay, P. Gouveia and C. Sernadas. *Analysis and Synthesis of and Synthesis of Logics*, 500pp. Springer, 2007.

86. Claudette Cayrol, Marie-Christine Lagasquie-Schiex: Graduality in Argumentation. *J. Artif. Intell. Res. (JAIR)* 23: 245-297, 2005.

87. Claudette Cayrol, Caroline Devred, and Marie-Christine Lagasquie-Schiex. Acceptability semantics accounting for strength of attacks in argumentation. *ECAI 2010*, pp. 995-996. [long version available at ftp://ftp.irit.fr/IRIT/ADRIA/rap-2010-13.pdf]

88. C. Cayrol and M.-C. Lagasquie-Schiex, 'Merging argumentation systems with weighted argumentation systems: a preliminary study', Technical Report RR-2011-18-FR, IRIT, (2011).

89. S. Cerrito. A linear axiomatization of negation as failure. *Journal of Logic Programming*, 12:1-24, 1992.

90. S. Cerrito. Negation and linear completion. In L. Farinas del Cerro and M. Penttonen, editors, *Intensional Logic for Programming*, pages 155-194. Clarendon Press, 1992.

91. M. Chalamish, D. M. Gabbay, and U. Schild. Intelligent Evaluation of Evidence using Wigmore Diagrams. In *PICAIL'11. Proceedings of the 13th International Conference on Artificial Intelligence and Law*, pp. 61–65. ACM New York, USA, 2011.

92. S. Chopra, A. Ghose, and T. Meyer, 'Social choice theory, belief merging, and strategy-proofness', *Information Fusion*, 7(1), 61–79, (March 2006).

93. G. C. Christie. *The Notion of an Ideal Audience in Legal Argument*. Kluwer, 2000.

94. Maria Laura Cobo, Diego C. Martinez, Guillermo Ricardo Simari: On Admissibility in Timed Abstract Argumentation Frameworks. *ECAI 2010*: 1007-1008

95. C. Cayrol and M. C. Lagasquie-Schiex. On the acceptability of arguments in bipolar argumentation frameworks. In L. Godo, ed., *ECSQARU*, pp. 378–389. Springer, Lecture Notes in Computer Science, vol 3571, 2005.

96. C. Cayrol and M. C. Lagasquie-Schiex. Coalitions of arguments: A tool for handling bipolar argumentation frameworks. *Int J Intell Syst*, 25(1):83–109, 2010.

97. C. I. Chesñevar, A. G. Maguitman, and R. P. Loui, *Logical Models of Argument. ACM Computing Surveys*, 32(4):337–383, December 2000.

98. Petr Cintula. The Ł $\prod$ and Ł $\prod \frac{1}{2}$ propositional and predicate logics. In *Fuzzy Sets and Systems*, **124**, 289–302, 2001.

99. K. Clark. *Negation as failure*. Originally published in 1978, and reproduced in *Readings in nonmonotonic reasoning*, Morgan Kaufmann Publishers, pages 311–325, 1987.

100. J. Coleman. *Risks and Wrongs*. Cambridge University Press, 1992.

101. S. Coste-Marquis, C. Devred, S. Konieczny, M.-C. Lagasquie-Schiex, and P. Marquis, 'On the merging of Dung's argumentation systems', *Artificial Intelligence*, **171**, 730–753, (July 2007).

102. Counterpart Theory. http://en.wikipedia.org/wiki/Counterpart_theory.

103. L. Couturat. *The Algebra of Logic*, Open Court, 1914

104. Sir Rupert Cross. *On Evidence*. Butterworth, 1999.

105. M. D'Agostino and D. M. Gabbay. *Depth-Bounded Approximations of Logical Systems*. Research Monograph in preparation.

106. A. Darwiche and P. Marquis. A knowledge compilation map. *Journal of Arti-cial Intelligence Research*, 17, 229–264, 2002.

107. H. Dennis. *Law of Evidence*, Sweet and Maxwell, 1992.

108. Deon Conferences. Journals.
    Please see: http://www.doc.ic.ac.uk/deon02/

http://www.cert.fr/deon00/
ARTIFICIAL INTELLIGENCE AND LAW
http://www.denniskennedy.com/ailaw.htm
http://www.iaail.org/
There are many useful Links to Evidence-Related Web Sites: at http://tillers.net

109. R. D'Inverno. *Introducing Einstein's Relativiety*, OUP, 1992.

110. P. Doherty, W. Łukasiewicz, and A. Szałas. A reduction result for circumscribed semi-Horn formulas. *Fndamenta Infomaticae*, 28(3–4), 261–271, 1996.

111. P. Doherty, W. Łukaszewicz, A. Skowron, and A. Szałas. *Knowledge representation techniques. A rough set approach*, vol. 202 of Studies in Fuziness and Soft Computing, Springer-Verlag, 2006.

112. P. Doherty, W. Łukasiewicz, and A. Szałas. Computing circumscription revisted. *Journal of Automated Reasoning*, 18(3), 297–336, 1997.

113. P. Doherty and A. Szałas. On the correspondence between approximations and similarity. In S. Tsumoto, R. Slowinski, J. Komorowski, and J.W. Grzymala-Busse, eds., *Proc. RSCTC'2004*, vol. 3066 of LNAI, pp. 143–152, 2004.

114. P. M. Dung. On the acceptability of arguments and its fundamental role in nonmonotonic reasoning, logic programming and n-person games. *Artificial Intelligence*, 77: 321–357, 1995.

115. P. M. Dung. An argumentation theoretic foundation for logic programming. *Journal of Logic Programming*, 22(2), 151–171, 1995.

116. P. M. Dung, P. Mancarella, and F. Toni. Computing ideal sceptical argumentation. *Artificial Intelligence*, 171(10-15):642–674, 2007.

117. P.E. Dunne, A. Hunter, P. McBurney, S. Parsons and M. Wooldridge. Inconsistency tolerance in weighted argument systems. In Proceedings of the 8th International Joint Conference on Autonomous Agents and multiagent Systems, pp. 851–858, 2009.

118. P.E. Dunne, A. Hunter, P. McBurney, S. Parsons and M. Wooldridge. Weighted argument systems: Basic definitions, algorithms and complexity results. *Artifical Intelligence* 175 (2): 457-486, 2011.

119. Wolfgang Dvorăk and Sara Alice Gaggl. Computation aspects of cf2 and stage 2 argumentation semantics. In *Proceedings of COMMA 2012*, pp. 273–284. IOS Press, 2012.

120. W. Dvorak and S. Woltran. On the intertranslatability of argumentation semantics. *JAIR*, 41: 445–475, 2011.

121. H.-D. Ebbinghaus and J. Flum. *Finite Model Theory*, Springer-Verlag, Heidelberg,1995.

122. R. Ebrahim. Fuzzy logic programming. *Fuzzy Sets and Systems*, 117:213–230, 2010.

123. K. Engesser, D. M. Gabbay and D. Lehmann. *A New Approach to Quantum Logic*, 300pp. College Publications, 2008.

124. T. Faulkner. *Projective Geometry*, Oliver and Boyd, 1949..

125. M. Fitting. Kripke–Kleene semantics for logic programs. *Journal of Logic Programming*, 2: 295–312, 1985.

126. M. Fitting. Negation As Refutation. In *Proceedings of the fourth annual symposium on logic in Computer Science*, R. Parikh editor, IEEE (1978), pp. 63-70.

127. D. M. Gabbay. *Semantical Investigations in Modal and Tense Logics*. Synthese, Volume 92, D. Reidel, 1976.

128. D. M. Gabbay and U. Reyle. N-Prolog: An Extension of Prolog with Hypothetical Implications I *Journal of Logic Programming*, 1, 319–355, 1984.

129. D. M. Gabbay. Theoretical foundations for nonmonotonic reasoning in expert systems. In *Logics and Models of Concurrent Systems*, K. Apt, ed., pp. 439–459. Springer-Verlag, 1985.

130. D. M. Gabbay. What is negation as failure? Manuscript, 1985.

131. D. M. Gabbay and M. Sergot. Negation as inconsistency. *Journal of Logic Programming*, 4(3): 1–35, 1986.

132. D. M. Gabbay. Modal Provability Foundations for Argumentation Networks *Studia Logica*, 93(2-3): 181–198, 2009,

133. D. M. Gabbay. N-Prolog: An Extension of Prolog with Hypothetical Implications 2 *Journal of Logic Programming*, 2, 251–283, 1986.

134. D. M. Gabbay. Modal provability foundations for negation by failure, in *Extensions of Logic Programming*, P. Schroeder-Heister, editor, pp. 179–222, Vol 475 of LNCS, Springer-verlag, 1990.

135. D. M. Gabbay and H. J. Ohlbach. Quantifier elimination in second-order predicate logic, *South African Computer Journal*, 7, 35–43, 1992.

136. D. M. Gabbay. *Labelled Deduction Systems*. Oxford University Press, 1996.

137. D. M. Gabbay. How to make your logic fuzzy (preliminary version). *Mathware and Soft Compting*, 3:5–16, 1996.

138. D. M. Gabbay. *Fibring Logics*. Oxford University Press, 1998.

139. D. M. Gabbay. How make your logic fuzzy. In *Fuzzy Sets, Logics and Reasoning about Knowledge*, D. Dubois, H. Prade, and E. P. Klement, eds., pp. 51–84. Kluwer, 1999.

140. D. M. Gabbay and N. Olivetti. *Goal Directed Algorithmic Proof Theory* (Monograph) Springer, 2000. 266 pp.

141. D. M. Gabbay and V. Shehtman. Flow products of modal logics. Draft, 2000. incorporated in [144]. Expanded version being prepared for a special issue in *Logica Universalis*, Springer, under the title: D. M. Gabbay, I. Shapirovsky and V. Shehtman, Flow Products.

142. D. M. Gabbay. Theory of hypermodal logics, *Journal of Philosophical Logic*, **31**, 211–243, 2002.

143. D. M. Gabbay. Sampling labelled deductive systems. In *A Companion to Philosophical Logic*, D. Jacquette, ed., pp. 742–869. Blackwell, 2002.

144. Dov M. Gabbay, A. Kurucz, F. Wolter, and M. Zakharyaschev. *Many-dimensional Modal Logics: Theory and Applications*. Number 148 in Studies in Logic and the Foundations of Mathematics. Elsevier, 2003.

145. D. M. Gabbay and J. Woods. Cooperate with your logic ancestors. *Journal of Logic, Language and Information*, **8**, iii–v, 1999.

146. D. M. Gabbay and J. Woods. *Agenda Relevance: A Study in Formal Pragmatics*, Elsevier, 2003, 521 pp.

147. D. M. Gabbay. Reactive Kripke semantics and arc accessibility. In Proceedings of CombLog04 (http://www.cs.manth ist.utl.pt/comblog04), W. Carnielli, F.M. Dionesio and P. Mateus, eds., Centre of Logic and Computation, University of Lisbon, pp.7–20, 2004. ftp://logica.cle.unicamp.br/pub/e-prints/comblog04/gabbay.pdf

148. D. M. Gabbay. Editorial to *Handbook of Philosophical Logic*, 2nd edition. Kluwer, 2002–2012.

149. D. M. Gabbay and A. D'Avila Garcez. Fibring neural networks. In *Proceedings of the 19th National Conference on Artificial Intelligence (AAA'04)*, San Jose, CA., pp. 342-347. AAAI Press, 2004.

150. D.M. Gabbay and J. Woods. *The Reach of Abduction: Insight and Trial*. Elsevier, 2005.

151. D. M. Gabbay, G. Pigozzi, and O. Rodrigues, 'Belief revision, belief merging and voting', in *Proceedings of the Seventh Conference on Logic and the Foundations of Games and Decision Theory (LOFT06)*, pp. 71–78. University of Liverpool, (2006).

152. D. M. Gabbay. *Logic for AI and Information Technology*, 500pp. College Publications, October 2007.

153. D. M. Gabbay. *The Leverhulme Lectures in Logic*, 200pp. Draft available.

154. D. M. Gabbay and A. Szałas. Second-order Quantifier Elimination in higher-order contexts with applications to the semantic analysis of conditionals. *Studia Logica*, 87:37–50, 2007.

155. D. M. Gabbay, R. Schmidt and A. Szałas. *Second-Order Quantifier Elimination*, 400pp. College Publications, 2008.

156. D. M. Gabbay. Reactive Kripke models and contrary to duty norms. *Proceedings of DEON 2008*, Deontic Logic in Computer Science. R. van der Meyden and L. van der Torre, eds. LNAI 5076, pp. 155–173. Springer, 2008.

157. D. M. Gabbay. Reactive Kripke Semantics and Arc Accessibility. In *Pillars of Computer Science: Essays dedicated to Boris(Boaz) Trakhtenbrot on the Occasion of his 85th Birthday*, Arnon Avron, Nachum Dershowitz and Alexander Rabinovich, eds., LNCS 4800, Springer, pp.292-341, 2008. Expanded version published in *Annals of Mathematics and Artificial Intelligence*, 661-4, 7–53, 2012. Earlier version published in

158. D. M. Gabbay. *Reactive Kripke Semantics, Theory and Applications*. 400pp. Monograph in preparation for Springer.

159. D. M. Gabbay and S. Modgil. Modal argumentation. Draft, May 2009.

160. D. M. Gabbay and J. Woods. Resource origins of non-monotonicity. *Studia Logica*, to appear, 2008.

161. D. M. Gabbay and A. S. d'Avila Garcez. Modes of attack in argumentation networks. *Studia Logica*, 93(2-3):199-230, 2009.

162. D. M. Gabbay. Modal provability foundations for argumentation networks. *Studia Logica*, 93(2-3):181-198, 2009.

163. D. M. Gabbay. Semantics for higher level attacks in extended argumentation frames. Part 1: Overview. *Studia Logica*, 93(2-3):355-379, 2009.

164. D. M. Gabbay. Fibring argumentation frames. *Studia Logica*, 93(2-3):231-295, 2009.

165. D. M. Gabbay. Dung's argumentation is equivalent to classical propositional logic with the Peirce-Quine dagger. *Logica Universalis*, 5(2), 255–318, 2011. DOI : 10.1007/s11787-011-0036-3.

166. D. M. Gabbay. An equational approach to argumentation networks. *Argument and Computation*, special issue on the equational approach to argumentation, 3(2-3), 87–142, 2012.

167. D. M. Gabbay. Introducing equational semantics for argumentation networks, In *Proc ECSQARU 2011*, W. Liu, ed., pp. 19–35. LNAI 6717, Springer, 2011.

168. D. M. Gabbay. Equational approach to default logic. In preparation, 90pp, 2012.

169. D. M. Gabbay. Bipolar argumentation frames and contrary to duty obligations, a position paper. In *Proceedings of CLIMA 2012*, M. Fisher *et al.*, eds. LNAI 7486, pp. 1–14, Springer, 2012.

170. D. M. Gabbay. The equational approach to CF2 semantics, short version. In *Proceedings COMMA 2012, Computational Models of Argument*, Edited by Bart Verheij, Stefan Szeider, and Stefan Woltran, IOS press, 2012, pp 141-153.

171. D. M. Gabbay. Logical foundations for bipolar argumentation networks. Submitted to Springer volume in Honour of Arnon Avron.

172. D. M. Gabbay Universal properties of two disjoint modalities — with application to deontic logic, products, argumentation, reactivity and conditionals. In preparation for *Logica Universalis*.

173. D. M. Gabbay, L. Gammaitoni and Xin Sun. The paradoxes of permission. An action based solution: Draft December 2012.

174. D. M. Gabbay and A. Szalas. Annotation theories over finite graphs. *Studia Logica*, 93, 147–180, 2009.

175. D. M. Gabbay. Logic and Networks. In preparation.

176. D. M. Gabbay and O. Rodrigues. An Equational approach to merging of argumentation networks. In M. Fisher and L. van der Torre, eds, *Computational Logic in Multi-Agent Systems, CLIMA 2012*, pp. 195–212. Lecture Notes in Computer Science Volume 7486, 2012. Expanded version submitted to special issue of JLC.

177. D. M. Gabbay and J. Woods. Non-cooperation in dialogue logic. *Synthese*, **127**, 161–180, 2001.

178. D. M. Gabbay and J. Woods. *Ad baculum* is not a fallacy. In *Proceedings of the Fourth Internatioal Conference of the International Society for the Study of Arguementation*, F.

H. van Eemeren, R. Grootendorst, J. A. Blair and C.A. Willard, eds. pp. 221–224, SicSat, Amsterdam, 1998.

179. D. M. Gabbay and J. Woods. More on non-cooperation in dialogue logic. *Logic Journal of the IGPL*, **9**, 305–324, 2001.

180. D. M. Gabbay. Dynamics of practical reasoning, a position paper. In *Advances in Modal Logic 2, Proceedings of Conference October 1999*, K. Segerberg *et al.*, eds, pp. 179–224. CSLI Publications, Cambridge University Press, 2001.

181. D. M. Gabbay and J. Woods. Formal approaches to practical reasoning. In *Handbook of the Logic of Argument and Inference: The Turn Towards the Practical*, D. M. Gabbay, R. H. Johnson, H. J. Ohlbach and J. Woods, eds. pp. 449–481. North-Holland, Amsterdam, 2002.

182. D. M. Gabbay and J. Woods. Normative models of Rational Agency. *Logic Journal of the IGPL*, Volume 11, Number 6, pp 597-613, 2003.

183. D. M. Gabbay and J. Woods. The law of evidence and labelled deductive systems. published in *Phi-News*, **4**, 5–46, October 2003, http//phinews.ruc.dk/phinews4.pdf. Also Chapter 15 in Gabbay D.M.; Canivez, P.; Rahman, S.; Thiercelin, A. (Eds.), *Approaches to Legal Rationality, Logic, Epistemology, and the Unity of Science* 20, pp. 295–331, 1st Edition, Springer, 2010. DOI 10.1007/978-90-481-9588-6_15.

184. D. M. Gabbay, A. Perotti, T. Rienstra, L. van der Torre and S. Villata. Simulating Proofs in Pure Argumentation Networks, Draft April 2011.

185. A. S. D'Avila Garcez, D. M. Gabbay and L. Lamb. Value based argumentation frameworks as neural networks. *Journal of Logic and Computation*, 15(6):1041-1058, Dec. 2005.

186. A.S. D'Avila Garcez, L.C. Lamb and D. M. Gabbay. Connectionist Non-classical Logics: Distributed Reasoning & Learning in Neural Networks (Monograph). Springer-Verlag, 2008.

187. D. M. Gabbay. The equational approach to logic programs. In LNCS 7265 Festschrift for Vladimir Lifschitz, *Correct Reasoning — Essays on Logic-based AI in Honour of Vladimir Lifschitz*, E. Erdem, J. Lee, Y. Lierler and D. Pearce, eds., pp. 279–296, Springer, 2012.

188. D. M. Gabbay. Temporal Deontic logic for the Generalised Chisholm set of contrary to duty obligations. In T. Agotnes, J. Broersen, and D. Elgesem, eds. , *DEON 2012*, LNAI 7393, pp. 91–107. Springer, Heidelberg, 2012.

189. D. M. Gabbay. Reactive Kripke Models and Contrary-to-duty Obligations Expanded version, original version 2008, revised 2012 into two parts, Part A Semantics, to appear in *Journal of Applied Logic*, http://dx.doi.org/10.1016/j.jal.2012.08.001. Part B Proof Theory, to be submitted to *Journal of Applied Logic*.

190. D. M. Gabbay. The equational approach to defeasible argumentation with applications to ASPIC, in preparation.

191. S. A. Gaggl and Stefan Woltran. Strong Equivalence for Argumentation Semantics Based on Conflict-Free Sets, *ECSQARU*, pp. 38–49, W. Liu, ed. LNAI 6717, Springer, 2011.

192. S. A. Gaggl and S. Woltran. cf2 Semantics revisited. In Baroni, P., Cerutti, F., Giacomin, M., and Simari, G. R., eds., *COMMA 2010*, volume 216, pp. 243–254. IOS Press, 2010.

193. Sarah Alice Gaggl and Stefan Woltran. The cf2 Argumentation Semantics Revisited. To appear in *Journal of Logic and Computation*, 2012.

194. L. T. F. Gamut. *Logic, Language, and Meaning* (University of Chicago Press, 1991).

195. A. S. d'Avila Garcez, D. M. Gabbay and L. C. Lamb. Argumentation Neural Networks. In *Proceedings of 11th International Conference on Neural Information Processing (ICONIP'04)*, Calcutta, India, Lecture Notes in Computer Science LNCS, Springer-Verlag, November 2004.

196. A. S. d'Avila Garcez, D.. Gabbay, and L.C. Lamb. Value-based argumentation frameworks as neural-symbolic learning systems. *JLC*, 15(6):1041–1058, 2005.

197. A. S. d'Avila Garcez, K. Broda, and D. M. Gabbay. *Neural-Symbolic Learning Systems: Foundations and Applications*. Springer, 2002.

198. A. J. García and G. R. Simari. Defeasible logic programming: An argumentative approach. *Theory and Practice of Logic Programming*, 4(1):95–138, 2004.

199. M. Gelfond. Answer sets. In *Handbook of Knowledge Representation*, Frank van Harmelen, Vladimir Lifschitz, and Bruce Porter, eds., pp. 285–316. Elsevier, 2008.

200. M. Gelfond and V. Lifschitz. The stable model semantics for logic programming. In *Proceedings of the Fifth International Conference on Logic Programming(ICLP)*, pp. 1070–1080, 1988.

201. L. Giordano and N. Olivetti. Negation as failure in intuitionistic logic programming. In *Proceedings of JICSLP* 1992. pp.431–445

202. J. Y. Girard, Y. Lafont and P. Taylor. *Proofs and Types*. Cambridge University Press, 1989.

203. S. A. Gomez and C. I. Chesñevar. Integrating Defeasible Argumentation with fuzzy art neural networks for pattern classification. *Journal of Computer Science and Technology*, 4(1):45–57, 2004.

204. V. Goranko. Temporal logics with reference pointers and computation tree logics, *Journal of Applied Non-Classical Logics*, 10, 3–4, 2000.

205. N. Gorogiannis and A. Hunter. Instantiating abstract argumentation with classical logic arguments: Postulates and properties. Elsevier DOI, 2010.

206. C. Gratie and A. Maqda Florea. Fuzzy labelling for argumentation frameworks. In *ArgMas2011, 8th International Workshop on Argumentation in Multi-Agent Systems*, Taipei, Taiwan, May 2011, pp. 18–25.

207. Davide Grossi. Doing argumentation in modal logic. 2009.

208. D. Grossi. On the logic of argumentation theory. In *Proceedings of AAMAS-2010*, 2010.

209. J. W. Guan and D. A. Bell. *Evidence Theory*, 2 volumes. Elsevier, 1991.

210. Rolf Haenni: Probabilistic argumentation. *J. Applied Logic* 7(2): 155-176, 2009.

211. P. Hajek, T. Havranek, and R. Jirousek. *Uncertain Information Processing in Expert Systems*. CRC Press, 1992.

212. P. Hajek and J. Valdes. An analysis of MYCIN-like expert systems. *Mathware and Soft Computing*, 1, 45–68, 1994.

213. J. Halpern. An analysis of first-order logics of probability. *Artificial Intelligence*, **46**, 311-350, 1990.

214. Do Duc Hanh, Pha Minh Dung and Phan Minh Thang. Inductive defence for sceptical semantics of extended argumentation. To appear in *Journal of Logic and Computation*, 2010.

215. Do Duc Hanh, Phan Minh Dung and Phan Minh Thang. Inductive Defense for Modgil's Extended Argumentation Framework. Submitted to *Journal of Logic and Computation*.

216. J. Harland. A Kripke-like model for negation as failure. In *Proceedings of the North American Conference on Logic Programming (NACLP)*, pages 626–642, Cleveland, Ohio, October 16-20 1989.

217. S. Haykin. *Neural Networks: A Comprehensive Foundation*. Prentice Hall, 1999.

218. I. Horrocks and U. Sattler. Ontology reasoning in the shoq(d) descriptionlogic'. In *Proc. of the 17th Int. Joint Conf. on Articial Intelligence (IJCAI 2001*, pp. 199–204, 2001.

219. A. Hunter. Ramification analysis with structured news reports using temporal argumentation. Proc. of the Adventures in Argumentation Workshop (at 6th ECSQARU), 2001. http://www.cs.ucl.ac.uk/staff/a.hunter/papers/ra2.ps

220. A. Hunter. Base Logics in Argumentation. In *Proceedings of COMMA 2010*, pp 275–286.

221. N. Immerman. *Descriptive Complexity*, Springer-Verlag, New York, Berlin, 1998.

222. J. Janssen, M. De Cock and D. Vermeir. Fuzzy argumentation frameworks. In *Proc. 12th Conference on Information Processing and Management of Uncertainty in Knowledge-based Systems (IPMU 2008)*, pp. 513–520, 2008.

223. H. Jakobovits and D. Vermeir. Robust semantics for argumenation frameworks. *Journal of Logic and Computation*, 9: 215–261, 1999.

224. F. V. Jensen. *an introduction to Bayesian Networks*, UCL Press, 1996.

225. R. H. Johnson, H. J. Ohlbach, D. M. Gabbay and J. Woods, eds. *Handbook of the Logic of Argument and Inference: The Turn Towards the Practical*. Studies in Logic and Practical Reasoning, Elsevier, 2002.

226. J. Kachniarz and A. Szałas. On some extensions of second-order quantier techniques. Unpublished manuscript, 2001.

227. A. Keane. *The Modern Law of Evidence*, Butterworth, 2000.

228. J. M. Keynes. *A Treatise on Probability*. Macmillan Press for the Royal Economic Society, 1973 (1st edition, 1921; paperback edition, Cambridge University Press, 1988).

229. L. M. Kirousis and P. G. Kolaitis. A dichotomy in the complexity of propositionalcircumscription. In *Proceedings of the 16th Annual IEEE Symposium on Logic in Computer Science, LICS'01*, pp. 71–80, 2001.

230. E. P. Klement, R. Mesiar and E. Pap. *Triangular Norms*. Dordrecht: Kluwer, 2000.

231. P. G. Kolaitis and C. H. Papadimitriou. Some computational aspects of circumscription. *Journal of the ACM*, 37(1), 1–14, 1990.

232. S. Konieczny and R. Pino-Pérez, *Proceedings of KR'98*, chapter On the logic of merging, 488–498, Morgan Kaufmann, 1998.

233. S. Konieczny and R. Pino-Pérez, 'Logic based merging', *Journal of Philosophical Logic*, **40**(2), 239–270, (2011).

234. R. A. Kowalski and F. Toni. Abstract argumentation. *AI and Law*, 4(3-4):275–296, 1996.

235. K. Kunen. Negation in logic programming. *Journal of Logic Programming*, 4: 289–308, 1987.

236. A. C. Leisenring. *Mathematical Logic and Hilbert's Epsilon-Symbol*, London: Macdonald, 1969.

237. J. Leite and J. Martins, Social Abstract Argumentation. In T. Walsh (Ed.) *Proceedings of the Twenty-Second International Joint Conference on Artificial Intelligence 2011*. [Available at the author's webpage]

238. S. A. Levin, ed. *Studies in Mathematical Biology*, Part II, *Populations and Communities*. Mathematical Association of America, 1978.

239. V. Lifschitz. Computing circumscription. In *Readings in Nonmonotonic Reason-ing*, pp. 121–127. Morgan Kaufmann, 1985.

240. V. Lifschitz. On the satisfiability of circumscription. *Artificial Intelligence J.*, 28, 17–27, 1986.

241. V. Lifschitz. Circumscription. In *Handbook of Articial Intelligence and Logic Programming*, vol. 3, D. M. Gabbay, C. J. Hogger, and J. A. Robinson, eds., pp. 297–352. Oxford University Press, 1991.

242. V. Lifschitz. What is answer set programming?, 2008, http://www.cs.utexas.edu/~vl/papers/wiasp.pdf

243. J. W. Lloyd. *Foundations of Logic Programming* (2nd edition). Springer-Verlag 1987

244. A. J. Lotka. *Elements of physical biology*. Baltimore: Williams & Wilkins Co, 1925.

245. M. MacCrimmon and P. Tillers, eds. *The Dynamics of Judicial Proof*. Physica-Verlag, 2002.

246. S. Maclane. *Categories for the working mathematician*. Springer verlag, 1971.

247. J. McCarthy. Circumscription: A form of non-monotonic reasoning. *Articial Intelligence J.*, 13, 27–39, 1980.

248. W. Makiguchi and H. Sawamura. A hybrid argumentation of symbolic and neural net argumentation (part 1). In *ArgMAS 2007*, pp 197–215, Honolulu, USA, 2007. Springer, LNCS.

249. David Makinson and Leendert van der Torre. Constraints for input/output logics. *Journal of Philosophical Logic*, 30: 155-185, 2001.

250. D. Makinson. On a fundamental problem of deontic logic. In *Norms, Logics and Information Systems*, P. McNamara and H. Prakken, eds;, 29–54. IOS Press, 1999.

251. N. Mann, A. Hunter. Argumentation Using Temporal Knowledge. COMMA 2008, Toulouse, France, May 28-30: 204-215. IOS Press, 2008.

252. D. Martinez, A. Garcia and G. Simari. An abstract argumentation framework with varied-strength attacks. In Proceedings of the 11th International Conference on Principles of Knowledge Representation and Reasoning (KR'08), 2008.

253. Prabhaker Mateti and Narsingh Deo. On Algorithms for Enumerating All Circuits of a Graph *Siam Journal on Computing - SIAMCOMP*, vol. 5, no. 1, pp. 90-99, 1976 DOI: 10.1137/0205007

254. Paul-Amaury Matt and Francesca Toni. A Game-Theoretic Measure of Argument Strength for Abstract Argumentation. *JELIA 2008*, pp. 285-297.

255. R. M. C. May. Simple mathematical models with very complicated dynamics. Nature, Vol. 261, pp. 459–475, 1976.

256. K. Mellouli, ed. *Symbolic and Quantitative Approaches to Reasoning with Uncertainty, 9th European Conference, ECSQARU 2007*, Hammamet, Tunisia, October 31 - November 2, 2007, Proceedings, volume 4724 of Lecture Notes in Computer Science. Springer, 2007.

257. G. Metcalfe, N. Olivetti and D. M. Gabbay. *Proof Theory for Fuzzy Logics*, Springer, 2008.

258. Dale Miller. A Survey of Linear Logic Programming. *Computational Logic: The Newsletter of the European Network in Computational Logic*, Volume 2, No. 2, December 1995, pp. 63–67

259. G. Mints. Complete Calculus for Pure Prolog (in Russian), *Proc. Acad. Sci. Estonian SSR*, 35:367-380 (1986).

260. S. Modgil. Reasoning about preferences in argumentation frameworks. *Artificial Intelligence*, 173(9–10): 901–93, 2009.

261. S. Modgil. An abstract theory of argumentation that accommodates defeasible reasoning about preferences. In [256, pp.648–659], 2007.

262. Sanjay Modgil and Henry Prakken. A General Account of Argumentation with Preferences http://www.dcs.kcl.ac.uk/staff/smodgil/GAP.pdf

263. S. Modgil and T. J. M. Bench-Capon. Integrating object and meta-level value based argumentation. In P. Besnard, S. Doutre, and A. Hunter, editors, *COMMA*, volume 172 of Frontiers in Artificial Intelligence and Applications, pages 240–251. IOS Press, 2008.

264. S. Modgil and T.J.M. Bench-Capon. Metalevel Argumentation. *Journal of Logic and Computation*, doi: 10.1093/logcom/exq054 First published online: September 28, 2010.

265. Bamshad Mobasher, Jacek Leszczylowski, Giora Slutzki and Don Pigozzi. Negation as Partial Failure, LPNMR (1993), p. 244-262.

266. M. Mozina, J. Zabkar, and I. Bratko. Argument based machine learning. *Artificial Intelligence*, 171(10- 15):922–937, 2007.

267. J. D. Murray. *Mathematical Biology*, Volume 1, Springer-Verlag, 2001.

268. F. S. Nawwab, T. Bench-Capon, and P. E. Dunne. Exploring the role of emotions in rational decision making. In *COMMA'10*, Desenzano del Garda, Italy, 2010.

269. Hung T. Nguyen and Elbert A. Walker. *First Course in Fuzzy Logic*, Third Edition, Chapman and Hall, 2006.

270. S. H. Nielsen and S. Parsons. A generalisation of Dung's abstract framework for argumentation: arguing with sets of attacking arguments. *LNCS Vol. 4766*, pp. 54–73. Springer, 2007.

271. S. H. Nielsen and S. Parsons. Computing preferred extensions for argumentation systems with sets of attacking arguments. In em Proceedings of the 2006 Conference on Computational Models of Argument: Proceedings of COMMA 2007, pp. 97–108, 2007.

272. A. Nonnengart and A. Szałas. A fixpoint approach to second-order quantifier elimination with applications to correspondence theory. In *Logic at Work: Essays Dedicated to the Memory of Helena Rasiowa*, E. Orłowska, ed., pp. 307–328. Vol 24 of *Studies in Fuzziness and Soft Computing*, Springer Physica-Verlag, 1998.

273. E. Oikarinen and S. Woltran. Characterizing strong equivalence for argumentation frameworks. *Artificial Intelligence*, to appear. PII: S0004-3702(11)00075-0DOI: 10.1016/j.artint.2011.06.003

274. Nicola Olivetti and Lea Terracini. N-Prolog and Equivalence of Logic Programs. *Journal of Logic, Language and Information* 1 (4), 1992.

275. N. Oren, C. Reed, and M. Luck. Moving Between Argumentation Frameworks. In *Proceedings of the 2010 conference on Computational Models of Argument: COMMA 2010*, Pietro Baroni, Federico Cerutti, Massimiliano Giacomin, Guillermo R. Simari, eds., pp. 379–390. IOS Press Amsterdam, The Netherlands, 2010.

276. Z. Pawlak. *Rough Sets. Theoretical Aspects of Reasoning about Data*, Kluwer AcademicPublishers, Dordrecht, 1991.

277. C. S. Peirce. A Boolean Algebra with One Constant. In Hartshorne, C, and Weiss, P., eds., *Collected Papers of Charles Sanders Peirce*, Vol. 4: 12-20. Harvard University Press, 1931-35.

278. Judea Pearl, 'Fusion, propagation, and structuring in belief networks', *Artificial Intelligence*, **29**(3), 241–288, (September 1986).

279. J. Pearl. *Probabilistic reasoning in intelligent systems: networks of plausible inference.* Morgan Kaufmann, 1988.

280. Ch. Perelman. *Justice, Law and Argument*. Reidel, 1980.

281. J. Pollock. Self-defeating arguments. *Minds and Machines*, 1(4):367–392, 1991.

282. J. L. Pollock. Rational Cognition in OSCAR. In *Proc. of the 6th International Workshop on Intelligent Agents VI, Agent Theories, Architectures, and Languages (ATAL 1999)*, volume 1757 of Lecture Notes in Computer Science, pp 71–90, 1999.

283. J. Pollock. Justification and defeat. *Artificial Intelligence* 67: 377–408, 1994.

284. J. L. Pollock. *Cognitive Carpentry. A Blueprint for How to Build a Person.* MIT Press, Cambridge, MA, 1995.

285. H. Prakken. A logical framework for modelling legal argument. In *Proc. of the 4th Int. Conf. on Artificial intelligence and Law, ICAIL*, ACM, pp 1–9, 1993.

286. H. Prakken. *Logical Tools for Modelling Legal Argument*, Kluwer, 1997.

287. H. Prakken, C. Reed and D. Walton. Argumentation schemes and generalisation in reasoning about evidence. *ICAIL-03*, June 24–28, 2003.

288. H. Prakken. An abstract framework for argumentation with structured arguments. *Argument and Computation*, 1:93–124, 2010.

289. H. Prakken and S. J. Modgil. Clarifying some misconceptions on the ASPIC+ framework. In B. Verheij, S. Szeider and S. Woltran, eds., *Computational Models of Argument. Proceedings of COMMA 2012*, pp. 442-453. Amsterdam: IOS Press, 2012.

290. H. Prakken and G. Sartor. Argument-based extended logic programming with defeasible priorities. *Journal of Applied Non-Classical Logics*, 7:25–75, 1997.

291. H. Prakken and G.A.W. Vreeswijk. Logical systems for defeasible argumentation. In D. M. Gabbay and F. Guenthner, eds., *Handbook of Phil. Logic*, second ed., Kluwer Academic Publishers, Dordrecht, Volume 4, pp. 219–319, 2002.

292. R. Price. The Stroke Function in natural deduction. *Zeitsehr. f.math. Logik und Grundlagen d. Math.*, 7, 117-128 (1961).

293. T. Przymusinski. Well founded semantics coincides with three valued stable semantics. Special issue of *Fundamenta Informatica* on Non-monotonic reasoning. 13(4), W. Marek, ed. 445–464, 1990.

294. I. Rahwan and G. Simari, eds. *Argumentation in Artificial Intelligence.* Springer, 2009.

295. I. Rahwan, F. Zablith, and C. Reed. Laying the foundations for a world wide argument web. *Artificial Intelligence*, 171(10-15):897–921, 2007.

296. I. Rahwan and F. Tohmé, 'Collective argument evaluation as judgement aggregation', in *Proceedings of the 9th International Conference on Autonomous Agents and Multiagent Systems: volume 1 - Volume 1*, AAMAS '10, pp. 417–424, Richland, SC, (2010). International Foundation for Autonomous Agents and Multiagent Systems.

297. Report of the Royal Commission of Criminal Justice, M 2263, 1993. London: HMSO, Ch 8, para 26.

298. Nicols D. Rotstein, Martn O. Moguillansky, Alejandro Javier Garca, and Guillermo Ricardo Simari. A Dynamic Argumentation Framework. *COMMA 2010*, 427-438.

299. G. Rowe and C. Reed. Translating Wigmore Diagrams. In *Proceedings of COMMA 2006*, pp. 171–182, 2006.

300. T. J. Schaefer. The complexity of satisability problems. In *STOC '78: Proceed-ings of the 10th Annual ACM symposium on Theory of Computing*, pp. 216–226, 1978.

301. E. Schröder. Vorlesungen über die Algebra die Logik, 3 vols. B. G. Tuebner, Leipzig, 1890–1904. Reprints, Chelsea, 1966; Thoemmes Press, 2000.

302. B. Selman and H. Kautz. Knowledge compilation using Horn approximations. In *Proceedings of AAAI-91*, pp. 904–909, 1991.

303. B. Selman and H. Kautz. Knowledge compilation and theory approximation. *Joural of the ACM*, 43, 193–224, 1996.

304. M. J. Sergot, R. A. Kowalski *et al.*. British Nationality Act. *Communications of the ACM*, **29**, 370–386, 1986.

305. Y. Seto. Proofs of Some Axioms by Stroke Function. In *Proc. Japan Acad.*, 44 , pp 1024-1026, 1968.

306. G. Shafer. *A Mathematical Theory of Evidence*. Princeton University Press, 1976.

307. H. M. Sheffer. A set of five independent postulates for Boolean algebras, with application to logical constants, *Transactions of the American Mathematical Society*, 14: 481-488, 1913.

308. J. C. Shepherdson. Negation as failure 2. *The Journal of Logic Programming*, Volume 2, Issue 3, October 1985, Pages 185-202.

309. J. C. Shepherdson. Negation as failure: a comparison of Clark's completed data base and Reiter's closed world assumption, *The Journal of Logic Programming*, Volume 1, Issue 1, June 1984, Pages 51-79

310. J. C. Shepherdson. A Sound and Complete Semantics for a Version of Negation as Failure, *Theoret. Comput. Sci.* 65(3):343-371 (1989).

311. J. C. Shepherdson and G. Mints. Type Calculi for Logic Programming, Report PM-88-01, School of Mathematics, Univ. of Bristol, 1988.

312. G. R. Simari and R.P. Loui. A mathematical treatment of defeasible reasoning and its implementation. *Artificial Intelligence*, 53:125–157, 1992.

313. Craig Smorynski. Modal logic and self-reference. In Dov M. Gabbay and F. Guenthner, editors, *Handbook of Philosophical Logic, 1st edition*, volume 2, pp. 441–495. Kluwer, 1984.

314. R. Solovay. Provability interpretations of modal logic. *Israel Journal of Mathematics*, 25, 1976, pp. 287-304.

315. Tran Cao Son. Answer Set Programming Tutorial, October 2005. http://www.cs.nmsu.edu/~tson/tutorials/asp-tutorial.pdf

316. R. Staerk. A Transformation of Propositional Prolog Programs into Classical Logic. In *LPNMR '95 Proceedings of the Third International Conference on Logic Programming and Nonmonotonic Reasoning*, Editors: V. Wiktor Marek Anil NerodeSpringer-Verlag London, 1995, pp 302-315

317. R. Staerk. Cut property and negation as failure. *International Journal of Foundations of Computer Science (IJFCS)*, 5:2, 129-164, 1994.

318. R. Staerk. A complete axiomatization of the three-valued completion of logic programs. *J. of Logic and Computation*, 1(6):811–834, 1991.

319. S. Strasser. Towards the Proof-Theoretic Unification of Dung's Argumentation Framework: an Adaptive Logic Approach. *Journal of Logic and Computation*, 21:133–156, 2011.

320. A. Szałas. On the correspondence between modal and classical logic: An automated approach. *Journal of Logic and Computation*, 3, 605–620, 1993.

321. Y. Tang, T. J. Norman, and S. Parsons. A model for integrating dialogue and the execution of joint plans. In *AAMAS*, pages 883–890, 2009.
322. Silvano Colombo Tosatto, Guido Boella, Leendert van der Torre and Serena Villata. Abstract Normative Systems: Semantics and Proof Theory, KR 2012.
323. S. Toulmin. *The Uses of Argument*. Cambridge University Press, 1958.
324. P. Turchin. *Complex Population Dynamics*. Princeton University Press, 2003.
325. S. Uglow. *Textbook on Evidence*, Sweet and Maxwell, 1997.
326. J. van Benthem. Minimal predicates, xed-points, and denability. *Journal of Symbolic Logic*, 70(3), 696–712, 2005.
327. J. Vauzeilles. Negation as Failure and Intuitionistic Three-Valued Logic. In *Proceedings of FAIR '91, the International Workshop on Fundamentals of Artificial Intelligence Research*, pp 228-241.
328. B. Verheij. Two approaches to dialectical argumentation: admissible sets and argumentation stages. In *Proceedings of the Eighth Dutch Conference on Artificial Intelligence (NAIC-96)*, pp. 357–368. Utrecht the Netherlands, 1996.
329. B. Verheij. *Rules, Reasons, Arguments: formal studies of argumentation and defeat*. PhD thesis, Maastricht University, The Netherlands, 1996.
330. Bart Verheij. A labeling approach to the computation of credulous acceptance in argumentation. In Manuela M. Veloso, editor, *Proceedings of the 20th International Joint Conference on Artificial Intelligence, Hyderabad, India*, pages 623–628, 2007.
331. S. Villata, G. Boella, D. M. Gabbay, and L. van der Torre. Modelling defeasible and prioritized support in bipolar arguentation. *AMAI, Annals of Mathematics in Artificial Intelligence*, speical issue: New Developments in Reactive Semntics 66(1–4), 163–197, 2012.
332. V. Volterra. Fluctuations in the abundance of a species considered mathematically. Nature, Vol. 118, pp. 558–560, 1926.
333. G. A. W. Vreeswijk. Studies in defeasible argumentation. *PhD thesis at Free University of Amsterdam*, 1993.
334. G.A.W. Vreeswijk. An algorithm to compute minimally grounded and admissible defence sets in argument systems. In P.E. Dunne and TJ.M. Bench-Capon, editors, *Computational Models of Argument; Proceedings of COMMA 2006*, pages 109–120. IOS, 2006.
335. T. Wakaki. Preference-based Argumentation Capturing Prioritized Logic Programming, *ArgMAS*, 2010.
336. J. Wang, G. Luo, and B. Wang, 'Argumentation framework with weighted argument structure', in *10th IEEE International Conference on Cognitive Informatics Cognitive Computing (ICCI*CC)*, pp. 385–391, (2011).
337. J. Williamson and D. M. Gabbay. Recursive Bayesian Networks and self-fibring logics. In *Laws and models of Science*, D. Gillies, editor, College Publications, 2004, pp 173-247.
338. J. Woods, R. H. Johnson, D. M. Gabbay and H. J. Ohlbach. Logic and the Practical turn. In *Handbook of the Logic of Argumentation and Inference: The Turn Towards the Practical*, D. M. Gabbay, R. H. Johnson and H. J. Ohlbach, eds. pp. 1–39. Volume 1 of *Studies in Logic and Practical Reasoning*, North-Holland, 2002.
339. J. Woods. *The Death of Argument: Fallacies in Agent-Based Reasoning*. Kluwer, Dordrecht and Boston, Applied Logic Series, 2004.
340. Yining Wu, M. Caminada and D. M. Gabbay. Object level reduction of logic programs to argumentation networks. *Studia Logica*, 93(2-3):355-379, 2009.

Index